D0855315

KINGDOMS & DOMAINS

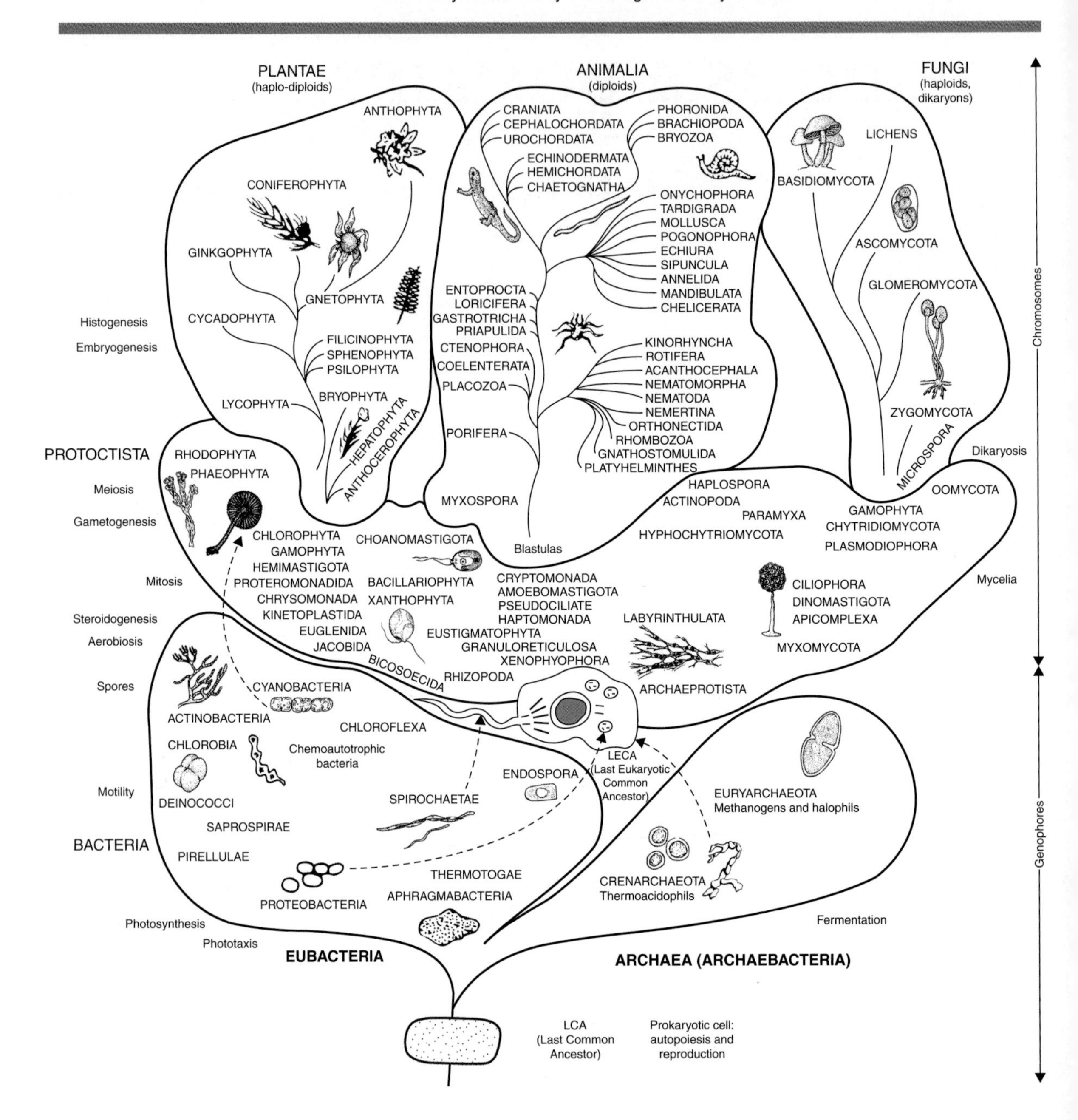

The phyla of life on Earth based on our modification of the Whittaker five-kingdom system and the symbiotic theory of the origin of eukaryotic cells.

KINGDOMS & DOMAINS
An Illustrated Guide to the Phyla of Life on Earth

Lynn Margulis
University of Massachusetts at Amherst

Michael J. Chapman
Marine Biological Laboratory
Woods Hole, Massachusetts, USA

AMSTERDAM • BOSTON • HEIDELBERG • LONDON •
NEW YORK • OXFORD • PARIS • SAN DIEGO •
SAN FRANCISCO • SINGAPORE • SYDNEY • TOKYO

Academic Press is an imprint of Elsevier

Development Editors: Janet Tannenbaum, Kendra Clark
Project Editor: Georgia Lee Hadler
Cover and Text Designer: Diana Blume
Illustration Coordinator: Susan Wein
Production Coordinators: Maura Studley, Mani Prabakaran
Composition: Electronic Publishing Center and Progressive Information Technologies
Manufacturing: RR Donnelley

Library of Congress Cataloging-in-Publication Data
Margulis, Lynn 1938– and Michael J. Chapman 1961–
Kingdoms & Domains: An Illustrated Guide to the Phyla of Life on Earth/Lynn Margulis, Michael J. Chapman — 4th ed.
p. cm.
Includes bibliographical references and index.
ISBN 0-7167-3026-X (hardcover: alk. paper).—ISBN 0-7167-3027-8 (pbk.: alk. paper).—
ISBN 0-7167-3183-5 (pbk.: alk paper/ref. booklet).
ISBN: 978-0-12-373621-5
1. Biology—Classification, Evolution
QH83.M36 1998
570′.1′2—dc21 97-21338
CIP

Printed in the United States of America

First Printing, 1982

COVER IMAGE—Classification schemes help us comprehend life on this blue and green planet. But classification schemes are an invention; the human hand attempting to sort, group, and rank the types of life that share Earth with us. Because no person witnessed the more than 3000 million years of the history of life, our domains, kingdoms, phyla, classes, and genera are approximations.

In the metaphor of the hand, the lines within the hand outline and separate the kingdoms. The thumb represents the earliest kingdom of bacteria (the Prokaryotae), which includes the Archaea (Archaeabacteria). The fingers, more like one another, represent the living forms composed of nucleated cells. The back of the hand and the baby finger are continuous; they form a loosely allied, ancient group of microbes and their descendants: members of kingdom Protoctista—seaweeds, water molds, ciliates, slime nets, and a multitude of other water dwellers. The ring and middle fingers stand together: The molds and mushrooms of kingdom Fungi and the green plants of kingdom Plantae made possible the habitation of the land. Members of kingdom Animalia, the most recent kingdom to venture onto dry land, are on the index finger.

No matter how we care to divide the phenomenon of life, regardless of the names that we choose to give to species or the topologies devised for family trees, the multifarious forms of life envelop our planet and, over eons, gradually but profoundly change its surface. Life and Earth become a unity, intertwined where each alters the other. A graphic depiction of our taxonomic hypothesis, the hand and globe image, conveys the intricate mergers, fusions and anastomoses that comprise the web of life. [Illustration based on a design by Dorion Sagan.]

This fourth edition is dedicated to Donald I. Williamson, Port Erin Marine Station, United Kingdom (who changed our view of the origins of animals and their larvae by recognition of the importance of evolutionary mergers) and to all other scientists, artists, teachers, and students who aided this labor of love of life on Earth (see Acknowledgments, page xxi*).*

CONTENTS

APPENDIX

List of Figures

Introduction

Chapter 1

Chapter 3

Chapter 4

Chapter 5

List of Tables

Introduction

Kingdom Prokaryotae (Bacteria, Monera, Prokarya)

FOREWORD

This lavishly illustrated book is an extravaganza. But then it is true that life, its subject, is also an extravaganza. It is further true that every species of organism that exists or ever existed is worthy of a book all its own, with striking art to illustrate it. The more we learn of the diversity of life, the more prodigious seem these products of over three thousand million years of evolution. Such is the "Creation," our living inheritance, to whose understanding and careful preservation we should feel uniquely committed. No one before the present era of biology could have imagined more than a small part of its true full efflorescence – not the scribes of Abrahamic sacred literature, not Aristotle who tried to encompass it all, not Linnaeus who devised the means to name and classify it systematically, and not even those biologists who have explored it so rigorously through the twentieth century.

About 1.8 million species, our rough estimates tell us (in 2009, serendipitously the bicentennial of Darwin's birth and the sesquicentennial of *The Origin of Species*), have been discovered and described. That includes perhaps three-fourths of the extant hundred thousand or so vertebrates, and, at a guess, ninety percent of the quarter million species of flowering plants thought to exist. But the sixty thousand known fungi are fewer than five percent of the estimated total, and the fewer than twenty thousand named nematode worms, the most abundant animals on Earth, are probably an even smaller fraction of the whole. Moreover, all this ignorance shrinks in the dismaying presence of the "dark matter" of the prokaryotic universe – or if you prefer, the Subkingdoms (Domains) of Archaea and Eubacteria. The exploration of what could turn out to be tens of millions or even hundreds of millions of well differentiated strains of these subvisible organisms has scarcely begun.

Kingdoms and Domains, this book, fearlessly enters the latter world and provides a guide through the rapidly shifting, most inclusive classification necessitated by new information pouring in about it. The technological breakthrough that is accelerating progress in this sector today is comparative genomics, more and more of it, learned faster and faster, and steadily falling in price per DNA base pair. Even so, the phenotypic traits must be added into the calculations as the reconstructed phylogeneticists push forward. It should be kept in mind that a small number of genes, or one step in symbiosis, can sometimes alter the structure and biochemistry of an organism in profound ways.

Meanwhile, the exploration of all of Earth's biodiversity, including the more familiar eukaryotic phyla, is about to be accelerated by the *Encyclopedia of Life*, an online database launched in 2008 into which complete knowledge of all organisms down to the species level will be compiled, organized and made available to anyone, anywhere, and at any time.

Our fascination with the possibility of life on other worlds is entirely understandable. Yet it is passing strange that we pay so little attention to the largely unknown world all around us. The strange yet lovely biosphere is our only harbor in the vastness of space. Perhaps this attractive and comprehensive book will help us to redirect our gaze, closer to home.

E. O. Wilson
University Professor Emeritus
Honorary Curator in Entomology
Harvard University
Cambridge, Massachusetts
USA

FOREWORD
To 1st–3rd editions

Like bureaucracy, knowledge has an inexorable tendency to ramify as it grows. In the early nineteenth century, the great French zoologist Georges Cuvier classified all "animals"—moving beings, both microscopic and visible—into just four great groups, or phyla. A century earlier, Linnaeus himself, the father of modern taxonomy, had lumped all "simple" animals into the single category "Vermes"—or worms.

Cuvier's four animal phyla have expanded to more than forty, distributed in two kingdoms, the Protoctista (for microscopic forms and their descendants) and the Animalia (for those that develop from embryos)—and remember, we have said nothing of plants, fungi, other protoctists, and bacteria as yet. The very names of these groups are imposing enough—kinorhynchs, priapulids, onychophorans, and gnathostomulids. Some biologists can spit out these names with a certain virtuosity, but most of us know rather few of the animals behind the names. This ignorance arises for two primary reasons: the names are simply now too many, and modern training in zoology is now so full of abstract theory that old-fashioned knowledge of organic diversity has, unfortunately, taken a backseat.

Margulis and Schwartz have generated here that rarest of intellectual treasures—something truly original and useful. If the originality comes before us partly as a "picture book," it should not be downgraded for that reason—for primates are visual animals, and the surest instruction in a myriad of unknown creatures must be a set of figures with concise instruction about their meaning—all done so admirably in this volume. It is remarkable that no one had previously thought of producing such a comprehensive, obvious, and valuable document.

My comments thus far have been disgracefully zoocentric. I have spoken only of animals, almost as if life were a ladder with animals on the top rungs and everything else inconspicuously and unimportantly below. The old taxonomies included two kingdoms (plants and animals, with unicells placed, in procrustean fashion, into one or the other camp), or at most three kingdoms (animals, plants, and unicells). With this work, and its 96 phyla distributed among five kingdoms, we place animals (including ourselves) into proper perspective on the tree of life—we are a branch (albeit a large one) of a massive and ramifying tree. The greatest division is not even between plants and animals, but *within* the once-ignored microorganisms—the prokaryotic Bacteria and the eukaryotic Protoctista. The five kingdoms are arrayed as three great levels of life: the prokaryotes, the eukaryotic microorganisms and their derivatives (Protoctista), and the eukaryotic larger forms (Plantae, Animalia, and Fungi). These last three familiar kingdoms represent the three great ecological strategies for larger organisms: production (plants), absorption (fungi), and ingestion (animals).

Some people dismiss taxonomies and their revisions as mere exercises in abstract ordering—a kind of glorified stamp collecting of no scientific merit and fit only for small minds that need to categorize their results. No view could be more false and more inappropriately arrogant. Taxonomies are reflections of human thought; they express our most fundamental concepts about the objects of our universe. Each taxonomy is a theory about the creatures that it classifies.

The preceding material is a slightly shortened and lightly altered version of the preface that I wrote for the original edition of this book. As I reread my words and consider the remarkable changes in this field during the past 15 years—a growth of knowledge and development of thinking that, for once, justly deserves the overused designation of "revolutionary"—I am particularly struck by the wisdom and discernment of Margulis and Schwartz in their original, and now even more compelling, choice of the Five Kingdom system for ordering the diversity of life.

Molecular sequencing of nucleic acids has provoked the enormous gain in our understanding during the past 15 years. We can now obtain a much more accurate picture of the branching pattern on the tree of life through time by measuring the detailed similarities among organisms for the fine structure of genes held in common by all: as a general rule, the greater the differences between any two kinds of organisms, the longer they have been evolving on separate paths since their divergence from a common ancestor.

The system advocated here—five great kingdoms of life divided into two great domains (the Prokarya with their simple unicellular architecture lacking nuclei and other organelles and forming the kingdom of Bacteria, versus the Eukarya made of more complex cells and including the other four kingdoms of Protoctista, Animalia, Fungi, and Plantae)—might seem to be challenged by the discovery by Carl Woese and others that the genealogical tree of life has only three great branches, including two among the Prokarya (the Archaea and the Eubacteria), with all Eukarya on a third branch, and the three great multicellular kingdoms of plants, animals, and fungi as twigs at the tips of this branch.

But classification must consider more than the timing of branching. Woese's surprising discovery makes excellent sense when we realize that life is at least 3.5 billion years old on Earth, and that only Prokarya lived during the first 2 billion years or so. Since Eukarya arose so much later, they are confined to a single branch on a system that records time of branching alone.

Classification must also record degree and amount of diversity and complexity (while never violating the primary signal of phylogeny, or order of branching), as well as the timing of branch points. When these criteria are added, the breaking of the enormous eukaryote branch into four kingdoms, and the compression of the two prokaryote branches into one kingdom of Bacteria seems fully justified, if only for our legitimately parochial interest in the astonishing diversity of organisms in our visible range of size and complexity.

Still, as the authors duly and happily note, and from an enlarged and less human-centered perspective, bacteria really are the dominant form of life on Earth—and always have been and probably always will be. They are more abundant, more indestructible, more diverse in biochemistry (if not in complexity and outward form), and inhabit a greater range of environments than all the other four kingdoms combined. But we cannot grasp this fundamental fact, and so much else about evolution, until we abandon our biased view of life as a linear chain leading to human complexity at a pinnacle, and focus instead upon the rich range of diversity itself as the primary phenomenon of life's spread and meaning. And we cannot grasp life's full diversity without such excellent works as this book, dedicated to presenting the full story of life's vastness—from the "humble" and invisible (to us!) bacteria that really dominate life's history to the arrogant, fragile single species, *Homo sapiens*—a true upstart and weakling, but the Earth's first creature endowed with the great evolutionary invention of language, a device that may only lead to our self-destruction, but that also yields all our distinctive glories, including our ability to understand by classifying.

Foreword to 1st–3rd editions
Stephen Jay Gould (1943–2002)
Museum of Comparative Zoology
Harvard University

PREFACE

This book provides an illustrated guide to the diversity of life on Earth. As a comprehensive reference to both microbes and their larger descendants, it serves as a guidebook to living organisms based on their evolution. What do they look like, where do they dwell, how are they related to one another, how best do scientists group them? We try to answer such questions by photographs, simplified text and drawings. Brief essays introduce the broad outlines of the "higher" (largest, most inclusive) taxa. If curiosity leads, references and further reading are included. NASA scientists opened our eyes to the need for an illustrated guide to the diversity of life on Earth to inform their search for extraterrestrial life. *Kingdoms and Domains* includes diagrams and photographs of whole organisms that should enable recognition of life-forms even in outer space. We write for all students of biology at any level of expertise—whether participants in biology, biodiversity, zoology, botany, mycology, systematics, evolution, ecology, genetics, and geomicrobiology—or curious naturalists, geologists, park rangers, space scientists, and armchair explorers. What characteristics distinguish the members of one taxon? What broad view of evolutionary relationships is most valid?

The reader is encouraged to seek evidence of the history of life by firsthand visits to fossil sites, prehistoric dioramas, and museum exhibits. The extraordinary history of life, documented in fossils including the molecular fossil record, inspires us to look appreciatively at present-day life.

Tips for reading this book include major habitats illustrated for each phylum (the seven ecostrips are explained on page 25). Colophons (page 28) indicate how the organism is viewed in the photographs. A chronology of the past four thousand million years of Earth history is summarized in Figure I-1, and Table I-1 summarizes the classification scheme.

Our frontispiece illustrates differences between the archaebacteria and the eubacteria as well as distinctions among the eukaryotes—animals, plants, fungi, and protoctists. These differences and distinctions are summarized in the Introduction and fleshed out in the introductory essays for each kingdom. Concepts of kingdoms and phyla originate in the classification proposals of scientists of the twentieth century—including Robert Whittaker and Herbert Copeland—built on earlier attempts of Linnaeus, Jussieu, Cuvier, and Haeckel to order the biota. We have extended these proposals and present a five-kingdom system consistent with both the fossil record and the most recent molecular data.

The molecular data have perhaps most profoundly affected our view of bacteria and protoctists. In accord with changes in bacterial classification that recognize differences in ribosomal RNA molecules, we have classified 16 phyla in superkingdom Prokarya (Prokaryotae, Monera): 2 phyla in subkingdom Archaea and 14 phyla in subkingdom Eubacteria. We have incorporated protoctist reclassification based on the *Handbook of Protoctista* second edition with up-to-date ultrastructural, ecological, and molecular information for more than 30 protoctist phyla.

In kingdom Fungi, the three phyla of our earlier edition have been expanded to six. We incorporate both phycobiont and mycobiont into the symbiogenetic unity phylum Lichenes (≡phylum Mycophycophyta). Like all other eukaryotic taxa, only more conspicuously, lichens evolved by symbiogenesis. We now include the deuteromycotes—fungi that have lost differentiated asci or basidia—in phylum Ascomycota; some of these fungi will be moved to Basidiomycota and when their sexual (meiotic spore-forming) tissue is discovered.

Plant classification has changed in accord with recognition of the morphological and biochemical differences between the nonvascular plants and mosses, liverworts, and hornworts. These bryophytes probably evolved independent of one another from algal ancestors. Each—formerly a class in phylum Bryophyta—is now afforded phylum status: Bryophyta (mosses), Hepatophyta (liverworts), and Anthocerophyta (hornworts). Research from the frontiers of botany that has enlightened our understanding of plant relationships includes the discovery by ethnobotanists of diverse drugs, details of nitrogen gas and nitrogenous compounds in forests, elucidation of the evolutionary origin of the seed plants, and advances in understanding frequency and modes of plant hybridization.

Concepts of animal taxa also change rapidly. Relationships within and among the phyla continue to be modified as phylogenetic information from ultrastructural, developmental, morphological, and molecular sources continues to abound. Ribosomal RNA sequences and evidence from proteins tell us that the closest relatives of animals are fungi; plants are more distant relatives. As Haeckel observed in 1874, animals and choanomastigotes (unicellular protoctists) share common ancestry. Molecular data support his inference. Sponges and comb jellies form a lineage within the animal kingdom; the placozoan *Trichoplax* and Cnidaria share common ancestry. The more complex bilaterally symmetrical animal phyla (A-23 to craniates like us A-37) as well as mandibulates (insects and crustaceans, A-21) and mollusks (A-26), probably evolved by merges of phyletic lineages as Donald I. Williamson argues. Adult rotifers (A-14) and onychophora (A-28) became larvae of holometabolous (metamorphosing) animals through anatomosis of phylogenies, probably by "forbidden fertilizations" (hybridogenesis), see Box A-i, p. 238 on larval transfer).

The rhombozoans and orthonectids, formerly together in phylum Mesozoa, are now separate phyla because of their unique characteristics. The five hooked-mouthed worms (pentastomes), now a class within Crustacea, led us to abandon phylum Pentastomida. We have moved arthropods into three phyla: Chelicerata, Mandibulata, and Crustacea. Because some recent Burgess shale fossil evidence suggests arthropods are monophyletic, these three arthropod taxa may eventually be reunited. As fragmentary molecular data become more complete, all animal phylogenetic relationships will be refined.

Why do classification schemes change? Every taxon—class, order, phylum, domain, kingdom—based on the study of relationship—is artificial. We recognize only eukaryotic species as natural taxa. A case in point is the phylum Loricifera—a group of minute marine organisms first described by R. P. Higgins in 1983. Because loriciferans could not be placed in any previously known inclusive taxon without stretching the phylum concept, he established a new phylum just for them. Phylum Loricifera was a hypothesis to be tested, as are all new taxa. After more than two dozen years of evidence gathering on the biology of loriciferans, Loricifera persists as a phylum. Priapulids appear to be closest to them, and the kinorhynchs are more distant relatives. In 1995, a new species, *Symbion pandora*, was reported and a new phylum, Cycliophora, was proposed to accommodate this single species. As is the case for loriciferans, the relationships of *Symbion* to other phyla will be tested—possible relatives are entoprocts and bryozoans (ectoprocts). All newly suggested hypothetical life cycles and morphologies, and metabolic pathways require rigorous scrutiny with the goal of more accurate classification. We do not yet accept much new taxonomy, especially claims based on molecular data but rather we await firsthand biological confirmations of evidence.

As in the earlier editions, a handy reference Appendix expanded in a list of c. 4000 genera includes all those mentioned in this book and many others. The vernacular or common names by which these genera are known are given, as is the phylum for each genus. The glossary provides definitions of terms. So that readers may move easily from glossary to chapter essays, we frequently indicate the phylum and kingdom to which a term applies. Again an Appendix lists all kingdoms and phyla.

Six sets of 35-mm color transparencies of the five kingdoms including phylogenetic drawings are available from Ward's Natural Science Establishment,

Inc., 5100 West Henrietta Road, P.O. Box 92912, Rochester NY 14692-9012 (1-800-962-2660). Five of these transparency sets depict a member of each phylum of bacteria, protoctists, fungi, plants, and animals; the sixth set introduces the general features of each kingdom and prokaryote–eukaryote distinctions. A printed teacher's guide describes each slide.

Five Kingdoms is available in translation from the following publishers: Spanish: 2008, Tusquets, Barcelona; Japanese: Nikeii Science, Tokyo; German: Spektrum der Wissenschaft, Heidelberg; Portuguese: 2004, Guanabara Koogan, SA Rio de Janeiro, Brazil.

Our colleagues generously continue to provide photographs, drawings, manuscripts, publications, and constructive criticism for the complete rewriting of this fourth edition. We thank them for their great kindness and invite them to continue their contributions. We welcome additional relevant new material. Please send corrections, critical comment, new illustration possibilities, etc. to either coauthor at University of Massachusetts-Amherst.

This book grew out of our own need and that of our students for a single-volume reference to the diversity of life on Earth. Your response tells us that our passion for our astonishing planetmates is shared.

ACKNOWLEDGMENTS

The artists' ways of viewing organisms enhance our own appreciation immensely. We thank the artists who gave assistance in the preparation of our earlier editions: Laszlo Meszoly, Michael Lowe, Peter Brady, Christie Lyons, and Robert Golder. Artists who rendered illustrations new for this edition include Christie Lyons, Kathryn Delisle and James MacAllister. Lowell M. Schwartz created our chapter-opening background patterns and the tiny organism-studded text borders that distinguish the phylum essay pages for each kingdom. For their generous provision of photographs, drawings and corrections we thank, George Bean, Ray Evert, Robert P. Higgins, Eugene Kozloff, Carl Shuster, Jr. and David K. Smith.

We thank the librarians at the University of Massachusetts (both Boston and Amherst) and Duke University for their dedicated aid. We acknowledge the continuing encouragement and critical comments of our colleagues at the University of Massachusetts in Amherst and in Boston. We are grateful to Stephanie Hiebert, Donna Reppard, Celeste Asikainen, Dorion Sagan, Kendra Clark and Idalia Rodriguez for aid in manuscript preparation.

LM acknowledges the support of the Geosciences Department, the Provost's Office, and the Graduate School at the University of Massachusetts at Amherst as well as the Tauber Fund, the Alexander von Humboldt, Stiftung and Balliol College, University of Oxford, UK. MJC is deeply grateful to Kendra Clark and Idalia Rodriguez, editorial assistants par excellence. He also wishes to thank Dr. Elizabeth Connor and artist Kathy Delisle for help with images, Mark and Mary Anne Alliegro for their patience and support, and Susan Elizabeth Sweeney for tending the Peaceable Kingdom during preparation of this volume.

For urging us forward in countless ways, we thank our families: Susan Elizabeth Sweeney; Sophie, Reuben and Menina; Dorion, Jeremy, and Tonio Sagan; Jennifer Margulis di Properzio, Zachary Margulis-Ohnuma and Ricardo Guerrero.

We extend our thanks to the instructors and many researchers who provided new reviews during information and preparation of this fourth edition: O. Roger Anderson, John Archibald, Celeste Asikainen, Jerri Bartholomew, Sharon Bartholomew-Began, Gordon Beakes, Samuel Bowser, J.P. Braselton, Martin Brasier, Sandy Carlson, Allen Collins, Margery Coombs, Peter Del Tredici, Michael Dolan, Stephen Fairclough, Mark Farmer, Victor Fet, Wilhelm Foissner, Rebecca Gast, Andrew Gooday, Peter Hayward, Robert Higgins Jr., Hiroshi Kawai, Bryce Kendrick, Lafayette Fredericks, Fayett, J.J. Lee, Joyce Longcore, James MacAllister, Sandra McInnes, Michael Melkonian, Daniel Miracle, Sandy Paracer, John Pilger, John Pojeta, H.-R. Preisig, Dennis Richardson, Margaret Riley, Karlene V. Schwartz, Pat Gonzalez, Bruce Scofield, Stanley Shostak, Douglas G. Smith, William Stein, Wolfgang Sterrer, Erik V. Thuesen, Leleng To, Keith Vickerman, Donald I. Williamson and Jeremy Young.

Without the superb professional aid of Kendra Clark and Idalia Rodriguez in all aspects of manuscript preparation, this fourth edition would have never been completed.

Our heartfelt gratitude to Pat Gonzalez, Andrew Richford, Nancy Maragioglio at Academic Press, and to ETI BioInformatics (University of Amsterdam) for their help with the CD and its translation.

KINGDOMS & DOMAINS

INTRODUCTION

Great contributors to our concepts of kingdoms of phyla

Carolus Linnaeus (Carl von Linné, Swedish) originated the concept of binomial nomenclature and a comprehensive scheme for all nature.
Antoine-Laurent de Jussieu (French) established the major subdivision of kingdom Plantae.
Georges Leopold Cuvier (French) established the major "embranchements" (phyla) of kingdom Animalia.
Ernst Haeckel (German) was the innovator of kingdom Monera for many organisms—neither animal nor plant.
Herbert F. Copeland (Sacramento City College, California) reclassified all the microorganisms in recognition of Englishman John Hogg's (1860) Kingdom Primogenium (members Protoctista) for the nucleated organisms: all eukaryotes except plants and animals.
Robert H. Whittaker (Cornell University, New York) founded the five-kingdom system, by recognition of the kingdom Fungi.

Carolus Linnaeus
1707–1778
[The Bettmann Archive.]

Antoine-Laurent de Jussieu
1748–1836
[The Bettmann Archive.]

Georges Leopold Cuvier
1769–1832
[Bibl. Museum Hist. Nat. Paris.]

Herbert F. Copeland
1902–1968
[Courtesy of Mrs. H. F. Copeland.]

Robert H. Whittaker
1924–1980
[Courtesy of R. Geyer.]

Ernst Haeckel
1834–1919
[The Bettmann Archive.]

Box I-i The wisdom of Darwin and Gould

Charles Darwin (1802–1889) never actually used the word "evolution;" rather, he wrote of "descent with modification." As keenly interested in how living forms change with time as in biblical genealogies, he proposed: "All true classification is genealogical; that community of descent is the hidden bond which naturalists have been unconsciously seeking, and not some unknown plan of creation, or the enunciation of general propositions, and the mere putting together and separating objects more or less alike."

Genealogies, Darwin inferred, are in essence classification systems. Our goal in this book on *Kingdoms and Domains* is to organize knowledge about life genealogically, that is, in a manner that reflects the history of living forms based on the widest possible information about genetics, physiology, chemistry, and fossil history. But our classification must be accessible and useful to all those who study nonhuman life-forms. Therefore, albeit informed by current scientific data, we simplify our presentation for accessibility to a general readership including students, teachers, professors, naturalists, and professional scientists. This simplification is most marked in the bacterial chapter, where we have confined our presentation to an artificially small but manageable number of 15 representative phyla (inclusive groups). We direct the reader to specific literature on the *animals* (zoology, entomology, herpetology, ichthyology, helminthology, etc.), *plants* (botany, dendrology, horticulture, agriculture), *fungi* (mycology, plant pathology), *protoctista* (protozoology, phycology, parasitology, veterinary medicine, mycology) and *bacteria* (microbiology, bacteriology, virology). The organisms have been objects of study in traditional disciplines that often fail to describe them properly. As our esteemed colleague Prof. James Walker quipped, when biologists discarded the two-kingdom (plant versus animal) system "we botanists got four out of five." Of course, the old system suffered from the still-popular misconception that all organisms other than animals must be plants!

Prof. Stephen J. Gould (1941–2002), recently called "the greatest evolution writer of his time" in the new book of his works edited by Ed. Barber and Niles Eldredge (Norton 2007), celebrated Darwin's genius when he too admonished us to study oddities and peculiarities: "Darwin answers that we must look for imperfections and oddities, because any perfection in organic design or ecology obliterates the paths of history and might have been created as we find it. This principle of imperfections became Darwin's most common guide." Our classification of microbes, organisms whose bodies require magnification to be seen, follows the Darwinian–Gouldian principle: we comprehend oddities and peculiarities in the microbial world as the best possible clues to their evolutionary history. Astounded by the prodigious diversity of life, we endorse Gould's sentiment that classification of 10–100 million different kinds of extant life on Earth is no mere "stamp collecting." Rather, it is the application of Darwin's "descent with modification" theory to the evolutionary history of life on Earth. Taxonomy, therefore, is the most important enterprise in biology, truly both "Queen" and "King" of the biological sciences. May Darwin be given the last word in this introductory note to our tome: as he wrote in the 1859 edition of *Origin*, "Anyone whose disposition leads him to attach more weight to unexplained difficulties than to the explanation of a certain number of facts will certainly reject my theory." Our dispositions attach less weight to unexplained molecular minutiae than to the explanatory power of the concept of common descent of all life on Earth.

INTRODUCTION

CLASSIFICATION OF LIFE, NAMES OF ORGANISMS

This book is about the biota, the cover of living beings at the surface of planet Earth. A catalog of life's diversity and virtuosities, *Kingdoms and Domains* provides a manageable ordering of life-forms. Here an internally consistent, complete system of higher (most inclusive) taxa is presented. We judge it robust, valid, and current, given the fragmented and often inconsistent professional literature from which our information is drawn.

Biologists, whether in the field or in the laboratory, study the parts of individual organisms or their populations, communities, or ecosystems. Organisms are classified on the basis of body form (morphology), genetic similarity, metabolism (body chemistry), developmental pattern, behavior, and, in principle, all of their characteristics. They are grouped together with similar organisms into genera and species. At least 3 million and perhaps 30 million species of living organisms now exist. A far greater number, greater than 99.9 percent of all that ever lived, are estimated to be extinct. This book can only mention the greatest diversity of all: that of past life as documented in the fossil record. Only living groups are depicted here.

The science of systematics, the classification of the living, represents the international scientific effort to provide order to this incredible variety. Modern systematists group closely related species into genera (singular: genus), genera into families, families into orders, orders into classes, classes into phyla (singular: phylum), and phyla into kingdoms (or domains) (Table I-1). This conceptual hierarchy grew gradually, in the course of over a century, from a solid base established by the Swedish botanist Carolus Linnaeus (1707–1778), who began the practice of binomial nomenclature for c. 10,000 species of plants and animals. Every known organism has a unique two-part name, usually Latin in form. The first part is the same for all organisms in its genus. The second part denominates the species within the genus. For example, *Acer saccharum*, *Acer nigrum*, and *Acer rubrum* are the scientific names of the sugar maple, the black maple, and the red maple, respectively.

Groups of all sizes, from species on up to kingdom, are called taxa (singular: taxon). Taxonomy is the analysis of an organism's characteristics for the purpose of assignment to a taxon. Since the time of Linnaeus, the growth of biological knowledge has greatly extended the range of

Table I-1 The Classification of Two Organisms

Taxonomic level	Humans	Garlic
Kingdom	Animalia	Plantae
Phylum (division)*	Chordata	(Angiospermatophyta)
Subphylum†	Vertebrata	—
Class	Mammalia	Monocotyledoneae
Order	Primates	Liliales
Family	Hominoidea	Liliaceae
Genus	*Homo*‡	*Allium*‡
Species	*sapiens*§	*sativum*

*Botanists use the term "division" instead of "phylum."

†Intermediate taxonomic levels can be created by adding the prefixes "sub" or "super" to the name of any taxonomic level.

‡Genera names are capitalized and italicized.

§Species names take lower case and are also italicized.

characteristics used in taxonomy. Linnaeus based his classification on visible structures of living organisms. Later, extinct organisms and their traces—fossils—were named and classified. In the nineteenth century, the discoveries of paleontologists and Charles Darwin's revelation of descent with modification (evolution by natural selection) encouraged systematists to hope that their classifications reflected the history of life. Classifications were converted into phylogenies, family trees of species or higher (more inclusive) taxa. To this day, very few lineages from fossil organisms to living ones have been traced, yet the truest classification is still held to be the one that best reflects the evidence for relationship by common ancestry.

In the twentieth century, advances in developmental biology and biochemistry gave the taxonomist new tools. Phylogenies, for example, can be based on patterns of development, on the linear sequence of amino acids that compose proteins, or on gene sequences—the sequences of nucleotides in the genes (nucleic acids). Techniques of electron microscopy and optical microscopy have greatly improved. Their results enable scientists to document internal structural details in the smallest life-forms and those of constituent cells of larger organisms. Computers that can organize and compare massive quantities of sequence data allow scientists to measure the relatedness of life-forms by comparison of their gene sequences. This procedure underlies the fields of genomics and proteomics (molecular systematics).

The field of biology called molecular systematics or molecular evolution stems from two innovations. The first deciphers the linear, genetically determined sequence of component monomers (amino acid residues or nucleotides) in macromolecules, primarily proteins and nucleic acids (DNA, RNA or protein). The second is the computer-based comparisons of immense quantities of data. Each cell of each life-form contains 500–50,000 genes. The average size of a gene exceeds 1000 nucleotides. Hence the molecular data for only the linear order of the protein and nucleic acid components in a single cell can range from half a million (0.5×10^6) to 50 million pieces of relevant information. Only high-speed computers, properly programmed, can organize and meaningfully compare such quantities of sequence data.

Fine structure, genetics, metabolism, behavior, development, and natural history—the biology of organisms—are refractory to quantitative study. These features (semes) often defy uniform description required for statistical measure of relatedness. Molecular systematics which compares widely distributed molecules with conserved functions revolutionizes our understanding of evolution, especially microbial evolution. All life, because it evolved from common ancestors, must at all times make and use certain long-chain DNA and protein molecules to maintain itself, to grow and reproduce. The number and order of components (amino acids in proteins and nucleotides in DNA and RNA) in these macromolecules are called their sequences. Molecular sequences, for example, of ribosomal RNA or ATPase enzymes essential for life, retain identical functions and overall structure, change only slowly through time, and so make useful standards that contribute to systematics. In our classification system we use insights from molecular systematics. But we integrate molecular data with whole organism biology to provide a classification system consistent with widely scattered information.

THE CELL AS A UNIT; THE KINGDOMS OF LIFE

From Aristotle to the middle of the twentieth century, and in many cultural traditions, members of the living world were assigned to one of two kingdoms: plant or animal. Toward the end of the nineteenth century, however, many scientists noted that certain organisms, such as bacteria and slime molds, differ from plants and animals more than plants and animals differ from each other. Third and fourth kingdoms to accommodate anomalous organisms were proposed from time to time. Ernst Haeckel (1834–1919), the German naturalist who disseminated Darwin's ideas of "descent with modification" and "natural selection," made several proposals for a third kingdom of organisms. The boundaries of Haeckel's new kingdom, Protista, fluctuated in the course of his long career, but his consistent aim was to set the most "primitive" and ambiguous organisms apart from the plants and animals, with the implication that larger organisms evolved from protist ancestors. Haeckel recognized the bacteria and "blue-green algae" (which are really bacteria in their own right, not algae) as a major group—the Monera, distinguished by their lack of a cell nucleus—within the protist kingdom. However, most biologists either ignored proposals for additional kingdoms beyond plants and animals or considered them unimportant curiosities, the special pleading of eccentrics.

The climate of opinion regarding the kingdoms of life began to change in the 1960s, largely because of the knowledge gained by new biochemical and electron-microscopic techniques. These techniques revealed fundamental affinities and differences on the subcellular level that encouraged a spate of new proposals for multiple-kingdom systems. Among these proposals, a system of five kingdoms (plants, animals, fungi, protists, and bacteria), first advanced by Robert H. Whittaker in 1959, was greatly indebted to the earlier and highly original four-kingdom (plants, animals, protoctists, and bacteria) work of Herbert Copeland. The five-kingdom scheme steadily gained support for more than three decades. With modifications necessitated by recent molecular insights we use the Whittaker system. Briefly, our five kingdoms are Prokaryotae with its two domains Archaea and Eubacteria, Protoctista (algae, protozoa, slime molds, and other less-known aquatic and symbiotic organisms), Animalia (animals that develop from embryos formed by fertilization of eggs by sperm), Fungi (mushrooms, molds, and yeasts), and Plantae (mosses, ferns, and other spore- and seed-bearing plants). For the plant kingdom, we follow James Walker of the University of Massachusetts, Amherst (personal communication). We distribute 12 plant phyla among two broad groups. The subkingdom Bryata includes the nonvascular plants (mosses, liverworts, and hornworts) and subkingdom Tracheata all others (the vascular plants). Although Walker and other botanists use "Anthocerophyta" for the nonvascular group, we simplify name, and call only the hornworts Anthocerophyta.

Our five kingdoms group into two superkingdoms: (1) Prokarya, containing the sole prokaryote kingdom, Bacteria, and (2) Eukarya, containing the symbiogenetically generated four kingdoms, all the eukaryotes. Sociopolitical terms such as "kingdom," "domain", "class", and "order" are anachronisms that we hope eventually will be replaced. Yet their current widespread acceptance makes convenient their continued use.

The most serious challenge to five-kingdom schemes is the three-domain system of the microbiologists led by Carl Woese of the University of Illinois. By use of molecular criteria, especially small ribosomal RNA nucleotide sequences, microbiologists argue for three major groups: two domains (Archaea and Bacteria) of organisms that consist of prokaryotic cells and one domain (Eukarya) that contains all other organisms (Figure I-1). Fungi, plants, and animals are three of the kingdoms of the Eukarya

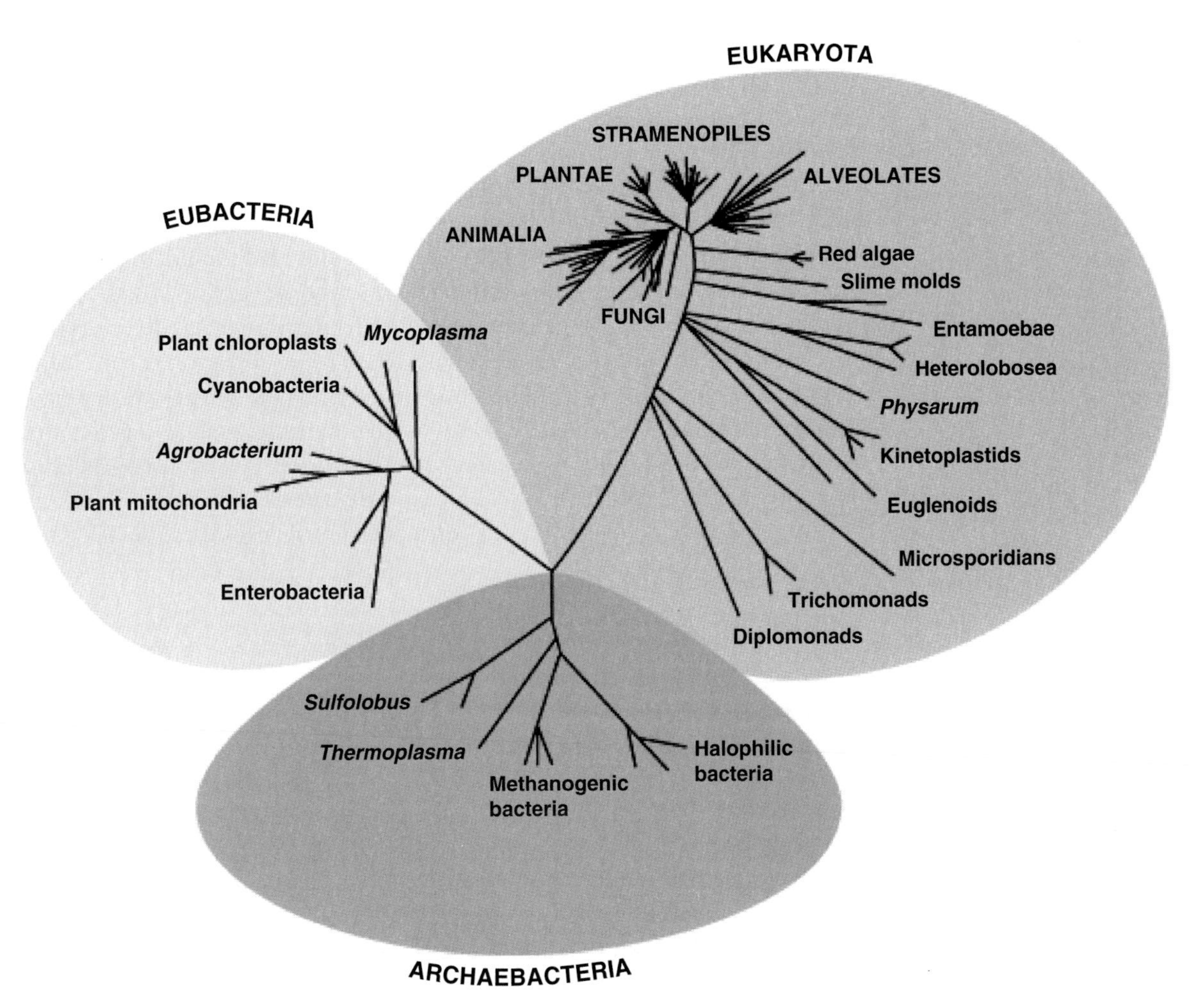

Figure I-1 Relations between eukaryotic higher taxa based on a single important criterion: nucleotide sequences in the genes for small-subunit ribosomal RNAs. The lengths of the lines are proportional to the number of differences in the nucleotide sequences. The "crown group" (Fungi, Animalia, Plantae, Stramenopiles) is envisioned to be those more recently evolved eukaryotes most closely related to large organisms. The main difference between this scheme, based solely on molecular biology criteria, and ours is that we try to take into account all the biology of the living organisms. This single measure, useful to compare all extant life, was developed by George Fox and Carl Woese (1977). Since then human awareness of the importance, diversity, and vastness of the distribution of prokaryotes has developed everywhere. We have begun to understand how profound is our ignorance to the prokaryotic world that sustains us. [Modified from G. Hinkle and M. Sogin, unpublished, with permission.]

domain, as they are in our five-kingdom scheme. However, within each of the three domains are numerous additional kingdoms—many that correspond to our Protoctista phyla in the five kingdoms.

Although we are deeply indebted to Carl Woese (University of Illinois), Mitchell Sogin (Marine Biological Laboratory at Woods Hole), and other molecular sequence analysts for their unprecedented contributions to the reorganization of the living world, we reject the bacteriocentric three-domain scheme on biological, evolutionary, and pedagogical grounds. Biologically, this trifurcation fails to recognize symbiogenesis, fusion of former bacteria to generate new individuals, as the major source of innovation in the evolution of eukaryotes. Furthermore, its three domains and multiple kingdoms are established solely by the criteria of molecular sequence comparisons, whereas each kingdom in our five-kingdom scheme is uniquely defined by the use of myriad semes of whole organisms. Molecular, morphological, developmental, metabolic, and other criteria are used. Pedagogically, proliferation of so many kingdoms in the three-domain system defeats the purpose of manageable classification. The diversity of our planetmates requires a classification system from which information is retrieved by teachers, naturalists, and other nonspecialists. For these reasons, although we make extensive use of molecular sequence data, we reject Woese's concept that privileges these data.

PLANT OR ANIMAL? HISTORY OF THE HIGHEST TAXA

In 1927, the French marine biologist Edouard Chatton (1883–1947) wrote a paper for an obscure journal published in Sète, in southern France, that used the term *procariotique* (from Greek *pro*, meaning before, and *karyon*, meaning seed, kernel, or nucleus) to describe bacteria and "blue-green algae," organisms that lack a nucleus. The term *eucariotique* (from Greek *eu*, meaning true) described nucleated animal and plant cells. The distinction between Prokarya and Eukarya had already been clearly recognized and explained by the Russian botanist from Voronezh, Boris Mikhailovich Kozo-Polyansky in 1924. Alas, no one in the west read Russian. In the past four decades, the insight of Kozo-Polyansky and Chatton into the nature of life was abundantly verified. Virtually all biologists now agree that this basic divergence in cellular structure, which separates the bacteria and the "blue-green algae" from all other cellular organisms, probably represents the greatest single evolutionary discontinuity to be found in the present-day world.[1]

This distinction we retain at the superkingdom level: Superkingdom Prokarya contains prokaryotes and only prokaryotes, the great diversity of organisms with bacterial cell organization. Members of the other four

[1] Stanier, R. Y., Adelberg, E. A., and Doudoroff, M. *The microbial world*, 3d ed. Prentice-Hall, Englewood Cliffs, NJ; 1963.

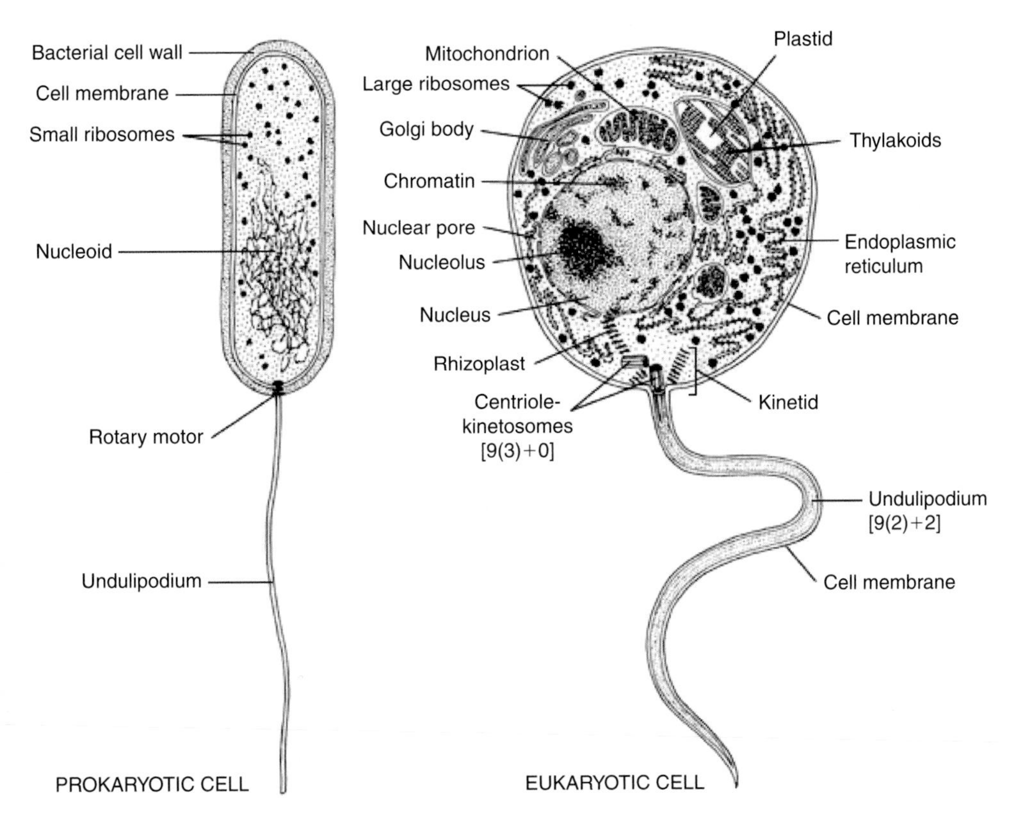

Figure I-2 Typical organism cells, based on electron microscopy. Not all prokaryotic or eukaryotic organisms have every feature shown here. Note that these cells are not drawn to scale; the eukaryote should be two to ten times larger in diameter than the prokaryote. "[9(3)+0]" and "[9(2)+2]" refer to the microtubule arrangement in cross section of kinetosomes and undulipodia, respectively (Figure I-3). [Drawing by K. Delisle.]

kingdoms are classified in superkingdom Eukarya. They are all eukaryotes that evolved symbiogenetically just as Kozo-Polyansky (1890–1957) insisted in his prescient book *Symbiogenesis: A new principle of evolution* written in 1924 when he was in his early thirties.

Both in structure and in biochemistry, eukaryotes and prokaryotes differ by far more than the presence or absence of a membrane-bounded nucleus (Figure I-2; Table I-2 summarizes their major differences). By comparison of illustrations of bacteria with those of protoctists, one notes that prokaryotic cells are usually simpler in structure (but not necessarily in chemistry) and smaller than eukaryotic cells. The distinction between prokaryote and eukaryote is immediate and definitive from electron micrographs. Multiple structures occur within eukaryotic cells. Any visible structure inside a cell is an organelle. Prokaryotes have simple organelles (carboxysomes or gas vacuoles, for example), but eukaryotic cells contain

Table I-2 Comparisons of Prokaryotes and Eukaryotes

Prokaryotes	Eukaryotes
DNA (the genophore*) seen in electron micrographs as nucleoid (not membrane bounded). Genophores, plasmids, not chromosomes. DNA not coated with protein.	Membrane-bounded nucleus containing chromosomes made of DNA and proteins.
Evolved by duplication and mutation of DNA.	Evolved by duplication and mutation of DNA and by symbiogenesis: permanent associations between at least two different kinds of prokaryotes.
Cell division direct, by binary or multiple fission or budding. Centriole-kinetosomes, mitotic spindle, microtubules lacking.	Cell division by various forms of mitosis; formation of mitotic spindles (or at least some arrangement of microtubules).
In sexual recombination, genetic material (as plasmid, virus, genophore, or other replicon) is transferred from donor to recipient. Reversible transfer of genes (DNA) without any cytoplasmic fusion.	Sexual systems in which two partners (often male and female) participate in fertilization. Alternation of diploid and haploid forms by fertilization and meiosis. Reversible formation of hybrid (fused) nuclei (karyogamy) at fertilization by mating. Gamontogamy,† syngamy,‡ conjugation, etc. usually including cytoplasmic fusion, called cytogamy.
All lack tissue development.	Some have extensive development of tissues and organs.
Strict anaerobes (which are killed by oxygen), facultative anaerobes, microaerophiles, aerotolerant, and aerobic organisms.	Almost all are aerobes, needing oxygen to live; exceptions are either archaeprotists (Phylum Pr-1) or organisms evolved from aerobes.
Highly diverse modes of metabolism: vary in sources of energy (light, organic or inorganic molecules), sources of electrons (organic or inorganic molecules), and sources of carbon (organic or CO, CO_2, CH_4).	Same metabolic patterns of oxidation within the group (Embden–Meyerhof glucose metabolism, Krebs-cycle oxidations, cytochrome-electron transport chains). They are either organoheterochemotrophs§ (most) or photolithoautotrophs¶ (most plants and algae).
Bacterial flagella, composed of flagellin protein. Flagella rotate.	Complex [9(2)+2] undulipodia composed of tubulin and hundreds of other proteins (Figure I-3).
Most are small cells (1–10 μm). All are microbes; many, if not most, are multicellular (colonial) in nature.	Most are large cells (10–100 μm) or composed of large cells. Some are microbes; most are large multicellular organisms or colonial.
Mitochondria absent; cofactors and enzymes for oxidation of organic molecules, if present, are bound to cell membranes (not packaged separately).	Enzymes for oxidation of three-carbon organic acids are packaged in mitochondria [except for archaeprotists (Phylum Pr-1)].
In photosynthetic species, enzymes for photosynthesis are bound as chromatophores to cell membranes, not packaged separately. Various patterns of anaerobic and aerobic photosynthesis, including the formation of end products such as sulfur, sulfate, and oxygen.	In photosynthetic species, enzymes for photosynthesis are packaged in membrane-bounded plastids. All photosynthesizers produce oxygen.

*"Bacterial chromosome," a term used by molecular biologists to refer to the genophore, is confusing and its use should be avoided.

†Gamontogamy: in protoctists, such as foraminiferans, fusion of two or more gamonts (reproducing cells or organisms) followed by gametogamy (fusion of gametes). Examples: Copulation, conjugation.

‡Syngamy: process by which two haploid cells fuse to form a diploid zygote; fertilization.

§Organoheterochemotroph: refers to metabolic mode in which organism uses organic compounds (from other organisms, living or dead) as sources of energy, carbon, and electrons (for example, *Escherichia coli*).

¶Photolithoautotroph: refers to metabolic mode in which organism uses light as a source of energy and uses inorganic compounds (such as CO_2 and H_2S) as sources of carbon and electrons (for example, *Chlorobium*).

unique heritable organelles, some of them separated by their own membranes from the cytoplasm (Figure I-2). Mitochondria (singular: mitochondrion), ovoid organelles that specialize in production of energy by enzymatic oxidation of simple organic compounds, are found in nearly all eukaryotes. Prokaryotes lack mitochondria, but enzymes bound in their membranes may catalyze equivalent oxidations.

Green plant cells and algal cells contain one or several plastids—membrane-enclosed organelles in which complex structures made of membranes, chlorophyll, and other biochemicals photosynthesize. In photosynthetic bacteria, however, chlorophyll and other photosynthetic components are seen as granules in and on membranes rather than discrete organelles.

UNDULIPODIA, CENTRIOLES, AND KINETOSOMES

Many eukaryotic cells—plant sperm, those of most protoctists and animal sperm—at some stage in their life history have flexible, long, protruding organelles called undulipodia (singular: undulipodium) misnamed in much of the published literature as "flagella". All cilia, eukaryotic flagella, and most sperm tails are examples of undulipodia. Undulipodia are composed of thin cytoplasmic tubules called microtubules. In undulipodia the microtubules are arranged in a canonical array. In transverse (cross) section, they show a characteristic ninefold symmetry. The undulipodium is enclosed in a membrane that is simply an extension of the cell membrane. The ninefold symmetry characteristic of undulipodia is also found in the centriole kinetosomes (often called eukaryotic basal bodies) from which the shafts (axonemes) grow. The centrioles by themselves are small barrel-shaped bodies that, in many cells, appear at the poles of the spindle during mitosis. Undulipodia are composed of more than 400 different proteins. The main type of protein which makes up the microtubules is called microtubule protein, or tubulin.

Swimming prokaryotic cells bear long thin moveable extensions, called flagella (singular: flagellum; Figure I-2). In much of the published literature, undulipodia are called "flagella;" however, these extensions are not undulipodia. Flagella are far smaller, and they are not composed of microtubules. Neither are flagella covered by membrane, rather, their naked shaft is extracellular. It protrudes through the cell membrane and is composed of a single globular protein that belongs to a class of proteins called flagellins. The beating motion of an undulipodium is caused by the transformation of chemical into mechanical energy by motor proteins arrayed along the full length of the microtubules in this organelle. The flagellum's motion results from the rotation at its rotary motor, a basal attachment that is embedded in the cell membrane.

SEX AND REPRODUCTION

Reproduction is the increase in number of cells or organisms, whether unicellular or multicellular. Growth is increase in size. All species of organisms grow and reproduce, although the details of how they do it vary. Even though fusion of parental gametes accompanies reproduction in humans and in the animals we best know, biologically, sex is entirely distinguishable from reproduction. Sex is defined as the formation of an organism whose genes come from more than a single individual. Sex, the recombining of genes from two or more individuals, does occur in prokaryotes, but prokaryotic sex is not directly required for reproduction.

Prokaryotic cells do not open their membranes and fuse their contents. Rather, genes from the fluid medium, from other prokaryotes, from viruses, or from elsewhere unidirectionally enter prokaryotic cells. A prokaryote that carries some of its original genes and some new genes is called a recombinant. This propensity for gene uptake, along with the lack of a nucleus and the other features listed in Table I-2, defines one of the two highest taxa, or superkingdoms: Prokarya, organisms composed of bacterial cells. All other organisms are Eukarya, organisms composed of nucleated cells, that evolved by symbiogenesis (Table I-2).

Eukaryotic cells reproduce by mitosis. They form chromosomes—tightly coiled gene packages bound together by proteins and attached to the inner membrane of the nucleus. At least two chromosomes are located in the nucleus of every eukaryotic cell; some protoctists have more than 16,000 chromosomes in a single nucleus at certain stages. Although all cells and species of organisms made of cells must either reproduce or die, the way that eukaryotes make more eukaryotic cells or organisms made of cells is highly peculiar to each of the eukaryotic kingdoms and forms the basis of our classification system.

KINGDOMS AND DOMAIN CRITERIA

All animals reproduce by fertilization of an egg by a smaller sperm to form the fertile egg—zygote—that divides by mitosis to make a rudimentary embryo. The first embryonic stage is called the blastula (Figure 3-1).

Plants form spores that, by themselves, grow into one of two kinds of gametophyte (a plant that forms gametes). Either they are male and form sperm (or pollen) or they are female and form the female gametophyte that develops the egg. The egg remains on the mother plant and is fertilized by fusion with sperm nuclei. The fertilized egg—with a chromosome set from each parent—then develops into a plant embryo, a young multicellular stage common to all plant groups. The embryo stays in the mother's (maternal) tissue at first. Eventually it grows into the adult plant, capable again of making spores with only one set of chromosomes (Figure 5-3).

Fungi reproduce by means of fungal spores, propagules that are capable of generation—from a single parent—the entire fungus again. Some fungi enter their sexual phase only when the environment no longer favors uniparental reproduction. At such times, genetically novel spores, but no embryos, are produced. Fungal spores are usually more resistant to water loss, starvation, and other adverse conditions than is the growing fungus. Fungi lack undulipodia at all stages in their life history.

Protoctists display a huge range of variation in life history features—but none fits the description of animal, plant, or fungus. The protoctist kingdom includes the microbial (few- or single-celled) eukaryotes and their immediate multicellular descendants. Because protoctists are grouped together as the microbial symbiotic complexes from which animals, plants, and fungi were removed, it is not too surprising that their life stories are extraordinarily varied. "Protist" refers to the smaller protoctists, but some people use the term for all of them. Some 250,000 species are estimated to exist now.

The differences between members of the kingdoms are further explained in the opening sections of each of the five chapters. They are summarized here in Table I-3.

VIRUSES

Antony van Leeuwenhoek (1632–1723), the discoverer of the microbial world, called the microorganisms that he found everywhere in vast numbers "very many little animalcules." For more than a century after his discoveries, it was commonly held that these little animals arose spontaneously from inanimate matter. The chemist Louis Pasteur (1822–1895) and the physicist John Tyndall (1820–1893) showed conclusively that, like large organisms, microbes are produced only by other microbes. Microbes are simply life-forms best visualized and understood when seen with a microscope. Whether bacteria, tiny protoctists, or fungi, microbes are either cells or composed of cells, or they are viruses.

All organisms classified in one of the five kingdoms (Prokaryotae, Protoctista, Animalia, Fungi or Plantae) either are cells or are composed of cells. The arguably living forms that do not fit this description are viruses (Figure I-3). Composed of DNA (deoxyribonucleic acid) or RNA (ribonucleic acid) but not both, viruses lack membrane-bounded cell organization; most are enclosed in a protein coat. Viruses are much smaller than

Table I-3 Taxonomic Summary*

Nonsymbiogenetic origin Superkingdom PROKARYA (PROKARYOTAE)

Nonnucleated (prokaryotic) cells. Chromonemal genetic organization ultrastructurally visible as nucleoids. Cell-to-cell transfer of genophores, that is, of the chromoneme (large replicons) and of plasmids (and other small replicons). Ether- (isoprenoid-derivative) or ester-linked membrane lipids, without steroids, cytoplasmic fusion absent. Flagellar rotary motor motility. Concept of ploidy inapplicable. Photo-, chemoautotrophs, vast diversity of metabolic modes.

Kingdom BACTERIA (PROKARYOTAE, PROCARYOTAE, MONERA)

Bacterial cell organization.

DOMAIN ARCHAEA

Methanogens, thermoacidophiles, halophiles, and probably some Gram-positive bacteria.

DOMAIN EUBACTERIA

Gram-negative and most other bacteria. Variable metabolic modes. Two surrounding lipoprotein membranes define the periplasmic space.

Symbiogenetic origin Superkingdom EUKARYA (EUKARYOTAE)

Nucleated (eukaryotic) cells, all evolved from integrated bacterial symbioses. Membrane-bounded hereditary organelles. Chromosomal genetic organization. Intracellular, microfilament- and microtubule-based motility (actin, myosin, tubulin–dynein–kinesin). Microtubule organizing centers. Whole-cell and nuclear fusion (karyogamy). Flexible steroid-containing (for example, cholesterol, cycloartenol, and ergosterol) membranes. Meiosis and fertilization cycles underlie Mendelian genetic systems. Levels of ploidy vary.

Kingdom PROTOCTISTA (Hogg, 1860)

Mitotic organisms capable of internal cell motility (that is, cyclosis, phagocytosis, pinocytosis). Many motile by undulipodia. Binary or multiple fusion. Meiosis and fertilization cycles absent or details unique to phylum. Photoautotrophs, ingestive and absorptive heterotrophs.

Kingdom ANIMALIA

Embryo called a blastula (diploid) formed after fertilization of egg by sperm (fusion of haploid anisogametes—karyogamic cells that differ in size). Females deliver mitochondria to the zygote in cytogamy. Meiosis produces gametes. Diploids. Most are ingestive heterotrophs; some are absorptive heterotrophs.

Kingdom FUNGI

Hyphal or cell fusion. Zygotic meiosis to form resistant propagules (spores). Lack undulipodia at all stages. Haploids. Absorptive heterotrophs.

Kingdom PLANTAE

Maternally retained diploid embryo formed from fusion of mitotically produced gamete nuclei. Sporogenic meiosis produces male (antheridium; sperm-producing haploid plant organ) or female (archegonium; egg-producing haploid plant organ). Gametes formed in antheridium and archegonium and fertilized in archegonium. Alternating generations of haploid and diploid organisms. Most are oxygenic photoautotrophs.

*For the major higher taxa: Prokarya (Archaebacteria, Eubacteria); Eukarya (Protoctista, Fungi, Animalia, and Plantae), brief technical descriptions accompany the introduction of each of their sections.

Box I-ii Life is growth

What is the difference between a live bull one minute, and the same bull lying dead in the ring in the second minute? No doubt the composition does not change for the first few minutes after death: the cells with their DNA, various RNAs, ribosomes, proteins, and lipid membranes are still intact. But not for long: the difference is that energy flow and matter transformation are stopping, identity will soon be lost, and all the processes that build it up will soon cease.

Little progress has been made on the question of "origins of life" since the many famed experiments begun by Miller and Urey that produced organic compounds from simple gases thought to be present on the Archaean Earth. Why, if organic compounds are made with ease in the laboratory from mixtures of common gases, have we seen so little progress in "origin-of-life" experiments? Our colleague William Day (2002) claims that the "origin-of-life" problem ought not be more difficult than other experimental science. But since members of the scientific community, especially biologists, do not understand what life *is*, we cannot explain its origin. Day posits that life is a *growth system* that intrinsically involves chemical change (metabolism) and energy flow. Without the incessant flow, cells and bodies made of cells cannot sustain themselves. This flow of energy and matter (including water), and growth of necessity, has not wavered since the earliest life on Earth. Living matter has always been a cyclical process of growth, since life began. In this way, living matter differs from both dead matter and chemical systems that have never been alive.

Scientists have assumed that life's properties result from the precise composition and organization of its components. We have hypothesized that life originates from prebiotic substances that must be assembled into bodies of sufficient complexity to function. This is the "cake mixture" idea: if we assemble the correct components in the right proportions and supply appropriate energy or "spark" the mixture, life will appear. But this concept is fatally flawed. No single cell, no living body is ever static; all dynamically regenerate themselves by breakdown of components with lytic enzymes and resynthesis of new parts, as shown by Schoenheimer and colleagues (1938, 1942). Biological turnover is not a chemical equilibrium, rather, it is a steady state of metabolic reactions that continually transform matter and chemical energy: biosynthesis of worn-out membranes and cytoplasm, physical work, thermoregulation, digestion and excretion all consume some sort of fuel, without which organisms die. Autopoietic (self-maintenance) cycles cease, and one-way degradation reactions continue until thermodynamically unstable structures collapse and life is irreversibly lost.

Mere composition, then, does not make a living system. Life is a matter and energy flow system of sustained growth, and still requires autopoiesis following containment of its continuous growth in mature organisms. Like a waterfall, life is a process that starts at the beginning and continues as it forms; fundamentally a *growth process*. The essence of living cells and multicellular organisms is inseparable from the energy and material flow that drives it. As a process, life cannot be created from chemicals *in toto*, but rather it formed and grew from the beginning in historical order.

Growth is life's principle. Conditions intrinsic to growth have prevailed since the start of Earth's living system. Constituents (by-products of synthesis and growth) and reproduction (a repetition of the growth process) require incessant influx of energy. External energy sources, either light or oxidation of biochemical fuels, are only temporarily dispensable to life in propagules (for example, spores, cysts, seeds). These bits of living matter have evolved to tolerate near-total desiccation or absolute zero temperatures. Return to permissive environmental conditions always requires energy and material flow to restart life's cyclical growth process.

Problems associated with the "origins-of-life experiments" may be resolved with adherence to Day's concept. Metabolism, recombination, reproduction, and evolution are aspects of a continuous repetitive growth process. No preexisting store of matter or information accumulated prior to life's creation, nor was one necessary. As life emerged as a process of matter and energy flow, it grew step by step and accumulated ordered information bit by bit. Prior to evolution of today's complex genetic-protein synthetic system, compositional information lay in the structures of the metabolites and the sequential order of their reactions. The indispensable requirement for the origin and continuity of life was energy flow to sustain growth.

Growth followed a reproducible course: each energy-transforming constituent was synthesized *in situ* at the right time. Modern cells share common ancestry and common biochemical history: the growth process was honed by natural selection, the winnowing process intrinsic to evolution. If, then, life never simply assembled from parts, the "origin-of-life" problem narrows to a straightforward question: "How did life, with no preexisting mechanism, start a lipid-membrane-bounded energy–material flow system that sustained the growth process?" Many attempts have been made to address this question (Deamer and Fleischaker, 1994).

Carbon dioxide, nitrogen, and hydrogen sulfide (H_2S), as well as phosphoric acid (H_3PO_4), present in the early Archaean eon (c. 3500 mya), in our view, were reduced as hydrogen sulfide (H_2S) was oxidized in the formation of certain organic compounds. This oxidation/reduction reaction occurs spontaneously in sunlight, which energized a flow of electrons between the reactants. Life began, we hypothesize, with a simple form of what became the active site of the small iron–sulfur (Fe–S) complex of today's electron carrier protein, ferredoxin. In the formative stage of life, a primordial forerunner of ferredoxin, probably a mineral cluster, transduced sunlight directly into electromotive energy. After abiotic membrane-lipids formed and closed, the concentration of light-capturing and energy-transducing molecules within cells permitted evolution of an extended, more elaborate photosynthetic apparatus. In the laboratory, illuminated FeS_2 particles suspended in solution with an electron donor and carbon dioxide behave like photoelectric cells. When this or similar source of photochemical energy in membrane-bounded cells converted carbon dioxide and nitrogen to organic compounds, the biological revolution began.

cells. Although viruses replicate, they can do so only by entering a cell and using its living metabolism for their perpetuation. Outside cells, viruses cannot replicate, feed, or grow. Some viruses can even be crystallized, like minerals. In this state, viruses can survive for years unchanged—until they contact the specific living tissues they require for their perpetuation.

Viruses are probably more closely related to the cells in which they replicate than to one another. They may have originated as replicating nucleic acids that escaped from cells—they must always return to living tissue, to actively metabolizing cells, to use the complex chemicals and structures they require for replication. Thus, the polio and flu viruses are probably more closely related to people, and the tobacco mosaic virus (TMV) to tobacco, than polio virus and TMV are to each other. Although not cellular, viruses have been essential to the flow of genes through the biosphere.

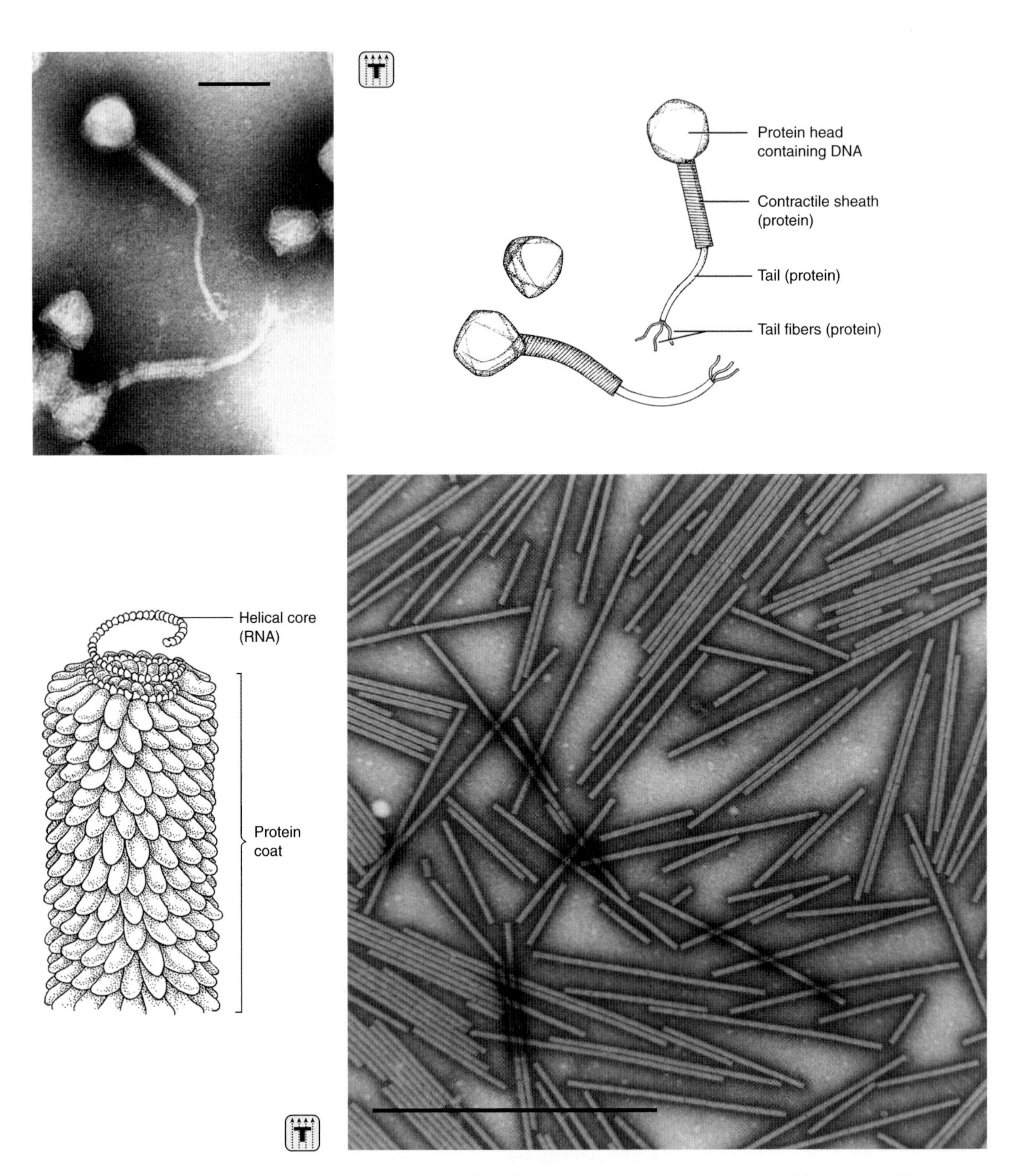

Figure I-3 (Top) A DNA virus, *Botulinum* ϕ, which attacks *Clostridium botulinum*; TEM, bar = 0.1 μm. (Bottom) An RNA virus, TMV, which causes a blight of tobacco plants; TEM, bar = 1 μm. [Photographs courtesy of E. Boatman; botulinum drawing by R. Golder: TMV drawing by M. Lowe.]

They are crucial to the maintenance of cyclical symbioses, those that predictably reform and dissolve as the environment changes.

THE ENVIRONMENT

EARTH HISTORY: THE GEOLOGIC RECORD

Planet Earth is almost 5000 million years old. The oldest fossils yet discovered, bacterium-like filaments from rocks of the Pilbara gold fields near North Pole in Western Australia, are at least 3500 million years old. Bacterium-like spheroids also have been found in rocks of that age from the Swaziland rock system of southern Africa. Thus, for most of its history, Earth has supported life. Almost everything that we know about the history of the planet and of life comes from evidence in the rock record. Geologists have made a chronology of this history from the composition of rocks, the order of their formation, and the fossils in them (Figure I-4). The upper layers of an undisturbed sequence of sediments are younger than the lower ones. This geological rule has been called the "law of superposition." Rock layers at different places on Earth's surface are matched by examining the fossils in them, whereas absolute ages are determined by radioactive-dating methods.

The longest divisions of geological time are called eons. The sequence of rocks in the latest, the Phanerozoic, is known in such detail that this eon is divided into eras, eras into periods, and the periods into epochs (not shown). Of the other eons, only the Proterozoic is known well enough to allow generally accepted subdivisions (eras, epochs). The subdivisions of the Phanerozoic are so well known because of the worldwide abundance of its fossils. The fossil record of eukaryotic organisms is Phanerozoic, except for microfossils of protoctist cysts and many Proterozoic trails, burrows, and body fossils of unknown origin. Evidence of pre-Phanerozoic eukaryotes ("Ediacaran biota") is known from more than 20 localities. Often spherical, microfossils of unidentified eukaryotes are called "acritarchs."

The most abundant fossils from before the Phanerozoic eon are rocks called stromatolites. A typical stromatolite is a column or dome of rock a few centimeters wide made of horizontal layers. The layers are apparently the remains of sediment trapped or precipitated and bound by growing communities of bacteria, primarily cyanobacteria. Fossil stromatolites often extend over hundreds of meters laterally and several meters in height. Comparable bacterial communities exist today, but in only a few isolated, extreme environments, such as salt ponds, do they lithify to become stromatolites. During the Archaean and Proterozoic eons, more than half of

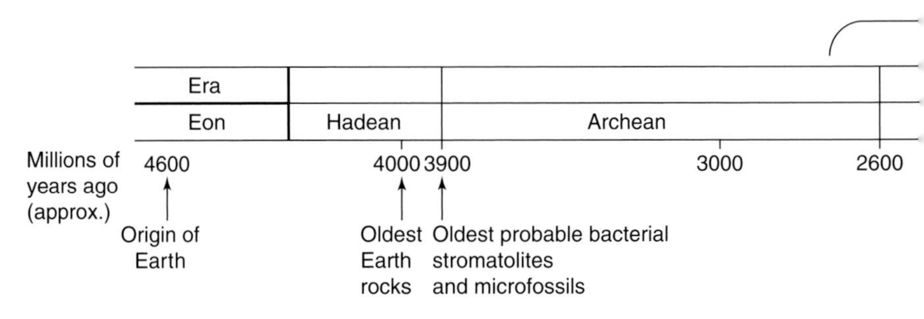

Figure I-4 Time line of Earth history. Eons (time-rock divisions) in which unambiguous fossils first appear: bacteria—early Archaean; Protoctista—middle Proterozoic; animals—late Proterozoic [Ediacaran (Vendian) era]; plants and fungi—early Phanerozoic (Paleozoic era, Silurian period). See for time-rock units on the standard international stratigraphic column.

Earth's existence, the planet was the uncontested territory of kingdom Bacteria. The stromatolite communities recorded to modern places of refuge only after the rise of protoctists and animals. Some of the new organisms must have grazed voraciously in the lush bacterial pastures.

SEVEN "ECOSTRIPS:" ECOSYSTEMS AND THEIR HABITATS

At the top of the right-hand page of each phylum essay is a scene with one or more arrows pointing to the typical habitats of the members of the phylum. Seven different environments and their habitats: temperate seashore—rocky, sandy, and muddy; temperate forests, lakes, and rivers; deserts and high mountains; tropical forest; tropical seas, reefs, continental shelves and slopes, and seashore; tectonically active anoxic environment; and the abyss at a tectonically active ocean rift zone are shown in Figure I-5. Tectonically (Greek *tekton*, carpenter) active pertains to changes in the structure of Earth's crust. The tropical seashore, which is not shown as a separate environment, will be indicated by arrows pointing toward the top

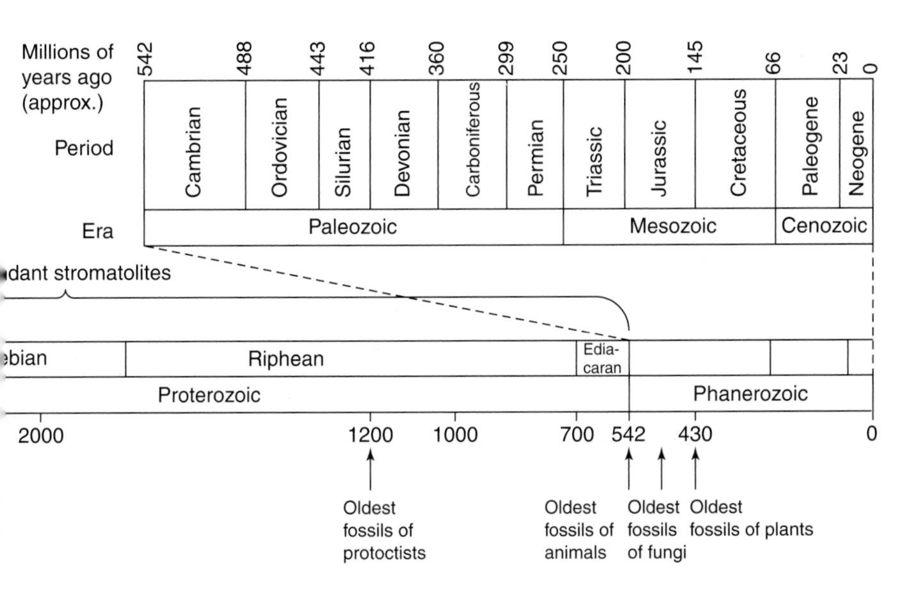

of the tropical seas scene when appropriate. Our depictions are certainly not complete, because members of a phylum may be found in habitats not illustrated, yet our scenes include habitats in which life frequently and abundantly abides.

EVOLUTION AND TAXA

We recognize and describe two superkingdoms: Prokarya (all bacteria) and Eukarya (all nucleated organisms). In these two highest, most inclusive taxa are the kingdoms. Superkingdom Prokarya contains kingdom Prokaryotae, equivalent to Bacteria, Procaryotae, or Monera in other schemes. Its two domains (≡subkingdoms), Archaea and Eubacteria, we infer evolved prior to the eukaryotes. The subkingdom Archaea contains two phyla: B-1, the Euryarchaeota, or the methanogens and extreme halophils, and B-2, the Crenarchaeota, primarily extreme thermophils. The subkingdom Eubacteria contains 12 phyla.

Products of symbiogenesis and prokaryotes, the Eukarya kingdoms (Protoctista, Animalia, Fungi, and Plantae) comprise a total of 91 phyla.

Temperate seashore – rocky, sandy, and muddy; sunlit surface

Temperate forests, lakes, and rivers; sunlit surface

Deserts and high mountains; sunlit surface

Tropical forest; sunlit canopy

Tropical seas, reefs, continental shelves and slopes, and seashore; sunlit to depths of 200 or fewer meters

Tectonically active anoxic environment; sunlit surface

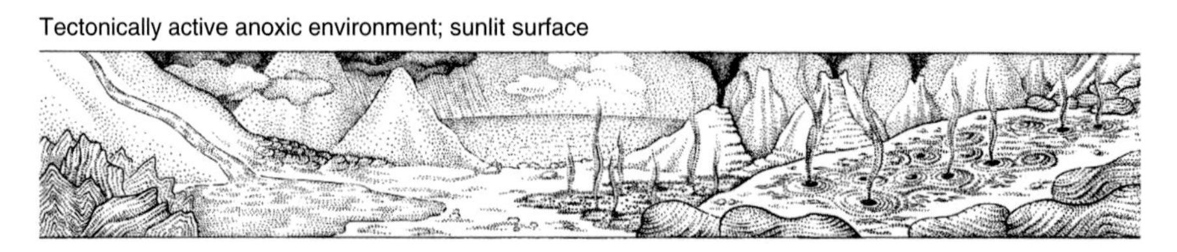

Abyss at tectonically active ocean rift zone; sunlight absent

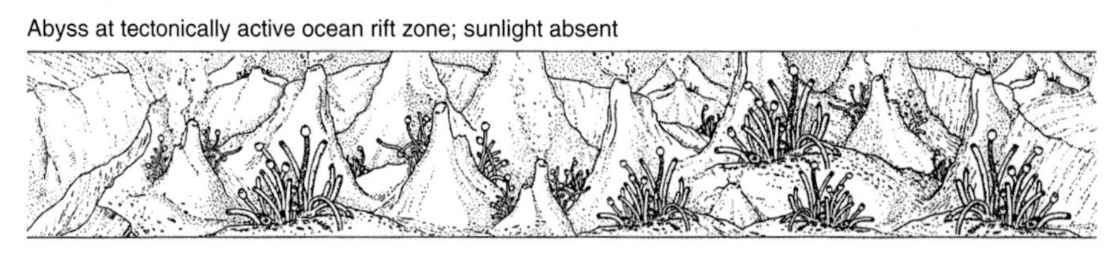

Figure I-5 Environments: the seven scenes used to designate typical habitats.

We now recognize 36 protoctist phyla, 37 animal phyla, 6 fungal phyla, and 12 plant phyla. Within each Eukarya kingdom, the phyla are arranged in an approximate order from the simplest (and presumably earliest) to have evolved to the more complex, presumably more recent forms. The highest (most inclusive taxa) within a kingdom (phyla and class) represent the most ancient evolutionary divergences; the lowest, least inclusive taxa (genera and species), represent the most recent. However, because the evolutionary relationships of many groups are unknown, this is not an absolute rule. Organisms are grouped in the same taxon for now only because they have clearly distinguishable semes (traits in common; for example, the rays of the actinopods), whether or not their common ancestry has been documented.

We begin each chapter with a phylogenetic diagram that shows the likely evolutionary relationships among the phyla of that kingdom. Each phylogeny is a branching structure evolving through time from a single ancestral group to the extant phyla. A time line along one axis shows relevant geological eras. Solid lines indicate accepted lineages; dashed lines are lineages that are provisional. Each extant phylum is illustrated with a thumbnail sketch of a member organism.

The introduction to each chapter defines and describes the evolutionary features of the entire kingdom. It is followed by essays, each describing one phylum in that kingdom. Phyla differ enormously in size. Some have only a single species; others have millions. Each phylum description begins with a list of examples of its genera, not all of which are mentioned in the text. We have been selective: some genera are significant research models, some have vast economic importance, and some are new genera. The lists give experienced students a firm idea of each phylum and offer clues to further reading.

ENVIRONMENTAL EVOLUTION AND GAIA THEORY

Signposts of *Kingdoms and Domains*

In this book unfamiliar terms are defined in the General Glossary beginning on page 001; unfamiliar life forms are listed in the Organism Glossary beginning on page 002.

A blue strip studded with tiny drawings of representatives runs across the top of the text on phylum description pages. Distinctive organisms for each indicate the kingdom to which the phylum belongs: budding bacteria for Prokaryotae, *Mastigamoebae* for Protoctista, salamanders for Animalia, mushrooms for Fungi, and tulip tree leaves for Plantae.

The text is accompanied by photographs and drawings of representative species; the species illustrated are among those listed in the Table of Contents

COLOPHON	 (eye)	 (hand lens)	 (microscope)	 S	 T
OPTICAL EQUIPMENT	Unaided eye	Hand lens	Light microscope	Scanning electron microscope	Transmission electron microscope
(Approximate) SIZE OF SUBJECT IN METERS	10^{-3}–10^{-1}	10^{-4}–10^{-2}	10^{-6}–10^{-4}	10^{-8}–10^{-2}	10^{-9}–10^{-5}

Figure I-6 Key to photograph colophons.

(pp. vii–x). For most phyla, the main representative photograph is interpreted in the two-page spread that contains a labeled anatomical drawing.

Most organisms were photographed alive. The chief exceptions are those visualized by electron microscopy. Transmission electron microscopy requires dead samples that have been chemically fixed, embedded in a transparent matrix, and cut extremely thin. Even scanning electron microscopy, which generates in black and white a more three-dimensional view of the organism, usually requires treatment with deadly fixation techniques. Since samples must be viewed in a vacuum the organism by then is dead.

The legend of each states the type, if any, of microscopy used to take the photograph: LM stands for light (optical) microscopy, TEM and SEM for transmission and scanning electron microscopy, respectively. The legend also gives the organism's length represented by the scale bar in the photograph. A colophon with each photograph indicates the kind of optical equipment needed to see the subject of the photograph (Figure I-6).

CHAPTER, FIGURES, AND TABLE NUMBERS IN THIS BOOK

For easy location of all text and figures, we classify the illustration in this book by I (Introduction), B (kingdom Prokaryote, Bacteria), Pr (kingdom Protoctista), A (kingdom Animalia), F (kingdom Fungi), and Pl (kingdom Plantae). Thus a callout such as "Table Pl-3, Figure A-7 or I-10" signals the third table in Chapter 5 (Plants), the seventh figure in Chapter 3 (Animals), or the 10th figure in the Introduction, respectively. A callout such as "Phylum A-37, Table 1; Figure A refers to the first table in the 37th phylum (Craniata) in Chapter 3 (Animalia), Figure A *Ambystoma* on page 359. Term definitions, except names of organisms and groups of organisms (taxa glossary page 561) are in the general glossary that begins on page 463.

Selected recommendations for further reading appear at the end of each kingdom chapter. The Appendix includes a list of kingdoms and an alphabetical list of genera with the phylum and common names (if any).

References

Day, W., *How life began*. The Foundation for New Directions; Cambridge, MA; pp. 154–155. 2002.

Deamer, D. L., and G. Fleischaker, *Origin of life: The central concepts*. Jones and Bartlett; Sudbury, MA; 1994.

Schoenheimer, R., and D. Rittenberg, "The application of isotopes to the study of intermediary metabolism." *Science* 87:221–226; 1938.

Schoenheimer, R., *The dynamic state of body constituents*. Harvard University Press; Cambridge, MA; 1942.

Bibliography

Bengtson, S., *Early life on Earth*. Columbia University Press; New York; 1994.

Calder, N., *Timescale: An atlas of the fourth dimension*. Viking Press; New York; 1983.

Copeland, H. F., *The classification of lower organisms*. Pacific Books; Palo Alto, CA; 1956.

Corliss, J. O., "The protista kingdom and its 45 phyla." *Biosystems* 17:87–126; 1984.

DeDuve, C., "The birth of complex cells." *Scientific American* 274(4):56–57; April 1996.

Eldredge, N., *Fossils*. Abrams; New York; 1993.

Erwin, D. H., *The great Paleozoic crisis: Life and death in the Permian*. Columbia University Press; New York; 1993.

Gould, S. J., *Wonderful life: The Burgess shale and the nature of history*. Norton; New York; 1989.

Hillis, D. M., and C. Moritz, *Molecular systematics*. Sinauer; Sunderland, MA; 1990.

Hogg, J., "On the distinctions of a plant and an animal, and on a fourth kingdom of nature." *Edinburgh New Philosophical Journal* 12:216–225; 1860.

Johnson, K. R., and R. K. Stucky, *Prehistoric journey: A history of life on Earth*. Denver Museum of Natural History (Roberts Rinehart Publishers); Boulder, CO; 1995.

Lipps, J., *Fossil prokaryotes and protists*. Blackwell; New York; 1995.

Margulis, L., *Symbiosis in cell evolution*, 2d ed. W. H. Freeman and Company; New York; 1993.

Margulis, L., "Archaeal–eubacterial mergers in the origin of Eukarya." *Proceedings of the National Academy of Sciences, USA* 93:1071–1076; 1996.

Margulis, L., and D. Sagan, *Microcosmos: Four billion years of evolution from our microbial ancestors*. Simon and Schuster; New York; 1986.

Margulis, L., and D. Sagan, *Origins of sex: Three billion years of genetic recombination*. Yale University Press; New Haven and London; 1991.

Margulis, L., and D. Sagan, *What is life?* Simon and Schuster; New York; 1995.

Margulis, L., and D. Sagan, *What is sex?* Simon and Schuster; New York; 1997.

McMenamin, M. A., and D. L. S. McMenamin, *Emergence of animals*. Columbia University Press; New York; 1990.

Raff, R. A., *The shape of life: Genes, development, and the evolution of animal form*. University of Chicago Press; Chicago; 1996.

Schopf, J. W., *Major events in the history of life*. Jones and Bartlett; Sudbury, MA; 1992.

Sogin, M. L., H. G. Morrison, G. Hinkle, and J. D. Silberman, "Ancestral relationships of the major eukaryotic lineages." *Microbiologia SEM* 12:17–28; 1996.

Taylor, T. N., and E. L. Taylor, *Biology and evolution of fossil plants*. Prentice-Hall; Englewood Cliffs, NJ; 1993.

Ward, P. D., *On Methuselah's trail: Living fossils and the great extinctions*. W. H. Freeman and Company; New York; 1992.

Whittaker, R. H., "On the broad classification of organisms." *Quarterly Review of Biology* 34:210–226; 1959.
Whittington, H. B., *The Burgess shale*. Yale University Press; New Haven, CT; 1985.
Woese, C., "Microbiology in transition." *Proceedings of the National Academy of Sciences, USA* 91:1601–1603; 1994.
Woese, C. R., O. Kandler, and M. L. Wheelis, "Towards a natural system of organisms: Proposal for the domains Archaea, Bacteria and Eucarya." *Proceedings of the National Academy of Sciences, USA* 87:4576–4579; 1990.

SUPERKINGDOM **PROKARYA**

Origins not by symbiogenesis

Single membrane–bounded genetic systems all composed of prokaryotic cells that contain genophores, often visible as nucleoids by electron microscopy. Protein synthesis occurs on 16S rRNA-28S rRNA two-component small ribosomes. Only DNA-level unidirectional recombination is present. There is no cell fusion; lack of nuclear and cytoplasmic fusion (that is, fertilization) implies non-Mendelian genetics. Unidirectional gene transfer occurs by conjugation and various forms of small genome transfer (plasmid, viral, and other transduction). Microscopic observation reveals that they lack visible intracellular motility. Reproduction is by binary fission, budding, budding of filaments, fission of stalked sessile parent to produce flagellated offspring, polar (end-to-end) growth, or multiple fission. Propagules include resistant spores, motile filaments, cystic forms, and radiation resistant walled cells.

B-6A2 *Anabaena* [Courtesy of N. J. Lang.]

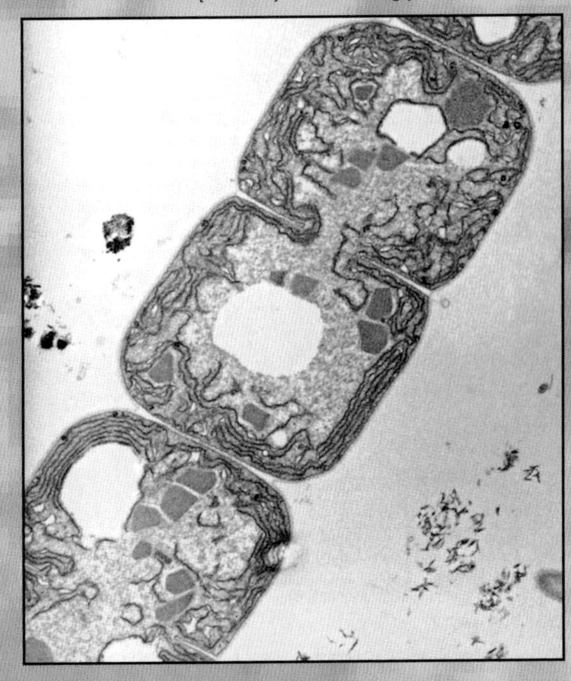

Superkingdom Prokarya
- Kingdom Bacteria
  - Subkingdom (Domain) Archaea (Archaebacteria)
  - Subkingdom (Domain) Eubacteria

CHAPTER ONE

KINGDOM PROKARYOTAE
(BACTERIA, MONERA, PROKARYA)

B-3 *E. Nitrobacter winogradskyi*. [Courtesy of S. W. Watson, *International Journal of Systematic Bacteriology* 21:261 (1971).]

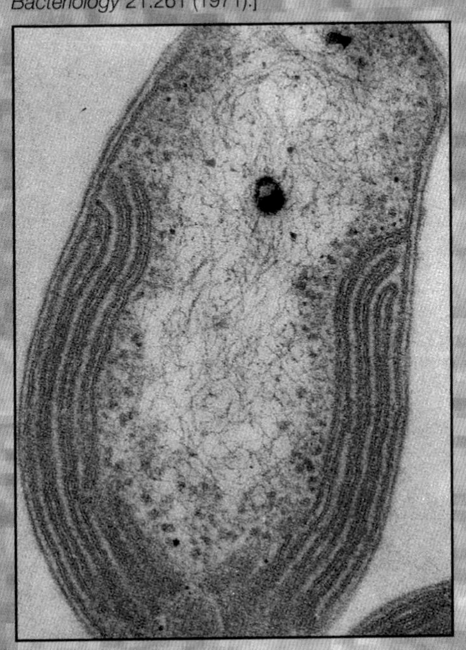

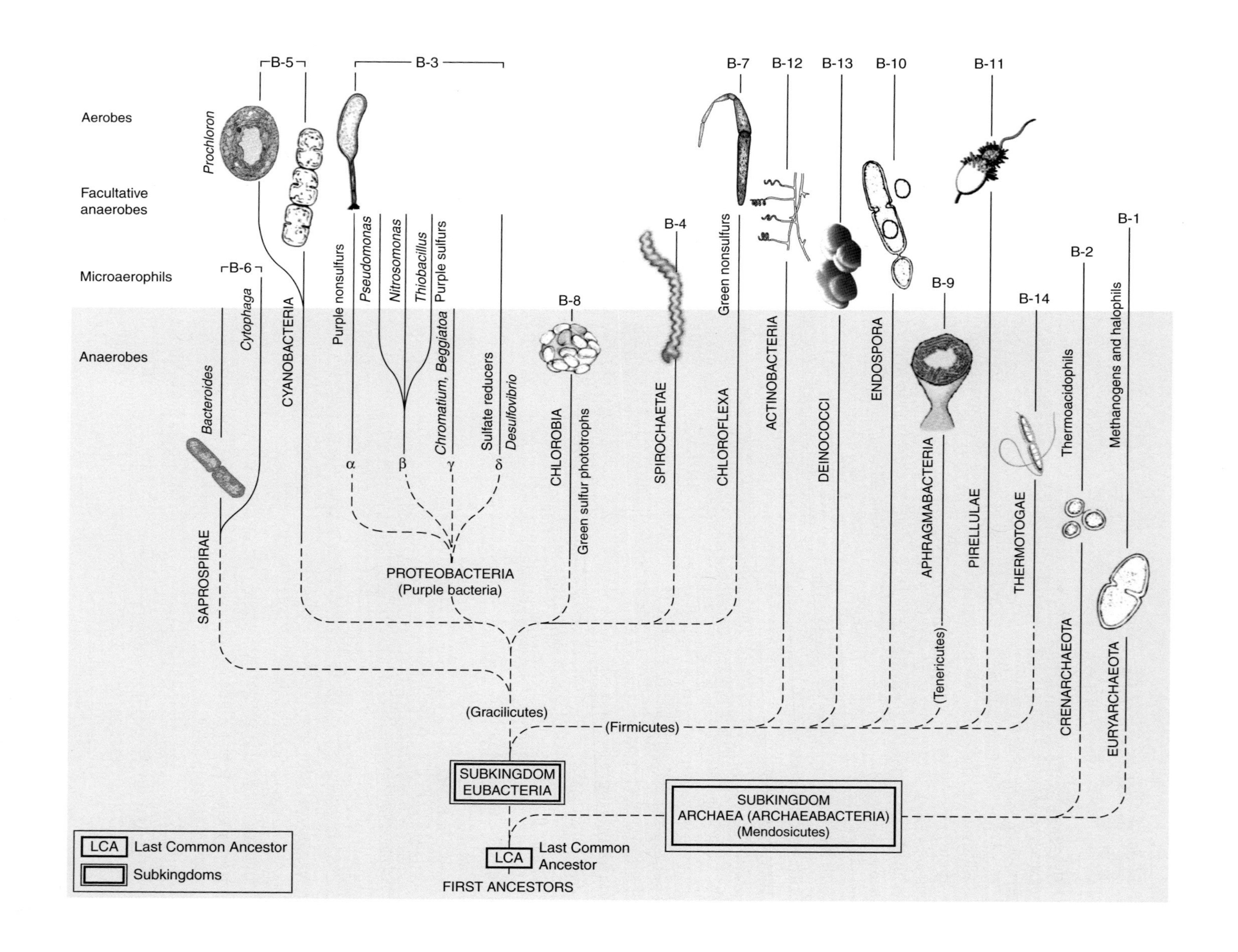

B-5
B-3
B-7
B-12
B-13
B-10
B-11
Aerobes
Facultative anaerobes
Microaerophils
Anaerobes
Prochloron
B-6
Cytophaga
Bacteroides
CYANOBACTERIA
Purple nonsulfurs
Pseudomonas
Nitrosomonas
Thiobacillus
Purple sulfurs
Chromatium, Beggiatoa
Sulfate reducers
Desulfovibrio
α
β
γ
δ
B-4
B-8
CHLOROBIA
Green sulfur phototrophs
SPIROCHAETAE
Green nonsulfurs
CHLOROFLEXA
ACTINOBACTERIA
DEINOCOCCI
ENDOSPORA
B-9
APHRAGMABACTERIA
(Tenericutes)
PIRELLULAE
B-14
THERMOTOGAE
B-2
Thermoacidophils
CRENARCHAEOTA
B-1
Methanogens and halophils
EURYARCHAEOTA
SAPROSPIRAE
PROTEOBACTERIA
(Purple bacteria)
(Gracilicutes)
(Firmicutes)
SUBKINGDOM
EUBACTERIA
SUBKINGDOM
ARCHAEA (ARCHAEABACTERIA)
(Mendosicutes)
LCA
Last Common Ancestor
FIRST ANCESTORS
LCA Last Common Ancestor
Subkingdoms

KINGDOM PROKARYOTAE (BACTERIA, MONERA, PROKARYA)

Greek *pro*, before; *karyon*, seed, kernel, nucleus
Greek *moneres*, single, solitary
Greek *bakterion*, little stick (diminutive of *baktron*, stick, staff, rod)

Bacterial cell structure characterizes all prokaryotes whether single- or multicellular. Sexuality produces genetic recombinants temporally and spatially independent of reproduction. Branched filaments with terminally and/or cyclically differentiated cells (for example, heterocysts, endospores) are among the most structurally complex. All major modes of metabolism are represented in the group (Table B1-1). Maximum metabolic diversity and lithospheric (geologic) and atmospheric interaction are relative to eukaryotes. The fossil record of bacterial communities extends from the lower Archean eon (Figure I-5) to the present. A comparison of gene sequence in ribosomal RNA (rRNA) molecules is useful for the identification and distinction of modern lineages.

Kingdom Bacteria comprises all organisms with prokaryotic cell structure: they have small ribosomes surrounding their nucleoids, but all lack membrane-bounded nuclei. In activity and potential for rapid, unchecked growth, bacteria are unrivaled among living organisms. About 10,000 different forms have been described as "species"; most are cyanobacteria (B-6). Their genes are easily passed from one to another; the set of genes, or genophore, is organized into thin (25 Å) fibrils that, when visible as a light region in electron micrographs, are called the "nucleoid." The distinguishing traits of all the prokaryotes (and therefore of bacteria) are listed and compared with those of the eukaryotes in Table I-2. Prokaryotes, unlike all members of the superkingdom Eukarya, lack pore-studded nuclei that contain chromosomes. Prokaryotes also lack membrane-bounded organelles such as mitochondria and plastids; they did not evolve by cell symbiosis.

In this fourth edition, we have changed the kingdom name (from Monera or Bacteria) to kingdom Prokaryotae. *Bergey's Manual of Systematic Bacteriology* provides information about identifying and classifying bacteria. It unites all bacteria on the basis of their "prokaryotic" nature, under the name Prokaryotae. We support the bacteriologists who bring together all living organisms with genophores, not chromosomes, small ribosomes, and unidirectional gene flow, under the term "Prokaryotae" (*pro*, before; *karyon*, seed, nucleus). Our readers who are not professional microbiologists prefer the more familiar term "Bacteria". However, many microbiologists now use "Bacteria" to distinguish Eubacteria from Archaea. We accept the distinction between these two domains of prokaryotes, but at the subkingdom and not at the kingdom level. Today's kingdom of prokaryotes corresponds to the "Moneres" of the German biologist Ernst Haeckel. Haeckel used the term "Monera" (from the Greek *moneres* for single, solitary) to refer to bacteria

Table B1-1 Metabolic Modes in Prokarya

Energy	Nutrition: Electrons (Hydrogen)*	Nutrition: Carbon	Examples†	Phylum
Light	Inorganic compounds and compounds with one carbon atom	Carbon dioxide	Photoautotrophs:	
			Chlorobium (H_2S)	B-8
			Chromatium (H_2S)	B-3
			Cyanobacteria (H_2O)	B-6
			Rhodospirillum (H_2)	B-3
	Organic compounds	Acetate, lactate, pyruvate	Photoheterotrophs:	
			Chromatium (some)	B-3
			Chloroflexus	B-7
			Halobacterium	B-1
			Heliobacterium	B-10
			Rhodomicrobium (some)	B-3
			Rhodospirillum (some)	B-3
Chemical compounds	Inorganic compounds	Carbon dioxide	Methanogens:	
			Methanococcus (H_2)	B-1
			Hydrogen-oxidizing pseudomonads (H_2)	B-3
			Nitrosomonas (NH_3)	B-3
			Methylotrophs:	
			Methylosinus (CH_4)	B-3
		Organic compounds	Manganese oxidizers (Mn^{2+})	B-10
			Sulfide oxidizers:	
			Beggiatoa (H_2S)	B-3
			Desulfovibrio (SO_4^{2-})	B-3
	Organic compounds	Organic compounds	Heterotrophs:	
			Escherichia coli	B-3
			Bacillus	B-10

*Source of electrons for reduction of carbon to synthesize cell material.

†Examples include genera that have more than a single possible physiological mode. In parentheses is the most common source of electrons for the example in question.

as a group within his "Protista" kingdom in his three-kingdom scheme. Our acceptance of Prokaryotae as the only kingdom of life that did not evolve by symbiosis supports the use of the term "Prokaryotae" to refer to kingdom Bacteria and kingdom Prokarya rather than the older Monera.

Since taxonomy including all species designations in kingdom Bacteria are subject to continual revision for accessibility and in accordance with our focus on whole-organism criteria for classification, we present all prokaryotae

A

C

B

Figure B-1 Bacterial structures: living stromatolites (A, B). The living stromatolites are microbial mats that have hardened and turned to stone (lithified) (C). Found today in Hamelin Pool, Shark Bay, Western Australia, such limestone structures are made by communities of bacteria. The dominant stromatolite-builder here is a coccoid (spherical) cyanobacterium called *Entophysalis*. Besides *Entophysalis* many other bacteria are present. Stromatolites, which may be thought of as petrified microbial mats, are important clues to interpreting the fossil record of prokaryotes. Unlithified microbial mats, here in Baja California Norte, Mexico (B) may be precursor to stromatolites (C) or laminated cherts, if they preserve. In (C) the Cambrian carbonate stromatolites that outcrop in Colorado are indicators of a bygone >500 million year-old tropical shallow sea. Although living stromatolites are rare today such limestone layered rocks were widespread and abundant through the Proterozoic eon from 2500 to 542 million years ago—before the evolution of fungi, animals, and plants.

in the single kingdom Bacteria organized into 2 subkingdoms—Archaebacteria and Eubacteria—and 14 Phyla. Our taxa of the prokaryotes, consistent with *Bergey's Manual*, reflect not only molecular sequence data but also physiology, biochemistry, ultrastructure and habit, ecological niche, and symbiotic association.

Given the flux and flurry that molecular sequence measurement instigated with its vast expansion of information from nature as new microbes, refractory to laboratory cultivation, are identified, we remain conservative. The well-documented taxa are assigned to the 14 prokaryotic phyla, and many new lineages await consensus about their physiological range. Powerful, new tools of molecular biology continue to reveal astonishing diversity and metabolic virtuosity on this bacterial planet Earth.

The term "Prokaryotae" simultaneously recognizes the bacterial, prokaryotic nature of all Archaea (Archaebacteria). Prokaryotes are the hardiest of living beings. Some survive very low temperatures, well below freezing point, for years; others thrive in boiling hot springs. Still others even grow well in very hot acid or live by deriving hydrogen as energy and carbon dioxide (CO_2) as a source of self-made body parts directly from rocks. By forming propagules, or spores—traveling particles of life that contain at least one copy of all the genes of a bacterium—many tolerate boiling water or total desiccation. Prokaryotes are the first to invade and populate new habitats such as land that has been burned or newly emerged volcanic islands.

We dimly recognize the activities of thriving communities of bacteria (usually supported by photosynthesizers) as "scum," "slime," "gloop," "microbial mats," "floc," "nepheloid," and other derogatory terms. Prokaryotic communities of different kinds of bacteria living together survive in an extraordinary range of habitats inhospitable to protoctists, plants, animals, and fungi. The absolute requirements for growth of all of them are liquid water and sources of energy and matter (elements including carbon, hydrogen, nitrogen, sulfur, phosphorus, oxygen, calcium, magnesium, sodium, potassium, zinc, and a few others) in the appropriate form and amounts. Some bacteria survive and grow at great oceanic depths or even inside granites or carbonate rocks. Others have been captured in nets from above the atmosphere by stratospheric airplanes. Yet no organism—not even the hardiest bacterium—is known to complete its life history suspended in the air or any other gas.

Some other activities of bacteria are still only poorly known. The incorporation of soluble metal ions such as those of manganese and iron into rocks—nodules on lake and ocean floors—is accelerated by bacterial action. Layered chalk deposits called stromatolites (Figure B-1) are produced at the seashore by the trapping and binding of calcium carbonate–rich sediments by growing communities of bacteria, especially cyanobacteria. Gold in South African mines is found with rocks rich in organic carbon, associated with fossil bacteria and probably of microbial origin. In Witwatersrand, the miners find the gold, deposited apparently more than 2500 mya, by following the "carbon leader." Copper, zinc, lead, iron, silver, manganese, and sulfur all seem to have been concentrated into ore deposits by biogeochemical processes that include bacterial growth and metabolism. Bacteria are the only organisms that, in a process called "nitrogen fixation," convert nitrogen (N_2; the air's most abundant gas) into organic nitrogen and carbon–hydrogen–nitrogen compounds usable as food.

Because of their limited morphology and the paucity of their fossil record, bacteria have evolutionary relationships that have been exceedingly difficult to ascertain. However, in recent years, advances in molecular biology have enhanced our understanding of the evolutionary relationships among the tiny but highly diverse prokaryotes. Great insights have emerged from comparative studies of the long-chain ribonucleic acid (RNA) molecules that are components of the ribosomes of all organisms. The assumption, probably valid, on which this work is based is that changes in the sequence of nucleotides in RNA molecules reveal evolutionary histories of the modern bacteria. Because ribosomes are universally found in all organisms and are crucial for the same cell function, rRNA molecules are thought to have changed very slowly through evolutionary time. Under this assumption, Carl Woese of the University of Illinois, Otto Kandler and Wolfram Zillig of the University of Munich, and many other colleagues have concluded that bacteria assort into two fundamentally different, and therefore ancient, groups: the Archaebacteria (Archaea) and the Eubacteria. We agree and therefore recognize these groups as subkingdoms. However, we cannot accept the trifurcation of all life into three "primary kingdoms" [(1) Archaebacteria = Archaea, (2) Eubacteria = Bacteria, and (3) Eukarya], as these authors urge, because eukaryotes, as composite beings that evolved through symbiogenesis, are different in principle from eubacteria and archaebacteria, both of which are prokaryotes.

Small though they are, bacteria, which are extremely numerous and fast growing, are crucial to the health of our digestive system. They are responsible for soil production and soil maintenance in agriculture and forestry and for the existence of the air we breathe. Modern food processing began with an awareness of the nature of bacteria. Canning, preserving, drying, salting, and pasteurization are techniques that prevent the entry of even a single bacterium or growth of the few that remain. The success of these techniques is remarkable in view of the ubiquity of bacteria. Every spoonful of garden soil contains some 10^{10} bacteria; a small scraping of film from your gums might reveal some 10^9 bacteria per square centimeter of film—the total number of bacteria in your mouth is greater than the number of people who have ever lived. Bacteria make up some 10 percent of the dry weight of mammals. They normally cover our skin, especially on damp surfaces such as under the arms, between the toes, and in the vagina. They line nasal, otic, and oral passages and live on the surface of and in pockets in the gums and between the teeth. Most pack the digestive tract, especially the large intestine. Only a very few freaky bacterial associates of humans are pathogenic.

Pathogens are simply bacteria (or, occasionally, protoctists or fungi) capable of causing infectious diseases in animals or plants. The word "germ," like the word "microbe," has no precise or specific meaning. A germ is a small living organism capable of growth by using another organism as food. A microbe, or microorganism, is an organism so small that one needs a microscope to see it well [for example, some cyanobacteria that superficially look like plants (B-6) or myxobacteria that superficially resemble slime molds (B-3) are multicellular microbes]. Thus the smaller fungi, most protoctists, and all but the largest bacteria are also called microbes.

Bacteria can cure as well as cause disease. Many of our most useful antibiotics (a kind of allelochemical, a compound made by one form of life that inhibits the growth of a different, usually microbial, life-form) come from microbes. Among the best-known antibiotics, streptomycin, erythromycin, chloromycetin, and kanamycin come from specific kinds of bacteria, whereas penicillin and ampicillin come from certain species of fungi.

Bacteria tend to have a rather simple morphology: spherical (cocci), rod shaped, comma or spiral shaped. The most complex undergo developmental changes in form: a single bacterium may reproduce and give rise to populations that metamorphose into stalked structures; grow long, branched filaments; or form tall bodies that release resistant spore-like cysts. Some produce highly motile, swimming or gliding colonies. Knowledge of bacterial structure, unless complex and on the level of ultrastructure (the electron microscope), is seldom a source of insight into function. In this respect, bacteria differ from protoctists, animals, fungi, and plants. Because their differences lie chiefly in their internal chemical metabolism, many kinds of bacteria can be distinguished only by the chemical composition or the chemical transformations that they cause.

Bacteria are grouped by cell wall properties that are distinguished by a color-staining procedure. A universally applied diagnostic test of bacteria is whether they stain purple or pink with the Gram test, a staining method developed by the Danish physician Hans Christian Gram (1853–1938). Gram-positive organisms (which stain deep purple) differ from Gram-negative ones (which stain light pink). The chemistry of the bacterial cell walls—the presence of an extra membrane in Gram-negative bacteria—is the basis for classification.

Bacteria, especially in highly ordered but flexible communities, can effect a large number of different chemical transformations. A summary of metabolic patterns of strains of bacteria growing in pure culture is presented in Table B-1. The range of metabolic capabilities is far greater than that of all the eukaryotes. Metabolically speaking, bacteria represent the extreme range of biological diversity. Molecular biological detection of unique DNA sequences in natural habitats alerts us to the fact that the full extent of bacterial diversity has yet to be appreciated.

Although some very complicated molecules are made by certain plants and fungi, the biosynthetic and degradative patterns—the chemistry of food use and energy generation—in all plant and fungal cells are remarkably similar. Animals and protoctists exhibit even less variation in their chemical repertoires. The metabolism of eukaryotes, in short, is rather uniform; its patterns of photosynthesis, respiration, glucose breakdown, and synthesis of nucleic acids and proteins are fundamentally the same in all nucleated organisms. Bacteria, on the other hand, differ not only metabolically from eukaryotes, but also from each other.

The work of most bacteriologists (microbiologists) concerns the role of bacteria in health and disease. Bacterial activities in our environment have been much less studied, but are even more significant. Bacteria release and remove from the Earth's atmosphere all the major reactive gases, such as nitrogen, nitrous oxide, oxygen, carbon dioxide, carbon monoxide, several sulfur-containing gases, hydrogen, methane, and ammonia, among others. Protoctists and plants also make substantial contributions to atmospheric gases, such as oxygen, and ruminant animals contribute methane, but few, if any, differ from those gases produced by bacteria. Yet many important reactions are limited to the prokaryote repertoire.

The soil of Earth and the regolith—the loose, rocky covering of any planet—on the surface of Mars and the Moon differ enormously. Mars and the Moon are very dry and lack atmosphere relative to Earth, but the differences extend far beyond just moisture content. The surface of Earth—its regolith, sediments, and waters—is rich not only in living bacteria, small animals, protists, yeasts, and other fungi, but also in the complex organic (carbon plus hydrogen) compounds that they produce. The less tractable biological products such as tannic acids, lignin, and cellulose tend to accumulate, whereas much more actively metabolized organics, such as sugars, starches, organic phosphorus compounds, and proteins, are produced and removed more rapidly. All these organic compounds are—directly or indirectly—the products of chemoautotrophy or photoautotrophy, processes that use chemical oxidation energy or sunlight, respectively, to convert the carbon dioxide of the air into the organic compounds of the biosphere and, ultimately, into the organic-rich sediments from which we obtain oil, gas, and coal. In fact, the soil and rocks of Earth, mostly the biogenic limestone sediments, contain more than 100,000 times as much carbon as Earth's living forms do.

Chemoautotrophy is limited to certain groups of bacteria. Photosynthesis, which often is incorrectly attributed only to algae and plants, is carried out by many groups of bacteria. Chemoautotrophy and photoautotrophy are often, but not always, correlated with processes that use inorganic chemicals or light, respectively, to generate energy to make organic compounds. Both types of organic compound production are forms of strict autotrophic nutrition, the synthesis of all food and derivation of energy exclusively from inorganic sources. Heterotrophy, the alternative mode of nutrition, refers to obtaining food and energy from preformed organic compounds—from either live or dead sources. Like algae and plants, most photosynthetic bacteria convert atmospheric carbon dioxide and water into organic matter and oxygen; unlike them, many bacteria are also capable of very different modes of photosynthesis—for example, the use of hydrogen sulfide instead of water as the source of hydrogen atoms to attach to carbon dioxide in the making of organic compounds and the elimination of sulfur but not oxygen. Bacterial photoautotrophy and chemoautotrophy are essential for environmental chemical element and compound cycling through the biosphere on which all life depends. Bacteriologists refer to photoautotrophy as the mode of nutrition for organisms (at the top of Table B-1) that nourish themselves by light reactions that drive CO_2 fixation. Photosynthesis refers to any process of living tissue in which light energy is used to build organic matter.

Probably the most important evolutionary innovation on Earth, if not in the solar system and the galaxy, was photoautotrophy, the transformation of the energy of sunlight into usable form: the chemical energy of food or energy-storage molecules (such as carbohydrates, lipids, and proteins). Photoautotrophy, the process, began in anaerobic bacteria more than 3000 mya. Bacteria that derive their energy from sunlight, carbon from CO_2 of the air, and electrons from H_2, H_2S, H_2O, or other inorganic sources are called "photolithoautotrophic bacteria." They "feed" on rocks (lithos) and sunlight.

Photoautotrophy is an essentially anaerobic process, and none of the proteobacteria (such as *Chlorobium* or *Chromatium*, B-3) carry it out when they are exposed to oxygen. Except for the Chloroflexa (B-7), green sulfur bacteria are hypersensitive to oxygen; they grow only photoautotrophically in the absence of oxygen. Some purple nonsulfur bacteria (B-3) can grow microaerophilically—that is, under oxygen concentrations less than the modern norm—or even aerobically, but only in total darkness. In that case, they derive their energy not from photoautotrophy, but heterotrophically from the breakdown of food.

Oxygen release, even though it is characteristic of photosynthesis in plants, algae, and cyanobacteria, is not an essential property of photosynthesis. The essential properties are the incorporation of carbon dioxide from the air into organic compounds needed for the growth of the photosynthesizer and the conversion of the energy of visible light into chemical energy in a form useful to cells. The conversion of light energy requires chlorophyll and other light-absorbing (pigment) molecules, although not always precisely the same chlorophyll molecules. The chemical energy currency produced is adenosine triphosphate (ATP), a nucleotide used in energy transformation reactions of all cells. Although details of the enzymatic pathways of photolithotrophy are still being worked out, it is clear that the five types of phototrophic bacteria (purple sulfur, purple nonsulfur, green nonsulfur, green sulfur, and oxygenic) differ in details of metabolism, in their source of electrons for CO_2 reduction, and in other ways.

To reduce the carbon dioxide in the air to organic compounds, cells need a source of electrons, which, as a rule, are carried by hydrogen atoms. The source of these electrons varies with the organism. In green sulfur bacteria (chlorobia of B-8), the electrons come from hydrogen sulfide (H_2S), although they may also come from hydrogen gas (H_2). The purple sulfur bacteria (B-3) also use H_2S as the hydrogen donor. In purple nonsulfur bacteria, such as *Rhodospirillum* and *Rhodopseudomonas* (B-3), the hydrogen donor is a small organic molecule such as lactate, pyruvate, or ethanol. Thus, the general phototrophic equation can be written as

$$2n\, H_2X + n\, CO_2 \xrightarrow{\text{light}} n\, H_2O + n\, CH_2O + 2nX$$

in which X varies according to the species. The molecule H_2X is the hydrogen donor. In proteobacteria (B-3), the hydrogen donor is never water; thus, oxygen is not a by-product of their photosynthesis. In cyanobacteria (B-6), algae (Pr-5, Pr-12, Pr-15 through Pr-18, Pr-25 through Pr-28, Pr-32, and Pr-33), and plants (Pl-1 through Pl-12), on the other hand, oxygen is released because water is the hydrogen donor. When H_2S is the hydrogen donor, the by-product of photosynthesis is sulfur, which may be excreted, stored as elemental sulfur, or further oxidized as a sulfur compound such as thiosulfate ($S_2O_3^{2-}$) or sulfate, and then excreted. No gas such as oxygen is released; rather, the form of the sulfur product depends on environmental conditions. When buried in sulfur-rich muds full of high concentrations of sodium sulfide, for example, intracellular sulfur globules are made even by some cyanobacteria (B-6). These same cyanobacteria generate oxygen from water in aerated, sulfur-poor conditions.

Genera of photosynthesizers that are placed in bacterial phyla include the archaeabacterium, *Halobacterium* (B-1), which uses rhodopsin rather than chlorophyll in its processing of light energy, and several groups that contain some kind of bacterial chlorophyll (that is, cyanobacteria and some proteobacteria). A new type of photosynthesizer, *Heliobacterium*, which by 16S rRNA criteria is related to Gram-positive low-GC (guanosine plus cytosine) eubacteria (B-10), grows either heterotrophically or phototrophically; this option, rare in prokaryotes, is common in algae (kingdom Protoctista). *Heliobacterium* uses bacteriochlorophyll ***g***, whereas the green sulfur (B-8) and nonsulfur (B-7) bacteria use still other chlorophylls to capture visible light and generate cell energy in the form of ATP. All cyanobacteria (B-6) use at least chlorophyll ***a***. Photosynthesis evidently evolved many times separately (polyphyletically) in prokaryotes, although members of only one group, cyanobacteria, were ancestral to the plastids of protoctists and plants.

The notion that the food web starts with the plants, followed by the herbivores, and ends with carnivorous animals is shortsighted. Zooplankton of the seas feed on protoctists; nonphotosynthetic protoctists feed on bacteria; bacteria (and fungi and animal scavengers) break down the carcasses of animals, plants, and algae, releasing back into solution such elements as nitrogen and phosphorus required by the phytoplankton. Because *phyto* means plants and because no plants float in the open ocean, we prefer the term "photoplankton" to refer to floating bacteria and algal photoautotrophs at sea and in lakes. Bacteria, because they are eaten by others, facilitate entire food webs. The ways in which we and other forms of life depend on bacteria, and evolved from them, will be explained in the descriptions of the phyla. Life on Earth would die out far faster if organisms in the superkingdom Prokarya became extinct than if any of the other life-forms disappeared. We believe that bacterial life on our planet thrived long before the large organisms, those that evolved by symbioses from communities of bacteria, ever appeared.

Bacteria have an ancient and noble history. They were probably the first living organisms and, with respect to everything but size, have dominated life on Earth throughout the ages. The oldest fossil evidence for bacteria dates to 3400 mya, whereas the oldest evidence for organisms belonging to any of the eukaryotic kingdoms is 1200 million years or so.

Biologists and geologists agree that, some 2000 mya, the cyanobacteria (the oxygen-releasing photosynthetic prokaryotes that used to be called "blue-green algae"; B-6) began one of the greatest changes known in the history of this planet: the increase in the concentration of atmospheric oxygen from far less than 1 part per thousand (<0.001) to its current level of about 200 parts per thousand, or 20 percent. Without this high concentration of oxygen, plants, people, and other animals would not have evolved.

Bacteria reproduce uniparentally by binary fission (one cell divides, giving rise to two similar offspring cells), by budding (a parent cell produces a small bud that later reaches the size of its parent) or by spores. Although bacteria participate in sexual donation of DNA from one (the donor) to another (the recipient) in the process of conjugation, this bacterial sexuality is not associated in time or space with reproduction. The extent of bacterial sexuality in nature is not well known, partly because—as with most other organisms—the "sex life" of bacteria is elusive. Any process that leads to the formation of a bacterium with genes from more than a single parental source is bacterial sexuality. If the source is a second bacterium in contact with the first, the bacterial sexual process is conjugation. If the second source of genes is a plasmid (circular fragment of DNA), virus (protein-coated DNA or RNA), or linear piece of DNA that carries bacterial genes, the bacterial sexual process has another name: transformation. When genes are transferred to recipient bacteria by viruses that infect bacteria, the process (a special case of transformation) is called transduction. Many bacteria excrete DNA. In the laboratory and in nature, DNA excreted by one bacterium is taken up and incorporated by another to form genetically recombinant prokaryotes.

No location anywhere on Earth lacks bacteria, but only a few places today are dominated by them. Some exclusively bacterial habitats, most often found in intemperate climates, are the bare rocks of cliffs, the interior of certain carbonate rocks, and muds lacking oxygen. Perhaps the most spectacular are the boiling hot springs and muds such as those in Yellowstone National Park, in Wyoming, or the brightly colored salt flats and shallow embayments of tropical and subtropical areas. Many such thermal springs, flats, and bays are dominated by microbial mats—cohesive, domed, or flat structures on soil, in air, or in shallow water that are caused by the growth and metabolism of bacteria, primarily filamentous cyanobacteria. By entrapping bits of sand, carbonate, and other sediment, such microbial communities grow to be quite conspicuous manifestations of biological activity.

The habitat scenes are notably arbitrary in this chapter because so many bacteria can be found in or on the bodies of eukaryotes: protoctist, animal, fungal, and plant. They abound as well in soil, air, and water samples, in vastly different habitats and locations. Bacterial communities in the intestines of mammals (Figure B-2) have been studied disproportionately.

Except for those rather extreme environments where microbial mats or thermal springs abound, eukaryotes seem to dominate our landscape. However, microscopic examination of a sample from any forest, tide pool, riverbed, chaparral, or other habitat reveals bacteria in abundance. When a specific type of bacterium is removed from nature for growth on its own

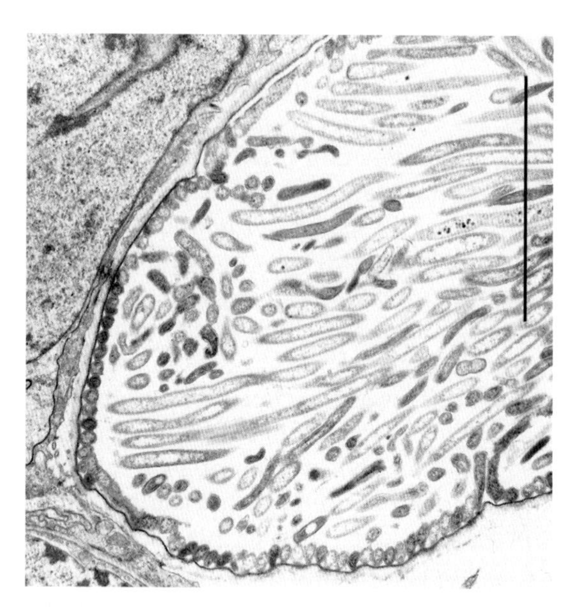

Figure B-2 An intact bacterial community from a pocket in the hindgut wall of the Sonoran desert termite *Pterotermes occidentis* (A-21). More than 10 thousand million bacteria per milliliter have been counted in these hindgut communities. Many are unknown. All survive anoxia. In our studies, 28–30 strains isolated were facultative aerobes that metabolize oxygen when available. Most are motile, Gram-negative heterotrophs, and thus most likely proteobacteria. Notice that some of the bacteria line the wall of the gut, whereas others float freely in the lumen. TEM, bar = 5 μm. [Courtesy of D. Chase.]

(pure culture) or with other microbes (mixed culture), that type, called a "strain," is given a name or identifying number. When environmentalists mourn the destruction of habitats by pollution, they are usually thinking of the loss of fish, fowl, and fellow mammals, and not strains of bacteria. If our sympathies were with the cyanobacteria and other bacteria instead, we would recognize the pollution of green scummy lakes, for example, as a sign of flourishing life.

Much of bacterial nomenclature is in dispute; there is no consensus among scientists on how to name and group the thousands of strains or how to relate them to bacteria in nature or in the literature. Most bacteria are still not identified; microbiologists, who study bacteria (as well as the smaller fungi and anaerobic protoctists), assert that the vast majority of bacteria have not yet been carefully studied and described. Microbiologists lack standard nomenclatural and taxonomic practices. Many believe that morphology in natural habitats alone is inadequate to name new bacterial strains, yet with some refractory to growth in the laboratory there is no option. Ironically, prokaryotes changed by captivity are often far better known than their wild relatives. Those who study animals (zoologists) or plants (botanists) in natural habitats have strict rules for giving names to new organisms. By contrast, the terminology and taxonomic practices of microbiologists are inconsistent and not directly comparable. Thus, inevitably, our groupings differ from those found in *Bergey's Manual of Systematic Bacteriology* or the four-volume work *The Prokaryotes* (see Balows *et al.*, 1991). We aim to make the taxonomic level of phylum conceptually comparable throughout all life and to avoid confusion and contradiction. We recognize 14 major prokaryotic phyla, which group the bacteria by clearly distinguishable morphological and metabolic traits. However, unlike the phyla in the eukaryotic kingdoms, these prokaryotic phyla are not all-inclusive. Many small groups of bacteria that are difficult to classify have been omitted from the descriptions of members of phyla for reasons of clarity and lack of space. As the amount of molecular systematic data increases and 16S rRNA and other detailed information becomes available, we employ it, keeping members of morphologically distinctive metabolic groups together as well as we can.

Because bacteria that differ in nearly every measurable trait can receive and permanently incorporate any number of genes from each other or from the environment, Sorin Sonea (University of Montreal) and others have argued that bacteria form a single worldwide web of relations. Although strain names are easily applied, the concept of "species," applicable to named eukaryotes, seems to us inappropriate for the Prokarya. Because prokaryotes can change their genetic properties so quickly and easily, we agree with Sonea and Mathieu's analysis (2002). We doubt that any single prokaryote with its simple genome evolved to form the distinctive definable species characteristic in all polygenomic Eukarya. Flexibility must be tolerated in prokaryotic taxonomy. One of our goals is to be maximally informative and useful in our treatment of these versatile, abundant, and totally indispensable planetmates.

Box Prokaryotae-i: Whole-organism criteria, not partial phylogenies

Taxonomy—the naming, identification, and classification of myriad forms and manifestations of all life on Earth—remains a daunting enterprise. We evolutionists follow Darwin's dictum: our classification systems attempt to reflect evolutionary history. This book groups whole living organisms, not their component or isolated parts.

The rise of two modern subdisciplines, especially since the third edition of *Five Kingdoms* (1998), has imposed quantitative rigor on the task of cataloguing life. Cladistics is a taxonomic subdiscipline that attempts to show evolutionary relationships between organisms using shared derived traits (synapomorphies). This practice insists on monophyly in all recognized lineages. Molecular evolution—comparisons of amino acid sequences in proteins and nucleotide sequences in nucleic acids (DNA and RNA)—has uncovered hitherto unknown phylogenetic relationships. Taxonomists have been greatly aided in recent decades by the advent of comparative molecular criteria that are measurable in all cells, whether bacteria or bees. The depiction of cladistic trees for ascertaining relatedness of representative organisms, and to chart "phylogenetic" branch points, introduces standardized rules into the analysis. Carl Woese and coworkers (since Woese and Fox, 1977) have made an invaluable contribution to taxonomy in their delineation of three great domains of life (Archaea, Bacteria, and Eukarya) based on ribosomal DNA sequences. We heartily acknowledge the value of such studies. Not only are data in principle more accessible to students and scientists, but standardization makes it more conceptually accurate. However, molecular phylogenies are partial; like all partial phylogenies, they are flawed in that they measure only a few criteria of relatedness: nucleotide sequence identity for some gene or genes. But organisms constantly exchange DNA through lateral gene transfer. Such gene exchange is rampant not only in bacteria but also in eukaryotes, where it is intrinsic to both sexuality and symbiogenesis. More problematic still, although cladistic trees can tell us at what point in evolutionary history a branching occurred, their representation never includes anastomoses (fusion of branches reflecting merged lineages). The branches that fuse to mark the symbiotic mergers are routinely omitted; thus these molecular phylogenies have a systematic error.

A robust framework for taxonomy must include not only recent, often-revised molecular measures of relatedness, but *all that we know* about the organisms under study: life histories, ecology, symbioses, morphology, physiology, and other data handed down by our illustrious predecessors, the giants on whose shoulders we stand.

Accordingly, our taxonomy acknowledges recent molecular insights but derives fundamentally from microbiology, the bacterial "elements of life," and from these elements in symbiotic association. Symbiosis is a quintessentially eukaryotic phenomenon: the mitochondria, chloroplasts, and even, as new evidence suggests, the kinetosome, undulipodium, Golgi, and nucleus (collectively, the *karyomastigont*) are all of symbiotic origin. Meiotic sexuality is likewise a fundamentally eukaryotic strategy to shuffle chromosomes and present a "moving target" to predators, environmental pathogens, and other threats. To illustrate our approach, consider the subdivision of kingdom Protoctista into four subkingdoms, groups based on the presence or absence of undulipodia and/or meiotic sex (meiosis and fertilization).

mode I phyla include organisms whose natural history (for example, symbiotrophy to necrotrophy, saprotrophy) favored secondary loss of the undulipodium. They either never had or lost complex sexual life histories: for example, rhizopod amoebae and haplosporidians.

mode II phyla include organisms that exhibit complex sexual life histories, but whose undulipodia have been lost: for example, red algae, conjugating green algae, and cellular slime molds.

Cells of organisms in **mode III** phyla reversibly form undulipodia for swimming but retract their undulipodia before cell division and reassign kinetosomes as mitotic spindle MTOCs (microtubule organizing centers): for example, amoebomastigotes, prymnesiophytes, and euglenids.

And finally, members of **mode IV** phyla, by far the largest group of organisms, have retained both reversible formation of undulipodia and complex sexual life histories. Not surprisingly, this group includes the most cosmopolitan and diverse protists as well as direct ancestors of crown taxa (for example, chytrids, green and brown algae, foraminiferans, and dinomastigotes).

Although molecular data confirm the bulk of our classification scheme, strict cladists may raise objections at certain points: for example, our assignment of the chytrids to the protoctist rather than the fungal kingdom. To this we respond that redesignation of chytrids as fungi was based on studies of only a few genes; hundreds of more genes comprise the chytrid undulipodium, which is completely absent in all fungi, probably because most are terrestrial. (Cilia and other undulipodia require liquid to function.) The molecular data confirm direct descent of fungi from chytrids, and of land plants from green algae; however, a truly biological classification of these lineages must consider more than one, or even a few, gene sequences: it must account for as many living phenomena as possible, in the tradition of Darwin as ascertained through countless, meticulous observations of a group of organisms, and the species and individuals taken as revwpresentatives of the group.

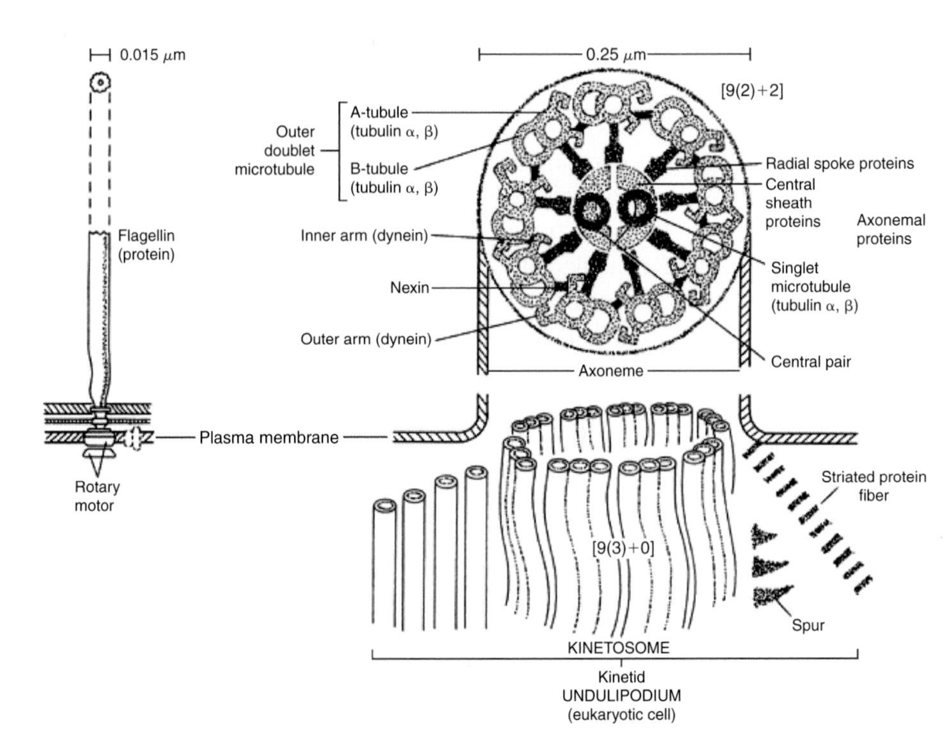

Figure Prokaryotae-i-1 A bacterial flagellum (left) compared with the undulipodium of eukaryotes (right). Kinetosomes, which always underlie axonemes, are associated with fibers, tubules, and possibly other structures. The organelle system, the kinetosome with its associated structures (e.g., fibers, microtubules, spurs) is called the kinetid. nm, nanometer; μm, micrometer. See Figure Pr-1, P 120. [Drawingvw by K. Delisle.]

Box Prokaryotae-ii: Two superkingdoms, not three domains

Modern classification systems are hypotheses about the genetic relationships among species and the evolutionary history of life. When the traditional "Plant versus Animal" dichotomy was supplanted by discoveries in microbiology, most scientists accepted the classification of all life into five kingdoms: Bacteria (or Monera), Protoctists (protists and their macroscopic relatives), Animals, Plants, and Fungi (Figure Prokaryotae-ii-1). Each kingdom unambiguously belongs in one of two "superkingdoms": prokaryotes are cells that lack membrane-bounded nuclei and protein-rich chromosomes, but synthesize proteins on small ribosomes; eukaryotes are organisms composed of nucleated cells.

Carl Woese and his colleagues, in 1990, proposed the three-domain classification based on DNA sequence data. Woese's analysis assigns all organisms to one of the three fundamental groups, or domains: Archaea, Bacteria, and Eukarya. The Archaea, of which relatively few have been studied, are prokaryotes: they resemble bacteria and are classified with bacteria in the five-kingdom scheme. The Eukarya (meaning "true nucleus") include protoctists, animals, plants, and fungi. Woese coined the term *progenote* to denote the last universal common ancestor (LCA) at the root of his three-domain tree.

The five-kingdom and three-domain schemes are contradictory, especially in the basis for classification. The five-kingdom classification reflects studies of whole live organisms; no single criterion is adequate. Each of the five kingdoms is uniquely defined by use of inherited characters (semes) shared by all of its members. The life history, genetic structure, ecology, symbiotic relationships, morphology, and development are all considered. The division of life into three domains, by contrast, is based on the DNA sequence of one or a few genes.

The assignment of life into any of the three domains requires molecular sequence comparison of only a few inherited traits. Traditionally, "family trees" (phylogenetic diagrams based on a limited number of traits) are called "partial phylogenies."

The five-kingdom classification system is vastly different from the Woesian three-domain scheme in its view of the origin and nature of cellular life. According to Woese, the three modern cell types (Bacteria, Archaea, and Eukaryotes) emerged very early in Earth history from a now-extinct progenote. The three-domain ancestor was not any one species, rather a composite "progenote lineage" that evolved as a whole. The two-superkingdom, five-kingdom classification posits

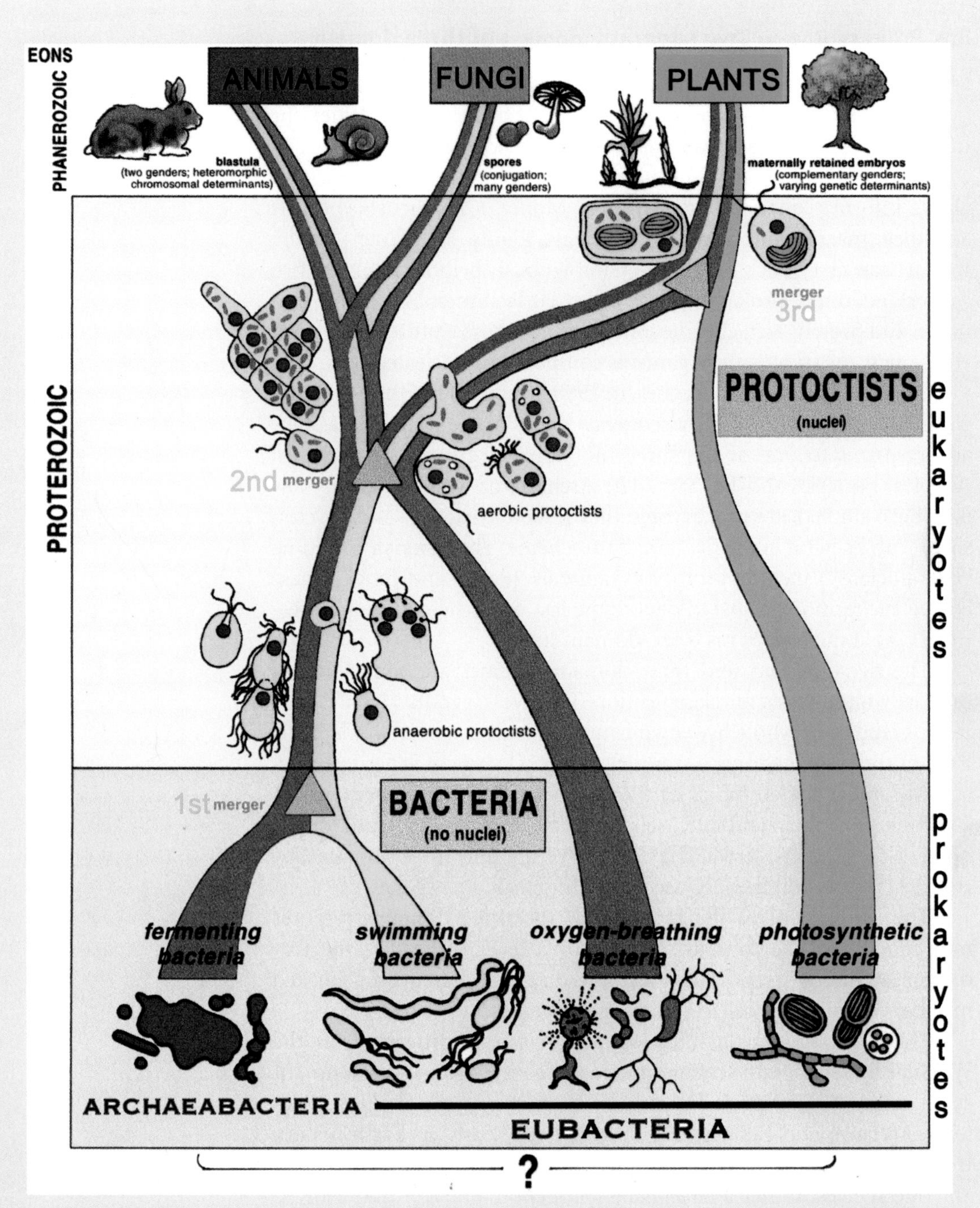

Figure Prokaryotae-ii-1 Five-kingdom, two super kingdom classification of life on Earth. [Illustration by K. Delisle.]

that prokaryotes evolved from a common, bacterial cell origin before the symbiogenetic origin of the earliest eukaryotes.

Archaeabacteria and eubacteria were present in the Earth's Archaean eon (3800–2500 mya). Nucleated cells evolved from symbioses between at least two and as many as four different kinds of prokaryotic cells. The recognition that bacteria (prokaryotes) were the earliest forms of life but nucleated cells were symbiotic combinations of bacteria was first published by Boris Kozo-Polyansky in 1924. The nomenclature for this fundamental distinction (*procaryotíque/eucaryotíque*) originated with Edouard Chatton in 1925 (Sapp, 2005).

Box Prokaryotae-iii: Modes of multicellularity

Multicellularity occurs in all lineages of life. All plants and animals develop from multicellular embryos and consequently are unicellular only in their reproductive propagules [for example, plant spores; animal eggs, sperm; and specialized propagules such as gemmae of sponges (A-3) and statoblasts of ectoprocts (A-29)]. The fungal, bacterial, and protoctist kingdoms are assemblages of single-celled and multicellular taxa. Studies of the world's most ancient rocks have shown bacterial cells in organized filaments, suggesting that multicellularity is a very ancient evolutionary phenomenon. The multicellular habit most likely began when cell division was blocked or impaired, and then evolved under selection pressure for cell differentiation, that is, division of labor: whereas a single cell must be generalized in function, multicellular organisms have some cells specialized for motility, others for phagocytosis, still others for reproductive functions, and so on. Multicellular organisms acquire clear advantages of size over their unicellular predecessors. They may outcompete their ancestors for resources in a given niche or deter predators more effectively.

Evolutionists have advanced three models for the origin of multicellularity. In the symbiogenetic theory, unicells from different species could have evolved the habit of living together (symbiosis) that became more and more obligate with time, leading ultimately to permanent association of cells into tissues and organs. Many extant examples of symbiogenetic multicellularity, such as the well-studied incorporation of *Trebouxia* algae into lichens (F-6), support this theory; however, one would then expect to find genetic evidence of distinction between

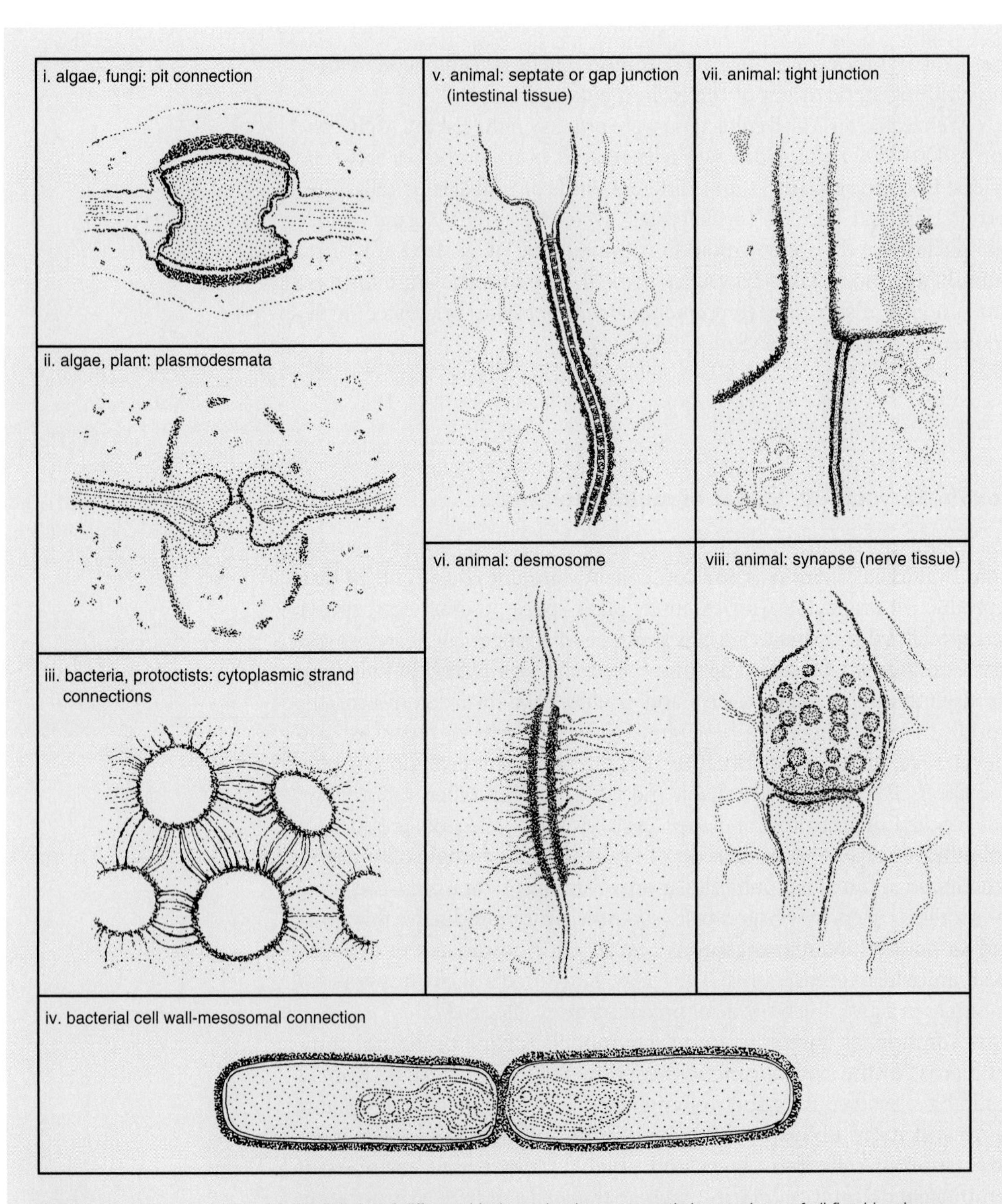

Figure Prokaryotae-iii-1 Multicellularity of different kinds evolved convergently in members of all five kingdoms. Animal tissue-cell multicellularity is most elaborate, distinctive and kingdom-specific (v-viii). Plants and green algae tend to have cytoplasmic strands that extend through gaps in their cellulosic walls (ii). Here only major trends are depicted. We recognize that many variations exist on cell junction patterns especially in multicelluar heterotrophs: bacteria, protoctists and animals.

diverse tissues in other multicellular organisms, analogous to the relict bacterial genomes in mitochondria and chloroplasts. Because no such elements of distinction have been reported for most taxa, multicellularity through symbiogenesis was probably the exception rather than the rule.

In the cellularization (syncytial) model, partitioning of multinucleate cells may have led to multicellularity. Many protoctists, such as ciliates, hypermastigotes, and slime molds, form syncytia that sometimes contain hundreds of nuclei. Some animal, plant or fungal tissues, moreover, such as mammalian muscle and flowering plant endosperm, form through partitioning of embryonic syncytia. Here again, however, it seems impossible to generalize the syncytial mechanism to all tissues in all phyla.

Finally, the colonial model holds that individual cells of one species retained their connections to one another following cell division and formed colonies, which ultimately led to division of labor and organization into tissues. The morphological identity of choanomastigotes (Pr-36) with choanocytes of sponges (A-3), as well as the volvocine series in green algae (Pr-28) in which examples of the same cell type are found singly (*Chlamydomonas*), in small groups (*Pandorina*, *Eudorina*), and in colonies of thousands of cells, differentiated for different functions (*Volvox*), lends credence to this mechanism. Multicellularity must have evolved independently at least several times in evolution of eukaryotes in these different ways, as suggested by the large number of distinctive unicellular taxa that have peculiar multicellular relatives (for example, holotrichous ciliates and the stalked multicellular slime mold look-alike, *Sorogena stoianovitchii*, multicellular euglenas or diatoms). The diversity of cell–cell junctions in nature also supports many independent evolutionary origins of the multicellular condition.

Multicellularity in prokaryotes is best developed in myxobacteria and cyanobacteria, some species of which form straight, branching filaments, bulbous multicellular cysts ("fruiting bodies"), nets, multiple trichomes in common sheaths, and so on. Pit connections between walled cells occur in algae and fungi; they are also found in plants, sometimes including plasma membranous connections between adjacent cells (plasmodesmata). Because animals have evolved the greatest diversity of cell types in a multicellular body (100–150 different cell types, compared with 10–20 in plants, fungi, and protoctists), animals have also evolved the most complex variety of cell–cell junctions, including desmosomes, gap junctions, septate junctions, and synapses (Figure Prokaryotae-iii-1).

SUBKINGDOM (DOMAIN) ARCHAEA

The distinguishing characteristics of archaebacteria concern primarily the gene (DNA) sequences that determine the sequence of the small-subunit rRNA. A 16S rRNA, about 1540 nucleotides long, from small ribosomal subunits, is comparable in function in the superkingdom Prokarya. The nucleotide sequences in both 16S and 5S RNA, constituents of the 30S rRNA small subunit, show that archaebacteria are more closely related to each other than they are to eubacteria. Archaebacterial ribosomes in ultrastructure resemble ribosomes of Eukarya more than they do with those of eubacteria (Figure B-3). The major lipids of archaebacteria are ether linked with phytanol side chains (C_{20}). In other bacteria and in eukaryotes, most lipids are ester-linked. Archaebacteria lack the peptidoglycan layer typical of cell walls of eubacteria. A single DNA-dependent RNA polymerase enzyme with complex structure—more than six subunits—is present in archaebacteria. Archaebacteria include methanogenic, halophilic, and thermoacidophilic bacteria. All other—that is, most—bacteria are eubacteria.

We emphasize the ancient "extreme" environments in which members of two archaebacterial phyla—Euryarchaeota and Crenarchaeota—tend to be found. The habitats of archaebacteria, tectonically active environments, were considered far more abundant on the surface of Earth during the Archaean eon more than 3000 mya. Archaeabacteria were prematurely classified as "extremophiles." Further field study has shown archaebacteria to be widespread in seawater, lakes, soils, and other environments not subject to extremes. The distribution of archaeabacteria in nature is under intense investigation. Here we depict them thriving as methanogenic (methane-producing), halophilic (salt-loving), and thermoacidophilic (heat- and acid-loving) bacteria in settings of oxygen-depleted (anoxic) muds, soils, or comparable places: geysers; hot springs; places where sea vents spew water vapor, sulfide, hydrogen, and other oxygen-depleted gases; salty seashores; boiling muds; or landscapes of lava and ash-ejecting volcanoes. The extreme environments that dominated the early Earth certainly harbored archaebacteria, but not exclusively. Direct investigation of the distinctive 16S rRNA sequences in nature, in microbes that cannot be grown in the laboratory, reveals astonishing microbial diversity. We expect research to continue to reveal new taxa of archaebacteria with unsuspected physiologies, morphologies, and community relationships. They underscore the depth of our ignorance of Prokarya on Earth, especially archaebacteria with their unknown morphology and physiology. All of the archaebacteria (Archaea of Carl Woese and his colleagues) are formally grouped in the division Mendosicutes (≡Superphylum) by professional bacteriologists.

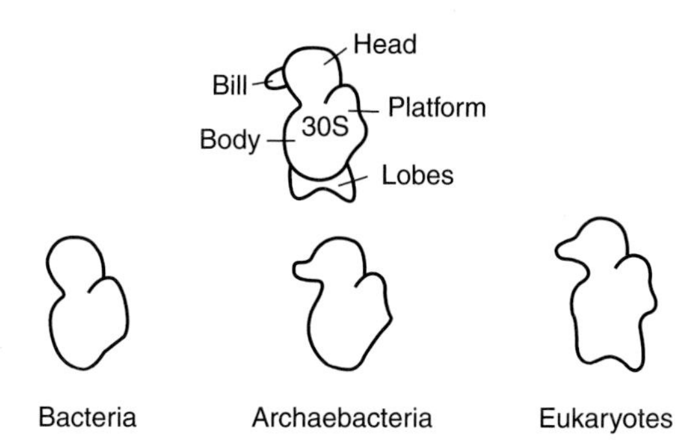

Figure B-3 Shapes of the smaller portion of ribosomes, 30S subunits, are compared. "S" refers to number of "Svedbergs", a measurement of the rate of descent of the portions in a standardized centrifuge. As a universal organelle of protein synthesis intact ribosomes are required for autopoiesis (organismic self-maintenance). In live cells small subunits (30–40S) bound to larger ones (the 50–70S) comprise each ribosome. By comparison of small subunits in the three domains (eubacteria, archaebacteria and eukarya) a greater ribosomal resemblance of the archaebacteria to the eukarya ribosomes, is apparent. [Based on information from George Fox and publication in 1983 of J.A. Lake's "Ribosomal evolution: The structural basis for protein synthesis in archaebacteria, eubacteria, and eukaryotes," *Cell* 33:318–319.]

Division: Mendosicutes

B-1 Euryarchaeota

(Methanogens and halophils)

Greek *eury*, broad, wide; *archae*, old; *-otes*, quality

GENERA

Haloarcula
Halobacter
Halococcus
Haloferax
Methanobacillus
Methanobacterium
Methanobrevibacter
Methanococcoides
Methanococcus
Methanocorpusculum
Methanoculleus
Methanogenium
Methanohalophilus
Methanolobus
Methanomicrobium
Methanoplanus
Methanopyrus
Methanosaeta
Methanosarcina
Methanosphaera
Methanospirillum
Methanothermu
Methanothrix
Natronobacteri
Natronococcus

Phylum Euryarchaeota comprises two classes: very strange and different groups of archaebacteria, methanogens and halophils. They are grouped together on the basis of rRNA sequence similarities. Methanogens can be Gram-positive or Gram-negative, motile by means of undulipodia or immotile. All three classic bacterial shapes—rod, spirillum, and coccus—are represented: methanogens vary from short rods and irregular cocci to spirilla, large cocci in packets, and filaments. So far, no methanogens with branched filaments or internal periplasmic undulipodia have been discovered. Most tolerate moderate or high temperatures. Methanogens are found worldwide in sewage, marine and freshwater sediments, and the intestinal tracts of animals—both ruminants and wood-eating (cellulose-ingesting) insects. Formally recognized genera are distinguished by morphology and physiology. They have names prefixed by "methano-": *Methanosarcina*, *Methanobacterium*, *Methanococcus*, *Methanobacillus*, and *Methanothrix*.

Halophils live in extremely salty or alkaline environments. They do not form specialized tolerant propagules. Before the use of molecular phylogenetic techniques, the close relationship between halophils and methanogens was unknown.

Methanogenic bacteria do not use sugars, proteins, or carbohydrates as sources of carbon and energy. Most can use only three smaller carbon compounds as food sources: formate, methanol, and acetate. Methanogens are known for their extraordinary way of gaining energy: they form methane (CH_4) by reducing CO_2 and oxidizing H_2, a scarce commodity. Because they obtain both these gases from the air around them and they cannot tolerate oxygen, their distribution is limited. In addition to using H_2, some use formate, methanol, or acetate as a source of electrons for reducing CO_2. The overall metabolic reaction is

$$CO_2 + 4H_2 \rightarrow CH_4 + 2H_2O$$

Methanobacillus omelianski, which was thought for many years to be a methane-producing bacterium that requires ethanol, is now recognized as a symbiosis between two similarly shaped but metabolically distinct bacteria. One, known simply as "organism S," is a eubacterium: an anaerobic, Gram-negative fermenting bacterium that produces hydrogen from ethanol. It forms no CH_4 whatsoever. At a certain concentration, the H_2 produced by *M. omelianski* is toxic to the fermenting bacterium. The other bacterium, called "strain MOH," is a methanogen that combines the H_2 provided by its associate with atmospheric CO_2 to form CH_4 in the autotrophic reaction described earlier.

Methanogenic bacteria are the source of "marsh gas" and, indeed, of most natural gas. More than 90 percent of natural gas is methane. Methane is produced in swamps, estuaries, bogs, and sewage treatment plants by these archaebacteria. If methanogenic bacteria did not move organic carbon from the sediments—in the form of acetate, formate, methanol, and CO_2 derived by other organisms from sugars, starch, and other photosynthate—to the atmosphere in its gaseous form (as CH_4), the carbon produced by photosynthesis would be irretrievable; it would remain buried in the ground. Methane may be made from CO_2 and H_2 or formaldehyde (H_2CO) (as in *Methanobacterium* or the sulfur-reducing methanogen *Methanothermus*), or it may be made from acetate (as in *Methanothrix* or *Methanosaeta*).

For every atom of carbon buried as carbohydrate or other reduced compound produced by oxygenic photosynthesis, a molecule of oxygen (O_2) has been released. Without the intervention of methanogens, the amount of buried carbon and the amount of O_2 in the atmosphere would increase. An excess of atmospheric O_2 would lead to spontaneous fires that would threaten the entire biosphere. However, most methane released by methanogens reacts with O_2 spontaneously or, in methylotrophic eubacteria (B-3), to form cell material or CO_2. Balance is thereby maintained.

Methanogenic bacteria produce some 2000 million tons of methane per year—a quantity equivalent to several percent of the total annual production of photosynthesis on the entire Earth. Thus carbon is spared from burial and returned to the atmosphere, where it can be used again by plants and other life-forms. The methane in today's Earth's atmosphere (more than one part per million) is produced by methanogenic bacteria. Although some bacteria emit methane as an end product of carbohydrate fermentation, no eukaryotes (or prokaryotes other than methanogenic bacteria) form methane from CO_2 and H_2. A good part of the world's methanogenesis, perhaps some 30 percent, comes from "animal fermentation tanks on four legs" that we recognize as cows, elephants, and other mammals that enjoy a high-cellulose diet (Figure A). Wood-eating termites, roaches, beetles, and other cellulose eaters also harbor methanogens. The rumen (a specialized cow "stomach") could not function, just as our sewage treatment plants could not function, without methanogenesis.

Enzymes connected in series, including the autofluorescing flavin-derivative coenzyme F_{420} (so named because it absorbs light at 420 nm), are essential for generating methane. Therefore, they are present in cells that are actively methanogenic. When ultraviolet light is shone on these often-ordinary-looking bacteria, they fluoresce blue green, betraying their methanogenic identity. Another coenzyme, which is unique to methanogenic bacteria and catalyzes the last step of the pathway before methane is vented to the air, is a yellow tetrapyrrhole (like chlorophyll and heme) that (unlike iron-containing heme) contains a nickel atom in its carbon ring system. Called coenzyme F_{430}, this compound does not fluoresce. The presence of additional unique coenzymes, such as methanofuran, tetrahydromethanopterin, and coenzyme M, further supports the idea that methanogenic anaerobes are unique.

Halophils are aerobes. Unlike oxygen-intolerant methanogens, they respire oxygen. The salt-requiring bacteria are incapable of methane production, but like methanogens, they live in extreme environmental conditions and do not form spores. Halophils abide in a variety of newly discovered hypersaline environments, feeding on a variety of carbon compounds. Some examples include rod-shaped bacteria such as *Halobacterium salinarium*, which lives on salted fish; *H. sodomense*, which requires high concentrations of magnesium; and the salt-tolerant sugar eater *H. saccharovorum*. Flattened, disk- or

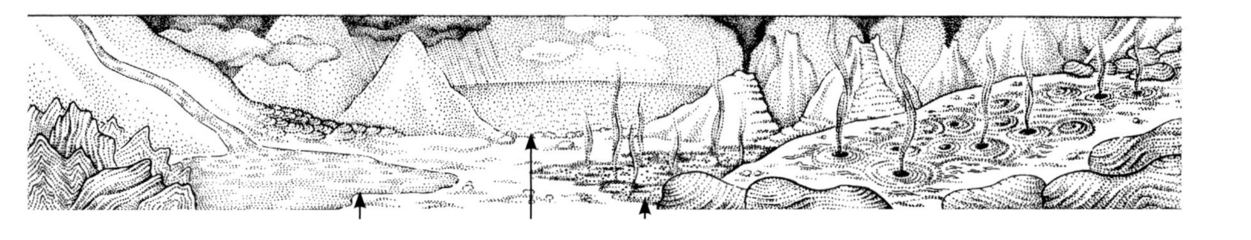

cup-shaped cells are members of the genus *Haloferax*, whereas triangular and rectangular halophils are usually assigned to the genus *Haloarcula*. Most halophils have obligate aerobic metabolism and are motile.

Halophils live in environments of high ionic strength (Figure B). Notably, halophils abide in saturated solutions of sodium chloride in salt works and brine all over the world. They produce bright pink carotenoids and can even be spotted from airplanes and orbiting satellites as pink scum on salt flats. Many of their proteins have modifications that allow them to function only at high salt concentrations. Their cell walls are quite different from those of other bacteria in that they lack derivatives of diaminopimelic and muramic acids. Ordinary lipoprotein membranes burst or fall apart at high salt concentrations, but the halobacters' special lipids include derivatives of glycerol diether that stabilize the membranes in high concentrations of salt.

Natronobacterium (including *N. gregori*, *N. magadii*, and *N. pharaonis*) has been found in extremely alkaline "soda lakes." It is a rod-shaped organism that grows best at pH 9.5. The related spherical archaeabacterium, *Natronococcus occultus*, was isolated from a highly alkaline soda lake.

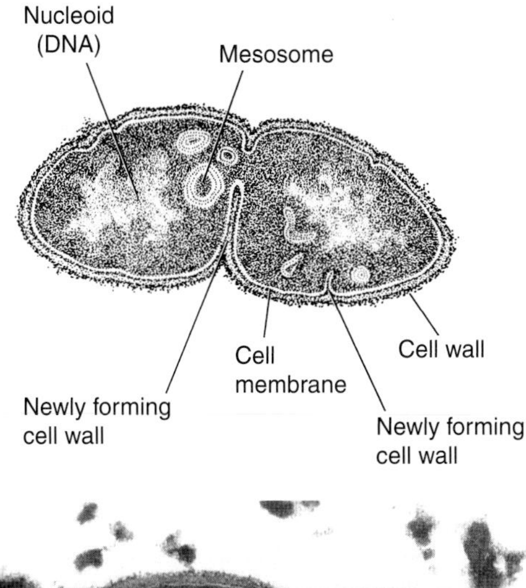

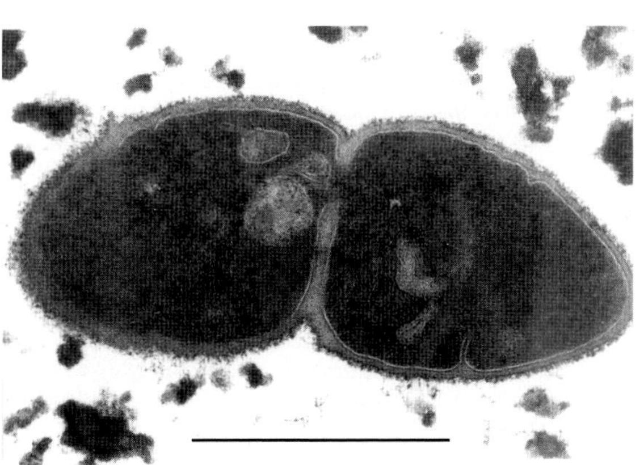

Figure A *Methanobacterium ruminantium*, a methanogenic bacterium taken from a cow rumen. The bacterium has nearly finished dividing: a new cell wall is almost complete. Notice that a second new cell wall is beginning to form in the right-hand cell. TEM, bar = 1 μm. [Photograph courtesy of J. G. Zeikus and V. G. Bowen, "Comparative ultrastructure of methanogenic bacteria;" reproduced by permission of J. G. Zeikus and the National Research Council of Canada from the *Canadian Journal of Microbiology* 21: 121–129 (1975); drawing by I. Atema.]

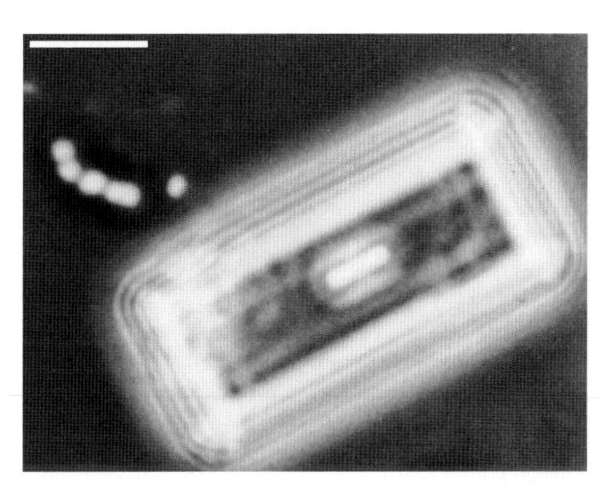

Figure B Halophilic bacteria in saturated salt solution. A string of five spherical bacteria (*Halococcus* sp.) are shown near a salt (sodium chloride) crystal. A rod-shaped bacterium (probably *Halobacter* sp.) is on the surface of the crystal. These salt-loving archaebacteria are tiny; the fuzzy rings around the three-dimensional salt crystal are due to the microscopic imaging. LM, bar = 5 μm.

Division: Mendosicutes

B-2 Crenarchaeota
(Eocytes, Thermoacidophils)

Greek *krene*, spring; *archae*, old; *-otes*, quality

GENERA
Acidianus
Desulfurococcus
Pyrobaculum
Pyrodictium
Pyrolobus
Sulfolobus
Thermococcus
Thermofilum
Thermoplasma
Thermoproteus

Recently, some 20 percent of the Earth's microbiota has been estimated to live deep in rocks, in hot vent waters, in sulfur pools, and elsewhere in total darkness. They are autotrophs that "fix" or assimilate CO_2 into their bodies using inorganic chemicals as energy sources: H_2, CH_4, NH_3, and other reduced (H-rich) compounds. More and more of these autotrophs have been identified by molecular techniques as Crenarchaeota archaebacteria.

The archaebacteria in Crenarchaeota are encountered in sulfurous hot springs all over the world. They inhabit the geothermal sources of Iceland, the geysers of Yellowstone National Park, submarine volcanic eruption fluids, and other habitats with conditions far too hot, too acidic, and too sulfur rich and oxygen poor for the far more familiar eubacteria. Thermoacidophils studied, such as *Thermoproteus* and *Sulfolobus*, have strong, acid-resistant cell walls composed of a glycoprotein material arranged in a hexagonal subunit pattern. *Thermoplasma*, like the mycoplasmas of Phylum B-9, to which it is not related, lacks walls entirely.

Cells of the genus *Sulfolobus*, which thrive in waters at 90°C and pH values of 1–2 (the acidity of concentrated sulfuric acid), were first isolated in culture in 1972. Some die of cold at temperatures below 55°C. Growing in environments ranging in pH from 0.9 to 5.8, preferring acid waters at a pH ranging from 2 to 3, *Sulfolobus acidocaldarius* is well named (Figure A). These archaebacteria have cell walls that lack peptidoglycan, and they are facultatively autotrophic. They use elemental sulfur as their energy source and fix CO_2, but they may also use glutamate, yeast extract, ribose, and other organic compounds. *Sulfolobus* generally is an aerobe or a microaerophil (colonizer of low-oxygen habitats) that oxidizes organic matter. Some strains can aerobically reduce iron (Fe^{3+} to Fe^{2+}). Live *Sulfolobus* cells can be seen tightly adhering to the surface of elemental sulfur crystals when they are viewed by fluorescence microscopy.

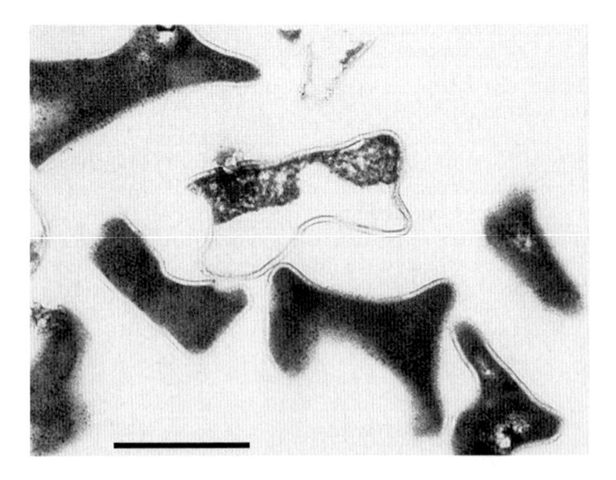

Figure A *Sulfolobus acidocaldarius*, although pleiomorphic like *Thermoplasma*, has well-bounded cells. TEM (negative stain), bar = 1 μm. [Photograph courtesy of D. W. Grogan.]

Pyrobaculum islandicum fixes CO_2 and grows with H_2 in acid water that is nearly boiling (95–98°C, pH 5–6). But it can also switch to heterotrophic metabolism metabolism and grow on acetate in H_2-rich anoxic environments. *Acidianus brierleyi* grows in even more acid water (pH 2.2) but at lower temperatures (70°C). Both these heat-loving crenarchaeota convert CO_2 into their bodies but use enzymatic pathways that differ from each other and that are entirely unlike the typical CO_2-fixing ribulose 1,5-bisphosphate (RuBP) carboxylase pathways of cyanobacteria, algae, plants, and sulfide-oxidizing bacteria.

Thermoplasma is a genus with one well-known species—*Thermoplasma acidophilum*—and about seven newly isolated strains (Figure B). The new *Thermoplasma* isolates are under investigation in Germany and Japan. The best-studied *T. acidophilum* comes from hot coal-refuse piles (waste tailings from coal mines) or from hot springs in Yellowstone National Park. First described in the 1970s, thermoplasmas are ecologically distinctive because of the extremely hot and acidic conditions under which they live, thriving at nearly 60°C and pH values of 1–2. Having no competition under such conditions—because no other organisms tolerate the hot acid so dangerous to their DNA and proteins—thermoplasmas can easily be grown in pure culture (that is, the extreme conditions favored by thermoplasmas exclude potential contaminating bacteria from the thermoplasma laboratory culture). However, observation of these live cells is very difficult because, at 37°C (human body temperature) or cooler and at pH 3 or greater, these thermoplasmas die. (Microscopes are ill equipped to maintain samples at high temperatures.)

Thermoplasmas are the only prokaryotes known to contain DNA coated with basic proteins similar to histones, the chromosomal proteins of most eukaryotes. The protein coating is believed to protect their DNA from destruction in hot acid. *Methanothermus*, a methanogen, also has this protein coating on its DNA. *Thermoplasma* may be related to the ancestor of the nucleocytoplasm of eukaryotes.

In recent years, submarine vents have yielded new genera of thermoacidophils such as *Pyrodictium*, a strict anaerobe that forms an anchorlike structure, a flagellin-like fibrous network for attachment. Growing at temperatures as high as 110°C, it rivals the bacterium growing at the hottest temperature known (113°C): *Pyrolobus*.

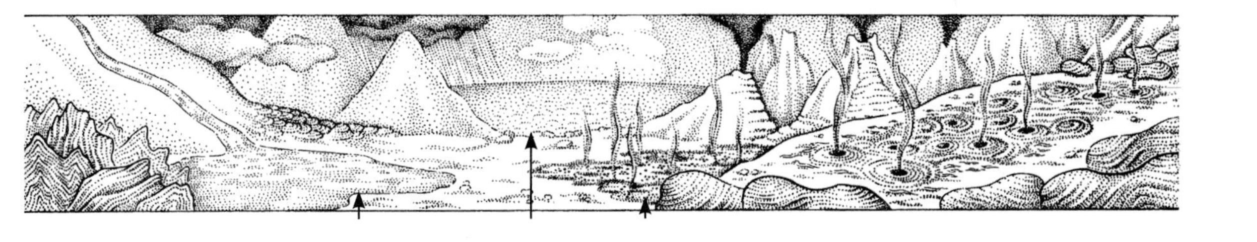

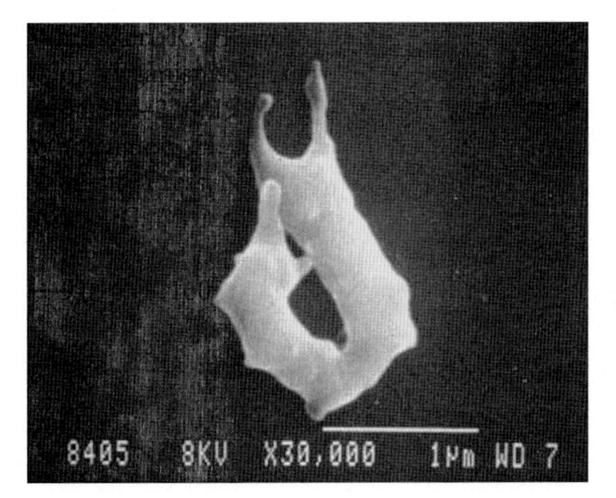

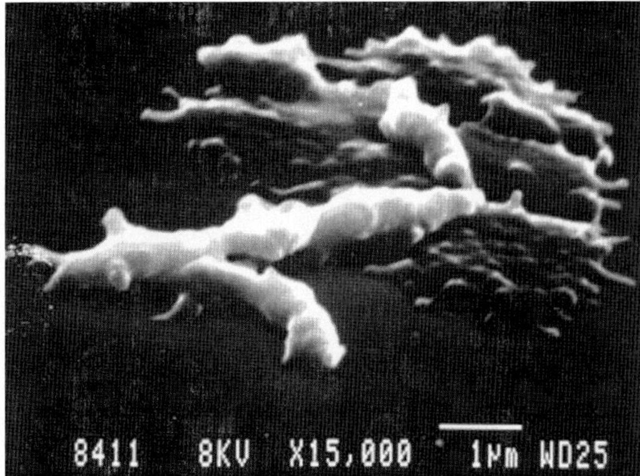

Figure B *Thermoplasma acidophilum* from a culture at high temperature, less than 50 percent oxygen, and low pH. Scanning electron microscopy reveals a great variety of morphologies in a single culture of *Thermoplasma*. When these same organisms are grown with particles of elemental sulfur, they flatten and adhere. SEM, bar = 0.5 µm. [Photograph courtesy of D. G. Searcy.]

SUBKINGDOM (DOMAIN) EUBACTERIA

All bacteria, all members of the superkingdom Prokarya that are not archaebacteria are "eubacteria": "true," or "good," bacteria. This latter group, comprising the vast majority of microbes described and named in the literature, is exceedingly diverse. Morphology and metabolism vary enormously. Multicellular bacteria with complex structures or capacity for cell differentiation are eubacteria. So are bacteria that derive energy directly from sunlight and make their food from carbon dioxide from the air (that is, photoautotrophic bacteria). All cells, including all bacteria, maintain an intact cell membrane at all times. The cell wall, if present, is always outside this intact cell (plasma) membrane (also called the "cell envelope"). Prokaryotes are grouped by bacteriologists according to their cell walls, a classification scheme not necessarily consistent with that based on 16S rRNA. The comprehensive and consistent classification in this book has reconciled the views and results published by as many scientists as possible in a form feasible for students to follow.

The various classifications for cell walls of eubacteria are Gram-negative walls (B-3 through B-8), no wall outside the cell membrane (B-9), or Gram-positive or proteinaceous walls (B-10 through B-14). The Gram-negative stain reaction correlates with the presence of an outer lipoprotein layer of the cell wall and a thin inner peptidoglycan (peptide units attached to nitrogenous sugars) layer bounded on the inside by the plasma membrane. Gram-positive walls may be associated with certain wall-bound proteins.

Bergey's Manual recognizes three major groups of eubacteria on the basis of their cell walls. The eubacteria of the archaebacterial division Mendosicutes (Latin *mendosus*, having faults; *cutis*, skin), by contrast, lack conventional peptidoglycan walls and show a great diversity in wall structure. This means that archaebacteria may or may not retain the violet-colored Gram stain. The three eubacterial divisions (superphyla) are Gracilicutes, Firmicutes, and Tenericutes.

Division Gracilicutes (Latin *gracilis*, slender, thin) includes all Gram-negative eubacteria that, even though they have an outer lipoprotein membrane, have generally thinner cell walls. Very few spore-forming eubacteria are Gram-negative, but some, such as *Sporomusa*, do exist.

Division Tenericutes (Latin *tener*, soft, tender) comprises eubacteria that lack cell walls. They correspond to Aphragmabacteria (B-9) and are genetically incapable of the synthesis of precursors of peptidoglycan. On the basis of their RNA molecules and other features, most scientists believe that wall-less eubacteria evolved by loss of walls from members of the Firmicutes, the Gram-positive eubacteria.

Division Firmicutes (Latin *firmus*, strong, durable) includes the Gram-positive eubacteria, which tend to have thick conspicuous peptidoglycan walls that all lack the outer, membranous lipoprotein layer. The organisms in this division may be

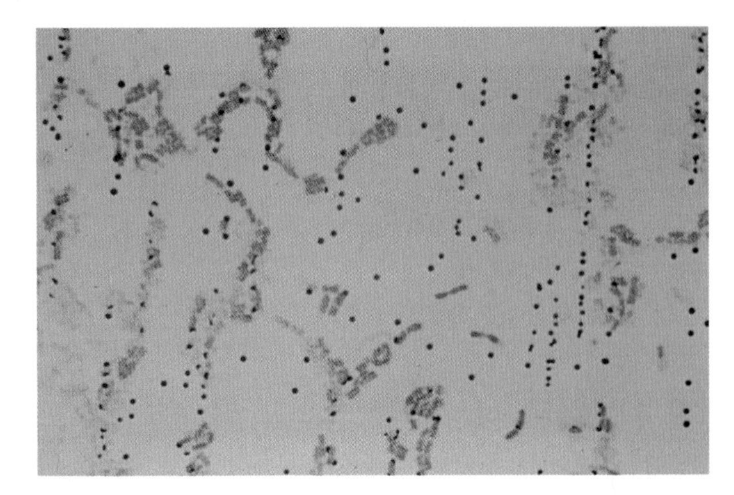

Figure A Eubacteria, Gram-negative stained rods (pink) and Gram-positive stained cocci (purple). [LM; courtesy of R. Guerrero].

spherical, rod shaped, filamentous, or composed of branching filaments. No Gram-positive eubacteria are phototrophic, but many form resistant propagules, spores, inside their cells or at the cell tips.

Members of both subkingdoms Archaebacteria (Mendosicutes: B-1 and B-2) and Eubacteria (Gracilicutes: B-3 through B-8; Tenericutes: B-9; and Firmicutes: B-10 through B-14) are diagrammed on the phylogenetic tree at the beginning of this chapter.

Division: Gracilicutes

B-3 Proteobacteria
(Purple Bacteria)

GENERA

Acetobacter	*Alginobacter*	*Bdellovibrio*
Actinobacillus	*Amoebobacter*	*Beggiatoa*
Aerobacter	*Ancalomicrobium*	*Beijerinckia*
Aeromonas	*Aquaspirillum*	*Beneckea*
Agrobacterium	*Archangium*	*Bradyrhizobium*
Alcaligenes (*Achromobacter*)	*Asticcacaulis*	*Caedibacter*
	Azotobacter	(continued)

The nearly 2000 species of this great group of eubacteria are classified by 16S rRNA data, morphology, and metabolism rather than by any single method. Because these ways of grouping the proteobacteria are mutually exclusive, we try to summarize the basic biology of the group members by briefly mentioning distinguishing characteristics.

Comparisons of the nucleotide sequences of 16S rRNA of thousands of bacterial isolates have led to reorganization of this vast group of bacteria into four nameless major lineages, referred to by Greek letters: α, β, γ, and δ (see the table on the facing page). These groups correspond to no consistent morphology or metabolism. Only representative genera are listed in this book. New genera are discovered frequently.

Large, slime-producing, multicellular gliding bacteria (for example, *Archangium*, *Myxococcus*, *Chondromyces*, and *Stigmatella*) traditionally grouped together as myxobacteria have been placed with other genera in the δ group of proteobacteria.

Among the metabolic variations displayed in this enormous and extremely diverse phylum, respiration is aerobic for many genera, reducing O_2 to H_2O. In the absence of oxygen, they do not stop growing, as obligate aerobes must; rather, as facultative aerobes, they continue to respire, using compounds and ions such as sulfate (SO_4^{2-}), nitrate (NO_3^-), or nitrogen (N_2) as the terminal electron acceptor and reducing them to sulfide (S^{2-}) or elemental sulfur (S^0), nitrite (NO_2^-), and nitrous oxide (N_2O), respectively. Cytochrome electron-transport pathways are used in these reductions, which are called "respiration." The pathways used with oxygen are the same as those used with nitrate. In facultatively aerobic species, two respiration products can be excreted by the same eubacterium, depending on physiological and ecological conditions. Many members of the phylum are chemoheterotrophic; that is, they require reduced organic compounds both for energy and for growth. But at least two genera of oxygen-respiring proteobacteria, *Bdellovibrio* and *Daptobacter*, are predaceous. They attack and live off other members, *Chromatium*, by reproducing in the cytoplasm of their prey.

Morphologically, these organisms range from solitary, simple unicells, on the one hand, to several classes of complex morphological types, such as stalked, budding, and aggregated bacteria, on the other hand.

Enterics, Gram-negative eubacteria that inhabit intestines, have long been associated with human, plant, and other animal diseases; many have been isolated from intestinal tissue or from diseased plants. The enterics include many rod-shaped microbes (Figure A), most of which have flagella distributed all around the cell (peritrichous): *Escherichia*, *Edwardsiella*, *Citrobacter*, *Salmonella*, *Shigella*, *Klebsiella*, *Enterobacter*, *Serratia*, *Proteus*, *Yersinia* (*Pasteurella*), and *Erwinia*.

The enterics are distinguished from one another by the carbohydrates that they use (lactose, glutamic acid, arabinose, sugar, alcohols, citrate, tartrate, and other fairly small organic compounds) and by chemical abilities. Some hydrolyze urea, produce gas from glucose, or break down gelatin. They are also distinguished by their sensitivity to specific bacteriophages, by their surface antigens, by their attraction to certain hosts, by their pathogenicity, and by their morphological traits, such as mode of motility and distribution of flagella.

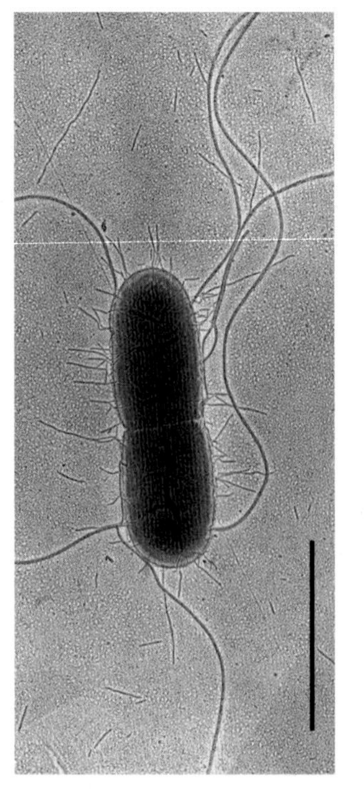

Figure A Peritrichously (uniformly distributed) mastigoted *Escherichia coli.* A new cell wall has formed and the bacterium is about to divide. The smaller appendages, called "pili," are known to make contact with other cells in bacterial conjugation. However, even many strains that do not conjugate have pili. TEM (shadowed with platinum), bar = 1 μm. [Courtesy of D. Chase.]

These Gram-negative, rod-shaped, unicellular heterotrophs grow rapidly and well. Although none produce spores, they seem to have remarkable persistence, as do some vibrios, waiting things out under conditions of adversity and vigorously taking advantage of new food sources. Water samples everywhere yield enterics when incubated under the proper conditions of growth. It is not too extreme to assert that most life on Earth takes the form of facultatively aerobic, Gram-negative, unicellular, rod-shaped bacteria. At the base of microbial food webs, they provide food for innumerable protoctists and other organisms. As the chief object of study by molecular biologists, *Escherichia coli* is better known than any other single organism on Earth.

A second group of enterics comprises mainly comma-shaped organisms, most of which have a single polar flagellum. Bacteria of this shape, called "vibrios," include at least seven genera: *Vibrio*, *Beneckea*, *Aeromonas*, *Pleisiomonas*, *Photobacterium*, *Xenorhabditis*, and *Zymomonas*. Members of the genus *Vibrio* are associated with cholera. They ferment carbohydrates into mixed products, including acids, but do not give off carbon dioxide and hydrogen. *Beneckea*, a marine vibrio that requires salt, is capable of fermentative or aerobic respiring growth on a broad range of carbon sources. Some vibrios, *Photobacterium* and *Xenorhabditis*, are bioluminescent. Photobacteria associate with certain tropical, marine fish, which culture the bacteria

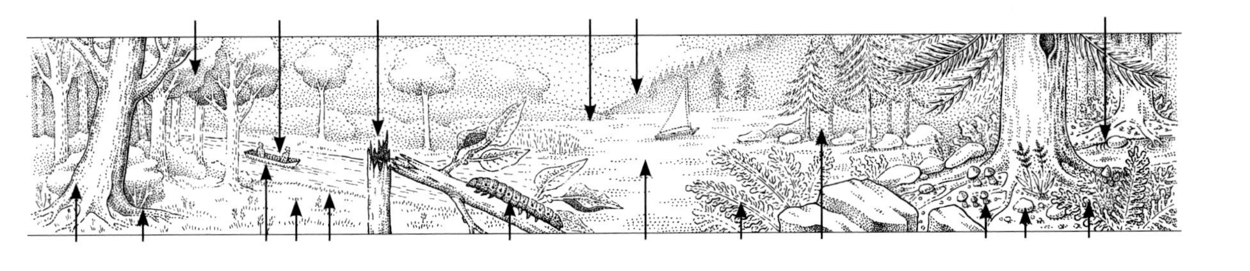

Best-Known Proteobacteria

Group	Genus	Phenotypes: Morphology	Phenotypes: Metabolism
Alpha (α) group	*Agrobacterium*	Motile rod*	Crown gall former, heterotroph
	Aquaspirillum	Motile, spirillum-shaped bacterium	Heterotroph
	Beijerinckia	Motile rod	Nitrogen fixer
	Hyphomicrobium	Budding bacterium	Heterotroph
	Nitrobacter	Short, budding rod	Ammonia oxidizer, nitrate oxidizer (marine)
	Paracoccus	Motile rod	Heterotroph
	Pseudomonas†	Motile rod, polar flagella	Heterotroph
	Rhizobium	Symbiotic with root nodules	Nitrogen fixer
	Rhodobacter	Flagellated rod	Purple nonsulfur phototroph
	Rhodomicrobium	Budding bacterium	Purple nonsulfur phototroph
	Rhodopseudomonas	Flagellated rod	Purple nonsulfur phototroph
	Rhodospirillum	Spirillum-shaped bacterium	Purple nonsulfur phototroph
Beta (β) group	*Alcaligenes*	Motile rod	Hydrogen oxidizer
	Neisseria	Nonmotile coccus	Heterotroph
	Nitrosomonas	Rod	Ammonia oxidizer, nitrite oxidizer
	Pseudomonas†	Motile rod, polar flagella	Heterotroph
	Sphaerotilus	Filament	Iron oxidizer
	Spirillum	Motile, spirillum-shaped bacterium	Heterotroph
	Thiobacillus	Motile rod	Sulfide oxidizer
Gamma (γ) group	*Azotobacter*	Rod	Nitrogen fixer
	Beggiatoa	Filaments form rosettes	Sulfide oxidizer
	Chromatium	Motile rod, sulfur globules in cells	Purple sulfur phototroph
	Escherichia	Motile rod, peritrichous flagella	Heterotroph
	Legionella	Motile rod	Heterotroph
	Leucothrix	Filament	Sulfide oxidizer
	Pseudomonas† (fluorescent)	Motile rod, polar flagella	Heterotroph
	Salmonella	Motile rod	Heterotroph
	Shigella	Motile rod	Heterotroph
	Thiocapsa	Gelatinous colonies of cocci	Purple sulfur phototroph
	Thiospirillum	Motile, spirillum-shaped bacterium	Purple sulfur phototroph
	Vibrio	Motile, comma-shaped bacterium	Heterotroph
Delta (δ) group	*Archangium*	Myxospores stalkless, rod cell ends tapered	Aerobic heterotroph
	Bdellovibrio	Motile, tiny curved rod	Predator
	Chondromyces	Single–branched, myxospore-bearing stalks	Aerobic heterotroph
	Desulfotomaculatum	Spore-forming rod	Sulfate reducer
	Desulfovibrio	Motile rod, polar flagella	Sulfate reducer
	Desulfuromonas	Small rod	Sulfate reducer, obligate anaerobe, obligate sulfur reducer
	Myxococcus	Glider	Heterotroph
	Polyangium	Multiple myxospore-bearing stalks	Aerobic heterotroph
	Stigmatella	Myxospores stalkless, rod cell ends rounded	Aerobic heterotroph

*Motility may be by flagella (swimming) or by unknown means, that is, slow movement in contact with a surface, called "gliding motility."

†Pseudomonads are aerobic, straight, or slightly curved, rod-shaped cells that swim by polar flagella. They are incapable of fermentation but oxidize a wide range of organic substances. Analysis of their 16S rRNA genes shows that any given strain may belong to any one of the three groups of Proteobacteria: α, β, and γ.

GENERA
Calymmatobacterium
Cardiobacterium
Caulobacter
Chloropseudomonas
Chondromyces
Chromatium
Chromobacterium
Citrobacter
Coxiella
Daptobacter
Desulfacinum
Desulfonema
Desulfotomaculum
Desulfovibrio
Desulfuromonas
Edwardsiella
Enterobacter
Erwinia
Escherichia
Ferrobacillus
Flavobacterium
Gluconobacter
Haemophilus
Hyphomicrobium

in special pockets called "light organs." The growth requirements of *Photobacterium* are generally far more restricted than those of the free-living marine genus *Beneckea*. *Xenorhabditis* is known from its associations with nematodes and nematode-eating insects.

Members of the genus *Aeromonas*, informally called "aeromonads," are common in ponds, lakes, and soils. They are coccoids or straight rods with rounded ends; most are motile by means of polar flagella. When growing anaerobically, they reduce nitrate to nitrite. Most of them contain cytochrome *c*, an electron-transporting protein whose activity is due to its inclusion of a porphyrin molecule, and catalase, an enzyme that decomposes hydrogen peroxide into water and oxygen. Very eclectic in their tastes, they use a wide variety of food sources, especially plant materials such as starch, casein, gelatin, dextrin, glucose, fructose, and maltose.

Similar unicellular, Gram-negative bacteria include *Chromobacterium*, *Haemophilus*, *Actinobacillus*, *Cardiobacterium*, *Streptobacillus*, *Calymmatobacterium*, and several symbionts of ciliates in Phylum Pr-6. Certain *Paramecium* symbionts, originally called kappa, lambda, sigma, and mu particles, were revealed to be proteobacteria with complex requirements for growth. Once thought to be cytoplasmic genes of the ciliate itself, they are now known to be members of the genus *Caedibacter*.

Many enterics produce colorful pigments: the violet, ethanol-soluble violacein of *Chromobacterium*; the red prodigiosin of *Serratia* and *Beneckea*; and the red yellow, orange, and brown pigments, some of which are carotenoids and some not, of *Flavobacterium* and *Aeromonas*. No one yet understands the functions of these pigments.

Other proteobacteria are extremely heterogeneous. In one group, the prosthecate bacteria, appendages called "prosthecae" (stalks) made of living material protrude from the cells. *Caulobacter* and *Asticcacaulis* have single polar or subpolar prosthecae (Figure B). Their life histories superficially resemble those of some marine animals: a stalked sessile form divides to produce a motile form, which swims away; its offspring, in turn, is a sessile form.

The budding bacteria, or hyphomicrobia, reproduce by the outgrowth of buds that eventually swell to parent size. Hyphomicrobial colonies may form quite complex networks that resemble the mycelia of fungi. This sort of budding is found both in photosynthetic microbes, such as *Rhodomicrobium* (Figures C and D), and in heterotrophs, such as *Hyphomicrobium*.

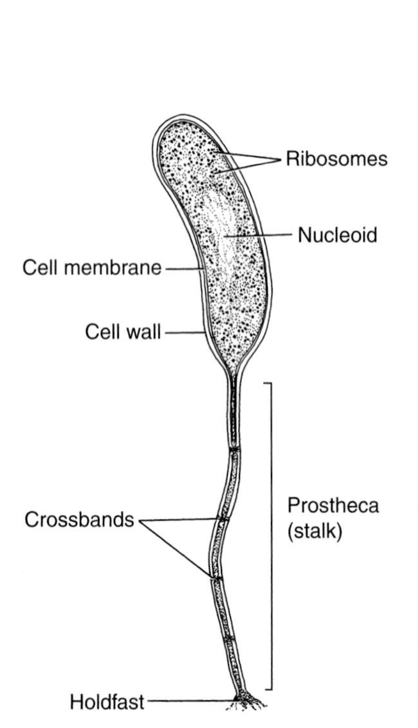

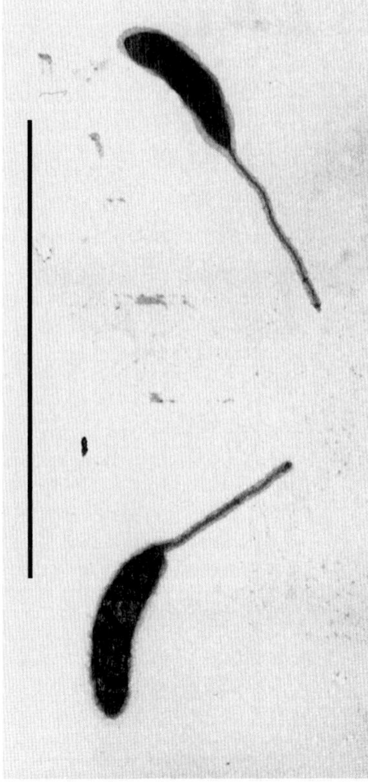

Figure B Stalked cell of *Caulobacter crescentua*, which in nature would be attached to plants, rocks, or other solid surfaces. This cell divides to form swarmer cells. TEM (negative stain, whole mount), bar = 5 μm. [Photograph courtesy of J. Staley; drawing by I. Atema.]

Klebsiella	*Neisseria*	*Nitrosomonas*	*Proteus*	*Rickettsia*	*Stigmatella*
Legionella	*Neorickettsia*	*Nitrosospira*	*Pseudomonas*	*Rickettsiella*	*Streptobacillus*
Leucothrix	*Nitrobacter*	*Nitrospira*	*Rhizobium*	*Rubrivivax*	*Thermodesulfo-rhabdus*
Macromonas	*Nitrococcus*	*Oceanospirillum*	*Rhodobacter*	*Salmonella*	*Thiobacillus*
Methylococcus	*Nitrocystis*	*Paracoccus*	*Rhodoferax*	*Serratia*	*Thiobacterium*
Methylomonas	*Nitrosococcus*	*Photobacterium*	*Rhodomicrobium*	*Shigella*	*Thiocapsa*
Methylosinus	*Nitrosogloea*	*Pleisiomonas*	*Rhodopseudomonas*	*Sphaerotilus*	(continued)
Myxococcus	*Nitrosolobus*	*Polyangium*	*Rhodospirillum*	*Spirillum*	

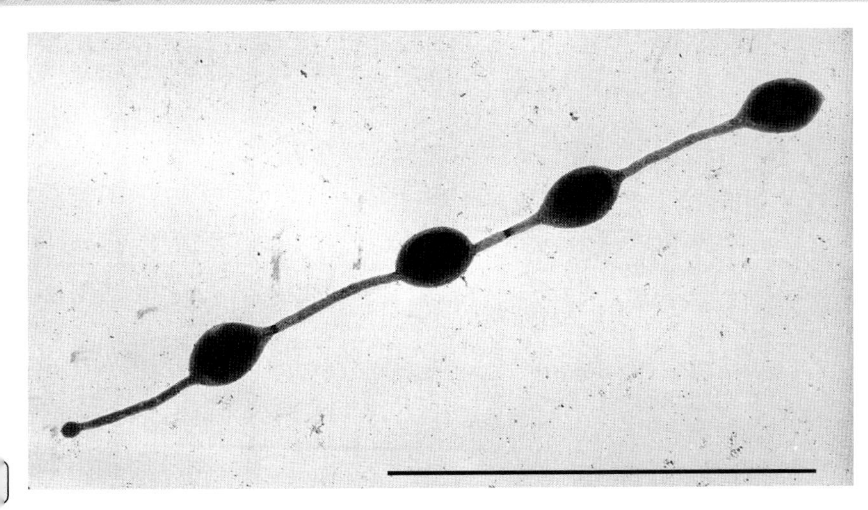
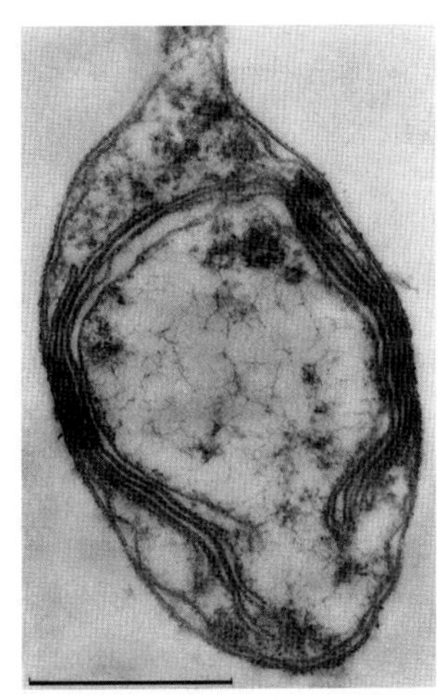

Figure C *Rhodomicrobium vannielii*, a phototrophic, purple nonsulfur bacterium that lives in ponds and grows by budding. (Left) A new bud is forming at lower left. TEM, bar = 1 μm. (Right) Layers of thylakoids (photosynthetic membranes) are visible around the periphery of this *R. vannielii* cell. TEM, bar = 0.5 μm. [Courtesy of E. Boatman.]

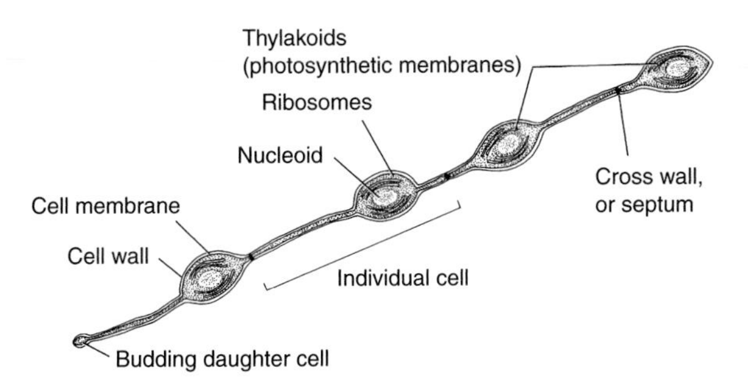

Figure D *Rhodomicrobium vannielii.* [Drawing by I. Atema.]

The aggregated bacteria, which include *Sphaerotilus*, can be recognized in the light microscope by their distinctive metal-rich clumps of cells. These microbes oxidize iron or manganese, which they then deposit around themselves as manganese and iron oxides. The bacteria are thought to gain some energy from these oxidations; however, there is still no proof that any of them are obligate lithotrophs. All prefer fixed nitrogen and carbon compounds and grow faster if supplied with organic food.

Nonmastigoted bacteria of the genus *Neisseria* are infamous as the cause of gonorrhea and one form of bacterial meningitis. They can grow on their own.

The acetic acid bacteria, *Gluconobacter* and *Acetobacter*, oxidize ethanol to acetic acid—wine to vinegar. They form sheaths having a rectangular cross section.

The rickettsias, unable ever to grow on their own, have an obligate intracellular existence in animals. They have residual Gram-negative walls, cell walls reduced in size, and their metabolism is limited. Several strains cause Rocky Mountain spotted fever.

The best-known chemolithoautotrophs belong to the α, β, or γ group of Proteobacteria. They probably evolved from phototrophic ancestors. Chemolithoautotrophy is a metabolism that functions without sunlight and without preformed organic compounds—not a single vitamin, sugar, or amino acid. Thus, chemolithoautotrophic bacteria represent the pinnacle of metabolic achievement. They live on air, salts, water, and on inorganic source of energy. Provided with nitrogenous salts, oxygen, carbon dioxide (CO_2), and an appropriate reduced compound to use as an energy source, they make all of their own nucleic acids, proteins, and carbohydrates and derive their energy from oxidation of the reduced compound. Some are capable of using organic compounds as food, but all can do without them. Chemolithoautotrophic bacteria are crucial to the cycling of nitrogen, carbon, and sulfur throughout the world because they convert gases and salts unusable by animals and plants into usable organic compounds. The maintenance of the biosphere depends on such metabolic virtuosity, yet chemolithoautotrophy is strictly limited to bacteria.

Chemolithoautotrophic bacteria are grouped by the compounds that they oxidize to gain energy. At least three types are distinguished: oxidizers of nitrogen compounds, of sulfur compounds, and of methane.

Chemolithoautotrophs that oxidize reduced nitrogen compounds include morphologically distinct organisms that oxidize nitrite to nitrate: *Nitrobacter*, *Nitrospira*, *Nitrocystis*, and *Nitrococcus.* Nitrobacters are short rods, many of them pear-shaped or wedge-shaped (Figure E). Elaborate internal membranes extend

B-3 Proteobacteria
(continued)

GENERA

Thiocystis
Thiodictyon
Thiomicrospira
Thiopedia
Thiosarcina
Thiospira
Thiospirillum
Thiothece
Thiovulum
Vibrio
Xenorhabditis
Yersinia (Pasteurella)
Zymomonas

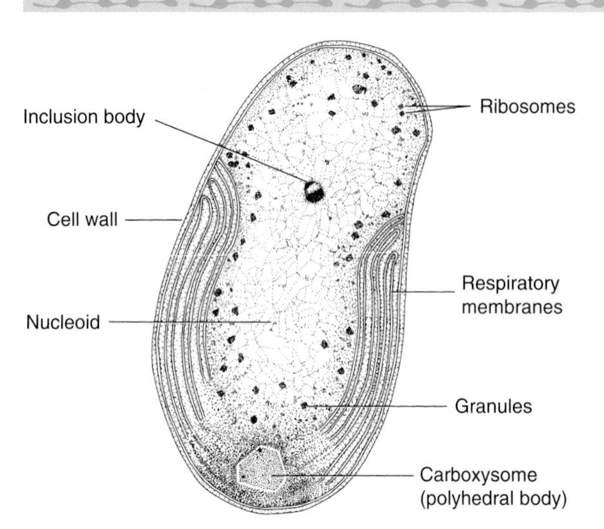

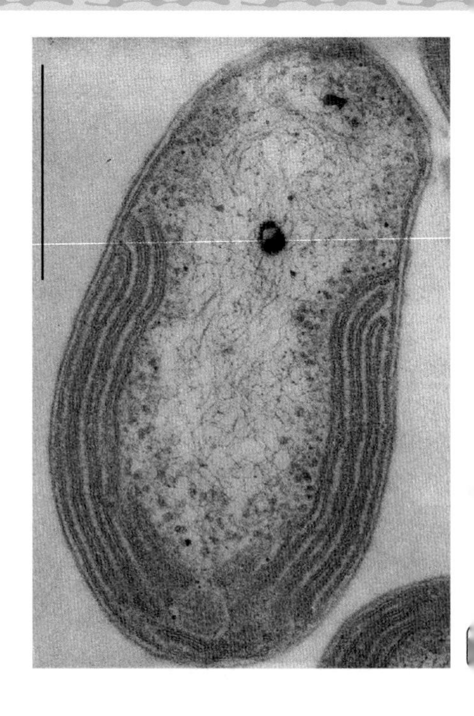

Figure E *Nitrobacter winogradskyi.* This specimen is young and thus lacks a prominent sheath. Carboxysomes are bodies in which are concentrated enzymes for fixing atmospheric CO_2. This species is named for the Russian Sergius Winogradsky, who pioneered the field of microbial ecology. TEM, bar = 0.5 μm. [Photograph courtesy of S. W. Watson, *International Journal of Systematic Bacteriology* 21:261 (1971); drawing by I. Atema.]

along the periphery of one end of the cell. Old cultures of *Nitrobacter winogradskyi*, a widely distributed soil microbe, form a flocculent sediment made of gelatinous sheaths produced by the bacteria. Organic compounds and even ammonium salts inhibit the growth of this nitrobacter. *Nitrospira* are long slender rods that lack elaborate internal membranes. They are marine bacteria, strict aerobes, and strict chemolithoautotrophs. Nitrococci are spherical cells containing distinctive internal membranes that form a branched or tubular network in the cytoplasm.

The other group of nitrogen-compound oxidizers contains the chemolithoautotrophs that oxidize ammonia (NH_3) to nitrite (NO_2^-) for energy: *Nitrosomonas*, *Nitrosospira*, *Nitrosococcus*, and *Nitrosolobus*. They live in environments containing both oxygen and ammonia, such as at the edges of the anaerobic zone at the sedimentation interface where the solid surface contacts seawater or freshwater of soil or of lakes and rivers. *Nitrosomonas* species are either ellipsoidal or rod-shaped; they may be single, in pairs, or in short chains. They are rich in cytochromes, which impart a yellowish or reddish hue to laboratory cultures. Internal membranes extend along the cell periphery. They grow at temperatures between 5°C and 30°C. *Nitrosospira* is a genus of spiral-shaped freshwater microbes that lack internal membranes. *Nitrosococcus* cells are spherical; they grow singly or in pairs and often form an extracellular slime. Aggregates of cells attach to surfaces or become suspended in liquid. *Nitrosolobus* cells are variously shaped, lobed cells that are motile by means of peritrichous flagella. They divide by binary fission (cell division producing two equal offspring cells).

There are at least five genera of organisms currently recognized that grow by oxidizing inorganic sulfur compounds. Their cells contain sulfur globules (products of oxidizing sulfide to elemental sulfur), and they live in high concentrations of hydrogen sulfide or other oxidizable sulfur compounds. The genera are of four distinct morphological types: nonmotile rods embedded in a gelatinous matrix (*Thiobacterium*), cylindrical cells having polar flagella (*Macromonas*), ovoid cells having peritrichous flagella (*Thiovulum*), and spiral cells having polar flagella (*Thiospira* and *Thiobacillus*).

Thiobacillus is the best-known genus of sulfur oxidizers and has been grown in culture. Most are Gram-negative rods and motile by means of a single polar flagellum. They derive energy from the oxidation of sulfur or its compounds, such as sulfide, thiosulfate, polythionate, and sulfite. The final oxidation product is sulfate, but other sulfur compounds may accumulate under certain conditions. One species, *Thiobacillus ferrooxidans*, also oxidizes ferrous compounds. *Thiobacilli* will grow on strictly inorganic media, fixing CO_2 to produce cell material.

Proteobacteria that oxidize the reduced single-carbon compounds, CH_4 or CH_3OH (methanol), are methylomonads. Two genera may be distinguished by morphology: *Methylomonas* (various Gram-negative rods) and *Methylococcus* (spherical cells usually appearing in pairs). Methylomonads cannot grow on complex organic compounds; rather, they use CH_4 or CH_3OH as their sole source of both energy and carbon. In fact, the growth of many methylomonads is inhibited by the presence of organic matter.

Beside these chemolithoautotrophs are organic carbon–requiring chemotrophs that respire sulfur compounds. The sulfate reducers of the δ group of proteobacteria are obligate anaerobes quickly poisoned by exposure to oxygen. They require SO_4^{2-}, just as we require oxygen, for respiration. In this energy-yielding process, electrons from food molecules are transferred to inorganic compounds; by this transfer, the food molecules are oxidized and the inorganic compounds are reduced in the oxidation state. All reduce SO_4^{2-} to some other sulfur compound, such as elemental sulfur or hydrogen sulfide (H_2S), and synthesize cytochromes. Although these chemotrophs obtain energy from SO_4^{2-}, they also take in organic compounds, usually the three-carbon compound lactate or pyruvate, as a source of carbon, electrons, and energy. Thus, they are not autotrophs.

More than 18 genera of sulfate-reducing bacteria are known. Some, such as *Desulfovibrio* and the motile rod *Desulfotomaculum*, cannot oxidize acetate but can use other carbon sources for energy, such as lactate or pyruvate. Others, strict anaerobes, oxidize acetate to CO_2 by using this carbon compound as an electron source. Genera of acetate-using sulfate reducers include the large filament *Desulfonema* and the thermophils *Desulfacinum* and *Thermodesulforhabdus*.

The δ group of sulfate reducers—members of the best-known genus *Desulfovibrio*—are unicellular bacteria widely distributed in marine muds, estuarine brines, and freshwater muds. Either single polar flagella or bundles of them (lophotrichous flagella) provide motility. The genera that require sodium chloride for growth are considered marine bacteria. Sulfate reducers contain cytochrome c_3 and a pigment called desulfoviridin, which gives them a characteristic red fluorescence. Many also synthesize hydrogenases, enzymes that generate hydrogen, which protects the organisms from the hostile aerobic world.

Sulfate reducers release gaseous sulfur compounds, including H_2S, into the sediments, thus playing a crucial role in cycling sulfur—a constituent of all proteins throughout the world. In iron-rich water, the H_2S formed by these bacteria reacts with iron, leading to the deposition of pyrite (iron sulfide, also known as "fool's gold"). It is thought that Archaean and Proterozoic iron deposits may be due, at least partly, to the activity of sulfate-reducing bacteria. No symbiotic forms of sulfate-reducing bacteria have been reported; members of all groups are free-living.

Desulfotomaculum is a genus of unicellular, straight or curved, rod bacteria that are motile by means of peritrichous flagella. The genus is distinguished by the formation of resistant endospores and is commonly found in marine and freshwater muds, in the soil of geothermal regions, in the intestines of insects and bovine animals, and in certain spoiled foods.

Also in the δ group of proteobacteria are large, heterotrophic, multicellular bacteria—the myxobacteria. The myxobacteria represent, with the cyanobacteria (B-6), the acme of morphological complexity among the prokaryotes that form upright, propagule-dispersing, multicellular bodies. Individual myxobacteria are obligately aerobic, unicellular, Gram-negative rods that may be as long as 5 μm. Some aggregate into complex colonies that show distinctive behavior and form. The cells are typically embedded in slime consisting of polysaccharides of the cells' own making.

When soil nutrients or water are depleted, members of certain genera of myxobacteria (for example, *Stigmatella* and *Chondromyces*) aggregate and form upright structures composed of extracellular excretion products and many cells (Figures F and G). Bacterial cells within these reproductive bodies enter a resting stage; the resting cells are called "myxospores." In some

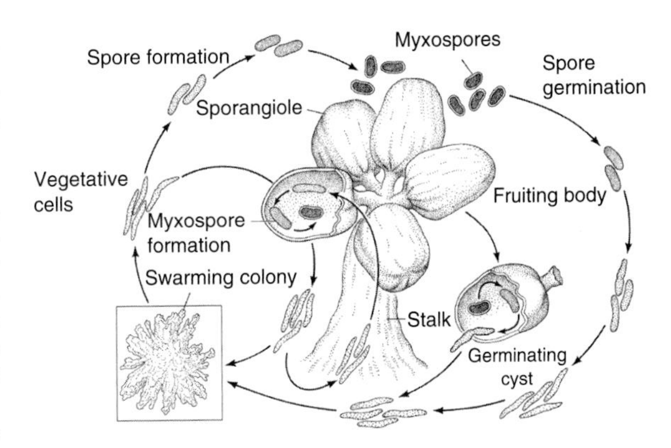

Figure F Life cycle of *Stigmatella aurantiaca*. [Drawing by L. Meszoly; labeled by M. Dworkin.]

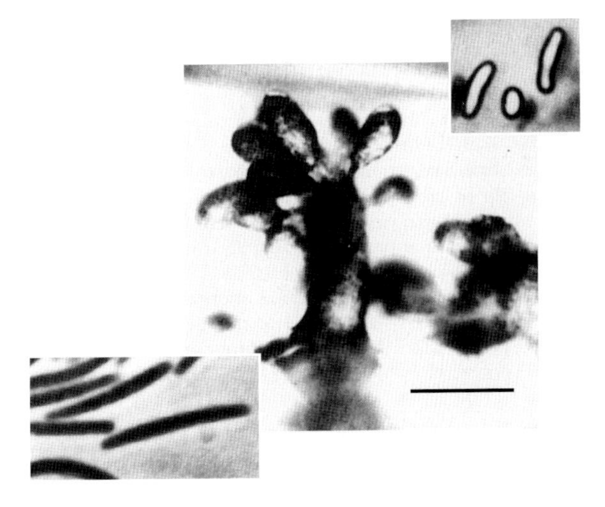

Figure G The reproductive body of *Stigmatella aurantiaca*, which grows on the remains of vegetation in soil. LM, bar = 100 μm. (Inset, bottom left) Growing cells, which glide in contact with solid surfaces. (Inset, top right) Myxospores. [Photographs courtesy of H. Reichenbach and M. Dworkin, in *The prokaryotes*, M. Starr *et al.*, eds. (Springer-Verlag; New York; 1981).]

taxa, these resting cells may become encapsulated, thick walled, and shiny; in others, they seem to be quite like growing bacteria. Some, such as *Polyangium violaceum*, form brightly colored reproductive structures. Others form branched stalks; these tiny "trees" may be barely visible to the unaided eye.

Myxococcus xanthus cells glide by retraction of pili. The cells make outer membrane lipoproteins required to form the pili. By transient membrane fusion and specific pili protein transfer from one *Mixanthus* cell to another, colony members "talk" to one another and correct one another's genetic defects. They change modes of foraging on other victim bacteria that become their food.

Some myxobacteria form thick-walled, darkly colored, spore-filled cysts, called "sporangioles," that open when wetted to release huge numbers of individual gliding bacteria; the gliding cells move together to form migrating colonies. The entire history of these myxobacteria is uncannily analogous to that of the slime molds (Pr-2).

Myxobacters, multicellular colonial proteobacteria, construct stalks with swollen ends (myxocysts, erroneously called "fruiting bodies" because these organisms were classified as "plants"). Inside the swellings, whether on the ground or borne aloft, they form propagules ("myxospores"). The propagules, thousands of which may be released at once, may be discharged through breached cysts or be dispersed through air. When they fall on moist soil, logs, or pond water surfaces, each survivor single cell capable of germination begins to grow and divide by bacterial cell division. Wherever it lands, as long as the habitat provides moisture and food, growth continues. The individual cells in a colony communicate with each other. Myxobacteria are often called "social bacteria." This group produces a range of toxic compounds made of protein: "bacteriocins" that regulate their social interactions. Bacteriocins may inhibit the growth or even kill members of their own population. The bacterial colonies may comprise millions of individual cells that glide in complex ways. Some generate branched, complex, emergent structures (Figure G). This group of proteobacteria (named after Proteus the Greek God, who could change his shape) has been studied. The diverse class of bacteriocin proteins have been identified in all prokaryotes studied so far, including Archaebacteria.

Phototrophic proteobacteria are morphologically diverse. Many are single cells and motile or immotile. Some grow as packets or as stalked budding structures, extensive filaments, or sheets of cells in which the spaces between the cells are filled with coverings, called "sheaths," composed of mucous material. Some contain gas vacuoles, giving them buoyancy and a sparkling appearance. In anoxic environments, most purple sulfur bacteria convert hydrogen sulfide into elemental sulfur, which they deposit inside their cells in tiny but visible granules; the presence, distribution, and shape of these granules can be used to distinguish them.

The phototrophic bacteria are delightfully colored in an astonishing range of pinks and greens, although in the bright sunlight, in the top layers of anaerobic muds, they become very dark, nearly black. Because each species has an optimum growth at a given acidity, oxygen tension, sulfide and salt concentration, moisture content, and so on, they often grow in layers—each in its appropriate niche. Well-lit anoxic sediments become layered communities of phototrophic and other bacteria.

Although many of the major photosynthetic bacteria belong here in the proteobacteria—for example, the purple nonsulfur bacteria such as *Rhodospirillum*, *Rhodomicrobium*, *Rhodoferax*, and *Rubrivivax* are in either the α or the β lineage, and the purple sulfur bacteria such as *Chromatium* or *Amoebobacter* are genera of photosynthesizers grouped together as δ purples—many others do not (Figure H).

Among the many different species of phototrophs, some are tolerant of extremely high or extremely low temperatures or salinities. In each group, some kinds are capable of fixing atmospheric nitrogen. The ability to fix atmospheric nitrogen, the conversion of nitrogen into organic compounds, such as amino acids, that include nitrogen in their structure, is conspicuously present in many members of this huge phylum. This important process is entirely limited to bacteria. Most notable are the free-living, aerobic (oxygen-respiring), soil nitrogen fixers, *Azotobacter* (Figure I) and *Beijerinckia* among them. Other close relatives of these free-living bacteria include some soil bacteria that can also live as plant root symbionts, such as the motile rods *Rhizobium* and *Bradyrhizobium.* All atmospheric nitrogen fixers contain nitrogenase, a large enzyme complex. Nitrogenase is composed of azo- and molybdoferredoxins—proteins containing iron and molybdenum and that are absolutely necessary for the reduction of nitrogen to organic nitrogen compounds such as glutamine. The genes for the entire process may be borne on plasmids, relatively small pieces of DNA, that transfer from one bacteria to another kind of bacteria. For this reason, organisms are not classified according to this ability (Figure I).

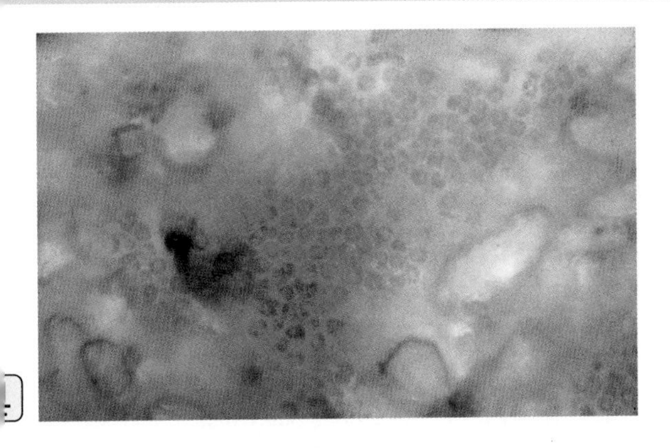

Figure H *Thiocapsa* sp., from Laguna Figueroa, Baja California Norte, Mexico. This multicellular, sulfide-oxidizing, non-oxygenic phototrophic purple sulfur bacterium commonly dwells in microbial mats and scums.

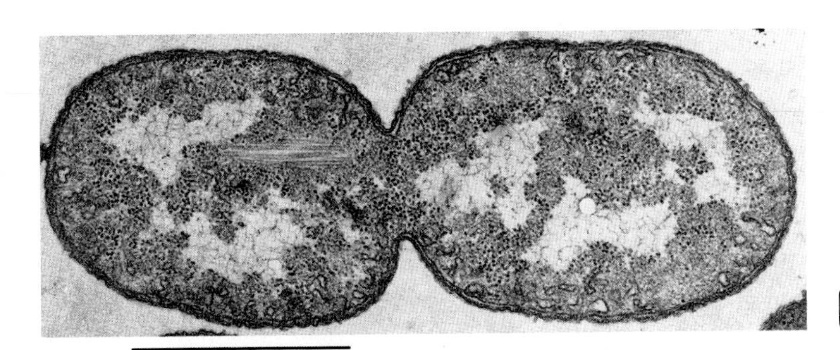

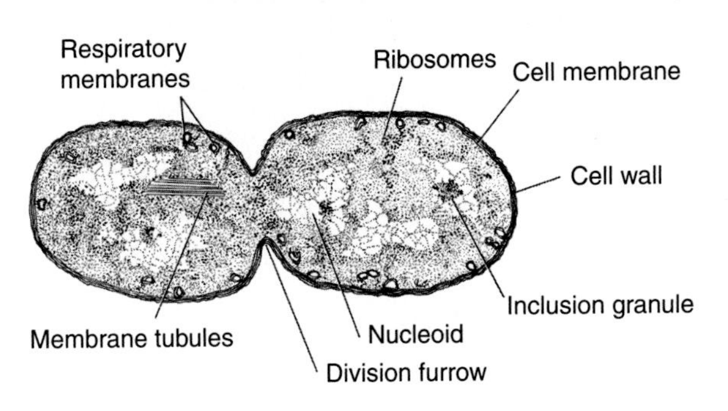

Figure I *Azotobacter vinelandii*, commonly found in garden soils. In this photograph, division into two cells is nearly complete. TEM, bar = 1 μm. [Photograph courtesy of W. J. Brill; drawing by I. Atema.]

Division: Gracilicutes

B-4 Spirochaetae

Latin *spira*, coil; Greek *khaite*, long hair

GENERA

Borrelia	*Hollandina*	*Pillotina*
Canaleparolina	*Leptonema*	*Spirochaeta*
Clevelandina	*Leptospira*	*Spirosymplokos*
Cristispira	*Mobilifilum*	*Treponema*
Diplocalyx	*Perfiel'eva*	

Spirochetes look like coiled snakes. Unlike other motile bacteria, they have from two to more than two hundred internal flagella (axial filaments or endoflagella) in the space between the inner (plasma) membrane and the outer cell membrane of the Gram-negative cell wall; that is, they have "periplasmic flagella" that are in the cell wall. The range of numbers is specific to the genus. Spirochetes are found in marine waters and freshwaters, deep muddy sediments, the gastrointestinal tracts of several different kinds of animals, and elsewhere. Their long, slender, corkscrew shape enables them to move flexibly through thick, viscous liquids with great speed and ease. In more dilute environments, many swim quickly with complex movements—rotation, torsion, flexion, and quivering. The unusual flagellar arrangement, probably responsible for their corkscrew shape and characteristic movements, distinguishes spirochetes from other bacteria. Each flagellum originates near an end of the cell and extends along the body. Thus, flagella anchored at opposite ends of the cell often overlap, like the fingers of loosely folded hands. The internal rotation of these flagella is thought to be responsible for the motility of spirochetes, just as is external ciliary rotation for other motile bacteria.

Major groups of spirochetes include leptospires (*Leptospira* and *Leptonema*), spirochaetas (*Spirochaeta* and related genera), and pillotinas (*Pillotina*, *Hollandina*, and *Diplocalyx*). Leptospires require gaseous oxygen, whereas most other spirochetes are quickly poisoned by its slightest trace. Certain leptospires (members of *Leptospira*) live in the kidney tubules of mammals. Often, they are carried with the urine into water supplies and can enter the human bloodstream through cuts in the skin, causing the disease leptospirosis. Because leptospires are so thin and hard to see in the microscope, the disease is often misdiagnosed.

The spirochaetas include at least three genera—*Borrelia*, *Treponema*, and *Spirochaeta*— some species of which are internal symbiotrophs of animals. Infamous as the cause of syphilis, *Treponema pallidum* is also responsible for yaws, a debilitating and unsightly tropical eye disease. *Treponema* have one to four flagella at each end of the cell. The genus *Spirochaeta* (Figure A) contains free-living marine water and freshwater spirochetes that are less than a micrometer wide and have a small number of overlapping flagella. They resemble *Treponema* but most are free-living.

Perfiel'eva spirochetes are free-living mud dwellers in both freshwater and saline water. They were first discovered as the motile, sulfide-oxidizer component of a seaside consortium thought to be a single bacterium named *Thiodendron latens* ("sulfur-tree lazy") with a morphologically complex life history. Associated with sulfidogens (*Desulfomaculatum*, *Desulfothiovibrio*, for example, B-3), at least six localities in Russia and the Pacific, Perfiel'eva are not only oxygen tolerant but their metabolism changes in the presence of small amounts (<2%) of ambient oxygen. They grow more rapidly, produce more acetate and less formate and possibly vary in other ways when grown microaerophilically. They die at great oxygen pressures (~2–20%), whereas they thrive under strictly anoxic conditions. The great microbial ecologist Perfiel'ev died before his former student Galina Dubinina, microbiologist at the Moscow Academy of Sciences, recognized the spirochete partner in this geochemically important cosmopolitan consortium.

The pillotinas, all symbionts of animals, are much larger than other spirochetes (Figures B and C). Some may be 3 μm

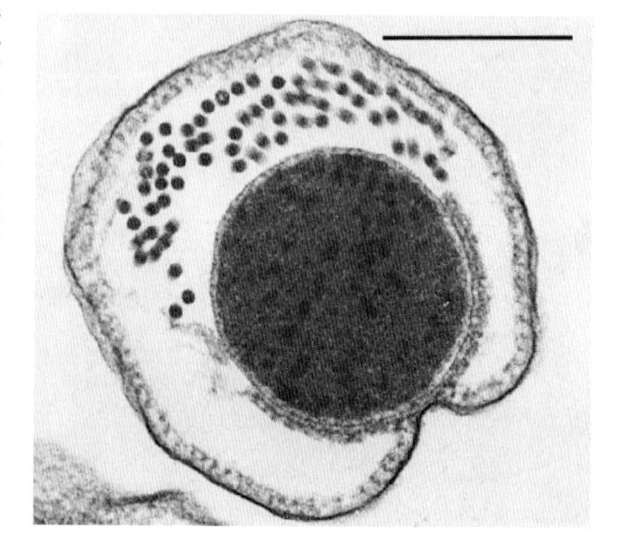

Figure B *Diplocalyx* sp., in cross section. These large spirochetes, which belong to the family Pillotaceae (the pillotinas), have many flagella. The several genera of Pillotaceae all live in the hindguts of wood-eating cockroaches and termites. This specimen was found in the common North American subterranean termite *Reticulitermes flavipes* (A-21). TEM, bar = 1 μm. [Courtesy of H. S. Pankratz and J. Breznak.]

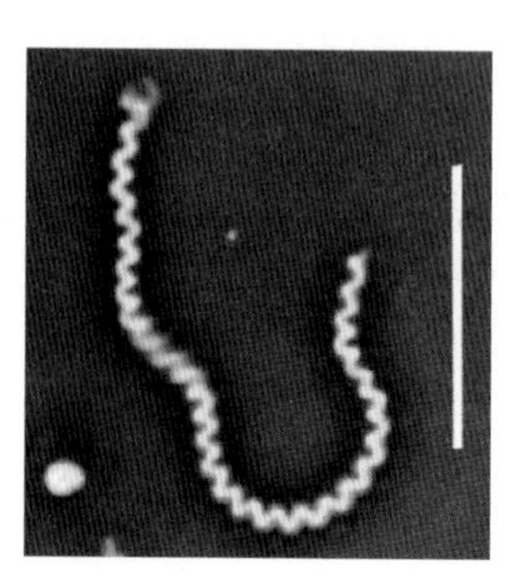

Figure A *Spirochaeta plicatilis* from the Fens, Boston. LM, bar = 10 μm. [Courtesy of W. Ormerod.]

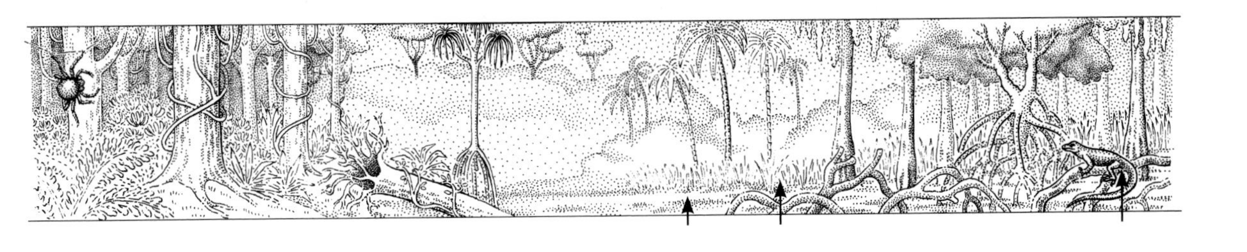

wide and hundreds of micrometers long. All have been found in symbiotic relations with animals. Members of the genus *Cristispira* inhabit the crystalline style of clams and oysters; the style is an organ that helps these molluscs (A-26) grind their algal food. The presence of the cristispires is contingent on environment and does not seem to influence the mollusc or its style. The other pillotina spirochetes—*Pillotina*, *Hollandina*, *Diplocalyx*, and *Clevelandina*—are found associated with other bacteria and protoctists in the hindgut of wood-eating cockroaches and dry-wood, damp-wood, and subterranean termites. The animals ingest wood, but the microbes inhabiting their intestines digest it. One of these microbes, *Spirosymplokos deltaeiberi* from

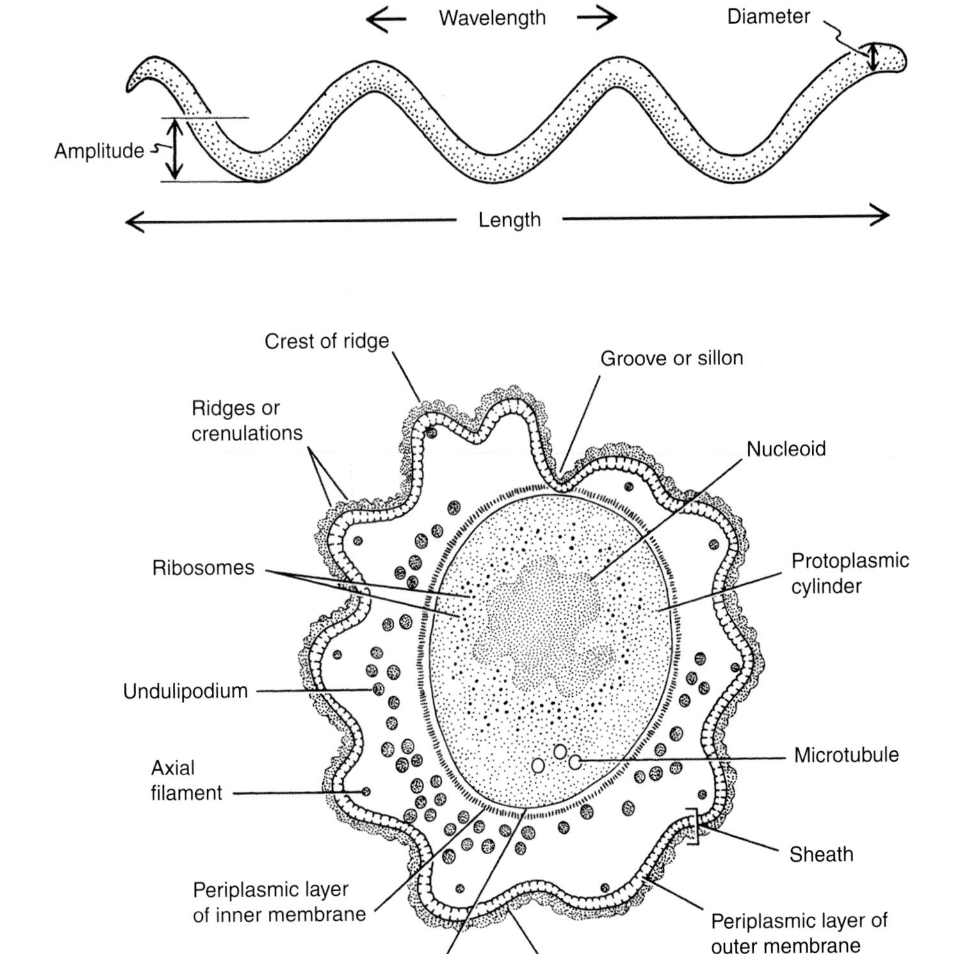

Figure C (Top) Features, in principle, measurable in all spirochetes. (Bottom) Cross section of a generalized pillotina spirochete. No single member of the group has all these features. [Drawing by K. Delisle.]

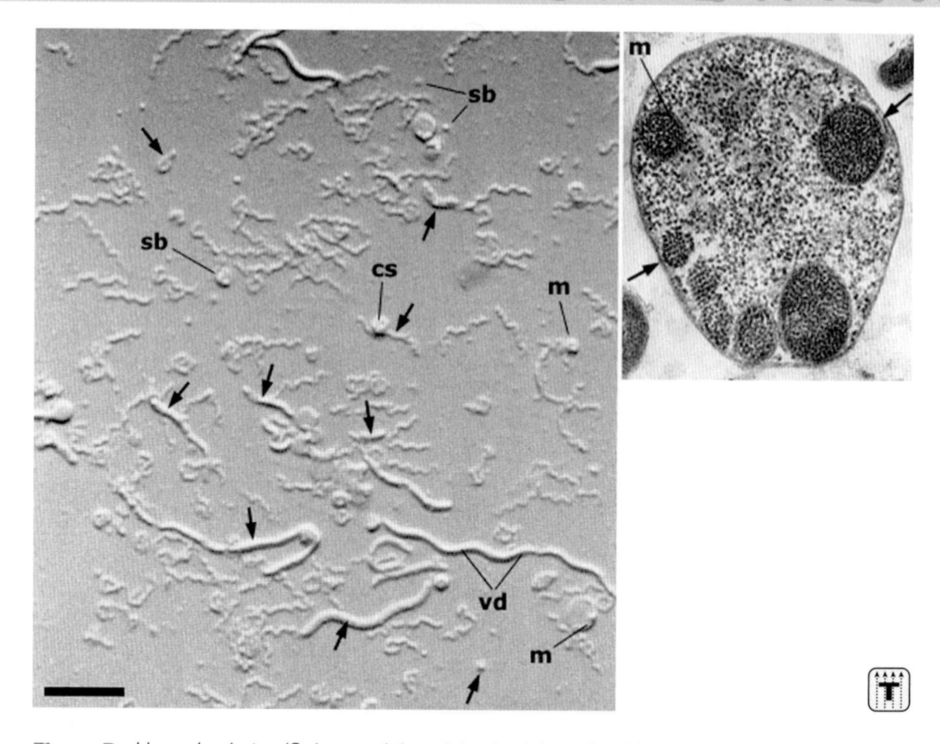

Figure D Live spirochetes (*Spirosymplokos deltaeiberi*) from the delta of the Ebro River, northeastern Spain. Variable diameter (vd), spherical bodies (sb), internal membranous structures (m), and probably composite structure (cs) can be inferred. TEM, bar = 10 μm. (Inset) Transverse section of internal development of composite structure as the membranes form around the internal offspring (arrows). TEM, bar = 1 μm.

microbial mats, is unique. Not only is its diameter tapered (variable), but it seems to produce "babies" by releasing small spirochete and membranous bodies that may function as spores (Figure D).

The spirochetes themselves probably lack cellulases, enzymes that initiate the breakdown of wood, but do have enzymes for digesting the products of the initial breakdown. In fact, the termite spirochetes are often in intimate contact with parabasalid protoctists (Pr-1) that contain cellulases. Many spirochetes, mostly unidentified, are observed in environments where active breakdown of algal or plant cellulose is taking place.

Most spirochetes are difficult to study in the laboratory. Some require nutritional media containing large, complex fatty acids. Requirements in general are not known, because only a few spirochetes have been cultured and none of them are pillotinas.

Division: Gracilicutes

B-5 Bacteroides–Saprospirae (Fermenting gliders)

Greek *sapros*, rotten; *spira*, coil, twist

GENERA

Alysiella	*Flexibacter*	*Saprospira*
Bacteroides	*Flexithrix*	*Simonsiella*
Capnocytophaga	*Fusobacterium*	*Sporocytophaga*
Cytophaga	*Herpetosiphon*	*Synergistes*
Flavobacterium	*Microscilla*	*Symbiothrix*

Gene sequencing of the 16S small-subunit rRNA molecule has revealed this lineage of Gram-negative eubacteria. *Bacteroides* and its relatives, anaerobic fermenters, form one subgroup (Figure A); the other, the flavobacterium subgroup, unites a set of oxygen respirers that move by gliding motility (Figure B): *Capnocytophaga, Cytophaga, Flexibacter, Microscilla, Saprospira,* and *Sporocytophaga.* There is another group of gliders—not related to Saprospirae by 16S rRNA criteria—that belong to Proteobacteria (B-3).

Members of the *Bacteroides* group, like us, are nutritionally organoheterochemotrophs: their sources of energy, carbon, and electrons are all organic (food) compounds. Unlike us, however, they are anaerobic fermenters restricted to anoxic environments. Members of the genus *Bacteroides* inhabit our intestinal tracts in large numbers.

Relatives of *Bacteroides* in insect guts and the rumen in large numbers form thick coatings on amitochondriate protists. *Symbiothrix dinenymphae, Tammella caduceiae* (termites), and *Synergistes jonesii* (rumen) are examples.

Fermentation is a metabolic process that uses organic compounds to produce energy; the end products are a different set of organic compounds from which some chemical energy has been extracted. Fermenting bacteria are distinguished by their inability to synthesize porphyrins—the metal- and nitrogen-containing ring compounds—which all photosynthetic and respiring organisms are capable of synthesizing.

Saprospira and its relatives are oxygen-respiring aerobes that tend to live in organic-rich environments such as decaying vegetation, rotting seaweed, and the like. Members of the flavobacterium subgroup are unified by their ability to glide at some stage in their life history and by heterotrophic (parasitic or saprobic) growth.

Glider cells—characteristic of these organisms but also of the δ group of Proteobacteria (B-3)—typically embedded in slime (polysaccharides of their own making), require contact with a solid surface to move; they cannot swim. Although they unmistakably move by gliding along surfaces, the part of the bacterium in contact with the surface shows no organelles of motility, such as flagella. Numerous tiny intracellular fibrils seen

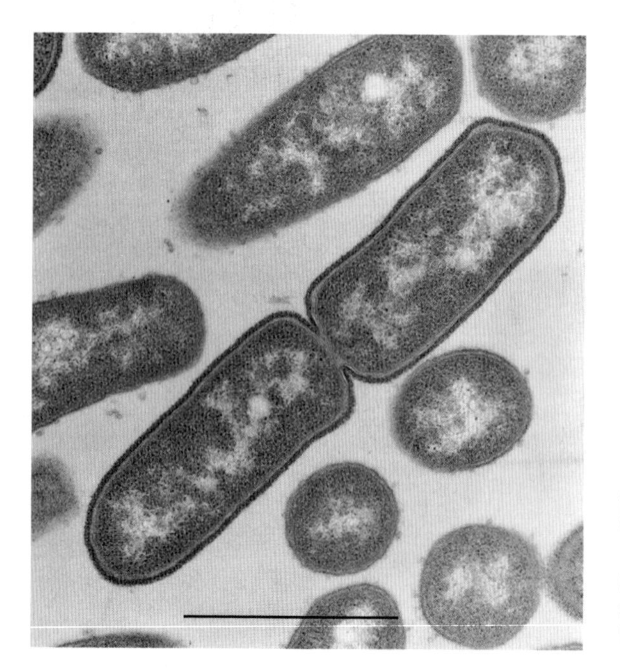

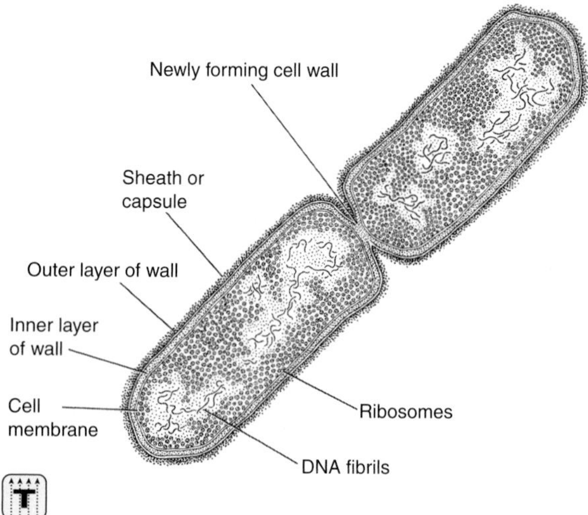

Figure A *Bacteroides fragilis*, an obligate anaerobe found in animal gut tissue, just prior to cell division. TEM, bar = 1 μm. [Photograph courtesy of D. Chase; drawing by L. Meszoly.]

in electron micrographs of some species have been correlated with motility, but the means of motility in these organisms is really not understood.

Cytophaga and its relatives (*Capnocytophaga*, *Saprospira*, and *Sporocytophaga*) constitute a natural group of straight rods or rigid helical filaments, gliding forms that produce orange pigments (carotenoids). *Cytophaga* breaks down agar, cellulose, or chitin; *Flexibacter* metabolizes less-tough carbohydrates, such as starch and glycogen; and *Herpetosiphon*, a filamentous member of this group with a sheath around its cells, breaks down cellulose, but not agar or chitin. Three filamentous genera are *Flexithrix* and helical *Saprospira*, neither of which forms myxocysts (cases from which myxobacteria emerge), and *Sporocytophaga*, which forms small, resistant cells called "microcysts."

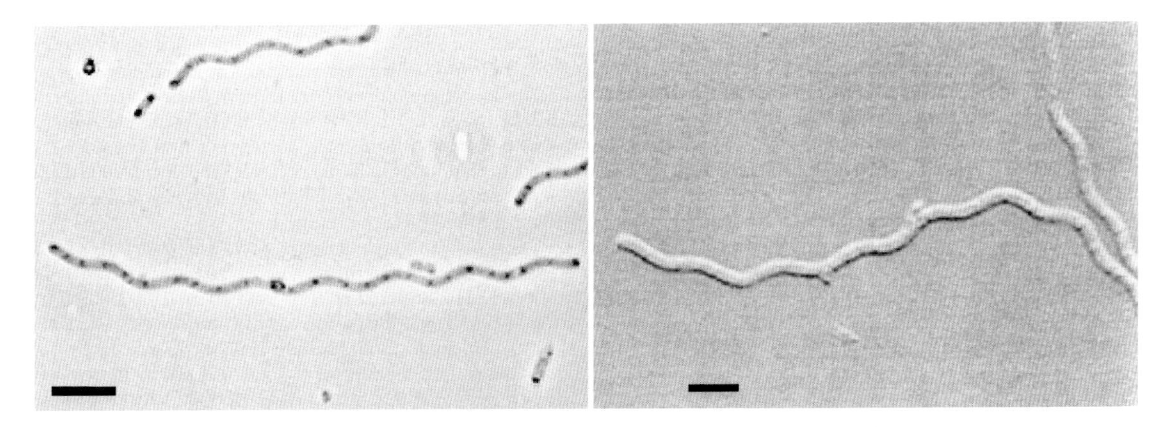

Figure B *Saprospira* sp., live from a microbial mat from Laguna Figueroa, Mexico. (Left) Internal polyphosphate granules (dark spots) are visible in this gliding cell. LM (phase contrast), bar = 5 μm. (Right) The surface of these helical rigid gliders, as seen by using Nomarski phase-contrast optics. LM, bar = 5 μm.

Division: Gracilicutes

B-6 Cyanobacteria

(Blue-green bacteria and chloroxybacteria, grass green)

Greek *kyano*, dark blue

GENERA

Anabaena	*Gloeocapsa*	*Prochlorococcus*
Anacystis	*Gloeothece*	*Prochloron*
Chamaesiphon	*Hyella*	*Prochlorothrix*
Chroococcus	*Lyngbya*	*Spirulina*
Dermocarpa	*Microcystis*	*Stigonema*
Entophysalis	*Nostoc*	*Synechococcus*
Fischerella	*Oscillatoria*	*Synechocystis*

These two groups of oxygenic photosynthetic bacteria differ in color and pigmentation. By far the largest group, with perhaps a thousand genera, is the blue-greens. Only a few recently described genera of the green, or chloroxybacteria, photosynthesizers are known: *Prochlorococcus*, *Prochloron*, and *Prochlorothrix*. Because studies of 16S rRNA show these chloroxybacteria to be more closely related to certain cyanobacteria than they are to each other, the group has been united. They are called "chloroxybacteria" for their grass-green color and oxygenic habit.

Until fewer than two decades ago, cyanobacteria were called blue-green algae or cyanophyta and were considered plants. Physiologically, cyanobacteria are remarkably similar to algae and plants. As the ultimate producers, these three kinds of life feed and energize all of the others—with the minor exception of some obscure chemolithoautotrophic bacteria (Table B-1). Together with anoxygenic phototrophic bacteria, algae, plants, and cyanobacteria sustain other life on Earth by converting solar energy and carbon dioxide into organic matter that provides food and energy for the rest. Algae, plants, and cyanobacteria photosynthesize according to the same rules, by use of water as the hydrogen donor to reduce carbon dioxide:

$$12H_2O + 6CO_2 \xrightarrow[\text{chlorophyll}]{\text{light}} C_6H_{12}O_6 + 6O_2 + 6H_2O$$

Like algae and plants, cyanobacteria and their grass-green relatives respire the oxygen that they produce by photosynthesis in the dark. Furthermore, like certain anoxygenic phototrophs (B-3), many cyanobacteria can use hydrogen sulfide instead of water as the hydrogen donor. Thus, under high-sulfide conditions, their photoautotrophic metabolism deposits sulfur instead of releasing oxygen:

$$2H_2S + CO_2 \xrightarrow{\text{light}} CH_2O + 2S + H_2O$$

Stained preparations viewed in the light microscope, confirmed by detailed electron microscopic observations, show that the blue-greens and chloroxybacteria have Gram-negative walls (p. 42). The similarity between cyanobacteria, on the one hand, and nucleated algae and plants, on the other, is now understood to apply only to the plastids (photosynthetic organelles and their nonphotosynthetic derivatives) of algal and plant cells: the plastid evolved when cyanobacteria became permanent symbionts inside protoctista cells.

Many cyanobacteria in high-sulfide conditions photosynthesize as the green and purple sulfur bacteria do. They deposit sulfur in their cells. Yet all can produce oxygen gas. The production of oxygen gas distinguishes the cyanobacteria and the grass-green chloroxybacteria from other photoautotrophic Prokarya.

Cyanobacteria, like many other bacteria, contain organelles: distinct subcellular structures visible with the light microscope in live cells. Many have nucleoids and carboxysomes. The former are fibrils of DNA; the latter stores the most abundant enzyme in the biosphere—ribulose bisphosphate carboxylase (RuBisCO), the universal CO_2-fixing enzyme of cyanobacteria, algae, and plants. Like most other photosynthetic organisms, cyanobacteria contain membranes called "thylakoids," often at the best-lit periphery of the cells. In most blue-greens, thylakoids are associated with spherical structures called "phycobilisomes." Photosynthetic pigments called "phycobilins" are bound to proteins in phycobiliprotein complexes that are embedded in the phycobilisomes. Chlorophyll ***a*** and two similar lipid-soluble pigments, phycocyanin and allophycocyanin, give cyanobacteria their bluish tinge. Many have an additional reddish pigment called "phycoerythrin." These pigments, which resemble nitrogen-containing porphyrin whose ring has been opened up to form a chain, are biosynthesized by the same reactions that produce porphyrins. Bile pigments of animals also contain porphyrin rings. The grass-green oxygenic phototrophic bacteria (*Prochloron*, *Prochlorothrix*, and *Prochlorococcus*) lack phycobiliprotein and phycobilisomes but contain a chlorophyll: chlorophyll ***b***.

Unobtrusive though they are, there are thousands of living types of cyanobacteria; in the ancient past, they dominated the landscape. The Proterozoic eon, from about 2500 million until about 600 mya, was the golden age of cyanobacteria. Remains of their ancient communities include trace fossils called "stromatolites," and layered sedimentary rocks produced by the metabolic activities of microorganisms, especially of filamentous cyanobacteria. Certain stromatolitic communities still live in salt flats and shallow embayments of the Persian Gulf, the west coast of Mexico, the Bahamas, western Australia, and even under the ice in Antarctica. In the Proterozoic, however, such communities extended to all the continents. Stromatolites formed in fairly deep open waters at the lowest end of the photic zone, the surface waters into which light penetrates (about 200 m in clear seawater). Cyanobacteria occupied the kinds of environments that coral reefs do today.

Prochlorococcus, a marine bacterium resembling *Synechococcus* but with green pigmentation, may be one of the most common bacteria in the world. They were thriving at the base of the photic zone all over the ocean and for years they were called as "unidentified green coccoid."

There are two great classes of cyanobacteria:

Class Coccogoneae: coccoid (spherical) cyanobacteria

- Order Chroococcales: coccoids that reproduce by binary fission. Some genera are *Gloeocapsa*, *Chroococcus*, *Anacystis*, *Prochloron*, *Synechocystis*, *Synechococcus*, and the stromatolite-building coccoid *Entophysalis*.
- Order Chamaesiphonales: coccoids that reproduce by releasing exospores. Unlike bacillus spores, exospores are not necessarily resistant to heat and desiccation. *Chamaesiphon* and *Dermocarpa* are two genera in this group.
- Order Pleurocapsales: coccoids that reproduce by forming propagules (baeocytes). The parent organism disintegrates when the propagules are released. Unlike the endospores of other bacteria, these propagules are not resistant to desiccation and high heat.

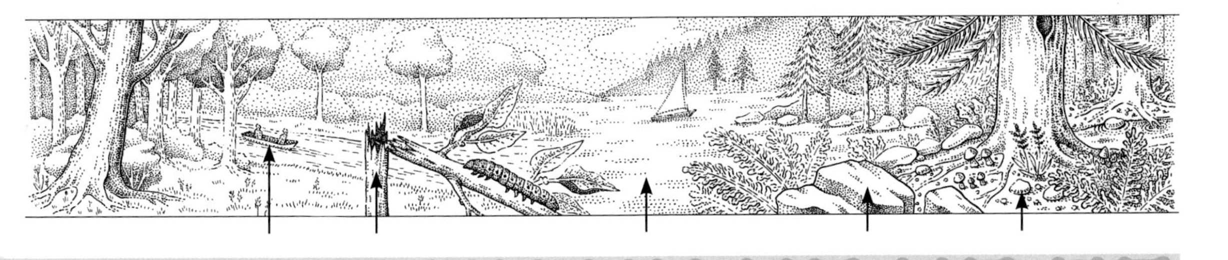

Class Hormogoneae: filamentous cyanobacteria

In many members of this class, filament fragments, called "hormogonia," containing as many as several dozen cells break off, glide away, and begin new growth.

Order Nostocales: filaments that either do not branch or exhibit false branching—that is, a branch formed not only by a single growing cell but also by slippage of a row of cells. Examples include *Anabaena* (Figure A), *Nostoc*, *Oscillatoria*, and *Prochlorothrix*.

Order Stigonematales: filaments that exhibit true branching. A single cell dividing by binary fission in a multicellular body may have two places of growth on it and thus leading to the formation of two new filaments from the same cell. These cyanobacteria are among the morphologically most complex of the prokaryotes. Examples include *Fischerella* and *Stigonema*. The *Stigonema*, shown in Figure B, is from the Nufunen Pass at about 2500 m in the Austrian Alps, where, like moss (Pl-1), it grows on granite as ground cover as a terrestrial cyanobacterium.

Many cyanobacteria take in nitrogen from the air and incorporate (fix) it into organic compounds. Some of the species fix nitrogen in cells that specialize by losing their chlorophyll-containing thylakoids and turning colorless. The photosynthetic cell changes into a generally larger and more spherical nonphotosynthetic cell called a "heterocyst." Several genera of nitrogen-fixing cyanobacteria that lack heterocysts or any other morphological clue to their ability have been described. Early claims that cyanobacteria do not inhabit the open ocean are incorrect. The tiny marine cyanobacterium *Synechococcus*, abundant in seawater, was ignored for years. Species of *Oscillatoria* that are capable of nitrogen fixation, but lacking heterocysts, have also been reported.

Prochloron was seen before it was scientifically described in the late 1960s, but it was assumed to be a green alga. When they were carefully studied with the electron microscope, however, it became clear that these green coccoids are oxygenic phototrophic prokaryotes. Morphologically, they are cyanobacteria. They were named "prochlorophytes" to indicate their

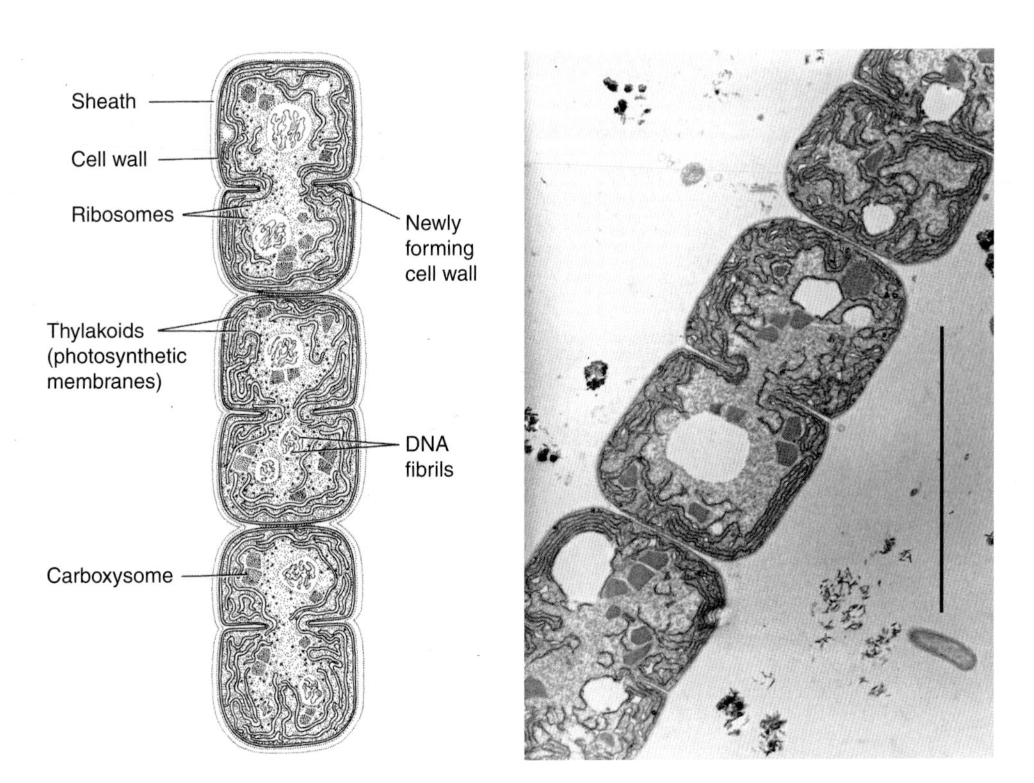

Figure A *Anabaena.* This common filamentous cyanobacterium grows in freshwater ponds and lakes. Within the sheath, the cells divide by forming cross walls. TEM, bar = 5 μm. [Photograph courtesy of N. J. Lang; drawing by R. Golder.]

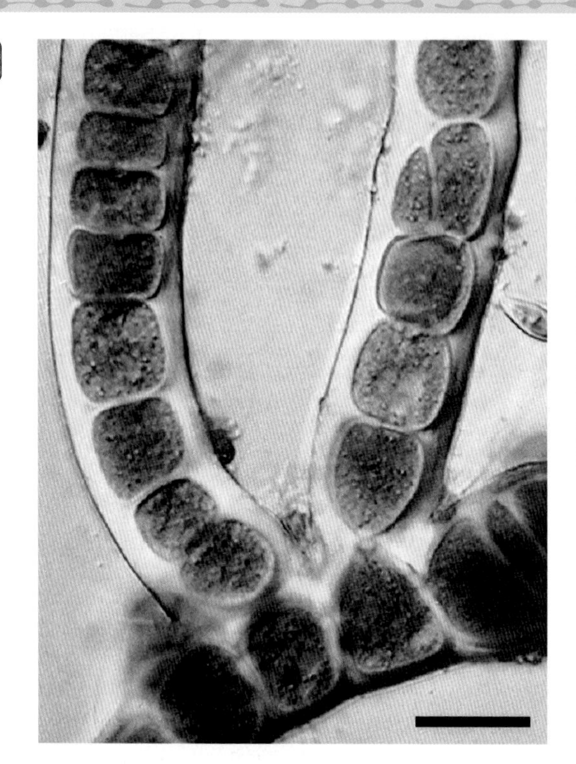

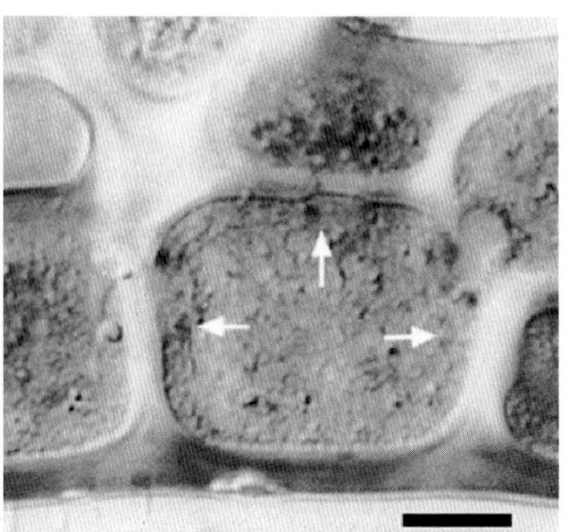

Figure B (Left) *Stigonema informe*, a multicellular, terrestrial cyanobacterium that grows luxuriantly in the high Alps, showing true branching. (Right) Close-up view of true branching, showing three growth points (arrows) on a single cell. LMs, bars = 10 μm. [Courtesy of S. Golubic.]

relationship with the green plants (Greek *phyton*, plant). They do resemble the plastids of plants and green algae.

Several strains, simple nonmotile coccoids primarily from the South Pacific, belong to the genus *Prochloron.* All form associations with tropical and subtropical marine animals, primarily tunicates (A-35). *Prochloron* has been found on two species of *Didemnum*—as gray or white surface colonies on *Didemnum carneolentum* and as internal but not intracellular cloacal colonies in *D. ternatanum.* They also reside as colonies in the cloaca of *Diplosoma virens* (Figure C), *Lissoclinum molle*, *L. patella* (Figure D), and *Trididemnum cyclops.* Why *Prochloron* grows only on the surface or in the cloaca of these tunicate hosts is a mystery.

Although they lack the phycobiliproteins of other cyanobacteria, chloroxybacteria contain carotenoid pigments. A large percentage of the total carotenoid is beta-carotene, the pigment that is found in carrots, green algae, and many other plants.

After *Prochloron*, a second filamentous green oxygenic photosynthesizer was found free living in the Loosdrecht Lakes, The Netherlands, in 1984. It resembled blue-greens such as *Oscillatoria* but had other features of *Prochloron.* It was named as *Prochlorothrix hollandica*, after its form and place of discovery and is the only grass-green cyanobacterium that grows in pure culture.

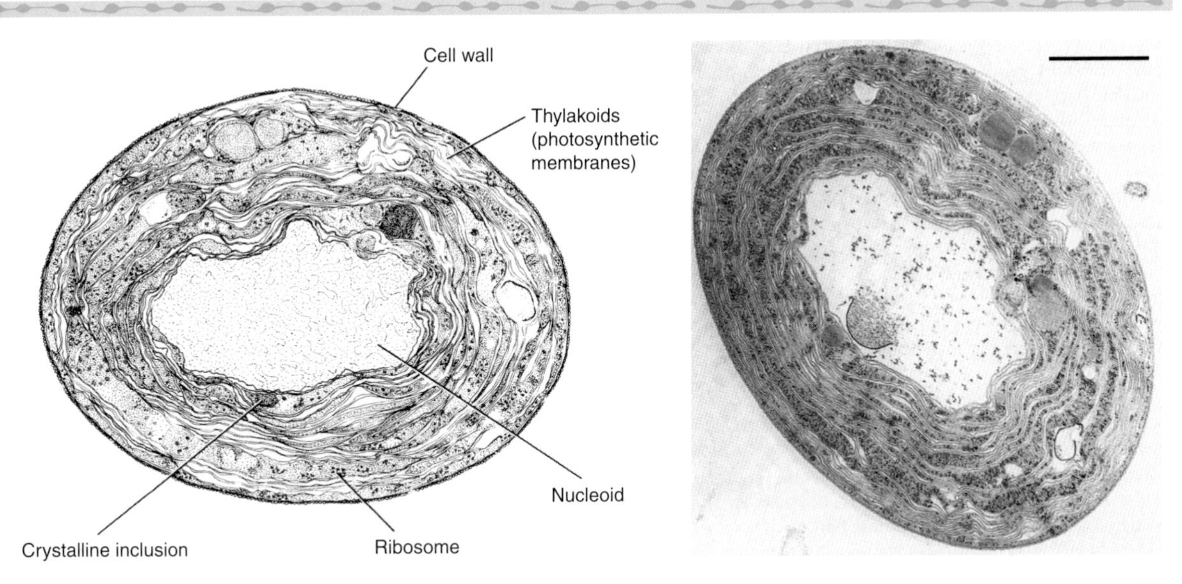

Figure C Thin section of *Prochloron* from the tunicate *Diplosoma virens* (A-35). TEM, bar = 2 μm. [Photograph courtesy of J. Whatley and R. A. Lewin, *New Phytologist* 79:309–313 (1977); drawing by E. Hoffman.]

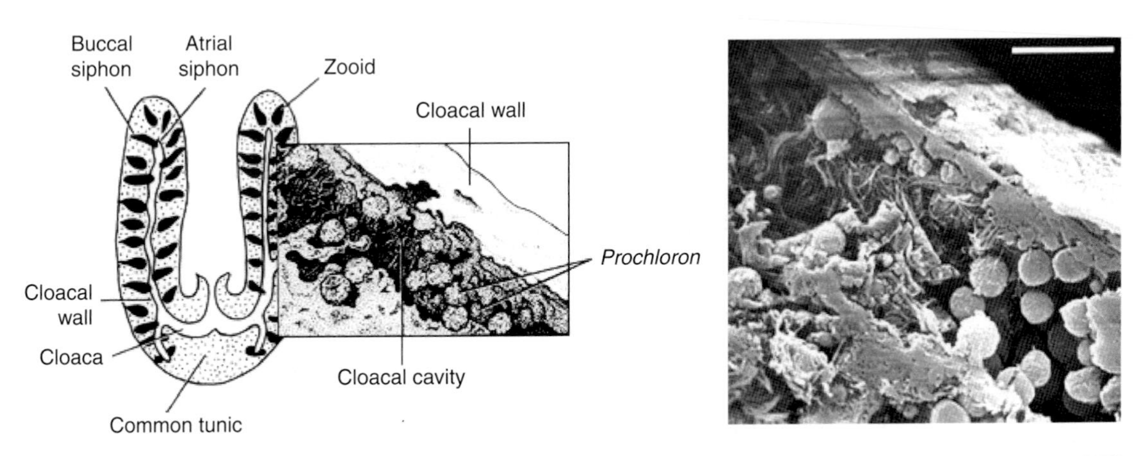

Figure D Cloacal wall of *Lissoclinum patella* (A-35) with embedded small spheres of *Prochloron.* The tunicate *L. patella* is native to the South Pacific. SEM, bar = 20 μm. [Photograph courtesy of J. Whatley; drawing by E. Hoffman.]

Division: Gracilicutes

B-7 Chloroflexa

(Green nonsulfur phototrophs)

GENERA
Chloroflexus
Heliothrix
Oscillochloris

Greek *chloro*, green; Latin *flexus*, bend

New knowledge, especially of the sequence of nucleotides in their 16S rRNA genes, has led to the separation of these phototrophic bacteria in their entirety from other green sulfur phototrophic bacteria, with which they had been grouped. Since they were discovered in the 1960s in hot springs and microbial mats, there have been hints that *Chloroflexus* were not members of Chlorobia (B-8), the standard green sulfur phototrophic bacteria that they superficially resemble. These hints include tolerance for oxygen gas and lack—at all times—of a need for a sulfur-rich environment. In contrast, Chlorobia are obligate phototrophs and obligate anaerobes that are extraordinarily intolerant to free oxygen and, because they require sulfur in their photosynthetic metabolism, dwell only in sulfide-rich, anoxic environments.

Chloroflexus is grouped now not on the basis of its phototrophic metabolism but with two other genera, *Heliothrix* and *Oscillochloris*. Like *Heliothrix* and *Oscillochloris*, *Chloroflexus* is a filamentous, gliding, nonsulfur bacterium (Figure A). *Oscillochloris* is such a large filament, with cells as large as 5 μm in diameter and a conspicuous holdfast that anchors it to surfaces, that it superficially resembles cyanobacteria such as *Oscillatoria* (B-6). [In contrast, the gliding filaments of one of only a few recognized species of *Chloroflexus* (*C. aurantiacus*) are 0.8 μm in diameter.] However, both the physiology and the 16S rRNA gene sequence of *Oscillochloris* group it here with the phylum Chloroflexa.

Chloroflexus, a typical bacterial cell with tendencies to form filaments, can be grown in quantity in the laboratory. The presence of chlorosomes—cigar-shaped, membrane-bounded structures (Figure B)—rather than the flat-membraned thylakoids, further distinguishes them. The lack of CO_2 fixation through the Calvin–Benson cycle will probably be observed in the heretofore unstudied members of this small phylum, with its few species (at least two of *Chloroflexus* and one each of the other two genera). On 16S rRNA phylogenies, the Chloroflexa green nonsulfur bacteria are closer to the spirochetes (B-4) and the Saprospirae (in the *Bacteroides–Saprospirae* group, B-5) than they are to the green sulfur phototrophs (Chlorobia, B-8).

Because of its cultivability, nearly all the detailed information on members of this phylum comes from studies of *Chloroflexus*. Although large populations of cells of *Chloroflexus*, like those of *Chlorobium*, appear green, the photosynthetic apparatus of *Chloroflexus* is organized into chlorosomes (Figure B). Chlorosomes, like thylakoids of cyanobacteria, algae, and plants, are repositories of the chlorophylls and their binding proteins. The single membrane is composed of light-harvesting bacteriochlorophylls and the straight-chain (aliphatic) carotenoids alpha- and beta-carotene. (These carotenoids resemble similar pigments of cyanobacteria, algae, and plants, although oxygenic photosynthesis is unknown in any Chloroflexa.)

Chloroflexus may grow well photoautotrophically with CO_2 as its carbon source and with hydrogen or hydrogen sulfide as its hydrogen donor. These properties, including the assumption that the complex and demanding ability to photosynthesize is fundamental, led to the classification of this green sulfur phototroph with the Chlorobia (B-8). But the 16S rRNA gene sequence data are not the only evidence that leads us to place *Chloroflexus* in its own phylum. Another characteristic that

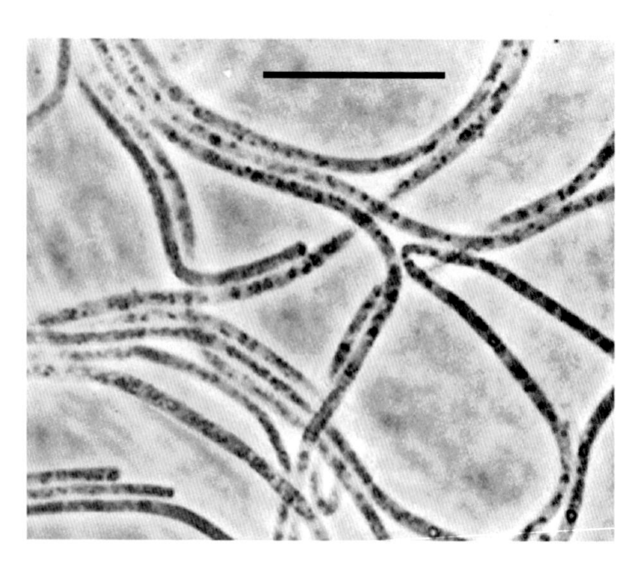

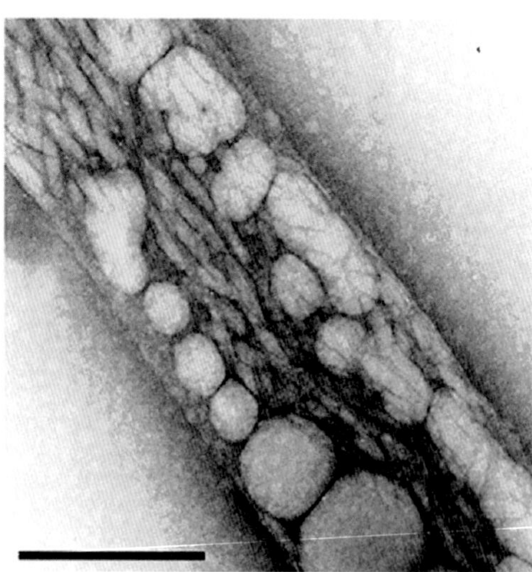

Figure A (Left) Live photosynthetic gliding filamentous cells, 1 μm in diameter, of *Chloroflexus* from hot springs at Kahneeta, Oregon. LM (phase contrast), bar = 5 μm. [Courtesy of B. Pierson and R. Castenholz, *Archives of Microbiology* 100:5–24 (1975).] (Right) Magnified view showing the typical membranous phototrophic vesicles that contain the enzymes and pigments for photosynthesis. EM (negative stain), bar = 1 μm. [Courtesy of R. Castenholz.]

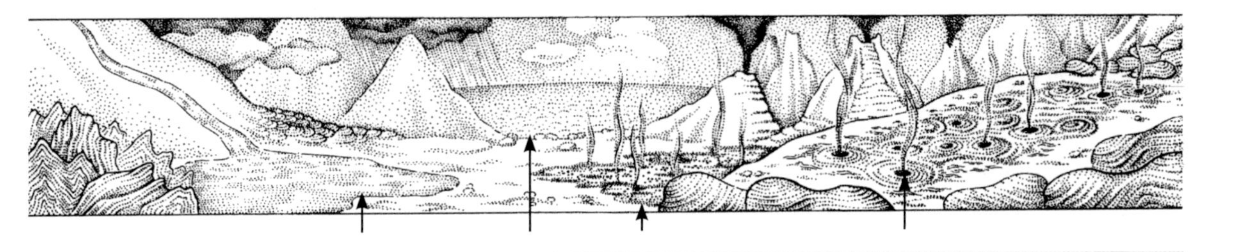

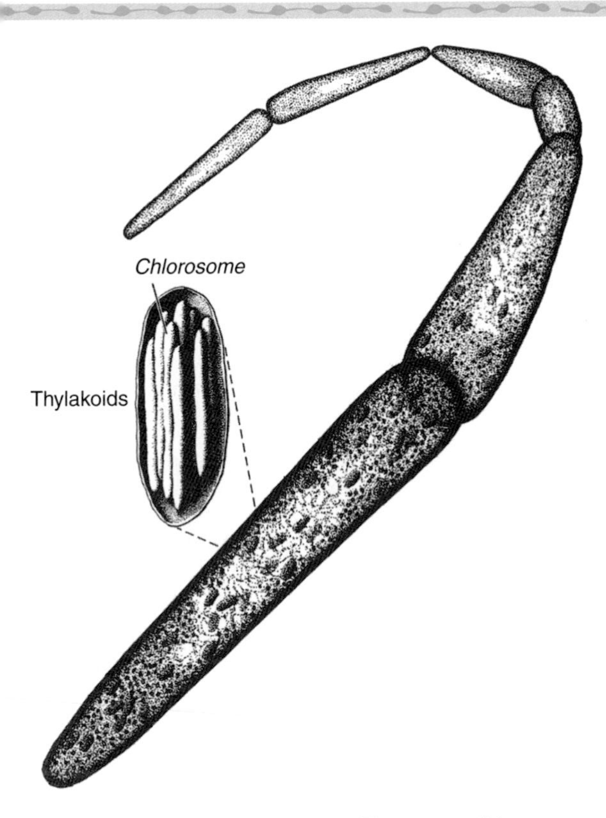

Figure B *Chloroflexus aurantiacus.* Filamentous, thin photosynthesizers showing distribution of their chlorosomes as seen by light microscopy. (Inset) The entire chlorosome as reconstructed from electron micrographs. The membranous plates are the sites of the bacterial chlorophylls and their bound proteins. [Drawings by C. Lyons.]

Figure C *Chloroflexa* habitat. Laguna Figueroa, Baja California Norte, recolonizing microbial mat.

distinguishes *Chloroflexus* is that, atypically for other green phototrophs, it grows well heterotrophically in the dark. *Chloroflexus* thrives on organic foods, including a wide variety of sugars, amino acids, or other small organic acids, as sources of carbon.

The electron acceptor–reduction potential of *Chloroflexus,* during phototrophic growth, resembles that of the phototrophic purple bacteria (B-3), which are more tolerant of oxygen than are *Chloroflexus* and *Chlorobium,* much more than it does with that of Chlorobia (B-8). This measurable reduction potential, a property that predicts the conditions in which cells absorb light and generate chemical energy for cell reactions, is useful for understanding both the habitat distribution and the physiological limitations of photosynthetic and respiring microbes.

Chloroflexus and many purple bacteria enjoy a reduction potential of about −0.15V (compared with the more negative and thus more reduced reduction potential of −0.5 in the Chlorobia). Like all the purple and the other green phototrophs, *Chloroflexus* is not oxygenic; nevertheless, in spite of its oxygen tolerance and pigment composition (green color and plant-like carotenoids), it is not like purple bacteria or cyanobacteria either. *Chloroflexus* is also unique in the way that it handles carbon dioxide. Two atmospheric CO_2 molecules, one at a time, are bound to acetyl coenzyme A thus producing hydroxypropionyl CoA molecules. Twice-carboxylated acetyl CoA yields methylmalonyl CoA; this molecule forms acetyl CoA and glyoxylate when it is rearranged. Glyoxylate, probably through an amino acid pathway (serine or glycine), is converted into cell material in a series of reactions which reuses the acetyl CoA.

Thus, neither the ribulosebisphosphocarboxylase cycle (Calvin–Benson cycle, in which CO_2 is fixed by a 5-carbon ribulose 1,5-bisphosphate molecule and ultimately yields two 3-carbon glyceraldehyde and 3-phosphate molecules) nor the reverse Krebs cycle (TCA, or tricarboxylic acid, cycle) of Chlorobia reduces carbon dioxide into cell material in *Chloroflexus.* No other group of organisms uses such a pathway for carbon dioxide removal from the air and incorporation into the cell material of the biosphere. The *Chloroflexus* hydroxypropionate pathway is one of about half a dozen distinct and independently evolved CO_2-fixing metabolic schemes in all of life. From this perspective, the metabolism of the cyanobacteria, algae, and plants, all of which use only the Calvin–Benson cycle, is extremely uniform.

Chloroflexus has been discovered associated with heat-tolerant filamentous cyanobacteria in hot springs at temperatures between 40°C and 70°C. *Heliothrix* was found at the surface of microbial mats replete with *Chloroflexus* (Figure C). *Heliothrix* requires high light intensities for growth, but is not a *Chloroflexus* because it lacks chlorosomes and contains only a single bacteriochlorophyll (bacteriochlorophyll ***a***). Its gliding filamentous structure, its 16S rRNA sequences, and the fact that it contains carotenoids nearly identical with those in *Chloroflexus* support its placement in this phylum.

Division: Gracilicutes

B-8 Chlorobia

(Anoxygenic green sulfur bacteria)

Greek *chloro*, green; *bios*, life

GENERA
Ancalochloris
Chlorobium
Chlorochromatium[1]
Chloroherpeton
Chloronema
Pelodictyon
Prosthecochloris
Thiomargarita namibiensis

[1]Only the phototrophic member of the symbiotic complex.

Not all photosynthetic organisms are phototrophic. Any life-form that uses light for the synthesis of cell material is photosynthetic, but only those that do not need fixed organic compounds in the medium are phototrophs. The most independently productive photosynthesizers are the photolithoautotrophs: their energy comes from visible light, their carbon comes from atmospheric CO_2, and their electrons for the reduction of CO_2 to cell material comes from inorganic compounds, such as hydrogen sulfide, thiosulfate, sodium sulfide (Na_2S), hydrogen gas, and water (H_2O).

To reduce the CO_2 in the air to organic compounds, any phototroph needs a source of electrons; members of *Chlorobia* depend on sulfur for this source. These light-requiring eubacteria dwell in sulfide-rich, sunlit habitats. In such obligately anaerobic green sulfur bacteria, photosynthetic pigments are organized on chlorosomes, as they are in *Chlorochromatium* (Figure A) here and in *Chloroflexus* (B-7, Figure B). When H_2S or Na_2S is the hydrogen donor, the by-product of photosynthesis is sulfur, which may be stored or excreted by cells as elemental sulfur or as more oxidized sulfur compounds.

Chlorobia are inevitably found in anaerobic muds—those dark, sulfur-smelling beach areas at the edge of the ocean or in ponds and lakes over sulfur-rich rock. In today's oxygenated world, the greenish, brown, or yellow Chlorobia are banished beneath the blue and green cyanobacteria (B-6), the red, pink, and purple phototrophs (B-3), and the Chloroflexa (B-7). Chlorobia tend to be at the bottom of the photic zone. So oxygen-intolerant and light-requiring on their own, Chlorobia escape the need to dwell at the bottom of the photic zone by establishing associations with facultatively oxygen-respiring heterotrophic bacteria (Figure B).

All these phototrophic bacteria, delightfully colored in an astonishing range of yellows, pinks, and greens, are cosmopolitan in sunlit muds, scums, and on most well-lit limestones. In the bright sunlight, in the top layers of anaerobic muds, absorbing all incident light, they become very dark, nearly black. Because each species has an optimum growth at a given acidity, oxygen tension, sulfide and salt concentration, moisture content, and so on, they grow in layers—each in its appropriate niche—which we see as bands of differing color of sediment. Layered communities of phototrophs support with their productive mode of life many other types of bacteria; these form sedimentary structures such as microbial mats and living stromatolites (Figure B-1).

Chlorobium is a large genus; some strains are tolerant of extremely high or extremely low temperatures or salinities. The organisms of this species tend to be small nonmotile coccoids deeply embedded in their anoxic communities. Most of these green photosynthesizers are refractory to propagation in pure culture under laboratory conditions. Members of the genus *Chlorobium* lack gas vesicles; membranous intracellular inclusions are used in floating. Less well known genera of Chlorobia, anaerobic phototrophs, include only one other genus whose members lack gas vesicles: *Prosthecochloris*. Gas vesicles in all the other genera of Chlorobia aid in regulation of position in the water column. Both of the two known species

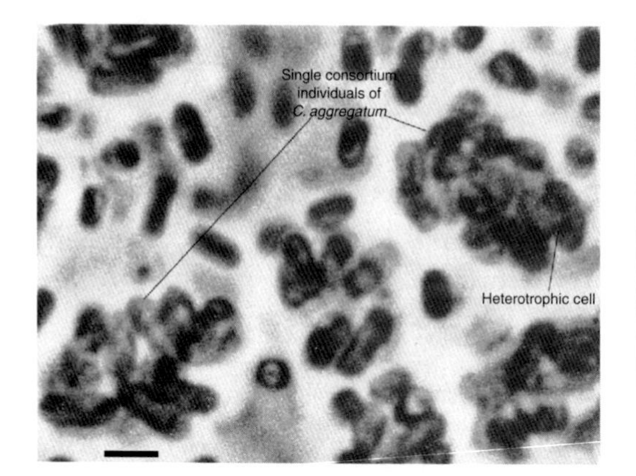

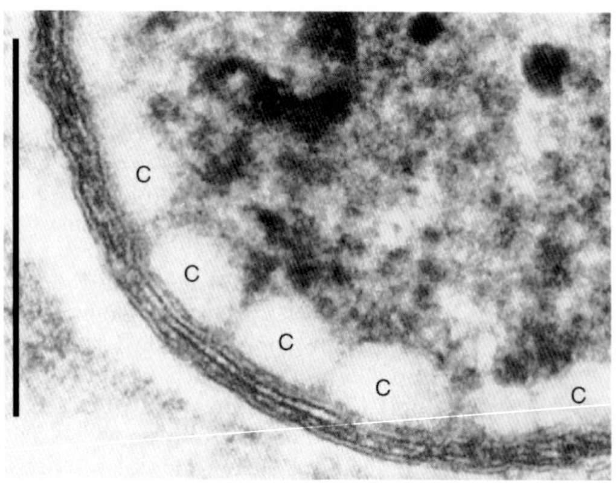

Figure A *Chlorochromatium aggregatum.* TEM (above, left; bar = 1 μm) consortium bacterium, in which a single heterotroph (facing page, left; bar = 1 μm) is surrounded by the several pigmented phototrophs with their chlorosomes (c), seen here as peripheral vesicles (above, right; bar = 0.5 μm). From Lake Washington, near Seattle.

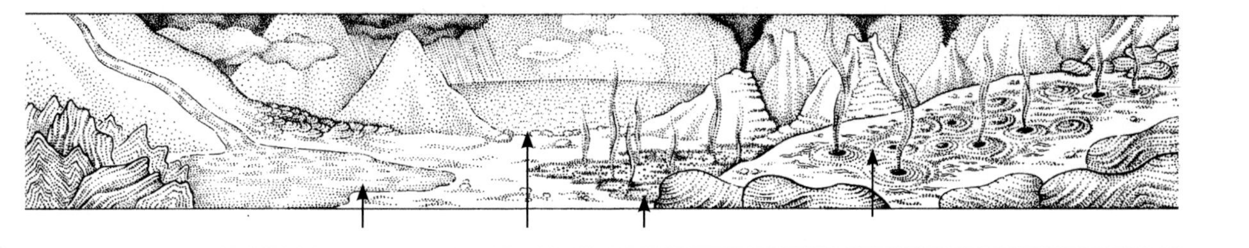

of *Prosthecochloris* form cell protrusions from their ellipsoid or spherical cells. As wall appendages that are not filled with cytoplasm, the protrusions are called "prosthecae." None swim; no Chlorobia are mastigoted.

The remaining four gas-vacuolate genera of Chlorobia are less familiar to us. They can be morphologically distinguished. *Pelodictyon* cells, which are branched, nonmotile rods, form a loose, irregular network. The spherical cells of *Ancalochloris* bear prosthecae. Whereas *Chloroherpeton* is a gliding rod, *Chloronema* forms large-diameter (2.0–2.5 μm) gliding filaments. No doubt a great variety of other Chlorobia reside in nature, awaiting students of anoxic, sulfurous, well-lit habitats to discover them.

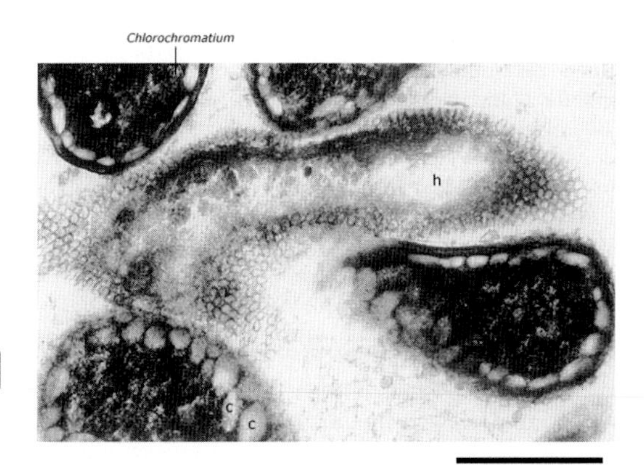

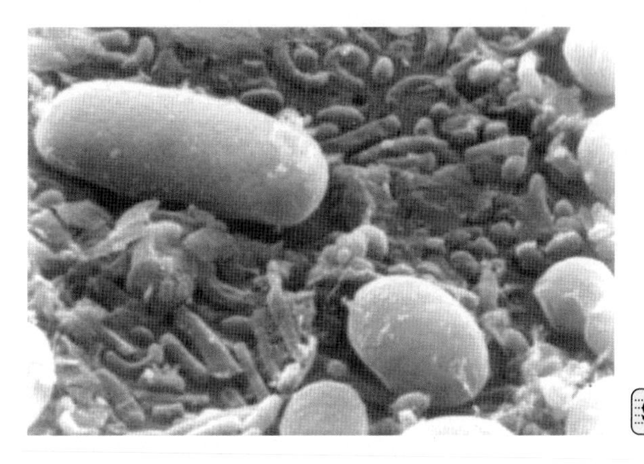

Figure B Anoxygenic layer of photosynthesizer. [SEM, Lake Ciso, Banyoles. Isabel Esteve.]

The photosynthetic cells responsible for the productivity of the consortium are *Chlorobium*, whereas the motility needed to approach the light but flee from oxygen gas is due to the central heterotroph (h). [Bar = 0.5 μm Micrographs by D. Caldwell and others.]

Division: Tenericutes

B-9 Aphragmabacteria
(Mycoplasmas)

Greek *a*, without; *phragma*, fence

GENERA

Acholeplasma
Anaplasma
Bartonella
Cowdria
Ehrlichia
Mycoplasma
Spiroplasma
Wolbachia

All cells, including those of wall-less eubacteria, the mycoplasmas, are bounded by a cell membrane of their own making that permits passage to water, salt ions, and small organic compounds. Outside the ubiquitous lipid–protein bilayer (double appearing in electron micrographs), called the unit membrane in most other bacteria, cells are bounded by a rigid cell wall that, although varied in composition, contains polysaccharides attached to short polypeptide molecules. This peptidoglycan wall is absent in all aphragmabacteria, although fuzzy extramembranous materials may coat the cells. Aphragmabacteria are bounded by the single, simple or decorated unit membrane because they are incapable of synthesizing certain polysaccharides (for example, diaminopimelic and muramic acids) that form the finished walls of most other bacteria. Lacking cell walls, they are resistant to penicillin and other drugs that inhibit wall growth.

Typically, bacteria range from 0.5 to 5.0 μm along their longest axis. Because the diameter of many aphragmabacteria is less than 0.2 μm, they are invisible even with the best light microscopes. Many are pleiomorphic; their shapes vary from irregular blobs, filaments, to even branched structures reminiscent of tiny fungal hyphae.

How wall-less bacteria reproduce is not clear. In some, tiny coccoid structures appear to form inside the cells, emerging when the "parent" organism breaks down. Others seem to form buds that become new organisms. Some apparently reproduce by binary fission (the common bacterial division mode where the cell divides into two roughly equal parts). On agar plates, the best-known mycoplasmas typically form tiny colonies, with a dark center and lighter periphery, resembling the shape of a fried egg.

Most well-known kinds are studied because they are seen in large numbers in diseased mammals and birds. They live in profusion inside the cells of animal tissues, where as symbiotrophs they derive their carbon, energy, and electrons from the animals. Some, such as *Ehrlichia*, cause tick-borne diseases, but the presence of many, probably most, is invisible without high-power electron microscopy. Few have been cultivated outside the tissue in which they reside. Those that have, primarily *Mycoplasma*, require very complicated growth media that include steroids, such as cholesterol. These lipid compounds, produced by most and required by all eukaryotes, are seldom, if ever, found in prokaryotes. However, in aphragmabacteria of the genus *Mycoplasma*, cholesterol constitutes more than 35 percent of the membrane's lipid content. For a bacterium, this is an extremely large fraction and may be the legacy of a long biological association between *Mycoplasma* and animal tissue rich in complex ring compound lipids such as steroids. Through evolutionary time, long association of these bacteria with eukaryotes' lipids led to dependency. All strains so far cultured require long-chain fatty acids (a kind of lipid) for growth, and most ferment either glucose (a sugar) or arginine (an amino acid). The fermentation products are usually lactic acid and some pyruvic acid.

Mycoplasma are of economic and social importance because they cause certain types of pneumonia in humans and domestic animals (Figures A and B). Under conditions of animal cell debilitation, these common symbiotrophs that are normally benignly present can be responsible for the death of cells in whole mammals or in laboratory tissue cultures. Widespread in insects (A-21), vertebrates (A-37), and plant tissues, many are too small to identify as aphragmabacteria—to see whether cell walls are present—even with electron microscopes.

Often, bacteriologists have informally used the word "mycoplasma" for all members of the genus *Mycoplasma* and other wall-less organisms without formally raising them to phylum status. Other well-known genera include *Acholeplasma*, which do not require steroids, and *Spiroplasma* and others, which do require them (as *Mycoplasma* does).

Members of *Spiroplasma* were isolated from the leaves of citrus plants affected with a disease called "stubborn." Whether the spiroplasmas actually cause the disease has not been determined with certainty. Like other aphragmabacteria, the spiroplasmas are variable in form, lack cell walls, and form colonies that look like fried eggs. *Spiroplasma* cells are helical in shape and motile, showing a rapid screwing motion or a slower waving movement, yet they lack flagella or any other obvious organelles of motility. How do they move? We don't know.

An old hypothesis that the various aphragmabacteria separately (convergently) evolved from different Gram-positive bacteria by loss of walls is borne out by modern observations. Some *Mycoplasma* probably do represent minimal life on this planet and are truly primitive in the sense that their ancestors never had walls. Among them are the smallest organisms known. Most bacteria have 3000–6000 genes. The DNA of the strain *Mycoplasma genitalium* has only about 4.5×10^8 daltons of DNA, which is 10 times less than most bacteria. The complete sequence reveals this organism to have fewer than 500 genes and to make fewer than 500 proteins. Figure C shows a similar mycoplasma.

Wolbachia, a once free-living bacterium, has coevolved with insects: wasps, beetles, flies, and many others. This bacterium has given the animals many new traits *Wolbachia* genes have been passed into the nucleus of the insect.

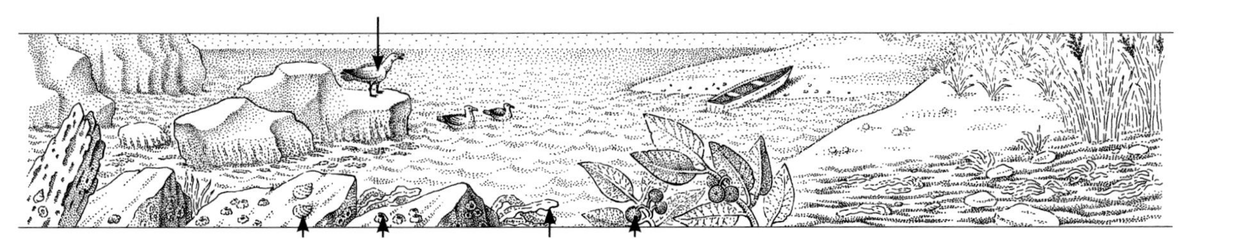

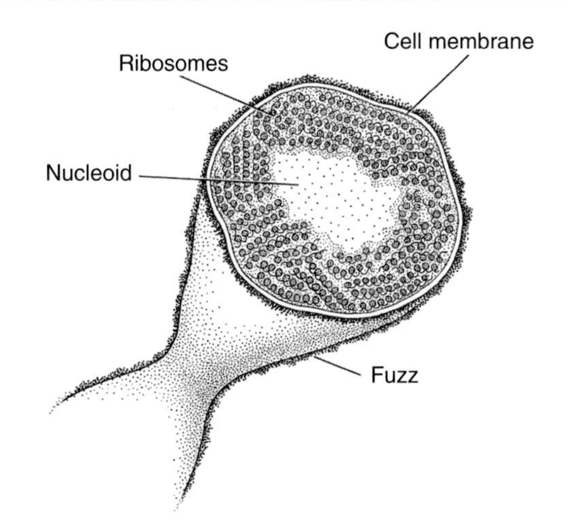

Figure A A generalized mycoplasma. [Drawing by L. Meszoly.]

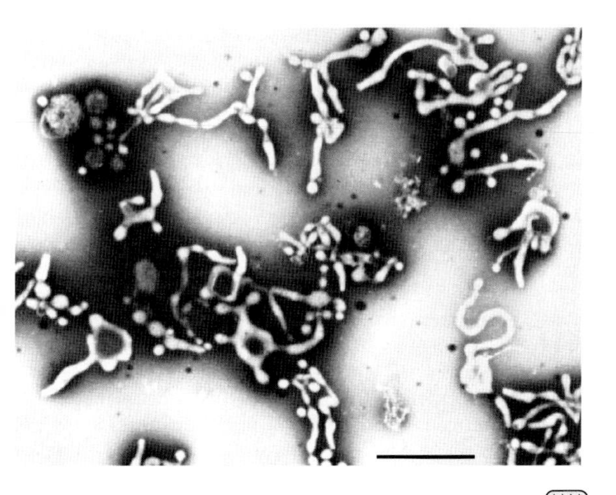

Figure B *Mycoplasma pneumoniae,* which lives in human cells and causes a type of pneumonia. TEM (negative stain), bar = 1 μm. [Courtesy of E. Boatman.]

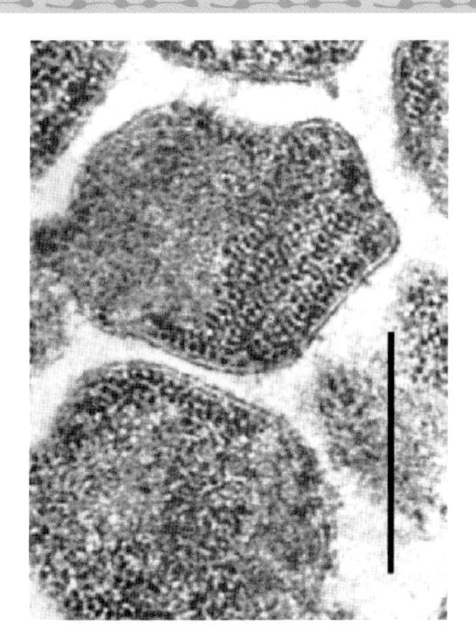

Figure C *Mycoplasma gallisepticum*, symbiotroph in chicken cells. TEM, bar = 0.5 μm. [Courtesy of J. Maniloff, *Journal of Cell Biology* 25:139–150, 1965.]

Division: Firmicutes

B-10 Endospora

(Endospore-forming and related low–G + C Gram-positive bacteria)

Greek *endos*, within; *spora*, seed

GENERA

Bacillus
Clostridium
Coleomitus
Heliobacterium
Lactobacillus
Leuconostoc
Lineola
Peptococcus
Peptostreptococcus
Ruminococcus
Sporolactobacillus
Sporomusa
Sporosarcina
Streptococcus
Thermoanaerobacter
Thermoanaerobacterium

Endospora include fermenting obligate anaerobes and obligate or facultative aerobes, most of which stain as Gram-positive walled cells and all of which belong to the 16S rRNA low-GC group of eubacteria. This great group of heterotrophs that comprises over 2500 species is the largest (most speciose) of the bacterial phyla. Many form endospores within their cells. Endospores, specialized propagules formed within the parent ("mother") cell, are resistant to conditions such as heat and desiccation that harm the trophic (growing) cells.

Most bacteria belonging to this phylum are motile by means of polar, peritrichous, or laterally inserted flagella. At least three genera are capable of oxygen respiration: *Bacillus*, a huge and important genus; *Sporosarcina*; and *Sporolactobacillus*. Only members of the genera *Clostridium* and *Sporomusa* are obligate anaerobes. Whereas *Clostridium* and all the other thousands of bacterial strains in this phylum are Gram-positive, *Sporomusa*—insect symbionts—surprisingly have Gram-negative cell walls.

Endospora comprises thousands of strains distributed in a multitude of habitats all over the world. Growing cells are rod or sphere shaped, as a rule, and their spores either elliptical or spherical. Each parent cell produces only a single, waterborne or airborne, spore, which may land anywhere. A special compound, calcium dipicolinate, constitutes 5–15 percent of the dry weight of the spore. The core of the spore contains one copy of the parent bacterium's genetic material (the nucleoid in Figure A). It is enclosed in a cortex made of peptidoglycan cell wall material and surrounded by an outer layer, called the "spore coat." Spore development is quite complex. In this dormant stage, endospora cells survive for years without water and nutrients. If nutrients and moisture are plentiful, the spore germinates and growth ensues until nutrient or water is depleted. As conditions become unfavorable for growth, sporulation sets in; the spore's position in the cell may be either central or terminal. Finally, the parent cell shrivels or disintegrates, releasing the spore.

Many strains of endospora produce antibiotics during the active growth stage. Most species produce acid, and all can metabolize glucose. Several produce gas or acetone as products of glucose catabolism. Many species can hydrolyze starch to glucose, and some also produce pigments such as the reddish brown or orange pulcherrinin and the brown or black melanin.

Some endospore formers break down tough plant substances such as pectin, polysaccharides, and even the lignin and cellulose of wood. The ability to degrade lignin and cellulose is unusual in the living world, which accounts in part for the persistence of forest litter, petroleum, peat, coal, and other organic-rich sediments. Lignin- and cellulose-degrading bacteria tend to associate in communities. Because no population of a single bacterial species totally degrades any of these refractory materials—lignin and cellulose—all the way to CO_2 and H_2O in days or weeks, in nature the breakdown requires communities of differently metabolizing bacteria.

The requirements for growth vary widely among endospora. Some have a high tolerance or even a requirement for salt; others do not. Vitamins and other complex growth factors are needed by certain strains, but not by others. Some clostridia can fix molecular nitrogen from the air, but most have more complex organic nitrogen requirements. Some may grow at a temperature as low as −5°C; some strains, called "psychrophils," grow optimally at −3°C. Some endospore formers, even members of *Bacillus*, are thermophilic; found in hot springs, they grow at temperatures above 45°C.

The genus *Bacillus* has so many members that information about it requires entire books. *Arthromitus*, first described as a "plant" (because it was not an animal) rooted to the intestines of beetles, termites, and other animals, is a symbiotic bacillus. Joseph Leidy (1823–1891), founder of the Philadelphia Academy of Natural Sciences, and his successors recognized the ubiquity of animal-associated, large–filamentous, spore-forming bacteria before the bacteriologists had developed a vocabulary for their description. Hence, many animal-associated endospore-forming filamentous bacilli have obsolete names such as *Lineola longa*, *Coleomitus*, and *Arthromitus*. Some of these animal symbionts are morphologically indistinguishable from *Bacillus anthracis* (some strains of which are associated with the disease anthrax); they deserve further study.

Entire colonies of *B. circulans* are motile, rotating as a unit for unknown reasons and by unknown mechanisms. Many *Bacillus* grow as filaments or chains and form colonies characterized by distinctive growth patterns.

Sporosarcina ureae are composed of spherical cells 1–2 μm in diameter, form tetrads (packets of four), and occasionally grow in distinctive cubical bundles that give the genus its name (Latin *sarcina*, packet). Urea, the metabolic product excreted by many animals, is converted by *S. ureae* into ammonium carbonate.

In appearance and metabolism, *Sporolactobacillus* resembles *Lactobacillus*—a well-known lactic acid–producing bacterium that grows vigorously on milk products—except that *Sporolactobacillus* produces endospores and respires in the presence of oxygen.

Miscellaneous important Gram-positive (nonspore forming) bacteria, many grouping with lactic acid bacteria (*Peptococcus*, *Leuconostoc*), including some obligate anaerobes, are discussed here. Their classification is in flux. Lactic acid bacteria do not use oxygen in their metabolism, and many do not tolerate oxygen at all: it inhibits their growth or quickly kills them. Nutritionally, they are heterotrophs of the kind called "chemoorganotrophs"—they require a mixed set of organic compounds to grow and reproduce.

Lactic acid bacteria, such as *Lactobacillus*, *Streptococcus* (the bacterium associated with strep throat infections), and *Leuconostoc*, are rod-shaped organisms famous for their ability to ferment sugar, in particular that in milk, and to produce lactic acid as well as acetate, formate, succinate, carbon dioxide, and ethanol. Their complex nutritional requirements include amino acids, vitamins, fatty acids, and other compounds depending on the species and strain. A 5–10 percent solution of carbon dioxide enhances their growth. They cannot form spores. Some may tolerate oxygen, but none use it for metabolism.

Peptococcaceae, spherical bacteria that lack flagella and spore formation, range in diameter from 0.5 to 2.5 μm. They are found singly, in pairs, in irregular masses, as three-dimensional packets, and as long and short chains of cocci. Many produce

gases such as CO_2 and H_2 when they ferment carbohydrates, amino acids, and other organics. Like most lactic acid bacteria, they have complex growth requirements. Many have been isolated from the mouths or intestines of animals. Three genera are recognized: *Peptococcus*; *Peptostreptococcus*, which uses protein or its breakdown products for energy; and *Ruminococcus*, which can break down cellulose.

Eapulipiscium, in the intestines of surgeon fishes, inhabits coral reefs. Since it may be up to 100 μm in length × 20 μm in width and "babies" have been seen inside it, was first published as a protist. But, no, it is another genus of Firmicutes that undergoes multiple fission and releases some 30 offspring. The parent cell is one of the largest prokaryotes known.

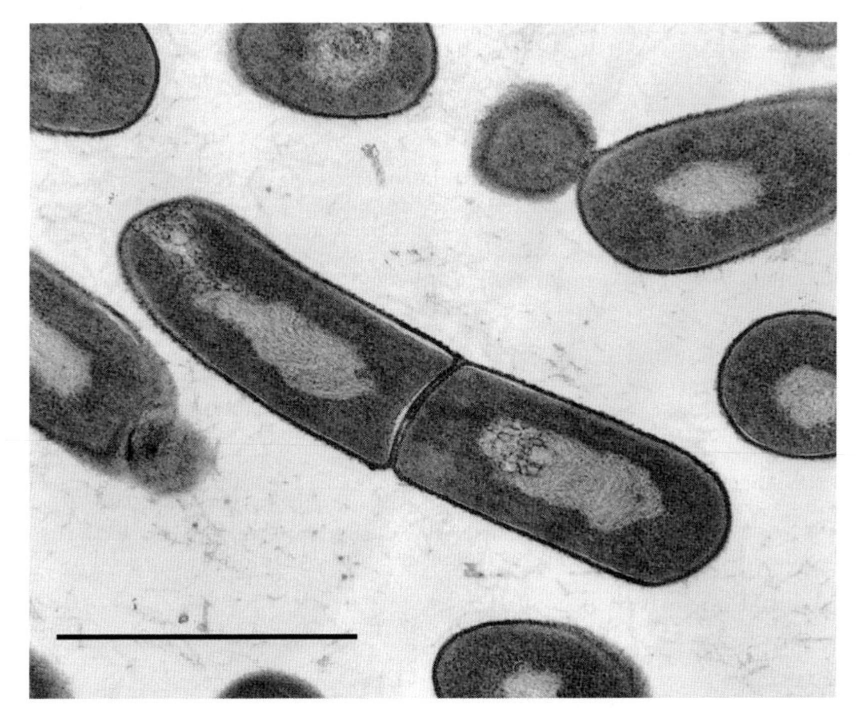

Figure A This unidentified *Bacillus* has just completed division into two offspring cells. Such spore-forming rods are common both in water and on land. TEM, bar = 1 μm. [Photograph courtesy of E. Boatman; drawings by I. Atema.]

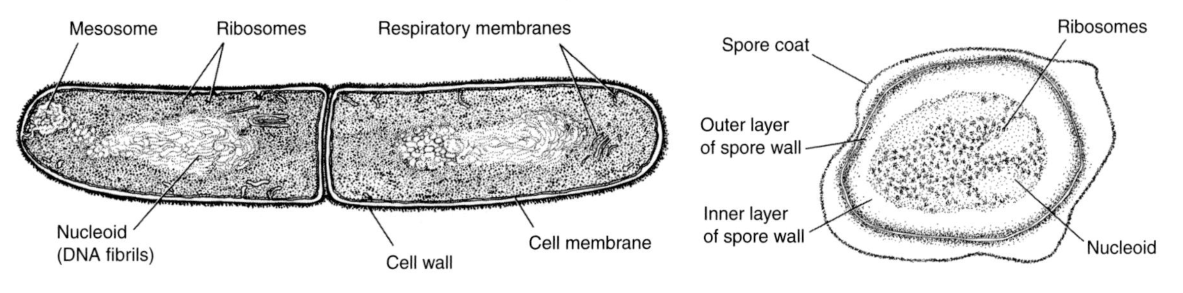

Division: Firmicutes

B-11 Pirellulae

(Proteinaceous-walled bacteria and their relatives)

Named after the genus *Pirellula* (Latin *pirus*, pear)

GENERA
Blastobacter
Chlamydia
Gemmata
Isosphaera
Pirellula (formerly Pirella)
Planctomyces

This new diverse group was revealed by 16S rRNA sequence information. At least two of these genera—*Pirellula* and *Planctomyces* (and probably *Gemmata*)—have unique, proteinaceous, nonpeptidoglycan cell walls. They do have bacterial cell walls, and although they superficially resemble the rickettsias, *Chlamydia* are classified, from sequence data, on their own branch, far closer to the pirellas than to the rickettsias. *Pirellula* and *Planctomyces* contain large quantities of two amino acids: proline and the sulfur-rich amino acid cysteine. Their lipids—palmitic, oleic, and palmitoleic—are strange for bacteria, although common in eukaryotes. Their aberrant walls render them insensitive, as are eukaryotes, to cell wall–inhibiting antibiotics such as cycloserine, cephalosporin, and penicillin.

Obligate aerobes, *Gemmata*, *Pirellula*, and *Planctomyces* are found in freshwater. They are heterotrophs that grow in extremely dilute solutions of salts and food. Some attach to surfaces of rock or vegetation by a kind of stalk (holdfast). Those that have prosthecae (appendages) differ from the ordinary prosthecate bacteria of Phylum B-3 (Proteobacteria)—for example, *Ancalochloris*, a green phototroph, or the heterotrophs *Asticcacaulis* and *Ancalomicrobium*—by the nature of their cell walls and RNA sequences.

Pirellula, *Blastobacter*, and *Planctomyces* form buds in their reproductive process: a small cell appears as a protrusion on its parent (Figures A and B). This cell grows by polar growth: a new cell wall is formed from a single point instead of the typical (nonpolar) growth by intercalation of new peptidoglycan units all over the surface of the cell. In polar growth, internal membranous and other organelles, such as carboxysomes or gas vesicles, do not take part in cell division, freeing up the parent cell for internal elaborations. This is most conspicuous and provocative in *Gemmata obscuriglobus* (Figures C and D), which, totally independently of eukaryotes, has evolved a nucleoid membrane. The double-layered unit membrane entirely encloses its bacteria-type DNA, producing an organelle that superficially resembles a true nucleus. The nucleoid lacks the pores, chromatin attachment sites, microtubules, and other characteristics of nuclei. Nor in any other way is *Gemmata* like a eukaryotic organism.

Planctomyces grow as pear-shaped or globular (large spherical) cells with long stalks. Because stalks contain no cytoplasm within the cell wall, they are not prosthecae. Although these stalks may be long and accompany undulipodia as protrusions on the surfaces of the cells, they are proteinaceous extensions as seen in Figure B.

Thought to be fungi when they were first described, *Planctomyces* still bear this misleading name (meaning "floating fungus"). Because they do float on lakes (they are planktic organisms), they should be renamed *Planctobacter* or something else more fitting of their prokaryotic nature.

Members of the genus *Chlamydia*, as obligate symbiotrophs refractory for study in pure culture, are enigmatic—well known from their medical context only. These bacteria, too, lack conventional peptidoglycan walls; hence, although they stain as Gram-negative bacteria, like all eukaryotes that also stain the telltale pink of the Gram-negative cell, they are better thought of as neither Gram-positive nor Gram-negative. *Chlamydia psittaci* is correlated with a parrot-transmitted disease of humans, and *C. trachomatis* is associated with the trachoma type of blindness. All dwell obligately inside animal cells (Figure E), and most work on these organisms has been directed at eliminating them from human cells.

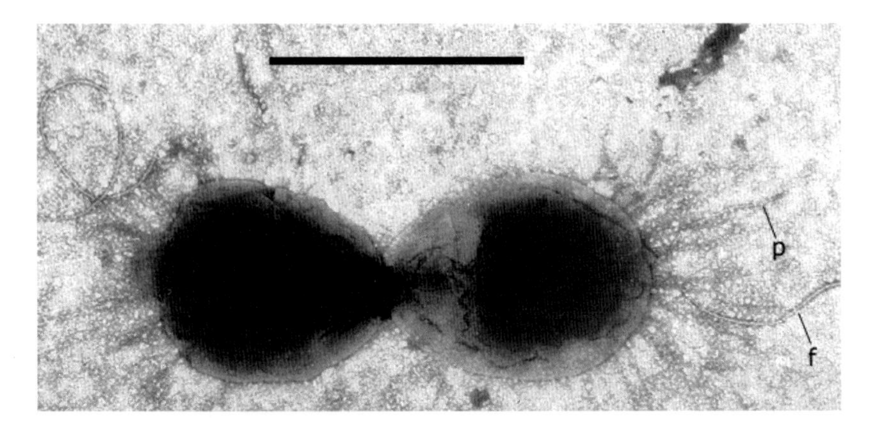

Figure A Dividing cells of *Pirellula staleyi* still attached to one another. Note pili (adhesive fibers; p) and polar undulipodia (f). TEM (negative stain, whole mount), bar = 1 μm. [Courtesy of J. Staley.]

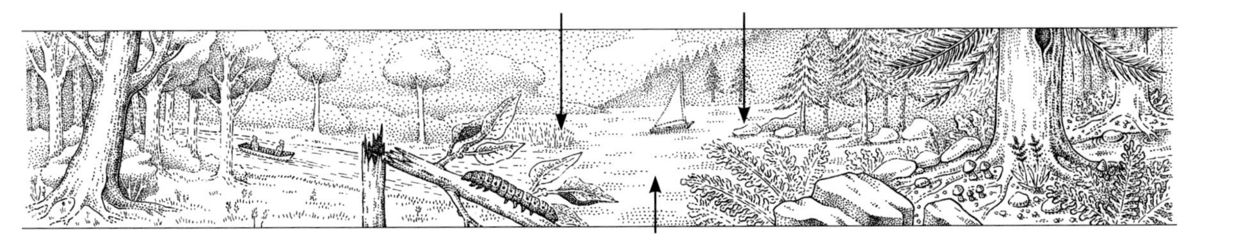

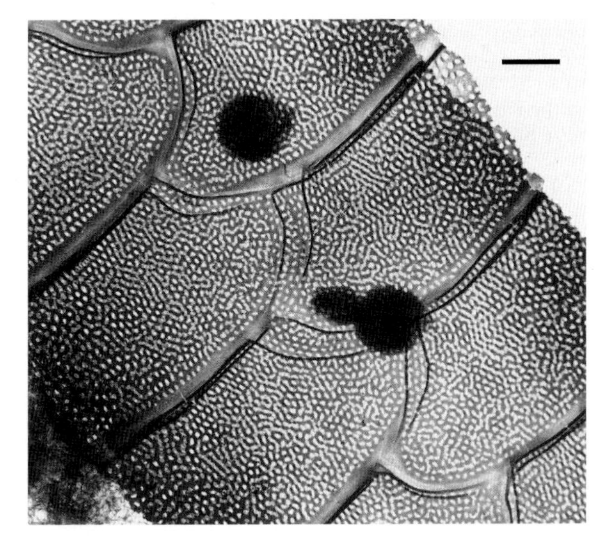

Figure B Pirellula sp. on a diatom. [Photograph courtesy of J. Staley.]

Figure C *Gemmata obscuriglobus*. Budding globular cells (arrowheads) as seen in a growing population. LM, bar = 10 μm. [Courtesy of J. Fuerst.]

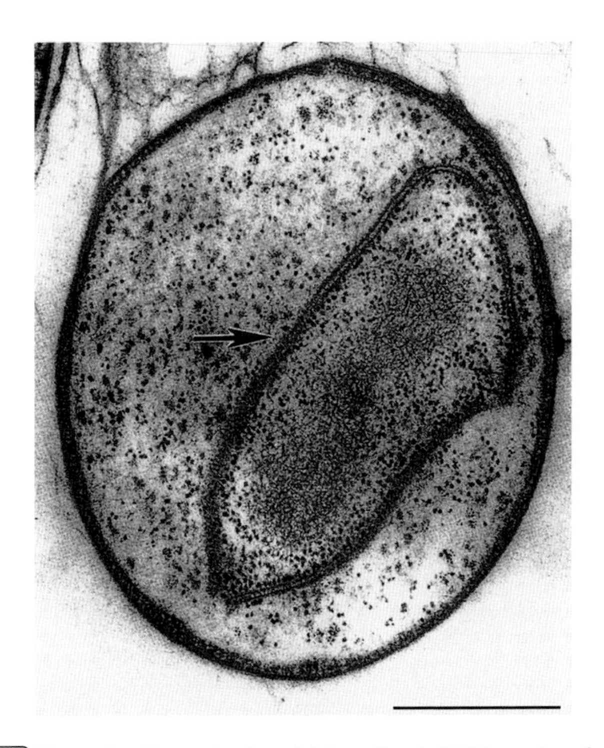

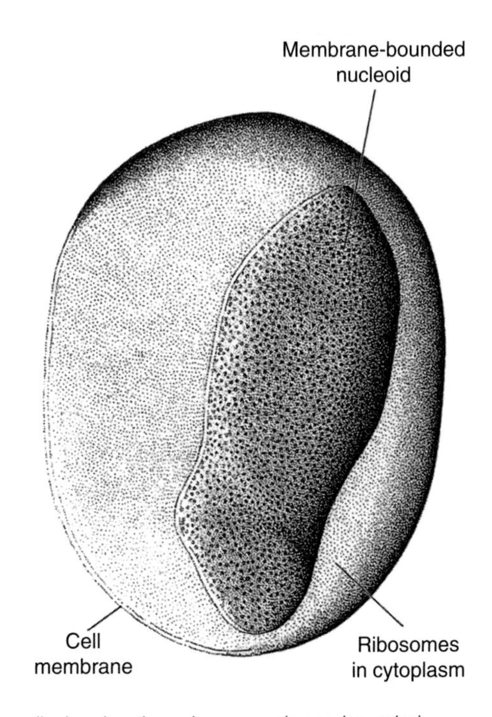

Figure D *Gemmata obscuriglobus.* Equatorial thin section of a single cell, showing the unique, membrane-bounded nucleoid (arrow). TEM, bar = 0.5 μm. [Photograph courtesy of J. Fuerst; drawing by C. Lyons.]

Although severely reduced in size relative to other bacteria by their obligate symbiotrophy, *Chlamydia* form propagules that are small, dense, and relatively resistant to desiccation. These structures, called "elementary bodies," are converted into reticulate structures that are capable of multiplication inside the animal cell. No known biochemistry for the metabolic production of energy molecules has been detected in purified preparations of *Chlamydia*, leading to the assertion that they are energy symbiotrophs. They are claimed to depend on ATP from the animal cells in which they reside.

The DNA content of these tiny propagules is only twice that of the largest viruses such as *Vaccinia*, since at least some of their RNA is standard ribosomal, apparently, in principle, they synthesize proteins and evolved from free-living bacteria with standard ribosomally based protein synthesis.

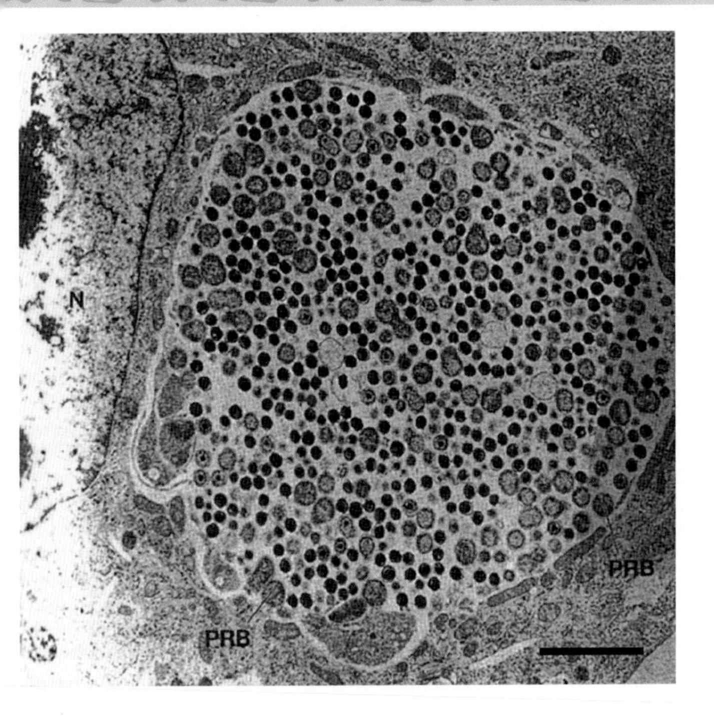

Figure E *Chlamydia psittaci.* Elementary bodies (dark small spheres) and progeny reticulate body (PRB) of *Chlamydia* in mammalian cells in tissue culture. The nucleus (N) of the animal cell is at left. TEM, bar = 1 μm. [Courtesy of P. B. Wyrick.]

Division: Firmicutes

B-12 Actinobacteria

(Actinomycetes, actinomycota; and related high–G + C Gram-positive bacteria)

GENERA

Actinomyces	*Dermatophilus*	*Nocardia*
Actinoplanes	*Frankia*	*Propionibacterium*
Arthrobacter	*Micromonospora*	*Streptomyces*
Cellulomonas	*Mycobacterium*	*Thermoactinomyces*
Corynebacterium	*Mycococcus*	*Thermomonospora*

Greek *aktis*, ray; *bakterion*, little stick

This phylum, uniting similar bacteria as determined by morphological, physiological, and 16S rRNA criteria, includes the coryneform bacteria and filamentous actinobacteria. The coryneforms are unicellular, Gram-positive organisms, straight or slightly curved rods with a tendency to form club-shaped swellings. Corynebacterial genera include *Corynebacterium* and the long or spiny *Arthrobacter*; members of the latter genus may absorb food from nearly pure water through as many as 20 spines (cytoplasmic projections) present in each cell. This phylum also contains the cellulose-attacking *Cellulomonas* and *Propionibacterium*, whose species produce propionic or acetic acids as products of sugar metabolism. The offspring cells of many fissioned coryneform bacteria or actinobacteria typically remain attached in a Y- or V-shaped configuration or remain attached to form hyphaelike structures.

The multicellular actinobacteria include filamentous prokaryotes that were originally mistaken for fungi. Unfortunately, even though they are prokaryotic in all of their features, they are still sometimes called "actinomycetes." Actinobacteria are distinguished by their stringy threads, superficially like fungal hyphae, and by the production of actinospores at their tips as resistant propagules. Actinobacteria probably evolved the "fungal" habit of growing hyphae (long strings) from cells that did not separate after binary fission. Hyphae in profusion form a visible mass of filaments, called a "mycelium." The actinobacterial mycelium probably evolved long before that of the fungi, all of which are eukaryotes.

This large and diverse group of eubacteria includes some that grow septate or nonseptate multicellular filaments. In *Mycobacterium*, the tuberculosis microbe, the filaments are short, whereas *Actinoplanes* and *Streptomyces* produce long and complex filaments that form and release actinospores.

Actinospores are sometimes inaccurately called conidia, but true conidia are the eukaryotic, haploid-propagating spores of the basidio- and ascomycotous fungi (F-5 and F-4). Actinospores also differ from bacterial endospores, which are formed within a parent cell and then released (B-10). In the development of all actinospores, the entire cell converts into a thick-walled, resistant spore. Thus, actinospores, endospores, and fungal conidia are all convergent structures—they represent a polyphyletic response to similar environmental pressures.

At least six of the actinobacteria groups form true actinospores, but only two enclose them in external structures. In these two (*Frankia* and *Actinoplanes*), actinospores are borne inside structures called "sporangia," by analogy with fungi. The many members of the genus *Frankia* are symbiotic in plants, where, like *Rhizobium* (B-3), they induce nodules that fix atmospheric nitrogen. The complex of threads of *Actinoplanes* forms dense mycelia that are still easily mistaken for fungal hyphae.

In *Dermatophilus*, the cells of the filaments of the mycelium divide transversely and in at least two longitudinal planes to form coccoid, motile bacteria that swim away and then lose flagellar motility, settle, and develop again into filaments. Some species form pathogenic lesions on human skin; others have been isolated from soil. The motile bacteria have been miscalled zoospores, a name restricted to motile eukaryotic propagules that swim by undulipodia and are capable of continued growth. Zoospores, for example, are commonly produced by oomycotes (Pr-21), chytridiomycotes (Pr-35), and plasmodiophorans (Pr-20). The term "zoospore" ought always, as it is in this book, to be restricted to eukaryotes. A multicellular mycelial habit alternating with motile stages (for example, flagellate bacteria or undulipodiated zoospores), in which there is no sexual encounter or fusion in either stage, has evolved independently in aquatic organisms: in several types of prokaryotes [for example, in actinobacteria and *Caulobacter* (B-3, Figure B) and in several distinctly different groups of eukaryotic microorganisms]. In microbes living on land, especially in soil, the tendency to form and release aerial propagules to be scattered by wind or by other organisms has evolved on repeated occasions: in myxobacteria (B-3), actinobacteria here, and several protoctist and most fungal groups.

The family Nocardiaceae includes the widely distributed genus *Nocardia*. Nocardias typically form mycelial filaments that fragment, yielding single nonmotile bacteria. They tend, especially in old cultures, to be Gram-variable (some Gram-negative cells) rather than clearly Gram-positive. Nocardias are very tenacious, and many survive, but do not grow, under noxious conditions of acidity, high salt, dryness, and other environmental extremes. If their entire developmental pattern is not observed, they can easily be mistaken for unicellular bacteria. Eventually, however, they betray their nocardial nature by forming filaments, mycelia, Y- and V-shaped cell groups, and actinospores. Some pathogenic strains and at least one capable of nitrogen fixation are known.

Members of the huge group constituting *Streptomyces* (more than 500 species have been described for this genus) form mycelia that tend to remain intact (Figure A). From the mycelia grow quite remarkable and well-developed chains of aerial actinospore-bearing structures (Figures B and C). Some are easily confused, at least superficially, with the smaller fungi. This group is justly well known for its synthetic versatility in producing streptomycin and other antibiotics.

The two genera grouped with *Micromonospora*—*Thermomonospora* and *Thermoactinomyces*—are found in soil, usually on plant debris. They form spores singly, in pairs, or in short chains on either aerial or subsurface mycelia. The mycelia are branched and septate, and the propagules are often brown. Some strains tolerate high temperatures.

Analysis of gene sequences of 16S rRNA confirms the validity of grouping together all the actinobacteria. This tight cluster of thousands of strains shows a range in guanosine–cytosine ratios from 51 to 79 and is called the high–G + C Gram-positive eubacteria. These diverse actinobacteria are distinguished from the genus *Bacillus* and its relative, which are low–G + C Gram-positive eubacteria (B-10). However, nature fails to cooperate with our tidy classifications: at least one strain of *Micromonospora* produces typical endospores instead of actinospores.

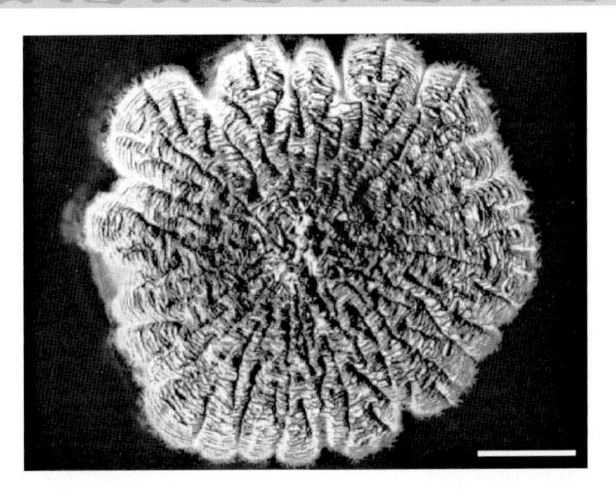

Figure A Colony of *Streptomyces rimosus* after a few days of growth on nutrient agar in petri plates. Bar = 10 μm. [Courtesy of L. H. Huang, Pfizer, Inc.]

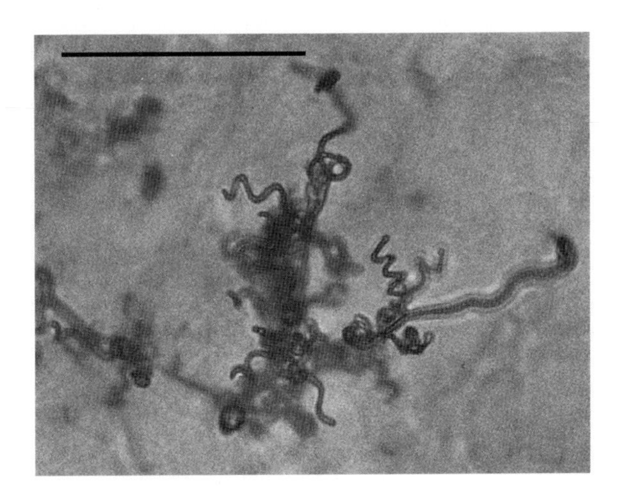

Figure B Aerial trichomes (filaments) bearing actinospores of *Streptomyces*. LM, bar = 50 μm. [Courtesy of L. H. Huang, Pfizer, Inc.]

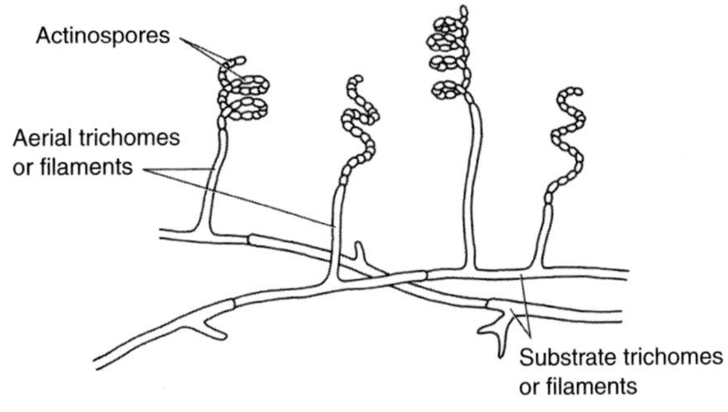

Figure C Part of a mycelium of *Streptomyces*. [Drawing by R. Golder.]

B-13 Division: Firmicutes
Deinococci
(Radiation-resistant or heat-resistant Gram-positive bacteria)

GENERA
Deinococcus
Thermus
Truepera

Greek *deino*, terrible, whirling; Latin *coccus*, berry

Deinococci, a new phylum based primarily on uniqueness as determined by 16S rRNA gene sequences, comprises Gram-positive, highly resistant, heterotrophic bacteria that require oxygen for growth. In the literature before the 1990s, all deinococci and many other aerobic spheres that do not form spores or other distinctive propagules were called "micrococci". Although at present only two well-studied genera (*Deinococcus* and *Thermus*) are formally assigned to this phylum, the colonization of radiation- and heat-resistant surfaces by coccoid bacteria suggests that more will be revealed by further study.

The spherical cells in *Deinococcus* are found singly or grouped in pairs and characteristically divide by binary fission in more than one plane to produce tetrads, irregular cubical packets of four cells (Figures A through C). Those micrococci that morphologically, but not by 16S rRNA criteria, resemble *Deinococcus* have been removed from this phylum. *Thermus* is heat resistant (60°C–80°C), whereas *Deinococcus* is radiation resistant.

Deinococci are strictly or facultatively aerobic—some can ferment, but all respire, using oxygen as the terminal electron acceptor. Because they tolerate as much as 3 million rads of ionizing radiation, they are easily isolated by bombardment with radiation, a procedure that kills everything else. (The human lethal dose is about 500 rads, and other bacteria can be killed by 100 rads.) Deinococci synthesize the respiratory pigments (cytochromes) and a class of quinones, also participants in respiration, called "menaquinones." Most species metabolize sugars: glucose, for example, is oxidized either to acetate or completely to carbon dioxide and water. They metabolize glucose by the hexose monophosphate pathway, rather than by the Embden–Meyerhof pathway used by many other heterotrophic bacteria and eukaryotes. Some species also oxidize smaller organic compounds, such as pyruvate, acetate, lactate, succinate, and glutamate, by the citric acid, or Krebs, cycle, characteristic of mitochondria. Some deinococci can grow in hypersaline environments, such as 5 percent NaCl in water (seawater is about 3.4 percent total salts). They break down hydrogen peroxide by using the enzyme catalase. *Deinococcus radiodurans* distinctively groups with *Thermus* by 16S rRNA analysis. Both *Deinococcus* and *Thermus* have tough cell walls of strange composition. The diaminopimelic acid of the peptidoglycan is absent in *Deinococcus*; instead, ornithine, a nonprotein amino acid, is a wall component. Unlike other Gram-positive bacteria, an extra outer-membrane layer surrounds the wall. The composition of the outer membrane in *Deinococcus* differs from that of Gram-negative bacteria, which also have an outer, as well as an inner (plasma), membrane.

Deinococcus cells are protected by multilayered walls. Most *Deinococcus* strains are colored pink or red, owing to the high concentration of carotenoid pigments made by their cells. Besides having resistance to ionizing radiation, many resist desiccation and ultraviolet light treatment even better than do endospores. The remarkable resistance of *Deinococcus* to mutagenic chemicals, radiation, and other treatments that generate genetic change in most bacteria has led to failure to induce stable mutations. The genetics of these organisms is poorly understood, although it is clear that excellent enzyme repair systems for DNA lead to highly efficient recovery from attempts to induce heritable damage.

Thermus aquaticus, discovered by Thomas Brock in his studies of natural hot springs bacteria of Yellowstone National Park, Wyoming, was catapulted to fame when it became the source of the Taq polymerase. This enzyme, a DNA polymerase that is so heat resistant that it survives the temperature cycling required for the PCR (polymerase chain reaction), is now in everyday use in thousands of laboratories that sequence DNA for scientific, legal, and industrial applications. The PCR technique permits small amounts of specific pieces of DNA to be replicated (amplified) many times for study. Kary Mullis was awarded the Nobel Prize in Chemistry in 1993 for his invention of PCR.

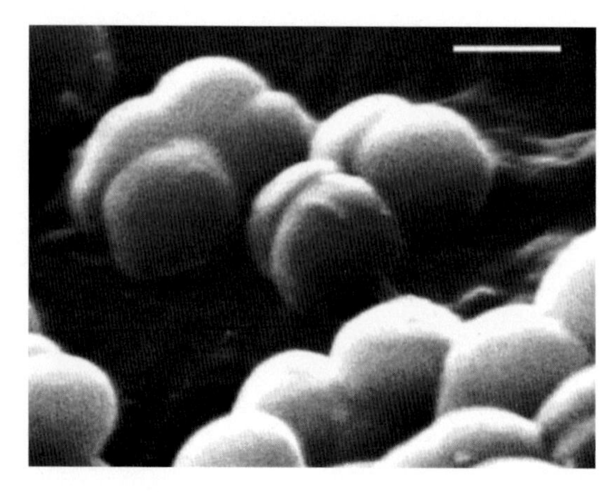

Figure A *Deinococcus radiodurans*. SEM (whole mount), bar = 1 μm. [Courtesy of J. Troughton.]

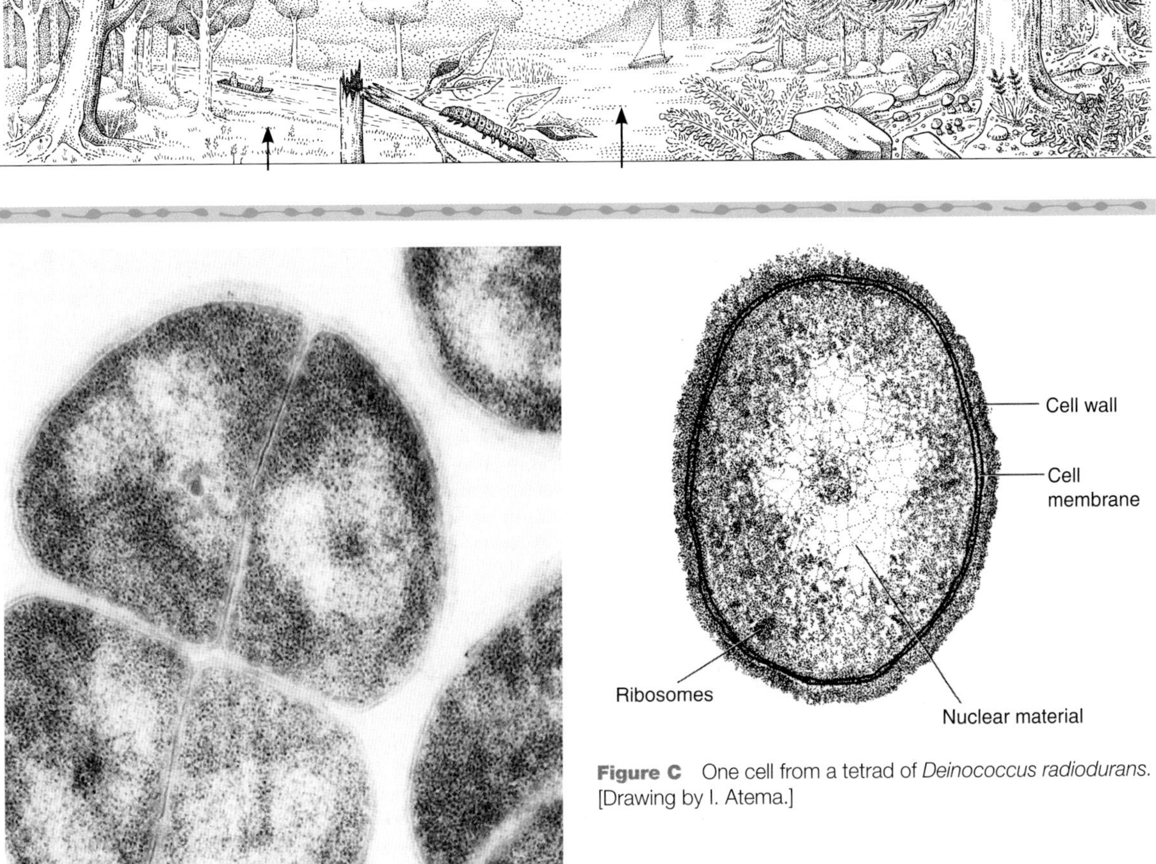

Figure B Transverse section of packet of four radiation-resistant *Deinococcus radiodurans* cells. TEM, bar = 1 μm. [Courtesy of A. D. Burrell and D. M. Parry.]

Figure C One cell from a tetrad of *Deinococcus radiodurans*. [Drawing by I. Atema.]

Division: Firmicutes

B-14 Thermotogae
(Thermophilic fermenters)

GENERA
Aquifex
Fervidobacterium
Thermosipho
Thermotoga

Named after the genus *Thermotoga* (Greek *thermo*, heat; Latin, *toga*, Roman citizen garment)

These newly discovered eubacteria are so strikingly different in their rRNA sequence from all others that they warrant a new phylum. Four genera have been described; no doubt many others exist. Both *Thermotoga* and *Thermosipho* are inhabitants of the hot vents at the bottom of the sea, where, as the continents separate, their spreading centers produce hot water and gases at temperatures that exceed 90°C. The differences between them are mainly morphological. *Thermotoga* is conspicuously covered by its "toga," several times thicker than and outside of its cell wall (Figures A and B). The guanosine–cytosine content of *Thermosipho* is lower than that of *Thermotoga*, another feature by which the isolates can be distinguished.

Thermotoga is exceedingly tolerant of high heat—it is one of the most heat-resistant eubacteria known. The cells have been cultured in the laboratory at temperatures ranging from 50°C to 90°C and grow optimally at 80°C. Although not an archaeabacterium, *Thermotoga* produces unique, lipidlike, extremely long-chain fatty acids. Nevertheless, like bacteria, these organisms do have peptidoglycan cell walls. As anaerobes, these cells have a strict requirement to avoid oxygen. Lacking a respiratory apparatus for any reduced inorganic compounds, they are obligate fermenters of sugars and other organic compounds, including complex protein digests.

The hot, deep-sea marine sediments have been under scientific study only since 1977, when the "dark gardens" replete with tubeworms, sulfur-rich clams, and other unique fauna were first discovered. Microbiologists require growth in "pure" culture (single bacterium) for the introduction of new bacterial names into the literature. Because high-temperature, high-pressure, anoxic conditions are not typically obtained in the laboratory, we suggest that it is highly likely that only a few thermotogas have been described; many other relatives of thermotogas lie waiting to be discovered in the hottest nether regions of the planet.

A new one, *Thermotoga subterranea*, another strict anaerobe, was isolated as recently as 1995 from a deep, continental oil reservoir in the East Paris Basin (France). It grew at temperatures between 50°C and 75°C, with optimum growth at 70°C. This *Thermotoga* was inhibited by elemental sulfur, but reduced cystine and thiosulfate to hydrogen sulfide. The guanosine–cytosine content, the presence of a lipid structure unique to the genus *Thermotoga*, and the 16S rRNA sequence put it with the other thermotogas but not as one of the recognized species: *T. maritima*, *T. neapolitana*, and *T. thermarum*.

Although *Aquifex*—another hot springs, submarine, volcanic eubacterium—belongs to this group, it lacks the ability to grow on sugar or protein. Rather, it oxidizes inorganic compounds—hydrogen (H_2), elemental sulfur (S^0), or thiosulfate ($S_2O_3^{2-}$)—by using either O_2 (under low oxygen tensions, hence microaerophilically) or NO^{3-} (anaerobically).

Fervidobacterium isolated from Iceland hot springs (hence *F. islandicum*) is also new to science. As extreme environments come under scrutiny, many new kinds of bacteria are expected to surprise us.

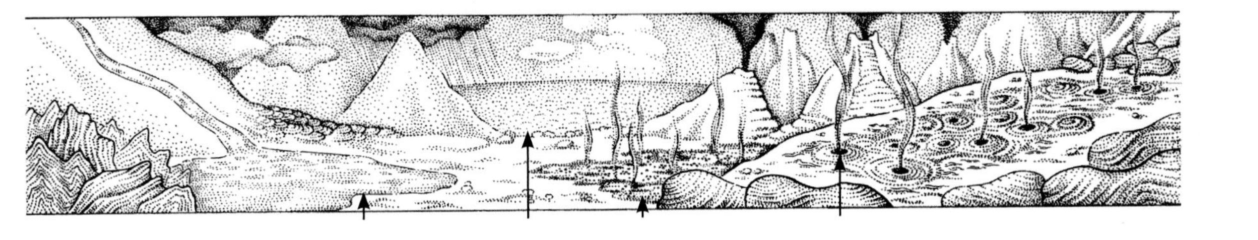

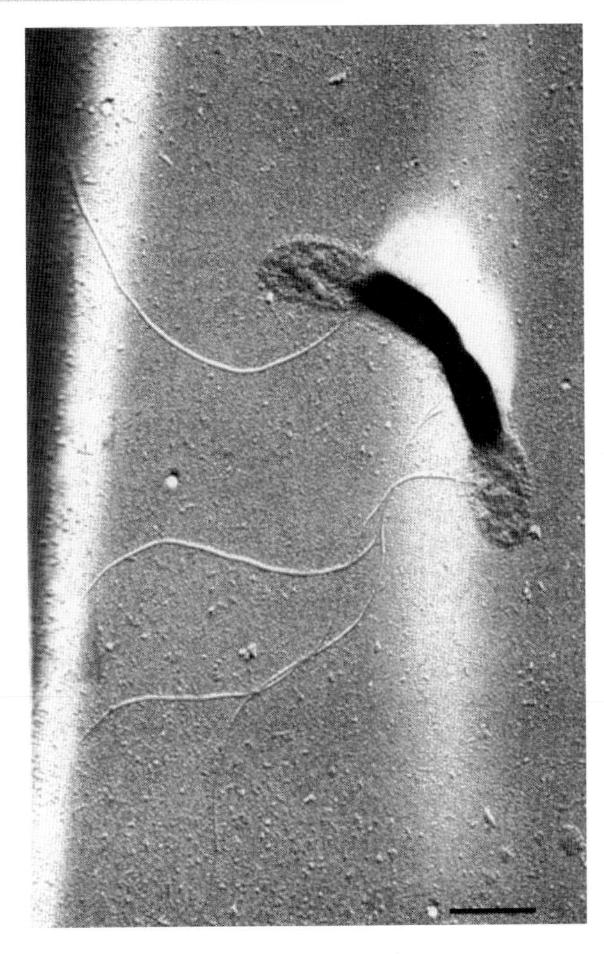

Figure A *Thermotoga thermarum.* The two cells are in division inside the thick toga. Here, the toga extensions can be seen by shadowcasting. TEM (negative stain), bar = 1 μm. [Courtesy of K. O. Stetter and R. Huber.]

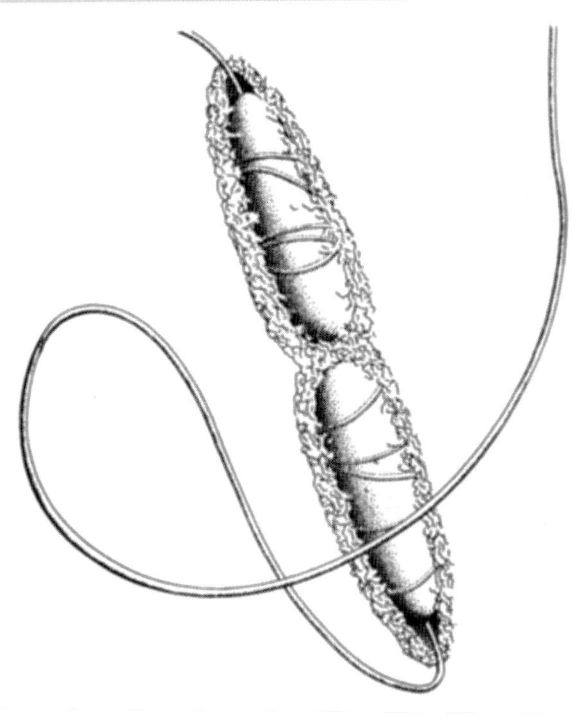

Figure B *Thermotoga* cell in division, entirely surrounded by the toga. The composition and function of the toga that surrounds the cell and the nature of the cell projections are not known. [Drawing, based on thin-section TEM images, by C. Lyons.]

Bibliography: Bacteria

General

Balows, A., H. G. Trüper, M. Dworkin, W. Harder, and K.-H. Schleifer, eds., *The prokaryotes: A handbook on the biology of bacteria: Ecophysiology, isolation, identification, applications, 4 vols.*, 2d ed. Springer-Verlag; New York; 1991.

Bengtson, S., ed., *Early life on Earth*. Columbia University Press; New York; 1994.

Dixon, B., *Power unseen: How microbes rule the world*. W. H. Freeman and Company; New York; 1994.

Dyer, B., *A field guide to bacteria*. Cornell University Press; New York; 2003.

Ferry, J. G., *Methanogenesis: Ecology, physiology, biochemistry and genetics*. Chapman & Hall; New York; 1993.

Gillies, R. R., and T. C. Dodds, *Bacteriology illustrated*, 4th ed. Churchill; New York; 1976.

Ingraham, J., and K. Ingraham, *Introduction to microbiology*. Wadsworth; Belmont, CA; 1995.

Krieg, N. R., and J. G. Holt, eds., *Bergey's manual of systematic bacteriology*. Williams & Wilkins; Baltimore and London; 1984–1989. Vol. 1, *Gram-negative bacteria of medical and commercial importance: Spirochetes, spiral and curved bacteria, Gram-negative aerobic and facultatively aerobic rods, Gram-negative obligate anaerobes, Gram-negative aerobic and anaerobic cocci, sulfate and sulfur-reducing bacteria, rickettsias and chlamydias, mycoplasmas*. Vol. 2, *Gram-positive bacteria of medical and commercial importance: Gram-positive cocci, Gram-positive endospore-forming and non-spore-forming rods, mycobacteria, nonfilamentous actinomycetes*. Vol. 3, *Remaining Gram-negative bacteria*: *Phototrophic, gliding, sheathed, budding, and appendaged bacteria, cyanobacteria, lithotrophic bacteria, and the archaeobacteria*. Vol. 4, *Filamentous actinomycetes and related bacteria*.

Madigan, M. T., *Brock's biology of microorganisms*, 8th ed. Prentice-Hall; Englewood Cliffs, NJ; 1996.

Margulis, L., *Early life*. Jones and Bartlett; Boston, MA; 1982.

Sagan, D., and L. Margulis, *Garden of microbial delights: A practical guide to the subvisible world*. Kendall-Hunt; Dubuque, IA; 1993.

Sapp, J. "The prokaryote–eukaryote dichotomy: Meaning and mythology." *Microbiology and Molecular Biology Reviews* 69:292–305; 2005.

Schleifer, K. H., and E. Stackebrandt, eds., *Evolution of prokaryotes*. Academic Press; London; 1985.

Sneath, P. H. A., ed., *International code of nomenclature of bacteria*. American Society for Microbiology; Washington, DC; 1992.

Sonea, S., and L. Mathieu, Prokaryotology. Montreal, CA: University of Montreal Press; 2002

Starr, M. P., H. Stolp, H. G. Trüper, A. Balows, and H. G. Schlegel, eds., *The prokaryotes: A handbook on habitats, isolation and identification of bacteria, 4 vols.*, 2d ed. Springer-Verlag; Heidelberg and New York; 1991.

Woese, C. R., and G. E. Fox, Phylogenetic structure of the prokaryotic domain: the primary kingdoms. PNAS 74(11):5088–5090; 1977.

Woese, C. R., and R. S. Wolfe, *The bacteria: A treatise on structure and function, Vol. 3: Archaebacteria*. Academic Press; New York; 1985.

Woese, C. R., O. Kandler, and M. L. Wheelis, "Towards a natural system of organisms: Proposal for the domains Archaea, Bacteria, and Eucarya." *Proceedings of the National Academy of Sciences, USA* 87:4576–4579; 1990.

Wolfe, S., *The biology of the cell*, 4th ed. Wadsworth; Belmont, CA; 1995.

B-1 Euryarchaeota

Bult, C. J., *et al.*, "Complete genome sequence of the methanogenic archaeon, *Methanococcus jannaschii*." *Science* 273:1058–1073; 1996.

Madigan, M. T., *Brock's biology of microorganisms*, 8th ed. Prentice-Hall; Englewood Cliffs, NJ; 1996.

Mah, R. A., D. M. Ward, L. Baresi, and L. Glass, "Biogenesis of methane." *Annual Review of Microbiology* 31:309–341; 1977.
Morell, V., "Life's last domain." *Science* 273:1043–1045; 1996.
Woese, C. R., "Archaebacteria." *Scientific American* 244(6):98–122; June 1981.

B-2 Crenarchaeota

Kandler, O., "Evolution of the systematics of bacteria." In: K. H. Schleifer and E. Stackebrandt, eds., *Evolution of prokaryotes.* FEMS Symposium 29. Academic Press; London; pp. 335–361; 1985.
Lanyi, J. K., "Physical chemistry and evolution of salt tolerance in halobacteria." In: C. Ponnamperuma and L. Margulis, eds., *Limits of life.* D. Reidel; Dordrecht, The Netherlands; pp. 61–68; 1980.

B-3 Proteobacteria

Barton, L. L., ed., *Sulfate-reducing bacteria.* Plenum Press; New York; 1995.
Bermudes, D., L. Margulis, and G. Hinkle, "Do prokaryotes have microtubules?" *Microbiological Reviews* 58:387–400; 1994.
Blankenship, R. E., M. T. Madigan, and C. E. Bauer, eds., *Anoxygenic photosynthetic bacteria.* Kluwer Academic; Dordrecht, The Netherlands; 1995.
De Ley, J., "Proteobacteria (purple *bacteria*)." In: A. Balows *et al.*, eds., *The prokaryotes,* Vol. 2, 2d ed. Springer-Verlag; New York; pp. 2111–2140; 1992.
Frederickson, J. K., and T. C. Onsott, "Microbes deep inside the earth." *Scientific American* 275:68–73; October 1996.
Nudleman, E., D. Wall, and D. Kaiser, "Cell-to-cell transfer of bacterial outer membrane lipoproteins." *Science* 309:125–127; 2005.
Schlegel, H. G., and H. W. Jannasch, "Prokaryotes and their habitats." In: A. Balows et al., eds., *The prokaryotes,* Vol. 1, 2d ed. Springer-Verlag; New York; pp. 75–125; 1992.
Shapiro, J. A., "Bacteria as multicellular organisms." *Scientific American* 258(6):82–89; June 1988.

B-4 Spirochaetae

Bermudes, D., D. Chase, and L. Margulis, "Morphology as a basis for taxonomy of large spirochetes symbiotic in wood-eating cockroaches and termites: *Pillotina* gen. nov., nom. rev.; *Pillotina calotermitidis* sp. nov., nom. rev.; *Diplocalyx* gen. nov., nom. rev.; *Diplocalyx calotermitidis* sp. nov., nom. rev.; *Hollandina* gen. nov., nom. rev.; *Hollandina pterotermitidis* sp. nov., nom. rev.; and *Clevelandina reticulitermitidis* gen. nov., sp. nov." *International Journal of Systematic Bacteriology* 38:291–302; 1988.
Breznak, J. A., "Genus II. *Cristispira* Gross, 1910, 44." In: N. R. Krieg and J. G. Holt, eds., *Bergey's manual of systematic bacteriology,* Vol. 1. Williams & Wilkins; Baltimore; pp. 46–49; 1984.
Canale-Parola, E., "The genus Spirochaeta." In: A. Balows *et al.*, eds., *The prokaryotes, Vol. 4,* 2d ed. Springer-Verlag; New York; pp. 3524–3536; 1992.
Holt, S. C., "Anatomy and chemistry of spirochetes." *Bacteriological Reviews* 42:114–160; 1978.
Margulis, L., and G. Hinkle, "Large symbiotic spirochetes: *Clevelandina, Cristispira, Diplocalyx, Hollandina,* and *Pillotina.*" In: A. Balows *et al.*, eds., *The prokaryotes,* Vol. 4, 2d ed. Springer-Verlag; New York; pp. 3965–3978; 1992.

B-5 Saprospirae

Reichenbach, H., and M. Dworkin, "Introduction to the gliding bacteria." In: M. P., Starr, ed. *The prokaryotes.* Springer-Verlag; Berlin; pp. 315–327; 1981.

B-6 Cyanobacteria

Carr, N., and B. Whitton, eds., *The biology of the blue-green algae.* Blackwell; Oxford, England; 1973.

Cohen, Y., and E. Rosenberg, eds., *Microbial mats.* American Society for Microbiology; Washington, DC; 1989.

Golubic, S., "Microbial mats of Abu Dhabi." In: L. Margulis and L. Olendzenski, eds., *Environmental evolution.* MIT Press; Cambridge, MA; pp. 103–130; 1992a.

Golubic, S., "Stromatolites of Shark Bay." In: L. Margulis and L. Olendzenski, eds., *Environmental evolution.* MIT Press; Cambridge, MA; pp. 131–148; 1992b.

Golubic, S., M. Hernandez-Marine, and L. Hoffman, "Developmental aspects of branching in filamentous Cyanophyta/Cyanobacteria." *Algological Studies* 83:303–329; 1996.

B-7 Chloroflexa

Fuller, C., ed., *Phototrophic bacteria.* Proceedings of the second international meeting. ASM Press; Washington, DC; 1991.

Stolz, J. F., ed., *Structure of photosynthetic bacteria.* CRC Press; Boca Raton, FL; 1990.

B-8 Chlorobia

Pearson, B., and R. Castenholz, "The family Chloroflexaceae." In: A. Balows *et al.*, eds., *The prokaryotes, Vol. 4,* 2d ed. Springer-Verlag; New York; pp. 3754–3774; 1992.

Stolz, J. F., ed., *Structure of photosynthetic bacteria.* CRC Press; Boca Raton, FL; 1990.

B-9 Aphragmabacteria

Fraser, C. M., *et al.*, "The minimal complement of *Mycoplasma genitalium.*" *Science* 270:397–403; 1995.

Gordon, R. E., W. C. Haynes, and C. H. N. Pang, *The genus Bacillus.* Handbook No. 427. U.S. Department of Agriculture; Washington, DC; 1973.

Razin, S., and E. A. Freundt, "Mycoplasmas." In: N. R. Krieg and J. G. Holt, eds., *Bergey's manual of systematic bacteriology,* Vol. 3. Williams & Wilkins; Baltimore; pp. 742–775; 1986.

Woese, C. R., E. Stackebrandt, and W. Ludwig, "What are mycoplasmas?: The relationship of tempo and mode in bacterial evolution." *Journal of Molecular Evolution* 21:305–316; 1985.

B-10 Endospora

Breznak, J. A., J. M. Switzer, and H.-J. Seitz, "*Sporomusa termitide* sp. nov., an H_2/CO_2-utilizing acetogen isolated from termites." *Archives of Microbiology* 150:282–288; 1988.

Clements, K. D., and S. Bullivant, "An unusual symbiont from the gut of surgeonfishes may be the largest known prokaryote." *Journal of Bacteriology* 173:5359–5362; 1991.

Slepecky, R. A., and H. E. Hemphill, "The genus *Bacillus*—nonmedical." In: A. Balows et al., eds., *The prokaryotes,* Vol. 2, 2d ed. Springer-Verlag; New York; pp. 1663–1696; 1992.

Sonenschein, A. L., J. A. Hoch, and R. Losick, *Bacillus subtilis and other Gram positive bacteria: Biochemistry, physiology and molecular genetics.* American Society of Microbiology; Washington, DC; 1993.

B-11 Pirellulae

Franzmann, P. D., and V. B. D. Skerman, "*Gemmata obscuriglobus,* a new genus and species of the budding bacteria." *Antonie van Leeuwenhoek* 50:261–268; 1984.

Fuerst, J. A., "The planctomycetes: Emerging models for microbial ecology, evolution and cell biology." *Microbiology* 141:1493–1506; 1995.

Fuerst, J. A., and R. I. Webb, "Membrane-bounded nucleoid in the eubacterium *Gemmata obscuriglobus.*" *Proceedings of the National Academy of Sciences, USA* 88:8184–8188; 1991.

Staley, J. T., J. A. Fuerst, S. Giovannoni, and H. Schlesner, "The order Planctomycetales and the genera *Planctomyces, Pirellula, Gemmata,* and *Isosphaera.*" In: A. Balows *et al.*, eds., *The prokaryotes, Vol. 4,* 2d ed. Springer-Verlag; New York; pp. 3710–3730; 1992.

B-12 Actinobacteria

Callwell, R., and D. Callwell, *Actinomycetes and streptomycetes.* American Society of Microbiology; Washington, DC; 1996.

Ensign, J., "Actinomycetes." In: A. Balows *et al.*, eds., *The prokaryotes,* Vol. 1, 2d ed. Springer-Verlag; New York; pp. 811–815; 1992.

B-13 Deinococci

Griffiths, E., and R. S. Gupta, "Identification of signature proteins that are distinctive of the *Deinococcus–Thermus* phylum." *International Microbiology* 10:201–208; 2007.

Kocur, M., W. E. Kloos, and K.-H. Schleifer, "The genus Micrococcus." In: A. Balows *et al.*, eds., *The prokaryotes,* Vol. 2, 2d ed. Springer-Verlag; New York; pp. 1300–1311; 1992.

B-14 Thermotogae

Jeanthon, C., A. L. Reysenbach, S. L'Haridon, A. Gambacorta, N. R. Pace, P. Glenat, and D. Prieur, "*Thermotoga subterranea* sp. nov., a new thermophilic bacterium isolated from a continental oil reservoir." *Archives of Microbiology* 164(2): 91–97; 1995.

Kristjansson, J. K., ed., *Thermophilic bacteria.* CRC Press; Boca Raton, FL; 1992.

Schliefer, B., ed., *Deep-sea bacteria.* CRC Press; Boca Raton, FL; 1996.

SUPERKINGDOM EUKARYA

Origins by symbiogenesis

Organisms in this superkingdom composed of cells that contain more than two protein-coated 100 Å unit fibril Feulgen-stainable chromosomes per cell. The cells reproduce by mitosis. In mitosis pore-studded membrane-bounded nuclei totally or partially dissolve and re-form as two offspring nuclei. Display viable cytoplasmic and nuclear fusion and their reciprocal processes (for example, cytoplasmic and chromosomal doubling with subsequent meiotic or equivalent reduction of cytoplasmic and nuclear content; hence, Mendelian genetics). Cells have tubulin–actin cytoskeletons capable of intracellular motility visible *in vivo*. Cells contain kinetosome-centriole [9(3)+0] microtubular organizing centers or their derivatives. Organisms lack unidirectional gene (naked DNA) transfer except in the laboratory. There are at least two genomic ancestors to eukaryotic cells, one eubacterial and one archaebacterial.

A-5A1 *Bolinopsis infundibulum*
[Courtesy of M. S. Laverack.]

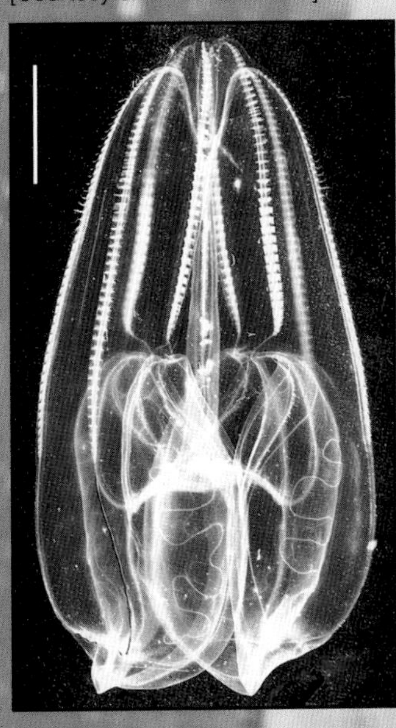

Superkingdom Eukarya
- **Kingdom Protoctista**
- **Kingdom Animalia**
- **Kingdom Fungi**
- **Kingdom Plantae**

Box Eukarya-i: Undulipodia from sulfur syntrophies

If the waste of one organism (sulfide, HS^- in this case) becomes food or fuel of a second (a different kind of microbe that converts HS^- to elemental sulfur, S) and then the sulfur can be used by the first organism to make more sulfide, we speak of a *syntrophy* (syn, together; trophy, feeding). Syntrophic partnerships are common in bacteria that tend to digest food outside of but near their bodies. The very first step in the origin of the first protoctists (and therefore the first nucleated cells) involved production of HS^- by the archaebacterial ancestor, a wall-less, heat-tolerant cell much like today's *Thermoplasma acidophila* (Searcy, 1980). We agree, but also posit that the partner in the syntrophy was an active swimming cell, a spirochete eubacterium that chemically converted low quantities of HS^- to elemental sulfur. Because the swimming cell could tolerate only low concentrations of oxygen ($<2\%$), the reaction HS^- + oxygen = S for oxygen removal was crucial for its survival. Both partners sought organic compounds as food. Stuck together as a motility symbiosis, they tolerated ranges of oxygen from 0.000001% to about 5%. They maintained the association as they traversed regions of abundant sulfur compounds in search of food. Strong selective advantages to this sulfur syntrophy soon made it obligate; archaebacterial–eubacterial genomes merged, and membrane systems fused. The swimmer, we hypothesize, became the undulipodium (cilium or eukaryotic undulipodium with the well-studied [9(2+)+2] pattern of microtubules). From this versatile, wide-ranging partnership evolved the earliest intracellular motility, including the nucleus with mitotic cell division ("dance of the chromosomes"). The same organelle (the *centriole* or *kinetosome*) that gave rise to the microtubules of the undulipodium also became the spindle pole in mitosis; thus, the nucleus, centriole-kinetosome, and undulipodium formed the quintessentially eukaryotic organellar system called the *karyomastigont*. Origin of the karyomastigont was the crucial first step in the evolution of eukaryotes from prokaryotes. Other types of cell movement (phagocytosis, exocytosis, cyclosis, and so on), on this reckoning, evolved under very low atmospheric oxygen pressures before the incorporation of oxygen-respiring bacteria that became mitochondria and later, photosynthetic bacteria that became plastids.

The cells of animal, plant, fungal, and protoctist bodies, all composed of nucleated ("eukaryotic") cells, tend to be larger and structurally more complex, sometimes vastly larger than those of any bacteria. However, the chemistry and fossil history of bacteria convince us that bacteria were the ultimate common ancestors of all extant life. Bacterial genes are made of DNA; bacterial cell membranes, invariably composed of fatty compounds linked to phosphoric acid and studded with proteins, are similar to those of eukaryotes.

What else took place during the evolutionary transition from bacterial cells to those of eukaryotes? Although some controversy persists over details, all scientists agree that nucleated cells (animal, plant, fungal, and protoctist) evolved not from any single hypertrophied ancestral bacterium, but from a mixed community of bacteria by way of mergers: literal incorporations of different kinds of bacteria. An α-proteobacterium that breathed oxygen became the mitochondrion, and photosynthetic cyanobacteria became the chloroplasts and all other plastids. At least two very different bacterial lineages had already fused in the formation of ancestors of all eukaryotes. We suspect that these were archaebacteria, which converted sulfur to sulfide, and eubacteria, common spirochetes that converted sulfide to sulfur for oxygen protection. Animal, plant, and other eukaryotic cells retain sulfur metabolism similar to today's archaebacteria. The cytoplasm of all eukaryotes investigated still produces HS^-. Hydrogen sulfide peaks just before mitosis and declines afterward as measured in some ciliates and even reversibly puts mice to sleep when given in tiny amounts. Our ancient legacy of sulfur syntrophy is still with us today.

References

Searcy, D. G., and R. J. Delange, Thermoplasma acidophilum histone like protein: Partial amino acid sequence suggestive of homology to eukaryotic histones. *Biochim. Biophys. Acta* 609:197–200; 1980a.

Searcy, D. G., and D. B. Stein, Nucleoprotein subunit structure in an unusual prokaryotic organism: Thermoplasma acidophilum. *Biochim. Biophys. Acta* 609:180–195; 1980b.

Box Eukarya-ii: Eukaryosis—Life histories, life cycles

The three major kingdoms of eukaryotes—fungi, plants, and animals—can be distinguished by the position of meiosis in their life cycles: fungi are haploid with zygotic meiosis, plants alternate haploidy (one set of chromosomes) and diploidy (two sets of chromosomes) with sporogenic meiosis, and animal bodies are diploid with

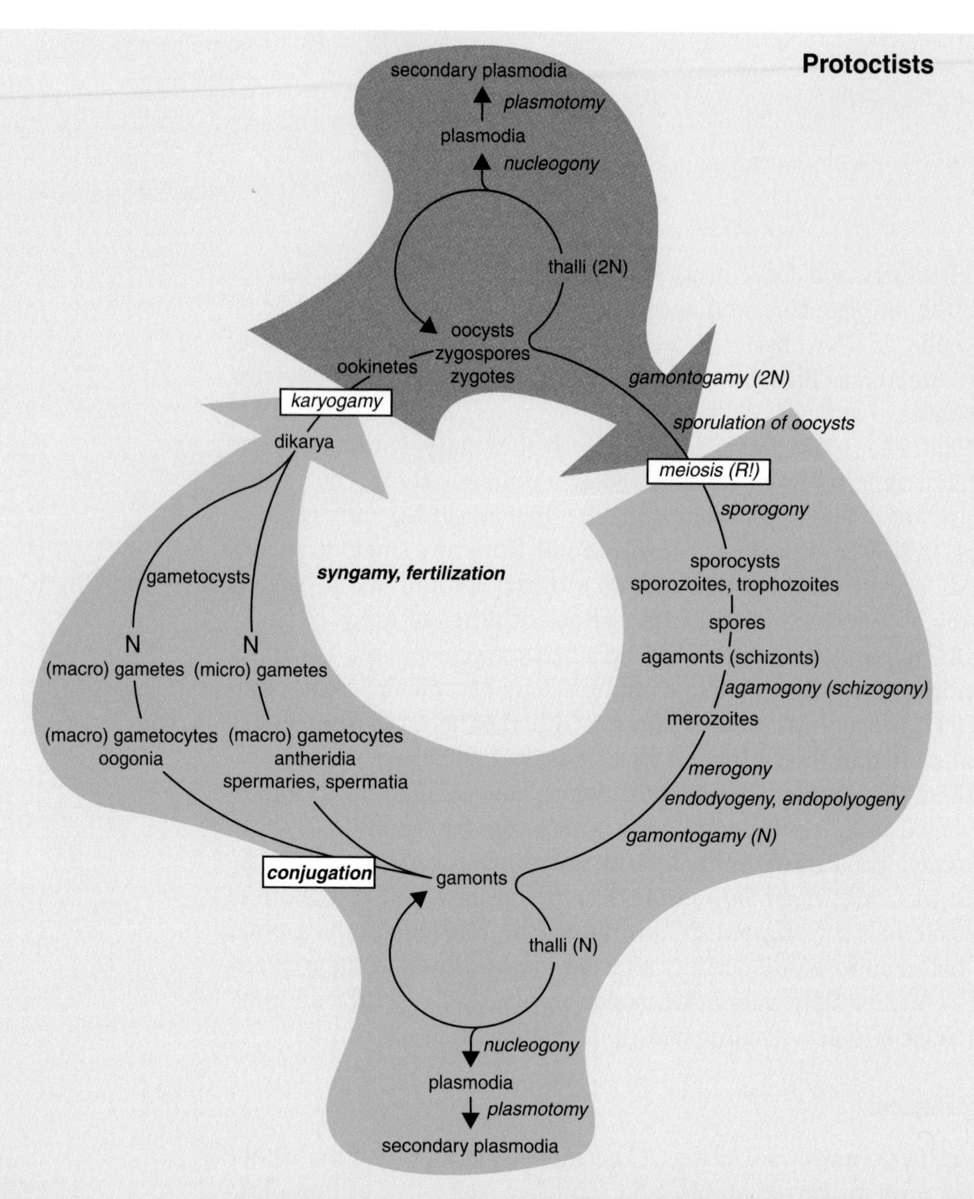

Figure Eukarya-ii-1 Generalized protoctist life cycle. Meiosis gives rise to haploid nuclei in cells of organisms. These occur e.g., in Apicomplexa (Pr-7) as resistant sporocysts, motile sporozoites or feeding trophozoites. Depending on environmental conditions a haploid cell or multicellular organism may remain in a uniparental, trophic or reproductive state as a haploid agamont (if it reproduces before it makes gametes). Or by mitotic growth and differentiation it may become a gamont. A gamont is an organism, either haploid or diploid, that by mitosis or meiosis respectively, makes gamete nuclei or gamete cells. The haploid organism may differentiate reproductive thalli, plasmodia, pseudoplasmodia or other structures without meiosis and remain an agamont. The haploid may form egg-producing oogonia, sperm-filled antheridia or develop isogametous (look-the-same) gametes in which case it changes, by definition, from an agamont to a gamont. Protoctist generative nuclei or cells may also remain in the diploid state and grow large and/or reproduce by multiple fission, hyphae, plasmodia, thalli, spores or other agamontic life history forms. Some diploid nuclei undergo meiosis in uni- or multicellular protoctists to produce more offspring as agamonts, gamonts or gametes. Gametes may be haploid nuclei only (as in some ciliates and foraminifera) or whole gamont bodies (as in many sexual algae or water molds. Gamontogamy, cytogamy and/or karyogamy (= conjugation, sexual fusion of cytoplasm of gaemete-formers or their gametes, nuclear fusion), spore-differentiation and other processes may regenerate diploids that quickly return, by meiosis, to haploidy. Or the diploid state, as in animals and flowering plants, may be protracted. Life cycles of the "crown taxa" (animals, fungi and plants) are limited specializations for ploidy levels and meiotic pathways. Sexuality (including gender differentiation) ranges from complete absence to such extravagant variation that the Protoctista Kingdom is the taxon in which Darwin's "imperfections and oddities" of meiosis-fertilization cycles must have evolved. Generalities in this figure (many described in Raikov, 1982 or Grell,1972) are well represented in foraminifera (Pr-3), ciliate (Pr-6) and red algal (Pr-33) protoctists.

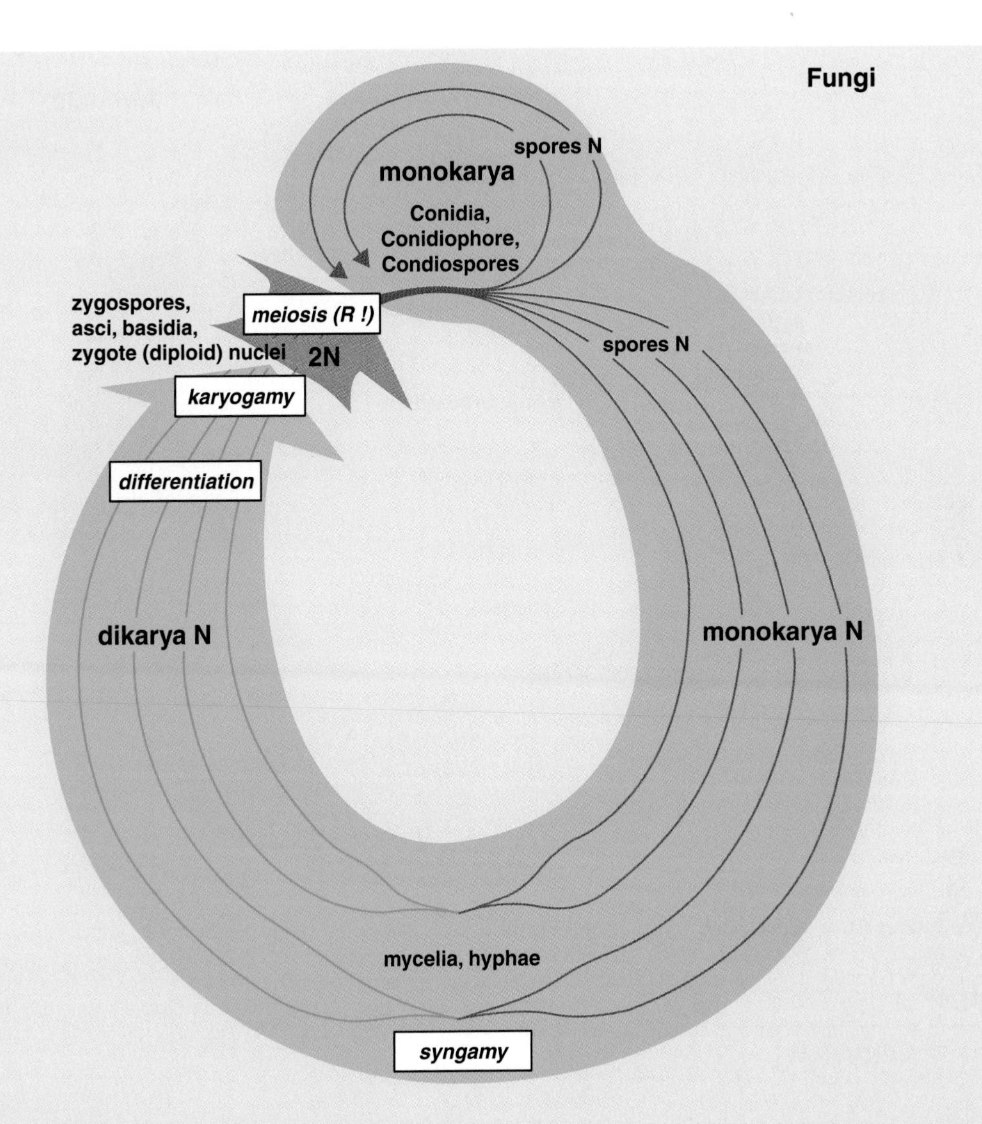

Figure Eukarya-ii-2 Generalized fungal life cycle. In the fungi, the haploid phase of the life cycle predominates. Haploid spores germinate to produce filamentous hyphae (collectively, a mycelium) in which haploid nuclei (monokarya) often occur syncytially, in absence of membranous cell boundaries. Two genetically distinct hyphae may fuse (syngamy) such that the syncytium now contains nuclei of two distinct genotypes (dikarya). Fusion of nuclei of such dikarya in fungal sporophytes or "fruiting bodies" (for example, asci, basidia; spore-bearing structures once construed as plants) is the fungal equivalent of fertilization. The highly reduced diploid phase of the life cycle consists only of the zygote fertilized nucleus or zygospore, in which meiosis occurs, to regenerate haploid spores. [Credit: James MacAllister.]

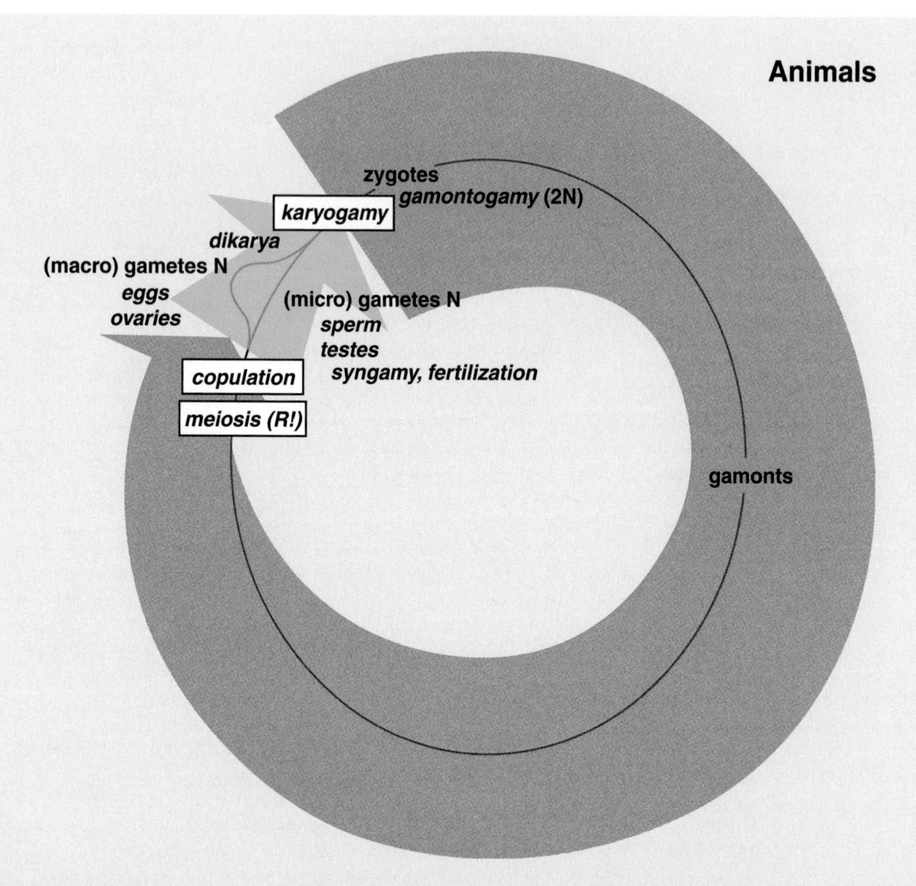

Figure Eukarya-ii-3 Generalized animal life cycle. In the animals, the diploid phase predominates. With a few insect and herpetological exceptions, all animals are multicellular diploids. A gamete-producing animal body (gamont) produces haploid eggs (females), sperm (males) or in many cases both, by meiosis. These gamete unicells represent the highly reduced haploid phase of the animal life cycle. Following copulation or external fertilization, the diploid zygote divides by mitosis to form the animal embryo called the blastula. This embryo further develops into a sexually mature diploid gamont. [Credit: James MacAllister.]

somatic meiosis. The "imperfections and oddities" Darwin cited as illustrative of "descent with modification" are in the protoctists, which show every sort of variation on life cycle pattern from no meiosis-fertilization cycle at all to the complexity of the foraminifera with every sort of possibility. We speak of *life cycle* when we know ploidy of the cells, but only of *life history* when our information is limited to morphological data without karyology (Figures Eukarya-ii-1 through Eukarya-ii-41,2,3,4)

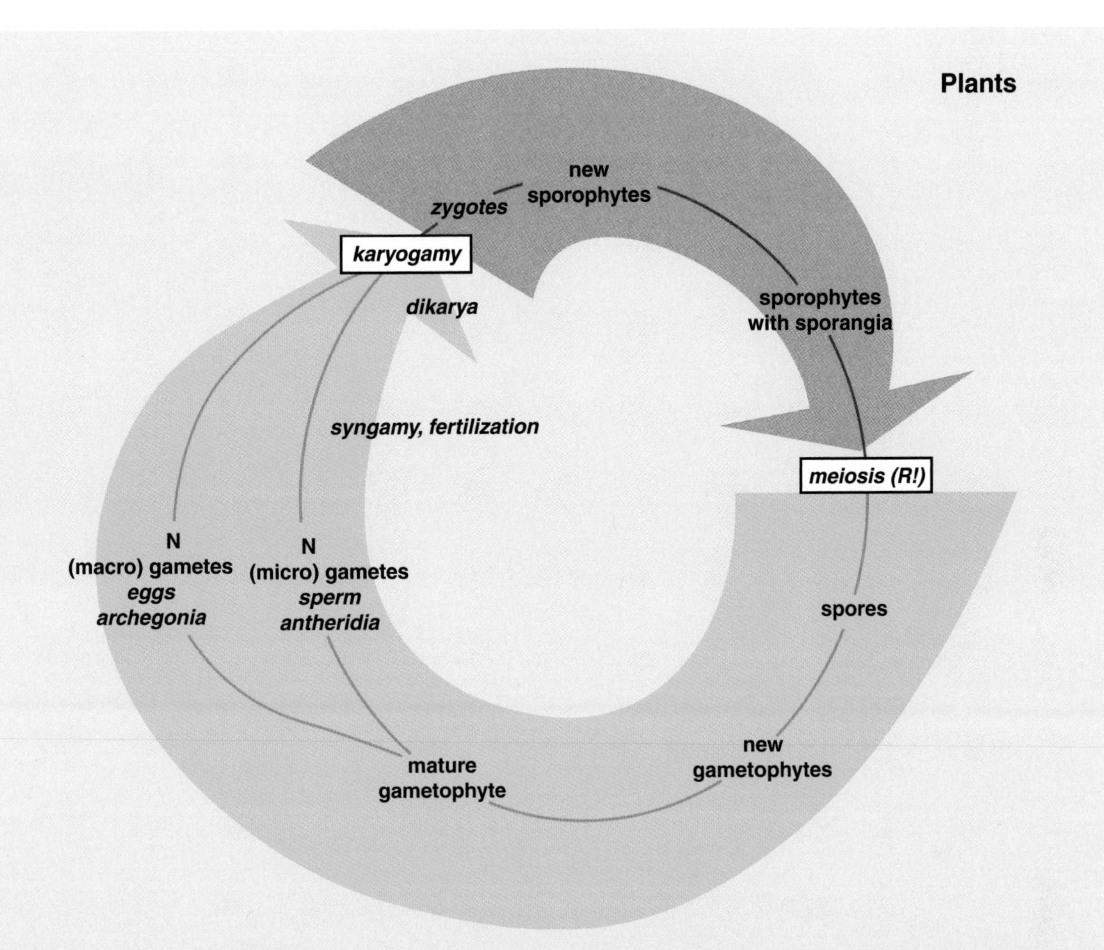

Figure Eukarya-ii-4 Generalized plant life cycle. Plants exhibit alternation of generations between the spore-producing, diploid sporophyte and the gamete-producing, haploid gametophyte. Depending on the plant group, either sporophyte or gametophyte may be more conspicuous, however, both phases of the life cycle are multicellular. Sporophyte plants produce sporangia organs in which sporogenic meiosis occurs to form single cells called spores. Plant spores are not necessarily resistant or hardy. Heterosporous plants produce two kinds of spores (smaller or larger) that divide by mitosis to produce gametophyte plants. The gametophytes differentiate egg- and sperm-producing organs (archegonia and antheridia, respectively) that by mitosis (not meiosis) produce gametes. Fertilization of egg nuclei by sperm nuclei (karyogamy) produces a zygote that divides by mitosis to regenerate the diploid sporophyte. [Credit: James MacAllister.]

CHAPTER TWO

KINGDOM PROTOCTISTA

1 **PR-18C** *Diploneis smithii*

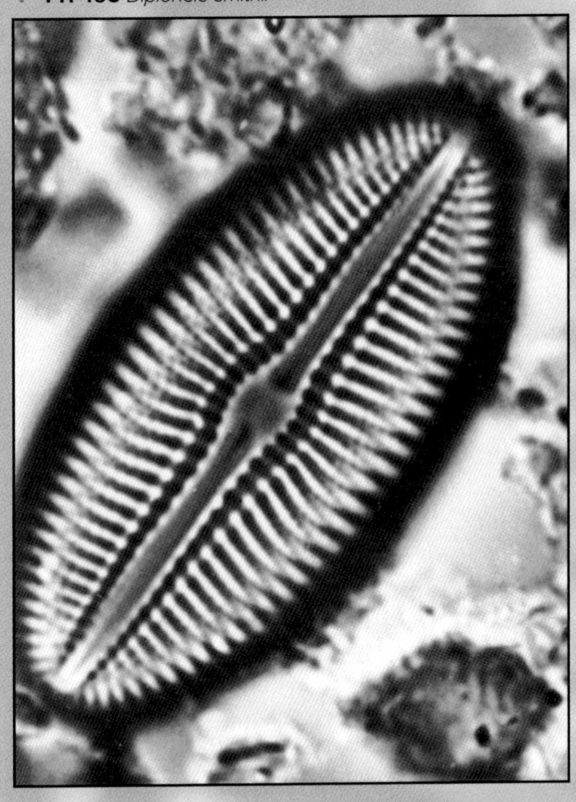

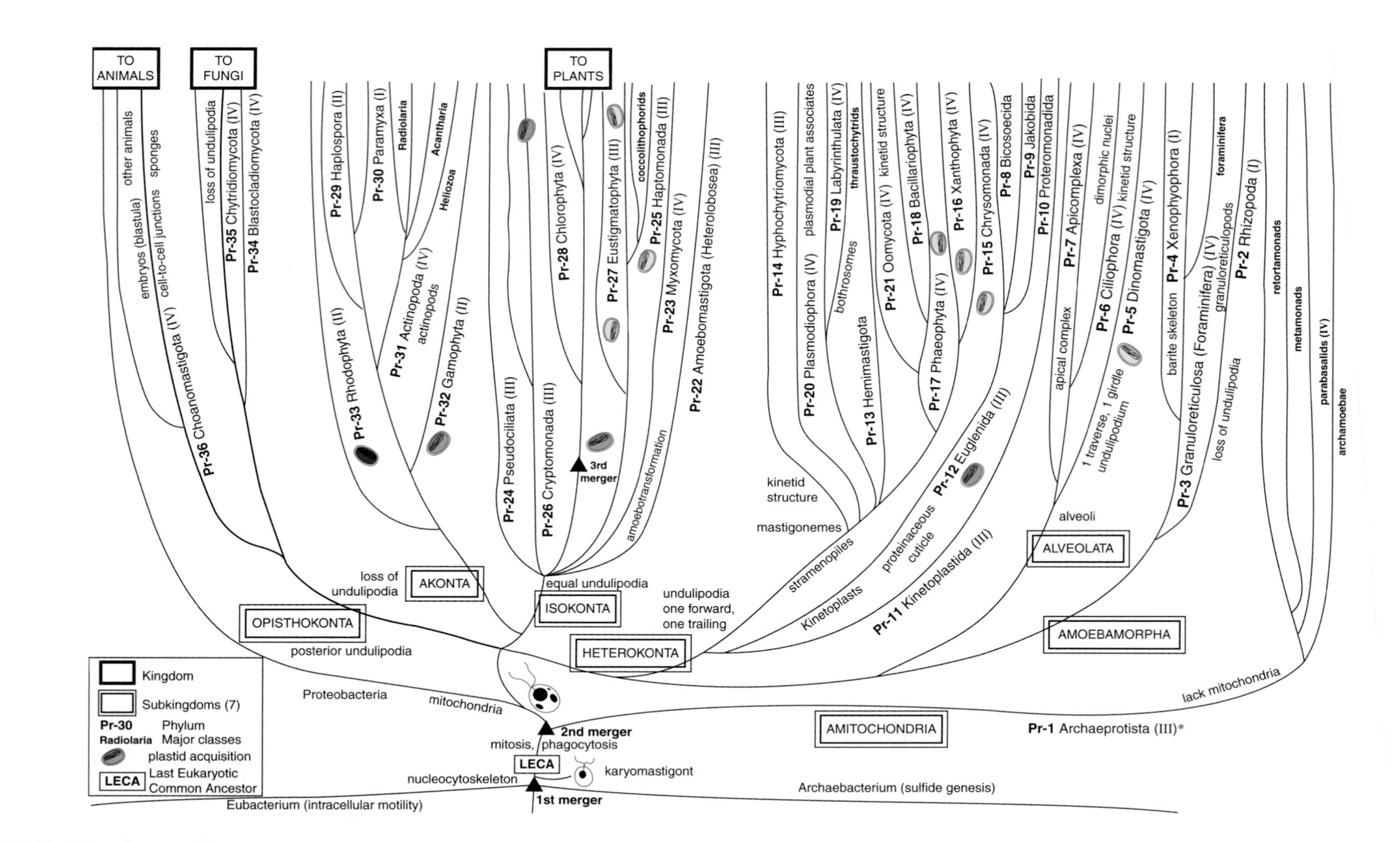

TO ANIMALS
TO FUNGI
TO PLANTS
other animals
embryos (blastula)
sponges
cell-to-cell junctions
Pr-36 Choanomastigota (IV)
loss of undulipodia
Pr-35 Chytridiomycota (IV)
Pr-34 Blastocladiomycota (IV)
Pr-29 Haplospora (II)
Pr-30 Paramyxa (I)
Radiolaria
Acantharia
Heliozoa
Pr-31 Actinopoda (IV)
actinopods
Pr-33 Rhodophyta (II)
Pr-32 Gamophyta (II)
Pr-24 Pseudociliata (III)
Pr-26 Cryptomonada (III)
Pr-28 Chlorophyta (IV)
3rd merger
Pr-27 Eustigmatophyta (III)
coccolithophorids
Pr-25 Haptomonada (III)
Pr-23 Myxomycota (IV)
amoebotransformation
Pr-22 Amoebomastigota (Heterolobosea) (III)
Pr-14 Hyphochytriomycota (III)
Pr-20 Plasmodiophora (IV)
plasmodial plant associates
Pr-19 Labyrinthulata (IV)
thraustochytrids
bothrosomes
Pr-13 Hemimastigota
Pr-21 Oomycota (IV)
kinetid structure
Pr-18 Bacillariophyta (IV)
Pr-17 Phaeophyta (IV)
Pr-16 Xanthophyta (IV)
Pr-15 Chrysomonada (IV)
Pr-8 Bicosoecida
Pr-9 Jakobida
Pr-10 Proteromonadida
Pr-7 Apicomplexa (IV)
apical complex
dimorphic nuclei
Pr-6 Ciliophora (IV)
kinetid structure
Pr-5 Dinomastigota (IV)
1 traverse, 1 girdle undulipodium
Pr-4 Xenophyophora (I)
barite skeleton
foraminifera
granuloreticulopods
Pr-3 Granuloreticulosa (Foraminifera) (IV)
loss of undulipodia
Pr-2 Rhizopoda (I)
retortamonads
metamonads
parabasalids (IV)
archamoebae
kinetid structure
mastigonemes
stramenopiles
Kinetoplasts
proteinaceous cuticle
Pr-12 Euglenida (III)
Pr-11 Kinetoplastida (III)
alveoli
ALVEOLATA
AMOEBAMORPHA
loss of undulipodia
AKONTA
equal undulipodia
ISOKONTA
undulipodia one forward, one trailing
HETEROKONTA
OPISTHOKONTA
posterior undulipodia
Proteobacteria
mitochondria
lack mitochondria
AMITOCHONDRIA
Pr-1 Archaeprotista (III)*
2nd merger
mitosis, phagocytosis
LECA
karyomastigont
nucleocytoskeleton
1st merger
Archaebacterium (sulfide genesis)
Eubacterium (intracellular motility)
Kingdom
Subkingdoms (7)
Pr-30 Phylum
Radiolaria Major classes
plastid acquisition
LECA Last Eukaryotic Common Ancestor

FOUR MODES

Protoctists, as Eukarya, due to their cytoskeletal proteins display intracellular motility visible in live cells at the level of light microscopy. The complex evolutionary history of the karyomastigont, the organellar system that evolved into the mitotic spindle, generated many higher (more inclusive) taxa. The karyomastigont: nucleus, nuclear connector, undulipodium (as kinetosome/centriole-axoneme, Figure Pr-1 (page 121) generated the highly variable mitotic apparatus of the protoctists that later became the cannonical apparatus of mitosis, and eventually meiosis in the larger protoctists, animals, fungi and plants.

The intimate relation between forms of intracellular motility (cyclosis, endocytosis, exocytosis, phagocytosis, pinocytosis) and sexuality (sperm motility, mitosis and its derivative process, meiosis) permits us recognition of four modes of developmental behavior in Protoctista phyla.

The four modes are phyla whose member species:

I. Lack both undulipodia and meiotic sex
II. Lack undulipodia but have meiotic sex
III. Have undulipodia but lack meiotic sexual life cycles
IV. Have undulipodia and meiotic sexual life cycles

KINGDOM PROTOCTISTA

Greek *protos*, very first; *ktistos*, to establish

Nucleated microorganisms and their descendants, exclusive of fungi, animals, and plants, evolved by integration of former microbial symbionts—nonmeiotic or sexuality meiotic with variations in the meiosis–fertilization cycle. Fossil record extends from the Lower Middle Proterozoic era (about 1200 million years ago) to the present. Evolved from aquatic motile ancestors with [9(3)+0] kinetosome-centriole microtubule cytoskeleton (the karyomastigont).

Kingdom Protoctista comprises the eukaryotic microorganisms and their immediate descendants: all algae, including the seaweeds; undulipodiated mastigote molds, water molds, the slime molds and slime nets; the traditional protozoa; and other even more obscure aquatic organisms. Its members are not animals (which develop from a blastula), plants (which develop from maternally retained plant embryos), or fungi (which lack undulipodia and develop from fungal spores). Nor are protoctists prokaryotes. Protoctist cells contain microtubules, nuclei, and other characteristically eukaryotic features (Table I-2). Many photosynthesize (have plastids), and most are aerobes (have mitochondria). Most have [9(2)+2] undulipodia with their kinetosome bases (Figure I-3) at some stage of the life history. All protoctists evolved from symbioses between at least two different kinds of bacteria—often many more than two. As the symbionts integrated, new levels of individuality appeared.

Modes I-IV for each Protoctist phylum are listed at the top of the page on the last page of that phylum. Animals and plants belong to mode IV since they display meiotic sexual cycles and are capable of formation of undulipodia (animal somatic, including sensory, cilia and most animal and plant sperm). Since members of the Kingdom Fungi are, by definition, amastigote they belong to mode II: Lack undulipodia but have meiotic cycles.

Many different combinations of ancient bacteria into symbiotic consortia did not pass the test of natural selection. But those that survived gave rise to the eight-subkingdom modern-day protoctist lineages. Protoctists, therefore, are grouped according to their karyomastigont-derived organellar system structures.

In the mitochondriates, the slowly evolving membranous structures include the cristae. They may be flat as in the stramenopiles (heterokonts and opisthokonts), tubular as in the alveolates, discoid as in the amoebamorphs, or altogether absent as in the archaeprotists (Phylum Pr-1). Photosynthetic pigment profiles, essential to chloroplast function, are major criteria similarly employed by taxonomists to resolve the bewildering diversity of kingdom Protoctista.

Undulipodia and their insertions, the kinetosomes always embedded in kinetids (Figure I-3 right and Figure Pr-1), because of their relationship to the origins of mitoses, are crucial to understanding protoctists. Undulipodia were present in common ancestors to all the phyla even before mitochondria, given that the anaerobic archaeprotists bear them. The kinetosome-centrioles as they move and reproduce are related to mitotic cell

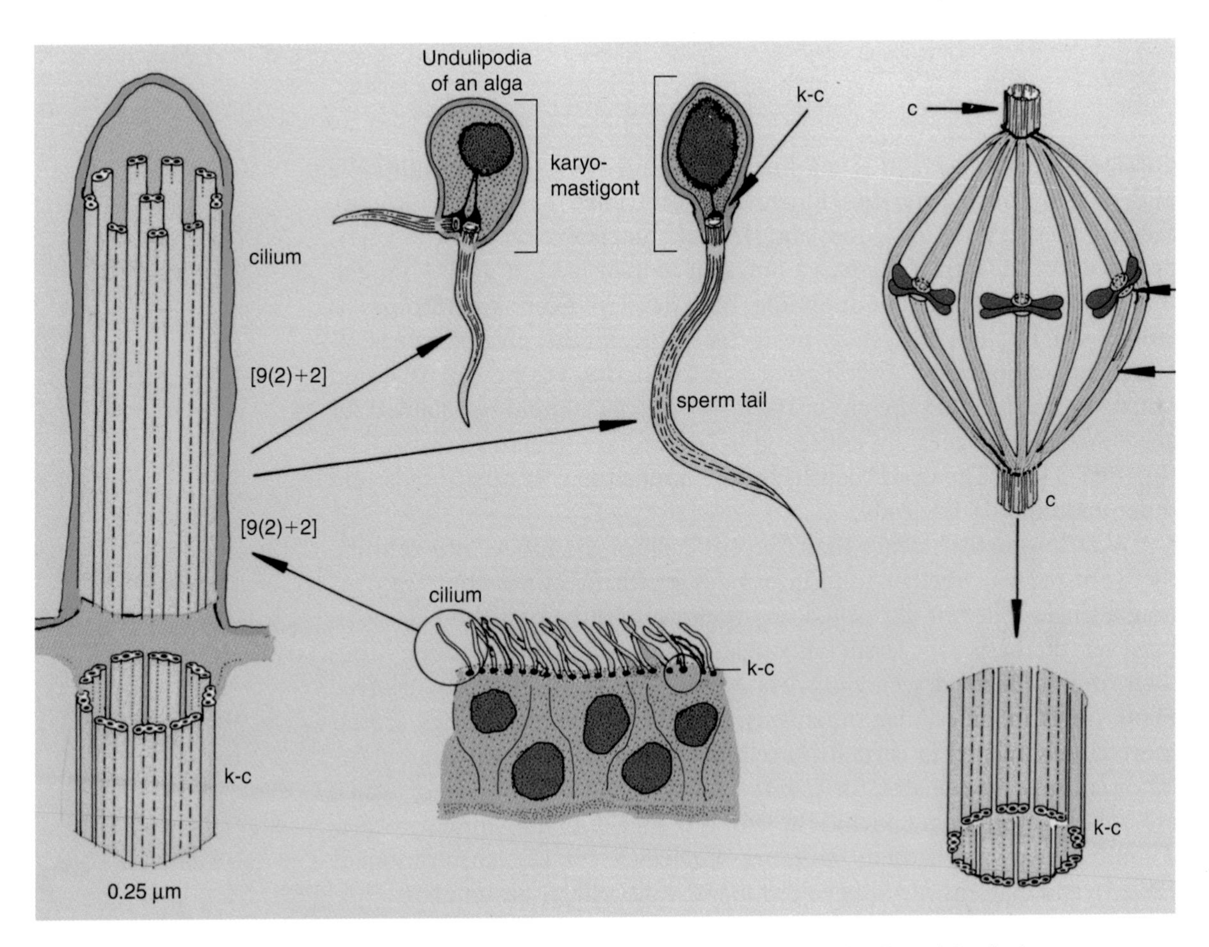

Figure Pr-1 Relation of microtubule cytoskeletal system to mitotic spindle (microtubules See Figure I-3 yellow). u = undulipodium, k-c = kinetosome-centre. [Credit: K. Delisle.]

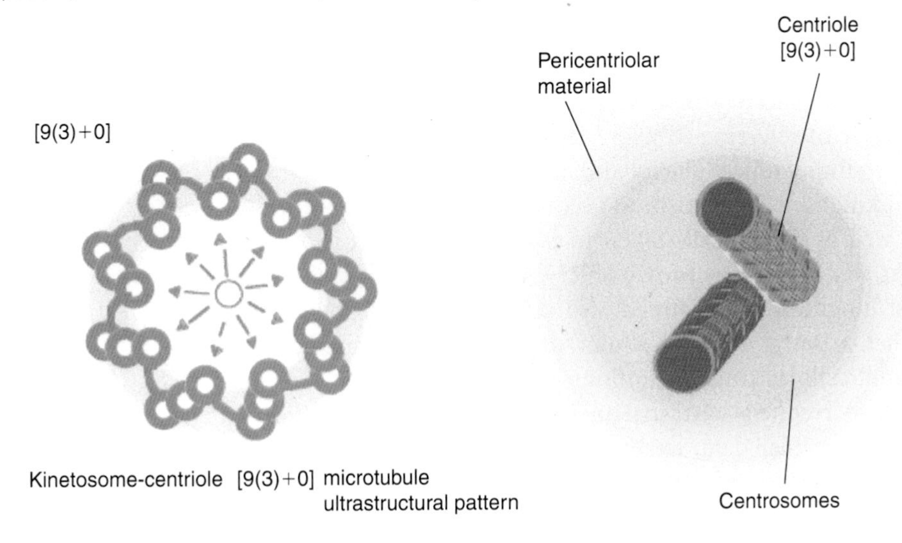

Figure Pr-2 Kinetosome-centriole. [Credit: M. Alliegro.]

division (Figure Pr-2). In some lineages, all members bear undulipodia underlain by their generative kinetosome-centrioles; in others (Akonta), they are absent; yet most (Alveolata, Heterokonta, Isokonta, Opisthokonta) produce, lose, or retract them as a function of their life histories. Although the importance of the undulipodia that develop from kinetosomes is emphasized by those who study protoctists—algologists, invertebrate zoologists, microbiologists, mycologists, parasitologists, protozoologists, and others—some feel that the use of the term "flagella" should be retained for eukaryotes. But flagella are entirely unrelated rotary structures of bacteria (Figure I-3 left). The word "flagella," when applied to cilia, sperm tails, and other undulipodia, misleads.

Why "protoctist" rather than "protist"? Since the nineteenth century, the word protist, whether used informally or formally, has come to connote a single-celled or few-celled tiny organism. In the past three decades, however, the basis for classifying single-celled organisms separately from their multicellular descendants has weakened. Multicellularity evolved many times in unicellular organisms. Many multicellular beings are far more closely related to certain unicells than they are to other multicellular organisms. The ciliates, for example (Pr-6, Ciliophora), most of which are unicellular microbes, include at least one species that forms a sorocarp, a multicellular cyst-bearing structure. Euglenids (Pr-12), chrysomonads (Pr-15), and diatoms (Pr-18) evolved many multicellular descendants.

Here we adopt the concept of protoctist propounded in modern times by Californian botanist Herbert F. Copeland in 1956. The word was introduced by English naturalist John Hogg in 1860 to designate "all the lower creatures, or the primary organic beings;—both *Protophyta, …* having more the nature of plants; and *Protozoa …* having rather the nature of animals." Copeland recognized, as had several scholars in the nineteenth century, the absurdity of referring to giant kelp by the word "protist," a term that had come to imply unicellularity and, thus, smallness. He proposed an amply defined kingdom Protoctista to accommodate certain multicellular organisms as well as the unicells taken to resemble their ancestors—kelp for example, as well as the tiny brownish cryptomonad alga, *Nephroselmis.* The protoctist kingdom thus defined also solves the problem of blurred boundaries that arises if the unicellular organisms are assigned to the intrinsically multicellular kingdoms: fungi, animals, and plants.

Attempting to reconcile ultrastructural and genetic information with newly acquired molecular data, we here propose 36 protoctist phyla. This number is more a matter of taste than tradition, because no rules that define protoctist phyla are enforced. Our groupings are debatable; for example, some argue that the cellular and plasmodial slime molds (in Pr-2 and Pr-23) should be united. Some believe that the oomycotes (Pr-21), hyphochytrids (Pr-14), and chytrids (Pr-35)—which we place in Protoctista—are really fungi, and that chlorophytes (Pr-28) are plants. Some insist that chaetophorales and prasinophytes, which here are within Chlorophyta (Pr-28), ought to be raised to phylum status. Most would reunite conjugating green algae (Pr-32, Gamophyta) with the others in Chlorophyta (Pr-28). Arguments for and against these views exist. Our system has the advantage of limiting the number of highest taxa in explicit recognition of protoctist lineages and precise definition of the three kingdoms of large organisms. Although it has the disadvantage that some eukaryotic taxa have little in common with one another, grouping together xenophyophores, cercomonads, water molds, and the others in the single taxon kingdom Protoctista is superior to the tradition of ignoring them entirely.

Protoctists are aquatic: some marine, some freshwater, some terrestrial in moist soil, and some symbiotic in moist tissues of others. Nearly all animals, fungi, and plants—perhaps all—have protoctist associates. Phyla such as Archaeprotists (Pr-1), Apicomplexa (Pr-7), Plasmodiophora (Pr-20), and Paramyxa (Pr-30) include myriad species, most of which live in the tissues of others.

No one knows the number of protoctist species. Although 60,000 extinct foraminifera alone are documented in the paleontological literature and more than 10,000 live protoctists are described in the biological literature, Georges Merinfeld (Dalhousie University, Halifax, Nova Scotia) estimates that there are more than 65,000 extant species, whereas John Corliss (University of Maryland) suggests that there are more than 250,000. Water molds and plant "parasites" are described in the literature on fungi, parasitic protozoa in the medical literature, algae by botanists, and free-living protozoa by zoologists. Contradictory practices of describing and naming species have led to contradictions that this book attempts to resolve. Another problem is that much protoctist diversity is in tropical regions where scientists are scarce. Furthermore, the documentation of new species of protoctists often requires time-consuming life-cycle and ultrastructural study. Most funding is limited to temperate-zone protoctists of economic interest. Many protoctists are sources of food, industrial products, or disease.

Remarkable variation in cell organization, cell division, life cycle, sexuality, and bio-mineralized skeletons or scales is evident in this diverse group of eukaryotic microbes and their relatives. Whereas the algae are oxygenic phototrophs, most others are heterotrophs that ingest or absorb their food. In many, the mode of nutrition varies with environmental conditions; many photosynthesize when light is plentiful and feed in the dark. Although protoctists are more diverse in life style and nutrition than are animals, fungi, or plants, metabolically they are far less diverse than bacteria.

Increasing knowledge about the ultrastructure, genetics, life cycle, developmental patterns, chromosomal organization, physiology, metabolism, fossil history, and especially the molecular systematics of protoctists has revealed the many differences between them and animals, fungi, and plants. The major protoctist groups, described here as phyla in one of eight subkingdoms, are so distinct as to deserve kingdom status in the minds of some authors. See the forthcoming second edition of *Handbook of Protoctista* (Margulis and Chapman; editors) and *Illustrated Glossary of Protoctista* (Margulis, McKhann, and Olendzenski; editors, 1992). Because even after years of study no two editors can master all the biological details of the protoctists, we expect animated discussion about their optimal taxonomy. With awe for protoctist diversity, a recognition of their common eukaryotic heritage, and a sense of humility in Nature's vastness, we present our 36 protoctist phyla.

Box Pr-i: Symbiogenesis and the origin of organelles

Molecular and morphological evidence clearly points to a symbiotic origin of two classes of eukaryotic organelles: mitochondria and chloroplasts. The hypothesis that the nucleus and the eukaryotic motility organelles (cilia and eukaryotic "flagella;" that is, undulipodia) emerged from an earlier symbiosis that generated the first eukaryotic cell is now under investigation in the Margulis laboratory. One line of evidence that supports this hypothesis is the connection between mitosis and motility; the microtubule-based cytoskeleton is required both to generate a mitotic spindle and to assemble an undulipodium (the 9[2]+2 microtubular motility organelle). The microtubules of spindle and undulipodium are identical in diameter (24 nm), structure, and composition (tubulin protein).

In the three-domain system, life began as a trifurcation when the archaea, eubacteria, and eukaryotes evolved from the progenote (Box Prokaryotae-ii). At the same time as the appearance of the eubacterial lineage (Eubacteria), ancestors of today's eukaryotes also split off from the Archaea (Archaebacteria). Therefore according to Woese, although members of the Archaea (thermoacidophils, halophils, and methanogens) physically resemble bacteria, they are actually more closely related to humans. Proponents of the three-domain system further suggest that because bacteria and archaea are two distinct lineages that arose independently, the term "prokaryote" should be abandoned as uninformative. But we who support five kingdoms in two superkingdoms (prokaryote and eukaryote) counter that cell structure, gene organization, and metabolism are far more similar in the two groups of prokaryotes than they are between any prokaryote (bacteria or archaebacteria) and eukaryote. The term "prokaryote," which refers to commonalities of structure, genetics, and biochemistry, should be retained. Both archaebacteria and eubacteria are naturally classified as subgroups of the superkingdom Bacteria. We regret the term "Archaea" that overemphasizes differences between the prokaryotic subgroups.

The ribosomal DNA (rDNA) sequences that generated the three-domain system are used to identify organisms and relate them to others in "family diagrams," that is, phylogenies. The rDNA analysis produces monophyletic phylogenies. All organisms stem from this single common ancestor and no others (Figure Pr-i-1). Unambiguous phylogenetic diagrams are impossible to establish, because transfer of genes or entire sets of genes (genomes) between different life-forms has occurred many times throughout history. Symbiotic mergers between different organisms give rise to new species. The neatly diverging branches of monophyletic phylogenies are abstractions of "misplaced concreteness." Tree diagrams based on comparative-sequence identity alone are deficient, because they lack fusions of branches denoting symbiotic mergers.

Chloroplasts as free-living cyanobacteria were engulfed but not digested by an early eukaryote. The symbiotic origin of chloroplasts, the photosynthetic organelles in plants and algae, is confirmed by many concurrent lines of evidence including molecular studies. Cells of a modern cyanobacterium and chloroplasts in algal and plant cells

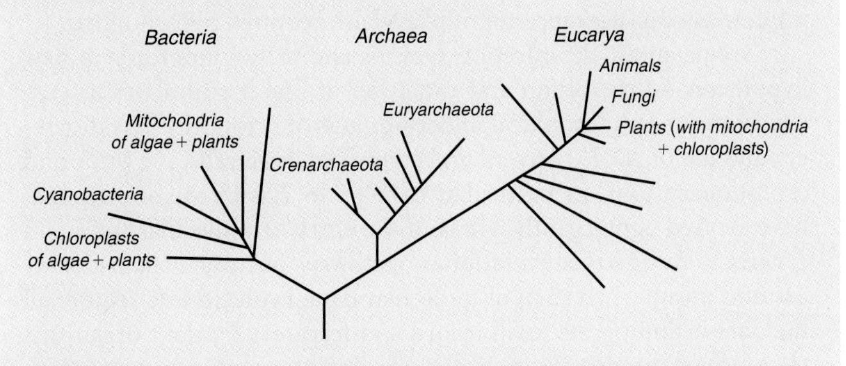

Figure Pr-i-1 "Tree of Life" based on ribosomal DNA (rDNA) sequence comparisons (Adapted from Sogin et al., 1993). Note absence of fusions between branches.

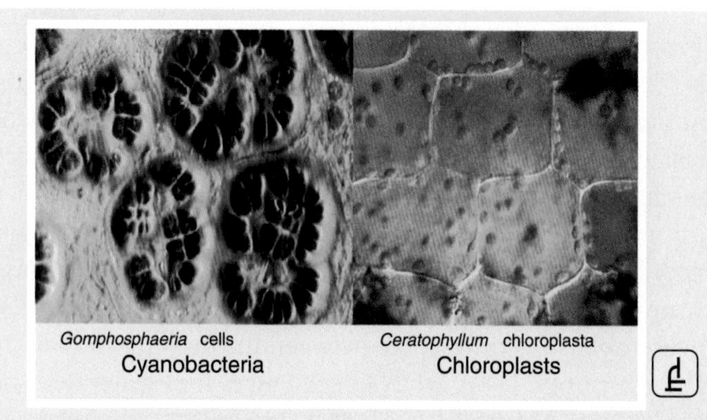

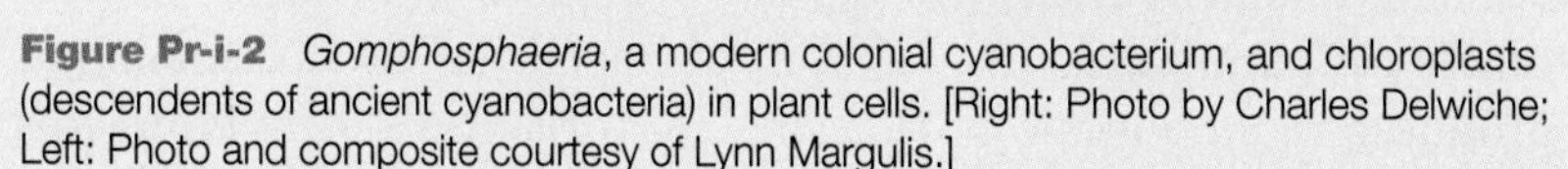

Figure Pr-i-2 *Gomphosphaeria*, a modern colonial cyanobacterium, and chloroplasts (descendents of ancient cyanobacteria) in plant cells. [Right: Photo by Charles Delwiche; Left: Photo and composite courtesy of Lynn Margulis.]

are depicted in Figure Pr-i-2. Plants and cyanobacteria are vastly different; their structures, life histories, modes of reproduction, metabolic capabilities, and molecular sequences squarely support their placement in separate domains or kingdoms. Yet the evolution of plants and cyanobacteria are inextricably linked; ancient cyanobacteria are now parts of plants. Note the positions of cyanobacteria and plants on the five-kingdom and three-domain phylogenies (Figures A-i-1 and Pr-i-1). The three-domain tree, more faithful to monophyly than to whole-organism biology, ignores symbiosis; plants and cyanobacteria lie on distant branches of separate domains. We prefer the five-kingdom scheme, in which all plants have at least three different ancestors: nucleocytoplasm, mitochondria, and chloroplasts. Symbiotic acquisition of an oxidative eubacterium, before the evolution of chloroplasts, implies that the algal ancestor of plants also acquired mitochondria.

Debate about classification systems and techniques leads to new hypotheses on the origin and evolution of life; it stimulates discussion about the philosophical underpinnings of taxonomy. The discovery and use of rDNA, present and conserved in all cells, is a profound evolutionary tool. In particular, it helps to classify organisms that have evolved convergently. We laud the entry of molecular tools and criteria to reconstruct evolutionary pathways but warn against overuse and misinterpretation of molecular data. Evolutionists require all the data including the fossil record and features of extant organisms to construct the best-fit, most plausible hypotheses for ancient events and consequent extant phylogenetic relationships.

Box Pr-ii: Mitosomes and hydrogenosomes

Mitosomes—synonym "cryptons"—are limited to Protoctista that live under anoxic conditions. They are tiny cell organelles (<0.05 μm to 0.15 μm), bounded by a double membrane. They lack the enzymes required to synthesize ATP. First discovered in *Entamoeba histolytica*—a pathogenic small encysting ameba especially abundant in fecal samples of diarrhea patients—these organelles do not coexist in any cells with mitochondria. Over 200 mitosomes per *E. histolytica* cell have been detected. Their function is not known. They have been reported in unicellular protists sensitive to oxygen gas that lack the metabolic ability to respire oxygen. These include *Blastocystis*, *Cryptosporidium*, *Giardia intestinalis*, *G. lamblia*, and several species of microsporidians (F-2). Mitosomes vary in morphology from simple double-membrane-bounded vesicles to diverse forms possessing cristaelike, goblet-shaped, branched, or filamentous invaginations of the inner membrane.

Mitosomes do not use oxygen as a terminal electron acceptor. They lack the urea cycle, oxidative degradation of fatty acids, heme protein biosynthesis, and other traits of mitochondria. Like mitochondria, however, they contain the enzymes crucial to iron–sulfur cluster assembly: cysteine desulfurase (IscS), and a scaffold protein (IscU). They also contain the "heat-shock protein" chaperonin 60, which leads some to conclude that mitosomes are actually highly reduced mitochondrial homologues. However, none of the enzymes in common between mitochondria and mitosomes actually react with oxygen, whereas hydrogen gas is produced (Meyer, 2007). They therefore may be derived from archaeprotistan hydrogenosomes (Pr-1; Dolezal *et al.*, 2005).

Hydrogenosomes are found in trichomonads, ciliates, and fungi. Larger than mitosomes, hydrogenosomes tend to be 1 μm in diameter. In some, an inner membrane with cristaelike folds has been seen on thin-section electron microscopy. In others (Figure Pr-ii-1), an electron-dense granular appearance with no apparent relationship to mitochondria is seen. No hydrogenosomal genome was detected in *Trichomonas*, *Neocallimastix*, *Piromyces*, or *Psalteriomonas*. However, in the ciliate *Nyctotherus ovalis*, gut symbiont of cockroach and millipede, a small organellar genome was reported (Akhmanova *et al.*, 1998; Hackstein, 2001). The defining features of hydrogenosomes are double-membrane-bounded cytoplasmic organelles that produce hydrogen gas via hydrogenase enzymes. From the biochemical details, it is clear that hydrogenosomes are polyphyletic. The fungi and some ciliates

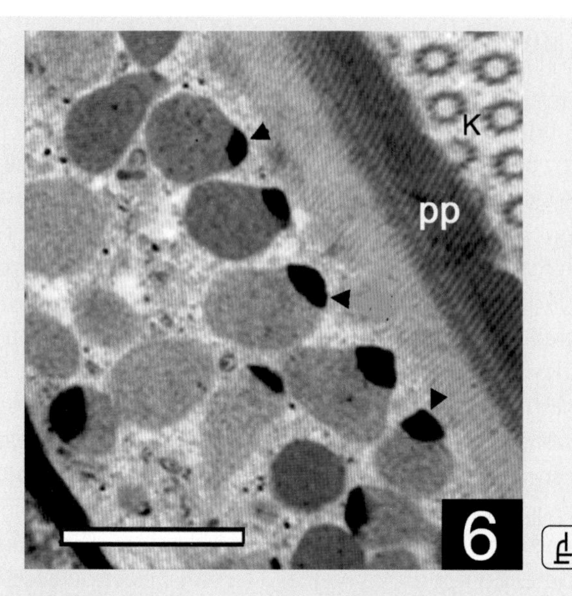

Figure Pr-ii-1 Hydrogenosomes of *Staurojoenina assimilis* bar = 2 μm (Wier *et al.*, 2004).

(for example, *Neocallimastix*, family Plagiopylidae) have hydrogenosomes that descended from mitochondria when aerobic ancestors of these organisms colonized anoxic niches. Hydrogenosomes in metamonads (for example, *Giardia*), parabasalids (Figure Pr-ii-1), and other entirely anaerobic protoctist taxa likely originated directly from bacterial (clostridial) ancestors. By contrast, no conclusive finding of a mitosome genome has yet been published. Although mitochondria appear monophyletic in their evolution from alpha-proteobacteria, hydrogenosomes arose independently several times in diverse taxa (Embley *et al.*, 1995, 1997; Finlay and Fenchel, 1989). By extension, then, the even more highly reduced mitosome has polyphyletic origins as well.

Those who research mitosomes hypothesize that mitosomes evolved from mitochondria that lost their oxygen metabolism upon return to anoxic habitats. Animal mitochondria have lost more than 80% of their original genes by transfer to the nucleus. Plants have lost somewhat fewer mitochondrial genes, perhaps 20–60%. If mitosomes really are degenerate mitochondria, they represent an endpoint of this trend. They lost all of their genes to the nuclei of cells in which they reside, and now depend entirely on nuclear genes for protein synthesis. If they are modified hydrogenosomes, they may have evolved from anaerobic bacteria such as *Clostridium* (B-9) under anoxic conditions.

References

Akhmanova, A., F. Voncken, T. van Alen, A. van Hoek, B. Boxma, G. Vogels, M. Veenhuiss, and J. H. P. Hackstein, "A hydrogenosome with a genome." *Nature* 396:527–528; 1998.

Dolezal, P., O. Smid, P. Rada, Z. Zubácová, D. Bursac, R. Suták, J. Nebesárová, T. Lithgow, and J. Tachezy, "*Giardia* mitosomes and trichomonad hydrogenosomes share a common mode of protein targeting." *Proceedings of the National Academy of Sciences USA* 102(31):10924–10929; 2005.

Embley, T. M., B. J. Finlay, P. L. Dyal, R. P. Hirt, M. Wilkinson, and A. G. Williams, "Multiple origins of anaerobic ciliates with hydrogenosomes within the radiation of aerobic ciliates." *Proceedings of the Royal Society London B, Biological Sciences* 262:87–93; 1995.

Embley, T. M., D. S. Horner, and R. P. Hirt, "Anaerobic eukaryote evolution: hydrogenosomes as biologically modified mitochondria?" *Trends in Ecology & Evolution* 12:437–441; 1997.

Hackstein, J. H. P., A. Akhmanova, F. Voncken, A. van Hoek, T. van Alen, B. Boxma, S. Y. Moon-van der Staay, G. van der Staay, J. Leunissen, M. Huynen, J. Rosenberg, and M. Veenhuis, "Hydrogenosomes: Convergent adaptations of mitochondria to anaerobic environments." *Zoology* 104:290–302; 2001.

Martin, W.F., and M. Müller, eds., *Origin of mitochondria and hydrogenosomes*. Springer-Verlag; Heidelberg; 2007.

Meyer, J., "[Fe-Fe] hydrogenases and their evolution: A genomic perspective." *Cellular and Molecular Life Sciences* 64:1063–1084; 2007.

Sogin, M.L., Hinkle, G. and Leipe, D.D., (1993). Universal tree of life. Nature 362, 795; doi:10.1038/362795a0.

Wier, A.M., M. F. Dolan, and L. Margulis, "Cortical symbionts and hydrogenosomes of the amitochondriate protist *Staurojoenina assimilis*." *Symbiosis* 36:153–168; 2004.

SUBKINGDOM (Division) AMITOCHONDRIA

Pr-1 Archaeprotista

Greek *karyon*, nucleus, kernel; *blastos*, bud, sprout

GENERA

Barbulanympha	***Coronympha***	***Foaina***
Blastocrithidia	***Deltotrichonympha***	***Giardia***
Calonympha	***Devescovina***	***Herpetomonas***
Chilomastix	***Dinenympha***	(continued)

Although the vast majority of nucleated organisms—in fact, all of them except those in this phylum and one other (F-1)—are aerobes with a mandate to respire atmospheric oxygen, the ancestors of eukaryotes were originally anaerobes, killed by oxygen. Living in habitats that recall ancient anoxic environments, members of this phylum are relicts of the eukaryotic life of those early days: all are anaerobes. Molecular phylogenies show that these anaerobes branch off very early from all other eukaryotes, and no evidence exists that they ever did have mitochondria. New knowledge, especially from coupling ultrastructure, physiology, and sequence studies of nucleotides in ribosomal RNA (rRNA), has led to impressive new insights into our ultimate eukaryotic ancestors. Because every member of every taxon here lacks mitochondria, this new information leads biologists to agree that these organisms were without mitochondria from the onset of their evolution. (It is more logical to argue that mitochondria were never in the ancestors than arguing that mitochondria disappeared independently in many aerobic species many argue that mitochondria constitute such a trace.) The former phylum Zoomastigina has therefore been dissolved, with members lacking mitochondria now placed in the present phylum (Archaeprotista) and those with mitochondria in phyla Bicosoecida (Pr-8), Euglenida (Pr-12), and Choanomastigota (Pr-36).

Phylum Archaeprotista and a second phylum, Microspora (F-1), are best grouped as the amitochondriates; other names for them in the literature are "hypochondriates" and "Archaeozoa." Members of phylum Microspora are easily unified as one of the old "sporozoan" groups: tiny amitochondriate intracellular symbiotrophs of animals. The grouping of organisms that creates phylum Archaeprotista, however, is new.

Archaeprotists were traditionally ignored or poorly known as tiny "micromastigotes" or "animal symbiotrophs." Despite this neglect, the work of some master biologists—Harold Kirby (1900–1952), Lemuel R. Cleveland (1892–1969), André Hollande (1910–1994), Guy Brugerolle, J.-P. Mignot, and others—now allows three classes to be well delineated: Archamoebae, Metamonada, and Parabasalia.

Class Archamoebae includes free-living, freshwater and marine organisms grouped into two subclasses: (1) Pelobiontae (or Caryoblastea) consists of the anaerobic amebas that lack undulipodia and swimming stages and (2) Mastigamoebae unites the anaerobic ameboids that bear undulipodia at some stage in their life histories, conferring the ability to swim.

Subclass Pelobiontae contains only one genus, *Pelomyxa* (Figure A). These giant cells, visible to the naked eye, may be relicts of the earliest living eukaryotes. They are classified as eukaryotes by definition—they have membrane-bounded nuclei. However, they lack nearly every other cell-inclusion characteristic of eukaryotes; they have no endoplasmic reticula, Golgi bodies, mitochondria, chromosomes, or centrioles. *Pelomyxa*

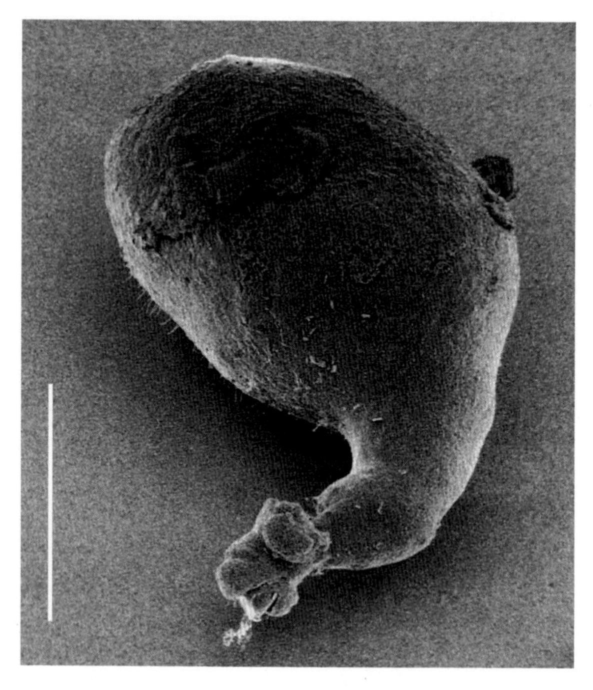

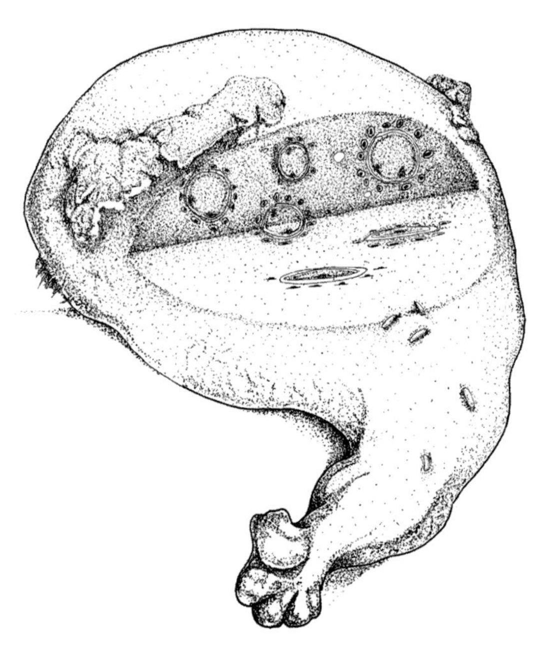

Figure A *Pelomyxa palustris*. SEM, bar = 100 μm. [Photograph courtesy of E. W. Daniels, in *The biology of amoeba*, K. W. Jeon, ed. (Academic Press, New York, 1973); drawings by R. Golder.]

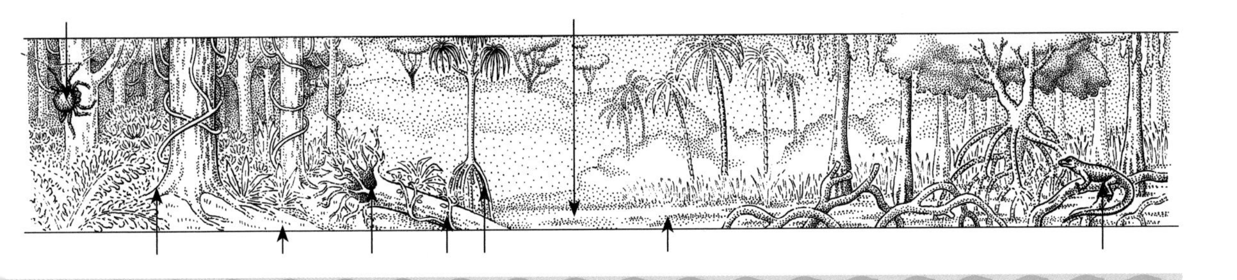

palustris, a large ameba with many nuclei, is the only well-documented species described in recent literature. Its nuclei do not divide by standard mitosis. Like bacterial nucleoids, the nucleus divides directly; new membranes form; and two nuclei appear where there had been only one.

In the late 1970s, some submembranous microtubules were seen in thin section, as was an intracellular, apparently nonfunctional undulipodium, suggesting that *Pelomyxa* evolved from undulipodiated ancestors. The microtubules and undulipodium do not take part in nuclear or cell division. No gametes are formed, and sexuality is absent.

Although it lacks mitochondria, *Pelomyxa* is microaerophilic, requiring lower concentrations of oxygen than do most eukaryotes. This giant ameba has three types of bacterial endosymbionts, functional analogues of mitochondria. At least two of the three endosymbiont types are methanogenic—that is, they produce methane gas instead of carbon dioxide. The endosymbionts of one type lie in a regular ring around each nucleus and are therefore called perinuclear bacteria. The others lie scattered in the cytoplasm; they are different species of bacteria having their own characteristic wall structures. *Pelomyxa* dies when treated with antibiotics to which its endosymbiotic bacteria are sensitive. Before the ameba dies, lactic acid and other metabolites accumulate in the cytoplasm; it is thought that the healthy symbiotic bacteria remove and metabolize the lactic acid from the cytoplasm. In any case, the ever-present bacterial partners seem to be required. *Pelomyxa* has vacuoles and stores glycogen, a polysaccharide that is also stored by many types of animal cells.

Pelomyxa palustris has been discovered in only one habitat: mud on the bottom of freshwater ponds in Europe, the United States, and probably North Africa. In fact, most studies were made on organisms taken from the former Elephant Pond at Oxford University, so named because, in the nineteenth century, the discarded carcasses of elephants were thrown there by taxidermists preparing museum exhibits. At the muddy bottom of such small ponds, including some in Massachusetts and Illinois, *Pelomyxa* feeds on algae and bacteria; there it grows and divides. It survives severe winters; *Pelomyxa* is best detected in autumn.

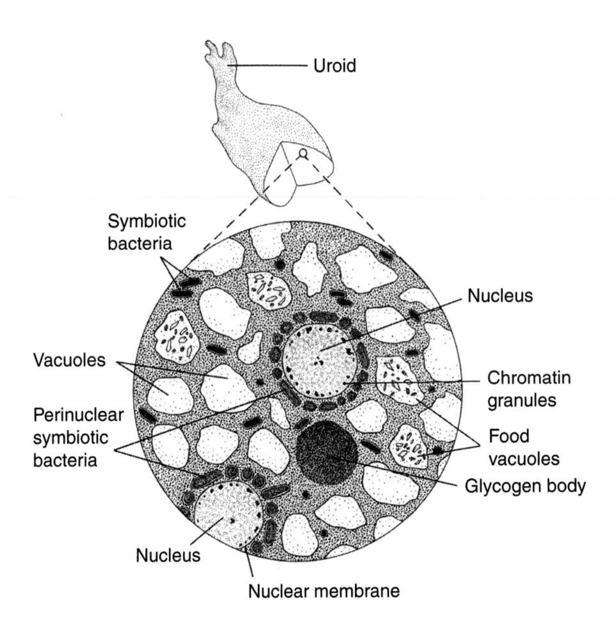

Mode III have undulipodia. but *most* lack meiotic sex. The observations filmed by L.R. Cleveland of sexuality in pyrsonymphids (*Notila*) and trichomonads (*Barbulanymphya, Trichonympha*) support I.B. Raikov's concept of multiple origins of meiotic fertilization cycles in early amitochondriate (Archaeprotists). Later sexual behavior was regularized in many lineages of Protoctista including choanomastigote ancestors of animals, chytrid ancestors of fungi and chlorophyte ancestors of plants. Margulis and Sagan, 2002, *Origins of Sex: Three billion years of genetic recombination*, University of California Press, Berkeley CA, paperback, details this concept.

Hexamastix	*Monocercomonas*	*Spironympha*
Hexamita	*Notila*	*Spirotrichonympha*
Histomonas	*Oxymonas*	*Staurojoenina*
Holomastigotoides	*Pelomyxa*	*Stephanonympha*
Joenia	*Pseudotrichonympha*	*Trichomitus*
Mastigamoeba	*Pyrsonympha*	*Trichomonas*
Mastigina	*Retortamonas*	*Trichonympha*
Metadevescovina	*Saccinobaculus*	
Microthopalodia	*Snyderella*	

when one ameba looks like a tiny, glistening droplet of water on dead leaves or submerged bark. Despite its hardiness, no scientist has been able to grow it in the laboratory. Those who study *P. palustris* must collect it from ponds.

Members of the Mastigamoeba subclass of class Archamoebae are unicellular, with each cell bearing at least one undulipodium. They may be either free living or parasitic, and all species known so far are asexual. They are heterotrophs that lack plastids. They are osmotrophs or phagotrophs. Two genera, *Mastigina* and *Mastigamoeba* (Greek *mastig*, whip), are single-celled heterotrophic protists known to protozoologists for the entire twentieth century.

The second class of phylum Archaeprotista, Metamonada, consists of cells in which the nuclei are attached to the undulipodia by thin fibers called nuclear connectors, or "rhizoplasts." (The nucleus with associated fibers is called a "karyomastigont.") The metamonads comprise three subclasses: Diplomonadida, Retortamonadida, and Oxymonadida—all referred to as "polymonads" (cells with a small number of undulipodia) in the old literature. As long ago as the 1940s, Kirby warned about the artificiality of using numbers of reproducing organelles, such as the number of nuclei and associated undulipodia, as a basis for taxonomy. Kinetid structure, the details of which are shared by close relatives but differ markedly in organisms only distantly related, makes it abundantly clear that cells with the same number of undulipodia are not necessarily related to each other. The number of kinetosomes per kinetid—usually one or two—is the same for thousands of unrelated organisms, but the elaborate, fine structure of the kinetosome arrangement, the kinetid, is common only to members of the same lower taxa (for example, families and genera). The approach of grouping organisms on the basis of the number of intrinsically reproducing organelles (for example, polymonads in one class) has been abandoned as electron microscopic techniques increasingly enable us deeper understanding of kinetids. *Giardia*, the most notorious of all metamonads, belongs to the subclass Diplomonadida. Like others in this subclass, *Giardia* has two karyomastigonts, as well as a Velcro-like adhesive ventral pad, allowing it to stick to our intestines and share our food. Members of the subclass Retortamonadida are small mastigotes with twisted cell bodies; this small mastigote has a trailing undulipodium that propels food into its mouth (cytostome). Most retortamonads (*Retortamonas* and *Chilomastix*) live as symbiotrophs in the digestive tracts of animals. *Pyrsonympha*, like all other members of the subclass Oxymonadida, live as symbiotrophs in the intestines of wood-eating cockroaches and termites. They have ribbon-shaped organelles called axostyles. These pulsating axostyles are composed of hundreds of microtubules, which are sometimes connected to each other by bridges and arranged in elaborate patterns. The other oxymonad genera are *Dinenympha* (the group has also been called Dinenymphida or Pyrsonymphida), *Notila*, *Oxymonas*, *Saccinobaculus*, and the single multinucleate genus, *Microthopalodia*.

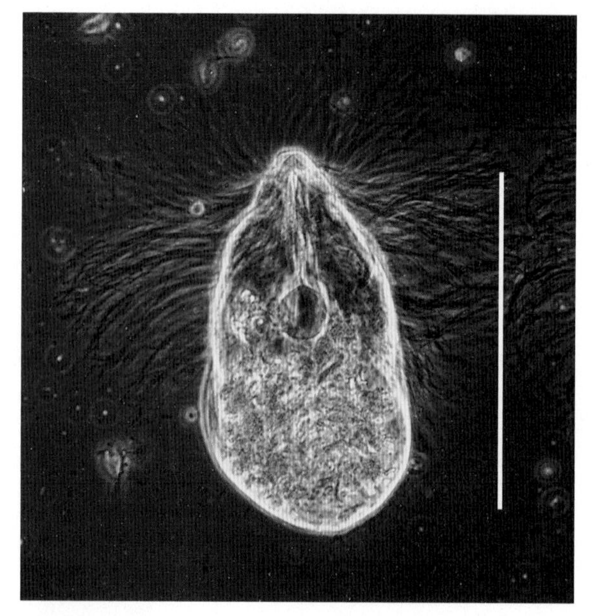

Figure B *Staurojoenina* sp., a wood-digesting hypermastigote from the hindgut of the dry-wood termite *Incisitermes* (*Kalotermes*) *minor* (A-21, Mandibulata). LM (stained preparation), bar = 50 μm.

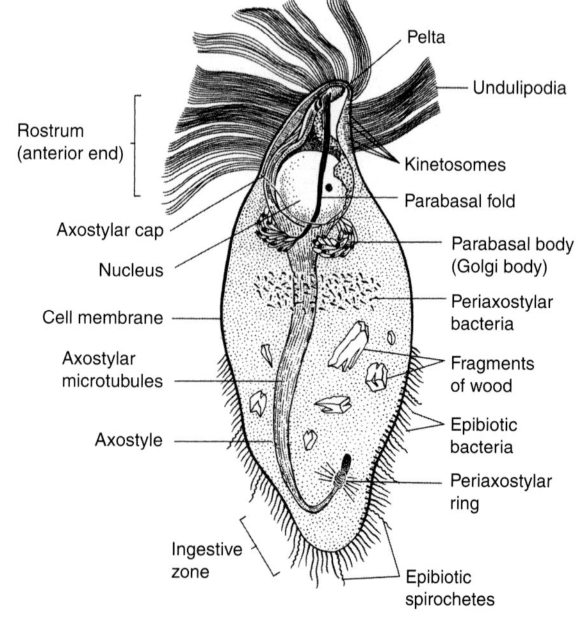

Figure C *Joenia annectens*, a hypermastigote that lives in the hindgut of a European dry-wood termite. *Joenia* is closely related to *Staurojoenina*. [Drawing by R. Golder.]

MODE III*

Members of the third class of phylum Archaeprotista, Parabasalia, also are symbiotic in the intestines of insects. Apparently, these microbes digest cellulose, from which they derive sugars both for themselves and for their hosts. Particles of wood are taken up through a sensitive posterior zone. In various stages of digestion, wood is often seen in the cytoplasm. A parabasalid bears at least four undulipodia, an axostyle, and conspicuous parabasal bodies. Well-defined homologues of the Golgi apparatus (dictyosomes) of animal cells and plant cells, parabasal bodies take part in the synthesis, storage, and transport of proteins. The presence of the membranous and granular parabasal body—often many parabasal bodies—distinguishes parabasalids from oxymonads. Sexuality is known, but because parabasalids are limited to the intestines of insects, no detailed study of it has been possible. In some species, the entire haploid adult is seen to transform into a sexually receptive gamete. Two gametes fuse into a gametocyst in which, it is thought, meiosis takes place—meiotic products, haploid adults, emerge from the cyst.

The class Parabasalia comprises two orders: Trichomonadida and Hypermastigida. Trichomonads typically bear 4 to 16 undulipodia. The undulipodia are often associated with supernumerary nuclei, two undulipodia per nucleus. On the basis of distinctive cell structure, in particular the manner of insertion of the undulipodia and its fibers in the cortical cell layer (just beneath the plasma membrane), four families of trichomonads have been described: Monocercomonadidae (*Hexamastix*, *Monocercomonas*, and *Histomonas*), Devescovinidae (*Devescovina* and *Metadevescovina*), Trichomonadidae (*Trichomonas* and *Trichomitus*), and Calonymphidae (*Calonympha* and *Snyderella*). Some members of the family Calonymphidae have more than a thousand nuclei. The order Hypermastigida, informally called hypermastigotes, have hundreds and even hundreds of thousands of undulipodia attached to special bands (Figures B through E) but generally only one or a few nuclei. The mitotic spindle, which grows out from these bands, is external to the nuclear membrane. Members of this order include *Staurojoenina*, *Joenia*, and *Trichonympha*.

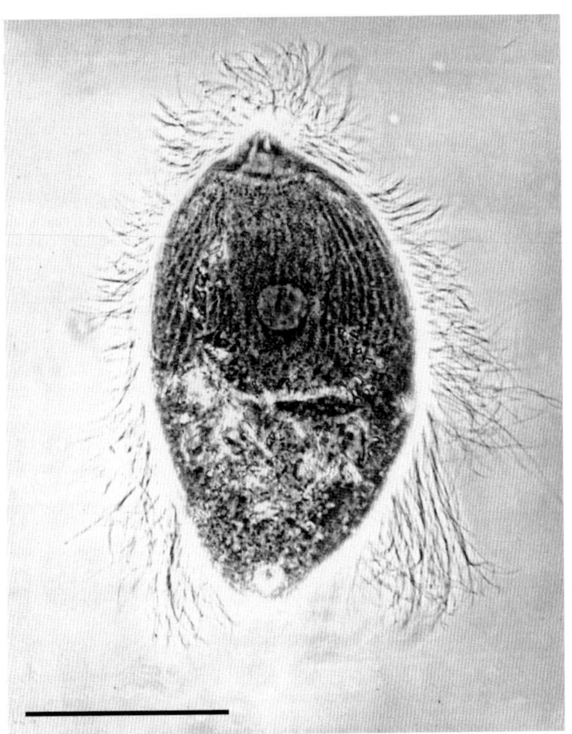

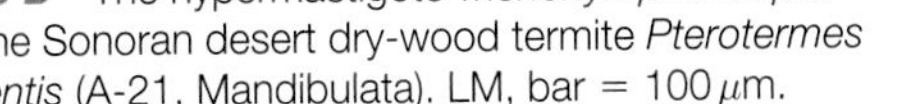

Figure D The hypermastigote *Trichonympha ampla* from the Sonoran desert dry-wood termite *Pterotermes occidentis* (A-21, Mandibulata). LM, bar = 100 μm. [Courtesy of D. Chase.]

Figure E Transverse section through the rostrum of a *Trichonympha* sp. from the termite *Incisitermes* (*Kalotermes*) *minor* from near San Diego, California, showing the attachment of undulipodia. TEM, bar = 5 μm. [Courtesy of D. Chase.]

* See note page 131.

SUBKINGDOM (Division) AMOEBAMORPHA

Pr-2 Rhizopoda

(Amastigote amoebae and cellular slime molds)

Greek *rhiza*, root; *pous*, foot

GENERA

Acanthamoeba
Acrasis
Acytostelium
Amoeba
Arcella
Centropyxis
Coenonia
Dictyostelium
Difflugia
Entamoeba
Guttulina
(continued)

As defined here, members of phylum Rhizopoda—amastigote amebas—are amebas that have mitochondria and lack undulipodia at all stages in their life histories. There are two classes. All the amebas in the first class, Lobosea, are single celled, either naked or with shells called tests. Some current classification systems place them in a supergroup Amoebozoa. In the second class, members of Acrasiomycota (Greek *acrasia*, bad mixture; *mykes*, fungus) are cellular slime molds—multicellular, land-dwelling derivatives of members of the first class. Although amastigote amebas lack undulipodia and in most cases the [9(3)+0] centrioles from which kinetosomes derive undulipodia, they are motile. However, centriole-like bodies have been observed in some species of the Acanthamoebidae. Defining features for phylum members are pseudopods (false feet; Figures A through D), flowing cytoplasmic processes used for forward locomotion and to surround and engulf food particles. Where studied, nonmuscle forms of contractile actomyosin proteins have been found to underlie pseudopodial movements. Like contraction of muscles, such movement is sensitive to variations in the concentration of calcium ion (Ca^{2+}).

Members of the first class are distributed worldwide in both freshwaters and marine waters, and they are especially common in soil. Many are symbiotrophic in animals; they may pass from host to host or from the soil/fodder to host. Although morphologically these amebas are among the most simple of the protoctists, from a molecular-evolution viewpoint, they are also among the most diverse. They are not monophyletic. All are microscopic, yet some are very large for single cells, hundreds of micrometers long. Lacking meiosis and any sort of sexuality, these amebas

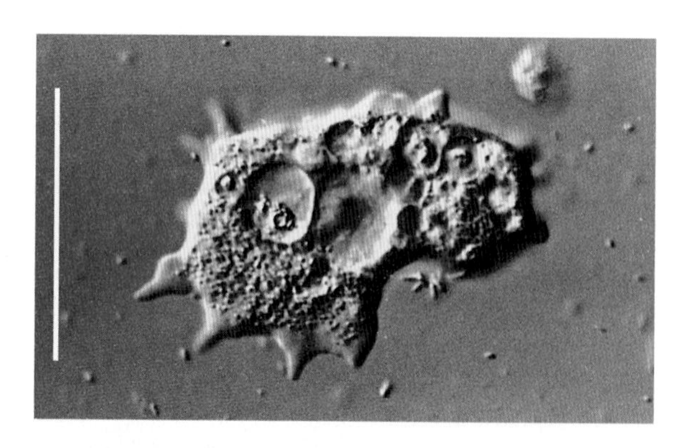

Figure A *Mayorella penardi*, a living, naked ameba from the Atlantic Ocean. LM (differential interference contrast microscopy), bar = 50 μm. [Courtesy of F. C. Page, in "An illustrated key to freshwater and soil amoebae," *Freshwater Biological Association Scientific Publication* 34 (1976).]

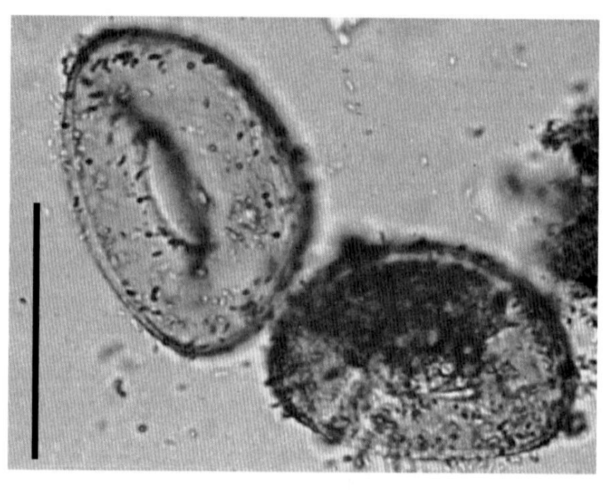

Figure C Two empty tests (shells) of the freshwater ameba *Arcella polypora*. LM, bar = 10 μm. [Courtesy of F. C. Page.]

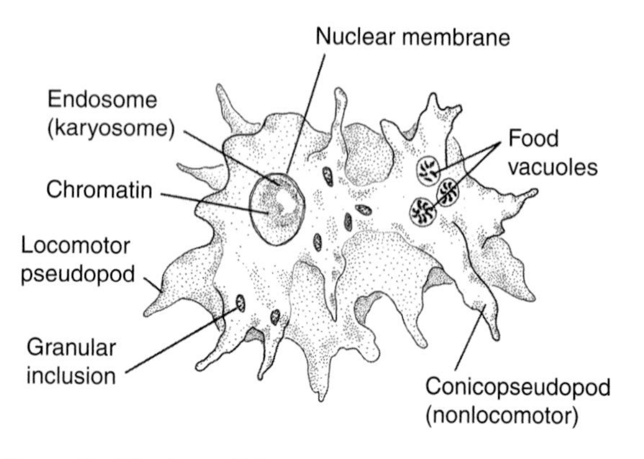

Figure B Structure of *Mayorella penardi* seen from above. [Drawing by E. Hoffman.]

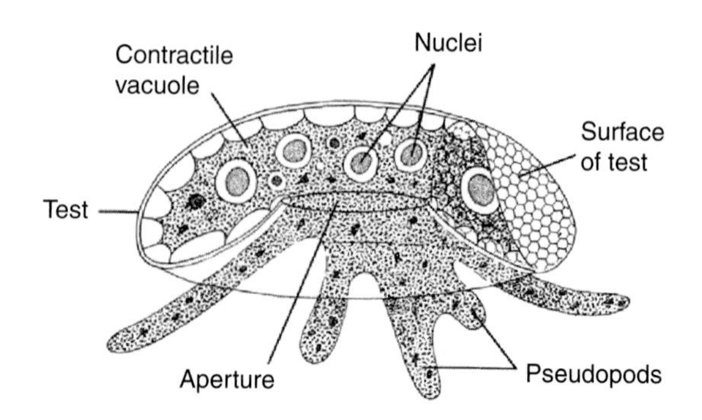

Figure D Structure of *Arcella polypora* showing the test composed of closely spaced, proteinaceous, hexagonal alveolae secreted from the cytoplasm. Cutaway view. [Drawing by R. Golder.]

reproduce by direct division into two offspring cells of equal volume. They have mitotic spindle microtubules and nuclear chromatin granules, which in some species form chromosomes. In these species, metaphase through telophase stages of cell division have been observed. The nuclear membrane persists well into the later stage of mitotic division; in some amebas, the nuclear membrane does not disperse at all during division.

The first subclass (naked amebas; subclass Gymnamoebia) in this phylum contains a single order (Amoebida) and five suborders: Tubulina, Thecina, Flabellina, Conopodina, and Acanthopodina.

The suborder Tubulina includes uninucleate, cylindrical, naked amebas. They are grouped into three families. The Amoebidae, which include the well-known *Amoeba proteus*, tend to be polypodial—an ameba has many feeding, changing, flowing pseudopods at one time. The Hartmannellidae, on the other hand, are monopodial—an ameba in this family forms one pseudopod at a time. Some form desiccation-tolerant resting cysts; the cell inside each cyst is binucleate. The Entamoebidae also are monopodial. The rhizopod nucleolus, which contains the ribosomal precursors, is organized into a conspicuous organelle (or several) called an endosome. These amebas are probably the most ancient to have endosomes, which are also found in the nuclei of other protoctists (for example, euglenids, Pr-12). That the Entamoebidae form cysts is of great importance because nearly all of them live in animals. Some, such as *Entamoeba histolytica*, are responsible for amebic dysenteries. The cysts enable the amebas to resist animal digestive enzymes. The amebal nuclei can divide inside the cysts without accompanying cytoplasmic division; this leads to four, eight, or even more nuclei per encysted cell. The cysts germinate in the animal digestive tract or they are transported to the soil in the host's feces.

The Thecina ameba seems to roll its wrinkled surface as it moves. Thecina amebas form a rather obscure group of free-living forms having various mitotic patterns.

Members of Flabellina form spatula-shaped pseudopods in which flowing endoplasm seems to erupt. Some, such as *Hyalodiscus* and *Vannella*, are fan shaped.

Members of Conopodina have pseudopodia that are shaped like fingers. When they move, they are longer than they are wide; some float on water, where they extend slender radiating pseudopods. *Mayorella* (Figures A and B) and *Paramoeba*, a mainly marine genus, belong to the family Paramoebidae, the only family in the suborder Conopodina. *Paramoeba eilhardi* contains two distinctive bodies called nebenkörper, which are packages of benign, omnipresent bacteria-like symbionts. *Paramoeba eilhardi* can be attacked and killed by certain other marine bacteria that are able to grow and divide only in its nucleus.

Members of the suborder Acanthopodina have finely tipped subpseudopodia; that is, each pseudopod extends smaller pseudopods of its own. The cell as a whole may be disk shaped. The suborder contains two families: Acanthamoebidae and Echinoamoebidae. *Acanthamoeba* forms a polyhedral or thickly biconvex cyst having a wall that contains cellulose. The many ubiquitous acanthamoebas aggressively devour bacteria, other amebas, and ciliates. Their populations achieve huge numbers, and tough cysts permit prolonged survival in soil—even in very dry soil. The Echinoamoebidae, members of the other family, are more or less flattened when they move; they have tiny, finely pointed pseudopods that look like spines.

Although naked amebas constantly change shape, the range of shapes that each takes on is genetically limited and species specific. Many naked amebas correspond to shelled forms (of which Testacealobosa, the second subclass within class Lobosea, consists) that are thought to have been derived from them. To construct their tests, amebas secrete intracellular mineralized particles or proteinaceous subunits (for example, *Arcella*) that are deposited on the cell surface to form the species-specific test. Others glue together sand grains, bits of carbonate particles, and other inorganic detritus, depending on what is available. In some species of testate amebas, only mineral particles of a particular size or chemical composition are selected. Indeed, certain species of ameba select only the siliceous shells of a particular species of diatom to construct their test. Some of these tests, such as those of *Arcella* (Figures C and D), are distinctive enough to be recognized in the fossil record. Such tests give the testate members of Rhizopoda a fossil record that extends well into the Paleozoic era. Some of the pre-Phanerozoic microfossils called acritarchs have been interpreted as tests of shelled amebas as well.

The organisms in the second class of this phylum, Acrasiomycota—cellular slime molds—are multicellular, land-dwelling, heterotrophic protoctists found in freshwater, in damp soil, and on rotting vegetation, especially on fallen logs. They enjoy a fascinating "dispersed" life history. In the course of their life history, independently feeding and dividing amebas aggregate into a slimy mass or slug that eventually transforms itself into a spore-forming reproductive body; the scattered spores germinate into amebas. Sexuality is rare or absent.

The taxonomy of slime molds has always been contested because slime molds have features commonly taken to be animal (they move; they ingest whole food by phagocytosis; and they metamorphose), plant (they form spores on upright reproductive bodies), or fungi (their spores have tough cell walls and germinate into colorless cells with absorptive nutrition—they live on dung and decaying plant material). The zoologists have called them "mycetozoa" (slime animals) and classified them with protozoa; the mycologists call them "myxomycetes." In some classifications, three of our phyla—Myxomycota (Pr-23), Labyrinthulata (Pr-19), and Plasmodiophora (Pr-20)—have been classified together with these acrasiomycotes as a single phylum—Gymnomycota or Gymnomyxa (naked fungi)—in kingdom Fungi. By 1868, Ernst Haeckel and John Hogg considered them neither plants nor animals but primitive forms that had not yet evolved to be members of either of the two great kingdoms. Hogg erected a new kingdom Protoctista or Primogenium (with protoctist members) to accommodate these organisms.

The class Acrasiomycota contains two subclasses: Acrasea and Dictyostelia. The members of both subclasses pass through a unicellular stage of ameboid cells that feed on bacteria. Later they form a multicellular, stalked reproductive structure, the sorocarp; the sorocarp that produces spores borne in a swelling called

Guttulinopsis
Hartmannella
Hyalodiscus
Mayorella
Minakatella
Paramoeba
Pocheina
Polysphondylium
Thecamoeba
Vannella

the sorus, which lies at the tip or just below the top of the stalk. In passing from the first stage to the second, the ameboid cells aggregate to form a pseudoplasmodium (slug). A true plasmodium, or syncytium, is a mass of protoplasm containing many nuclei formed by mitotic divisions but not separated by cell membranes. The acrasiomycote structure is called a pseudoplasmodium because it is made of mononucleate constituent cells that retain their cell membranes. It only superficially resembles the plasmodium of the true plasmodial slime molds (Pr-23).

Most acrasiomycotes will begin to aggregate if food is depleted and light is present. However, the exposure to light must be followed by a minimum period of darkness before development can continue.

The two subclasses differ in many ways and may not be directly related. In Acrasea, the stalk of the sorocarp consists of live cells that are capable of germination and lack cellulosic walls, whereas the stalk in Dictyostelia consists of a tube of cellulosic walls of dead cells. Dictyostelid ameboid cells are aggregated by their attraction to cyclic adenosine monophosphate (cAMP); the acrasids do not respond to cAMP.

In subclass Acrasea, the feeding stage consists of ameboid cells having broad, rounded pseudopods. The families of acrasids are distinguished primarily by the structure of the sorocarp. In some, the spore cells are different from the stalk cells; in others, all the cells are alike.

Acrasids, like all other rhizopods as grouped here, lack undulipodia. However, members of one genus of subclass Acrasea, *Pocheina*, exhibit an undulipodiated form. Molecular genetic evidence suggests that this genus, and perhaps all acrasids, belongs in the class Heterolobosea (amoebo-mastigotes) within an eclectic supergroup Excavata. This group includes undulipodiated protoctists with a feeding groove on the cell body, although some genetically related members have lost this feature. When spores of *Pocheina* germinate, they divide into motile, swimming cells, each bearing two undulipodia of equal length. Still poorly known, pocheinas have been reported from the former Soviet Union (Kazan) and North America; they live on conifer bark and lichenized dead wood.

Members of subclass Dictyostelia are far better known than members of subclass Acrasea. The amoeboid cells never form undulipodia, and molecular evidence suggests that they should be grouped with the amoebae in the supergroup Amoebozoa that includes other exclusively ameboid protoctists. Four dictyostelid genera have been described: *Acytostelium*, *Dictyostelium*, *Polysphondylium*, and *Coenonia*. Each has a number of species, at least 16 in *Dictyostelium*. The amoeboid cells of dictyostelids are usually uninucleate and haploid. However, cells having more than one nucleus and aneuploids, cells having an uneven number of chromosomes, have been reported. Some strains consist of stable diploid cells.

The typical dictyostelid life cycle is illustrated here for *Dictyostelium discoideum* (Figures E and F). The amebas have thin pseudopods and feed mainly on live bacteria. After the food supply is exhausted and the amebal population has reached a certain density, the cells cease feeding and dividing. Because of a pheromone (chemical attractant) called acrasin, cAMP, secreted by the amebas themselves, they begin to aggregate, streaming toward aggregation centers. The dispersed feeding stage of the life history terminates when the pseudoplasmodium forms. A thin slime sheath is produced around the mass of cells, forming a pseudoplasmodium that takes on the form of a slug. The slug begins to wander, leaving behind a slime track. As conditions become drier, the migration stops and the differentiation of the reproductive structure begins. In a complicated developmental sequence including differentiation, but not cell division, the sporophore (also called the sorocarp) with its cellulosic stalk forms. *Dictyostelium discoideum* is valued in developmental biology research because it grows rapidly and the separation in time of its trophic stage (feeding and growing stage of the ameba population) from its differentiation into stalk and propagule (spore) facilitates manipulation of its developmental stages.

Figure E The development of a reproductive body from a slug of *Dictyostelium discoideum.* Bar = 1 mm. [Courtesy of J. T. Bonner, from *The cellular slime molds*, © 1959, rev. ed. © 1967 by Princeton University Press; Plate III reprinted with permission.]

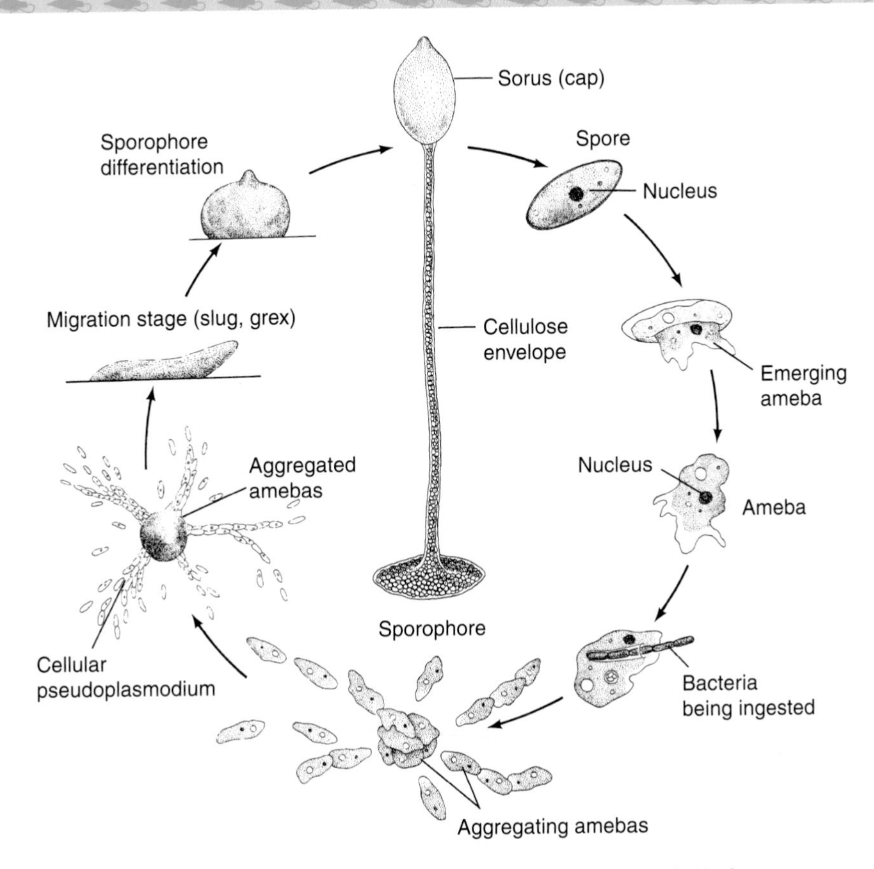

Figure F Life history of the cellular slime mold *Dictyostelium discoideum*. [Drawing by R. Golder.]

Pr-3 Granuloreticulosa

(Foraminifera and unshelled relatives)

Latin *foramen*, little hole, perforation; *ferre*, to bear

GENERA

Allogromia	*Elphidium*	*Miliola*
Bowseria	*Fusulina*	*Nodosaria*
Camerina	*Glabratella*	*Reticulomxya*
Discorbis	*Globigerina*	*Rotaliella*
Discospirina	*Iridia*	*Textularia*

Foraminifera (affectionately known as "forams") are easily defined: these organisms extend reticulopodia—thin, branching pseudopodia—that fuse to form networks in which bidirectional (two-way) streaming can be seen. This group can be divided into two major subdivisions: Monothalamia and Polythalamia. By far the better-known class, Polythalamia have hard, pore-studded, multichambered shells, or tests. In contrast, the early-evolving monothalamia include members that are "snot-shaped" slimy nets of messy, bactivorous masses that lack shells; others have soft leathery or hard agglutinated, single-chambered tests. Very few of these ancestral monothalamids have been studied in detail.

Forams are predominantly marine organisms, although freshwater and even terrestrial members are known. The smallest ones are some 10 μm in diameter, and the largest ones, visible to the naked eye, grow to several centimeters in diameter. The majority are tiny and live in sand or mud or attached to rocks, algae, or other organisms. Two groups of free-floating, modern planktic forams (Globigerinids and Globorotalids) are very important in the economy of the sea as predators and prey of zooplankton and, as major producers of calcium carbonate, in the carbon cycle.

The tests of forams are composed of organic materials, often reinforced with minerals. Some are made of sand grains; most are neatly cemented granules of calcium carbonate deposited from seawater. Some forams, by mechanisms that are poorly understood, choose echinoderm plates (A-34) or sponge spicules (A-3) to construct their tests. The test and the organism itself may be brilliantly colored—salmon, red, or yellow brown. A typical test looks like a clump of blobs of partial spheres (Figure A). One or more apertures in the test permit thin cytoplasmic projections, the microtubule-reinforced pseudopodia to emerge. The repeatedly branching and anastomosing pseudopodia form a network, or reticulopodium, which is used for feeding, crawling, and gathering material for tests. Forams are omnivorous: they eat algae, ciliates (Pr-6), actinopods (Pr-31), and even nematodes (A-11), and crustacean larvae (A-21). Many forams that live in shallow water, harbor photosynthetic symbionts—dinomastigotes (Pr-5), chrysomonads (Pr-15, planktic), and diatoms (Pr-18).

Figure A Adult agamont test of *Globigerina* sp., an Atlantic foraminiferan. SEM, bar = 10 μm. [Courtesy of G. Small.]

Although some foram species (for example, *Spiroloculina hyalina*) have been seen reproducing only by asexual budding or multiple fission, others that have been well studied—some dozen species—show a remarkably complex life cycle. The known cycles are variations on the theme of *Rotaliella* (Figure B). Meiosis takes place during reproduction in the agamont, a fully adult, diploid organism that releases small haploid juveniles called agametes. These agametes, which initially consist of just a single chamber called the proloculus, disperse and grow by adding new chambers to the test, ultimately becoming adults, called gamonts. The gamonts reproduce sexually, producing numerous gametes by mitosis. The fusion of haploid gametes forms diploid offspring, which grow to become agamonts.

The alternation of the diploid agamont and haploid gamont generations is obligatory in the species of *Rotaliella* but may be either obligatory or facultative in other foram species.. In fact, forams are the only heterotrophic protoctists that alternate morphologically distinct free-living adult generations. What complicates matters is that unlike other organisms except ciliates (Pr-6), forams show a striking nuclear dimorphism. The agamonts of *Rotaliella roscoffensis*, for example, contain four diploid nuclei. Three of these nuclei, the generative nuclei, reside in a chamber separate from that in which the larger somatic nucleus remains. The somatic nucleus never undergoes meiosis; it eventually becomes pycnotic (it stains heavily) and disintegrates. The three generative nuclei give rise to 12 haploid products by meiosis. These products become the nuclei of small haploid agametes. Later, in the gamonts, pairs of haploid nuclei, apparently of opposite sex, fuse to form diploid zygotes. In effect, these organisms show programmed cell death (selective "death" of the somatic nucleus), and each gamont fertilizes itself, although neither egg nor sperm is formed. The life cycle of *R. roscoffensis*, however, is highly derived and not typical of most forams. Nuclear dimorphism, for example, is known only in *Rotaliella* and its close relatives, and most forams reproduce sexually by liberating thousands of tiny, biundulipodiate gametes directly into the surrounding seawater where cross-fertilization occurs.

Foram tests have contributed greatly to the sediment on the bottom of marine basins, especially since the Triassic period. There are fossilized giant forams of great fame. Some, such as *Lepidocyclina elephantina*, had tests as thick as 1.5 cm. *Camerina laevigata* (also known as nummulites, the "coin stone") was a large (10 cm wide) foram that lived in warm, shallow waters during the Cenozoic era from the Eocene to the Miocene epoch (some 38 to 7 mya). Rocks bearing Eocene forams, many of them easily visible to the naked eye, abound on the shores of the Mediterranean. It is from such "nummulitic" limestone that the pyramids of Egypt were constructed.

The abundance of foram tests and their detailed architecture (the earliest ones appeared in the Cambrian) make them excellent stratigraphic markers. Geologists use the 40,000 or so fossil species to identify geographically separate sediment layers of the same age. Because the tests are often found in strata that cover oil deposits, recognition of foram morphology and knowledge of their distribution are helpful in petroleum exploration.

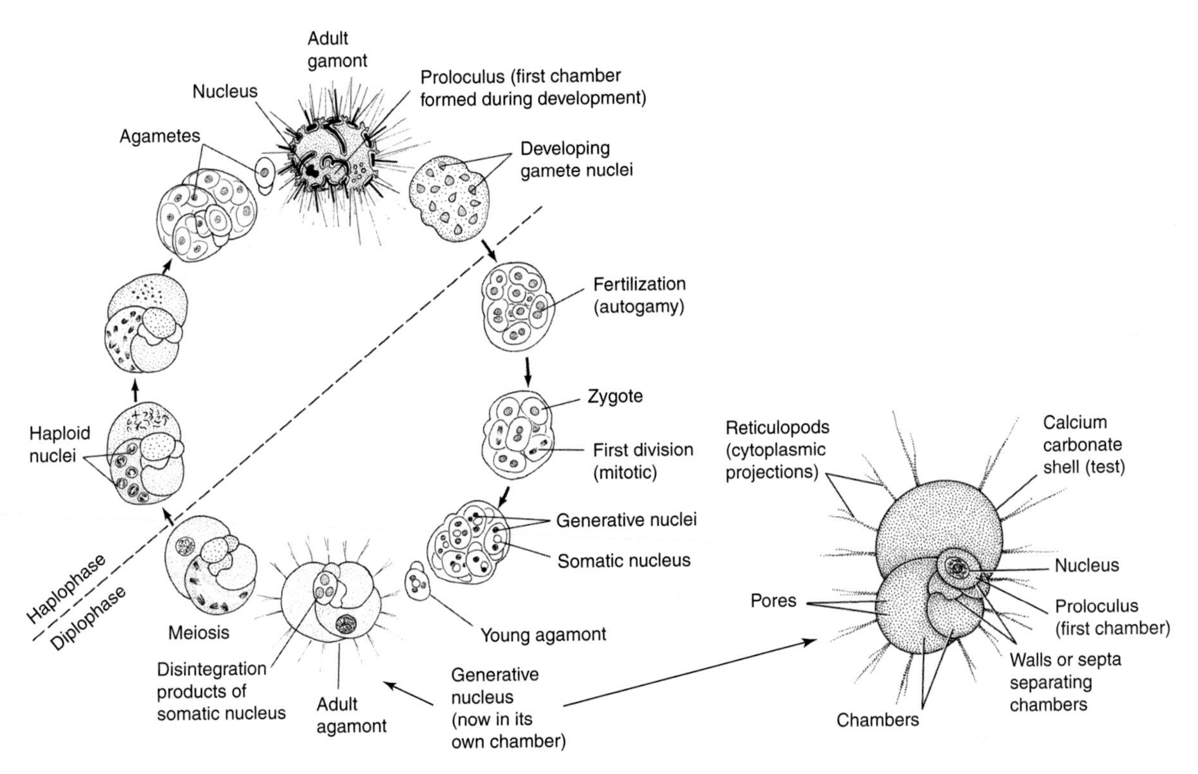

Figure B Life cycle of *Rotaliella roscoffensis* and adult gamont stage of *Rotaliella* sp. [Drawings by L. Meszoly; information from K. Grell.]

Box Pr-iii: Fossil record insights—Calibration of molecular clocks to the geologic record

A record of past life on Earth is preserved in rocks as fossils. The fossil record provides evidence of a minimum age for life-forms, what types of organisms lived to be preserved, how long they persisted, and how Earth has changed over the eons. With the discovery of DNA, a new kind of record for the history of life was possible.

Species change over time. According to the assumptions of molecular evolution, these changes are reflected in gene sequences. The frequency of change in a species' genes acts like a kind of molecular clock, which can trace the branches of a family tree back to the original root. Evolutionary biologists use molecular clocks to estimate rates of evolution, even when species have long since diverged from common ancestors.

Some dramatic conclusions are inferred from this technique. The estimate for the emergence of animals more than a thousand million years ago by molecular clock studies greatly exceeds those of paleontological estimates, which place their origin at about 600 mya. Why would inference from DNA sequence differ from the rock record? The discrepancies between dates inferred by molecular biologists and those derived from paleontological observations on the accuracy of the fossil record are due, at least, to the fallibility of built-in assumptions of molecular clocks.

Whereas the fossil record remains incomplete, molecular biologists, who generate thousands of molecular "cladograms" based on comparison of living species, may claim they work with time, but do not measure time directly. Paleontological information is absolutely necessary to calibrate molecular phylogenetic trees.

Use of the fossil record to provide well-constrained calibration points for molecular phylogeny depends on the completeness of the record and the reliability of paleontological ages. The accuracy of the fossil record depends on the quality of preservation of specimens, the proper identification of morphological characters, and accurate absolute geochronological data derived from magnetic and radiometric age studies. No ideal fossil record exists, but microfossils, such as foraminifera, provide a robust paleontological record for calibrating molecular clocks.

Several internal measures of accuracy make foraminifera ("forams") one of the best stratigraphic tools to calibrate the dates of the fossil record of life. Forams, protoctists in the phylum Granuloreticulosa (Pr-3), are abundant and widespread in marine sediment samples spanning salt marshes and estuaries to the deepest parts of the world oceans and from the tropical waters to polar seas. The preservation of their calcareous and agglutinated tests (shells) is often excellent. Their geological range begins in the earliest Cambrian period and extends to the present day. Varying patterns of evolution and extinction can be traced in detail. Micropaleontologists organize sedimentary layers into "biozones" based on the first and last occurrences of well-documented species (biostratigraphy). Biozones are used to establish relative ages (chronostratigraphy) of a sedimentary section based on the temporal succession of fossil species due to evolution. Over the past 60 years, fossil occurrences and biozones worldwide have been integrated with ever-improving paleomagnetic and radiometric age data to yield a robust geologic timescale (Figure Pr-iii-1). Foraminifera are widely used for calibration purposes by the oil industry and in marine geological research.

The quality of the foram fossil record is tested by independent molecular criteria. Molecular phylogenies based on rDNA sequences confirm many hypotheses derived from stratigraphic and morphologic information. This congruence lends confidence to the materials and methods of micropaleontology and supports the reliability of their phylogenetic reconstructions.

For ancient lineages that lack an adequate fossil record or for first appearances in the fossil record of complex animals or plants that imply an unpreserved ancestor, DNA researchers extrapolate a molecular rate of change that assumes that molecular clocks "tick" at a constant rate through time. However, this extrapolation produces inaccurate and even unreal results. For example, Samuel Bowser of the Wadsworth Center, Albany, New York, attempted to determine when the earliest "proto-foram" not preserved in the fossil record first appeared. The extrapolation produced a first appearance date of 10.8–4.3 thousand million years ago. If we assume that forams did not originate in outer space, the unconstrained molecular clock method that puts them before the formation of the Earth must be in error.

Molecular data provide a scaffold on which to place many features of evolution, whereas fossils provide a reality check: details of phylogenetic and ecologic history. Molecular studies, in principle, because they focus only on extant life, cannot answer questions about extinct lineages that abound in the rock record. Paleontology and molecular biology together complement the search for life's evolutionary history.

References

Bowser, S. "Forams." Lecture to Microbial Communities Seminar, University of Massachusetts at Amherst, 25 July 2007.

Culver, S. J., "Foraminifera." In: J. H. Lipps, ed., *Fossil Prokaryotes and Protist*. Blackwell Scientific Publications, Boston, MA; pp. 203–247; 1993.

Leckie, M., and K. St. John, "How old is it? Part 1 – Biostratigraphy." Joi Learning – School of Rock, n.p.; 2005.

Lipps, J. H., "Major features of protistan evolution: Controversies, problems and a few answers." *Anuário do Instituto de Geociências UFRJ* 29:55–80; 2006.

Lipps, J. H., "The furture of palenotology – The next 10 years." *Palaeontologia Electronica* v. 10(1); 2007

Mullen, L., "Synchronizing molecular clocks." *Astrobiology Magazine* 18 June 2003.

Pawlowski, J., and C. de Vargas, "The value of a good fossil record." *Nature Debates* 296(6709): 291–392; 1998.

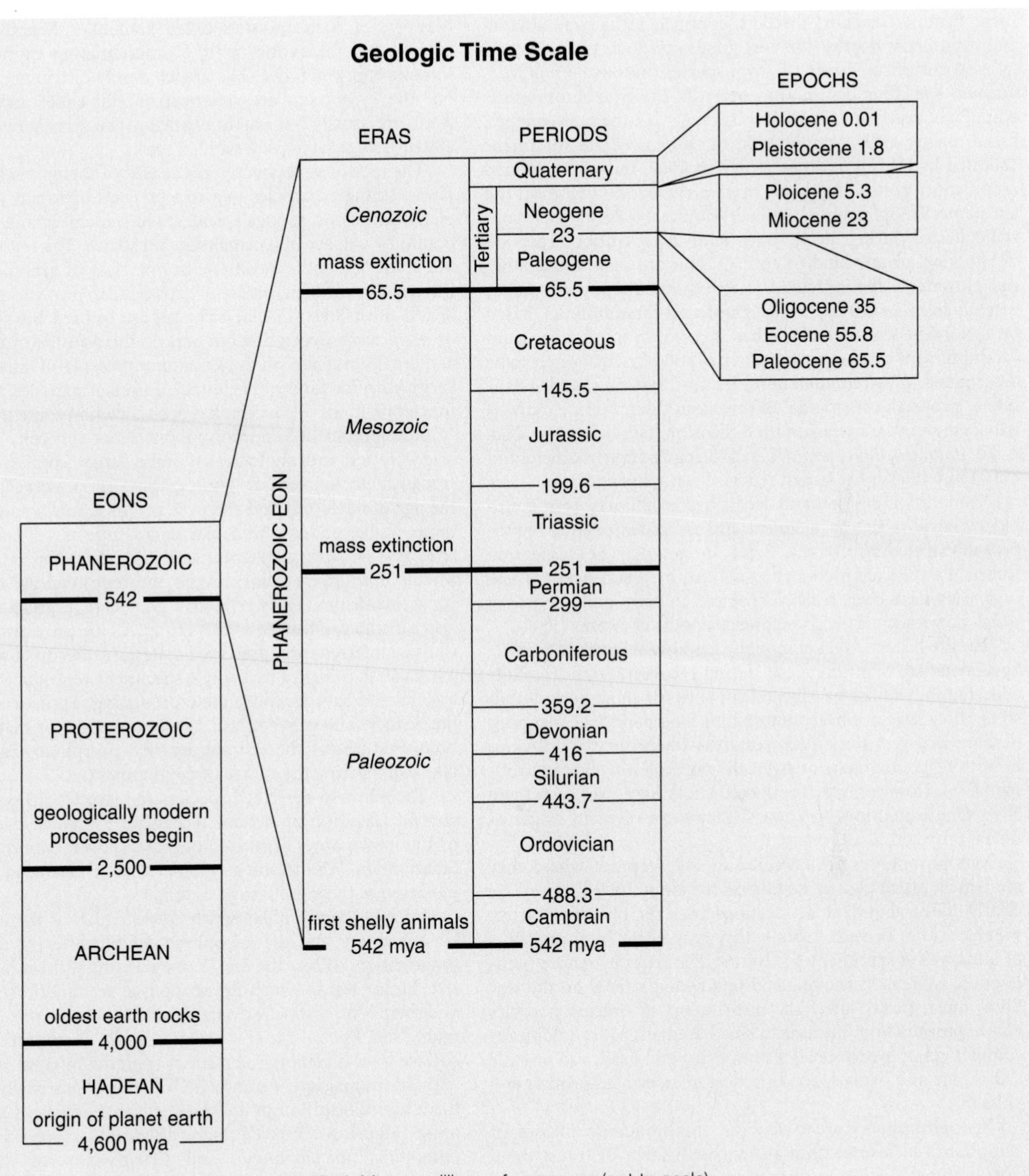

Figure Pr-iii-1 Geologic Time Scale, simplified. Mya = millions of years ago (not to scale).

Pr-4 Xenophyophora

Greek *xenos*, foreign (that is, the foreign particles used in test construction); *phyein*, to bring forth, beget; *pherein*, to bear

GENERA

Aschemonella	*Maudammina*	*Semipsammina*
Cerelasma	*Ocultammina*	*Spiculammina*
Cerelpemma	*Psammetta*	*Stannoma*
Galatheammina	*Psammina*	*Stannophyllum*
Homogammina	*Reticulammina*	*Syringammina*

Large benthic (seafloor) dwellers, confined to bathyal, abyssal, and even hadal depths, these enigmatic protoctists are objects of great curiosity. The first-known species (*Syringammina fragilissima*) was described in 1881 by H. B. Brady as a foraminiferan. Eight years later, Ernst Haeckel regarded them as sponges, based mainly on his examination of *Stannophyllum* specimens collected by HMS *Challenger*. Later, in 1907, they were defined as a distinct group of gigantic marine rhizopods and given the apt name Xenophyophora by F. E. Schultz, based on the material collected during the German Valdivia expedition. They are not sponges or any kind of animals. Nor are they testate amebas. However, a species related to *Syringammina fragilissima* has recently been shown to be an agglutinated foraminiferan, based on DNA gene sequences. Whether all xenophyophores are giant foraminiferans or whether the group is polyphyletic has yet to be determined. Whatever their affinities, some are huge and constitute a "protistan megafauna" of the ocean floor. Lumpy individuals 7 cm or more in maximum dimension have been described. A flat *Stannophyllum* specimen measuring 25 cm in diameter but only 1 mm thick is the largest reported xenophyophore.

Xenophyophores seem to feed phagotrophically (for example, by engulfment) on sediment and associated organic matter, probably including bacteria. Some species may be suspension feeders. Particles are presumably collected by pseudopodia. These organelles have been reliably observed in only one species, in which they resemble the granuloreticulopodia of forams (Pr-3).

The life-history stages of xenophyophores are largely unknown. Specimens are often damaged during recovery. Some carefully collected individuals have been kept alive for short periods, but so far, they have not been cultured for long periods. Large populations occur in some deep-sea areas where they are obvious in bottom photographs or from the portholes of manned submersibles. However, very few investigators have ever seen them alive. One would love to know if they form resistant stages or sexual propagules.

Xenophyophores are confined to the deep sea where they are widely distributed at depths below about 1000 m. They are clearly more abundant at locations overlain by a productive water column. In such regions, they may occur in abundances of hundreds of specimens per 100 m^2. The xenophyophore pseudopodial system, by moving and removing particles on the seafloor, must greatly affect the distribution of organic particles and organisms from bacteria to small animals. Morphologically complex xenophyophore tests, both living and dead, also provide a substrate and living space for a host of associated protists and animals.

Xenophyophores are plasmodia (multinucleate masses of cytoplasm) enclosed within a branched system of transparent, organic tubes, 30–90 μm diameter; together, the cytoplasm and tubes form the granellare system. Waste pellets (stercomata) accumulate outside the granellare in dark strings or masses bounded by an organic sheet. These masses are called stercomare. The wall of the granellare tubes is very thin, less than 0.5 μm. The pseudopodia presumably extend through the ends of the granellare branches. The strings of plasmodial cytoplasm running through the tubes contain numerous nuclei and huge numbers of barite crystals called granellare. Xenophyophores may be heterokaryotic, with a differentiation of nuclei into somatic and generative. The nuclei, evenly distributed throughout the cytoplasm, are spherical or ellipsoidal and measure 2–10 μm, usually 3–4 μm, in diameter. The granellare are generally the size of large bacteria (2–5 μm).

The tests of xenophyophores consist of foreign matter (*xenophyae*, stranger particles)—whole or parts of foram tests, radiolarian skeletons, sponge spicules, and mineral grains. These are bound by patches of a cementlike substance. The test surrounds the granellare and stercomare. In one class of xenophyophores, the test also contains a mass of extracellular, proteinaceous, 2- to 3-mm-thick fibers (linellae). The test can be hard, brittle, or more or less flexible, depending not only on the quantity of xenophyae and cement but also on the absence or presence of linellae. Color varies with the kind of agglutinated foreign particles. Selectivity for certain kinds of particles has been shown in some species.

Recognition of xenophyophores is not difficult. The characteristic test morphologies of some larger species are easily recognizable in seafloor photographs. In preserved material, the agglutinated test and the presence of granellare and stercomare, visible under a binocular microscope as yellowish, black, and grayish strings, respectively, are characteristic features. Microscopic investigation reveals the multinucleate cytoplasm packed with the highly refractive granellare, crystals which are also a distinctive feature of this group. All life activities, however, must be inferred from preserved material or in situ observations.

There is evidence that single specimens reproduce by gametogamy and have several gamete-producing rounds during their life history. The gametes bear two undulipodia or may be ameboid. Features of the test suggest that morphological changes take place during the course of development.

The 60 or so described species are placed in 15 genera. The current classification system recognizes a total of five families of xenophyophores organized into two classes, Psamminida and Stannomida. We include a complete list of genera in our enumeration of these groupings, as follows.

In class Psamminida, which consists of four families, linellae are absent and the xenophyae exhibit different degrees of organisation within the test. Family Psammettidae has a massive, lumpy test in which the xenophyae are randomly arranged and there is no distinct external layer: *Homogammina*, *Maudammina*, and *Psammetta* (Figure A). Family Psamminidae has a surface layer of xenophyae distinct from the internal xenophyae with additional layers sometimes developed internally; the test form varies from lumpy to branched or reticulate: *Galatheammina* (Figure B), *Reticulammina* (Figure C), *Psammina*, *Semipsammina*, *Spiculammina*, and *Cerelpemma*. In the family Syringamminidae, the test consists of a system of tubes, in some species very complex, in which the xenophyae are confined to the walls of the tubes, and the test is interiorly occupied only by granellare and stercomare: *Syringammina* (Figure D), *Ocultammina*, and *Aschemonella*. Family Cerelasmidae has xenophyae and large amounts of cement in no obvious order: *Cerelasma*.

In class Stannomida, which contains only one family (Stannomidae), the xenophyae are organized poorly or not at all, and

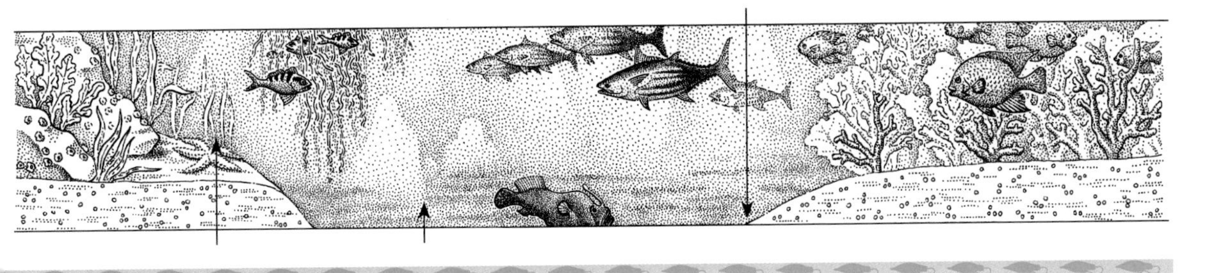

Figure A *Psammetta globosa* Schulze, 1906. "John Murray Expedition" St. 119. The specimen measures about 20 mm in diameter. Bar = 1 cm. [Photograph by O. S. Tendal.]

Figure C *Reticulammina lamellata* Tendal, 1972. NZOI "Taranui Expedition" St. F 881. Greatest dimension is about 30 mm. Bar = 1 cm. [Photograph by O. S. Tendal.]

Figure B *Galatheammina tetraedra* Tendal, 1972. "Galathea Expedition" St. 192. Greatest dimension from tip of arm to tip of arm is 18 mm. Bar = 2 cm. [Photograph by O. S. Tendal.]

Figure D *Syringammina fragillissima* Brady, 1883. "Triton Expedition" St. 11. Greatest dimension is about 40 mm. Bar = 1 cm. [Photograph by O. S. Tendal.]

linellae are present and the test is flexible with a branched, tree-like or flat morphology: *Stannoma* and *Stannophyllum*.

Xenophyophores have no proven fossil record. It has been suggested that *Palaeodictyon*, a trace fossil that forms a very regular network of tunnels, was formed by an *Occultammina*-like xenophyophore. However, this suggestion is not supported by any firm evidence.

SUBKINGDOM (Division) ALVEOLATA

Pr-5 Dinomastigota
(Dinoflagellata, Dinophyta)

Greek *dinos*, whirling, rotation, eddy; *mastigio*, whip

GENERA

Amphidinium	*Gonyaulax*	*Pfeisteria*
Ceratium	*Gymnodinium*	*Polykrikos*
Cystodinium	*Nematodinium*	*Prorocentrum*
Dinothrix	*Noctiluca*	*Protopsis*
Erythropsidium	*Peridinium*	*Warnowia*

Dinomastigota, Ciliophora (Pr-6), and Apicomplexa (Pr-7) are grouped together as alveolates because all have both alveolar sacs and common rRNA sequences. The characteristic cell envelope of dinomastigotes is a theca (or amphiesma) that consists of membranes including the plasma membrane, microtubules which form a cytoskeleton, and often thecal plates. The plates vary in thickness and in pattern of distribution (called tabulation) in different groups of dinomastigotes and are useful in identification from the subclass level through the species level in some groups. The upper half of the theca is called the epicone, and the lower half is called the hypocone. Some dinomastigotes do not have obvious surface plates and are called "naked" in contrast to those that do have them, the "armored" dinomastigotes.

The dinomastigote, often called "dino," has two undulipodia. One of them, the tranverse, bears fine hairs (mastigonemes) and rests in a characteristic groove, the girdle (cingulum), encircling the cell. The other, the longitudinal, is directed posteriorly along the sulcus, a groove in the surface of the theca. The undulipodia lie at right angles to each other. When they both move, the cell whirls. Specialized vacuole-like organelles (pustules), usually two per cell, open by canals into the kinetosomes and from there to the exterior of the cell. Stinging organelles (trichocysts) are found underlying the cell membrane of some dinos, a kind of extrusome; they are capable of sudden discharge to sting prey.

Of more than 4000 known species of dinomastigotes in 550 genera, most swim as members of marine plankton and are especially abundant in warm seas. Many genera have freshwater representatives. Photosynthetic dinos contain brownish plastids. The pigments of these plastids generally include chlorophylls ***a*** and c_2 (sometimes c_1 as well), beta- and gamma-carotenes peculiar to them, called peridinin. Photosynthetic dinomastigotes store starch.

Dinomastigotes are important primary producers, but approximately 50% are heterotrophs. When conditions are right, they often reproduce to form visually recognizable "blooms" called "red or brown tides." These tides can be environmentally significant because many blooming species (such as *Gonyaulax tamarensis*) produce paralytic or diarrheic shellfish poison toxins or fish-killing ichyotoxins (PSP or DSP), that are accumulated by fish and marine invertebrates.

Many dinomastigotes are bioluminescent and cause the twinkling of lights in the waves of the open ocean at night.

A large, unusual, bioluminescent dinomastigote *Noctiluca miliaris* is carnivorous, capturing small plankton with an immense feeding tentacle with which it sweeps through the water. Dinomastigotes are significant intracellular photosynthetic endosymbionts in marine corals and sea anemones (A-4), giant clams (A-26), and foraminifera (Pr-3) in shallow, well-illuminated tropical and semitropical seas. Much contemporary research on "global warming" is directed toward understanding the temperature-adaptive abilities of diverse members of the *Symbiodinium microadriaticum* complex, the endosymbionts in corals.

Although dinomastigotes are undoubtedly eukaryotic, their nuclear organization is so idiosyncratic that they have been called mesokaryotic (between prokaryotic and eukaryotic). The DNA of nearly all other eukaryotes is complexed with histone proteins to form fibrils that are 10 nm wide. The DNA of dinomastigotes is complexed with very tiny quantities of a peculiar alkaline protein rather than with the four or five common eukaryotic histones to form fibrils that are only 2.5 nm wide. In this sense, then, dinomastigote chromatin is organized like that of prokaryotes.

Furthermore, rather than condensing only during mitosis, the chromatin of dinomastigotes is always condensed into chromosomes (Figures A and B). This is particularly strange, because the condensed-chromosome stages in animal and plant cells are just those in which the genome is turned off: RNAs and proteins are not synthesized while chromatin is condensed. However, the genes in the condensed chromatin of dinomastigotes are not turned off—the genes in these condensed chromosomes continue to be expressed. The typical mitotic stages (interphase, prophase, metaphase, anaphase, and telophase) are absent in dinomastigotes. In some, microtubules penetrate the nucleus during division; the kinetochores, which in plants and animals are attached directly to chromosomes, in dinos are embedded in the nuclear membrane, which remains intact during the division cycle. Chromatin is segregated to offspring cells by attachment to the nuclear membrane. The pattern of cell division differs from one dinomastigote species to another as if, within the phylum, mitosis evolved in its own peculiar fashion. Many dinos also form hard, resistant cysts. Before cyst formation, they may engage in fusion, which seems to entail mating and gene exchange of some kind.

Dinomastigotes have speciated in some remarkable ways. Some, *Protopsis*, *Warnowia*, and *Nematodinium*, have eyespots consisting of a layer of light-sensitive bodies containing carotenoid pigments overlain by a clear zone. In the sedentary species *Erythropsidium pavillardii*, a complex cell eye, the ocellus, has evolved. The ocellus, which detects the approach of prey, includes a lens and a fluid-filled chamber underlaid by a light-sensitive pigment cup. The development of *E. pavillardii* reveals that the pigment cup has evolved from a plastid.

Many dinomastigotes have life cycles that involve a nonmotile, vegetative, stage and a motile, dinospore, stage. Some dinos have an ameboid stage. Perhaps the most impressive case of dino life cycles is the recently characterized species, *Pfeisteria piscicida*, which is of great—if negative—economic importance to the fishing industry.

Pfeisteria rapidly change from one stage to another. They produce terrible toxins, forcing many fish that feed on them—such as Atlantic menhaden and southern flounder—to turn belly up in a few minutes, before they settle into the sediment and transiently disappear. In the continued presence of fish flesh, the typical dinomastigote swimmer cell transforms into a stellate test-covered ameba, one of more than a dozen guises. In the absence of food, these shelled amebas—dinos in disguise—sink to become invisible benthic cysts lying in wait. The volatile toxins, inhaled or absorbed through the skin, caused illness in the scientists who first studied these symbiotrophs. Dinomastigotes also cause diseases of significant economic importance in crabs.

Dinomastigotes have left a significant fossil record extending back to the base of the Cambrian period; there is some evidence that they existed even earlier, in the late Proterozoic eon.

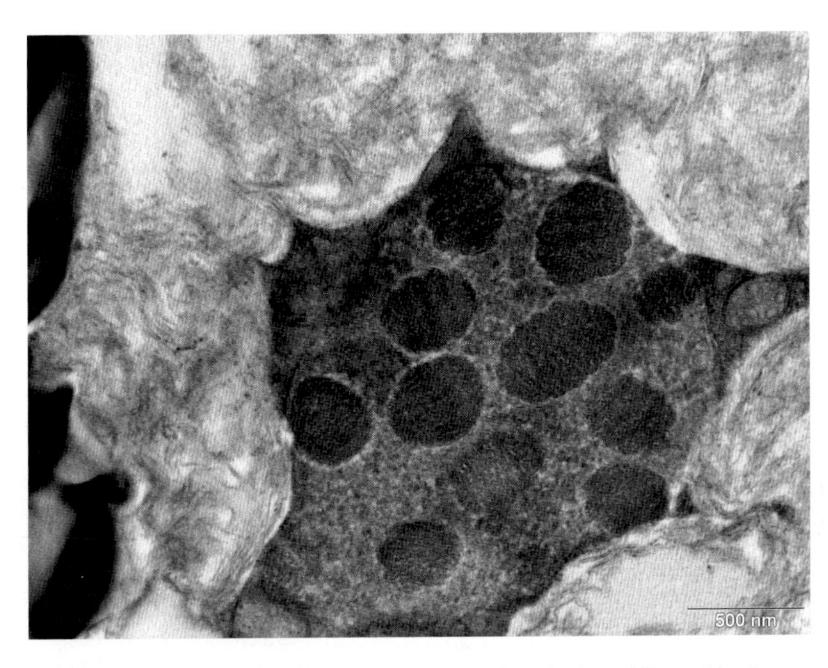

Figure A The nucleus of *Symbiodinium microadriaticum*, endosymbiont from the foraminifer an *Marginopora vertebralis*. Bar = 500 nm. [Courtesy of J. J. Lee.]

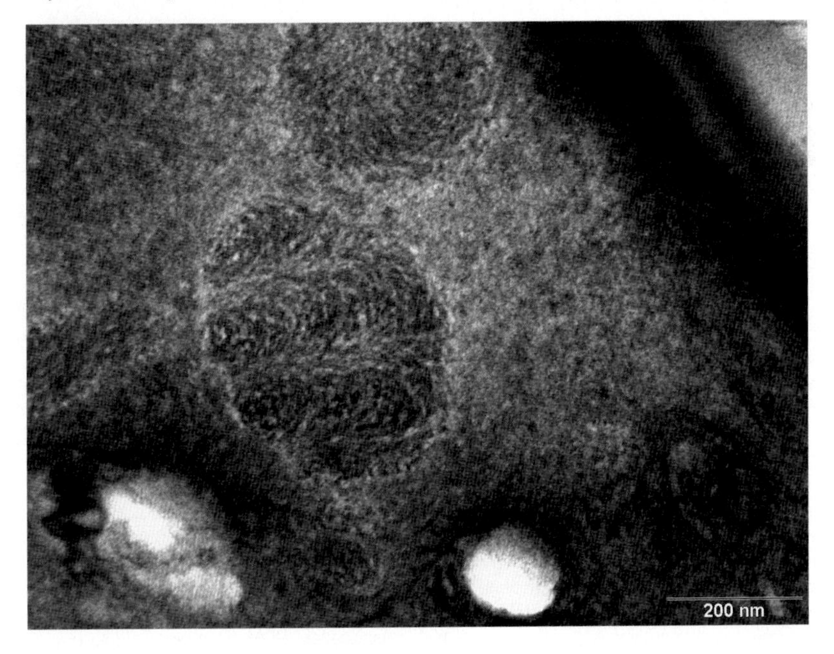

Figure B Chromosomes within the nucleus of *Symbiodinium microadriaticum*. The unusual structure of the chromosomes shows up only at high magnifications. Bar = 200 nm. [Courtesy of J. J. Lee.]

Pr-6 Ciliophora
(Ciliates)

Latin *cilium*, eyelash, lower eyelid; Greek *pherein*, to bear

GENERA
Balantidium
Blepharisma
Carchesium
Colpoda
Didinium
Dileptus
Ephelota
Euplotes
Gastrostyla
Halteria
Nyctotherus
Oxytricha
Paramecium
Pleurotricha
Prorodon
Spirostomum
Stentor
Stylonychia
Tetrahymena
Tokophrya
Vorticella
Uroleptus

Most ciliates, which are among the best-known protoctists, are bactivorous single cells. Ciliates are characteristically covered with cilia—short undulipodia with kinetosomes embedded in a tough, fibrillar outer cortex (proteinaceous cell layer) of the cell. Like the dinomastigotes (Pr-5) and apicomplexans (Pr-7), ciliates are alveolates, with pits embedded in their cortices. They possess two different types of nuclei, micronuclei and macronuclei, usually more than one of each kind. Nearly 10,000 freshwater and marine species have been described in biology literature. Probably many more exist in nature. Nearly all are phagotrophic, eating bacteria, tissue, or other protists, or they are osmotrophs, utilizing dissolved nutrients in rich waters.

The cilia of ciliates, like other undulipodia, including sperm tails, have the same ultrastructure, the ninefold symmetrical array of microtubules (the axoneme), with a kinetosome at its base. Cilia are modified to perform specialized locomotory and feeding functions. The most usual modification is the grouping of cilia and their underlying kinetosomes into cirri (bundles) or membranelles (sheets). Cirri or membranelles function as mouths, paddles, teeth, or feet. The ciliate undulipodia are embedded in an outer proteinaceous cell layer (the cortex) about 1 mm thick containing rows of kinetids (the kineties) comprising complex fibrous connections between them. Associated with each kinetid of the ciliate cortex is a parasomal sac, a small invagination of the plasma membrane used for osmotrophic nutrition.

Of the two types of nuclei in each ciliate, only the micronuclei, which apparently contain standard chromosomes, divide by mitosis. The macronuclei, which develop from precursor micronuclei in a series of complex steps, do not contain typical chromosomes. Instead, the DNA is broken into a great number of little chromatin bodies; each body contains hundreds or even thousands of copies of only one or two genes. Macronuclei are always required for growth and reproduction. They divide by elongating and constricting—not by standard mitosis. They take part in cellular functions such as the production of messenger RNA to direct protein synthesis. The micronuclei, not required for growth or reproduction, are dispensable, essential only for sexual processes unique to ciliates.

Most ciliates reproduce by transverse binary fission, dividing across the short axis of the cell to form two equal offspring. The anterior new cell is called the proter and the posterior one is the opisthe. However, certain stalked and sessile species, such as some suctorians, asexually bud off "larval" offspring. These offspring are "born": small rounded offspring, covered with cilia, emerge through "birth pores" of their entirely different-looking, stalked "mother."

Most ciliates undergo a sexual process called conjugation. The conjugants, two cells of compatible mating types ("sexes"), remain attached to each other for as long as many hours. Each conjugant retains some micronuclei and donates others to its partner. A series of nuclear fusions, divisions, and disintegrations follow, resulting in the two conjugants becoming "identical twins," as far as their micronuclei are concerned. The conjugants eventually separate and undergo a complex sequence of maturation steps. Although the micronuclei of the two exconjugants are now genetically identical (each conjugant having contributed equally), each new cell retains the cytoplasm and cortex of only one of the original conjugants. Because cytoplasmic and cortical inheritance in ciliates can be definitively distinguished from nuclear inheritance, these organisms are used in cell genetic analysis.

Ciliate classification has been revised dramatically in the past three decades because of the new information derived from rRNA sequencing studies and correlated with electron microscopy. "Holotrichs," ciliates with cilia over the entire surface, are not necessarily related—this formal name has been abandoned. Groups thought to be only very distantly related or unrelated, such as karyorelictans and stentors, are now known to be related, whereas organisms resembling each other superficially, such as *Euplotes* and *Stylonychia*, are more distantly related.

The most useful structure for the comparison of ciliates and the reconstruction of evolutionary history is the ciliate kinetid,

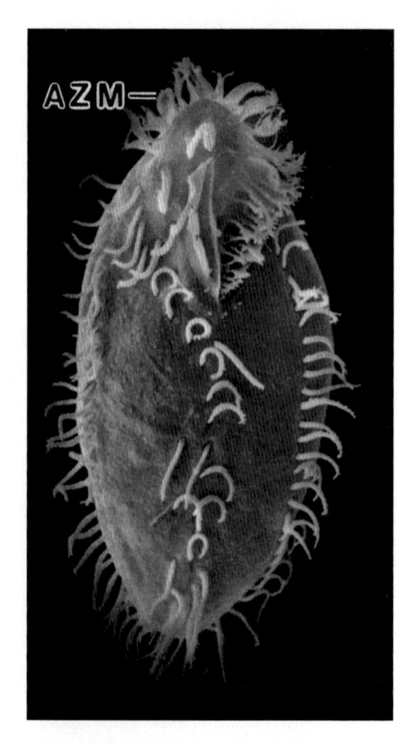

Figure A *Gastrostyla steinii*, a hypotrichous ciliate with a length of about 150 μm. The adoral zone of membranelles (AZM) is composed of ciliary plates each consisting of four ciliary rows. They sweep particulate food (bacteria and small ciliates) into the gullet. The cilia are condensed to bundles called cirri, whose arrangement is an important feature for classification. SEM. [Photograph courtesy of Foissner, W., Agatha, S., and Berger, H., Denisia 5:1–1459 (2002).]

S

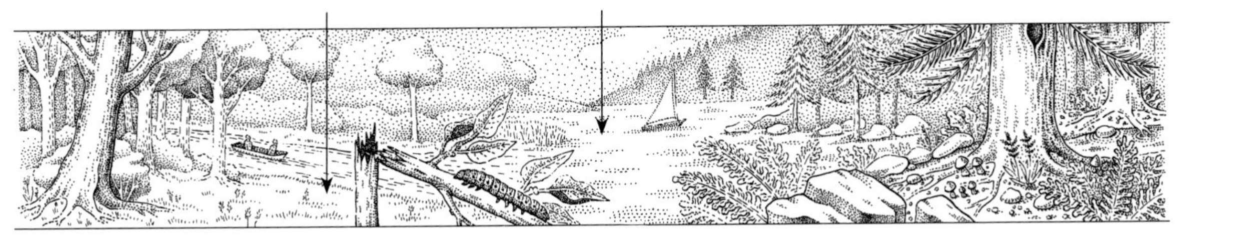

the structure consisting of one or more kinetosomes and their undulipodia, as well as the ribbons of microtubules and filaments—including kinetodesmal fibers—that surround them (Figure B). Kinetids are universal units of structure in all protoctist, animal, and plant cells that bear undulipodia. A kinetid with a single kinetosome is called a monokinetid, that with two kinetosomes is a dikinetid, and the rarer kinetid with many is called a polykinetid.

Presently, the Ciliophora are divided into two subphyla based on cortical ultrastructure and macronucleus division: the Postciliodesmatophora (classes Karyorelictea and Heterotrichea) and the Intramacronucleata (classes Litostomatea, Phyllopharyngea, Nassophorea, Colpodea, Spirotrichea, Armophorea, Plagiopylea, Prostomatea, and Oligohymenophorea).

Gastrostyla, an example of a spirotrich, is illustrated in Figure A. The subphylum Rhabdophora, ciliates that have kinetids with short kinetodesmal fibers and tangential transverse ribbons of microtubules, contains two classes: Prostomatea and Litostomea. Among many others, entodiniomorphs, bizarre-looking ciliates living as symbionts in the mammalian rumen, are classified in Litostomea. The third subphylum, Cyrtophora, contains four classes: Nassophorea, Phyllopharyngea, Colpodea, and Oligohymenophora. The cyrtophoran, unlike the rhabdophoran, disassembles its complex oral ciliature and makes two new ones in the process of cell division. Nearly all the well-known ciliates—*Colpoda* (class Colpodea: subclass Colpodida), *Tetrahymena* (class Oligohymenophora: subclass Scuticociliatida), *Vorticella* (class Oligohymenophora: subclass Peritrichia), *Paramecium* (class Nassophorea: subclass Peniculida), and *Stylonychia*, *Oxytricha*, and *Pleurotricha* (class Nassophorea: subclass Hypotrichia), as well as the subclass Suctoria (class Phyllopharyngea), belong in this great subphylum of ciliates.

Although many form spherical, resistant cysts, most ciliates lack hard parts and therefore do not fossilize; however, there is now good evidence for fossilized, soft-bodied ciliates in amber more than 100 million years old. The tintinnids, heterotrichs in subphylum Postciliodesmatophora, are exceptional marine ciliates that make shell-like structures from sand and organic cements. Their ancestors left evidence in the fossil record that they evolved before the Cretaceous period, some 100 mya.

Because of their various ciliary modifications, their rapid and controllable growth rates, and the ease with which they can be handled in the laboratory, ciliates are valuable for anatomical, genetic, and neurophysiological studies of single cells. Many ciliates are very harmful fish symbiotrophs, especially *Ichthyophthirius multifiliis* and therefore are of economic importance.

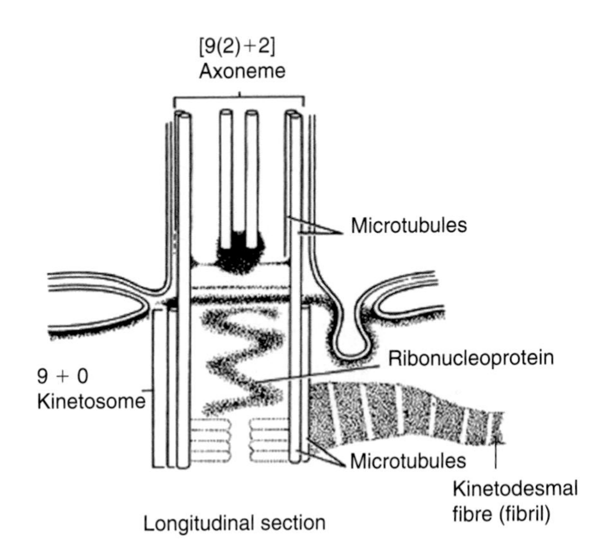

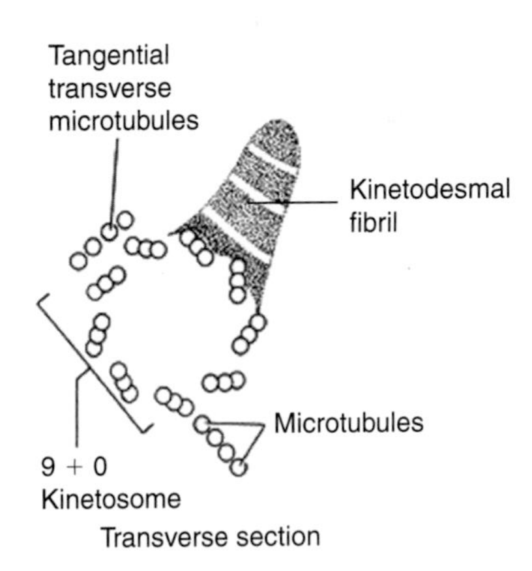

Figure B Kinetid reconstructed from electron micrographs. [Drawings by L. Meszoly.]

Pr-7 Apicomplexa
(Sporozoa, Telosporidea)

Latin *apex*, summit; *complexus*, an embrace, enfolding

GENERA
Babesia
Coccidia
Coelospora
Eimeria
Gregarina
Haemoproteus
Haplosporidium
Isospora
Plasmodium
Schizocystis
Selenidium
Toxoplasma

Members of Apicomplexa are single-celled symbiotrophs modified to penetrate tissue and obtain food from animals. All form spores. Unlike bacterial spores, apicomplexan spores are not heat- and desiccation-resistant cells; rather they are compact infective bodies that permit dissemination and transmission of the species from host to host. Along with phyla Pr-29, Pr-30, and A-2, apicomplexans have traditionally been grouped together as "sporozoa" because of their common habitat—animal tissue. Detailed studies of their structure, nucleic acid sequences, and proteins have firmly established the great differences that justify their separation into distinct phyla.

This phylum of alveolates is named for the "apical complex," a distinctive arrangement of fibrils, microtubules, vacuoles, and other cell organelles at one end of each cell. The apicomplexan group is probably monophyletic. Three classes are recognized: class Gregarinia, the gregarines, which includes *Gregarina*; class Coccidia, the coccidians, which includes *Eimeria* and *Isospora*; and class Hematozoa, the hemosporans and piroplasmids, which includes *Plasmodium*, *Haemoproteus*, and *Babesia*.

Apicomplexans reproduce sexually, with alternation of haploid and diploid generations. Both diploids and haploids can also undergo schizogony, a series of rapid cell divisions by mitosis that does not alternate with cell growth. Schizogony produces the small infective spores.

In fertilization, the undulipodiated male gamete (the microgamete) fertilizes a larger female gamete (the macrogamete) to produce a zygote (Figures A and B). The formation of the zygote is followed by the formation of a thick-walled oocyst (Figure C). Formation of the oocyst, rather than the infective spore, is the desiccation-, heat-, and radiation-resistant stage. The oocysts serve to transmit the microbes to new hosts. These cysts develop further by sporogony—rapid meiotic divisions inside the cyst produce infective haploid cells called sporozoites (Figures D and E).

The life cycles of apicomplexans may be complex and require various species of host. Many are bloodstream symbiotrophs. Many cause hypertrophy (gigantism) of the host cells in which they divide. The infection leads to duplication of host chromatin, causing a striking increase in the amount of host DNA, probably by polyploidization.

The coccidians are perhaps the best-known group of apicomplexans, because many of them cause serious and even fatal diseases of their animal hosts. *Isospora hominis* is the only coccidian that parasitizes humans, but others, such as *Eimeria*, affect livestock and fowl. Because these apicomplexans are generally acquired with food and thus find their way into the digestive tract, the major symptoms of coccidian disease are diarrhea and dysentery.

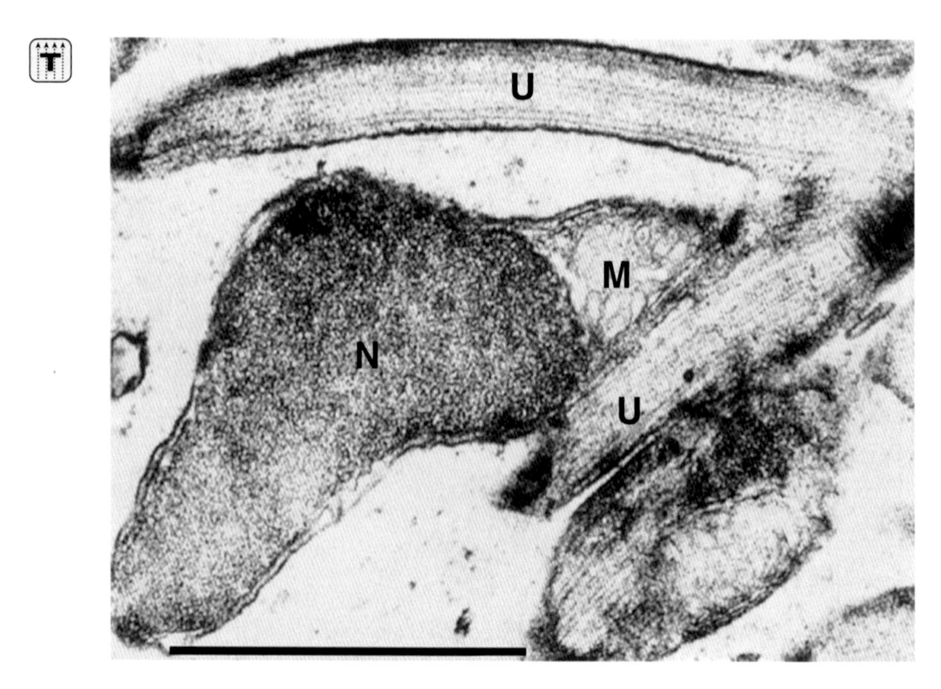

Figure A Microgamete ("sperm") kinetid of *Eimeria labbeana*, an intracellular symbiotroph of pigeons (A-37). N = nucleus; M = mitochondria; U = undulipodium; K = kinetosome. The structures above the nucleus are part of the apical complex. TEM, bar = 1 μm. [Courtesy of T. Varghese.]

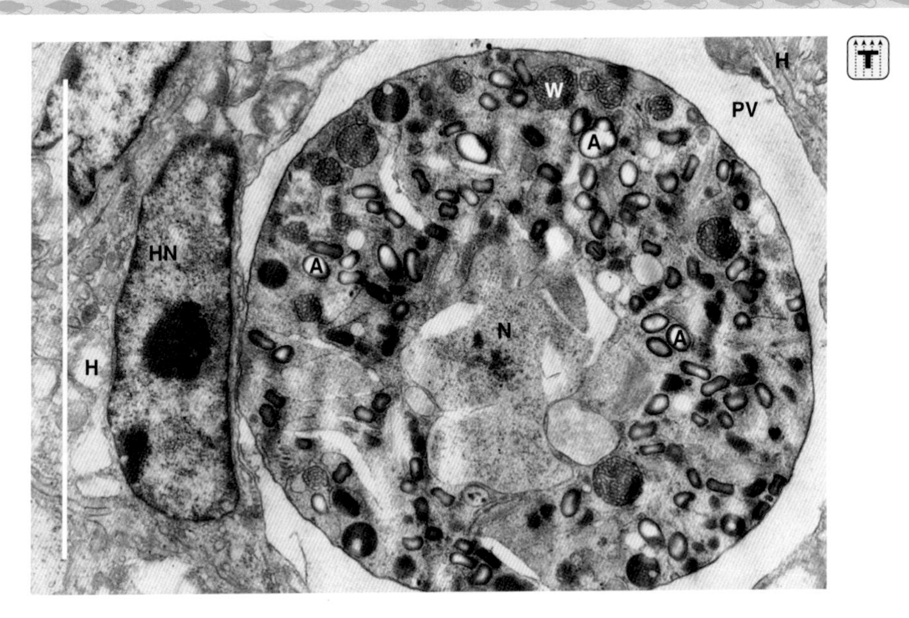

Figure B Macrogamete ("egg") of *Eimeria labbeana.* H = host cell; HN = host nucleus; PV = symbiotroph vacuole in host cell; N = macrogamete nucleus; A = amylopectin granule; W = wall-forming bodies, which later coalesce to form the wall of the oocyst. TEM, bar = 5 μm. [Courtesy of T. Varghese.]

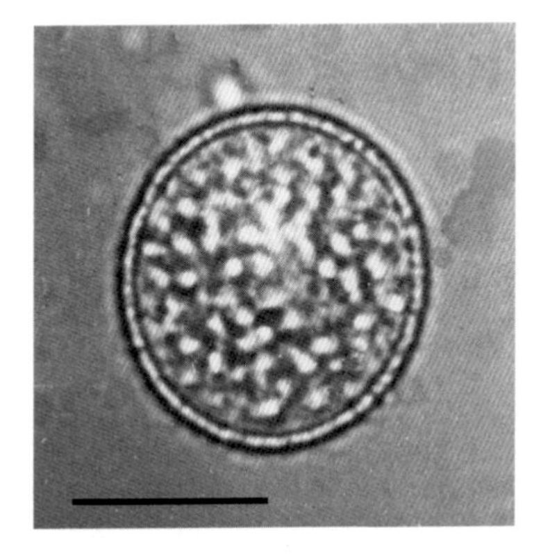

Figure C Unsporulated oocyst of *Eimeria falciformes*. LM, bar = 10 μm. [Courtesy of T. Joseph, *Journal of Protozoology* 21:12–15 (1974).]

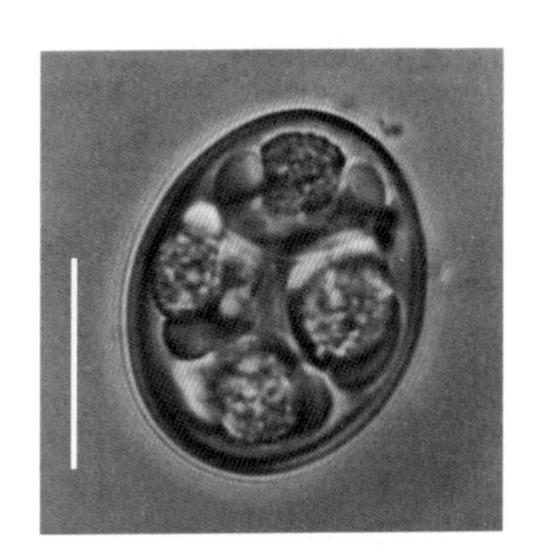

Figure D Four sporocysts of *Eimeria nieschulzi* in sporulated oocyst. LM, bar = 10 μm. [Courtesy of D. W. Duszynski.]

Pr-7 Apicomplexa

(continued)

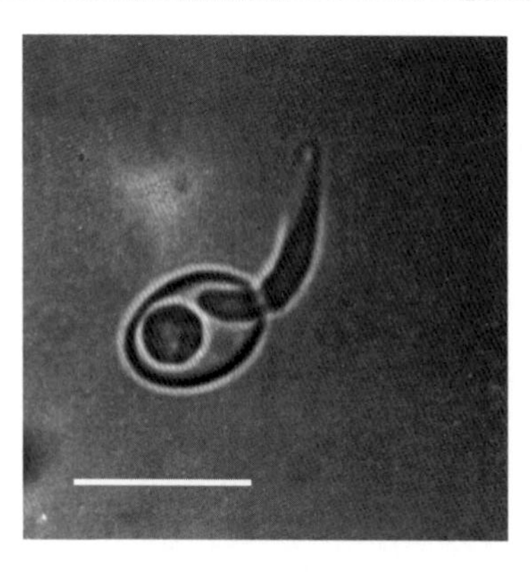

Figure E Sporozoite of *Eimeria indianensis* excysting from oocyst. LM, bar = 10 μm. [Courtesy of T. Joseph, *Journal of Protozoology* 21:12–15 (1974).]

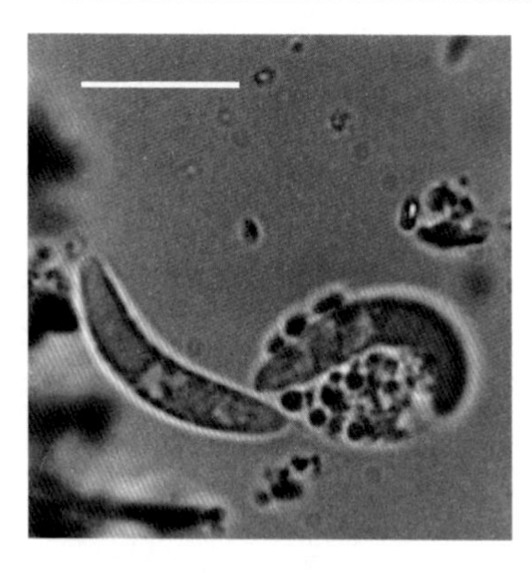

Figure F Free sporozoites of *Eimeria falciformes*. LM, bar = 10 μm. [Courtesy of D. W. Duszynski.]

An *Eimeria* infection begins when an oocyst is eaten. The oocyst germinates and produces sporozoites, which escape from the oocyst (Figures E and F) and enter the epithelial cells of the host, typically those of the gut lining, where they multiply by mitosis. Within the host cells, they develop into various forms called trophozoites, merozoites, and schizonts. The merozoites escape to infect more host cells. This cycle, called the schizogony cycle, can be repeated many times. Eventually, some of the merozoites within host cells develop into microgametes, which escape from the host cell. Other merozoites develop into macrogametes within host cells, where they remain. Fertilization takes place inside host cells. The resulting diploid zygotes undergo meiosis and develop into oocysts that typically exit from the anus with the feces (Figure G). Oocysts survive in soil until they are eaten by other potential hosts.

The most infamous apicomplexans important to human history are the malaria symbiotrophs, *Plasmodium* species. They are transmitted to humans by the female *Anopheles* mosquito (Phylum A-21, Mandibulata). Fertilization of *Plasmodium* takes place in the gut of the mosquito. The undulipodiated zygote embeds itself in the gut wall, where it transforms into a resistant oocyst. Within the oocyst, infective cells are formed by meiosis and sporogony (multiple fission). The sporozoites migrate to the salivary glands of the mosquito. With the bite of the mosquito, the sporozoites are injected into the human bloodstream, where they infect red blood cells. Inside a blood cell, a sporozoite develops further to become a feeding stage, the trophozoite, which grows at our expense—the sporozoan diet requires iron obtained from human hemoglobin. The trophozoite eventually undergoes schizogony to produce merozoites, small infective cells that escape from the ruined blood cell into the bloodstream. The merozoites attack and penetrate more blood cells, develop into trophozoites, and divide into more merozoites. After several such cycles, the merozoites differentiate into male and female gametes. These gametes must be taken up from the blood into the *Anopheles* mosquito gut—as they are when the mosquito draws in blood during its bite—to complete the fertilization stage of the *Plasmodium* life cycle. All the merozoites in a human host are produced and released more or less simultaneously—the pulse of formation and release of successive generations causes the characteristic periodic attacks of malarial fever.

There is a baffling variety of sexual life cycles and attack strategies in this group of protoctists; an enormous and contradictory terminology has made their study an arcane delight for specialists—especially for those interested in veterinary medicine.

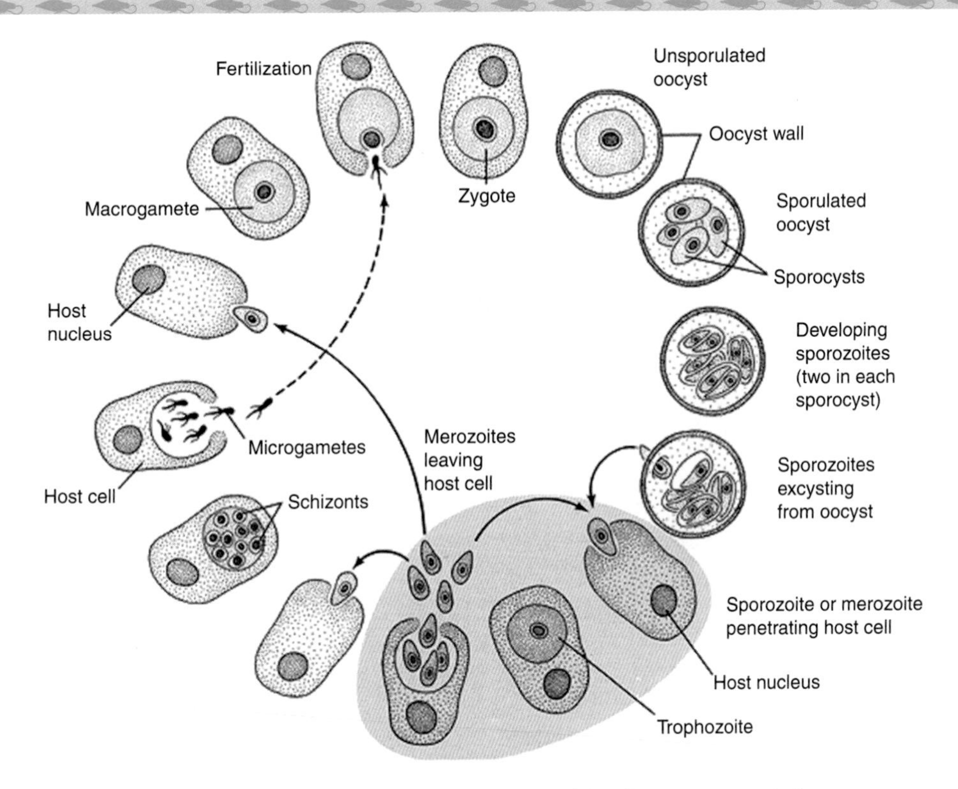

Figure G The life history of *Eimeria* sp. The shaded part of the diagram represents the schizogony cycle, which may repeat itself many times before some of the merozoites differentiate into gametes. [Drawing by L. Meszoly.]

SUBKINGDOM (Division) HETEROKONTA

Pr-8 Bicosoecida

GENERA
Acronema
Cafeteria
Pseudobodo

In the bicosoecid, a marine organism that bears a shell and two undulipodia, one undulipodium extends forward from the cell and is mastigonemate—it bears numerous tiny appendages called tinsel or flimmer. The other lies along the surface of the cell and is smooth. The bicosoecid kinetid is unmistakable; it has two rows of associated microtubules, one long (ten tubules) and one short (three tubules). Bicosoecids tend to form colonies. Sexuality has not been reported in this group. One newly described species, *Acronema sippewissettensis* (Figure A), is a tiny healthy eater that tolerates a few days of desiccating conditions by forming a walled stage. It has been grown in culture on bacterial food. Coming from a sulfurous mud flat, it easily survives anoxia. These colorless mastigotes resemble some of the chrysomonads (Pr-15, Figure D, page 170), and some chrysomonads may have evolved from bicosoecids by acquiring plastids. Alternatively, bicosoecids may have evolved from chrysomonads by loss of plastids.

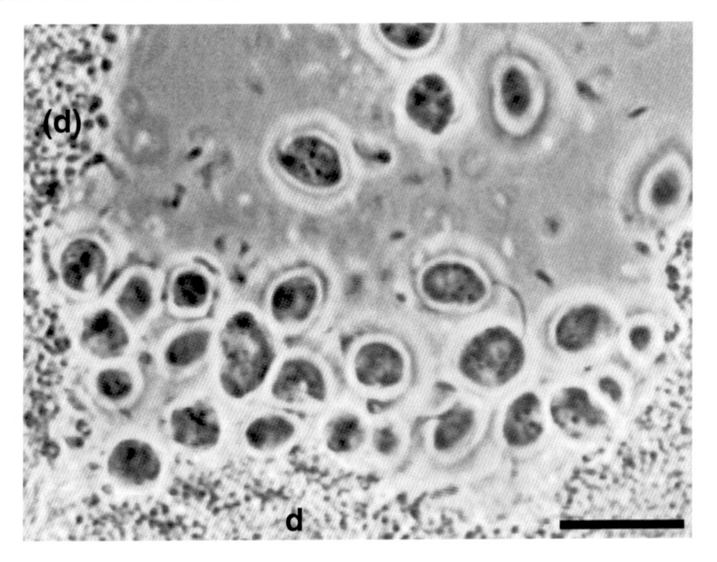

Figure A *Acronema sippewissettensis*. Lively mastigotes, recently emerged from weeks in their contracted desiccated (d) state. At the edge of the salt marsh, along with their food bacteria, the mastigote cells stop swimming as they lose water. They persist in clumps with bacterial spores (d) probably for at least a season. LM, bar = 10 μm.

Pr-9 Jakobida

GENERA
Andalucia
Histiona
Jakoba
Reclinomonas

Jakobids are small, and to visualize them electron microscopy is often required. Jakobids are mononucleate, biundulipodiated bacterivores. The anteriorly inserted undulipodia are equal in length. Although neither undulipodium bears hairs or scales, one undulipodium is directed forward and the other trails in typical heterokont style. The trailing undulipodium is pressed along, but not actually attached to, the ventral surface of the cell. A cytoplasmic flap, called the "lip," forms a groove on the right ventral side of the cell, and the posterior undulipodium lies in this groove. The groove serves as a mouth only in the sense that bacteria are ingested in it; there is no cytostome. Binary fission cell division is longitudinal as in mastigotes generally, and at least one offspring is an active swimmer with a reduced or no lip. Sexuality is unknown in the group. Only three genera have been reported. Cells of the genus *Jakoba* lack a lorica, or shell. An inhabitant of marine sediments, *Jakoba* may attach to the sediment surface by its anterior undulipodium. Freshwater mastigotes in the genus *Reclinomonas* have a hyaline lorica, which is covered with spiny scales. The cell resides dorsal side down inside the lorica—hence the name, *Reclinomonas. Histiona* also dwells in a lorica in freshwater habitats. The hyaline lorica of *Histiona* lacks scales; the cell lies upside down, with its posterior protruding from the lorica. The protruding "lip" on the posterior ventral side of the cell forms a sail-like extension (Figures A and B).

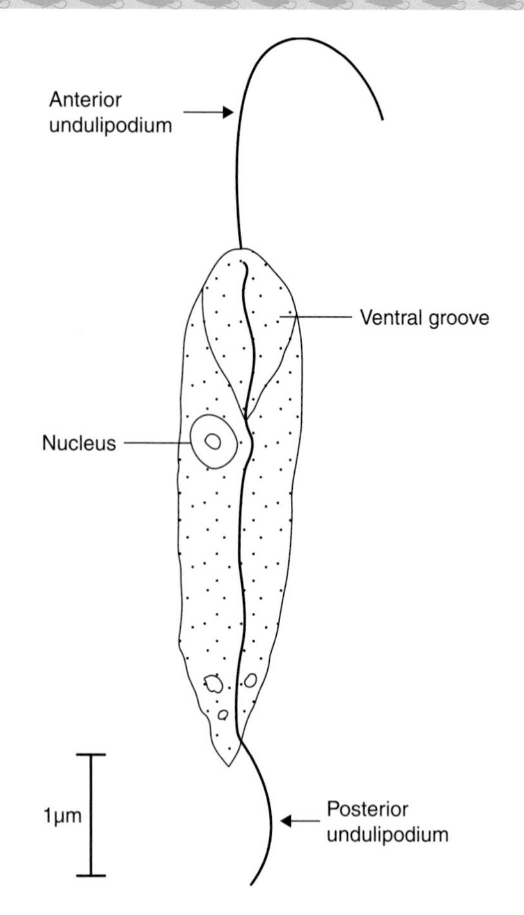

Figure A Structure of *Jakoba libera*. Bar = 1μm. [Drawing by M. Santiago.]

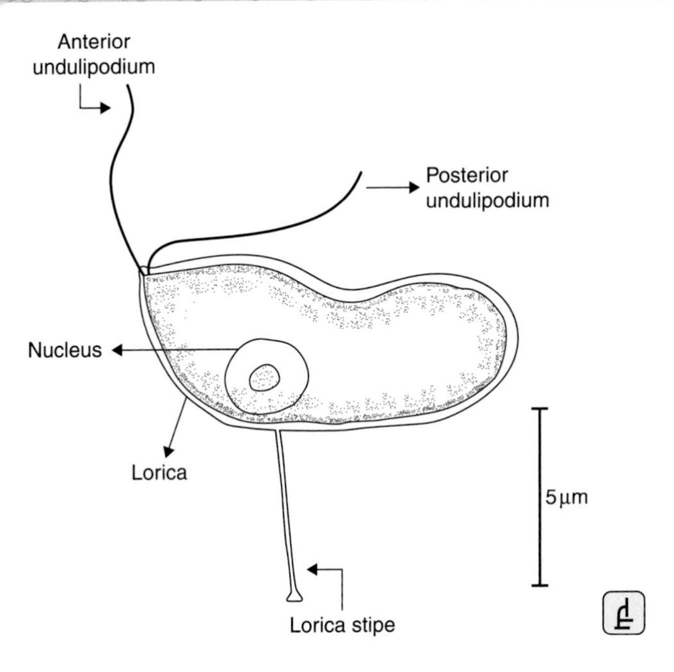

Figure B Structure of Reclinomonas Americana. Bar = 5 μm. [Drawing by M. Santiago.]

Pr-10 Proteromonadida

Greek *protos*, first; Latin *monas*, unit

GENERA
Proteromonas
Karotomorpha

Proteromonads, with two genera—*Proteromonas* and *Karotomorpha*—and five species, are small mastigotes limited to the posterior intestine of amphibians, reptiles, and some mammals, especially rodents. The proteromonad is a nonmastigonemate, heterokont cell with a rhizoplast that connects the nucleus and large Golgi body to the anterior kinetosomes. Proteromonads form resistant cysts that pass from the intestines to the soil. At least one species undergoes multiple fission (Figure A).

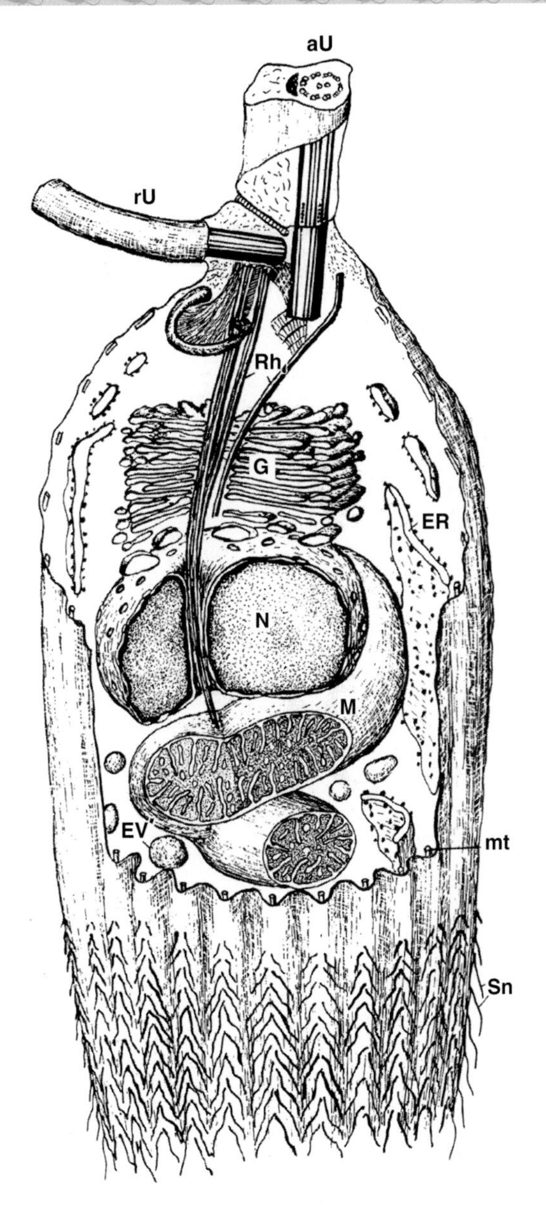

Figure A *Proteromonas*, diagrammatic reconstruction of its ultrastructure. In *Proteromonas*, the pair of kinetosomes is attached by a complex of fibers to the rhizoplast fiber (Rh) which traverses the golgi apparatus (G) and abuts on the mitochondrion (M) which lies under the nucleus (N). *Proteromonas* possess characteristic hairs, or somatonemes (Sn), covering the surface of the posterior part of the cell; they are inserted on the membrane in front of subpellicular microtubules (mt). The anteriorly directed undulipodium (aU) of *Proteromonas* has a dilated shaft containing microfibrils and a striated fiber parallel to the axoneme. Endoplasmic reticulum (ER); endocytotic vacuole (EV); recurrent undulipodium (rU).

Pr-11 Kinetoplastida

Greek *kineto*, pertaining to motion

GENERA
Blastocrithidia
Bodo
Crithidia
Cryptobia
Dimastigella
Herpetomonas
Ichthyobodo
Leishmania
Leptomonas
Phytomonas
Rhynchobodo
Rhynchomonas
Trypanoplasma
Trypanosoma

Kinetoplastid mastigotes are abundant in nature, either as free-living consumers of bacteria in freshwater, marine, and terrestrial environments or as symbiotrophs of animals, flowering plants, and, rarely, of other protists. Most kinetoplastids are fewer than 50 μm in length. The Bodonida (free-living kinetoplastids and some related symbiotrophs) have two heterodynamic undulipodia, one serving for forward propulsion, and the other trailing behind as a skid (Figure A). The majority of the symbiotrophic kinetoplastids belong to the Trypanosomatida and have a single undulipodium. In all kinetoplastids, the undulipodia

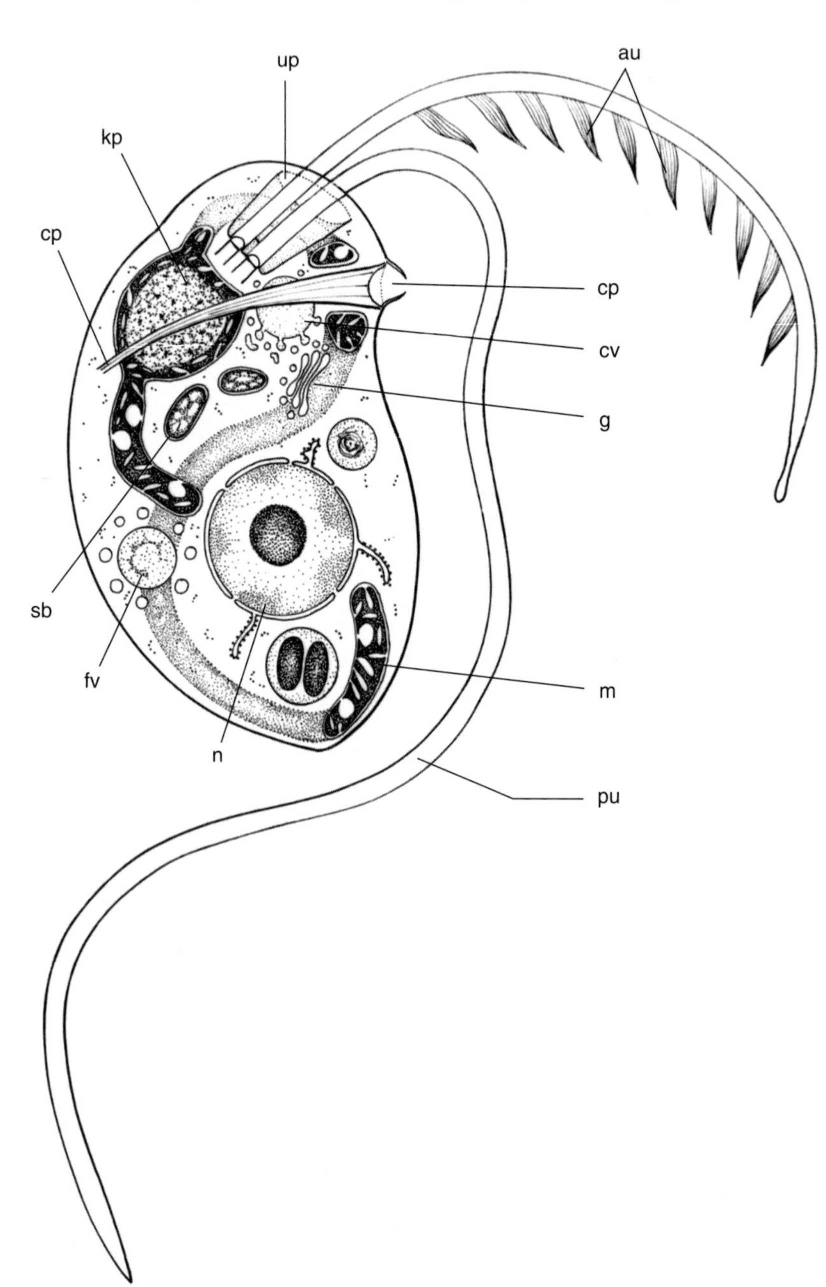

Figure A Structural features of *Bodo saltans* a common free-living kinetoplastid, based on electron microscopy. au = anterior undulipodium; cp = cytopharynx; cv = contractile vacuole; up = ciliary pocket; fv = food vacuole; g = Golgi; kp = kinetoplast; m = hooplike mitochondrion; n = nucleus; pu = posterior undulipodium; sb = symbiotic bacterium.

arise from a pocket, usually opening at the anterior end of the body, but in some trypanosomatid genera (*Trypanosoma* and *Blastocrithidia*), the pocket opening lies more toward the posterior end and the emergent undulipodium is attached along the body to the anterior end where it becomes free; beating of the attached undulipodium gives the appearance of an "undulating membrane" (Figure B). Kinetoplastids have been most studied as the causative agents of serious diseases transmitted by biting insects. In Africa, species of *Trypanosoma* transmitted by the tsetse fly cause sleeping sickness in human and nagana in cattle. In South America, *Trypanosoma cruzi* causes Chagas disease of humans, vectored by blood-sucking bugs (Hemiptera). Throughout the tropics and subtropics, species of *Leishmania* transmitted by sandflies (*Phlebotomus* spp.) cause dermal and visceral leishmaniasis in humans. Bug-transmitted *Phytomonas* species living in the phloem of plants cause fatal wilting in oil and coconut palms, also in coffee plants.

The kinetoplastids are set apart from other eukaryote organisms by distinctive organizational features. Most notable of these is the presence of a single, often branched mitochondrion, the kinetoplast, whose DNA is hugely amplified to form a stainable mass (or masses), usually located close to the base of the kinetid (Figure C, see page 147) . The circular DNA molecules composing the kinetoplast encode not only mitochondrial enzymes and mitochondrial rRNAs but also "guide RNAs" involved in editing defective transcripts of the mitochondrial enzyme genes. In trypanosomatids the circular molecules are catenated into a network linked via the mitochondrial membrane to the undulipodium base. In bodonids the kDNA molecules are not catenated. They are dispersed throughout the mitochondrion. The mitochondrion of kinetoplastids usually has discoid cristae, but in *Trypanosoma brucei*, causative agent of human sleeping sickness in Africa, the mitochondrion undergoes cyclical changes

Figure B Bloodstream form of *Trypanosoma brucei*, causative agent of human sleeping sickness. The undulipodium is attached to the body along most of its length and in beating deforms the body to give the appearance of an "undulating membrane." SEM, bar = 1 μm.

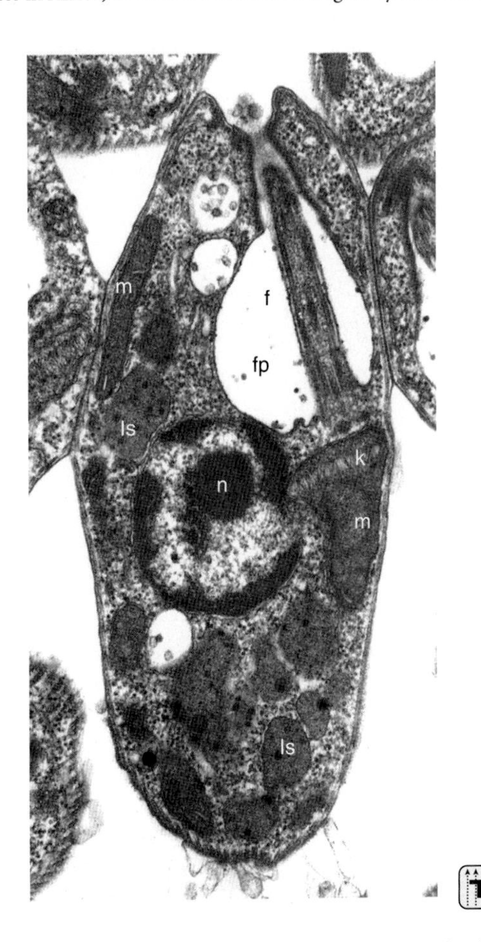

Figure C A longitudinal section through the ciliary pocket (fp), undulipodium (f), nucleus (n), and kinetoplast (k) of *Leishmania major*, causative agent of dermal leishmaniasis in humans. The kinetoplast consists of a network of interlocked circular DNA molecules and is embedded in a capsular region of the single reticular mitochondrion (m). ls = lysosome. TEM, bar = 0.5 μm.

in respiratory activity during the symbiotroph's life cycle through mammal and tsetse fly, and these changes are associated with switches from discoid to tubular cristae or loss of cristae altogether (Figure D). Another uniquely kinetoplastid feature is the segregation of enzymes of the glycolytic chain in membrane-bounded organelles, called glycosomes (in other eukaryotes glycolysis takes place in the cytoplasmic matrix). Storage carbohydrates are absent from kinetoplastids.

Other distinctive features of the Kinetoplastida include undulipodia which have a latticelike paraxial rod paralleling the axoneme, use of the undulipodium as an attachment organelle in many symbiotrophs, and a cytoskeleton composed of cortical microtubules. Endocytosis and exocytosis of macromolecules and insertion of new membrane into the body surface are confined to the ciliary pocket, although in phagotrophic bodonids and some trypanosomatids, a pocket-associated, microtubule-reinforced cytopharynx may also be responsible for ingestion.

The nucleus of kinetoplastids harbors a large number of chromosomes that do not condense during mitosis when they are partitioned on an intranuclear spindle not associated with the kinetosomes. There is evidence from molecular genetics of a sexual process in the life cycles of *Trypanosoma brucei* and *T. cruzi*. Meiosis and syngamy are believed to take place in the insect vector, just before the trypanosomes differentiate into the infective metacyclic stage (Figure D) that invades the mammal, but cytological evidence for these processes is still lacking. Moreover, unlike the malaria symbiotrophs (*Plasmodium* spp.), a sexual process in the vector is not necessary for completion of the trypanosome life cycle. With the exception of one bodonid (*Dimastigella trypaniformis*), sexuality remains to be discovered among the rest of the Kinetoplastida. Gene-sequencing studies and common structural features show that the Kinetoplastida are closely related to the Euglenoida and share a common ancestor that acquired a photosynthetic eukaryote symbiont that was later lost by the kinetoplastids, whose body size became reduced. Symbiotrophy has evolved several times in the evolution of the kinetoplastids.

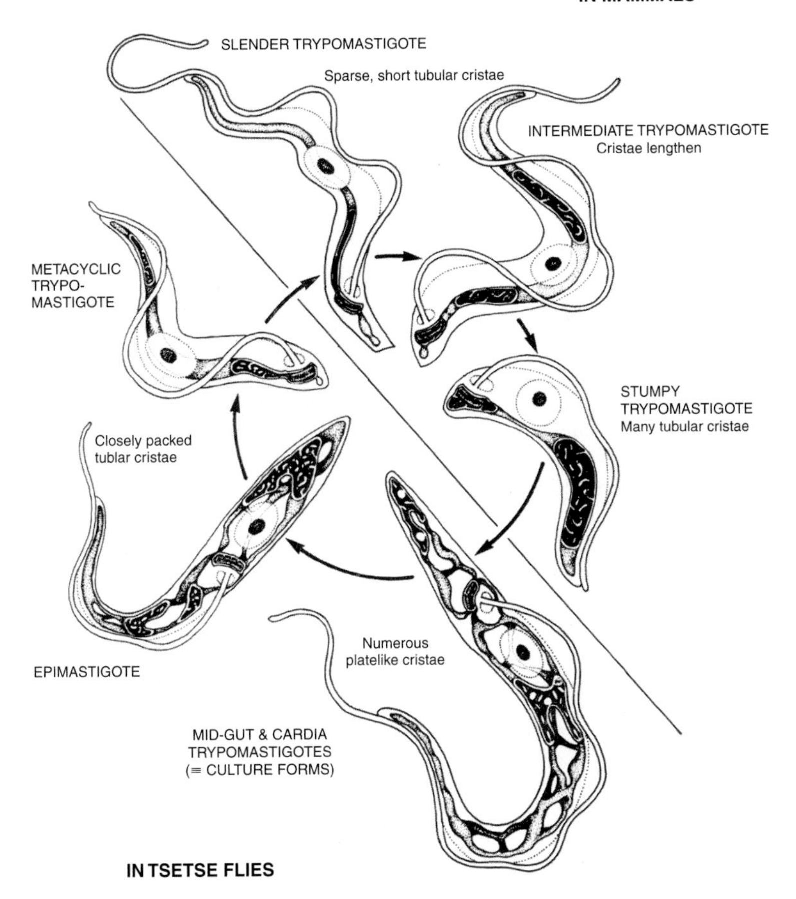

Figure D Diagram showing stages in the developmental cycle of *Trypanosoma brucei* in the mammalian host and in the tsetse fly (*Glossina* spp.) vector. The simple linear mitochondrion is inactive with few tubular cristae in the slender mammalian bloodstream trypanosome when the symbiotroph derives its energy from glucose by glycolysis. In the tsetse fly midgut, the mitochondrion becomes an active network with discoid cristae as the symbiotroph switches to utilizing the amino acid proline as a source of energy. Mitochondrial activation commences in the nondividing (stumpy) bloodstream trypanosome, whereas later stages in the development of the symbiotroph (epimastigote, metacyclic trypomastigote) in the vector's salivary glands show signs of progressive mitochondrial repression before being returned to the mammal as the metacyclic trypanosome when the fly bites a mammal, injecting trypanosomes in its saliva. [illustration courtesy of K. Vickerman.]

Pr-12 Euglenida

GENERA

Anisonema	*Dolium*	*Klebsiella*	***Ploeotia***
Ascoglena	*Dylakosoma*	*Lentomonas*	*Protaspis*
Astasia	***Entosiphon***	***Lepocinclis***	***Rhabdomonas***
Atraktomonas	***Euglena***	*Marsupiogaster*	*Rhizaspis*
Calkinsia	*Euglenamorpha*	***Menoidium***	*Rhynchopus*
Calycimonas	*Euglenopsis*	*Metanema*	*Scytomonas*
Clautriavia	***Eutreptia***	***Monomorphina***	*Sphenomonas*

Euglenids share a common ancestry with kinetoplastids and are thought to have evolved from free-living bacteriotrophs. One species, *Petalomonas catnuscygni*, has the characteristic feature of a euglenid (pellicle) and the defining feature of the kinetoplastids (a kinetoplast). The most likely explanation of the genesis of green, phototrophic, plastid-bearing euglenids is that a eukaryotrophic euglenid such as *Peranema* acquired a plastid secondarily by feeding on a green alga. This scenario is bolstered by the observation that an additional membrane surrounds the plastids of euglenids, likely the relict of the phagocytic event that entrapped them.

Although euglenids and green algae (Phylum Pr-28, Chlorophyta) both possess grass-green chloroplasts, they differ significantly from chlorophytes, even in their chloroplast pigmentation. The plastids of euglenids, like those of chlorophytes and plants, contain chlorophylls ***a*** and ***b*** only, as well as beta-carotene and the carotenoid derivatives alloxanthin, antheraxanthin, astaxanthin, canthoxanthin, echinenone, neoxanthin, and zeaxanthin. In addition, although euglenid plastids have the carotenoid derivatives diadinoxanthin and diatoxanthin, they are not present in chlorophytes (or in plants). Euglenids lack rigid cellulosic walls; instead, they have pellicles, finely sculpted outer structures made of protein strips underlain by microtubules. The pellicles are usually very flexible and many euglenids are capable of changing cell shape. Euglenids do not store starch; instead they store paramylon, a glucose polymer having the β-1-3 linkage of the monosaccharides.

Euglenid reproduction is not sexual; all attempts to find meiosis and gametogenesis in euglenids have failed. The nuclei of different individual organisms of the same species may have different amounts of DNA. The nuclei contain large karyosomes, also called endosomes, which are structures homologous to the nucleoli of other cells. Like nucleoli, the endosomes are composed of RNA and protein combined in bodies that are precursors to the ribosomes of the cytoplasm. Euglenids have a mitotic spindle composed of subspindles, each of which contains relatively few intranuclear microtubules. Each subspindle appears to function separately and attaches to and segregates a few of the chromosomes that have the appearance of granules. In euglenids, the nucleolus and the nuclear envelope remain intact throughout division. Many lack distinct, countable chromosomes; in cell division, their chromatin granules do not move in a single mass as in standard anaphases. The chromatin granules do not split at metaphase; rather, no metaphase plate is formed, and each granule autonomously proceeds to one of the nuclear poles, where the newly replicated undulipodia are located. Before nuclear division, the two kinetids move toward the anterior end of the cell, causing the cell to distort in a way characteristic of division as the cell divides lengthwise.

The species *Euglena gracilis* (Figure A) has proven to be a fine tool for analyzing cell organelles. Most species of *Euglena* possess chloroplasts, but in the case of the colorless species *Euglena longa* (syn. *Astasia longa*), a functional chloroplast has been permanently lost and only a remnant plastid with a highly reduced plastid genome remains. In other cases *Euglena gracilis* cells can be experimentally rendered colorless and later develop fully functional green plastids. Thus, the effect of light, chemical inhibitors, temperature, and many other agents on the development of chloroplasts and other organelles can be beautifully observed. For example, if the cells are placed in the dark, the chloroplasts regress; after a little growth, the euglenids turn

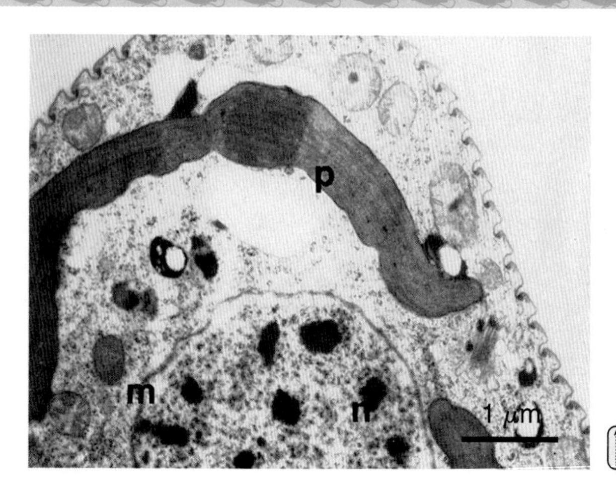

Figure A A thin section of *Euglena gracilis* grown in the light, showing the well-developed chloroplast (p). m = mitochondrion; n = nucleus. TEM, bar = 1 μm. [Courtesy of Y. Ben Shaul.]

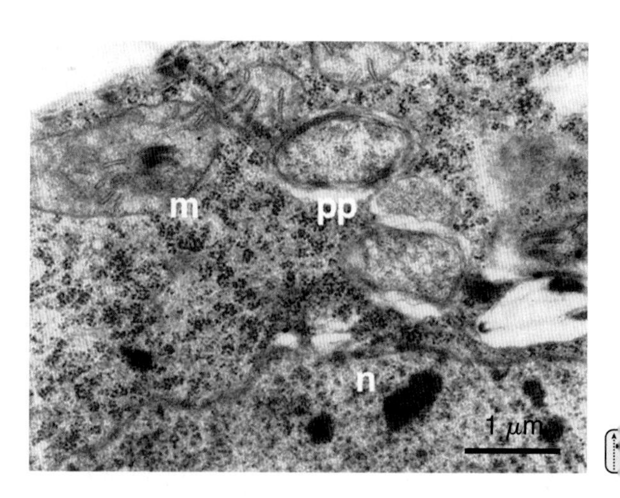

Figure B The same strain of *Euglena gracilis* as that shown in the previous figure, grown for about a week in the absence of light. The chloroplasts dedifferentiate into proplastids (pp). This process is reversible: proplastids regenerate and differentiate into mature chloroplasts after about 72 hours of incubation in the light. m = mitochondrion; n = nucleus. TEM, bar = 1 μm. [Courtesy of Y. Ben Shaul.]

Colacium	***Eutreptiella***	***Notosolenus***	***Strombomonas***
Cryptoglena	*Gyropaigne*	*Parmidium*	*Triangulomonas*
Cyclidiopsis	*Hegneria*	*Pentamonas*	***Trachelomonas***
Dinema	*Helikotropis*	***Peranema***	*Tetreutreptia*
Dinematomonas	*Heteronema*	***Petalomonas***	*Tropidoscyphus*
Discoplastis	*Jenningsia*	***Phacus***	***Urceolus***
Distigmopsis	***Khawkinea***		

white and become entirely dependent on external food suppliers for growth (Figure B). The "animal" cells can be reconverted into "plants." If dark-grown colorless euglenas are reilluminated, they turn light green within hours. Their chloroplasts go through a series of developmental changes induced by the light, and after around three days, they recover their bright green color.

Euglena gracilis is the only species known that can be genetically "cured" of chloroplasts without killing the organism. Although mutants of other photosynthetic protoctists and plants that lack the capacity for photosynthesis can be produced, *Euglena* is the only chloroplast-containing organism that survives and reproduces independently of chloroplast DNA. If *E. gracilis* cells are treated with ultraviolet light or with a number of other treatments to which the chloroplast genetic system is more sensitive than is the nucleocytoplasmic system, the genetic entities responsible for chloroplast development can be lost permanently. The euglenas then lose all their plantlike characteristics and become irreversibly dependent on food. By this treatment, the metabolism of the nucleocytoplasm can be studied in detail separately from that of the plastid.

Pr-13 Hemimastigota

Greek hemi, half; Greek mastigophora, bearing undulipodia

GENERA
Hemimastix
Paramastix
Spironema
Stereonema

The Hemimastigota or Hemimastigophora were recognized as a phylum only in 1988 when a new soil mastigote was described, viz., *Hemimastix amphikineta* (Figures D, F, G). Further research discovered some new taxa and revealed that several mastigotes, previously assigned to other protistan groups, belong to the Hemimastigota. Thus, the Hemimastigota now comprise four genera with a total of eight species. However, the wide ecological range indicates that there are more species waiting to be discovered.

The systematic home of the Hemimastigota is still in discussion, but two features indicate the euglenids as their nearest relatives. First, several Hemimastigota show euglenid metaboly, a special kind of movement not found in any other protistan group. Second, the Hemimastigota have, like the euglenids, a diagonal (rotational) symmetry of the cortical plates composing their pellicle (Figures D, E, G).

The Hemimastigota are small to middle-sized (10–60 μm), ellipsoidal to vermiform protists with a slight anterior constriction producing a head-like capitulum containing the transient mouth opening (Figures D, F, H). They have two rows of undulipodia, shorter or as long as the body, in more or less distinct furrows located laterally, where the cortical plates abut (Figures A–G). The basal bodies are monokinetids, each associated with a membranous sac, a short microtubule ribbon, a long microtubule ribbon, and nine filamentous arms (transitional fibres) forming a distinct basket. The cortex is unique, i.e., composed of two folded plates with diagonal (rotational) symmetry (Figure E); it is supported by microtubules with genus-specific arrangement and a thick granular layer (epiplasm), except in the regions along the undulipodia rows. There is a single nucleus with a prominent central nucleolus persisting throughout division. The contractile vacuole is near the posterior end (Figures A, C, D). The mitochondria have tubular to saccular cristae. The capitulum contains some complex, bottle-shaped extrusomes composed of a cylindroidal posterior and a rod-like anterior compartment. The Hemimastigota divide longitudinally and in free-swimming condition.

The Hemimastigota are not only remarkable for their morphology but also for their ecology and geographic distribution. There are species living in running and stagnant waters (*Stereonema geiseri, Spironema multiciliatum*), in the lake plankton (*Paramastix conifera*), and in soil (*Hemimastix amphikineta, Spironema terricola, S. goodeyi*). The most remarkable species is *Hemimastix amphikineta* (Figures D, F, G) because it is restricted to the southern hemisphere (Australia, Central and South America, Antarctica), where it is common and lives in a variety of habitats, such as mineral soil, leaf litter, moss, and the mud of bromeliad tanks. The same distribution pattern is known, inter alia, from several ciliates and many testate amoebae, showing that the split of the supercontinent Pangaea into Gondwana (~southern hemisphere) and Laurasia (~northern hemisphere) some 150 million years ago deeply influenced not only the distribution of vascular plants and higher animals, but also of single-celled organisms. Another species, *Spironema terricola* (Figure A), has been found as yet only in soil from the Grand Canyon, USA.

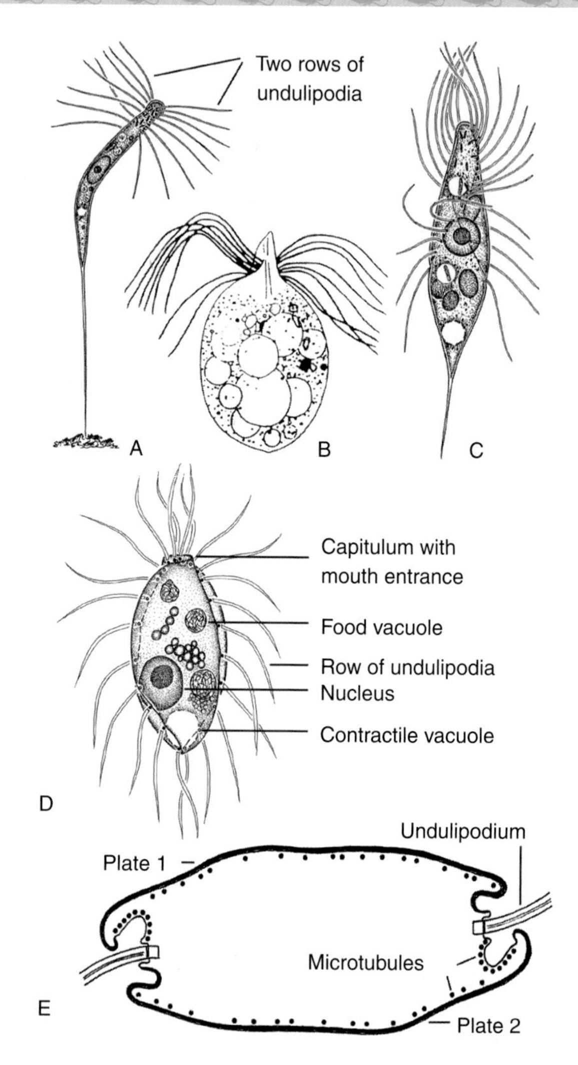

Figures A–E Hemimastigophoran mastigotes. A: *Spironema terricola*, length 40 μm. B: *Paramastix conifera*, length 15 μm. C: *Stereonema geiseri*, length 25 μm. D: *Hemimastix amphikineta*, length 17 μm. E: Schematized transverse section in the transmission electron microscope, showing that the cortex is composed of two plicate plates with diagonal (rotational) symmetry. [Courtesy of W. Foissner.]

Little is known on nutrition of the Hemimastigota, but the complex extrusomes in the capitulum and observations on *Hemimastix* indicate that they are rapacious carnivores feeding on minute, heterotrophic mastigotes. However, definite oral structures are not recognizable, and the mouth opening is a transient structure in the centre of the capitulum (Figure H).

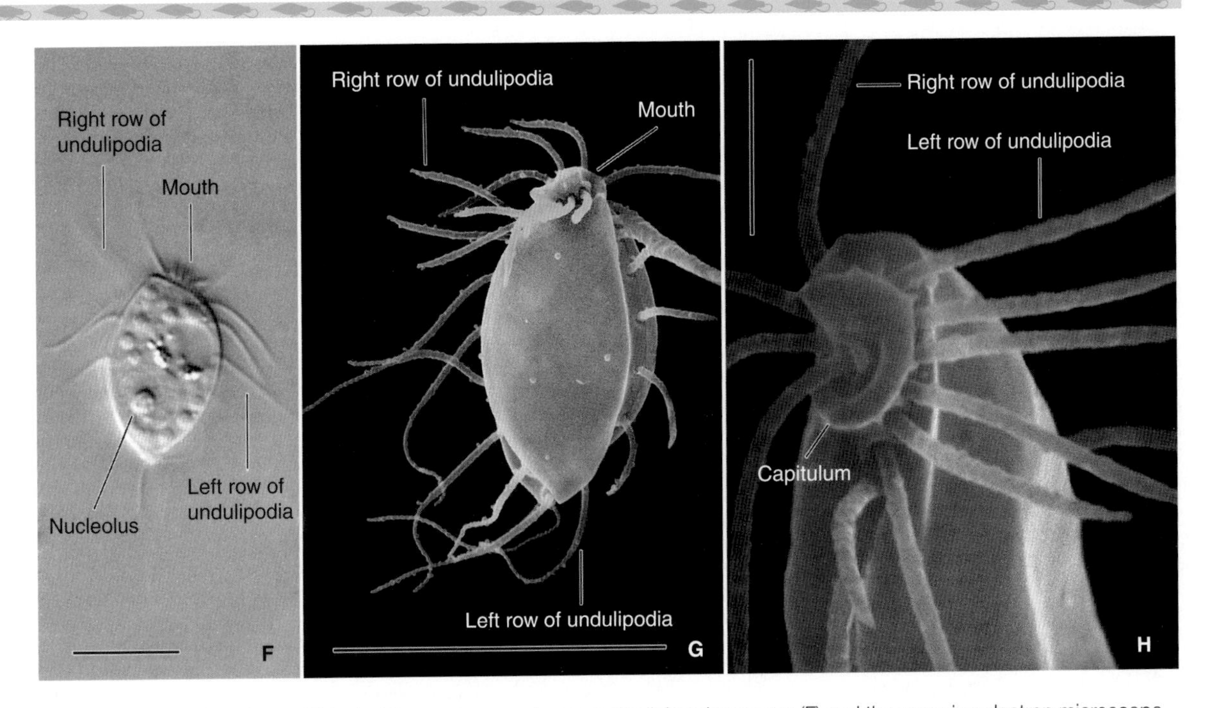

Figure F–H *Hemimastix amphikineta*, Venezuelan specimens in the light microscope (F) and the scanning electron microscope (G, H). F, G: Broad side views showing body shape and the two long rows of undulipodia, which make the organism looking like a ciliate. Bars 10 μm. H: Narrow side view of anterior body third showing the capitulum which contains the transient mouth. Bar 2 μm. [Courtesy of W. Foissner.]

Pr-14 Hyphochytriomycota

Greek *hyphos*, web; *chytra*, little earthen cooking pot; *mykes*, fungus

GENERA
Anisolpidium
Canteriomyces
Hyphochytrium
Latrostium
Rhizidiomyces

Hyphochytrids, with chytrids (Pr-35) and oomycotes (Pr-21), have traditionally been considered fungi. These aquatic osmotrophs do resemble fungi in their mode of nutrition, which may be either symbiotrophic or saprobic. Thin threads or filaments invade host tissue or dead organic debris where they release digestive enzymes and absorb the resulting nutrients (Figure A). However, they differ from fungi in that they produce undulipodiated cells. The kinetid structure, often inferred from the detailed description of motility, distinguishes these "mastigote molds" or "zoosporic fungi." The single anteriorly directed undulipodium that confers rapid swimming, quick changes in direction, and a wide spiral path is enough to distinguish hyphochytrids.

Hyphochytrids either are symbiotrophs on algae and fungi or live on insect carcasses and plant debris. The body, or thallus, of a hyphochytrid may be holocarpic (that is, where the entire body converts into a reproductive structure) or eucarpic (a part of the thallus develops into a reproductive structure while the remaining part continues its somatic function). In the holocarpic species, the thallus is formed inside the tissues of the host. In these species, the thallus consists only of a single reproductive organ bearing the branched rhizoids—rootlike tubes that penetrate the substrate—or hyphal feeding tubes that grow out of the substrate, in which cross walls, or septa, may or may not develop. The thallus in eucarpic species may reside on the surface rather than inside the host tissues.

From the reproductive organ, or zoosporangium, zoospores emerge through discharge tubes. Hyphochytrid zoospores are very active swimmers. True to their classification as stramenopiles, each bears one mastigonemate anterior undulipodium that moves in helical waves as well as in whiplash fashion (Figure B). The zoospores swim to new food sources. They stop swimming, encyst, withdrawing their undulipodia, and produce cell walls as they develop again into a thallus. Asexual reproduction by zoospores is the norm—and explains their need for aquatic habitats. Neither sexuality nor resistant spore formation has been confirmed.

All hyphochytrids live in freshwater and have been isolated primarily from soil and tropical freshwaters. However, they are probably distributed all over the world wherever their hosts are found. The five genera containing 23 species here are grouped into three classes: Anisolpidia, Rhizidiomycetae, and Hyphochytria.

The best-known hyphochytrid, *Rhizidiomyces apophysatus*, exploits water molds, such as *Saprolegnia* (Phylum Pr-21, Oomycota). After *Rhizidiomyces*, zoospores come to rest on their victims; they transform into spheres, withdrawing their undulipodia. The sphere does not divide but germinates, extending a germ tube into the host. With continued growth, the tube ramifies into a branching system of rhizoids in the host tissue. Between the rhizoids and the surface of the host, a swelling develops into a baglike structure called the apophysis of the sporangium. The sporangium itself grows at the outer end of the apophysis; it enlarges and forms an exit papilla—a raised bump with a hole in it—that becomes a discharge tube. Karyokineses (nuclear divisions) take place in the sporangium to form a plasmodial mass. Mitotic division figures in hyphochytrids are unmistakably distinguished from those in plasmodiophorans (Pr-20), oomycotes, and chytrids. The mass of multinucleate protoplasm passes through the discharge tube and emerges from it. The protoplasm then cleaves into a mass of individual zoospores that develop outward-directed undulipodia. The zoospores swim away to begin the cycle again (Figure C).

The details of the life cycles of most hyphochytrids are not known, partly because they are difficult to observe. *Hyphochytrium catenoides*, for example, grows mainly on the

Figure A Filamentous growth of *Hyphochytrium catenoides* on nutrient agar. LM, bar = 0.5 μm. [Courtesy of D. J. S. Barr.]

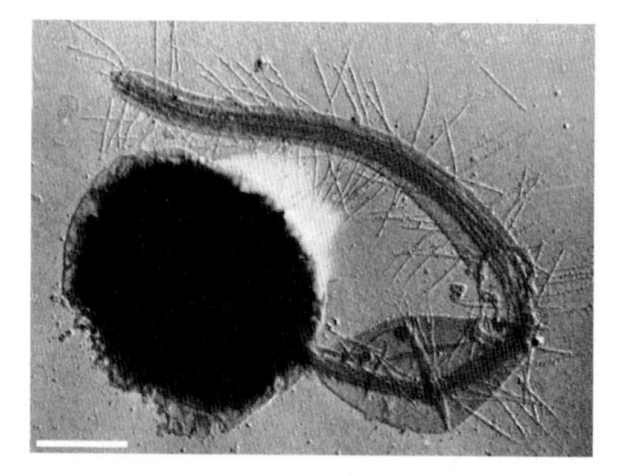

Figure B Zoospore of *Rhizidiomyces apophysatus*, showing mastigonemate undulipodium (right). TEM (negative stain), bar = 1 μm. [Courtesy of M. S. Fuller.]

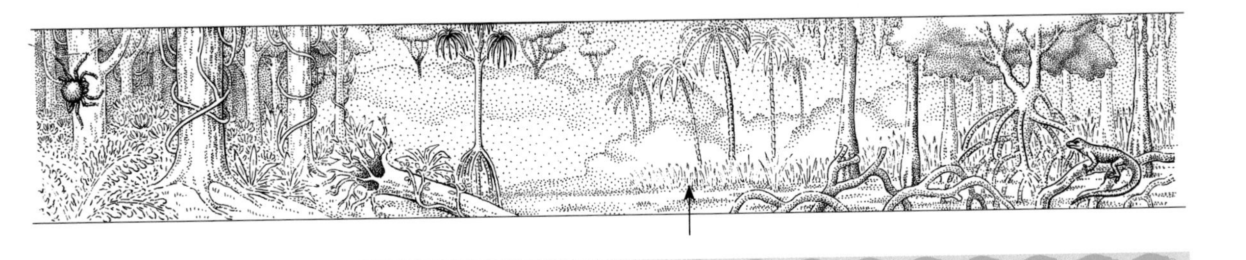

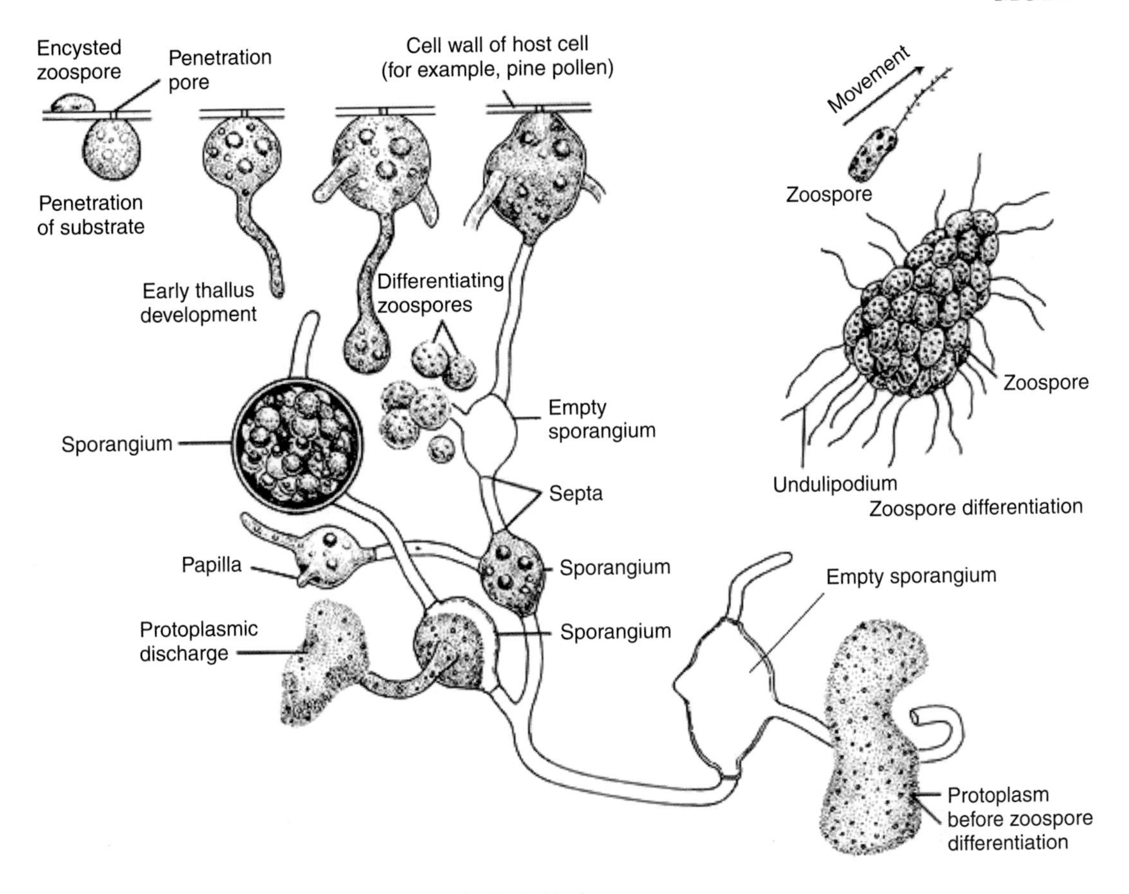

Figure C Life cycle of *Hyphochytrium* sp. [Drawing by R. Golder.]

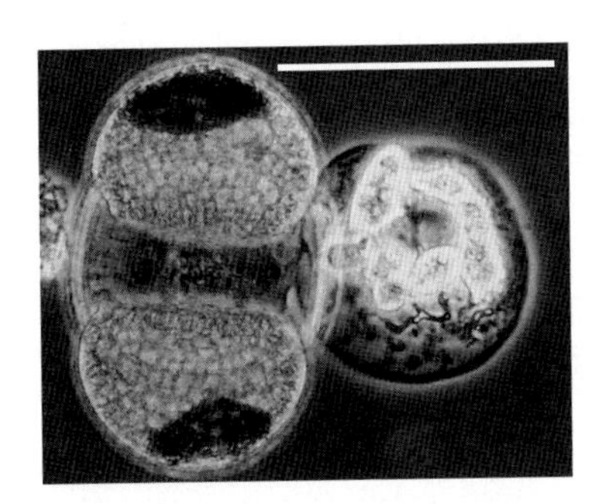

Figure D Sporangium (right) of *Hyphochytrium catenoides* on a ruptured pine pollen grain (left; PI-10). LM, bar = 0.5 μm. [Courtesy of D. J. S. Barr.]

pollen of conifers; the thallus stages develop inside empty pollen grains (Figure D). The cell walls of the hyphochytrids that have been studied are composed of chitin, but, in some species, they contain cellulose as well.

Pr-15 Chrysomonada

(Chrysophyta)

Greek *chrysos*, golden; *phyton*, plant

GENERA

Anthophysa
Chromulina
Chrysamoeba
Chrysocapsa
Chrysococcus
Chrysolepidomonas
Chrysosphaerella
Dinobryon
Epipyxis
Hibberdia
Hydrurus
Lagynion
Mallomonas
Ochromonas
Oikomonas
Paraphysomonas
Phaeoplaca
Spumella
Synura
Uroglena

The definition of the Chrysomonada (Chrysophyta) has changed markedly over time, especially in the last few decades when new results from electron microscopy, biochemistry, and molecular systematics became available. Whereas the Chrysomonada are now considered to represent a separate phylum within the larger grouping of the stramenopiles (Figure A), several taxa that were included in this phylum by earlier workers have now been transferred to other stramenopilan phyla. Thus, genera such as *Gonyostomum* and *Vacuolaria*, which in the previous editions of this book were included in the Chrysomonada, are now classified in the separate phylum Raphidomonada (Raphidophyta); the silicomastigotes (silicomastigotes) and pedinellids are now included in the Dictyochophyta; several coccoid and filamentous freshwater genera with distinct cell walls (for example, *Phaeothamnion*, *Phaeoschizochlamys*, and *Tetrasporopsis*) are now accommodated in the Phaeothamniophyta; some larger filamentous or saccate marine genera such as *Giraudyopsis* and *Phaeosaccion* were transferred to the Chrysomerophyta; some monadoid, coccoid, palmelloid, sarcinoid, and filamentous marine genera (for example, *Ankylochrysis*, *Aureococcus*, *Chrysocystis*, *Chrysonephos*, *Pelagococcus*, and *Sarcinochrysis*) to the Pelagophyta; and the marine unicellular loricate genus *Polypodochrysis* to the Pinguiophyta. Sometimes genera such as *Mallomonas* and *Synura* (Figures B and C) were also classified in a separate phylum, but as shown by molecular studies, they are most closely related to the Chrysomonada (Chrysophyta) and therefore are often retained in this phylum, as a separate class Synurophyceae, whereas the Chrysomonada *sensu stricto* represent the class Chrysophyceae.

Chrysomonads (both Chrysophyceae and Synurophyceae) occur in a wide range of freshwater habitats from polar regions to the tropics. Most commonly they are encountered in the plankton from lakes to small pools and in bogs. In temperate regions, many species have their main occurrence in spring when waters are cool, often just after the melting of the ice. There are also a few reports of chrysomonads growing on soil and snow. A few marine species are also known.

The class Chrysophyceae as conceived here comprise just over 100 described genera and some 800 species and the class Synurophyceae 4 genera and about 160 species. Additionally, a great number of infraspecific taxa have been described. A few further enigmatic organisms have also been tentatively assigned to this group (for example, Parmales, a group of silica-scaled oceanic organisms to the Chrysophyceae and genera such as *Conradiella*, *Jaoniella*, and *Pseudosyncrypta* to the Synurophyceae). Further studies must show if these really have a closer relationship with this group.

The phylum Chrysomonada mainly includes single-celled and colonial mastigotes ("flagellates"), but genera with primarily amoeboid (for example, *Chrysamoeba* and *Lagynion*), capsoid (*Chrysocapsa* and *Hydrurus*), coccoid (*Chrysosphaera*), or pseudoparenchymatous (*Phaeoplaca*) organization also occur. A few plasmodial genera (*Chlamydomyxa* and *Myxochrysis*) have also been tentatively classified in this phylum, but their taxonomic position remains uncertain.

Most members of the Chrysomonada are colored golden brown. They usually contain one or two parietal chloroplasts (rarely more). Several colorless genera also exist (for example, *Oikomonas*, *Paraphysomonas*, *Spumella* and *Anthophysa*); in some a leucoplast is present in place of the chloroplast. Chloroplast pigments usually include chlorophyll ***a*** and ***c*** (in Chrysophyceae chlorophyll c_1 and c_2, and in Synurophyceae chlorophyll c_1 only). A wide range of carotenes and xanthophylls is also present, with fucoxanthin normally as most dominant pigment giving the organisms the characteristic golden brown coloration. The chloroplast is surrounded by two membranes of endoplasmic reticulum that in typical Chrysophyceae (but not in Synurophyceae) is normally confluent with the outer membrane of the nuclear membrane. The characteristic cellular storage product is a beta-linked glucan (chrysolaminarin/leucosin). It is stored as a solution, usually in one or more large vacuoles near the posterior end of the cell (Figure D).

As typical for stramenopiles, the swimming cells normally have two anteriorly attached undulipodia of unequal sizes, one of which (the longer one) is bearing tripartite hairs (mastigonemes). This undulipodium is immature with respect to ciliary transformation. It is directed anteriorly and is beating in a sine wave. The shorter (mature) one is directed laterally in many chrysophyceans; in some genera (for example, *Chromulina*, *Hibberdia*, and *Oikomonas*), it is vestigial. In synurophycean genera it may also be vestigial (in most species of *Mallomonas*), whereas in *Synura* the second (smooth) undulipodium has almost the length of the hairy one, but has a stiffer beat. The two basal bodies form an angle of approximately 90° in most Chrysophyceae; in Synurophyceae they are more or less parallel. In many Chrysophyceae, the short undulipodium has a swelling at the base that is closely appressed to the cell surface underlain by the anterior end of the chloroplast and an eyespot (if present). The eyespot and the swelling at the base of the undulipodium are commonly referred to collectively as the photoreceptor

Figure A A new larger grouping including 20 phyla, from Chrysomonada (Chrysophyta) (Pr-15) through Hyphochytriomycota (Pr-14), has been established on the basis of similarity in gene sequences, which suggests that they have common ancestry. The most characteristic feature of the organisms of these phyla is the occurrence of cells with tripartite, hairy (mastigonemate) undulipodia in the heterokont style (anteriorly attached and of unequal lengths). These phyla are called stramenopiles, "straw bearers," referring to the hollow hairs that decorate their undulipodia. This larger grouping has also been formally described as kingdom Stramenopila (or sometimes including the Cryptomonada as kingdom Chromista). The Stramenopiles, as presently conceived, comprise 5 phyla of colorless organisms (Bicosoecida, Slopalinida, Labyrinthulata, Oomycota, and Hyphochytriomycota) and 15 phyla of pigmented organisms (Chrysomonada through Bolidophyta). The pigmented groups are sometimes collectively called Heterokontophyta or Chromophyta or Ochrophyta.

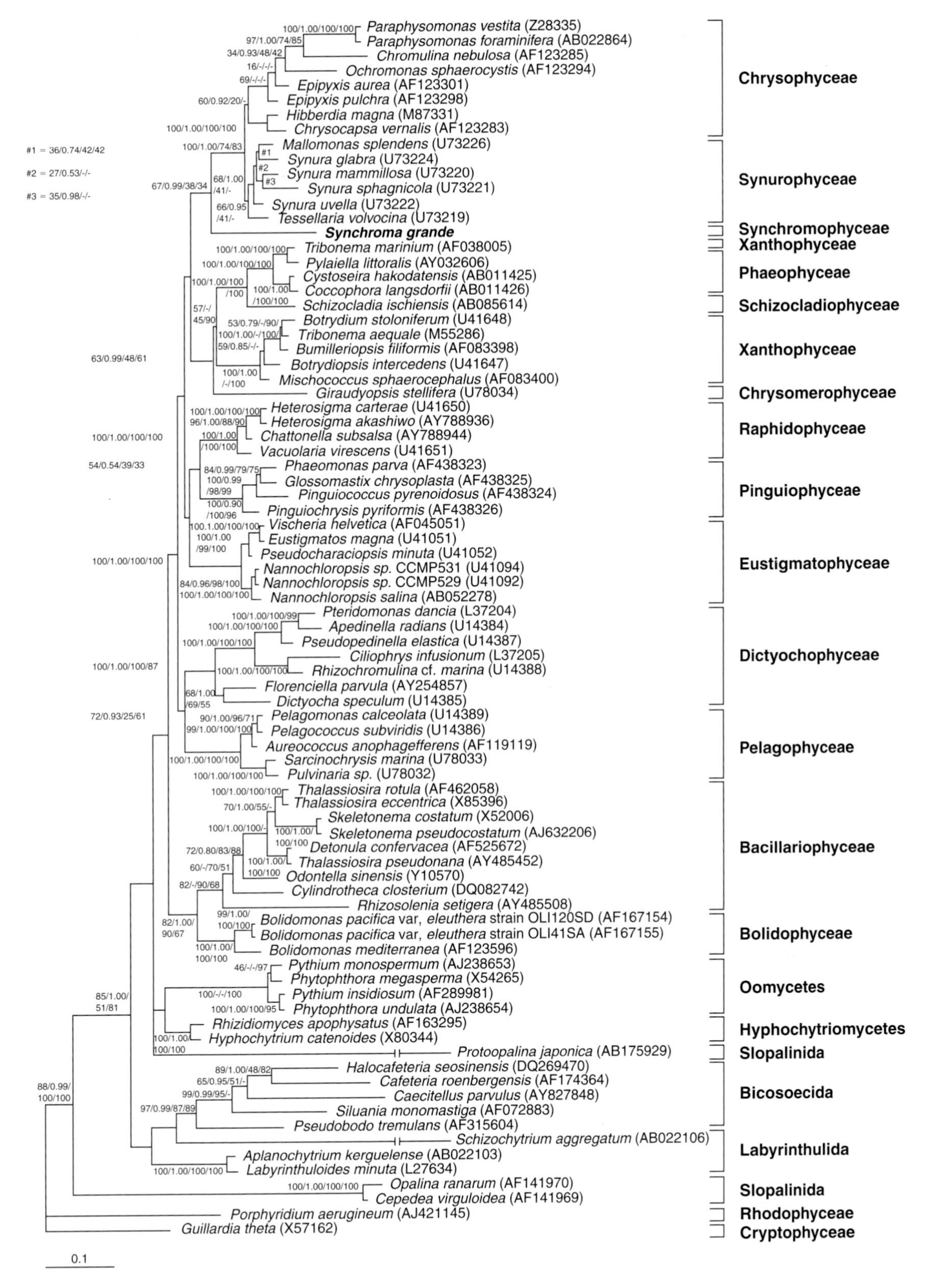

Phylogenetic tree of 18S rRNA gene analyses including the different phyla of Stramenopiles. Maximum likelihood tree based on a 1952 bp alignment (including gaps), rooted with a Cryptophyceae sequence. Bootstrap and Bayesian probability values at branches for maximum-likelihood (ML), Bayes, maximum parsimony (MP), and neighbor-joining (NJ) are separated by slashes. Accession numbers of species included in the analyses are shown in round brackets. Adapted from Horn et al. 2007 (Protist 158, page 280, Figure 2).

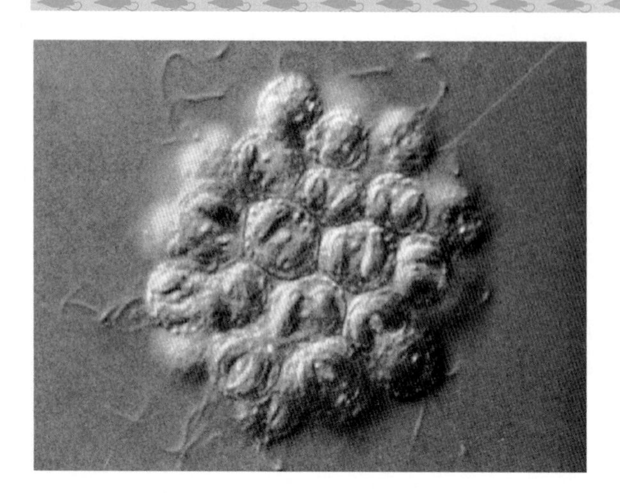

Figure B *Synura* sp., a living freshwater colonial chrysomonad from Massachusetts. LM; each cell is about 18 μm in diameter. [Courtesy of S. Golubic and S. Honjo.]

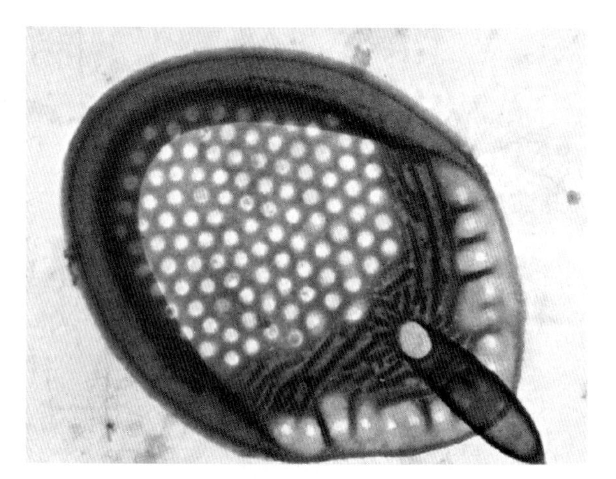

Figure C A siliceous surface scale from a member of the *Synura* colony shown in Figure B. SEM; greatest diameter is about 1 μm long. [Courtesy of S. Golubic and S. Honjo.]

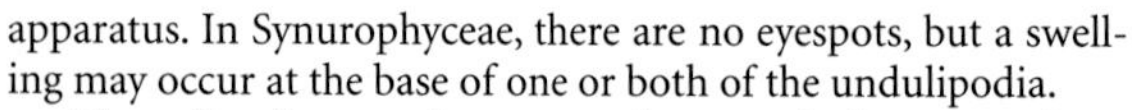

apparatus. In Synurophyceae, there are no eyespots, but a swelling may occur at the base of one or both of the undulipodia.

The cells of most chrysomonads are naked or sometimes embedded in a gelatinous matrix; in some genera, the cells are covered by organic or sometimes mineralized thecae or loricae (for example, *Chrysococcus*, *Dinobryon*, *Epipyxis*, and *Lagynion*), or rarely cell walls (*Chrysosphaera* and *Phaeoplaca*). Some genera are characterized by a cell-covering of organic scales (*Chrysolepidomonas*) or biradially symmetrical siliceous scales (*Chrysosphaerella*, *Paraphysomonas*, and *Spiniferomonas*). In Synurophyceae, presence of bilaterally symmetrical siliceous scales of complex species-specific structure is common.

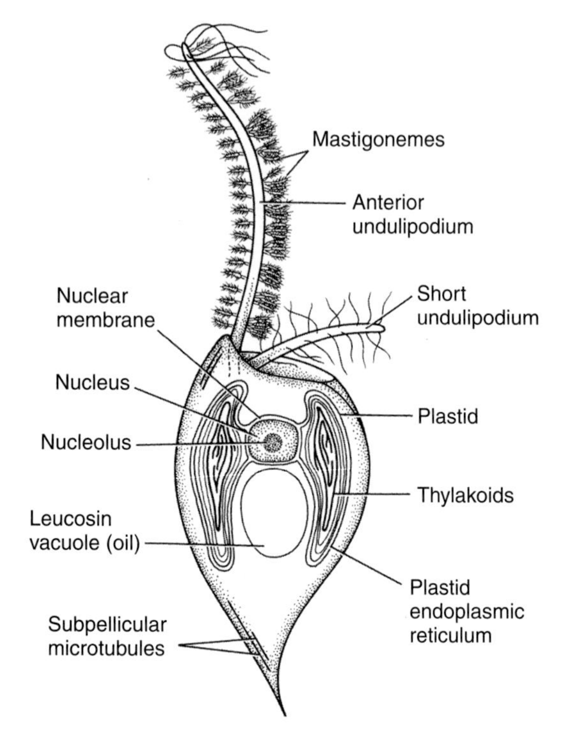

Figure D The freshwater, single-cell chrysomonad *Ochromonas danica;* the ultrastructures of *Ochromonas* cells and of single *Synura* cells are similar. [Drawing by M. Lowe.]

Another siliceous structure that is shared by both Chrysophyceae and Synurophyceae is the unique urn-shaped resting stage, called stomatocyst (or statospore), which can be produced both sexually and asexually. At the top of the stomatocyst is a pore, often surrounded by a tapering, conical collar and plugged with material that dissolves when the germinating ameboid chrysomonad emerges. There is great morphological variation between stomatocysts from different species. Their formation proceeds endogenously within an internal membrane system that is considered to be homologous to the silica deposition vesicle of diatoms. The species-specific siliceous cysts and scales preserved in sediments can be useful paleobiological indicators for reconstructing a wide variety of human influences on aquatic ecosystems, including acidification, eutrophication, and climate change. Siliceous scales have been identified from sediments as early as Middle Eocene, whereas putative stomatocysts have been recovered from marine sediments as old as Lower Cretaceous. This indicates that biosilicification within chrysomonads possibly originated in the Mesozoic.

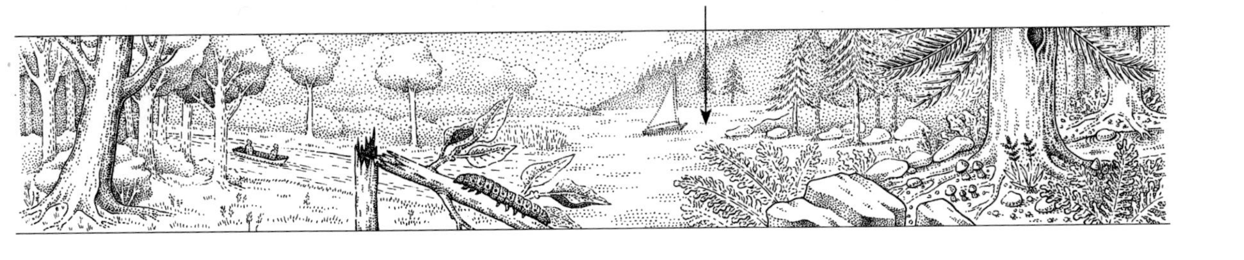

Nutrition varies from obligate phototrophy (for example, all synurophyceans) to obligate phagotrophy (colorless chrysomonads). Many species of Chrysophyceae are mixotrophic, that is, capable of photosynthesis as well as uptake of dissolved substances (osmotrophy) and food particles (phagotrophy).

Sexual reproduction has been observed in only few species. Undifferentiated cells may act as gametes, fusing apically or antapically and producing a short-lived binucleate planozygote with four undulipodia that subsequently undergoes encystment. Germination of sexual stomatocysts results in the formation of one, two, or four initial vegetative cells, depending on the species. The actual processes of karyogamy and meiosis have never been reliably documented.

Pr-16 Xanthophyta

Greek *xanthos*, yellow; *phyton*, plant

GENERA
Botrydiopsis
Botrydium
Botryococcus
Centritractus
Characiopsis
Chlamydomyxa
Chloramoeba
Chloridella
Heterodendron
Mischococcus
Myxochloris
Ophiocytium
Rhizochloris
Tribonema
Trypanochloris
Vaucheria

Xanthophytes, like eustigmatophytes (Pr-27), are yellow-green. However, the unique organization of their cells and their tendency to form strange colonies suggest that they are related to eustigs only by pigmentation. Their photosynthetic organelles, xanthoplasts, probably share ancestry with those of the eustigs, but the nonplastid part of the cell is far more like that of the chrysomonads (Pr-15). In fact, on the basis of a common morphology of the nonplastid part, some phycologists prefer to group xanthophytes with chrysomonads and phaeophytes (Pr-17)—excluding haptomonads (Pr-25)—in a single phylum called Heterokonta. In all members of this Heterokonta phylum, there are two anteriorly inserted undulipodia. One is directed forward and is mastigonemate—that is, hairy; the other, trailing, undulipodium is shorter and smooth. Although the nonplastid parts of the cells in these three Heterokonta groups probably do have common ancestry, the differences among these groups of protoctists and the uniformity within each group seem to us marked enough to justify the raising of each group to phylum status. Production of heterokont zoospores by xanthophytes, which resembles the pattern in certain mastigote molds, such as oomycotes (Pr-21) and plasmodiophorans (Pr-20), has led us to place phyla Pr-20 and Pr-21 near xanthophytes on our phylogeny.

The plastid pigmentation of the xanthophytes, like that of the eustigs, consists of chlorophylls ***a***, c_1, c_2, and ***e***. Several xanthins are found in the best-studied members: cryptoxanthin, eoxanthin, diadinoxanthin, and diatoxanthin; heteroxanthin and beta-carotene have been detected by spectroscopic methods. Oils are the storage products of photosynthesis—starch is absent. At least some of the glucose monomers in these oils are linked by β-1–3 linkages.

Xanthophytes typically have pectin-rich cellulosic walls made of overlapping discontinuous parts. Many cells are covered with scales characteristic of the species. In winter and under other adverse conditions, many species form cysts that are encrusted with iron or in which silica is embedded.

Xanthophytes populate freshwaters (Figure A). They are found in a variety of highly structured multicellular and syncytial (multinucleate) forms; some produce amebas or undulipodiated zoospores (Figures B and C). Although complex differentiation patterns are known, meiotic sexual cycles have not been reported.

This phylum contains four major subgroups, which we raise to the status of class. They are the Heterochloridales, Heterococcales, Heterotrichales, and Heterosiphonales. Each of the classes, traditionally reported as orders, has many genera.

Class Heterochloridales contains the morphologically least complex xanthophytes. It comprises two major groups, here called orders: Heterochlorineae (motile unicells) and Heterocapsineae (palm-shaped, flattened colonial forms). *Botryococcus*, a very common pond-scum organism, is perhaps the best-known member of the second group.

Class Heterococcales, which includes *Ophiocytium*, consists of coccoid cells. Their genera take various colonial forms—filamentous, branched, or bunched—reminiscent of the colonies formed by coccoid cyanobacteria (B-6), chrysomonads (Pr-15), and chlorophytes (Pr-28). However, the single coccoid unit cell has an internal organization characteristic of the xanthophytes. The well-known genus *Botrydiopsis* looks like a bunch of grapes.

Figure A Vegetative cells of *Ophiocytium arbuscula*, a freshwater xanthophyte from alkaline pools in England. LM (phase contrast), bar = 10 μm. [Courtesy of D. J. Hibberd.]

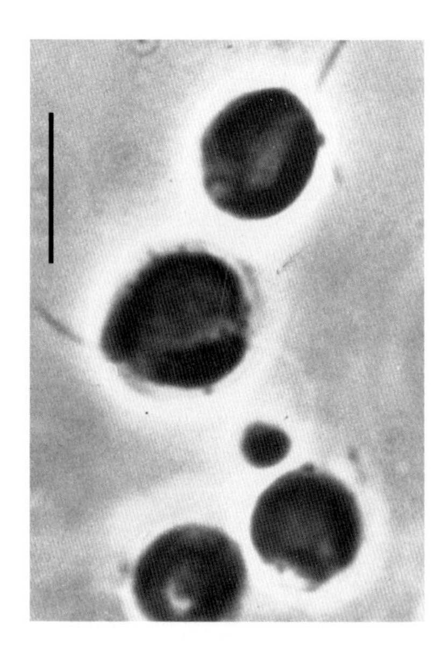

Figure B Living zoospores of *Ophiocytium majus*. LM, bar = 10 μm. [Courtesy of D. J. Hibberd and G. F. Leedale, *British Phycological Journal* 6:1–23 (1971).]

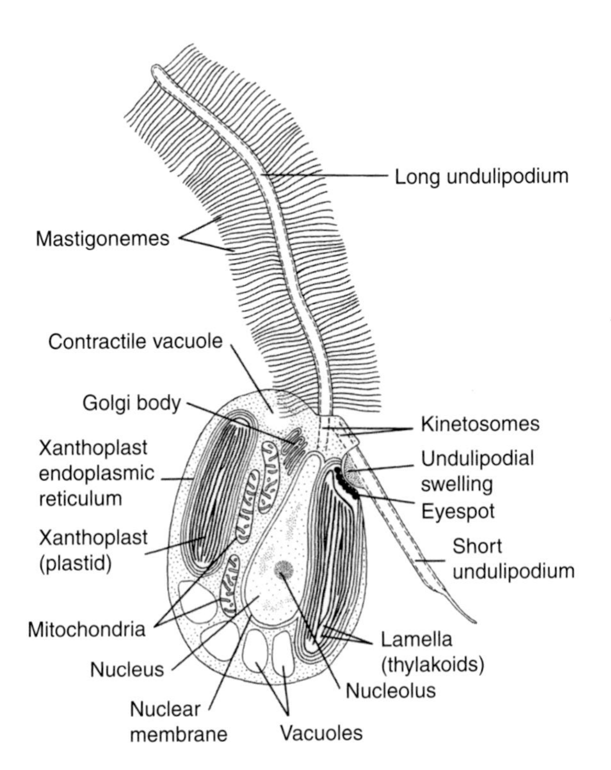

Figure C Zoospore of *Ophiocytium arbuscula*, showing typical heterokont undulipodia. [Drawing by R. Golder.]

Class Heterotrichales includes many complex multicellular organisms, most of them being variations on the theme of filaments. They include the highly branched, flaccid, tree-shaped alga called *Heterodendron*.

Class Heterosiphonales contains the most morphologically complex xanthophytes. Members of this class can be quite formidable in appearance. *Botrydium*, for example, develops a collapsible, balloonlike multicellular thallus in drying muds, superficially looking like a chytrid (Phylum Pr-35), which may become encrusted with calcium carbonate and may grow to nearly 1 m in length. It has an extensive system of branched rhizoids in which resistant, hard-shelled cysts develop. The cysts, when rehydrated, germinate into unicellular, heterokont zoospores typical of xanthophytes. The zoospores disperse, germinate, and grow into thalli, completing the life cycle. All zoospores are without sex and are competent by themselves for further development.

Although there are fewer than 100 well-documented species, this phylum of algae, best known as unsightly messes in muddy water, probably has many other members that have not yet come under scrutiny.

Pr-17 Phaeophyta

(Brown algae)

Greek *phaios*, brown; *phyton*, plant

GENERA

Alaria
Ascophyllum
Chorda
Chordaria
Cutleria
Desmarestia
Dictyosiphon
Dictyota
Durvillaea
Ectocarpus
Fucus
Ishige
Laminaria
Macrocystis
Pelvetia
Sargassum
Scytosiphon
Scytothamnus
Sphacelaria
Sporochnus
Tiloptetis
Undaria

The phaeophytes are brown seaweeds, a member of Heterokontophyta and often classified as Phaeophyceae; nearly all are marine. They are the largest protoctists. The giant kelps, for example, are as much as 40 m long. Some 2000 species in 300 genera have been described, all of them photosynthetic. Most phaeophytes live along rocky coastal seashores dominating the intertidal and upper subtidal zones, where they form seaweed beds. Communities of large brown seaweeds are referred to as "forests of the sea." They are crucial primary producers, providing habitat and food for other protoctists, marine animals, and microbes.

All known phaeophytes have multicellular thallus (algal "leaf"): filamentous, foliose, crustose, cushion shaped, or dendriform. Cell walls contain cellulose and alginates, and often the sulfated oligosaccharide fucoidan. The cytoplasmic connections between the cells (plasmodesmata), evolve into sieve tubes for conducting water and photosynthate, which in many cases is the sugar alcohol mannitol. The tubes are analogous, not homologous, to the sieve tubes of plants.

The phaeophytes principally have life histories alternating between haploid (gametophyte) and diploid (sporophyte) generations; however, in some members (for example, *Fucus* and *Sargassum*), superficially only one (diploid) generation dominates. When gametophytes and sporophytes are independent, depending on the differences in the gross morphology, isomorphic and heteromorphic life histories are recognized. Phylogenetically the isomorphic generation is primitive (plesiomorphic) in the phaeophytes, and perhaps as a response to the seasonal changes of environmental factors in temperate regions, heteromorphic life history evolved. Furthermore, by reductive evolution of the smaller generation, only one generation became dominant as in Fucales. The alternation of haploid and diploid generations in phaeophytes is analogous, but not homologous, to the alternation of generations in chlorophytes and plants.

Most reproductive cells (zoospores, gametes, and sperm) have two undulipodia of different morphology and function: the hairy anterior undulipodium provided with tubular mastigonemes on the surface and having locomotive function by its undulating ciliary motion, and the smooth posterior undulipodium having steering function as a rudder. Many mastigote cells show photo-orientation responses by the coordination of eyespot on the chloroplast and the photoreceptor on the posterior undulipodium. Male gametes and sperm show chemotaxis responding to sexual attractants (pheromones) from female reproductive cells (gamete and egg). About a dozen of volatile hydrocarbonic compounds (for example, ectocarpene and lamoxirene) are identified as sexual pheromones.

The morphology of phaeophyte mastigote cells resembles that of the heterokont phyla, the chrysomonads (Pr-15) and the xanthophytes (Pr-16). The phaeophytes have closest phylogenetic relationships with schizocladiophytes, phaeothamniophytes, and xanthophytes.

Sporangia (spore-producing structures) form on the sporophyte thallus and release zoospores (swarmer cells), not gametes. If meiosis precedes the release of the zoospores, they are haploid; otherwise, they are diploid. Although they resemble the phaeophyte sperm, zoospores are not gametes and do not require a fertilization step to develop and grow into the multicellular alga.

Brown algae contain a distinct set of photosynthetic pigments: their plastids (chloroplasts originated by secondary endosymbiosis) contain chlorophylls ***a*** and ***c***, but never chlorophyll ***b***. Fucoxanthin, the prominent carotenoid derivative, is responsible for the brown or olive-drab color of the thallus. The β-1,3 glucan carbohydrate stored as food is called laminarin (named after *Laminaria*, a well-known kelp genus). Phaeophyte cells also store lipids. Some members contain, in the cells, abundant osmiophilic globules called physodes. They include high concentrations of phlorotannins, characteristic bioactive substances of the phaeophytes, acting as defensive substances against herbivores and other organisms and ultraviolet irradiation.

Although most frequently found along seashores, some brown algae, such as the well-known *Sargassum* found in the Sargasso Sea, form immense floating masses far offshore. These algae provide food and shelter for unique communities of marine animals and microbes and lead to rates of primary productivity in the open oceans far higher than would be possible if the phaeophytes were absent.

Fucus, rockweed, is very common on temperate rocky seacoasts. As in other members of the order Fucales, the thallus is flattened; it branches in one plane (Figure A). The gametangia are contained in dark clotlike bodies, called conceptacles, scattered on the surface of heart-shaped swellings called receptacles (Figure B). Within a female gametangium, each oogonium produces one, two, four, or eight eggs. Oogonia containing ripe

Figure A Thallus of *Fucus vesiculosus* taken from rocks on the Atlantic seashore. Bar = 10 cm. [Courtesy of W. Ormerod.]

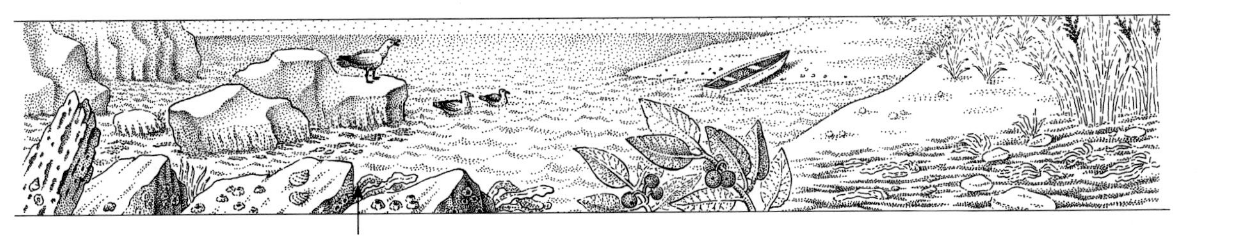

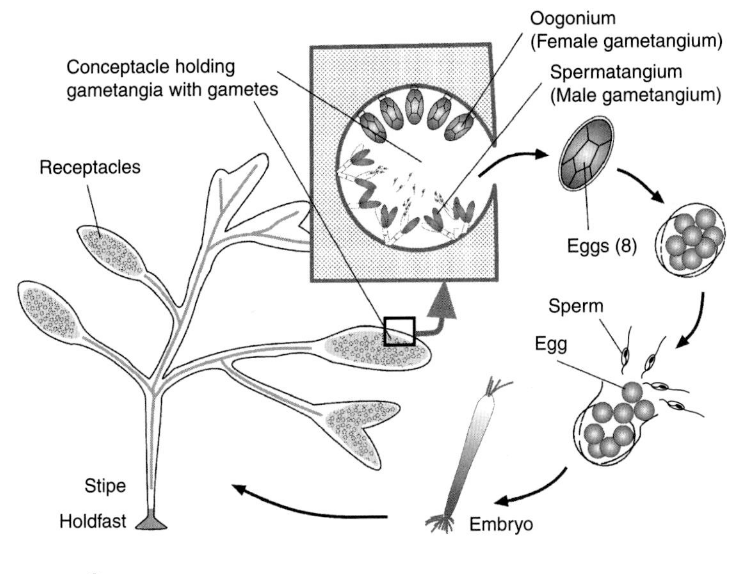

Figure B *Fucus*, showing fucalean-type life history without alteration of generations. [Courtesy of H. Kawai.]

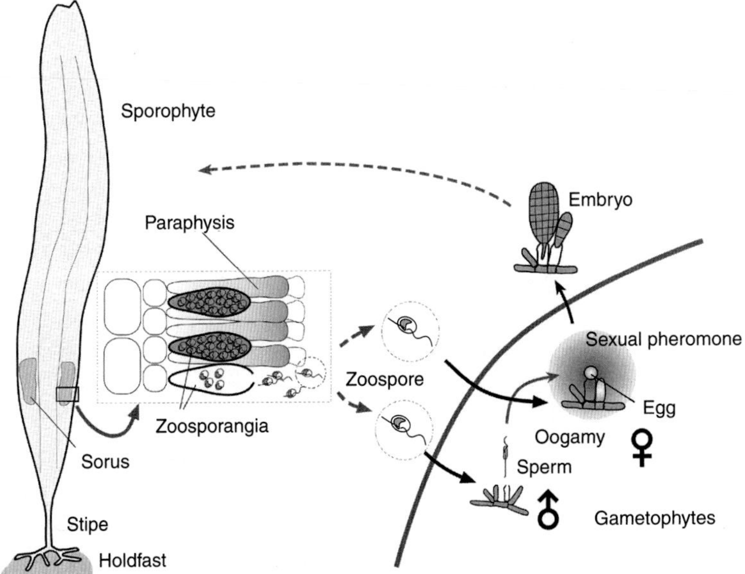

Figure C *Laminaria* showing heteromorphic life history alternating between large sporophyte and microscopic gametophytes. [Courtesy of H. Kawai.]

eggs are released into the water, where they are fertilized by sperm produced by antheridia in male gametangia. No embryo is formed: the thallus, unsupported by maternal tissue, develops directly. For these reasons alone, brown algae are not considered to be plants.

Laminaria, kelp, is also common on temperate rocky seacoasts, but its habitat is lower than *Fucus*. The *Laminaria* life history is heteromorphic, alternating between large sporophytes and microscopic gametophytes (Figure C). Zoosporangia are formed in sori on the thallus. Released and settled zoospores germinate to develop into male and female gametophytes, which are more tolerant of lower light intensity and higher temperature summer conditions than sporophytes. Gametophytes produce oogonia and spermatangia, and sexual reproduction occurs mediated by the sexual pheromones.

Pr-18 Bacillariophyta

Latin *bacillus*, little stick; Greek *phyton*, plant

GENERA

Achnanthes	*Diatoma*	*Nitzschia*
Amphipleura	*Diploneis*	*Pinnularia*
Arachnoidiscus	*Eunotia*	*Planktoniella*
Asterionella	*Fragilaria*	*Rhopalodia*
Biddulphia	*Hantschia*	*Surirella*
Coscinodiscus	*Melosira*	*Tabellaria*
Cyclotella	*Meridion*	*Thalassiosira*
Cymbella	*Navicula*	

Beautiful aquatic protists—perhaps 10,000 living species—diatoms are single cells or form simple filaments or colonies. Some 250 genera of these gorgeous protoctists are commonly described, and specialists recognize as many as 100,000 species, including those in 70 fossil genera. Valves, the two parts of the diatom test (shell), extend into the Mesozoic fossil record; the first ones appeared in the lower Cretaceous period.

Each valve of the diatom test is composed of pectic organic materials impregnated with silica (hydrated SiO_2), in an opaline state. The valves may be extremely elaborate and beautiful; their elegantly symmetrical patterns are used to test lenses for optical aberrations. Diatoms require dissolved silica for growth; they are so competent at the removal of silica from natural waters that they can reduce the concentration to less than one part per million, below the value detectable by chemical techniques.

Diatoms, important at the base of marine and freshwater food chains, are very widely distributed in the photic (illuminated) zones of the world. Some species are found in hypersaline ponds and lagoons, others in clear freshwater, and others in moist soil. Their cysts, empty tests, and dying cells can be found in unlighted regions of the ocean. Most species under study are obligate photosynthesizers, although some also require organic substances, such as vitamins, for growth. Some strains of *Nitzschia putrida* are saprobes.

Diatoms are generally brownish; for years, they were classified with the golden yellow algae, the chrysomonads (Pr-15). The xanthoplasts (plastids) of diatoms contain the pigments chlorophyll ***a***, chlorophyll ***c***, beta-carotene, and xanthophylls, including fucoxanthin, lutein, and diatoxanthin. The photosynthetic food reserve of diatoms, like that of chrysomonads, is the oil chrysolaminarin. Nevertheless, in life history, cell structure, and division, the diatoms differ greatly from the other golden yellow algae. The diatoms make up such an easily distinguished and large natural group that, in light of modern information, we provide them a phylum separate from the other organisms that have golden brown plastids.

The two great classes of diatoms are the Coscinopiscophyceae (Centrales) and the Bacillariophyceae (Pennales). The Coscinopiscophyceae, or centric diatoms, have radial symmetry, like that of *Thalassiosira* and *Melosira* (Figures A and B). The Bacillariophyceae, or pennate (featherlike) diatoms, have bilateral symmetry; many of them are boat-shaped (Figure C) or needle-shaped. The pennate diatom has a slit, called a raphe, between the valves. The raphe exudes cytoplasmically produced slime in which the diatom glides. The centric diatoms lack raphes and are never motile. In spite of the correlation of the raphe and movement, the detailed mechanism of gliding in the pennate diatoms is not known. The centric diatoms usually have numerous small plastids, whereas in the pennate diatoms, the plastids are fewer, as a rule, and larger.

The centric diatom class comprises nine major groups, recognized now as subclasses. Some examples are the discoid subclass, which includes *Thalassiosira*, a colonial diatom that extrudes threads, called setae, of chitin, which hold individual diatoms together in chains; the solenoids, elongate diatoms, cylindrical or subcylindrical in shape; the box-shaped biddulphioids, most with horns or other decorations on their tests; and the rutilarioids, with their naviculoid valves (that is, they depart from strict radial symmetry by having a boatlike shape with pointed ends that have radial or irregular markings).

The pennate diatoms are classified into two subclasses: small Eunophycidae and the enormous Bacillariophycidae. In classifying the four orders of subclass Bacillariophycidae, attention is given to the presence and development of the raphe. Araphids lack a true raphe; raphids show the beginnings of a raphe; monoraphids have a fully developed raphe but only on one of the valves; and biraphids have a fully differentiated raphe on each valve.

Diatoms are highly sexual organisms that, like animals, spend most of their life cycle in the diploid state. Meiosis occurs just before the formation of haploid gametes. After fertilization, the diploid zygote develops into the familiar diatom.

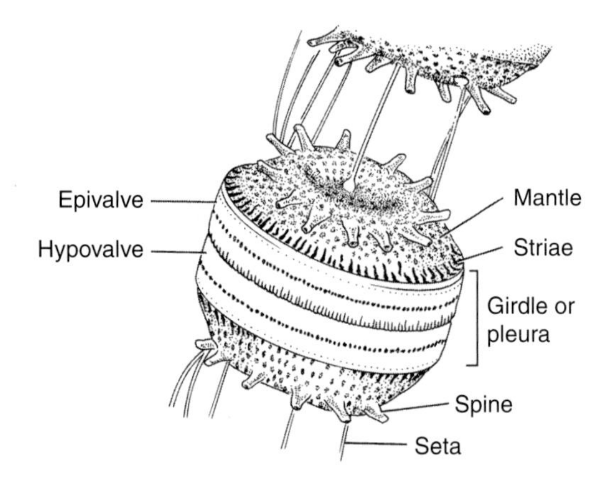

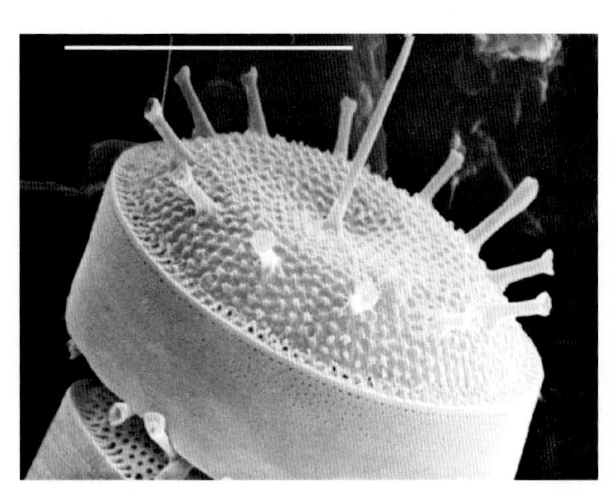

Figure A *Thalassiosira nordenskjøldii*, a marine diatom from the Atlantic Ocean. SEM, bar = 10 μm. [Drawing by E. Hoffman; photograph courtesy of S. Golubic.]

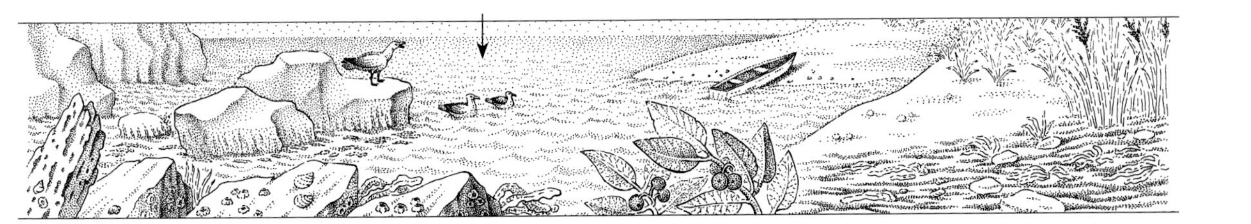

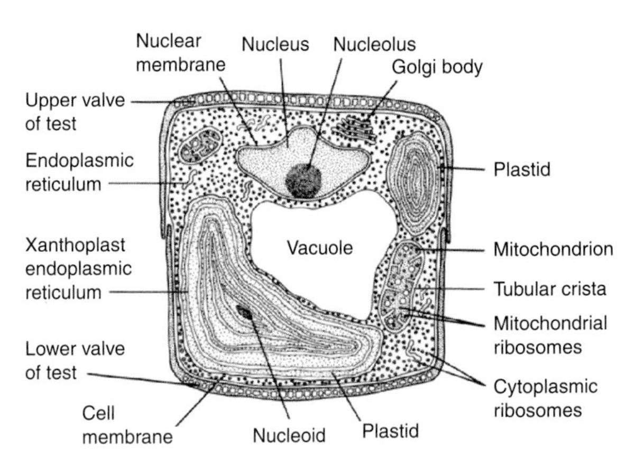

Figure B *Melosira* sp., a centric diatom. [Drawing by L. Meszoly.]

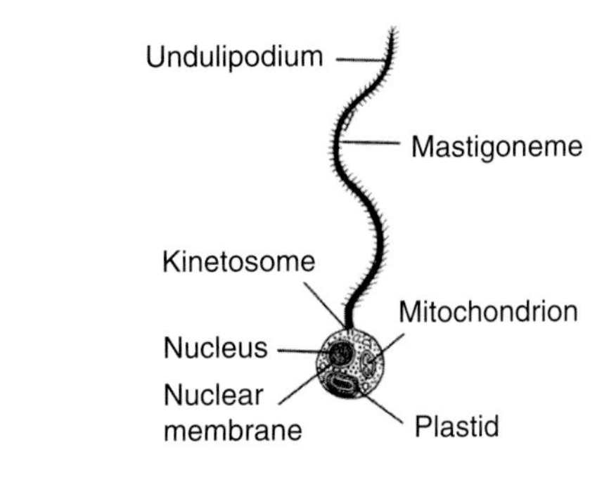

Figure D Sperm of *Melosira* sp. [Drawing by L. Meszoly.]

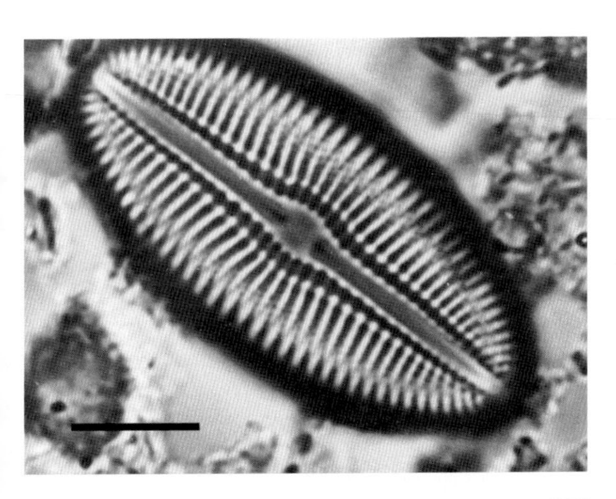

Figure C *Diploneis smithii*, a pennate (naviculate or boat-shaped) diatom from Baja, California. With the light microscope, only the silica test, which has been cleaned with nitric acid, is seen. LM, bar = 25 μm.

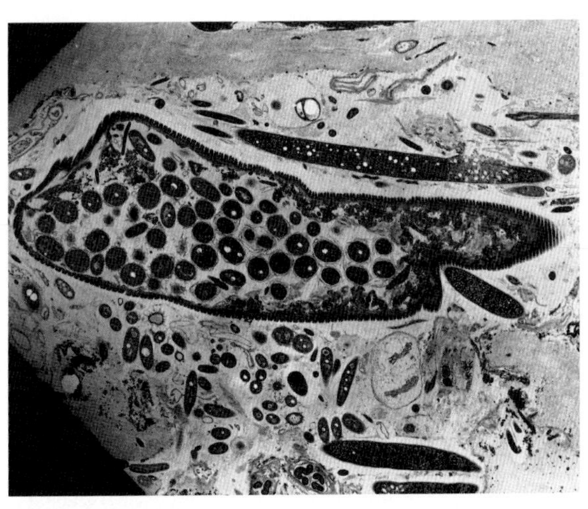

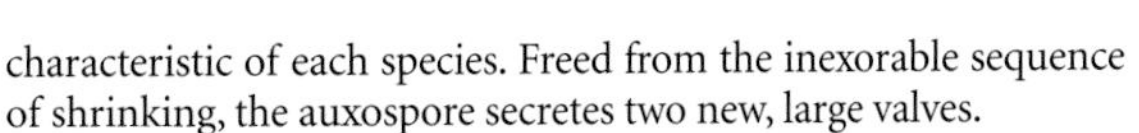

Figure E Diatom tests colonized probably by purple photosynthetic bacteria from young microbial mat, Laguna Figueroa, Baja California Norte, Mexico. TEM.

Diatoms reproduce by mitotic division into two offspring cells. In many, each offspring cell retains one valve of the parent shell and produces one new valve, which fits into the parental valve. Hence, the two offspring cells are each slightly smaller than the parent. This tendency for diatoms to decrease in size is counteracted by auxospore formation. An auxospore, not really a spore in the sense of being able to resist adverse conditions, is an enlarged, shell-less diatom formed when a protoplast (test contents) is released into the water column. The size of the auxospore is a characteristic of each species. Freed from the inexorable sequence of shrinking, the auxospore secretes two new, large valves.

In the pennate diatoms, auxospore formation is usually triggered by the sexual confrontation: it follows fertilization and zygote formation. Haploid male gametes (sperm), bearing single anterior undulipodia (Figure D), fertilize immotile haploid female protoplasts. The valves of the tests sometimes open for the purpose. The zygote becomes the large auxospore; reproduction by mitotic division follows. In the centric diatoms, auxospores may be formed without the intervention of the sexual processes of meiosis and fertilization.

Resilient siliceous diatom tests can be important structural components of microbial mat ecosystems (Figure E).

Pr-19 Labyrinthulata
(Slime nets and thraustochytrids)

Latin *labyrinthulum*, little labyrinth

GENERA
Althornia
Aplanochytrium
Japonochytrium
Labyrinthula
Labyrinthuloides
Schizochytrium
Thraustochytrium
Ulkenia

Molecular studies of rRNA have confirmed the relation between the two classes of this phylum: the thraustochytrids and labyrinthulids, or the slime nets. The phylum is defined as protoctists that use cellular organelles called bothrosomes (sagenogenosomes) to produce extracellular (ectoplasmic) slime matrix. The ectoplasmic matrix is continuous with the plasma membrane at the bothrosome.

The thraustochytrids, an obscure group of encysting marine protoctists that form swollen zoospore cases (sori) and superficially resemble chytrids (Pr-35), have traditionally been classified as fungi. Because they produce undulipodiated cells and also have other features that demonstrate an indisputable relation to labyrinthulids, these protoctists—seven genera with nearly three dozen species—belong as a class here. Their slime matrix does not surround each cell, as it does in the labyrinthulids; rather, this rich proteinaceous polysaccharide slime matrix emanates from the base of the developing sorus when it arises from single or clustered bothrosomes (sagenogens). The walls of the thick-walled sorus ("sporangium") contain large quantities of l-galactose; in some thraustochytrid walls, a high-sulfate galactan was detected.

All labyrinthulids have been united into a single genus, *Labyrinthula*, with at least eight distinguishable species. Labyrinthulids form colonies of individual cells that move and grow entirely within the slime matrix or slimeway, which is a slime track. This track (Figures A and B), probably mucopolysaccharides and actomyosin proteins, is laid down in front of the cells (Figures B and C). They have been called slime net amebas in some texts, but there is nothing ameboid about them.

The labyrinthulids form transparent colonies—they may be centimeters long. With the unaided eye, they look like a slimy mass on marine grass. Under the microscope, spindle-shaped cells can be seen migrating back and forth in tunnels within the slime, like little cars in tracks (Figure D and E). Labyrinthulid cells cannot move at all unless they are completely enclosed in the slime track. The mechanism of movement is unknown; it certainly does not require undulipodia or pseudopods. Calcium ions regulate contraction of actomyosin filaments in the slimeways. It seems that force is generated by the actomyosin musclelike movement along the surface of the slimeways, external to the cells in the slime layer, causing cells in the slimeway to travel at quite a rate—several micrometers per second.

The movements of trains of cells together back and forth in the track may look random. However, when a potential food source is sensed, such as a bacterial or yeast colony, the cells of the slime net move toward the source. Labyrinthulids are osmotrophic; they do not ingest the bacteria or yeast. Instead, extracellular digestive enzymes are released and the resulting small food molecules diffuse into and through the slime to nourish the colony. The labyrinthulid colony changes shape and size so extensively—that is, it is so pleiomorphic—that confusion

Figure A Live cells of *Labyrinthula* sp. traveling in their slimeway. LM, bar = 100 μm. [Courtesy of D. Porter.]

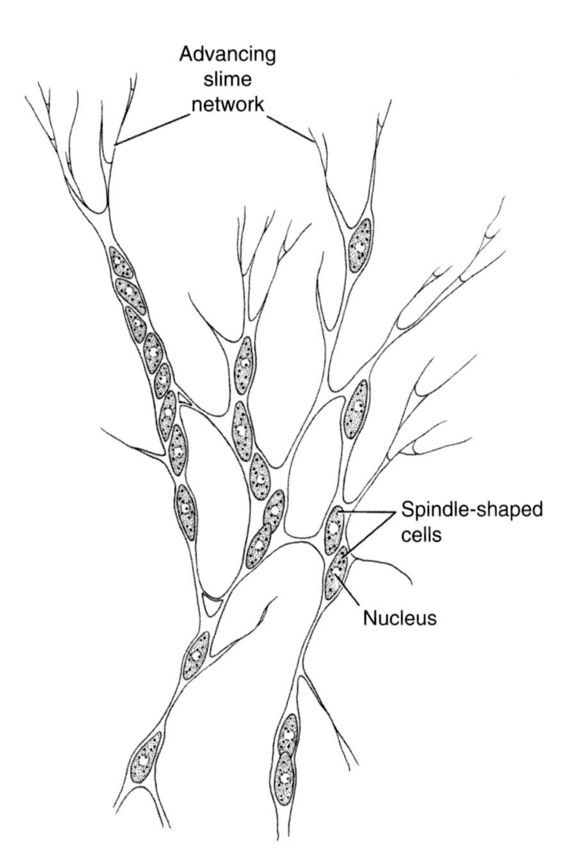

Figure B *Labyrinthula* cells in a slimeway. [Drawing by R. Golder.]

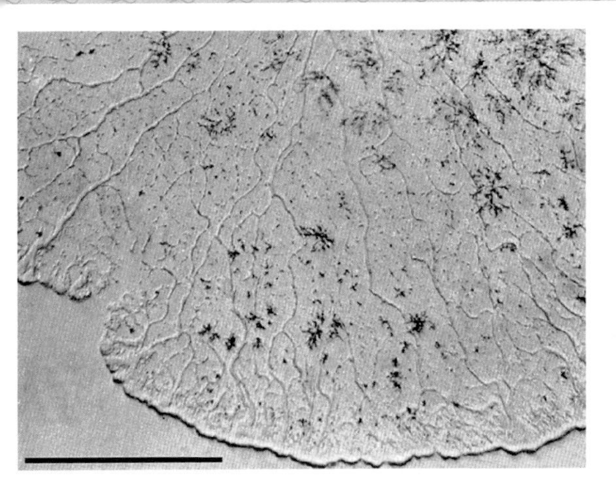

Figure C Edge of a *Labyrinthula* colony on an agar plate. Bar = 1 mm. [Courtesy of D. Porter.]

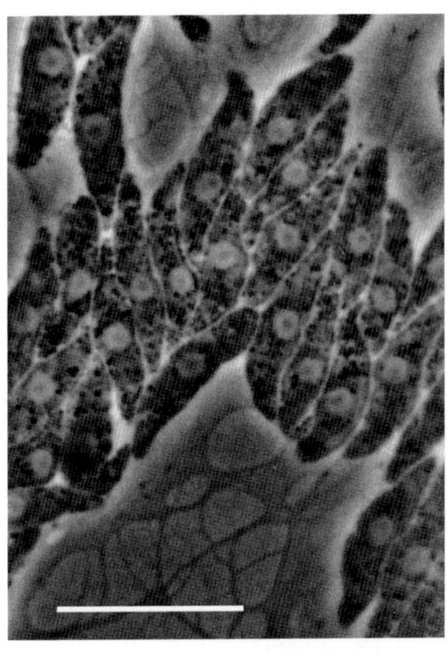

Figure D Live *Labyrinthula* cells in their slimeway. LM, bar = 10 μm. [Courtesy of D. Porter.]

surrounds its classification on the basis of morphological characteristics. Growth is by the mitotic division of cells inside the slime track, followed by separation of the offspring cells, capable of excreting slime.

When conditions become drier, labyrinthulids adopt a desiccation strategy. In the older parts of the colony, cells aggregate and form hardened, dark cystlike structures, clumps of cells surrounded by tough membrane. These cysts have no precise size or shape. The cysts are resistant to desiccation; they wait until moisture and food become more abundant and then rupture, liberating small spherical cells that grow again into spindle-shaped cells in a slime matrix that they produce. Undulipodiated

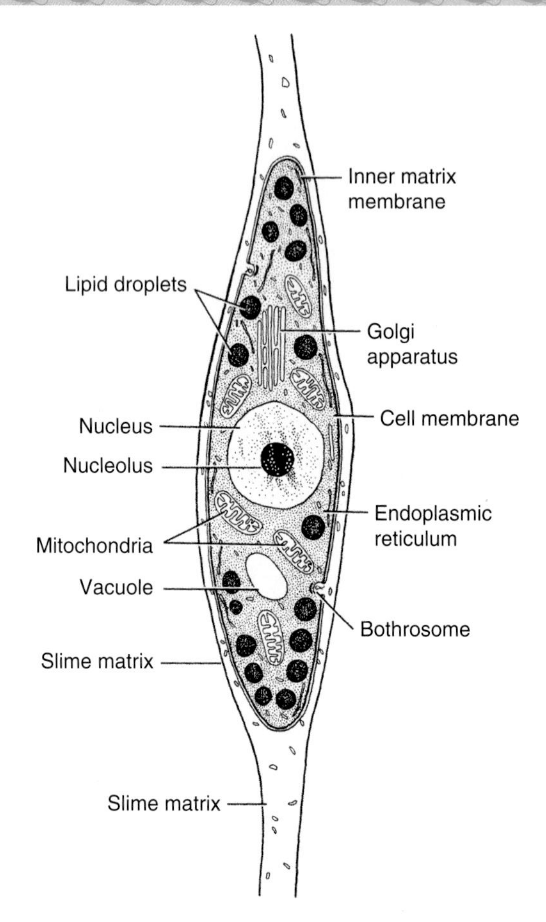

Figure E Structure of a single *Labyrinthula* cell. [Drawing by R. Golder.]

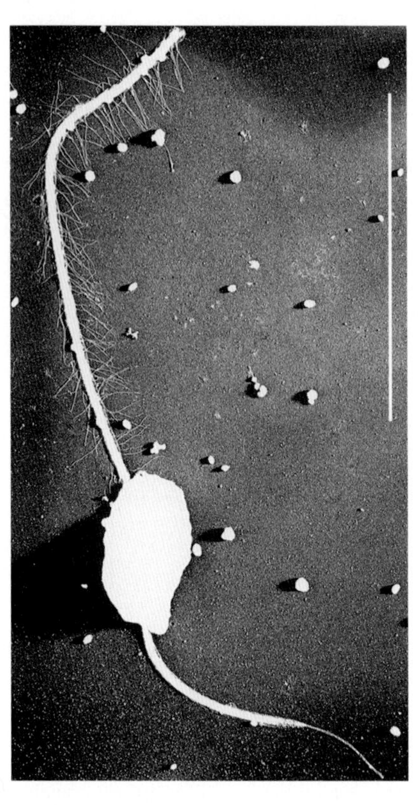

Figure F Zoospore of *Labyrinthula* sp., showing one anterior undulipodium with mastigonemes and one posterior undulipodium lacking them. SEM, bar = 10 μm. [Courtesy of F. O. Perkins, from J. P. Amon and F. O. Perkins, *Journal of Protozoology* 15:543–546 (1968).]

stages, zoospores, of *Labyrinthula marina*, *L. vitollina*, and *L. algeriensis* have been found (Figure F). Some seem to be products of meiosis that are isogametes. There is no difference between male and female gametes, which fuse to form a zygote. The zygote undergoes mitosis; its offspring apparently develop into the multicellular net.

Labyrinthula is best known because it grows on the eel grass *Zostera marina* (Pl-12, Anthophyta) and on many algae, such as *Ulva* (Pr-28, Chlorophyta). *Zostera* is very important to the clam and oyster (A-26, Mollusca) industries along Atlantic shores because it is the primary producer of ecosystems that include the clam and oyster beds. Blooms of *Labyrinthula* have been associated with diseased *Zostera*.

Pr-20 Plasmodiophora

New Latin *plasmodium*, multinucleate mass of protoplasm not divided into cells; Greek *pherein*, to bear

GENERA

Ligniera
Maullinia
Membranosorus
Octomyxa
Plasmodiophora
Polymyxa
Sorodiscus
Sorosphaera
Spongospora
Tetramyxa
Woronina

Plasmodiophorans, 11 genera and 30 species, are holocarpic, obligate symbiotrophs primarily in flowering plants (Pl-12); some species, however, infect members of Oomycota (Pr-21), Chlorophyta (Pr-28), or Phaeophyta (Pr-17). Several members induce hypertrophied tissues or galls on their hosts (Figure A). Economically significant members of the group include *Plasmodiophora brassicae*, *Spongospora subterranea*, *Polymyxa betae*, and *P. graminis*. *Plasmodiophora brassicae* causes clubroot of cabbage and other brassicaceous crops, and *S. subterranea* causes powdery scab of potatoes. *Spongospora subterranea* and the two species of *Polymyxa* are vectors for viruses that are pathogenic on a variety of crop plants.

The distinguishing characteristic of plasmodiophorans is a unique type of mitotic division known as cruciform division (Figure B). At metaphase, a persistent nucleolus aligns parallel to the spindle and perpendicularly to the equatorial plate of chromatin. When viewed from the side through both optical and transmission electron microscopy, the nucleus with its nucleolus at right angle to the chromatin appears as a cross, thus the term *cruciform division*. The nuclear envelope persists during cruciform division, constricting during telophase after the chromosomes have migrated to the poles in anaphase to form the two offspring nuclei. Other important features of plasmodiophorids include (1) centrioles paired in an end-to-end arrangement at the poles of nuclei; (2) multinucleate plasmodia as growth forms; (3) uninucleate zoospores with two, anterior, whiplash undulipodia of unequal lengths; and (4) uninucleate, single-celled, dense-walled, environmentally resistant resting spores.

Although the details of the life cycle of most species are not entirely understood, a generalized life cycle (Figure E) for the group can be given based on observations from several species, starting with a resting spore (upper left, Figure D). Resting spores, which generally occur in the soil, may occur singly as in *Plasmodiophora brassicae*, or in aggregates, the sporosori (Figure D). A resting spore germinates to release a primary zoospore that has two anterior, whiplash undulipodia of unequal lengths. When the zoospore swims, the shorter undulipodium is directed forward and the longer one is directed backward. The zoospore encysts on the exterior surface of a host cell and forms a cytoplasmic extension, the adhesorium. A dense organelle, the stachel, is within the adhesorium and is injected ahead of the contents of the encysted zoospore into the host cell through the host cell wall and cell membrane. Inside the host cell, the contents of the zoospore develop through synchronized cruciform divisions into a multinucleate, primary (sporangial) plasmodium. As the primary plasmodium matures, the last nuclear divisions are noncruciform because the nucleolus does not persist during division. These plasmodia with noncruciform divisions are referred to as transitional plasmodia. At maturity, after passing through the transitional stage, the primary plasmodium forms a thin wall and cleaves into several lobes to form sporangia, each of which in

Figure A Galls (brackets) caused by plasmodiophorids. On the left, a stem gall on *Veronica* sp. caused by *Sorosphaera veronicae*; on the right, a young root gall (clubroot) on Chinese cabbage caused by *Plasmodiophora brassicae*. [Courtesy of J. P. Braselton.]

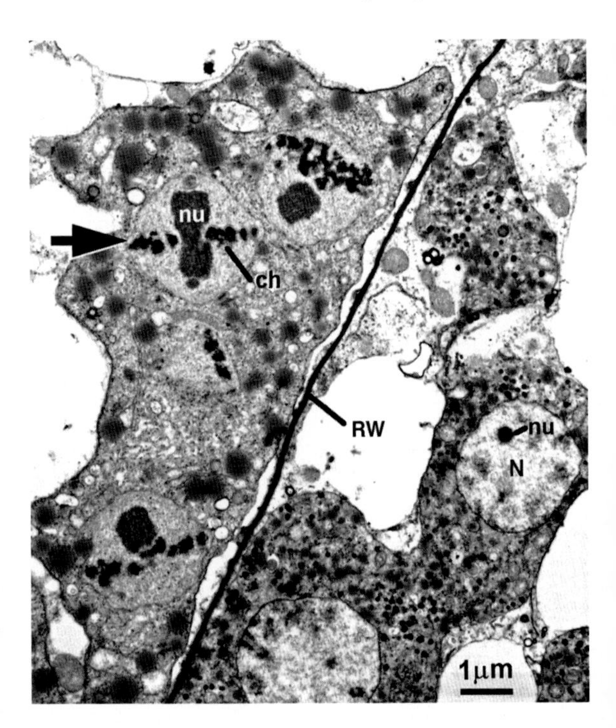

Figure B Portions of two shoot cells of a flowering aquatic plant, *Ruppia maritima*, which have been infected with secondary plasmodia of *Tetramyxa parasitica*. *Ruppias* cell wall (RW) separates the two cells. The plasmodium of *T. parasitica* in the left cell has cruciform divisions (arrow) with a persistent nucleolus (nu) perpendicular to the chromatin (ch) at metaphase, whereas the plasmodium in the right cell is in the transitional stage as indicated by the nucleus (N) with a smaller nucleolus (nu). TEM. [Courtesy of J. P. Braselton.]

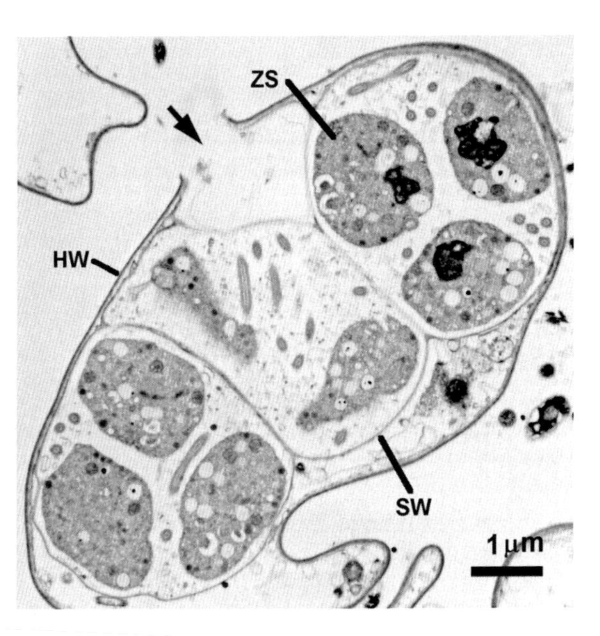

Figure C Portion of root hair of potato showing lobes of mature sporangia of *Spongospora subterranea*. Arrow indicates exit pore through one sporangial lobe. Also labeled are cell wall of the host (HW), walls of the sporangia (SW), and zoospores (ZS). TEM. [Courtesy of J. P. Braselton.]

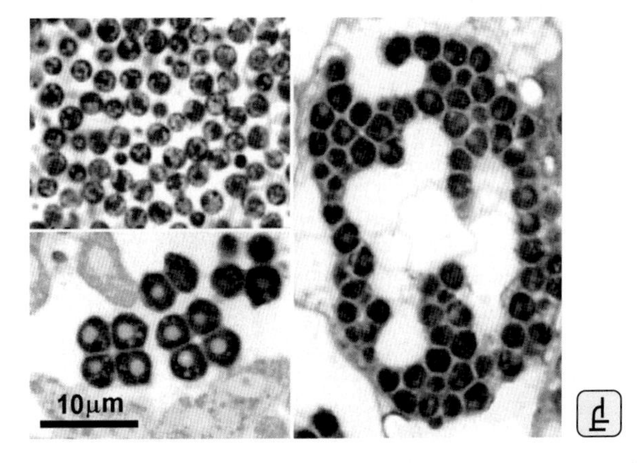

Figure D Resting spores of *Plasmodiophora brassicae* (upper left), *Tetramyxa parasitica* (lower left), and *Spongospora subterranea* (right). LM. [Courtesy of J. P. Braselton.]

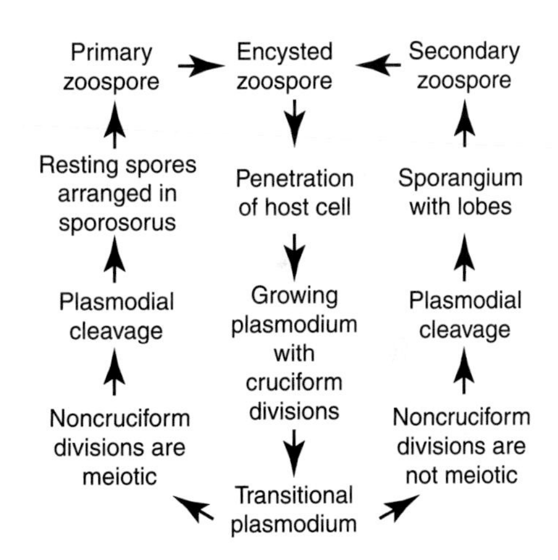

Figure E Generalized life history for plasmodiophorids based on several sources. [Courtesy of J. P. Braselton.]

turn cleaves to produce uninucleate, secondary zoospores (Figure C). Secondary zoospores are indistinguishable from primary zoospores. The secondary zoospores exit the sporangia through an opening in its wall or through an exit tube.

When secondary zoospores reinfect the cells of the host organism, they may develop into primary plasmodia and repeat the portion of the life cycle to form secondary zoospores described in the preceding paragraph; alternatively, under environmental conditions that are not understood, they may develop into secondary (sporogenic) plasmodia. Young secondary plasmodia have synchronous cruciform divisions as they grow. When a secondary plasmodium reaches its maximum size and ceases growth, its nuclei become less distinct in part because the nucleoli are reduced in size. These plasmodia, as with primary plasmodia, are referred to as transitional plasmodia. Noncruciform divisions in secondary transitional plasmodia are considered to be meiotic divisions because indicators of meiosis, synaptonemal complexes, are present in prophase nuclei. During the meiotic nuclear divisions, transitional sporogenic plasmodia cleave into uninucleate resting spores, each of which forms a dense cell wall. When the host dies and decays, the resting spores are released into the environment where they may remain viable for years before they encounter conditions for the cycle to begin again. Although meiosis is documented, it is not established where syngamy and karyogamy occur in the life cycle.

Plasmodiophorid genera are distinguished primarily by the arrangement of the resting spores in sporosori (Figure D). Examples of generic names based on shapes of sporosori include *Tetramyxa* (four resting spores), *Octomyxa* (eight resting spores), *Sorodiscus* (resting spores in a two-layered disk), *Membranosorus* (resting spores in a single-layered sheet), *Sorosphaera* (resting spores in a sphere), and *Spongospora* (resting spores in a spongy mass). Some species may be differentiated by the hosts they infect. Specific epithets based on generic names of hosts include *betae, brassicae, callitrichis, ectocarpii, graminis, heteranitherae, pythii, subterranea*, and *veronicae*.

Pr-21 Oomycota

(Oomycetes, oomycotes, Peronosporomycetidae)

Greek *oion*, egg; *mykes*, fungus

GENERA

Achlya
Albugo
Aphanomyces
Apodachlya
Dictyuchus
Eurychasma
Isoachlya
Lagenidium
Haliphthoros
Haptoglossa
Halophytophthora
Myzocytiopsis
Peronospora
Phytophthora
Plasmopara
Pythium
Rhipidium
Saprolegnia
Sapromyces

The oomycote zoospore morphology and molecular phylogeny (based on nuclear and mitochondrial genes) confirm these organisms as members of the stramenopiles and a sister clade, the photosynthetic Ochrophyta. The oomycote zoospore morphology and sequence of nucleotides in its rRNA gene confirm these organisms as stramenopiles ("straw bearers"). Oomycotes are variously called water molds, white rusts, and downy mildews. Some 50 genera and classes are recognized. Based on both morphology and molecular data, it is apparent that the majority of the genera fall into two major clades, the largely plant-pathogenic peronosporans (which also encompasses the lagenidialians that are mostly holocarpic animal pathogens such as *Lagenidium* and *Myzocytiopsis*) and the largely saprotrophic free-living saprolegnians (water molds, including familiar genera such as *Achlya*, *Dictyuchus*, and *Saprolegnia* as members of the leptomitales, such as *Apodachlya*). The white blister rusts (*Albugo*) are the earliest branching representative of the peronosporans, which includes the important plant-pathogenic orders, the pythiales and the peronosporales. The placement of the rhipidians (such as the sewage fungus, *Sapromyces*) is still uncertain because they have been associated with both the main clades, depending on the gene being analyzed. Recent molecular studies have also revealed that a number of holocarpic oomycote genera of uncertain affiliation, *Euryhasma*, *Haptoglossa*, and *Haliphthoros* all diverge before the two main clades. These genera either are exclusively marine or have marine representatives and are symbiotrophs of seaweeds, nematodes, and crustaceans, and indicate the marine origins of this lineage.

Like the related hyphochytrids (Pr-14) and true-fungal chytrids (Pr-35), oomycotes are either saprobes or symbiotrophs. A number of the more primitive genera such as *Eurychasma*, *Haptoglossa*, *Lagenidium*, and *Myzocytiopsis* are holocarpic, completely occupying their host cells or body cavities. However, most species feed by extending fungus threads, or hyphae, into invaded tissues, where they release digestive enzymes and absorb the resulting nutrients. These hyphae are usually coenocytic, although they may become plugged in response to injury or to delimit their reproductive organs. Unlike true fungi, the cell walls use cellulose to form the underlying microfibrillar skeleton. Although typically considered to be organisms of freshwater ponds and streams, oomycotes can be found in marine ecosystems, are widespread in soil, and are symbiotrophs of a wide range of plant and animal hosts. A few genera even attack vertebrates including fish (*Saprolegnia parasitica*) and mammals (*Pythium inosidium*). Most members of the pythians and peronosporans are plant pathogens, although one genus, *Halophytophthora*, plays a key saprotrophic role in breakdown and recycling of disgarded mangrove leaves.

The oomycotes are distinguished from the other fungus-like protists by the details of their oogonium and zoospore structure (Figures A and B) and by the nature of their sexual life cycle (Figure C). As with all heterokonts, the zoospores bear two undulipodia of unequal length. One of them is directed forward during swimming and is mastigonemate, whereas the other is smooth and trails behind. In most oomycotes, the zoospores are produced in an asexual reproductive organ, the zoosporangium that usually differentiates at the tips of nonextending hyphae. In the saprolegnians, a generation of pip-shaped primary zoospores is usually released directly from the sporangium (Figure D). These primary zoospores are weakly motile and serve to disperse the zoospores from the immediate vicinity of the parent mycelium. In contrast, in most peronosporans, an undifferentiated cytoplasmic mass (as in *Pythium* and *Lagenidium*) is released into an external vesicle, wherein the zoospores complete their differentiation or the differentiated zoospores are temporarily restrained by an evanescent vesicle (*Phytophthora* and Peronospora), which ruptures releasing reniform zoospores with lateral undulipodia (referred to variously as secondary-type or principal zoospores). These reniform zoospores eventually encyst and then germinate, usually to produce a new oomycote thallus.

Figure A Oogonium of *Saprolegnia ferax*, an oomycote from a freshwater pond. LM, bar = 50 μm. [Courtesy of I. B. Heath.]

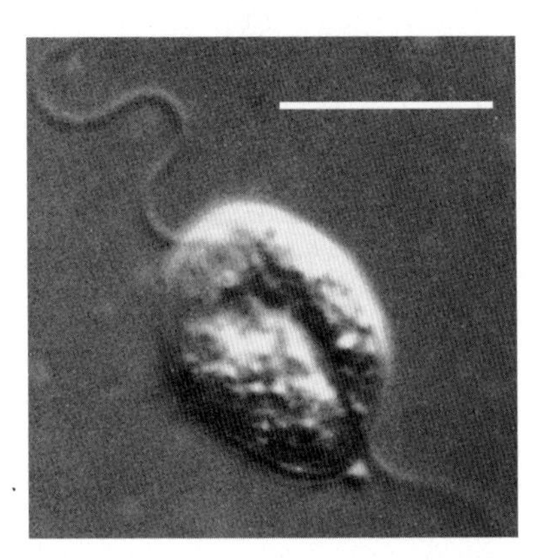

Figure B Zoospore of *Saprolegnia ferax*. LM, bar = 10 μm. [Photo courtesy of I. B. Heath; drawing by L. Meszoly.]

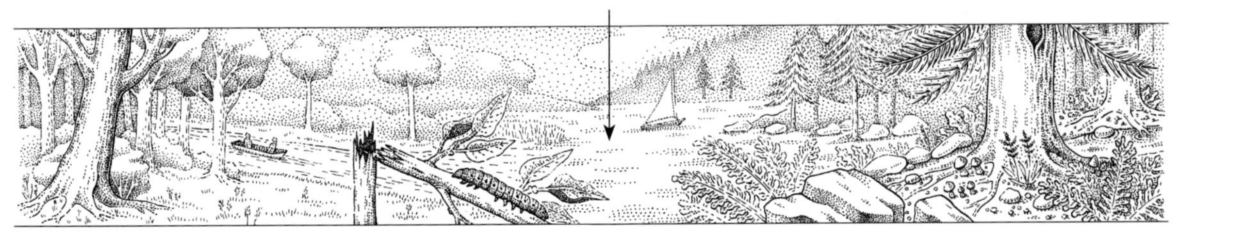

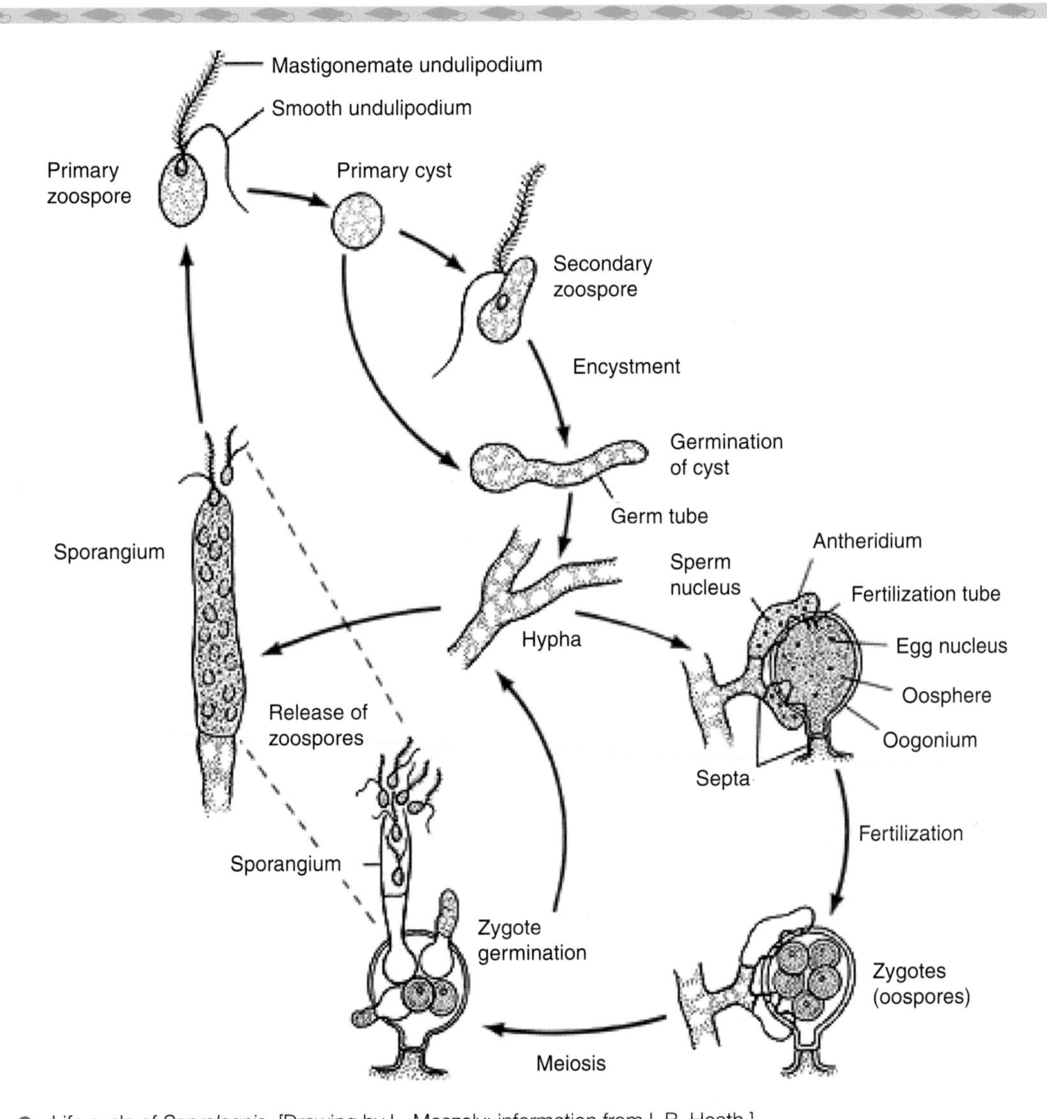

Figure C Life cycle of *Saprolegnia.* [Drawing by L. Meszoly; information from I. B. Heath.]

Sexuality is well developed in this phylum but with the exception of the marine genus *Lagenisma*, never includes motile gametes. The tips of the hyphae (or in case of holocarpic genera such as *Myzocytiopsis* and *Lagenidium*, neighbouring thallus compartments) swell to form specialized male and female gametangia, known by botanical analogy as antheridia and oogonia. The smaller antheridium transfers nonmotile male gametic nuclei via a fertilization tube to the differentiated egg cell(s), or oosphere. Unlike true fungi, oomycote thalli are normally diploid (or sometimes tetraploid), and the gametangial nuclei undergo synchronous meiosis in the antheridia and differentiating oogonium, to produce haploid gametic nuclei. Fertilization takes place inside the walled oogonium and once fertilized, the oospheres acquire a thick, multilayered resistant wall and acquire one or more refractile storage bodies (lipid bodies and a so-called vacuolar ooplast vacuole). The walls of some species become dark and sculpted and are often distinctive for a given species. Oospores often enter periods of considerable dormancy

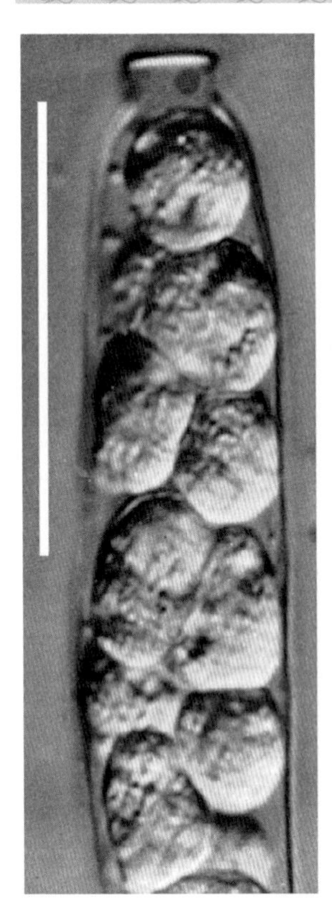

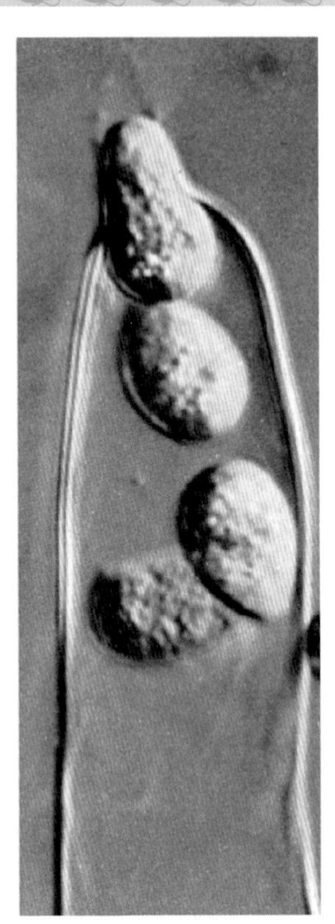

Figure D Zoospores in zoosporangium (left) and their release (right). LM, bar = 50 μm. [Courtesy of I. B. Heath.]

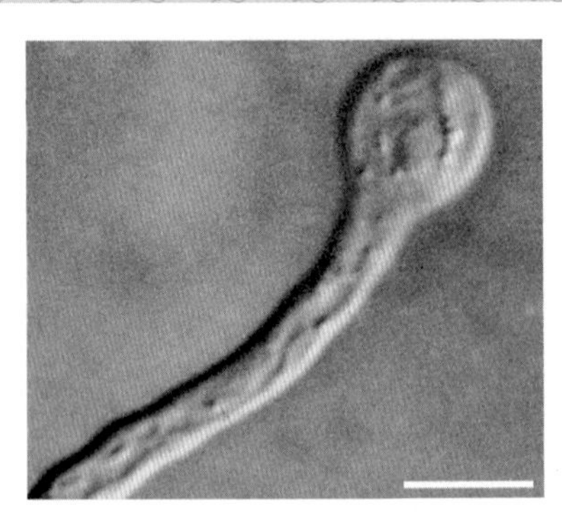

Figure E Germinating secondary cyst of *Saprolegnia ferax*. LM, bar 10 = μm. [Courtesy of I. B. Heath.]

before they germinate, usually by means of a small zoosporangium produced at the tip of a short germ tube (Figure E).

Hundreds of species have been reported. Nearly every crop plant (Pl-12) seems to have its own threatening oomycote species, such as *Phytophthora ramorum* on oak and *P. infestans* on potatoes. Some plant pathogens, such as *Albugo*, are dispersed in the air as undifferentiated sporangia, and it is not uncommon for sporangia to be dispersed and disseminated by wind. In plant pathogenic species, the zoospores respond to both chemical and electrical signals to help them home onto the roots of their hosts. Once the zoospores have settled (encysted) upon a suitable host they infect by producing germ tubes which directly penetrate the host surface. Members of the peronosporan order have been of utmost historical importance. The infamous *P. infestans*, as well as being the first funguslike organism identified as a pathogen of plants, was responsible for destroying the entire potato harvest throughout most of Europe in the mid-nineteenth century. The peasant farmers depended on the potato entirely for their food and thus this pathogen caused the mass migration of people from their homelands. Today other, root-infecting, *Phytophthora* species, such as *P. cinnamomi* and *P. ramorum* are causing massive destruction of forest trees and understory shrubs across the world.

The free-living class of oomycetes, the saprolegnians, is widely distributed in freshwater environments. Genera such as *Saprolegnia*, *Achlya*, *Isoachlya*, and *Dictyuchus* live on seeds, insect exuvae, submerged leaves, waterlogged twigs, and other decomposing organic substrates. *Saprolegnia parasitica* is very often encountered in freshwater aquaria as a white fuzz covering the fish fins. Whilst causing chronic infections of coarse fish, this species is a deadly pathogen of wild and cultured salmonids such as salmon and trout. The zoospores of this fish pathogen contain vesicles enclosing tightly coiled bundles of so-called boathook spines. Upon encystment these decorate the infective spores with bundles of long barbed spines that presumably help them grapple onto the surface of their prey. However, the most remarkable infection structures produced by any oomycote are the so-called gun cells of the nematode symbiotrophic genus *Haptoglossa*. The specialized infection cells produce an inverted tube enclosing a needle and they literally instantaneously inject themselves into the body cavity of a hapless nematode that passes too close. Each species comes equipped with its own distinctive weapon. Oomycotes appear to have evolved in the sea associated with marine crustaceans, nematodes, and seaweeds and appear to have been hardwired for symbiotrophy since their inception.

SUBKINGDOM (Division) ISOKONTA

Pr-22 Amoebomastigota (Amoebomastigotes, Heteroloboseans)

GENERA
Cercomonas
Gruberella
Mastigameba
Naegleria
Paratetramitus
Vahlkampfia
Willaertia

Amoebomastigota, or Schizopyrenida, transform from amebas into undulipodiated swimmers.

The amoebomastigotes, such as *Paratetramitus* (Figure A), are freshwater or symbiotrophic microbes distinguished by their ability to change from an undulipodiated to an ameboid stage and back again. This transformation, induced by the depletion of nutrients, is best studied in *Naegleria* because this genus can be cultured. When *Naegleria* amebas are suspended in distilled water, they develop kinetosomes, grow [9(2)+2] axonemes from them, and soon elongate into a mastigote form. They quickly swim in search of food bacteria. After they find it, they lose their undulipodia and return to an ameboid lifestyle.

That many amoebomastigotes lost the capacity to reversibly form kinetids including kinetosomes and their axonemes, has been frequently documented. The irreversible loss of undulipodia even occurs today in protists cultivated for extended periods under controlled conditions in the laboratory. We suspect that many, if not all, species of amastigote rhizopod amebas (Pr-2) evolved by loss of undulipodial motility. Therefore we support the suggestion, based primarily on molecular data and, in the case of desiccation resistent species, cyst and microcyst morphology, to erect a new phylum called Heterolobosea. We follow Frederick Page and Richard Blanton's proposal first presented in the mid-1980's to unite amebas with eruptive pseudopodia and, in some species, reversible ameba-mastigote transformations. The larger cohesive groups are vahlkampfids, gruberellids, their multinucleate (Willaertia) and their multicellular descendants: acrasids (sorocarp-forming ameboid slime molds). Most heteroloboseans show an ameba stage with cylindrical, not flattened pseudopods ("limax" amebas, that lack the fine extensions at the leading edge of the pseudopods called subpseudopodia). However not all species in this DNA-established group retain either the reversible ameba-mastigote transformation nor even the phagocytotic eruptive pseudopods. Most heterolobosean cells have single nuclei but the gruberellids tend to be multinucleate with from six to over thirty-five nuclei. Bactivorous feeding, and therefore survival requires retention of the amoeboid phagocytotic-pinocytotic morphology whereas other semes (mastigote-based rapid swimming, cyst formation) may be dispensable in stable habitats. The orphan genus *Stephanopogon*, although it does not form an ameba with eruptive pseudopodia at any stage and formerly was considered a ciliate, by molecular criteria apparently should be included in "Heterolobosea". We anticipate further refinements of this potential protoctist phylum. The reconstructed evolutionary history-taxonomy of amebas has challenged and delighted observers since the invention of the light microscope. With erection of Heterolobosea the end of classification schemes for amebas (heterogeneous, ubiquitous, free-living, symbiotic speciose groups of charming voracious, shape-changing protoctists whether crawling in corpses, compost, leaf-litter, soil, sand, stuck on frozen bull dung or swimming in pond water, marine pools or brain tissue) remains out of sight.

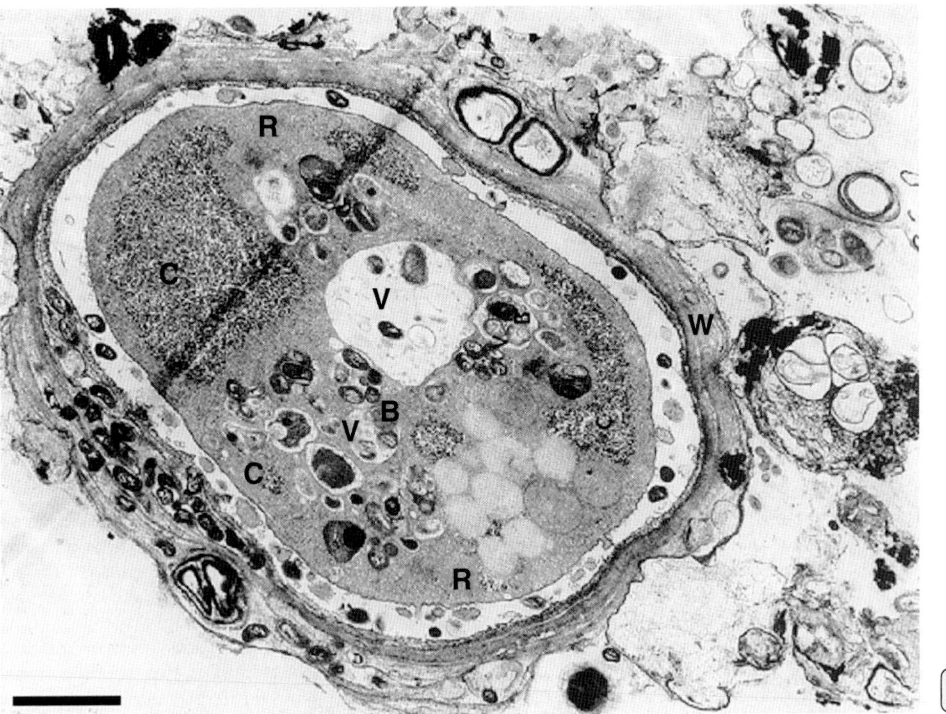

Figure A *Paratetramitus jugosus*, an amebomastigote that grows rampantly in microbial mats. From Baja California Norte, Laguna Figueroa, Mexico; these cysts and amebas are found with *Thiocapsa* (B-3) and other phototrophic bacteria. W = cyst wall; R = ribosome-studded cytoplasm; B = bacteria being digested in vacuoles (V); C = well-developed chromatin, source of chromidia (propagules). TEM, bar = 1 μm.

Pr-23 Myxomycota

(Myxogastria, plasmodial slime molds)

Greek *myxa*, mucus; *mykes*, fungus

GENERA

Arcyria	*Diderma*	*Metatrichia*
Barbeyella	*Didymium*	*Perichaena*
Cavostelium	*Echinostelium*	*Physarella*
Ceratiomyxa	*Fuligo*	*Physarum*
Cercomonas	*Hemitrichia*	*Protostelium*
Clastoderma	*Leocarpus*	*Sappinia*
Comatricha	*Licea*	*Stemonitis*
Dictydium	*Lycogala*	

The myxomycotes enjoy many names, including myxomycetes, mycetozoa, plasmodial (or acellular) slime molds, true slime molds, and Myxomycotina. Like the cellular slime molds—the acrasiomycotes (Pr-2)—these plasmodial slime molds form an ameboid stage that lacks cell walls and feeds on bacteria by engulfing them with pseudopodia—for example, phagocytosis. Also like the acrasiomycotes, the plasmodia can differentiate into reproductive structures that are stalked and funguslike in appearance. Unlike the acrasiomycotes, however, the myxomycotes are overtly sexual.

Like foraminifera (Pr-3) and plants (CH 5), some species of myxomycotes typically alternate free-living haploid and diploid phases. In the haploid phase, cells in this phylum bear two anterior undulipodia of unequal lengths. These cells convert into ameboid cells called myxamebas. The myxamebas reconvert just as readily into undulipodiated cells (mastigotes). Both the undulipodiated cells and the myxamebas may differ relative to mating types. Either two amebas or two mastigotes fuse to form a diploid zygote. Nuclei in the zygote undergo repeated mitoses to form a large mass of multinucleated cytoplasm, the plasmodium. Unlike comparable structures in acrasiomycotes, the plasmodium of myxomycotes is not cellular, because the diploid nuclei are not separated by cell membranes. The overall form of the plasmodium varies between orders, but, in some, the plasmodium takes on a veined or reticulated structure.

Plasmodia are found as a slimy wet scum on fallen logs, bark, and other surfaces. The plasmodia, although usually pigmented orange or yellow, do not photosynthesize. They are phagotrophs feeding on bacteria (and certain small protists) whose abundant populations develop on decaying vegetation. The size and shape of these slime molds are in no way predetermined; bits taken from a plasmodium form new individuals that can feed and grow independently. The plasmodium migrates in an ameboid manner. A pulsating back-and-forth movement of the plasmodium is obvious under the microscope, as an incessant intraplasmodial flow. The movement distributes metabolites and oxygen evenly.

If its surroundings become drier, the plasmodium may mature to the fruiting stage. Portions become concentrated into mounds, from which stalked, or sessile (nonstalked), spore-producing structures (sporangia, also called sporocarps) form. Meiosis takes place inside the maturing spores contained in the sporocarp. In some cases, three of the meiotic nuclei degenerate and the fourth persists as the nucleus of the mature spore, which germinates into either a haploid ameba or a haploid mastigote. Most of the life cycle of these organisms, then, is spent in the diploid stage. The direct development of haploid amebas and mastigotes into plasmodia, without fusion to form a diploid zygote, has also been reported, so the ploidy of any slime-mold mass must be ascertained in each case.

Some 400 to 500 species of myxomycotes are documented, with larger members of the phylum being best known. The color, shape, and size of the reproductive structure, the presence of a stalk, the presence of a sterile structure (the columella) at the top of the stalk, the presence of calcium carbonate ($CaCO_3$) granules or crystals in or on the reproductive body and the surface features of the spore wall distinguish species. Inside young sporocarps of some species, a system of threads develops, which are sterile in that they do not give rise to spores. This thread system, called the capillitium, differs between various myxomycote groups and is used as a taxonomic marker.

Although the mastigote ameba *Cercomonas* does not form plasmodia, we group it here with the myxomycotes and representative genus of the ancestral class, Cercomonadida. Insight into the many mastigotes of the subvisible world leads biologists to believe that these diverse, tiny, ubiquitous swimmers (classified in phyla Pr-11, Pr-12, Pr-22, Pr-23, Pr-24, or depending on detailed structure Pr-26) are ancestral to several lineages of larger heterotrophic protoctists (such as slime molds and water molds). The other two classes, both of which do form plasmodia, are Protostelida and Myxomycetes.

The major subgroups, here orders within class Myxomycetes, are Echinosteliales, Trichiales, Liceales, Stemonitales, and Physarales.

The spores of members of the first three orders—Echinosteliales, Trichiales, and Liceales—are pale, as a rule, and do not deposit $CaCO_3$. The sporocarps of members of order Echinosteliales are tiny (less than 1 mm high) and contain capillitia. The diploid feeding stage is a protoplasmodium—a microscopic amebalike plasmodium that lacks veins or reticulations. The order Echinosteliales comprises two families: Echinosteliidae, having only the genus *Echinostelium* (Figures A and B), and Clastodermidae, three tiny species belonging to the genera *Clastoderma* and *Barbeyella*. The sporocarps of members of order Trichiales contain sculptured capillitia. The diploid feeding plasmodium is midway between an aphanoplasmodium (a thin, inconspicuous plasmodium consisting of a reticulum, or network, of veins having fan-shaped leading fronts) and a phaneroplasmodium (a thick, more conspicuous structure whose veins and leading fronts are visible to the unaided eye). The genera include *Perichaena*, *Arcyria*, and *Metatrichia*. Members of order Liceales may be either protoplasmodial or phaneroplasmodial. Their sporocarps are of diverse shapes and lack capillitia. Genera include *Licea*, *Lycogala*, and *Dictydium*.

Members of the order Stemonitales, which typically form aphanoplasmodia, and of order Physarales, which typically form phaneroplasmodia, bear dark purplish to brown or black spores. They are the largest and best-known slime molds. Members of Stemonitales, which include *Stemonitis* and *Comatricha*, form dark fingerlike upright sporocarps. Members of order Physarales often have conspicuous $CaCO_3$ deposits in the capillitium and other parts of the sporocarp. Some 85 species of the genus *Physarum* are known. Their very active protoplasmic streaming and the ease with which some myxomycotes can be grown in the laboratory have made them useful in the study of proteins engaged in cell motility. The yellow plasmodia of *Physarum polycephalum* have nonmuscle actin and myosin proteins homologous to the actomyosin complexes of animal muscle.

MODE IV

Figure A Sporophore of the plasmodial slime mold *Echinostelium minutum.* LM, bar = 0.1 mm. [Photograph courtesy of E. F. Haskins.]

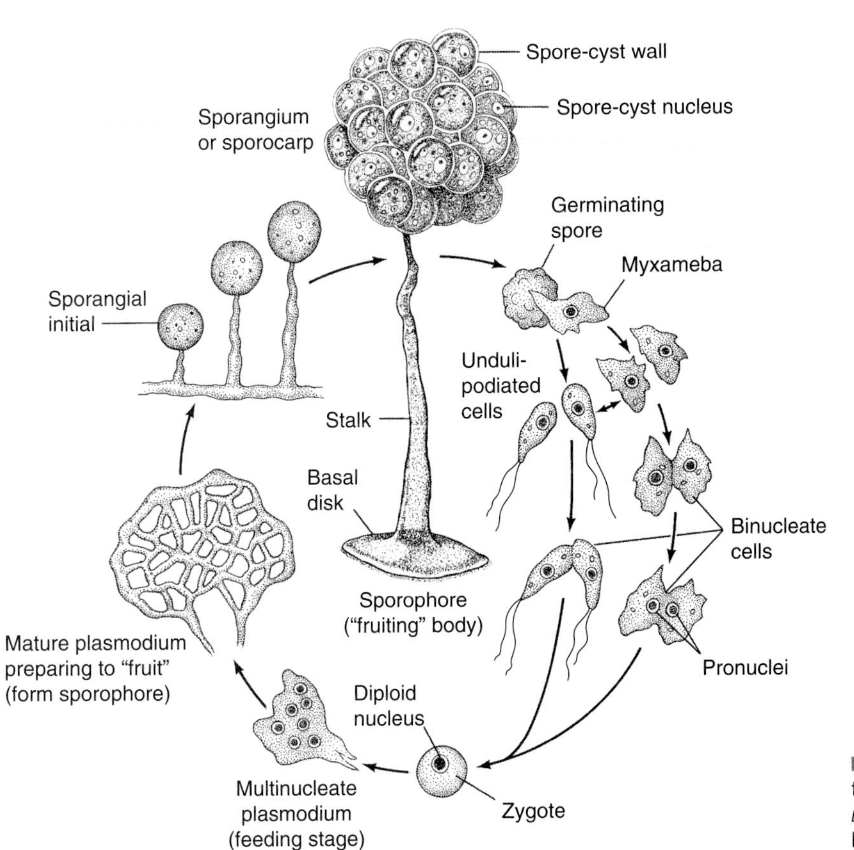

Figure B Life history of the plasmodial slime mold *Echinostelium minutum*. [Drawing by L. Meszoly.]

Pr-24 Pseudociliata

GENERA
Stephanopogon
Percolomonas
Psalteriomonas
Permina

The 16 nuclei per cell in all four genera of this provisional phylum are homokaryotic: all the same. Their name is derived from the fact that they are utterly unlike the ciliates that they superficially resemble. They lack dimorphic nuclei of ciliates (Pr-6 whose cells contain genetic diploid-micronuclei unable to synthesize messenger RNA and the physiological macronuclei that show all sorts of variation on the theme of ploidy). The fact that Pseudociliata can not be classified with ciliates is confirmed by the strange, unique structure of their kinetids (Figure Pr-24B) so different from those of the thousands of ciliates studied by electron microscopy. Pseudociliates are free-living marine benthic protoctists that feed on diatoms, other smaller protists and bacteria. Some are osmotrophic: they take in dissolved nutrients at the molecular level as proteins, amino acids or polysaccharides. They are undulipodiated; their undulipodia are aligned primarily in rows on the ventral surfaces of their single cells. Reproduction, in the absence of any signs of gender differences, meiosis or sexuality, occurs within a cyst as a form of palintomy. This process, a kind of uniparental multiple fission, yields many smaller offspring at once. They seem to have been cut from the larger parent although they do not differ much from it. The offspring pseudociliates emerge from the parental cyst and swim away as binucleates. *Stephanopogon* species have discoidal mitochondrial cristae (rather then the tubular mitochondrial cristae found in cells of ciliates and the dinomastigotes of Pr-5). Pseudociliates lack the typical stacked membranous cisternae of parabasalid (Pr-1) or other golgi bodies. The binucleate cells in three to four closed mitoses multiply to a total 12–16 nuclei per cell followed by encystment.

This phylum currently contains only four species all of which belong to the genus *Stephanopogon* depicted here (Figure Pr-24A): *S. colpoda* drawn from work of John Corliss, 1979; *S. mesnili* (based on a drawing by Andre Lwoff, 1923), *S. apogon* work of A. Borror, 1965 and *S. mobilensis*, work of Jones and Owen, 1974. See Margulis and Chapman, 2010.

Other genera of protists that may belong to Pseudociliata include *Percolomonas* sp. from anoxic sulfurous salt marshes. This free-living marine mastigote also lacks an amoeboid life history stage but on molecular criteria apparently can be related to *Stephanopogon. Percolomonas* cells have four anterior undulipodia, one of which, long and trailing, is used for adhesion to solid surfaces. The other three lie in a ventral groove also called the gutter. *Psalteriomonas* is an amoebomastigote whose cells contain four nuclei as four karyomastigonts but only during the swimming stage. During the alternative ameboid stage in its life history the *Psalteriomonas* cell is uninucleated. The genus has been collected from anoxic habitats and studied in the laboratory. Its members are apparently microaerophils whose cells harbor methanogenic symbionts and/or hydrogenosomes. *Permina* sp. with both amebal and mastigote stages, have encysting cells and mitochondria with flattened cristae.

Although these heterotrophic motile mitochondriate protist genera of Mode III are isolated from each other and from the other Protoctista and most, like *Stephanopogon,* lack any amebal life-history stages, they may be soon relatable to a new taxon: the amebomastigote Heteroloboseans of Phylum-22, see page 188. They exemplify our principles: the importance for taxonomy of the karyomastigont (especially the kinetid) and the convergent evolution of many semes: cyst formation, multinuclearity, multiple fission, multicellularity, organelle/bacterial symbiont acquisition and undulipodial loss. Problems of understanding taxonomy as reflected in the evolution of diversity are aided immensely by genomic and other molecular analyses but they are not definitively resolved without study of live communities of organisms in natural habitats through their seasonal and other temporal cycles.

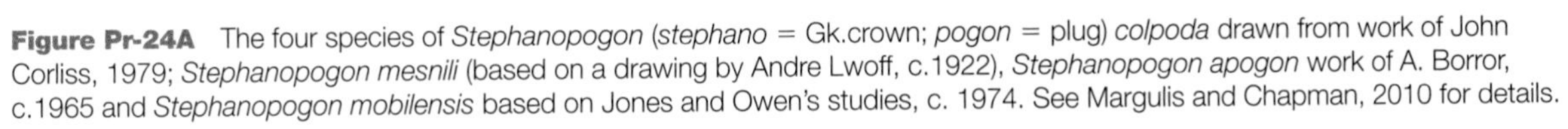

Figure Pr-24A The four species of *Stephanopogon* (*stephano* = Gk.crown; *pogon* = plug) *colpoda* drawn from work of John Corliss, 1979; *Stephanopogon mesnili* (based on a drawing by Andre Lwoff, c.1922), *Stephanopogon apogon* work of A. Borror, c.1965 and *Stephanopogon mobilensis* based on Jones and Owen's studies, c. 1974. See Margulis and Chapman, 2010 for details.

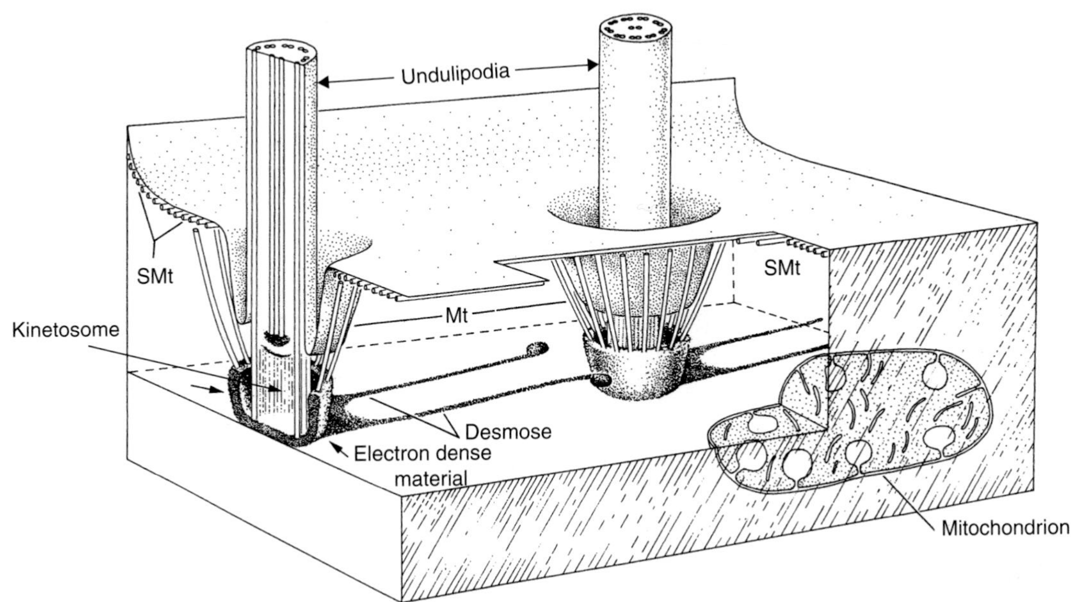

Figure Pr-24B Two kinetids with their emergent undulipodia are depicted in a three-dimension cut-away section of the cortex of a member of the genus *Stephanopogon* based on electron microscopy. Subpellicular microtubules (SMt) in a basket arrangement surround the kinetosome of each undulipodium and subpellicular microtubules (Smt) run longitudinally under the cell membrane. Dense material (arrows) from which extends the two-pronged desmose (pointers) that emanate from nodes in the cortex at each kinetosome (long arrow). Each linear array of kinetids forms a row, a kinety that is convergent, not homologous to a ciliate kinety (Pr-6). Work by Lipscomp and Corliss, references in Margulis and Chapman, 2010.

Pr-25 Haptomonada

(Prymnesiophyta, Haptophyta, coccolithophorids)

Greek *haptein*, fasten; *phyton*, plant

GENERA

Calcidiscus
Calciosolenia
Calyptrosphaera
Chrysochromulina
Coccolithus
Discosphaera
Emiliania
Gephyrocapsa
Hymenomonas
Phaeocystis
Pontosphaera
Prymnesium
Rhabdosphaera
Syracosphaera

The planktic, golden-brown algae, tiny, usually spherical, sometimes undulipodiated haptomonads have been viewed by marine biologists and paleontologists as two different types of organisms: (1) golden-brown, algae (Figure A) that resemble planktic chrysomonads (Pr-15) and (2) coccolith-covered coccolithophorids (Figure B). Coccoliths, the microscopic disklike calcium carbonate scales of renown to paleontologists, are produced by the cell, a coccolithophorids, which bears them as packed surface plates. Ultrastructural, developmental, and molecular biological studies have united the two photosynthetic chrysomonad-like alga and the coccolithophorid as the same organisms at different stages in their life histories.

Coccoliths first reported by Christian Ehrenberg in 1836 from the Cretaceous chalk and waters of Rugen Island in the Baltic Sea have been admired ever since for their beauty. Yet, study of the geochemical importance, scale biomineralization in relation to environmental variation including seasonal change, sexuality and developmental life history of the coccolith bearing haptomonads was only instigated by satellite observation at the end of the 20th century. Not until this 21st century, has the quantitative importance of this phylum of numerous, common but tiny algae been recognized.

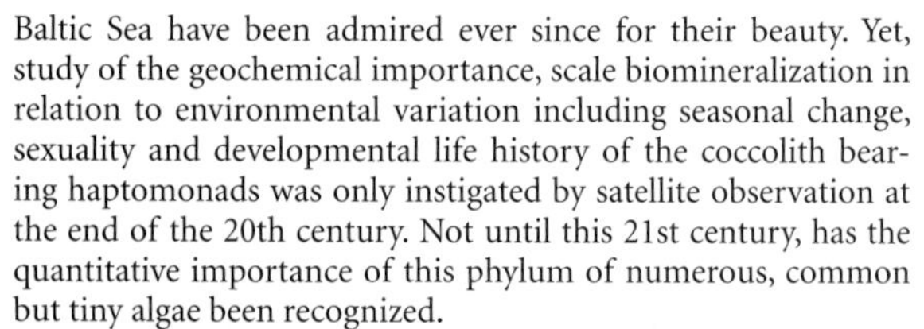

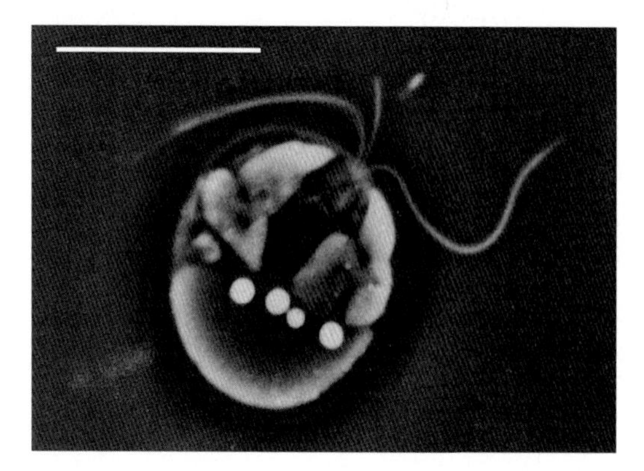

Figure A *Prymnesium parvum*, a living marine haptomonad, showing undulipodia and haptoneme. LM, bar = 10 μm. [Courtesy of I. Manton and G. F. Leedale, *Archiv für Microbiologie* 45:285–303 (1963).]

Figure B *Emiliania huxleyi*, a coccolithophorid from the Atlantic. It was not realized until the 1980's that Coccolithophorids are the resting stage of haptomonads. SEM, bar = 1 μm. [Courtesy of S. Honjo.]

Monothecate (one layer), Dithecate (2 layers) and Multilayered (3 or more) coccospheres (test of the coccolithophorid) are recognized. Each of these stages may have only one coccolith type, or 2 or more coccolith types on the same cell. When there are 2 layers the inside layer is called the endotheca and the outside layer is the exotheca. If any theca has only one kind of coccolith is is monomorphic, if 2 different types of coccoliths it is dimorphic (heteromorphic). If any layer has 3 or more kinds of layers it is polymorphic. Three forms of coccoliths include heterococcoliths (most common), holococcoliths and nannoliths. Multilayed coccospheres like *Emiliania* tend to have all the same kind of coccoliths (holococcolithophore). Often around the apical pole of the cell where the haptoneme and undulipodia emerge, circumundulipodial coccoliths often differ from the rest of the bodies coccolith. The use of scanning electron microscopy (SEM) to study the calcium carbonate "buttons" has revolutionized our understanding of this phylum.

Haptomonads, the larger taxon, includes scaleless algae and all those that bear calcium carbonate "buttons" (coccoliths) at any stage. They are primarily marine organisms, although some freshwater genera are known. Although they have golden yellow or brown plastids in common with chrysomonads, their cell structure differs enough to justify placing haptomonads in their own phylum. Unlike the chrysomonads, haptomonads have only a very weak tendency to become multicellular. Reproduction by uniparental binary fission is well known in all studied but details of complex developmental cycles that involve both miosis and syngamy (fertilization) are under study now. Cloned cells may display haploid motile "undulipodiated swarmers", or a haploid benthic naked (no scales) filamentous phase (*Apistonema*) that alternate with diploid heterococcolith stages. What is astounding is that a single cell of *Coccolithus pelagicus* a heterococcolithophore develops into *Crystallolithus hyalinus*, a holococcolithophore. What has been thought to be two different genera are developmental stages of a single cell (Figure C).

Haptomonads are distinguished by their haptonemes, scales, and in some, their coccoliths. The haptoneme is a thread, often coiled, which may be used as a holdfast to anchor the free-swimming protoctist to a stable object (Figure D). Each cell has one haptoneme, generally at its anterior end. The haptoneme, which may be a specialized modification of the ubiquitous [9(2)+2] undulipodium (Figure I-2), is a microtubular structure consisting of six doublets of microtubules arranged in a circle whose center is occupied by a single microtubule or none at all. With the haptoneme at the anterior end of the cell are two standard undulipodia and, generally, a membranous Golgi body.

In the transformation of the free-swimming stage into the resting, coccolithophorid stage, coccoliths and scales develop inside the cell: calcium carbonate crystals precipitate on

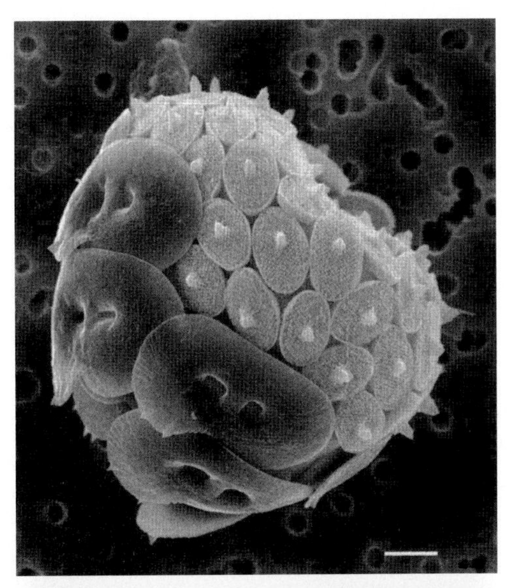

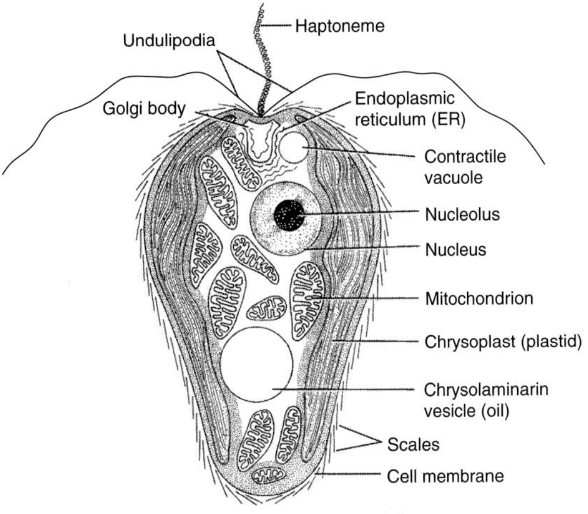

Figure D *Prymnesium parvum*, the free-swimming haptonemid stage of a haptomonad. The surface scales shown here are not cocoliths form. [Drawing by R. Golder.]

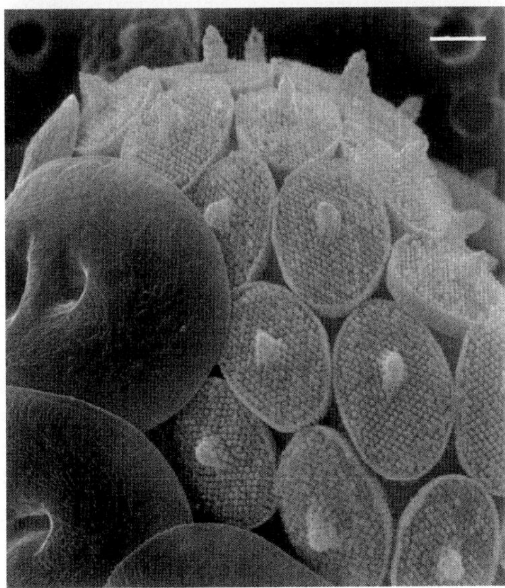

Figure C *Helicosphaera carteri*, (Wallich) Kamptner var. *carteri*: (A) a well-formed combination coccosphere of *H. carteri* (heterococcoliths) and the former *Syracolithus catilliferus* (holococcoliths). SEM, bar = 2 μm; (B) detail of A. SEM, bar = 1 μm. [Courtesy of L. Cros and J.-M. Fortuño, *SCI. MAR.*, 66 (Suppl 1):7-182 (2002).]

conspicuous scales, which are made of organic polymers inside the Golgi apparatus. The gradually assembling coccoliths are transported to the edge of the cell by microtubule-mediated processes. They are deposited, in some cases with exquisite regularity, on the cell surface. The scales and the coccoliths that form on them bear intricate patterns that are species or genus specific. These coccolithophorid stages are often resistant, permitting tolerance of conditions that would be prohibitive to the swimming, growing forms of haptomonads.

Coccolithophorids have continuously produced great quantities of particulate calcitic carbonate for 100 million years or so since the Cretaceous period; they have contributed significantly to the chalk deposits of the world. Because coccolithophorids are distinctive, they serve as stratigraphic markers; several hundred morphotypes or fossil species have been studied by geologists. For haptomonads, more is thus known about the morphology of fossils than of extant forms. The correlation between the haptonemid and the coccolithophorid stages has still not been made for many species. More work on haptomonad life cycles, including the development of calcium carbonate skeletal patterns, is needed in the context of the biology of the living cells.

Haptomonads typically have two golden yellow plastids (chrysoplasts) surrounded by a plastid endoplasmic reticulum that is continuous with the nuclear membrane. The plastids contain chlorophylls ***a***, c_1, and c_2 but lack chlorophylls ***b*** and *e*. In addition to beta-carotene, they have alpha- and gamma-carotenes. They have fucoxanthin, an oxidized isoprenoid derivative that is also found in diatoms (Pr-18) and brown algae (Pr-17). Fucoxanthin is probably the most important determinant of the brownish yellow color. Haptomonads do not store starch; rather, like euglenids (Pr-12), they form a glucose polymer having the β-1–3 linkage of the monosaccharides. This white storage material, called paramylon, is stored within pyrenoids (proteinaceous structures) between the thylakoids (photosynthetic membranes) of the plastids.

Populations of the naked haptomonad *Phaeocystis poucheti*, which lack coccoliths, are responsible for the production of large quantities of dimethyl sulfide, an atmospheric gas that helps form cloud condensation nuclei (and therefore rain over the open ocean).

Pr-26 Cryptomonada
(Cryptophyta)

Greek *kryptos*, hidden; Latin *monas*, unit; Greek *phyton*, plant

GENERA
Cyanomonas
Chroomonas
Cryptomonas
Hillea
Hemiselmis
Goniomonas
Guillardia
Proteomonas
Rhodomonas

Cryptomonads are flattened, elliptical swimming cells. Both heterotrophic and photosynthetic, they are found all over the world in moist places. Some commonly form blooms on beaches, whereas others have been found as intestinal symbiotrophs in domesticated animals. Palmelloid colonies (for example, nonmotile cells) embedded in the gel of their own making are known as well. These widely differing habitats have led differently trained scientists—such as marine botanists and parasitologists—to study them. Confusions in terminology and in taxonomy and general ignorance of their existence abounded, especially before electron microscopy and molecular analysis revealed them to be a clearly delineated group.

Like the euglenids of phylum Pr-12, cryptomonads may be pigmentless, animal-like "protozoa" or brightly pigmented and photosynthetic plantlike algae. Found primarily as free-living single cells, commonly in freshwater, they are unlike euglenids in details of cell structure and division. Their photosynthetic pigmentation, if present, also is unique.

The cryptomonad bears two anterior undulipodia inserted in a characteristic way along the gullet, also called the crypt. The colorless genus *Goniomonas*, for example, ingests particulate food through its gullet (Figures A through C). In the phagotrophic members of the group, which eat bacteria or other protoctists, the crypt is typically lined with trichocysts and bacteria-like bodies. Trichocysts expel poisons, which subdue and kill the microbial prey. Most members of the photosynthetic genera also have trichocysts.

Pigmented cryptomonads, as a rule, contain in their plastids chlorophyll $\boldsymbol{c_2}$ in addition to chlorophyll $\boldsymbol{a}$. Members of photosynthetic genera, such as *Cryptomonas* and *Chroomonas*, contain unique protein–pigment complexes called phycobiliproteins. Unlike most algae, they lack beta-carotene and zeaxanthin, but they contain alpha-carotene, cryptoxanthin, and alloxanthin. Many cryptomonads with those pigments are green or yellowish green. Others also contain phycocyanin or phycoerythrin and so tend to be deeper blue or deeper red. In general, phycocyanin pigments are strictly limited in nature: they are found in most cyanobacteria (B-6) and in the rhodoplasts of red algae (Rhodophyta, Pr-33), as well as in glaucophyte algae such as *Cyanophora paradoxa*, organisms whose plastids share features in common with the cyanobacteria from which they evolved. Cryptomonads acquired photosynthesis secondarily through the ingestion and retention of red algae. The symbiont nucleus persists in a miniaturized form called a nucleomorph, a feature shared with the chlorarachniophytes (Pr-3). Some members of the genus *Cryptomonas* are secondarily nonphotosynthetic and possess reduced plastids and nucleomorphs.

Meiotic sexuality and gametogenesis is virtually unknown in cryptomonads. Many have been grown and observed in the laboratory. They simply divide into two offspring cells. Just before cell division, new kinetosomes and undulipodia appear with a new crypt in proximity to the old one. The new oral structure then rotates, migrating to the opposite end of the cell. In the meantime, chromatin inside the closed nuclear membrane forms small knobby chromosomes that segregate into two bundles at opposite sides of the nucleus. The nucleus divides, cytokinesis ensues, and two offspring cells with a plane of mirror symmetry

Figure A *Goniomonas truncata*, a freshwater cryptomonad. SEM, bar = 5 μm [formally *Cyathomonas truncata* (Fresenius) Fisch, 1885].

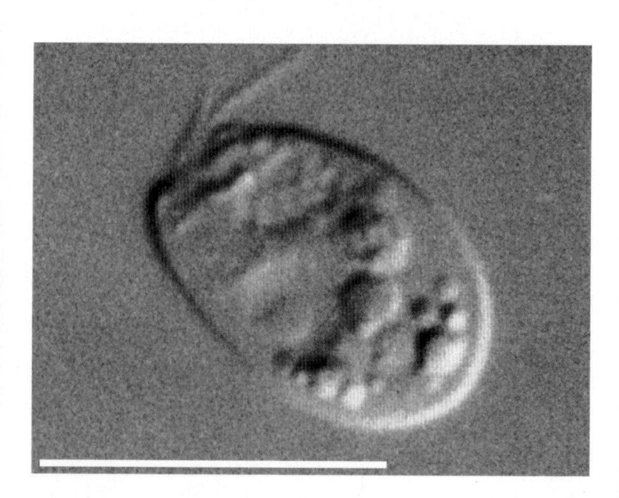

Figure B *Goniomonas truncata*, live cell. LM, bar = 5 μm. [Courtesy of F. L. Schuster.]

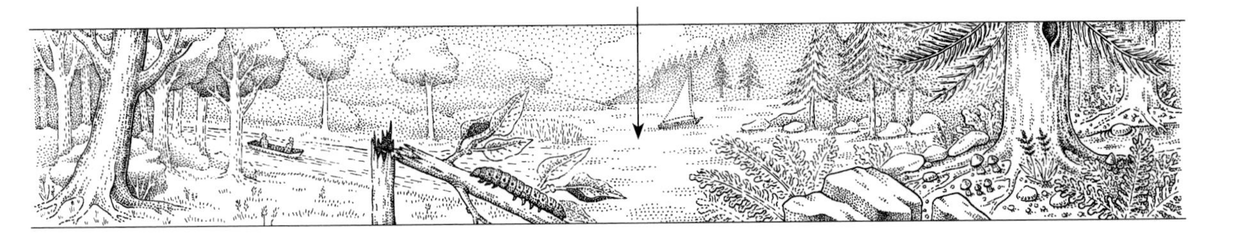

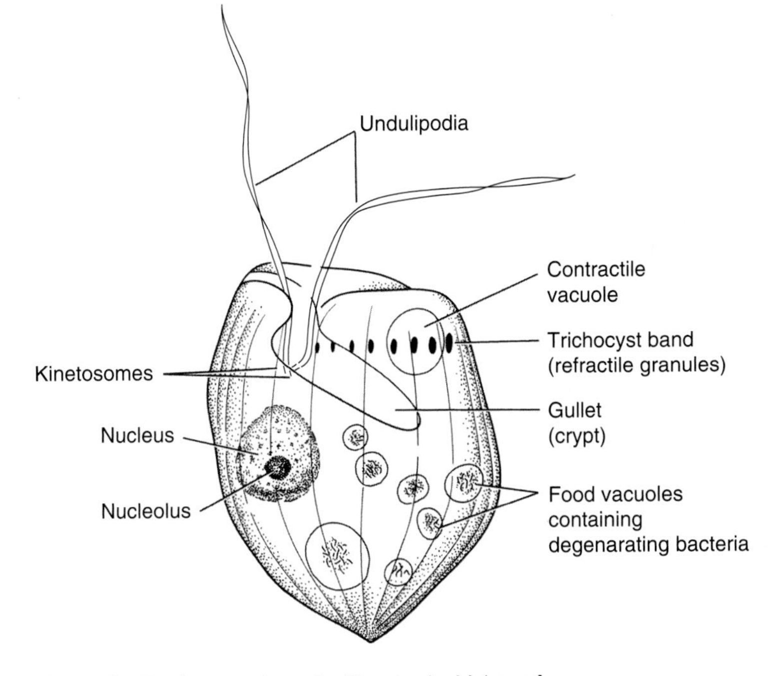

Figure C *Goniomonas truncata*. [Drawing by M. Lowe.]

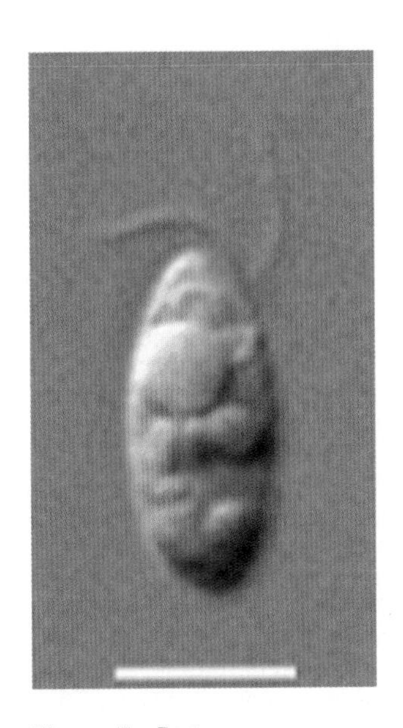

Figure E *Proteomonas sulcata*, a marine photosynthetic cryptomonad. LM, bar = 10 μm. [Courtesy of J. M. Archibald.]

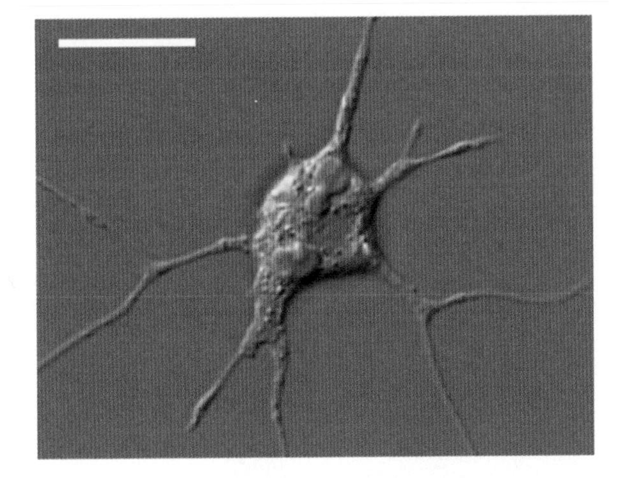

Figure D *Chlorarachnion reptans*, a chlorarachniophyte alga. LM, bar = 10 μm. [Courtesy of J. M. Archibald.]

between them separate. This type of reproduction by binary fission distinguishes cryptomonads, regardless of their nutritional mode, from other protoctists. It was first documented, beautifully, by Karl Belar in 1926.

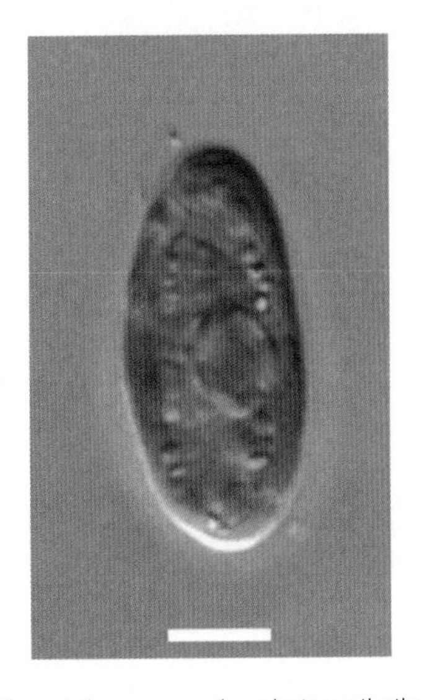

Figure F *Storeatula* sp., a marine photosynthetic cryptomonad. LM, bar = 10 μm. [Courtesy of J. M. Archibald.]

Pr-27 Eustigmatophyta

Greek *eu*, true, original, primitive; *stigma*, brand put on slave (as refers to eyespot), mark, spot; *phyton*, plant

GENERA

Chlorobotrys
Ellipsoidion
Eustigmatos
Monodopsis
Nanochloropsis
Pleurochloris
Polyedriella
Pseudocharaciopsis
Vischeria

Because they are yellowish green, form immotile coccoid vegetative cells, and propagate by motile, elongated asexual zoospores, eustigmatophytes were originally lumped together with the xanthophytes (Pr-16), which they resemble. Electron microscopic studies, however, reveal a distinctive eyespot and organization of the eustigmatophyte cell, distinguishing the morphology that justifies recognition of eustigmatophytes as a unique set of photosynthetic motile protoctists warranting their own phylum. Only the genera listed above are known to be in phylum Eustigmatophyta for sure, but this is due far more to lack of study at the ultrastructural level than to a paucity of these organisms. At present, *Pleurochloris*, *Polyedriella*, *Ellipsoidion*, and *Vischeria* (Figure A) are the major genera of "eustigs," as they are fondly called. Although multicellular eustigs exist—for example, colonies of *Chlorobotrys* cells surrounded by layered mucilage that form no zoospores—the majority are independent single cells. They live primarily in freshwater.

In the pigmentation of their plastids, eustigs are indeed very much like the true yellow-green xanthophytes. Their plastids, called xanthoplasts, contain chlorophyll ***a***, as all oxygen producing organisms do; in addition, they contain chlorophylls c_1, c_2, and *e*. They lack chlorophyll ***b***. They contain beta-carotene and several oxygenated carotenoids, depending on the genus. Violaxanthin is commonly present; epoxanthin, diadinoxanthin, and diatoxanthins may also be present. Eustigs store glucose not as starch but as a solid material (not yet chemically identified) that lies outside of the plastid. In some vegetative cells, a conspicuous polyhedral crystalline body constitutes the pyrenoid of the plastid; it is typically attached to the thylakoids by a thin stalk.

Although the pigments of eustigs are like those of xanthophytes (Pr-16), the cell organization is not. Most eustigs have only a single mastigonemate (hairy) anterior undulipodium, at the base of which is a conspicuous undulipodial swelling, T-shaped in transverse section (Figures B and C). An adjacent swelling filled with drops of carotenoids forms the eyespot; it probably communicates somehow with the undulipodium to direct the cell to optimally lighted environments. The eyespot is not associated with the plastid, nor is it membrane bounded. The xanthophyte lacks such a swelling on its anterior undulipodium but has a swelling on the posteriorly directed second undulipodium, which is apposed to a specialized part of the plastid. Some eustigs (for example, *Ellipsoidion*) have a second, smooth undulipodium.

The yellow-green eustig plastid is single and long; it lies in the center or at the posterior end of the cell and fills some two-thirds of its volume. The thylakoids are stacked inside, rather like the grana of plant plastids. Nearly all the endoplasmic reticulum (ER) of eustigs is associated with the plastid. The xanthoplast ER is not associated with the nuclear membrane, as it is in many other algae, and there is little developed free ER in the cytoplasm. This morphological arrangement suggests an integrated metabolic relation between the products of photosynthesis and the nucleocytoplasm-directed biosyntheses. The cell wall is entire—that is, it completely surrounds the cell—and in some

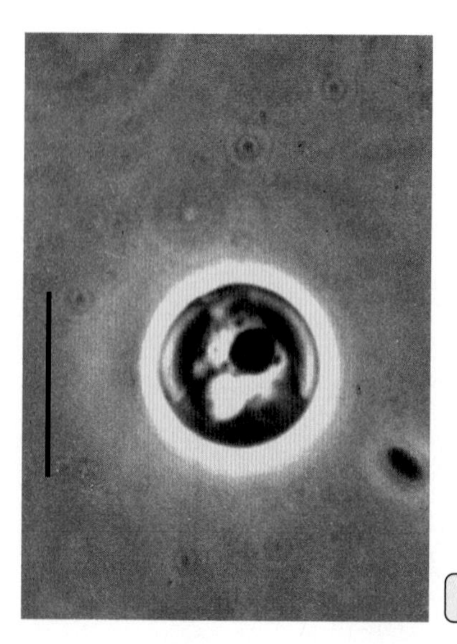

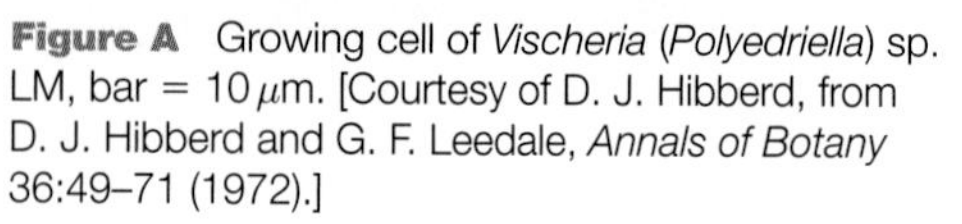

Figure A Growing cell of *Vischeria* (*Polyedriella*) sp. LM, bar = 10 μm. [Courtesy of D. J. Hibberd, from D. J. Hibberd and G. F. Leedale, *Annals of Botany* 36:49–71 (1972).]

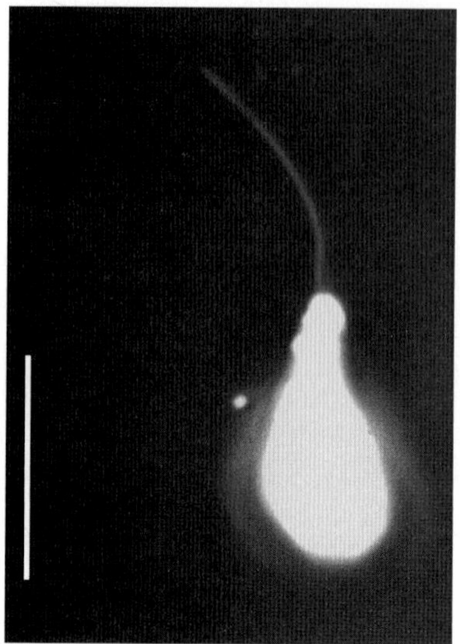

Figure B Zoospore of *Vischeria* sp. LM, bar = 10 μm. [Courtesy of D. J. Hibberd, from D. J. Hibberd and G. F. Leedale, *Annals of Botany* 36:46–71 (1972).]

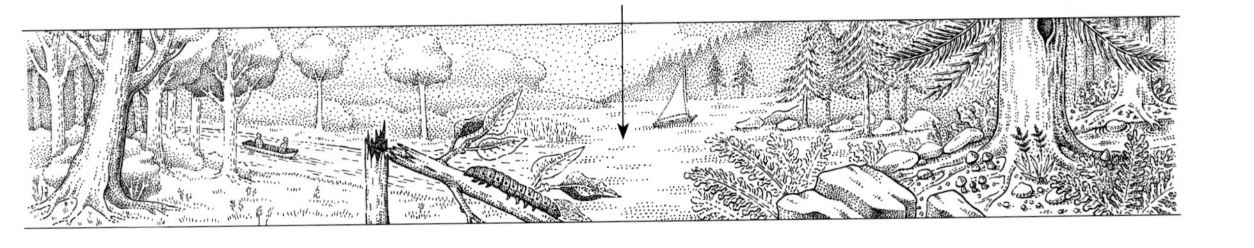

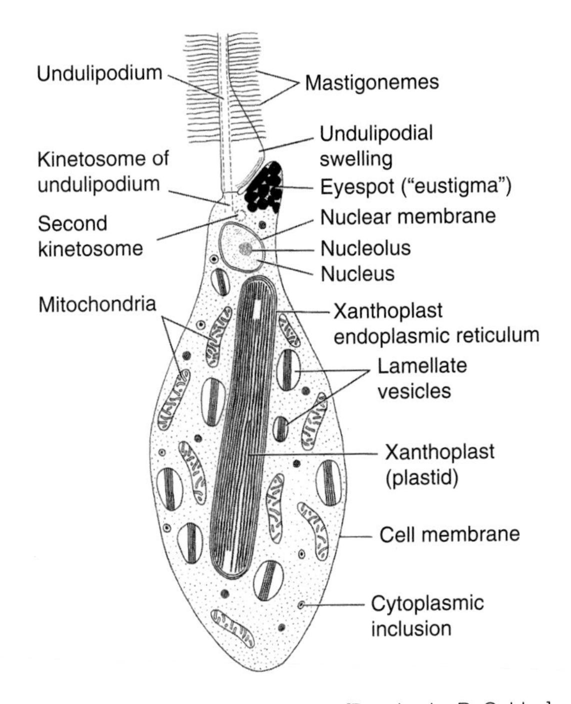

Figure C Zoospore of *Vischeria* sp. [Drawing by R. Golder.]

cases, it contains silica deposits. The cell divides directly into two offspring cells, and sexual processes are unknown in the group.

These planktic algae, at the base of aquatic food chains, are eaten by other protoctists and animals. Scientists have studied them only very little, however; so not much is known about their natural history.

Pr-28 Chlorophyta

(Green algae)

Greek *chloros*, yellow green; *phyton*, plant

GENERA

Acetabularia
Bryopsis
Caulerpa
Chaetomorpha
Chara
Chlamydomonas
Chlorella
Chlorococcum
Cladophora
Cylindrocapsa
Dunaliella
Enteromorpha
Fritschiella
Gonium
Halimeda
Klebsormidium
Lamprothamnium
Microspora
Nitella
Oedogonium
Penicillus
Platymonas
Pseudobryopsis
Pyramimonas
Spongomorpha
Stigeoclonium
Tetraselmis
Tetraspora
Tolypella
Trebouxia
Ulva
Urospora
Volvox

We highlight the three phyla likely to have hiding in their midst the ancestors of the other three kingdoms of eukaryotes: Chlorophyta, mastigote green algae and their relatives, from which green plants (embryophytes) arose; Chytridiomycota (Pr-35), which may have given rise, by loss of undulipodia, to the ancestors of the fungal lineage; and Choanomastigota (Pr-36), a collar-cell mastigote microbes, one lineage of which is likely to be ancestral to animals (Figure A). Current phylogenetic classification based on sequence comparisons divides the green plants (Viridiplantae) into two divisions: Chlorophyta and Streptophyta, the latter contain all green algae related to the embryophytes (bryophytes, ferns and seed plants), ie. genera such as *Mesostigma, Chlorokybus, Klebsormidium, Chaetosphaeridium, Entransia, Coleochaete, Chara, Nitella,* etc. plus the conjugating green algae (Zygnematophyceae). However, we accord embryophytes Kingdom status based on whole-organism criteria and semes held in common.

Chlorophytes, algae that contain grass-green chloroplasts surrounded by two envelope membranes and often form zoospores or gametes, generally have two or four undulipodia of equal length. About 500 genera with as many as 16,000 species have been described. Within the phylum, several evolutionary lines have led from unicellular forms to multicellular organisms. Their chloroplasts contain chlorophylls ***a*** and ***b*** as well as the carotenoids carotin, lutein, zeaxanthin, antheraxanthin, violaxanthin, and neoxanthin, which also occur in the leaves of green plants. In addition, special carotenoids occur in certain taxa or are synthesized under certain environmental conditions (for example, siphonaxanthin, prasinoxanthin, echinenon, canthaxanthin, loroxanthin, and astaxanthin). Starch, the α-1,4-linked glucose polymer, is the carbohydrate reserve synthesized and stored in the chloroplast. Although plant scientists agree that the ancestors of the plants were green algae, no consensus has emerged about the possible sister group of the embryophytes, the most likely candidates being the stoneworts (Charales).

Phylum Chlorophyta here excludes the gamophytes (Pr-32), which lack undulipodia, but unites the siphonales, charales, and prasinophytes with the chlorophytes in the strict sense because most are green algae that have undulipodia at some stage in their life history; a notable exception is the order Chlorococcales (for example, *Chlorella*). Whereas our scheme emphasizes the tendency of the unicellular, biundulipodiated algae to give rise to impressive and cohesive classes of reproductively and morphologically complex water "plants," current consensus among green algal researchers is that undulipodia have been independently lost on many occasions in the green algae, just as they have in other eukaryotes. Some recent phylogenetic analyses of plastid genomes, moreover, indicate that the Gamophyta (Pr-32) share at least some degree of common ancestry with the land plants. Moreover, like land plants, chlorophytes and gamophytes contain chlorophylls ***a*** and ***b***. Some members of both phyla are at least periodically resistant to desiccation; that one or several such algae were the progenitors of the land plants seems incontrovertible. Our classification scheme, however, lends more weight to whole-cell biology and life history of organisms than to molecular-sequence comparisons. The total absence of undulipodia at all life stages of the Gamophyta, including their sex cells, argues that loss of undulipodia was basal to this Mode II taxon. A later, more derived loss of undulipodia in the Chlorococcales is insufficient to displace the entire phylum Chlorophyta from Mode IV.

Chlorophytes are a major component of the phytoplankton in freshwater and seawater; it has been estimated that they fix more than a billion tons of carbon in the oceans and freshwater ponds every year. They also occur as macroalgae (seaweeds) along the rocky shores of the oceans.

The cell walls of green algae, like those of land plants, are composed of cellulose and pectins or of polymers of xylose (*Bryopsis* and *Caulerpa*) or mannose (*Acetabularia*, Figure A) linked to protein. The walls in many genera are encrusted with calcium carbonate, silica, and less frequently other minerals such as iron oxides.

Sexuality is rampant in this group; there is a trend from isogamy, in which two motile gametes of like size and shape conjugate and fuse, toward oogamy, in which a large immotile egg is fertilized by a small motile sperm. The sperm is very much like the individual adults (*Chlamydomonas*, Figure B), zoospores, or isogametes of many species in the phylum. In *Acetabularia*, a diploid zygote is the production of fertilization. It immediately undergoes meiosis to regenerate the haploid stage in the life cycle.

Within the Chlorophyta, as presented here, are four major classes and other groups of uncertain status. They are class Chlorophyceae, with 11 orders, including Volvocales, Oedogoniales, and Chaetophorales; class Ulvophyceae, which includes, among others, orders Ulotrichales, Siphonocladales, Ulvales, and Caulerpales; class Charophyceae; and class Prasinophyceae. Within each of these classes, except the prasinophytes, which are unicells, trends from unicellular forms to various types of complex colonies can be seen.

Figure A *Acetabularia mediterranea*, a living alga from the Mediterranean Sea. Bar = 1 cm. [Courtesy of S. Puiseux-Dao.]

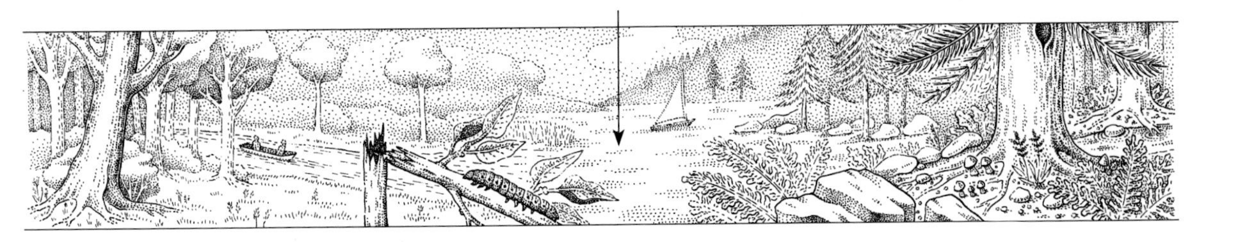

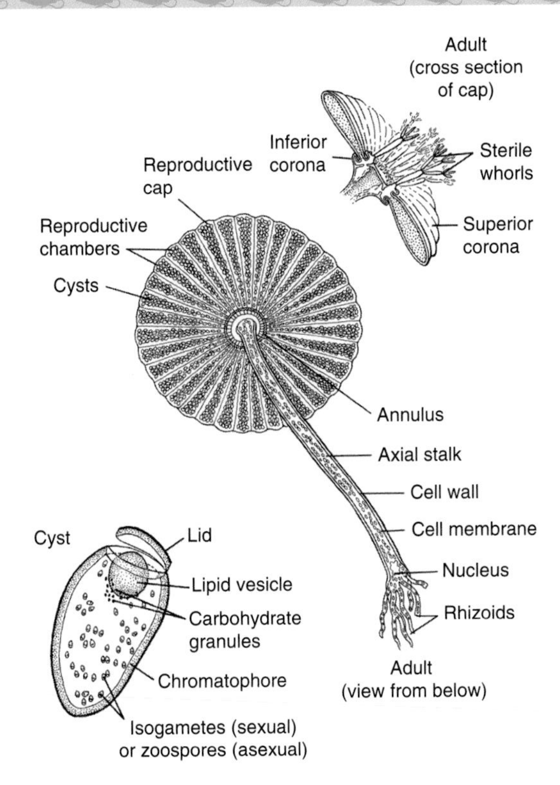

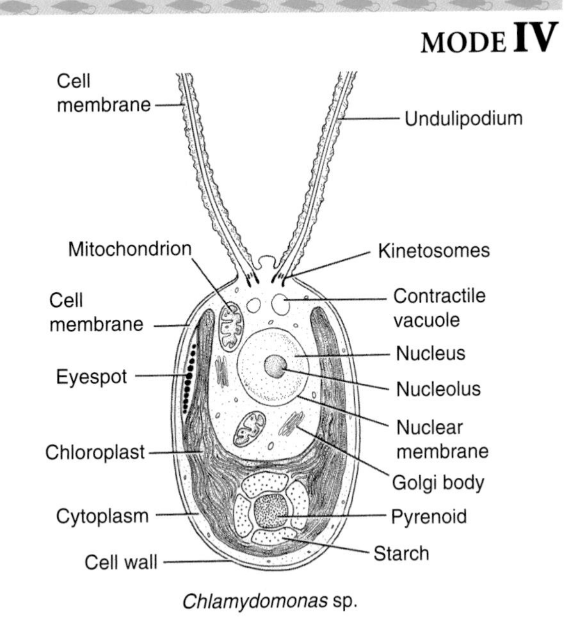

Figure B *Chlamydomonas* is similar in structure to the zoospores of *Acetabularia*. [Drawings by L. Meszoly.]

Class Chlorophyceae is very diverse and probably polyphyletic. Recently, some green algae were separated from other Chlorophyceae as a new class Trebouxiophyceae that includes the lichen symbiont *Trebouxia* and the well-known genus *Chlorella*.

As conceived here, the Chlorophyceae includes the ubiquitous tree-scum alga *Chlorococcum* and both symbiotic and free-living *Chlorella* species. In the laboratory, *Chlorella* sp. grow like weeds; during the 1950s, they were used to unravel the biochemistry of the "dark reactions" (Calvin-Bassham cycle). The "water nets," the Hydrodictyaceae, are another family in this class.

The Volvocales include *Chlamydomonas*. Probably more is known about the genetic control of mating, undulipodia, photosynthesis, and mitochondrial metabolism in *Chlamydomonas* than in any other protoctists.

The Oedogoniales produce zoospores having an unusual ring of many undulipodia; they have a unique method of cell division and an elaborate style of sexual conjugation. The relationship of *Oedogonium* and other members of the order Oedogoniales to other Chlorophyceae is not well understood.

The Chaetophorales (for example, *Stigeoclonium* and *Fritschiella*) are mainly branched multicellular algae; they are differentiated into prostrate and upright thalli.

In the class Ulvophyceae, the Ulotrichales are primarily filamentous (*Ulothrix*) or thalloid (*Ulva*, called sea lettuce, and the common estuarine form *Enteromorpha*).

Order Siphonocladales includes the family Cladophoraceae, green algae typically composed of multinucleate elongate cells. Algae of the Cladophoraceae may be unbranched (*Urospora* and *Chaetomorpha*) or branched (*Cladophora* and *Rhizoclonium*). Many species of common green seaweeds, such as *Codium* and *Acetabularia*, are in this group. Many are quite large, although all are syncytial: no cell membranes form, and so millions of nuclei and chloroplasts share the same cytoplasm.

Those in class Charophyceae, a paraphyletic lineage, are unicellular or multicellular and live in freshwater or brackish water. Their morphological diversity ranges from mastigotes (Mesostigma) to the highly complex multicellular stoneworts *Chara* and *Nitella*. The latter are favorite experimental organisms. It is from this class that the green plants (embryophytes) evolved.

The last class, Prasinophyceae, again a paraphyletic assemblage of early diverging Chlorophyta, whose members are unicellular, differs a good deal from all other green algae. Prasinophytes lack the typical chlorophyte gametes and sexual life cycle. Their cell structure differs from that of the standard chlorophyte. The typical prasinophyte has an anterior pit or groove, for example, from which emerge 1 to 16 undulipodia, and it bears scales on its cell surface including the undulipodia. Among the prasinophytes are some of the most abundant picoplanktic phototrophs in the oceans. The genus *Ostreococcus*, whose genome has been completely sequenced, is likely the smallest photoautotrophic eukaryote (1 μm). In the genus *Tetraselmis* (formerly *Platymonas*), the scales on the cell body are fused to a cell wall. This, together with the evolution of a microtubular system guiding cell division (the phycoplast), relates *Tetraselmis* and its relatives to the more advanced members of the Chlorophyceae. One *Tetraselmis* species is a regular tissue symbiont of the green photosynthetic flatworm *Convoluta roscoffensis* (A-7, Platyhelminthes). It is likely that several lineages of the prasinophytes will attain class status in the future.

SUBKINGDOM (Division) AKONTA

Pr-29 Haplospora

Greek *haplo*, single, simple; Latin *spora*, spore

GENERA
Haplosporidium
Minchinia
Urosporidium

Convergent evolution that led to small, dark, symbiotrophic structures in animal tissue unites this phylum, as well as Pr-30 (Paramyxa) F-1 (Microspora) and A-2 (Myxospora)—three phyla of propagule-forming (for example, cyst-forming) symbiotrophs—with an alveolate phylum, the Apicomplexa (Pr-7). Other "parasitic protozoa" were in the old "Sporozoa". The haplosporosome-forming symbiotrophs (Haplospora) are known mainly from fish and other marine animals; the paramyxans are nesting-cell symbiotrophs; and the multicellular myxosporans are the far better known traditional "sporozoa." Sporozoa, some of which are associated with serious diseases, were considered a class of animals in the phylum "Protozoa" when all the small heterotrophic protoctists were classified according to their importance to people. Fine-structure analyses with the electron microscope and complementary molecular studies of sequences of nucleotides, especially in rRNA, and of amino acids in proteins make it abundantly clear that "sporozoa" have little in common except their habitat. The investigation of these many very different organisms has thus been inappropriately restrained by ignorance of their great differences.

Clearly, the "sporozoan habit" convergently evolved in several free-living protist lineages: small protists invaded animal tissues and took up residence. The majority are now coevolved benign inhabitants with life histories tightly coupled to those of their hosts, but some remain necrotrophic (for example, they kill the host tissues in which they reside). Those evolving innovative modes of transfer from the bodies of their benefactors succeeded in leaving more offspring.

Only three genera with a total of 33 species are placed in Haplospora, a phylum of marine animal-tissue symbiotrophs. Because the phylum definition requires the propagule—spore—to have an anterior pore (opening or aperture) covered with a hinged cap or piece of wall material folded into the opening, other genera and species are suspected to exist in which this particular defining feature has not been seen. Knowledge of the group comes mainly from observations of live, infected marine animals coupled with electron microscopy of their tissues.

Haplosporans constitute a phylum of tissue and body-cavity symbiotrophs in marine invertebrate animals; the haplosporan is unicellular and begins its life history uninucleate. The cell contains tubulovesicular mitochondria with a small number of tubules and a distinctive organelle of unknown function, the haplosporosome, for which the group is named. Haplosporosomes, scattered throughout the cytoplasm, are electron-dense, generally spherical [but sometimes vermiform (wormlike), elliptical, or of other shape] objects that range in size from 0.07 to 0.25 μm in diameter. Both a unit membrane and an inner looser membrane delimit this organelle (Figure A).

Haplosporans lack walls and undulipodia at all stages in their life history. They have a paucity of ribosomes and ER but apparently do have Golgi bodies (Figure B). They have never been seen outside tissues of the animals in which they reside. Their spores seem to be shed into the water and, either there or in tissue, the spore lids probably open to release sporoplasm, but the entire life history has never been described for any species.

Figure A Haplosporosome of *Haplosporidium nelsoni* in which a limiting membrane (arrow) and internal membrane (double arrow) are visible. TEM, bar = 0.1 μm. [Micrograph by F. O. Perkins. Reprinted from *Handbook of Protoctista* (Jones and Bartlett, 1990).]

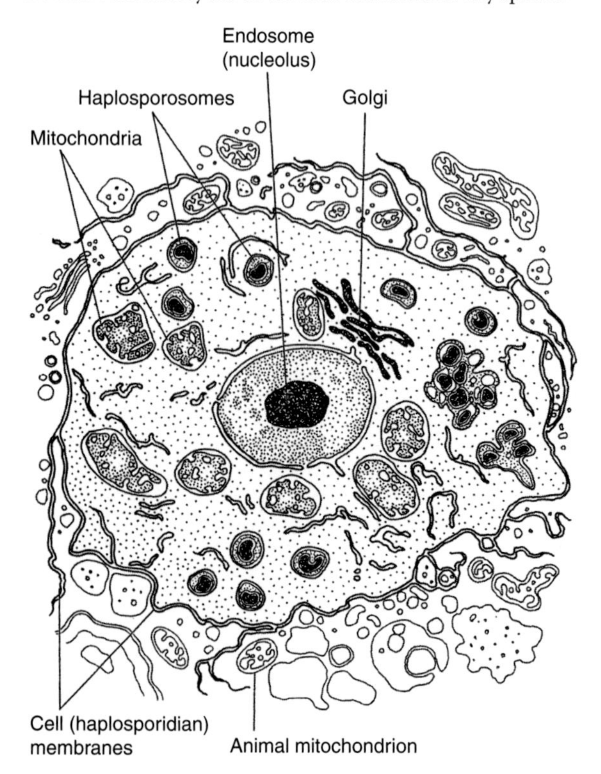

Figure B A generalized haplosporidian. Plasmodium with haplosporosomes in host tissue. [Drawing by K. Delisle.]

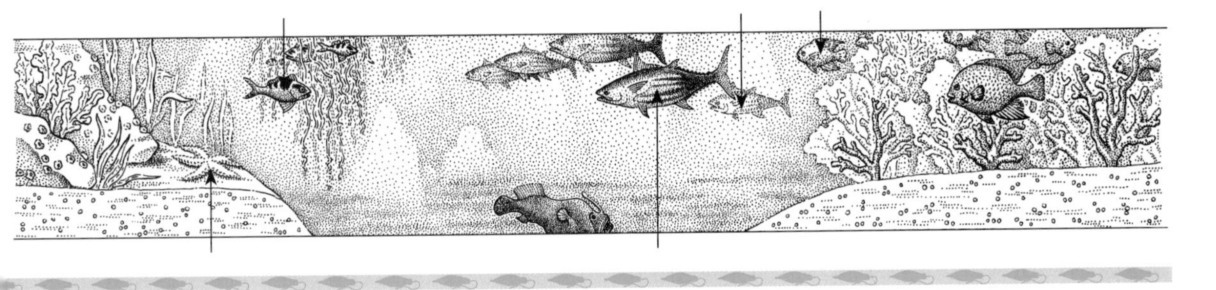

Before the formation of spores, haplosporans grow in tissue as uninucleate or multinucleate plasmodia (Figure C). The nuclear membrane does not disintegrate during karyokinesis (nuclear division; Figure D). The first sign of sporulation is the deposition of a thin wall around the larger multinucleate cells that then become, by definition, sporonts. The sporont undergoes cytokinesis (whole-cell division) in a manner that subdivides it into uninucleate cells called sporoblasts. A kind of sexual fusion then occurs in the development of the spores: pairs of uninucleate sporoblasts fuse to form binucleate sporoblasts. The nuclei of these dikarya fuse (karyogamy), and the resulting, presumably diploid, cell assumes the shape of an hourglass. A strange happening ensues: the anucleate half of the cell nearly entirely engulfs the nucleate half, forming the epispore cytoplasm (the former enucleate sporoblast) and the sporoplasm (the former nucleate part of the sporoblast). In the epispore cytoplasm, a new cup-shaped spore wall is formed that has an anterior constricted opening (aperture). The aperture is covered by a hinged lid or a tucked-in tongue of wall material. Decorations with distinguishing substructure are formed inside the epispore cytoplasm and transported to the outer surface of the spore wall. This distinctive ornamentation, prominent extensions of the spore, is only fuzzy at the level of light microscopy. Spore decorations are well resolved by electron microscopic analysis.

Animals in which these symbiotrophs thrive include limpet molluscs (A-26) and worms: nematodes (A-11), trematodes (A-7), and polychaetes (A-22). Haplosporans are easiest to find as "parasites of parasites", which are called "hypersymbiotrophs"—that is as indicated, symbiotrophs of symbiotrophs. For example, *Urosporidium* species are detected after they enter, presumably from the digestive tract, into the hepatopancreas of trematode worms. The worms themselves are symbiotrophs in bivalve molluscs such as oysters (A-26). After spore development begins, the worms change color to brown or black, indicating the presence of haplosporans. The worm-exploited soft tissue of the mollusc becomes watery; its greater transparency provides the investigator with a clue to the whereabouts of haplosporans. Infected tissue may be replete with haplosporan plasmodia having small, eccentrically placed nuclei with conspicuous endosomes (nucleoli, sites of ribosome synthesis). The Brownian movement of the haplosporosomes in the cytoplasm of the haplosporan plasmodia can be seen in live material, another way by which the existence of these obscure organisms is inferred.

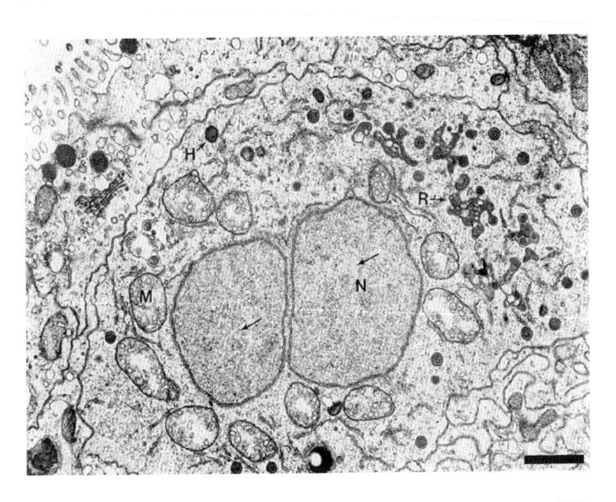

Figure C Plasmodium of *Haplosporidium nelsoni.* Nuclei (N), free haplosporosomes (H), mitochondria (M), microtubules (arrows) of the persistent mitotic apparatus, and membrane-bounded regions in which haplosporosomes are formed (R) are visible. TEM, bar = 1 μm. [Micrograph by F. O. Perkins. Reprinted from *Handbook of Protoctista* (Jones and Bartlett, 1990).]

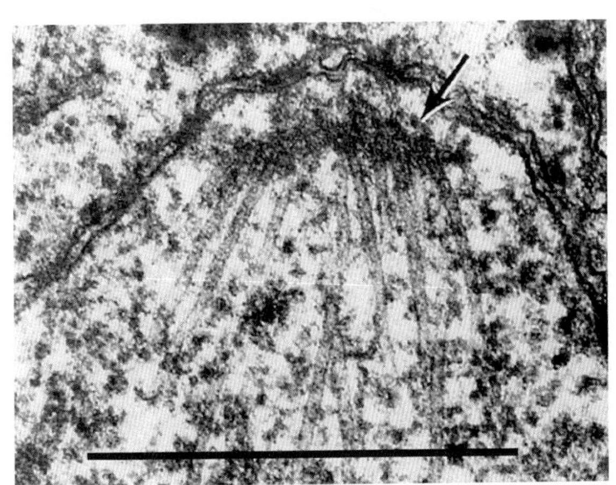

Figure D Fungal-like spindle pole body (arrow) of *Haplosporidium nelsoni* in mitotic nucleus with attached microtubules. TEM, bar = 1 μm. [Micrograph by F. O. Perkins. Reprinted from *Handbook of Protoctista* (Jones and Bartlett, 1990).]

Pr-30 Paramyxa

Greek, *para*, alongside of; *myxa*, mucus

GENERA
Marteilia
Paramarteilia
Paramyxa

Formerly ignored or lumped with "sporozoa," the symbiotrophic paramyxans are immediately distinguished from all other organisms by their "nesting-cell" behavior. Their propagules, called "spores" (as are other small spherical compact structures capable of further growth), consist of several cells enclosed inside one another that develop from "internal cleavage." A stem cell, ameboid in structure, divides, leading to an "endogenous bud," that is, an offspring cell fully inside the one that produced it. The parent cell is the sporont—the cell that bears the spore. Rather than the ubiquitous [9(3)+0] traditional centriole-kinetosomes, those of paramyxans have nine singlet microtubules, making their centriole-kinetosomes [9(1)+0]. Three genera, listed above, and six species are known.

The presence of haplosporosomes, organelles as described for other propagule-forming symbiotrophs, has led to the placement of paramyxans with *Haplosporidium* and other haplosporans (Pr-29). However, the unequivocal differences in the reproductive biology and other aspects of cell structure have led to the separation of this paramyxan phylum from haplosporans and other phyla of symbiotrophs that form protected propagating stages in animal tissue. Thus, they cannot be placed in other traditional "sporozoa" phyla, because they lack the apical complexes of apicomplexans (Pr-7), the polar filaments of microsporans (F-1), the capsules of myxosporans (A-2), and all other criteria for placement elsewhere.

The various genera of paramyxans are distinguished by two criteria: the number of spores and cells developing into spores in the sporont (such as the three illustrated in Figure A) and the taxonomic position of the animal on whose tissue the paramyxan depends. Paramyxans live in various invertebrate hosts: *Paramyxa* (Figure B) in the intestinal cells of annelids (A-22), *Paramarteilia* in the testes of crustaceans (A-21), and *Marteilia* in the hepatopancreas of bivalve molluscs (A-26).

The development of *Paramyxa paradoxa* in the cytoplasm of cells of a marine animal is shown in Figure C.

Although paramyxans have mitochondria and therefore probably some oxidative metabolism, their status as obligate intracellular symbiotrophs has precluded their cultivation. No metabolic or genetic studies are available, and no sexual behavior or motile stages of any kind have been reported.

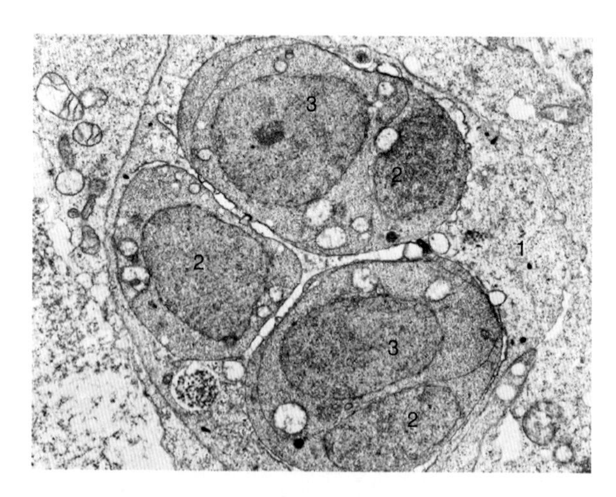

Figure A The stem cell of *Paramarteilia orchestiae* (1) containing three sporonts (2). In two of them, the tertiary cell (3) is already differentiated. This stage can be observed in all paramyxeans. TEM, bar = 1 μm. [From T. Ginsburger-Vogel and I. Desportes, "Étude ultrastructurale de la sporulation de *Paramarteilia orchestiae* gen. n., sp. n., symbiotroph de l'amphipode *Orchestia gammarellus* (Pallus)." *Journal of Protozoology* 26:390–403 (1979).]

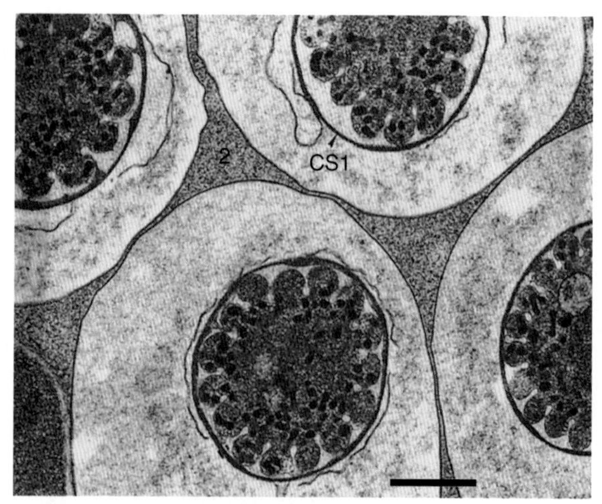

Figure B Transverse sections of four mature spores of *Paramyxa paradoxa.* The outer sporal cell (CS1) is reduced to a thin cytoplasmic layer (arrowhead). Infoldings and dense bodies of the secondary sporal cell can be seen. The light area around each spore results from its retraction in the sporont cytoplasm (2). TEM, bar = 1 μm. [From I. Desportes, "Étude ultrastructurale de la sporulation de *Paramyxa paradoxa* Chatton (Paramyxida) symbiotroph de l'annelide polychete *Poecilochaetus serpens.*" *Protistologica* 17:365–386 (1981).]

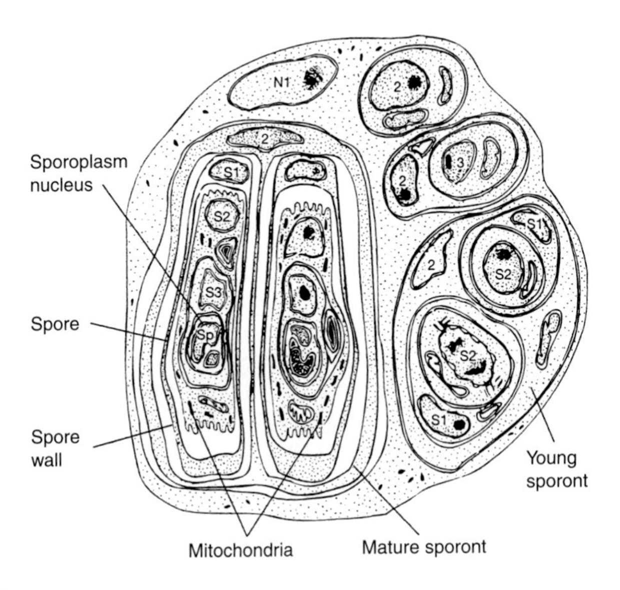

Figure C The development of *Paramyxa paradoxa* is shown here in the cytoplasm of cells of a marine animal. Only two of the four spores are shown in the young sporont and in the mature sporont. 2, Nucleus of secondary (stem) cell; 3, tertiary cell nucleus; N1, stem cell nucleus; S1, S2, S3, nuclei of sporal cells 1, 2, 3, respectively. [Adapted from I. Desportes, "The Paramyxa Levine 1979: An original example of evolution towards multicellularity. " *Origins of Life* 13: 343–352; 1984.]

Pr-31 Actinopoda

Greek *actinos*, ray; *pous*, foot

GENERA
Acantharia
Acanthocystis
Acanthometra
Actinophrys
Actinosphaera
Actinosphaerium (Echinosphaerium)
Aulacantha
Challengeron
Ciliophrys
Clathrulina
Collozoum
Heterophrys
Pipetta
Sticholonche
Thalassicola
Zygacanthidium

The marine protists that Ernst Haeckel traditionally called "radiolarians" along with other superficially similar protists with some radial symmetry are grouped as classes in the phylum Actinopoda for convenience and pedagogy. That actinopods represent convergently evolved lineages more related to other protists (Pr-8 through Pr-10) than they are to each other is likely, but in the absence of definitive taxonomic revision, we present the traditional actinopod grouping here, with its four classes. The first class is Heliozoa [Figure A(1)], commonly called freshwater sun animalcules even though there are marine forms. The second is the predominantly deep-dwelling Phaeodaria [Figure A(2)], and the third is the more open-ocean Polycystina [Figure A(3)]—these two classes constitute the traditional Phylum Radiolaria. The fourth class is Acantharia [Figure A(4)]; traditionally also grouped with Radiolaria), with their strontium sulfate skeletons.

Actinopods are distinguished by their long, slender, cytoplasmic axopods, also called axopodia [Figure A(5)]. These fine projections are stiffened by a bundle of microtubules running down the axis of the structure called an axoneme [Figure A(6)]. Each axoneme has an elaborate arrangement of microtubules characteristic of that actinopod group. Electron microscopic studies indicate that the classes Acantharia, Polycystina, and Phaeodaria, all considered marine "radiolarians," are products of evolutionary convergence and are only remotely related to one another. This has been further supported by molecular phylogenetic

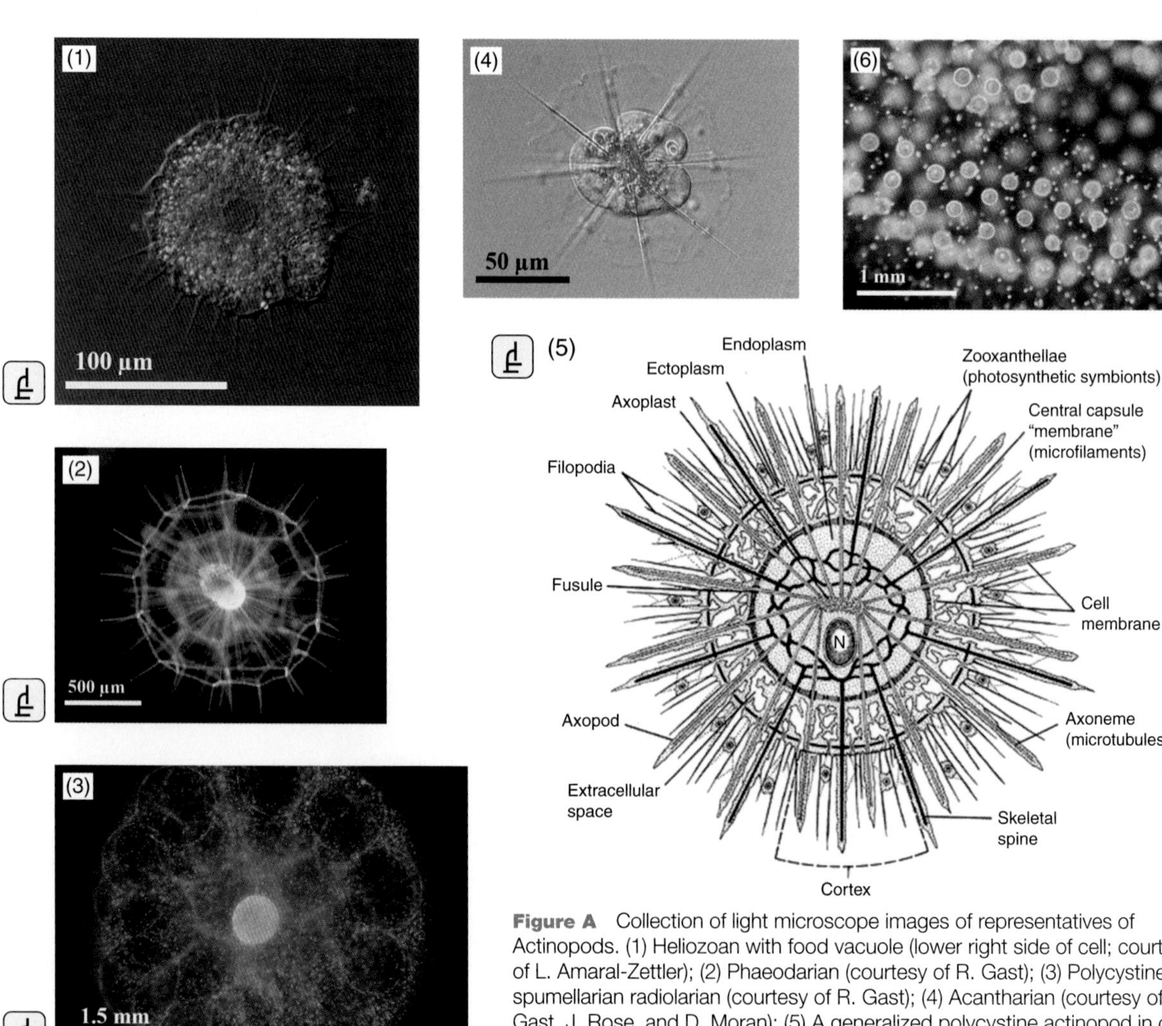

Figure A Collection of light microscope images of representatives of Actinopods. (1) Heliozoan with food vacuole (lower right side of cell; courtesy of L. Amaral-Zettler); (2) Phaeodarian (courtesy of R. Gast); (3) Polycystine spumellarian radiolarian (courtesy of R. Gast); (4) Acantharian (courtesy of R. Gast, J. Rose, and D. Moran); (5) A generalized polycystine actinopod in cross section; (6) Colonial radiolarian (courtesy of R. Gast).

analyses of small-subunit ribosomal gene sequence and protein-coding gene sequence (Amaral-Zettler *et al.*, 1997; Nikolaev *et al.*, 2004).

Heliozoans are primarily freshwater plankton, although estuarine, marine, and benthic (seafloor dwelling) species are known, comprising many genera and species. Most use their axopods to catch prey. Axopods radiate out into the water, surrounded along their length by plasma membrane. In some heliozoans, the axonemes grow out directly from the endoplasm; in others, each axoneme grows out from its own structure, the axoplast, located next to the nucleus. In a group called the centrohelidians, all the axonemes arise from a single axoplast, called a centroplast, whose center often contains a clearly defined organelle.

The rowing actinopod illustrated in Figure B(*i*), *Sticholonche zanclea* Hertwig, has been a particular enigma for taxonomists. Its peculiar skeleton, the placement of its axopods on the nuclear membrane (Figure B(*ii*)), and the hexagonal pattern of the axopods in cross section have justified its placement as the only species in the isolated order Sticholonchidea. This order was originally thought to be radiolarian (as suggested by A. Hollande, M. Cachon, and J. Valentin in 1967), but it has been recently placed in its own order (Taxopodida) based on molecular phylogenetic and ultrastructure information (Nikolaev *et al.*, 2004; Mikrjukov *et al.*, 2000). *Sticholonche* has microtubular oars and sets of moveable microfibrillar "oarlocks" and is found rowing in the Mediterranean with the splendor of a Roman galley. Unfortunately, it does not grow in the laboratory.

Many heliozoans have siliceous or organic surface scales or spines. In a few species, a spherical organic or siliceous cage encloses the entire cell. The cage has bars arranged in a repeating hexagonal pattern through which the axopods penetrate.

Reproduction in heliozoans by zoospores or swarmer cells is unknown except in the order Desmothoraca (for example, *Clathrulina elegans*). Sexual reproduction has rarely been seen, and most cells reproduce by binary or multiple fission or budding. In some multinucleate species, the nuclear and cytoplasmic divisions are not synchronized. In uninucleate forms, the axopods retract so that the organism does not move or feed during cell division.

A kind of autogamy (self-fertilization) has been reported in some heliozoans. A mature cell forms one or more cysts inside the cell. Meiosis apparently takes place in the cysts, and certain nuclei degenerate. Two of the final meiotic products in each cyst then fuse—their haploid nuclei form a new single diploid nucleus. The only surviving product of the two meiotic divisions and fusion emerges from the cyst as a mature heliozoan. Whether this inbred sort of reproduction is common is not known because of the paucity of study. Heterogamy (fusion of nuclei from different individuals) may also occur. In *Actinophrys*, two cells (but not their nuclei) may fuse just before they undergo autogamy. Gametes originating from one of the two cells have been seen to fuse with gametes originating from the other. Cell fusion is common in heliozoans, but whether it constitutes meiosis and fertilization is not known.

The polycystine and phaeodarian radiolaria are extremely common in tropical waters and often have strikingly beautiful skeletons made of silica. Of the more than 4000 actinopods

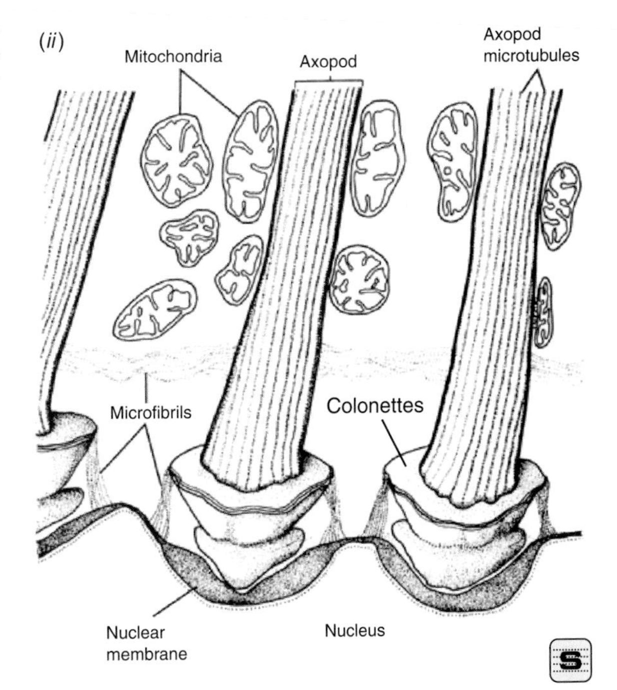

Figure B (*i*) A living *Sticholonche zanclea* Hertwig, taken from the Mediterranean off Ville Franche sur Mer Marine Station. LM, bar = 100 μm. (Courtesy of M. Cachon) (*ii*) The axopods of the oars (colonettes), of *Sticholonche*, showing their relation to the nucleus (central capsule) and the mitochondria. (Drawing by L.M. Reeves.)

described in the literature, some 500 are estimated to be polycystines. Along with diatoms, silicomastigotes, and sponges, they are responsible for the depletion of dissolved silica in surface waters.

Polycystines and phaeodarians differ in many ways. The polycystine skeleton is made of opal (hydrated amorphous silica) whereas the phaeodarian skeleton is made of silica and a large quantity of organic substances of unknown nature. The polycystine skeletal elements look solid under light microscope; however, electron microscopy reveals tiny canals and pores in their skeletons. The skeletal elements of phaeodarians look hollow even under the light microscope; their spines are tubular and the continuous shells of many species have a bubbly "Styrofoam" ultrastructure barely visible by light microscopy but conspicuous by electron microscopy. Crystals, but not skeletal components, of strontium sulfate ($SrSO_4$) are secreted by some adult polycystines in their endoplasm and perhaps by all of them in their undulipodiated swarmers, whereas $SrSO_4$ is unknown in phaeodarians. For a collection of images of radiolarian skeletons, see the work of Ernst Haeckel.

The capsule enclosing the central mass of cytoplasm in both polycystines and phaeodarians is not a flimsy microfibrillar open mesh net (as in acantharians) but is made of massive organic material. The polycystine capsule, probably composed of mucoproteins or mucopolysaccharides, is made of numerous juxtaposed plates, like the pieces of a jigsaw puzzle separated by narrow slits, whereas the phaeodarian capsule is a single continuous structure. The polycystine capsule grows in diameter during the life of the organism; the phaeodarian capsule cannot increase in diameter after it has formed—it can only thicken its wall.

The axonemes of the polycystine axopods studied so far are made of parallel microtubules aligned in geometrical arrays, with bridges between microtubules. Most species have many such axopods per cell. Polycystines usually have one axoplast, from which all axonemes originate, but some groups have other arrangements—for example, individual axoplasts, one per axoneme, are located near the nucleus. In phaeodarians, only two axonemes penetrate the capsule and the microtubules in the basal part of these axonemes are not linked by bridges. Light microscopy reveals a cortex of many thin peripheral pseudopods, which are perhaps branches of the two axopods. No polycystine axoneme is known to branch.

Polycystine orifices, called fusules, are complex mufflike structures, each filled with a dense plug that permits the passage of the axonemal microtubules, if they originate inside the endoplasm, but that hampers the circulation of cytoplasm between the endoplasm and the extracapsular pseudopodial network. The phaeodarian capsule normally has only three orifices of two kinds: a wide, complex astropyle, which is an opening that ensures exchange between the endoplasm and whatever cell parts that lie outside the capsule; and two, rarely more, parapyles. These openings, simpler than polycystine fusules, allow the passage of the two thick-cell axonemes. At each parapyle, there is a cup-shaped axoplast from which an axoneme originates. Outside of the capsule in front of the astropyle of many phaeodarians is a mass of predigested food called the phaeodium. The polycystines lack the phaeodium.

In the phaeodarian endoplasm are numerous strange tubes, called rodlets, about 200 nm wide, having a complex repeating ultrastructure. Their role is unknown (perhaps they take part in the secretion of the capsule). No such rodlets are known in the polycystines.

Polycystines supplement heterotrophy by photoautotrophy with symbiotic yellow or green algae (Figure A(3) and A(6); zooxanthellae or zoochlorellae; Phyla Pr-16 and Pr-28); phaeodarians lack algal symbionts.

Most polycystines and all phaeodarians have only one nucleus, large and polyploid. Only the phaeodarian nucleus undergoes an extraordinary equatorial division, superficially resembling classical mitosis, in which two monstrous "equatorial plates" are formed, each with more than 1000 chromosomes.

Class Polycystina is divided into the orders Spumellaria and Nassellaria. The spumellarian has fusules scattered all over its central capsule membrane; thus, its axopods radiate in all directions (Figure A(3)). The protist is usually spherical, ellipsoidal, or flattened, and so, naturally, is its skeleton. Some spumellarians form large colonies in which hundreds of individual organisms are embedded in a common mass of jelly (Figure A(6)). The fusules of nassellarians, which never form colonies, are clustered at one pole of the capsule membrane; their axopods are grouped in a conical bunch that leaves the cell at that pole.

Acantharians [Figure A(4)], generally spherical organisms, have a unique, radially symmetrical skeleton composed of rods of crystalline $SrSO_4$. The skeleton usually has 10 diametrical (20 radial) spines, called spicules, inserted according to a precise rule, known as the icosacanth law, described by Johannes Müller in 1859 (referenced within Wilcock *et al.*, 1988). The acantharian cell is a globe from whose center the spicules radiate and pierce the surface at fixed "latitudes" and "longitudes." Even in acantharians that do not have the general shape of a globe, these orientations are strictly observed, although some spicules are thicker and longer than the others. Some species have more than 20 spicules, as many as several hundred, but they are always grouped by some elaboration of Müller's law.

Acantharian cells are made of distinct layers. The innermost layer, coarsely granulated with many small nuclei, is the cell's central mass. Immediately surrounding the central mass is a perforated, flimsy network of microfilaments called the central capsule membrane. Through the central capsule membrane, the central mass extends several kinds of cytoplasmic outgrowths. There are cytoplasmic sheaths surrounding the skeletal spines, reticulopods, which are cross-connected netlike pseudopods lacking axonemes, filopods, which are thin pseudopods stiffened by one or very few microtubules, and a number of axopods (usually 54, but in some acantharians there may be several hundred) arising from axoplasts between the spines. At the periphery is the cortex, a thin, flexible layer of microfilaments, which may be arranged in intricate designs. The cortex is underlaid by a network of reticulopods, and where the $SrSO_4$ spines pass through, the cortex is pushed out, like a tent stretched out over tent poles [Figure A(4)]. At these points are filaments, the myonemes, which apparently control the tension of the cortex and bind it to the skeletal rods.

The delicate axopods increase the amount of cell surface exposed to the sea. They retard sinking and perhaps allow efficient scavenging of nutrients from the water. Prey, generally other protoctists and small animals, adheres to the axopods. Cytoplasm from the axopods then engulfs the prey and cytoplasmic flow transports it down the axopods toward the inner part of the cell, where it is digested.

Acantharians produce many small swarmer cells, each containing a drop of oil reserve and a crystal. They bear two [9(2)+2] undulipodia (Figure C). The undulipodia originate from kinetosomes in the anterior part of the swarmer cell. Some acantharians round up to form cysts in which they undergo mitotic divisions. Swarmers develop and are later released from these cysts, but meiosis has not been observed. Little about the development process is known because swarmers have been devilishly difficult to culture in the laboratory.

Most acantharians are effectively photoplankton as well because they harbor many haptomonad algae (Pr-25) that live and grow in them. The symbiotrophy permits the acantharians to obtain their energy and food by photosynthesis in the nutrient-poor open ocean. The acantharian wastes provide nitrogen and phosphorus for their haptomonad symbionts.

The acantharians and probably many other “actinopods” that elusively form cells that swim by use of undulipodia will be transfered out of subphylum Akonta when more is known about their life cycles, morphology, genetics and molecular biology.

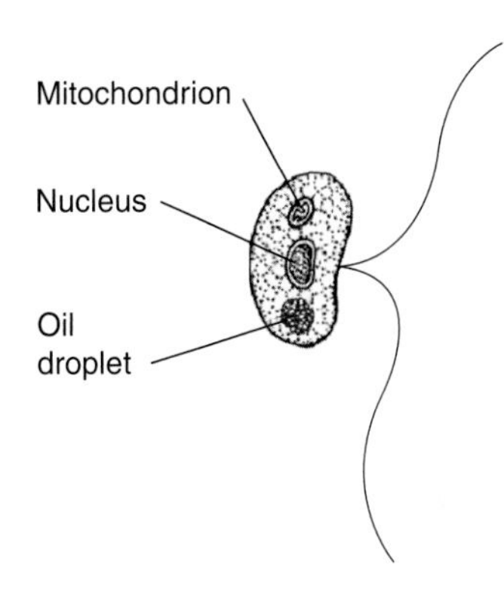

Figure C Generalized swarmer cell, as can be found in some acantharian actinopods.

Pr-32 Gamophyta
(Conjugaphyta, conjugating green algae)

Greek *gamons*, marriage; *phyton*, plant

GENERA
Bambusina
Closterium
Cosmarium
Cylindrocystis
Desmidium
Genicularia
Gonatozygon
Hyalotheca
Mesotaenium
Micrasterias
Mougeotia
Netrium
Penium
Spirogyra
Staurastrum
Temnogyra
Zygnema
Zygogonium

Gamophytes are green algae that lack undulipodia at all stages of their life history. Without motile sperm, any other sperm, or other means of locomotion, they engage regularly in sexual processes. They have symmetrical cells containing complex chloroplasts, which are usually aligned down the long axis of the cell. One large and conspicuous nucleus tends to be found in each cell. These conjugating green algae are found in ponds, lakes, and streams; no truly marine forms have been reported. To reproduce, the haploid growing cells either divide mitotically (uniparentally) or produce, by mitosis, amastigote ameboid cells that fuse to form the zygote. This usually develops into a resistant and conspicuous structure called a zygospore. Zygotic meiosis takes place within the zygospore, and haploid algal cells eventually emerge.

In their pigmentation, gamophytes are similar to all the other green algae: they have chlorophylls ***a*** and ***b***, and most are grass green in color. They are often classified with the chlorophytes (Pr-28).

Two classes are in the phylum Gamophyta as presented here, all oxygenic photoautotrophs: Euconjugatae (true conjugating algae) and Desmidales.

The euconjugates, which are generally filamentous forms, consist of one order (Zygnematales). In this order, two families are recognized: Mesotaeniaceae (for example, *Cylindrocystis*, *Mesotaenium*, *Netrium*), and Zygnemataceae. Most zygnemids, such as *Zygnema*, *Spirogyra*, and *Mougeotia*, form pond scums—stringy masses of long, unbranched filaments. In *Zygnema* and *Spirogyra*, the chloroplast is helically wound along the length of the long cylindrical cell; in *Mougeotia*, a single, large, flat, plate-shaped chloroplast extends the full length of the cell, as illustrated in Figures A and B. These organisms grow rapidly by mitosis, the filaments breaking off fragments that start new filaments, thus forming a bloom or scum in a few days.

During sexual union, two filaments, which are haploid, come to lie side by side. Protuberances grow and join to form conjugation tubes that link cells in opposite filaments. The cells of the "male" filament, with their chloroplasts, flow through the conjugation tube to fuse with the cells of the "female" filament. Each fusion results eventually in a dark, spiny diploid zygote in a chamber of the female filament. Because fertilization is often simultaneous, rows of such zygotes, seen as black zygospores, are common. After a period of dormancy, the zygotes are released into the water; they undergo meiosis and germinate to produce new haploid filaments.

The most speciose of the two classes are the desmids. In our classification, subfamilies of the desmids, each named for its best-known genus, are Penieae (*Penium*), Closterieae (*Closterium*), and Cosmarieae (*Cosmarium*). Several thousand species are known. Most are single cells—more precisely, they are pairs of cells whose cytoplasms are joined at an isthmus (Greek *desmos*, bond). The isthmus is the location of the single shared nucleus. Some desmids are colonial. In many desmids, the chloroplasts are lobed or have plates or processes that extend from the center toward the periphery of the cell. The outer layers of the cell wall form a shell, typically decorated with spines, knobs, granules, or other protrusions arranged in lovely designs. These outer layers are composed of cellulose and pectic substances and, in many cells, are impregnated with iron or silica; the inner layer, on the other hand, is composed of cellulose only and is structureless at the light microscopic level of magnification.

The outermost layer of the desmid cell is a mucilaginous sheath, sometimes thin and sometimes very thick and well developed. It is secreted through pores in the cell wall. The slow gliding movement of desmids is thought to be due to actin protein secretions in this mucilage.

In the typical desmid, the two "half cells" are mirror images of each other, and each has its own chloroplast. In uniparental reproduction, after the nucleus divides, the two partners simply

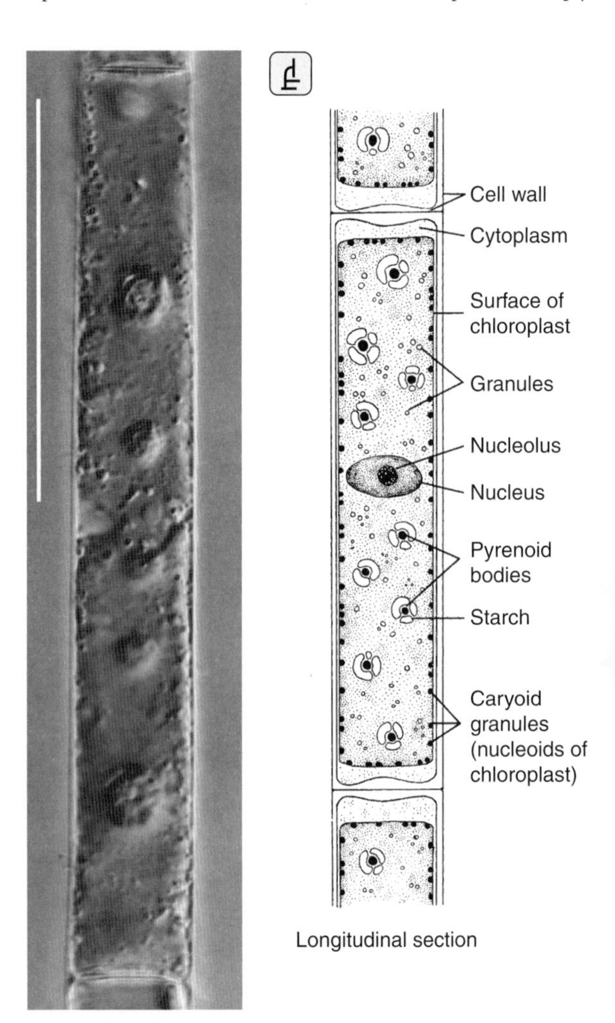

Figure A *Mougeotia* sp., a living freshwater green alga. LM (differential interference), bar = 100 μm. [Photograph courtesy of N. S. Allen; drawing by R. Golder.]

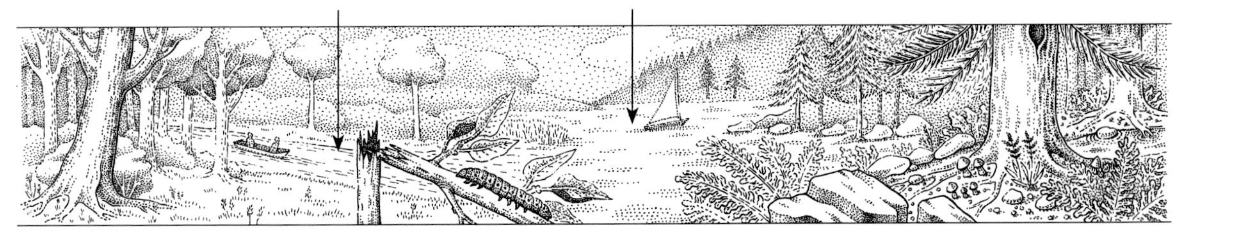

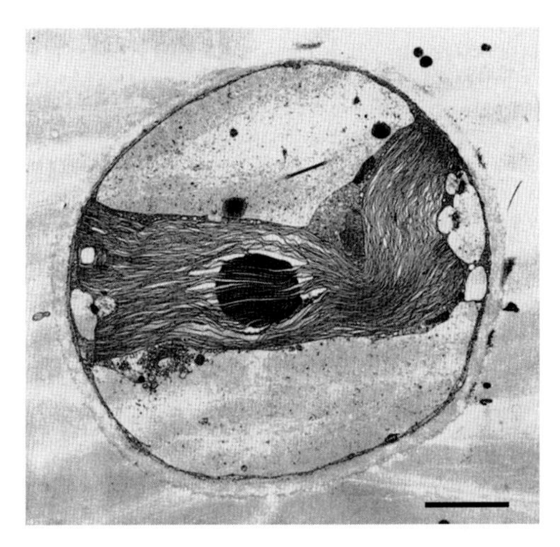

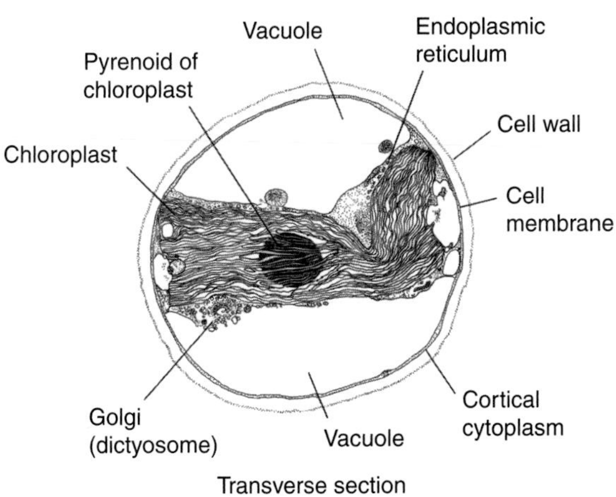

Figure B *Mougeotia* sp., a living freshwater green alga. TEM, bar = 5 μm. [Photograph courtesy of K. Klein and E. Wagner, *Photochemistry and Photobiology* 27:137–140 (1978); drawing by D. Salmon.]

separate. Each one grows a new half cell replacement. During sexual conjugation also, the two partners separate; both leave their shells and fuse outside, either with each other or with a liberated protoplast from another desmid, to form a dark, spiny zygote reminiscent of the zygotes of the Euconjugatae and of the zygomycotes (F-2). In *Desmidium cylindrium*, only one partner, the "male," leaves its shell; conjugation takes place inside the shell of the "female," and that is where the zygote is lodged.

Pr-33 Rhodophyta

(Red algae)

Greek *rhodos*, red; *phyta*, plant

GENERA

Agardhiella
Amphiroa
Bangia
Batrachospermum
Callophyllis
Chantransia
Chondrus
Corallina
Dasya
Erythrocladia
Erythrotrichia
Gelidium
Goniotrichum
Gracilaria
Hildebrandia
Lemanea
Lithothamnion
Nemalion
Polysiphonia
Porphyra
Porphyridium
Rhodymenia

The sexual organisms of this phylum and of the Gamophyta (Pr-32) probably evolved their peculiar conjugating mating systems independently of each other and independently of the fungi, most of which show similar behavior. Whereas the red seaweeds in phylum Rhodophyta are a huge and important group that deserves (and has received) many books of its own, the gamophytes—conjugating green algae, the desmids, and their kin—form a small group that has been removed from the great diverse phylum of other green algae (chlorophytes) because of a peculiar sexual system. Unlike sexual processes in other green algae, no undulipodiated gametes (motile sperm) are ever formed by any members of the phyla Rhodophyta and Gamophyta.

The red algae (rhodophytes), along with the phaeophytes (Pr-17), are the largest and most complex of the protoctists (Figure A). The largest rhodophytes are somewhat smaller and less complex than the largest phaeophytes, and some 50 genera, comprising about 100 different species of rhodophytes, grow symbiotrophically only on other red algae. Rhodophytes commonly inhabit the edges of the sea and are cosmopolitan in distribution. In the tropics, particularly, they abound on beaches and rocky shores. About 675 genera with 4100 species are known, the vast majority of which are marine. There are two subclasses, the Florideae and the Bangiales. (Only one class, Rhodophyceae, is recognized, and all species are placed in it.) Although rhodophytes are primarily marine organisms, some taxa are restricted to freshwater habitats and others live on land. Marine taxa live in littoral and benthic (seafloor) habitats where suitable substrata such as rocks and jetty pilings exist for attachment.

The red algae form a natural group—all species display several traits that characterize the phylum. They range from microscopic unicells and filaments (of single cells in rows or of multiple aligned rows of cells) to large (as much as 1 m), cell-packed, branched or unbranched, cylindrical, leaflike thalli (Figure B), including crustose (flat) and erect forms, some of which are calcified. Rhodophytes are distinguished by reddish plastids; rhodoplasts, with accessory, water-soluble pigments—allophycocyanin, phycocyanin, and phycoerythrin—localized in structures termed phycobilisomes found on the outer faces of the plastid photosynthetic lamellae (thylakoids). Other rhodoplast pigments include chlorophyll ***a***, alpha- and beta-carotene, lutein, and zeaxanthin. Thylakoids are present as single lamellae (that is, not stacked) in the rhodoplasts. Food reserves are stored as floridean starch (α-1–4-linked glucan) in granules outside the plastid.

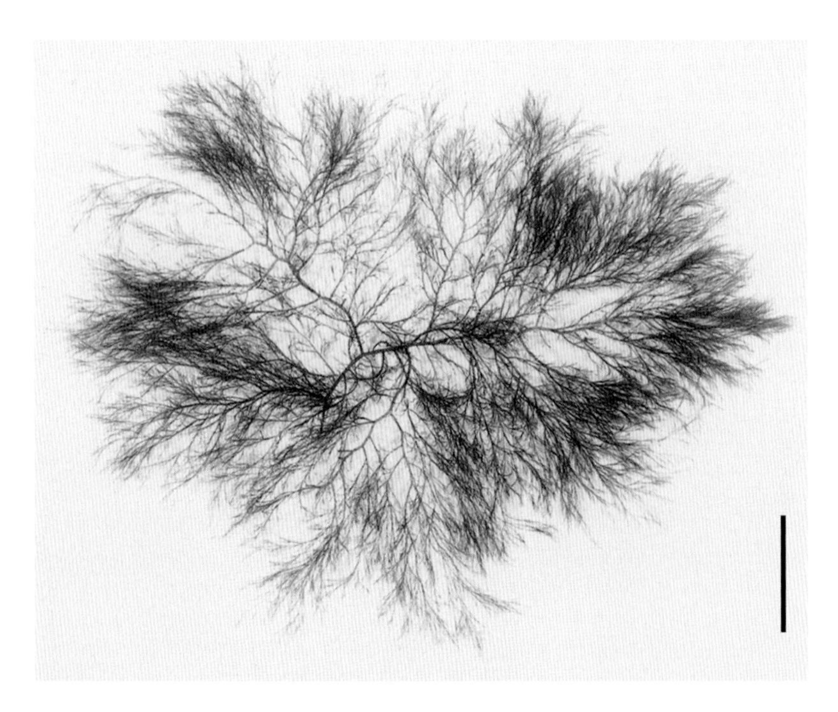

Figure A *Polysiphonia harveyi* from rocky shore, Atlantic Ocean. Bar = 1 cm. [Courtesy of G. Hansen.]

Characteristics of some red algae are "pit connections" between cells (a misnomer, because they do not connect cells but rather are proteinaceous plugs deposited in the pores that result from incomplete centripetal wall formation; Figure C); mitochondria associated with the forming faces of the membranous Golgi bodies; plastids surrounded by one or more encircling thylakoids; and a life history consisting of an alternation of two free-living and independent generations called gametophyte (Figure D) and tetrasporophyte, respectively. A third generation, the carposporophyte, is present on the female gametophyte.

Although none has undulipodia at any stage in the life history, all reproduce sexually. Reproduction is oogamous: a large egg cell is formed in a special female organ, the oogonium, which bears a long neck that is receptive to the male gamete. The male organ, the antheridium, produces a single male "sperm," which lacks an undulipodium and is incapable of locomotion. After male gametes are released near the female, at least one attaches to the neck of the female structure and moves down the neck to fertilize the egg. The physiology of this process is poorly known.

After fertilization, meiosis may take place immediately to form haploid spores. Alternatively, instead of meiosis taking place, diploid spores, called carpospores, may be formed. They are often formed in bunches of threads that grow out of the oogonium. In some of the more elaborate life cycles, the carpospores are formed in a special organ (cystocarp) that establishes a connection with the oogonium. The carpospores of some species develop into complex little "plants"—called carposporophytes

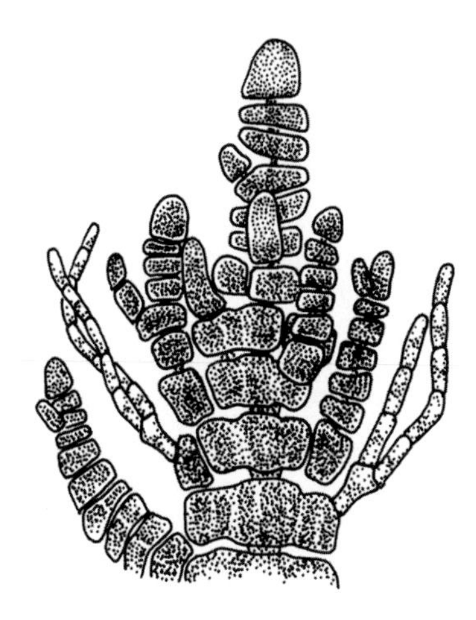

Figure B Apex of male thallus. [Drawing by R. Golder.]

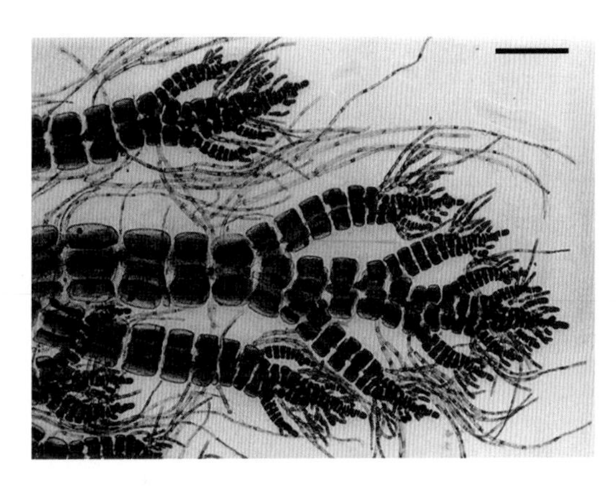

Figure C Apex of thallus, showing cells and pit connections. LM, bar = 0.1 μm. [Courtesy of G. Hansen.]

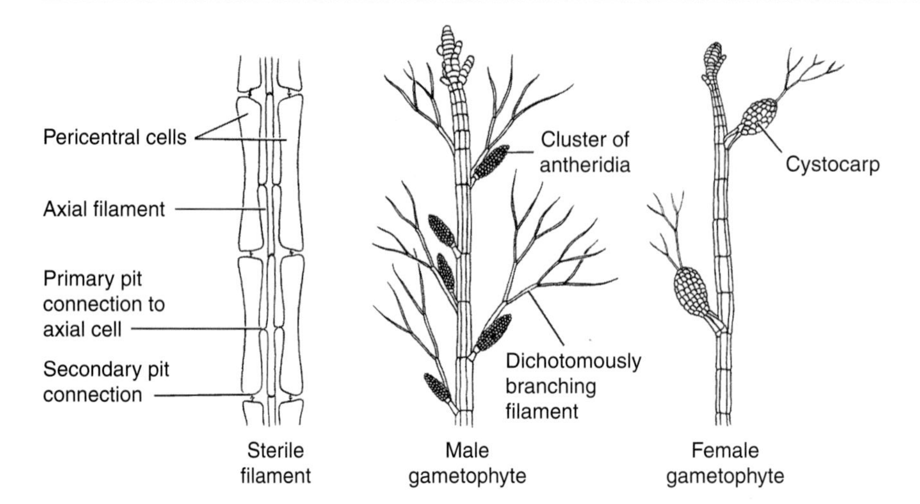

Figure D Sterile and sexually mature apices of thalli. [Drawings by R. Golder.]

(Figure E)—that bear organs called tetrasporangia. Meiosis takes place in the tetrasporangia; the four meiotic products, the tetraspores, are released into the sea. They germinate into haploid thalli that eventually produce oogonia or antheridia. Predominantly haploid or alternating haploid and diploid life cycles are common in the red algae; the details of most of the cycles have yet to be worked out.

When we note that many red algae calcify, we mean that they become encrusted with calcium carbonate. *Lithothamnion* looks like reddish circular crust on rocks, and *Corallina* looks like an encrusted tree. A single genus may have both calcifying and noncalcifying species. The propensity for calcification has produced a good fossil record for the phylum; mineralized forms of coralline algae first appear in the lower Paleozoic era. The first red algae in the fossil record are filaments dating to the late Proterozoic eon.

Rhodophytes show a marked parallelism of forms with other groups of algae—the chrysomonads (Pr-15), the chlorophytes (Pr-28), and the phaeophytes (Pr-17). As in these other phyla, there are heterotrichous filaments (*Chantransia*), prostrate disks (*Erythrocladia*), cushions (*Hildebrandia*), elaborate erect structures (*Batrachospermum*), compact tissuelike types (*Lemanea*), and delicate, many-branched forms (*Polysiphonia* and *Porphyra*).

Agar, the substance used to firm the broth on which colonies of microorganisms are grown so that they can be isolated and studied, is extracted from red algae. Agarose, so important to the gels of molecular biology and biochemistry, is extracted from *Gracilaria*, a member of the Gigantinales, one of the 11 orders in the Florideae subclass. Polysaccharides from these seaweeds are also used in the manufacture of ice cream and other food products, toothpaste, cosmetics, and pharmaceuticals. The leafy dulce of New England's rocky shore is dried and eaten whole.

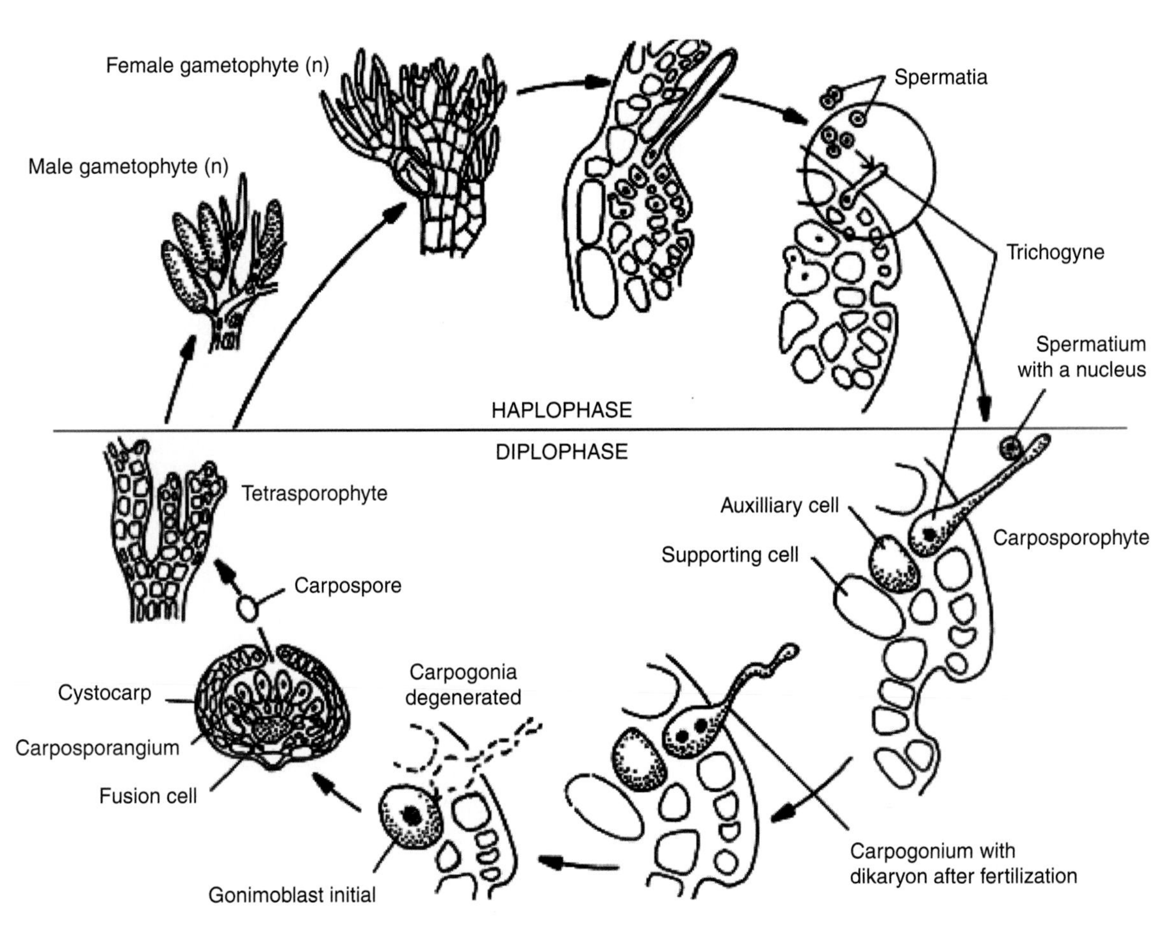

Figure E *Polysiphonia*: fertilization of carpogonia on the female gametophyte. [Drawing by K. Delisle.]

SUBKINGDOM (Division) OPISTHOKONTA

Pr-34 Blastocladiomycota

Greek *blastos*, germ; *klados*, twig or branch

GENERA

Allomyces
Blastocladiella
Coelomomyces
Physoderma
Catenaria
Catenophlyctis
Polycaryum
Sorochytrium

Recent molecular evidence, including a six-gene analysis, supports the Blastocladiomycota (blastoclades) as a lineage in the clade of true fungi, independent of the Chytridiomycota (Pr-35) where the order Blastocladiales was previously classified. Like the Chytridiomycota, blastoclades reproduce by the formation of posteriorly directed uni-undulipodiated reproductive cells.

Blastoclade zoospores (Figure A) have an unusual distinctive feature that can be seen by light microscopy. Virtually all the ribosomes of mastigote cells of members of this group are packed near the nucleus in a membrane-bounded structure called the nuclear cap. Transmission electron microscopy shows that the nucleus is angled toward the kinetosome and is sheathed by nine sets of three microtubules that extend from the proximal end of the kinetosome (Kinetid schematic; Figure A). The secondary centriole (nonmastigote centriole) is either at right angles to the kinetosome or absent. Like the Chytridiomycota, the blastoclades synthesize the amino acid lysine by the aminoadipic pathway and produce cell walls of chitin.

Blastoclades live in freshwater and soil; none have been reported from marine habitats. Like chytrids (Pr-35), hyphochytrids (Pr-14) and oomycetes (Pr-21) blastoclade thalli are coenocytic; walls form within the thallus primarily to separate reproductive structures from vegetative ones. The diversity of blastoclade morphology parallels that of the chytrids and ranges from holocarpy (the entire thallus converts to a reproductive structure) to monocentric, eucarpic (the thallus consists of rhizoidal system and a reproductive rudiment) to polycentric (the thallus, which may have determinate or indeterminate hyphal or rhizomycelial growth, bears many reproductive structures). Many of the Blastocladiomycota reproduce sexually and may have morphologically different or identical haploid and diploid stages. Sexual reproduction is by the fusion of undulipodiated cells called planogametes (Greek, *planos*, wandering), which look like asexual zoospores that are formed in zoosporangia.

Figure A Kinetid of Blastocladiomycota zoospores, the karyomastigont. K = kinetosome, nmc = nonmastigoted centriole, mt root = microtubule root. Props are found in the Blastocladiomycota and in most orders of the Chytridiomycota. [Courtesy: J. E. Longcore]

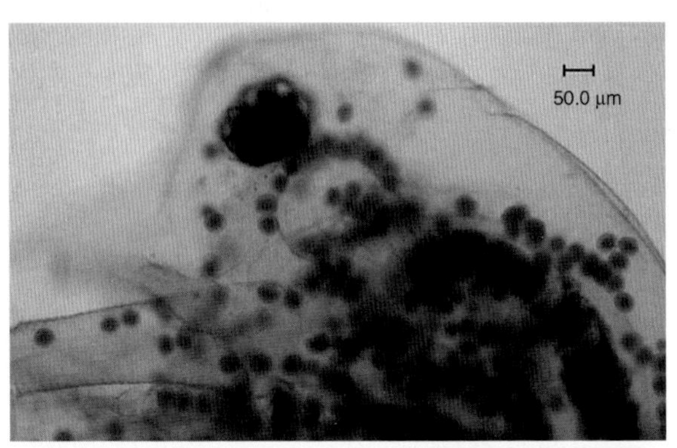

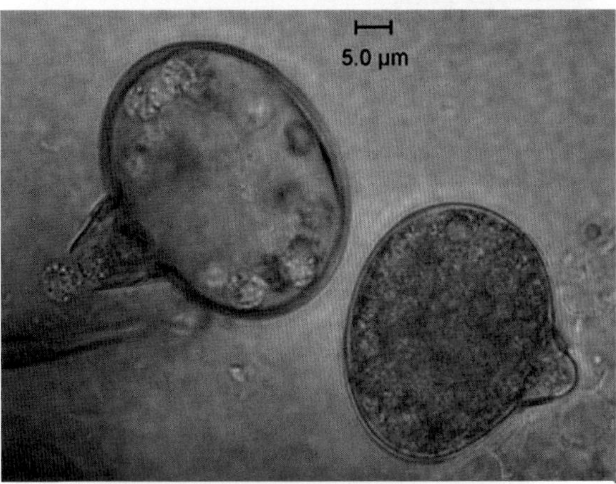

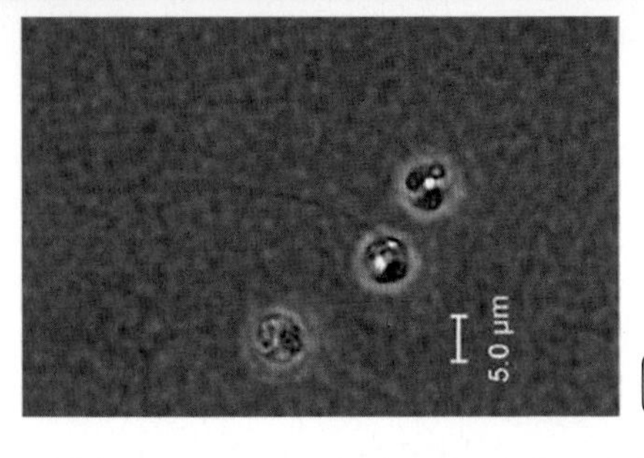

Figure B *Polycaryum laeve* in the hemocoel of *Daphnia pulicaria*. Monocentric, holocarpic (entire thallus forms the reproductive structure) thalli, motile spores leaving sporangium, and zoospores. [Photos Courtesy of J. E. Longcore]

The single class (Blastocladiomycetes) contains one order (Blastocladiales), five families, 13 genera, and about 175 species. Three of the families exclusively contain pathogens. *Physoderma zea-maydis* (Physodermataceae) causes brown-spot disease of corn, and other *Physoderma* spp. are obligate minor pathogens of various aquatic and semiaquatic plants. *Sorochytrium milnesiophthora* (Sorochytriaceae) infects tardigrades. *Polycaryum laeve* (family undetermined), once classified in the Haplosporidia (Pr-29), infects and can alter the peak density of *Daphnia* populations in lakes. Death and disruption of infected *Daphnia* individuals trigger germination of thick-walled spores (Figure B), but whether these motile spores are gametes or zoospores is yet to be determined. Species of *Coelomomyces* (Coelomomycetaceae) parasitize mosquito larvae but difficulties in growing this pathogen in pure culture have so far hindered its use in controlling mosquito populations. *Coelomomyces* requires alternate hosts (mosquito larvae and copepods) for the diploid and haploid stages of its life cycle. Species of *Catenaria* (Catenariaceae), although frequently pathogens of nematodes and other invertebrates, can live on decaying plant and animal matter and grow in pure culture; *Catenophlyctis* is saprobic. Most members of the Blastocladiaceae are saprobes; some members of this family, especially *Blastocladiella emersonii* and *Allomyces macrogynus*, serve as important research organisms for physiological, developmental and genetic studies. Some members of the Blastocladiaceae have well-developed, branching mycelia, and many have complex life cycles with several alternative developmental pathways. *Blastocladiella emersonii* (Figure C), for example, produces zoospores that have three distinct developmental options: a zoospore can form an ordinary colorless thallus, a thick-walled resistant thallus, or a tiny thallus that releases a single zoospore. Which option the organism takes depends on the quantity of food, moisture, and carbon dioxide available in the medium. These factors, in turn, are related to the degree of crowding.

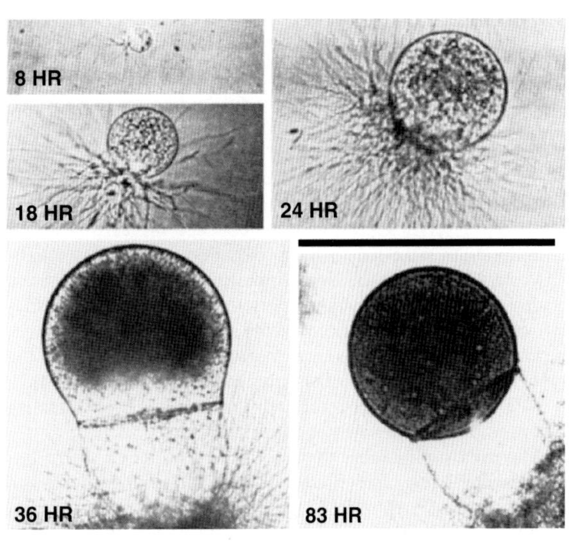

Figure C Development of the ordinary colorless sporangium of *Blastocladiella emersonii*. Hours are time elapsed after water was added to an initial small, dry sporangium. After 18 hours, rhizoids have proliferated. After 36 hours, the protoplasm has migrated into the anterior cell that becomes the sporangium. After 83 hours, the sporangium has thickened and zoospores have begun to differentiate from the coenocytic nuclei inside. LM, bar = 1 μm. [Photographs courtesy of E. C. Cantino and J. S. Lovett.]

Pr-35 Chytridiomycota

Greek *chytra*, little, earthen cooking pot; *mykes*, fungus

GENERA

Batrachochytrium
Chytriomyces
Lobulomyces
Monoblepharella
Monoblepharis
Nowakowskiella
Polychytrium
Rhizophlyctis
Rhizophydium
Rozella
Synchytrium
Zygohizidium

Recent analyses of gene sequence data have strengthened the hypothesis that the organisms in the Chytridiomycota, known as chytrids, are basal in the fungal lineage. Although the Chytridiomycota was thought to contain all fungal organisms that formed posteriorly uni-undulipodiated reproductive cells, recent analyses of multiple genes suggest that the possession of undulipodiated reproductive cells (zoospores) is a grade of development rather than being a unifying characteristic of a single group. Molecular analyses supported by ultrastructural features of the undulipodium have led to the segregation of the Blastocladiomycota (Pr-34) and the Neocallimastigota, which also possess posteriorly uni- or sometimes poly-undulipodiated reproductive cells, from the Chytridiomycota. Physiological, ultrastructural and molecular differences between the chytrids and the superficially similar protoctist stramenopiles ("straw bearers"), or heterokonts (Phyla Pr-8 through Pr-21) definitively establish that the stramenopiles are not ancestral to fungi.

Chytrids are microbes that live in fresh water or soil; a few are marine. Most, like Blastocladiomycota, (Phylum Pr-34), hyphochytrids (Phylum Pr-14) and oomycetes (Phylum Pr-21), grow and feed by extending threadlike hyphae or rhizoids into living hosts, recently dead biota, or other organic debris; some lack rhizoids and absorb nutrients through the wall of the thallus (the organismal body). They secrete extracellular digestive enzymes and absorb the resulting nutrients. The morphologically simplest chytrids grow and develop entirely within the cells of their hosts. The more complex produce reproductive structures on, or extending to, the host's surface, even though the vegetative and feeding parts of the chytrid thallus may be within the substrate. A few chytrids are necrotrophs (disease agents) of plants; and one species is an important pathogen of amphibians; *Batrachochytrium dendrobatidis* has been implicated in declines of amphibian populations on five continents.

The cell walls of all chytrids are composed of chitin; some contain cellulose as well. The chytrid thallus is coenocytic; that is, its many nuclei are not separated by cell walls. However, a septum, which is a plate composed of cell wall material, separates each reproductive organ from the thallus. Nearly all chytrids have motile stages that develop as zoospores inside a walled, sack-like structure known as a sporangium. The distinctive kinetids and the single posteriorly directed undulipodium of their zoospores distinguish chytrids from look-alike protoctists.

Although most chytrid reproduction is asexual, some chytrids reproduce sexually (Figure A). Many types of sexuality have been reported; the best documented include fusion of thalli in the Chytridiales and formation of oospheres (eggs) and antheridia (sperm) in the Monoblepharidales (Figure B). In both the zygote becomes thick-walled and serves as a resistant structure. In some species, the resistant structure germinates by producing an adnate zoosporangium, which releases zoospores that germinate into new chytrid thalli; in others, resistant sporangia germinate directly and grow into new thalli.

We recognize two classes (Chytridiomycetes and Monoblepharidomycetes) and six orders in Phylum Chytridiomycota: Monoblepharidales (*Monoblepharella*), Chytridiales (*Chytriomyces*), Spizellomycetales (*Spizellomyces*), Rhizophydiales (*Rhizophydium*), Rhizophlyctidales (*Rhizophlyctis*) and Lobulomycetales (*Lobulomyces*). The zoospore kinetid structure in each order is distinct and the characters in it are useful for defining orders and some families and genera. Ultrastructual characters, however, fail to resolve relationships between orders, and between families within the Chytridiales. Molecular analyses, particularly of the small and large subunits of the rRNA gene, have resolved the same clades as revealed by ultrastructural characters. About 825 species in 90 genera grouped into 25 families are currently recognized. The order is still in question to which some species belong that have not been studied by modern techniques, and many families and genera are known to be para- or polyphyletic.

In the Monoblepharidales, male planogametes are released from a specialized part of the thallus called the male gametangium or antheridium. Other specialized hyphae of the thallus produce a walled female gametangium or oogonium. Inside oogonia, protoplasm differentiates into uninucleate oospheres (eggs). The motile male gamete fuses entirely with the large, non-undulipodiated, nonmotile female gamete. The zygote wall thickens forming an oospore, in which meiosis takes place. The oospore germinates to begin growing the hyphae of a mycelium. The mycelium grows; eventually, either it produces sporangia that release asexual zoospores or it differentiates male and female gametangia. In *Monoblepharis polymorpha* the same thallus that produces asexual sporangia at certain temperatures is capable of producing the male and female gametangia and releasing compatible gametes at slightly higher temperatures.

Molecular and ultrastructural evidence also led to the description of the Rhizophlyctidales as a recent segregate of the Spizellomycetales where its position had been questionable. This order, containing four families, is based on *Rhizophlyctis rosea*, a species complex that is common on cellulosic substrates in soils.

The Chytridiales are unicellular, lacking well-developed mycelia (hyphae). Some species, however, produce a rhizomycelium, a system of branched hyphae with indeterminate growth on which are produced many zoosporangia. Other species may produce only root-like rhizoids that lack nuclei, or the thallus

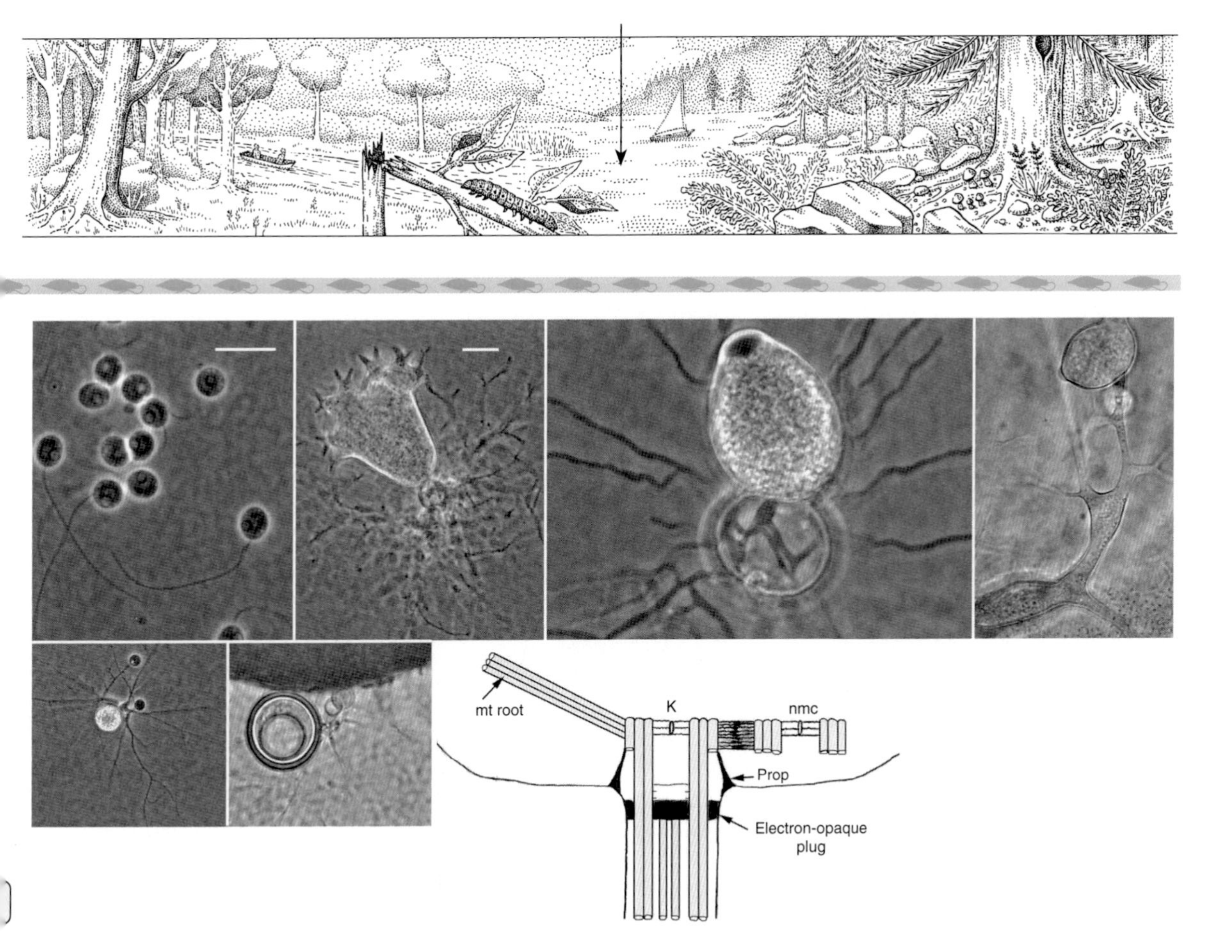

Figure A Chytridialean zoospores (bar=10 μm), monocentric, endogenously developed thallus of *Podochytrium dentatum* (Chytridiaceae) (bar=10 μm), exogenously developed thallus of *Chytridium lagenarium* (*Cladochytrium* clade), polycentric thallus of *Polychytrium aggregatum* (*Polychytrium* clade), sexual reproduction in the Chytridiaceae; resting spore formed after anastomosis of rhizoids from contributing thalli, mature, thick-walled, sexually produced resting spore, and schematic of kinetid of Chytridialean zoospore (K = kinetosome; nmc = nonmastigote centriole; mt root = microtubule root, which usually leads to the rumposome) [Courtesy: J.E. Longcore].

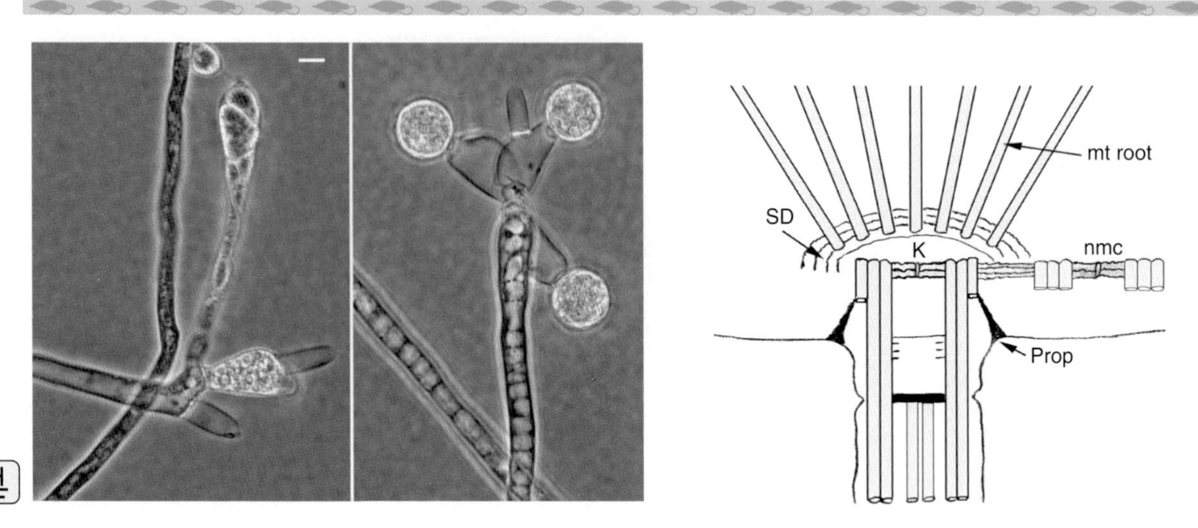

Figure B Zoospore discharge and sexual reproduction in *Monoblepharis polymorpha*; oospores of *M. polymorpha* (bar=10 μm). Schematic of *Monoblepharis* kinetid (K = kinetosome; nmc = nonmastigote centriole; mt root = microtubule root; SD = striated disk) [Courtesy: J.E. Longcore].

can be only a sporangium with no rhizoids. The Chytridiales is the largest and taxonomically the most difficult order of the Chytridiomycota. Ultrastructural and molecular characters have helped to refine phylogenetic hypotheses of the order resulting in new orders being segregated from it (e.g., Spizellomycetales, Rhizophydiales, Lobulomycetales); this process is ongoing and additional orders, now recognized as clades (related groups) within the Chytridiales, will no doubt be described.

The Order Spizellomycetales was separated from the Chytridiales on the basis of zoospore ultrastructure. In the Spizellomycetales the zoospore nucleus is connected to the kinetosome by microtubules; the nonmastigote centriole lies at an angle to the kinetosome and is connected to it only at the top. Rumposomes (fenestrated membrane cisternae that form honeycomb-like organelles) are absent, though common in zoospores of Chytridiales, Rhizophydiales and Monoblepharidales.

The Rhizophydiales is a recent segregate of the Chytridiales and is based on ultrastructural and molecular characters. The zoospore undulipodium of members of the new order does not contain an electron-opaque plug, which is present in most members of the Chytridiales. Most species consist of a thallus with a single rhizoidal axis, although exceptions exist. Sexual reproduction has been reported for only a few species of *Rhizophydium*. Most reports describe the beginning growth of a receiving thallus, followed by insertion of contents from a contributing thallus that either attaches directly to the female thallus or begins growth in proximity to the receiving thallus and then transfers its contents to the larger thallus through a fertilization tube. The zygote forms a resistant sporangium as in the Chytridiales. In contrast, many *Rhizophydium* species form asexual, thick-walled, resistant sporangia, also known as resting spores. Also segregated from the Chytridiales is the Lobulomycetales with a single described family. Its members are from soils and bogs but but metagenomic methods have yielded evidence of members of the order from marine as well as terrestrial ecosystems.

The Neocallimastigales are specialized inhabitants of the rumens and hindguts of mammalian herbivores; in this anoxic environment, the chytrids live on the cellulosic ingested food of the animal. The chytrids catabolize complex plant carbohydrates that would otherwise be indigestible by the mammal. Penetration of cellulosic material by the rhizoids of these anaerobic chytrids prepares the way for other biota within the rumen to further catabolize recalcitrant carbohydrates. Members of the Neocallimastigales are so well adapted to digestive systems that they lack mitochondria. Although thalli are similar to those of the Chytridiales, Spizellomycetales, and Rhizophydiales in being either monocentric or polycentric, the ultrastructural features of the zoospores of this group lack obvious homology to those of the aerobic orders. Zoospores in some genera are even polyundulipodiated.

Pr-36 Choanomastigota

Greek *choane*, funnel; *mastigio*, whip

GENERA
Codosiga
Proterospongia
Salpingoeca

Choanomastigotes are free-living, bactivorous eukaryotes found globally in marine, brackish, and freshwater environments. The organisms approximately are 10 μm in diameter, and their morphology is typified by a single undulipodium surrounded by a ring of 30–40 actin-filled microvilli that cap an ovoid cell body (Figures A and B). Internally, a kinetosome sits at the base of the apical undulipodium, and a second kinetosome/centriole is at a right angle to the kinetid. The nucleus occupies an apical-to-central position in the cell, and food vacuoles are positioned in the basal region of the cytoplasm (Karpov and Leadbeater, 1998; Leadbeater and Thomsen, 2000).

Choanomastigotes are found in the water column and adhering to substrates directly or through either a thin pedicel or periplast, if present. The life histories of choanomastigotes are poorly understood, and it is unclear whether there is a sexual phase to the life cycle. Nevertheless, choanomastigotes proliferate uniparentally, and observed division occurs by retraction of the undulipodium by longitudinal fission (Karpov and Leadbeater, 1998). Many species are thought to be solitary; however, coloniality seems to have arisen independently several times within the group. Colonial species retain a solitary stage (Leadbeater, 1983). Some choanomastigotes can undergo encystment, which involves the retraction of the undulipodium, its collar and encasement in an electron-dense fibrillar wall. Excystment occurs when cells are transferred to fresh media, although this process remains to be directly observed.

The hypothesized phylogenetic position of choanomastigotes sister to animals has been upheld by several independent analyses including phylogeny-based similarities in 18S rDNA, nuclear protein-coding genes, and mitochondrial genomes (Burger *et al.*, 2003; Lavrov *et al.*, 2005; Mendoza *et al.*, 2002; Steenkamp *et al.*, 2006; Wainright *et al.*, 1993). Finally, recent genome sequencing of the choanomastigote *Monosiga brevicollis* (Figure C) corroborates these phylogenetic analyses and has revealed that choanomastigotes encode homologues of metazoan signaling and adhesion genes (King and Carroll, 2001; King *et al.*, 2003; King *et al.*, in press). The phylogenetic position of choanomastigotes and their biology promise to provide insights into the earliest events in animal evolution.

Three orders of choanomastigotes, or their recent descendants, may soon be raised to phylum status. The Nuclearida

Figure A Immunofluorescent staining of *Monosiga brevicollis*. Undulation of the undulipodium generates water currents that propel free-swimming choanomastigotes through the water column and trap bacterial prey against the collar. Immunofluorescent staining of *Monosiga brevicollis* with anti-β-tubulin antibody (green) labels the cell body and undulipodium, DNA stained with DAPI (blue) highlights the nucleus and polymerized actin stained with phalloidin (red) marks the collar. [Illustration courtesy of S. R. Fairclough and image courtesy of M. Abedin.]

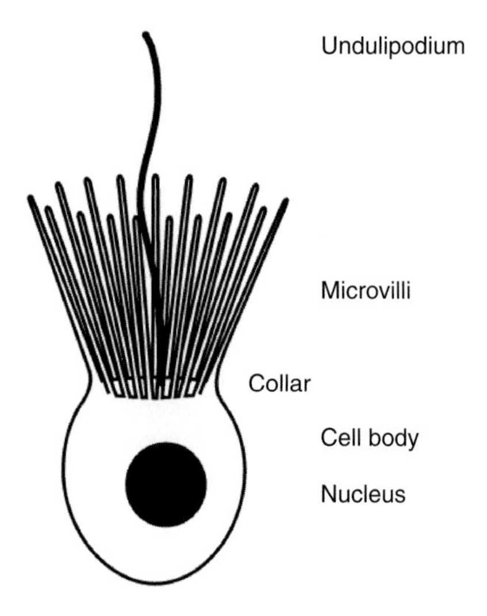

Figure B Choanomastigote morphology is typified by an ovoid cell body approximately 10 μm in diameter capped with a collar of actin-filled microvilli surrounding a single apical undulipodium. The undulipodium generate water currents that propel free-swimming choanomastigotes through the water column and trap bacterial prey against the collar. [Illustration courtesy of S. R. Fairclough.]

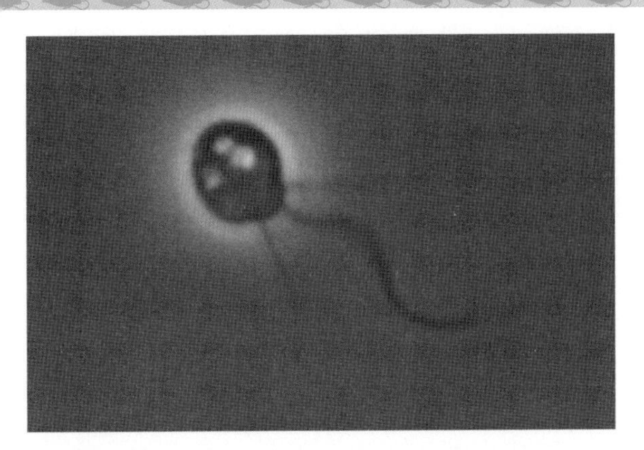

Figure C *Monosiga brevicollis* typifies choanomastigote morphology. An ovoid cell body approximately 10 μm in diameter capped with a collar of actin-filled microvilli surround a single apical undulipodium. Its beating generates water currents that propel free-swimming choanomastigotes through the water column and trap bacteria prey against the collar. Cell body approximately 5 μm in diameter. [Phase image courtesy of S. R. Fairclough.]

(*Neoparamoeba: Nuclearia*, and *Vampyrellidium*) (algal associates) and Ichthyophonida or "rosette agent" (*Amoebidium*, *Ichthyophonus*, and *Pseudoperkinsus*) (salmonid fish associates) have secondarily lost their undulipodia and assumed the ameboid habit. Dermocystida (for example, *Amphibiothecum* and *Dermocystidium*), the sister taxon of Ichthyophonida, are frog associates that alternate between opisthokont undulipodiated and ameboid stages. Most researchers currently place these taxa within class Mesomycetozoa of the phylum Choanomastigota and consider them ancestors of animals and fungi.

Bibliography: Protoctista

General

Abbott, I. A., and E. Y. Dawson, *How to know the seaweeds*, 2d ed. Brown; Dubuque, IA; 1978.

Copeland, H. F., *The classification of lower organisms*. Pacific Books; Palo Alto, CA; 1956.

Dyer, B. D., and R. A. Obar, *Tracing the history of eukaryotic cells*. Columbia University Press; New York; 1994.

Ebringer, L., and J. Krajcovic, *Cell origin and evolution*. Czechoslovak Society of Microbiology; Prague; 1995.

Fenchel, T., and B. J. Finlay, *Ecology and evolution in anoxic worlds*. Oxford University Press; Oxford; 1995.

Fritsch, F. E., *Structure and reproduction of the algae*, 2 vols, 2d ed. Cambridge University Press; Cambridge and New York; 1961.

Grell, K. G., *Protozoology*, 2d ed. Springer-Verlag; New York, Heidelberg, and Berlin; 1973.

Hogg, J., "On the distinctions of a plant and an animal, and on a fourth kingdom of nature." *Edinburgh New Philosophical Journal* 12:216–225; 1861.

Hülsmann, N., "Undulipodium: End of a useless discussion." *European Journal of Protistology* 28:253–257; 1981.

Lee, J. J., S. H. Hunter, and E. C. Bovee, *An illustrated guide to the protozoa*. Allen Press; Lawrence, KS; 1985.

Leedale, G. F., "How many are the kingdoms of organisms?." *Taxon* 23:261–270; 1974.

Lipps, J. H., *Fossil prokaryotes and protists*. Blackwell; New York; 1993.

Margulis, L., "Archaeal-eubacterial mergers in the origin of Eukarya: Phylogenetic classification of life." *Proceedings of the National Academy of Sciences, USA* 93:1071–1076; 1996.

Margulis, L., *Symbiosis in cell evolution*, 2d ed. W. H. Freeman and Company; New York; 1993.

Margulis, L., and D. Sagan, *Origins of sex: Three billion years of genetic recombination*, 2d ed. Yale University Press; New Haven; 1991.

Margulis, L., J. O. Corliss, M. Melkonian, and D. J. Chapman, eds., *Handbook of Protoctista: The structure, cultivation, habitats and life histories of the eukaryotic microorganisms and their descendants exclusive of animals, plants and fungi*. Jones and Bartlett; Boston, MA; 1990.

Margulis, L. and M. J. Chapman, eds. *Handbook of Protoctista*. The structure, cultivation, habitats and life histories of the eukaryotic microorganisms and their descendants exclusive of animals, plants & fungi. 2nd edition. Jones & Bartlett Publishing Company, Sudbury MA (in press); 2010.

Margulis, L., H. I. McKhann, and L. Olendzenski, *Illustrated glossary of Protoctista*. Jones and Bartlett; Boston, MA; 2009.

Margulis, L., and K. V. Schwartz, *Five kingdoms: Introduction to the phyla of life on Earth*. CD-ROM (IBM- or Macintosh-compatible). Ward's Natural History Establishment; Rochester, NY; 1994.

Margulis, L., K. V. Schwartz, and M. Dolan, *The illustrated five kingdoms: Guide to the diversity of life on Earth*. Addison-Wesley-Longman; Menlo Park, CA; 1994.

Raikov, I. B., *The protozoan nucleus: Morphology and evolution*. Springer-Verlag; Vienna; 1982.

Round, F. E., *The ecology of algae*. Cambridge University Press; Cambridge; 1981.

Sagan, D., and L. Margulis, *Garden of microbial delights: A practical guide to the subvisible world*. Kendall/Hunt; Dubuque, IA; 1993.

Sapp, J., *Evolution by association: A history of symbiosis*. Oxford University Press; New York; 1994.

Swofford, D. L., and G. O. Olson, "Phylogeny reconstruction." In: D. M. Hillis and C. Moritz, eds., *Molecular systematics*. Sinauer; Sunderland, MA; 1990.

Van De Graaff, K. M., S. R. Rushforth, and J. L. Crawley, *A photographic atlas for the botany laboratory*. Morton; Englewood, CO; 1995.

Pr-1 Archaeprotista

Daniels, E. W., and E. P. Breyer, "Ultrastructure of the giant amoeba *Pelomyxa palustris*." *Journal of Protozoology* 14:167–179; 1967.

Gunderson, J., G. Hinkle, D. Leipe, H. G. Morrison, S. K. Stickel, D. A. Odelson, J. A. Breznak, T. A. Nerad, M. Müller, and M. L. Sogin, "Phylogeny of trichomonads inferred from small-subunit rRNA sequences." *Journal of Eukaryotic Microbiology* 42:411–415; 1995.

Hollande, A., and J. Valentin, "Appareil de Golgi, pinocytose, lysomes, mitochondries, bactéries symbiontiques, attractophores et pleuromitose chez les hypermastigines du genre Joenia: Affinités entre Joeniides et Trichomonadines." *Protistologica* 5:39–86; 1969.

Jeon, K. W., ed., *The biology of Amoeba*. Academic Press; New York and London; 1973.

Lee, J. J., S. H. Hutner, and E. C. Bovee, *Illustrated guide to the protozoa*, 2d ed. Society of Protozoologists, Allen Press; Lawrence, KS; 1997.

Whatley, J. M., and C. Chapman-Andresen, "Phylum Karyoblastea." In: L. Margulis, J. O. Corliss, M. Melkonian, and D. J. Chapman, eds., *Handbook of Protoctista*. Jones and Bartlett; Boston, MA; 1990.

Pr-2 Rhizopoda

Bonner, J. T., *The cellular slime molds*, 2d ed. Princeton University Press; Princeton, NJ; 1967.

Grell, K. G., *Protozoology*, 2d ed. Springer-Verlag; New York, Heidelberg, and Berlin; 1973.

Martinez, J. A., *Free-living amebas: Natural history, prevention, diagnosis, pathology and treatment of disease*. CRC Press; Boca Raton, FL; 1985.

Olive, L. S., *The mycetozoans*. Academic Press; New York, San Francisco, London; 1975.

Page, F. C., *Marine gymnamoebae*. Institute of Terrestrial Ecology; Cambridge; 1983.

Page, F. C., *A new key to freshwater and soil gymnamoebae*. Institute of Terrestrial Ecology; Cambridge; 1988.

Pr-3 Granuloreticulosa

Cushman, J. A., *Foraminifera: Their classification and economic use*, 4th ed. Harvard University Press; Cambridge, MA; 1948.

Lee, J. J., "Living sands." *Bioscience* 45:252–261; 1995.

Lee, J. J., "Phylum Granuloreticulosa (Foraminifera)." In: L. Margulis, J. O. Corliss, M. Melkonian, and D. J. Chapman, eds., *Handbook of Protoctista*. Jones and Bartlett; Boston, MA; 1990.

Brasier, M., "Darwin's Lost World: The hidden history of animal life." Oxford University Press; Oxford, UK; 2009.

Pr-4 Xenophyophora

Gooday, A. J., B. J. Bett, and D. N. Pratt, "Direct observation of episodic growth in an abyssal xenophyophore (Protista)." *Deep Sea Research* 40:2132–2143; 1993.

Tendal, Ø. S., "Phylum Xenophyophora." In: L. Margulis, J. O. Corliss, M. Melkonian, and D. J. Chapman, eds., *Handbook of Protoctista*. Jones and Bartlett; Boston, MA; 1990.

Pr-5 Dinomastigota

Burkholder, J. M., and H. B. Glasgow, Jr., "Interactions of a toxic estuarine dinomastigote with microbial predators and prey." *Archiv für Protistenkunde* 145:177–188; 1995.

Burkholder, J. M., E. J. Noga, C. H. Hobbs, and H. B. Glasgow, Jr., "New 'phantom' dinomastigote

is the causative agent of major estuarine fish kills." *Nature* 358:407–410; 1992.
Fensome, R. A., F. J. R. Taylor, G. Norris, W. A. S. Sarjeant, D. I. Wharton, and G. L. Williams, *A classification of living and fossil dinomastigotes. Micropaleontology.* Special Publication No. 7. Sheridan Press; Hanover, PA; 1993.
Fritsch, F. E., *Structure and reproduction of the algae,* Vol. 1, 2d ed. Cambridge University Press; Cambridge and New York; 1961.
Hutner, S. H., and J. J. A. McLaughlin, "Poisonous tides." *Scientific American* 199:92–98; August 1958.
Scagel, R. F., R. J. Bandoni, G. E. Rouse, W. B. Schofield, J. R. Stein, and T. M. C. Taylor, *An evolutionary survey of the plant kingdom.* Wadsworth; Belmont, CA; 1966.
Taylor, F. J. R., "On dinomastigote evolution." *Biosystems* 13:65–108; 1980.

Pr-6 Ciliophora

Corliss, J. O., *The ciliated protozoa: Characterization, classification, and guide to the literature,* 2d ed. Pergamon Press; London and New York; 1979.
Curds, C. R., *British and other freshwater ciliated protozoa, Part 1, Ciliophora: Kinetofragminophora.* Cambridge University Press; Cambridge; 1982.
Curds, C. R., M. A. Gates, and D. M. Roberts, *British and other freshwater ciliated protozoa, Part 2, Oligohymenophora and Polyhymenophora.* Cambridge University Press; Cambridge; 1983.
Fenckel, T., *Ecology of protozoa: The biology of free-living phagotrophic protists.* Science Tech; Madison, WI; 1987.
Lynn, D. H., and E. B. Small, "Phylum Ciliophora." In: L. Margulis, J. O. Corliss, M. Melkonian, and D. J. Chapman, eds., *Handbook of Protoctista.* Jones and Bartlett; Boston, MA; 1990.
Lynn, D. H., "Morphology or molecules: How do we identify the major lineages of ciliates (Phylum Ciliophora)." *European Journal of Protistology* 3:356–364; 2003.
Olive, L. S., and R. L. Blanton, "Aerial sorocarp development by the aggregative ciliate, *Sorogena stoianovitchae.*" *Journal of Protozoology* 27:293–299; 1980.
Small, E. B., and D. H. Lynn, "Phylum Ciliophora Doflein, 1901." In: J. J. Lee, S. H. Hutner, and E. C. Bovee, eds., *Illustrated guide to the protozoa,* 2d ed. Society of Protozoologists, Allen Press; Lawrence, KS; 1997.

Pr-7 Apicomplexa

Hammond, D. M., and P. L. Long, *The coccidia: Eimeria, Isospora, Toxoplasma, and related genera.* University Park Press; Baltimore; 1973.
Kreier, J., *Parasitic Protozoa, Vol. 3, Gregarines, Haemogregarines, Coccidia, Plasmodia and Haemoproteids; Vol. 4, Theileria, Myxosporida, Microsporida, Bartonellaceae, Anaplasmataceae, Ehrlichia and Pneumocystis.* Academic Press; New York; 1977.
Noble, E. R., and G. A. Noble, *Parasitology,* 5th ed. Lea & Febiger; Philadelphia; 1982.

Pr-8 Bicosoecida

Dyer, B. D., "Phylum Zoomastigina, class Bicoecids." In: L. Margulis, J. O. Corliss, M. Melkonian, and D. J. Chapman, eds., *Handbook of Protoctista.* Jones and Bartlett; Boston, MA; 1990.

Pr-9 Jakobida

O'Kelley, C., "Jakoba." *Archives of Protistology* 1:2–100; 1994.

Pr-10 Proteromonadida

Brugerolle, G., and J. P. Mignot, "Phylum Zoomastigina, class Proteromonadida." In:

L. Margulis, J. O. Corliss, M. Melkonian, and D. J. Chapman, eds., *Handbook of Protoctista.* Jones and Bartlett; Boston, MA; 1990.

Pr-11 Kinetoplastida

Anderson, O. R., *Comparative protozoology: Ecology, physiology, life history.* Springer-Verlag; New York; 1987.

Pr-12 Euglenida

Anderson, O. R., *Comparative protozoology: Ecology, physiology, life history.* Springer-Verlag; New York; 1987.

Leedale, G. F., *Euglenoid mastigotes.* Prentice-Hall; Englewood Cliffs, NJ; 1967.

Pr-13 Hemimastigota

Foissner, I., and W. Foissner, "Revision of the family Spironemidae Doflein" (Protista, Hemimastigota), with description of two new species, *Spironema terricola* n. sp. and *Stereonema geiseri* n.g., n. sp.." *J. Euk. Microbiol.* 40:422–438; 1993.

Pr-14 Hyphochytriomycota

Fuller, M. S., "Phylum Hyphochytriomycota." In: L. Margulis, J. O. Corliss, M. Melkonian, and D. J. Chapman, eds., *Handbook of Protoctista.* Jones and Bartlett; Boston, MA; 1990.

Stevens, R. B., *Mycology guidebook.* University of Washington Press; Seattle; 1974.

Pr-15 Chrysomonada

Kristianse, J., "Phylum Chrysophyta." In: L. Margulis, J. O. Corliss, M. Melkonian, and D. J. Chapman, eds., *Handbook of Protoctista.* Jones and Bartlett; Boston, MA; 1990.

Round, F. E., "The Chrysophyta: A reassessment." In: J. Kristiansen and R. A. Andersen, eds., *Chrysophytes: Aspects and problems.* Cambridge University Press; New York; 1986.

Pr-16 Xanthophyta

Hibberd, D. J., "Phylum Xanthophyta." In: L. Margulis, J. O. Corliss, M. Melkonian, and D. J. Chapman, eds., *Handbook of Protoctista.* Jones and Bartlett; Boston, MA; 1990.

Pr-17 Phaeophyta

Clayton, M. N., "Phylum Phaeophyta." In: L. Margulis, J. O. Corliss, M. Melkonian, and D. J. Chapman, eds., *Handbook of Protoctista.* Jones and Bartlett; Boston, MA; 1990.

Pr-18 Bacillariophyta

Round, F. E., R. M. Crawford, and D. G. Mann, *The diatoms: Biology and morphology of the genera.* Cambridge University Press; New York; 1989.

Round, F. E., R. M. Crawford, and D. G. Mann, *A scanning electron microscope atlas of diatom structure and diatom genera.* Cambridge University Press; New York; 1993.

Pr-19 Labyrinthulata

Porter, D., "Phylum Labyrinthulomycota." In: L. Margulis, J. O. Corliss, M. Melkonian, and D. J. Chapman, eds., *Handbook of Protoctista.* Jones and Bartlett; Boston, MA; 1990.

Pr-20 Plasmodiophora

Dylewski, D., "Phylum Plasmodiophoromycota." In: L. Margulis, J. O. Corliss, M. Melkonian, and D. J. Chapman, eds., *Handbook of Protoctista.* Jones and Bartlett; Boston, MA; 1990.

Karling, J. S., *The Plasmodiophorales,* 2d rev. ed. Macmillan (Hafner Press); New York; 1968.

Sparrow, F. K., Jr., *Aquatic phycomycetes,* 2d ed. University of Michigan Press; Ann Arbor; 1960.

Pr-21 Oomycota

Dick, M. W., "Phylum Oomycota." In: L. Margulis, J. O. Corliss, M. Melkonian, and D. J. Chapman, eds., *Handbook of Protoctista.* Jones and Bartlett; Boston, MA; 1990.

Erwin, D. C., S. Bartnicki-Garcia, and P. H. Tsao, eds., *Phytophthora: Its biology, taxonomy, ecology, and pathology.* American Phytopathological Society; St. Paul, MN; 1983.

Pr-22 Amoebomastigota

Anderson, O. R., Comparative protozoology: Ecology, physiology, life history. Springer-Verlag; New York; 1987.

Hausmann, Klaus and Norbert Hulsmann, Protozoology. Thieme Verlag; New York, 1996.

Pr-23 Myxomycota

Alexopoulos, C. J., and C. W. Mims, *Introductory mycology,* 3d ed. Wiley; New York; 1979.

Frederick, L., "Phylum Plasmodial Slime Molds, Class Myxomycota." In: L. Margulis, J. O. Corliss, M. Melkonian, and D. J. Chapman, eds., *Handbook of Protoctista.* Jones and Bartlett; Boston, MA; 1990.

Hagelstein, R., *The mycetozoa of North America.* Hagelstein; Mineola, NY; 1944.

Olive, L. S., *The mycetozoans.* Academic Press; New York, San Francisco, and London; 1975.

Stephenson, S. L., and H. Stempen, *Myxomycetes: A handbook of slime molds.* Timber Press; Portland, OR; 1994.

Pr-24 Pseudociliata

Anderson, O. R., Comparative protozoology: Ecology, physiology, life history. Springer-Verlag; New York; 1987.

Hausmann, Klaus and Norbert Hulsmann, Protozoology. Thieme Verlag; New York, 1996.

Pr-25 Haptomonada

Green, J. C., and B. S. C. Leadbeater, eds., *The haptophyte algae.* Systematics Association, Clarendon Press; Oxford; 1994.

Green, J. C., K. Perch-Nielsen, and P. Westbroek, "Phylum Prymnesiophyta." In: L. Margulis, J. O. Corliss, M. Melkonian, and D. J. Chapman, eds., *Handbook of Protoctista.* Jones and Bartlett; Boston, MA; 1990.

Pr-26 Cryptomonada

Gillott, M., "Phylum Cryptophyta." In: L. Margulis, J. O. Corliss, M. Melkonian, and D. J. Chapman, eds., *Handbook of Protoctista.* Jones and Bartlett; Boston, MA; 1990.

Pr-27 Eustigmatophyta

Hibberd, D. J., "Phylum Eustigmatophyta." In: L. Margulis, J. O. Corliss, M. Melkonian, and D. J. Chapman, eds., *Handbook of Protoctista.* Jones and Bartlett; Boston, MA; 1990.

Hibberd, D. J., and G. F. Leedale, "Observations on the cytology and ultrastructure of the new algal class Eustigmatophyceae." *Annals of Botany* 36:49–71; 1972.

Pr-28 Chlorophyta

Graham, L., "The origin of the life cycle of land plants," *American Scientist* 73:178–186; March–April 1985.

Puiseux-Dao, S., *Acetabularia and cell biology.* Springer-Verlag; New York, Heidelberg, and Berlin; 1970.

Raven, P. H., R. F. Evert, and S. Eichhorn, *Biology of plants,* 6th ed. Worth; New York; 1998.

Pr-29 Haplospora

Perkins, F. O., ed., "Haplosporidian and haplosporidian-like diseases of shellfish." *Marine Fisheries Review* 41:25–37; 1979.

Perkins, F. O., "Phylum Haplosporidia." In: L. Margulis, J. O. Corliss, M. Melkonian, and D. J. Chapman, eds., *Handbook of Protoctista*. Jones and Bartlett; Boston, MA; 1990.

Pr-30 Paramyxa

Desportes, I., and F. O. Perkins, "Phylum Paramyxea." In: L. Margulis, J. O. Corliss, M. Melkonian, and D. J. Chapman, eds., *Handbook of Protoctista*. Jones and Bartlett; Boston, MA; 1990.

Pr-31 Actinopoda

Hollande, A., J. Cachon, and M. Cachon-Enjumet, "L'infrastructure des axopods chez les Radiolaires Sphaerellaires periaxoplastidies." *Comptes Rendu Hebdomedaire Seances Academie des Science (Séries D)* 261:1388–1391; 1965.

Pr-32 Gamophyta

Pickett-Heaps, J. D., *Green algae*. Sinauer; Sunderland, MA; 1975.

Pr-33 Rhodophyta

Gabrielson, P. W., and D. J. Garbary, M. R. Sommerfeld, R. A. Townsend, and P. L. Tyler, "Phylum Rhodophyta." In: L. Margulis, J. O. Corliss, M. Melkonian, and D. J. Chapman, eds., *Handbook of Protoctista*. Jones and Bartlett; Boston, MA; 1990.

Hoshaw, R. W., "Phylum Conjugaphyta." In: L. Margulis, J. O. Corliss, M. Melkonian, and D. J. Chapman, eds., *Handbook of Protoctista*. Jones and Bartlet; Boston, MA; 1990.

Pr-34 Blastocladiomycota

Barr, D. J. S., "Chytridiomycota." In: D. J. McLaughlin, E. G. McLaughlin, and P. A. Lemke, eds., *The Mycota VII Part A. Systematics and Evolution*. Springer-Verlag; Berlin Heidelberg; pp. 93–112; 2001.

Couch, J. N., and C. E. Bland, *The Genus Coelomomyces*. Orlando, FL; Academic Press, INC; 1985.

Hibbett, D. S., et al., "A higher-level phylogenetic classification of the fungi." *Mycological Research* 111(5):509–547; 2007.

Hoffman, Y., C. Aflalo, A. Zarka, J. Gutman, T. Y. James, and S. Boussiba, "Isolation and characterization of a novel chytrid species (phylum Blastocladiomycota), parasitic on the green alga *Haematococcus*." *Mycological Research* 112(1):70–81; 2007.

James, T. Y., P. M. Letcher, J. E. Longcore, S. E. Mozley-Standbridge, D. Porter, M. J. Powell, G. W. Griffith, and R. Vilgalys, "A molecular phylogeny of the mastigoted fungi (Chytridiomycota) and description of a new phylum (Blastocladiomycota)." *Mycologia* 98:860–871; 2006.

Johnson, P. T. J., J. E. Longcore, D. E. Stanton, R. B. Carnegie, J. D. Shields, and E. R. Preu, "Chytrid infections of *Daphnia pulicaria*: development, ecology, pathology and phylogeny of *Polycaryum laeve*." *Freshwater Biology* 51:634–648; 2006.

Olson, L. W., "*Allomyces*–a different fungus." *Opera Botanica* 73:1–96; 1984.

Roberson, R. W., and M. M. Vargas, "The tubulin cytoskeleton and its sites of nucleation in hyphal tips of *Allomyces macrogynus*." *Protoplasma* 182:19–31; 1984.

Sparrow, F. K., Jr., *Aquatic Phycomycetes*, 2d ed. University of Michigan Press; Ann Arbor; 1960.

Tambor, J. H. M., K. R. Ribichich, and S. L. Gomes, "The mitochondrial view of *Blastocladiella emersonii*." *Gene* 424:33–39; 2008.

Whisler, H. C., S. L. Zebold, and J. A. Shemanchuk, "Life history of *Coelomomyces psorophorae*." *Proceedings of the National Academy of Science, USA* 72(2):693–696; 1975.

Pr-35 Chytridiomycota

Barr, D. J. S., "Chytridiomycota." In: D. J. McLaughlin, E. G. McLaughlin, and P. A. Lemke, eds., *The Mycota VII Part A. Systematics and Evolution*. Springer-Verlag; Berlin Heidelberg; pp. 93–112; 2001.

James, T. Y., P. M. Letcher, J. E. Longcore, S. E. Mozley-Standbridge, D. Porter, M. J. Powell, G. W. Griffith, and R. Vilgalys, "A molecular phylogeny of the mastigoted fungi (Chytridiomycota) and description of a new phylum (Blastocladiomycota)." *Mycologia* 98:860–871; 2006.

James, T. Y., F. Kauff, et al., "Reconstructing the early evolution of fungi using a six-gene phylogeny." *Nature* 443:818–822; 2006.

Letcher, P. M., M. J. Powell, J. G. Chambers, J. E. Longcore, and P. M. Harris, "Ultrastructural and molecular delineation of the Chytridiaceae (Chytridiales)." *Canadian Journal of Botany* 83:1561–1573; 2005.

Letcher, P. M., M. J. Powell, P. F. Churchill, and J. G. Chambers, "Ultrastructural and molecular phylogenetic delineation of a new order, the Rhizophydiales (Chytridiomycota)." *Mycological Research* 110:898–915; 2006.

Letcher, P. M., M. J. Powell, D. J. S. Barr, P. F. Churchill, W. S. Wakefield, and K. T. Picard, "Rhizophlyctidales–a new order in Chytridiomycota." *Mycological Research* 112:1031–1048; 2008.

Longcore, J. E., A. P. Pessier, and D. K. Nichols, "Batrachochytrium dendrobatidis gen. et sp. nov., a chytrid pathogenic to amphibians." *Mycologia* 91(2):219–227; 1999.

Simmons, D. R., T. Y. James, A. Meyer and J. E. Longcore. "Lobulomycetales, a new order in the Chytridiomycota." *Mycological Research* (in press).

Sparrow, F. K., Jr., *Aquatic Phycomycetes*, 2d ed. University of Michigan Press; Ann Arbor; 1960.

Pr-36 Choanomastigota

Burger, G., L. Forget, Y. Zhu, M. Gray, and B. Lang, "Unique mitochondrial genome architecture in unicellular relatives of animals." *Proceedings of the National Academy of Sciences, USA* 100:892–897; 2003.

Karpov, S., and B. Leadbeater, "Cytoskeleton structure and composition in choanomastigotes." *Journal of Eukaryotic Microbiology* 45:361–367; 1998.

King, N., "Choanomastigotes." *Current Biology* 15:113–114; 2005.

King, N., "The unicellular ancestry of animal development." *Developmental Cell* 7:313–325; 2004.

King, N., and S. Carroll, "A receptor tyrosine kinase from choanomastigotes: Molecular insights into early animal evolution." *Proceedings of the National Academy of Science, USA*. 98:15032–15037; 2001.

King, N., H. Christopher, and S. Carroll, "Evolution of key cell signaling and adhesion protein families predates animal origins." *Science* 301:361–363; 2003.

Lavrov, D., L. Forget, M. Kelly, and B. Lang, "Mitochondrial genomes of two Demosponges provide insights into an early stage of animal evolution." *Molecular Biology and Evolution* 22:1231–1239; 2005.

Leadbeater, B., and S. Karpov, "Cyst Formation in a Freshwater Strain of the Choanomastigote Desmarella moniliformis Kent." *Journal of Eukaryotic Microbiology* 47:433–439; 2000.

Leadbeater, B., "Life-History and Ultrastrucutre of a New Marine Species of Proterospongia (Choanoflagellida)." *Journal of Marine Biology Association of the United Kingdom* 63:135–160; 1983.

Mendoza, L., J. Taylor, and L. Ajello, "The class mesomycetozoea: A heterogeneous group of microorganisms at the animal-fungal boundary." *Annual Review of Microbiology* 56:315–344; 2002.

Steenkamp, E., J. Wright, and S. Baldauf, "The Protistan Origins of Animals and Fungi." *Molecular Biology Evolution* 23:93–106; 2006.

Wainright, P., G. Hinkle, M. Sogin, and S. Stickle, "Monophyletic origins of the metazoa: An evolutionary link with fungi." *Science*. 260:340–342; 1993.

CHAPTER THREE

ANIMALIA

A-21E *Limenitis archippus* [Courtesy of P. Krombholz.]

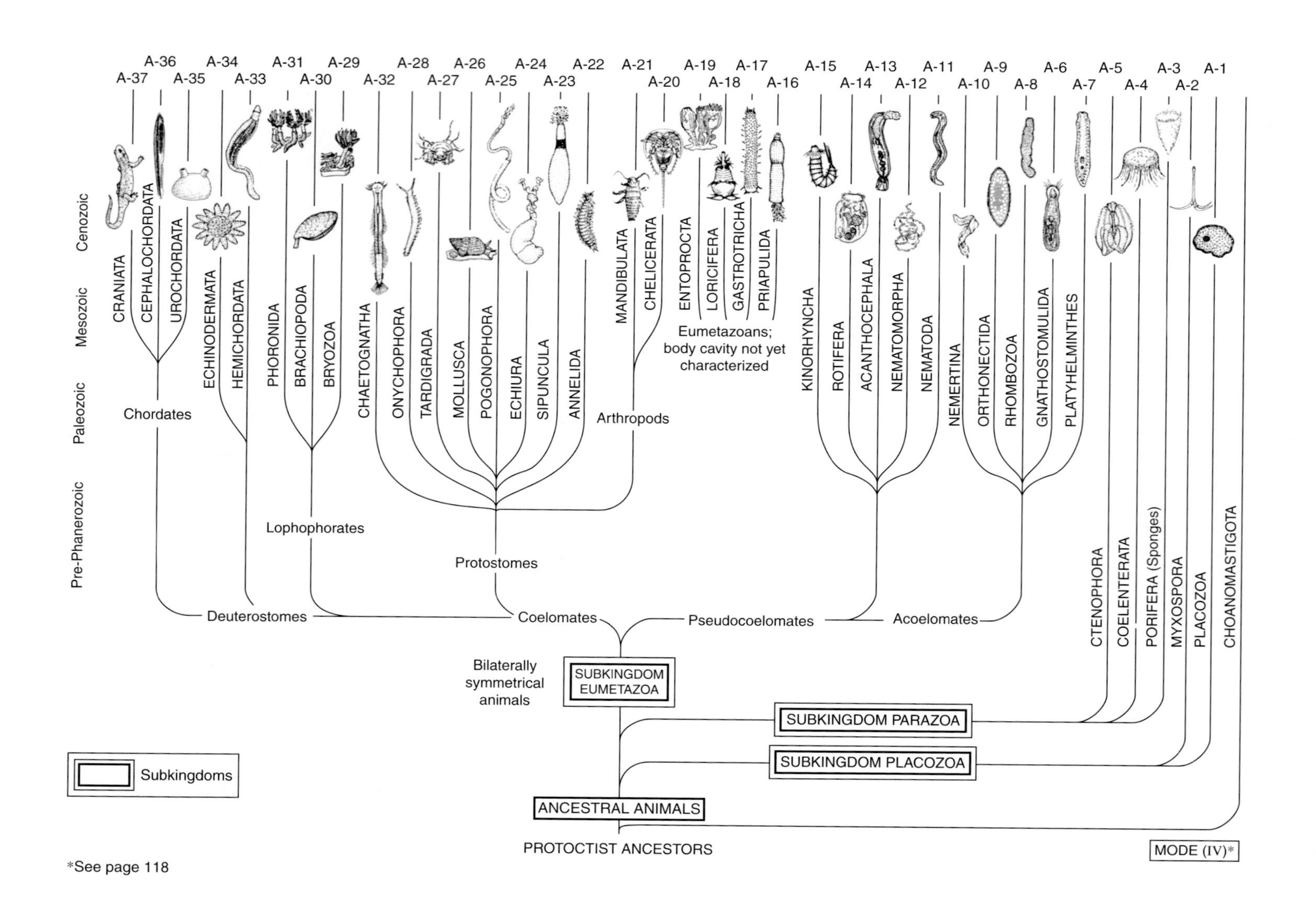

A-37
A-36
A-35
A-34
A-33
A-31
A-30
A-29
A-32
A-28
A-27
A-26
A-25
A-24
A-23
A-22
A-21
A-20
A-19
A-18
A-17
A-16
A-15
A-14
A-13
A-12
A-11
A-10
A-9
A-8
A-6
A-7
A-5
A-4
A-3
A-2
A-1
Cenozoic
Mesozoic
Paleozoic
Pre-Phanerozoic
CRANIATA
CEPHALOCHORDATA
UROCHORDATA
ECHINODERMATA
HEMICHORDATA
PHORONIDA
BRACHIOPODA
BRYOZOA
CHAETOGNATHA
ONYCHOPHORA
TARDIGRADA
MOLLUSCA
POGONOPHORA
ECHIURA
SIPUNCULA
ANNELIDA
MANDIBULATA
CHELICERATA
ENTOPROCTA
LORICIFERA
GASTROTRICHA
PRIAPULIDA
Eumetazoans; body cavity not yet characterized
KINORHYNCHA
ROTIFERA
ACANTHOCEPHALA
NEMATOMORPHA
NEMATODA
NEMERTINA
ORTHONECTIDA
RHOMBOZOA
GNATHOSTOMULIDA
PLATYHELMINTHES
Chordates
Arthropods
Lophophorates
Protostomes
Deuterostomes
Coelomates
Pseudocoelomates
Acoelomates
CTENOPHORA
COELENTERATA
PORIFERA (Sponges)
MYXOSPORA
PLACOZOA
CHOANOMASTIGOTA
Bilaterally symmetrical animals
SUBKINGDOM EUMETAZOA
SUBKINGDOM PARAZOA
SUBKINGDOM PLACOZOA
ANCESTRAL ANIMALS
PROTOCTIST ANCESTORS
MODE (IV)*
Subkingdoms

*See page 118

KINGDOM ANIMALIA

Latin *anima*, breath, soul, spirit

MODE IV*

Diploid organisms with gametic meiosis. The zygote develops by fusion (fertilization: cytogamy and karyogamy) of haploid egg and sperm (anisogametes) that form an embryo, the blastula. Meiosis, that occurs in the diploid mature female and male parent, yields anisogametes. Fossil record from 600 mya to present.

In the two-kingdom (animal vs. plant) classification—older and not used in this book—animals composed of many cells (multicellular) were considered Metazoa to distinguish them from Protozoa (one-celled animals). In our system, no one-celled animals exist; traditional protozoans are in the Protoctista kingdom. We define animals as heterotrophic, diploid, multicellular organisms that usually develop from embryos. The blastula, a multicellular embryo that forms from the diploid zygote produced by fertilization of a large haploid egg by a smaller haploid sperm, is unique to animals.

Because animal gametes—the egg and sperm—differ in size, they are called anisogametes. The diploid zygote produced by fertilization divides by mitotic cell divisions often a solid ball of cells that hollows out to become a blastula (Figure A-1). In many animals, the blastula develops an opening called the blastopore, which is the opening to the developing digestive tract. The blastopore will be the site of the mouth in some phyla or the anus in others. In some phyla young animals show neither of these two patterns. Some animals with spiral cleavage produce a blastula (stereoblastula) that is a solid ball of cells—their affinities remain unclear until more is known. Cephalopod molluscs (A-26), which have much yolk, lack blastocoels (embryonic cavities). Cell differentiation and cell migration transform blastula into a gastrula. Embryos have dead-end or tubular indentations that become the embryonic digestive tract in most.

The details of further embryonic development differ widely from phylum to phylum. Nevertheless, common developmental patterns provide clues to phyletic relationships. In many phyla, developmental details are known for a very few species in some phyla, for none. Because development is intricate and complex, it cannot be summarized in a few words. For similar reasons, concise, accurate definitions of the phyla cannot always be given. Our descriptions are more informal.

Multicellularity is not unique to animals; multicellular organisms abound in all the kingdoms. Examples include most Cyanobacteria (B-6) and Actinobacteria (B-12) in kingdom Prokaryotae; Phaeophyta (Pr-17), Oomycota (Pr-21), and Rhodophyta (Pr-33) in kingdom Protoctista; most members of kingdom Fungi; and all members of kingdom Plantae. However, multicellularity is most diverse in the animals; that is, many cells with highly specialized functions are grouped into tissues, tissues into organs. Complex junctions link cells into tissues in most phyla; two types of junctions unique to animals are desmosomes and gap junctions, which regulate communication and flow of materials between cells. Cell-to-cell connections are best seen in transmission electron microscopic thin section.

* see page 119

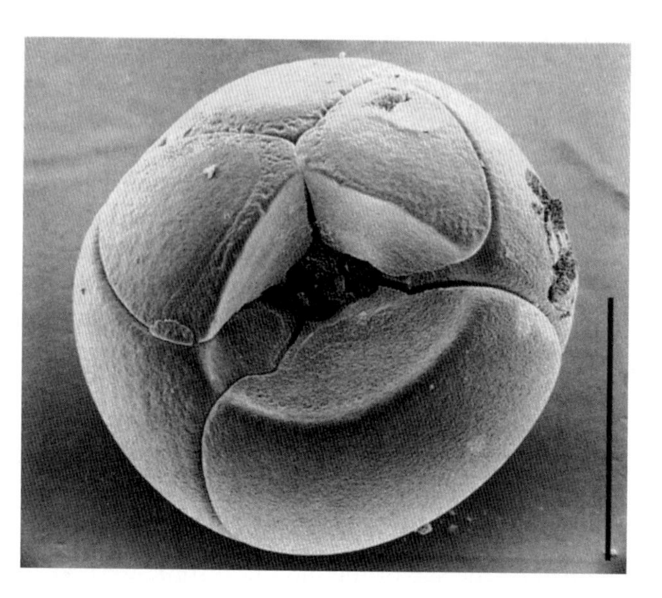

Figure A-1 Blastula, the embryo that results from cleavage of the zygote of *Xenopus*, the clawed frog. In many animal species, a sphere of cells surrounds a liquid-filled cavity, the blastocoel. One cell has been removed from this eight-cell embryo. SEM, bar = 0.5 mm. [Courtesy of E. J. Sanders, *Biophysical Journal* 15:383 (1975).]

S

Most animals ingest nutrients. Many animals take food into their bodies through an oral opening and then either engulf solid particles into digestive cells by phagocytosis ("cell eating") or, for liquid droplets, pinocytosis ("cell drinking") or absorb food molecules through cell membranes. Symbionts, such as the orthonectids (A-9) and gordian worms (A-12), often lack digestive systems. Solar-powered animals, such as *Convoluta paradoxa* (a platyhelminth, A-7) and *Elysia* (a mollusc, A-26), acquired photosynthetic symbionts, just as did the protoctists with closest common ancestry to green algae (Pr-28).

Animals that inhabit deep-sea black smokers (hydrothermal vents) and cold seeps (cold water rising through the seafloor) do not directly depend on sunlight for energy. Rather, the energy that powers their symbioses comes from inorganic compounds such as sulfides and methane that seep up through vents in the seafloor. Tube worms, clams, and other vent and cold-seep animals are nourished by symbioses with internal chemolithoautotrophic bacteria. A chemolithoautotroph is a self-feeding bacterium that uses energy released by inorganic chemical oxidations as the source of energy for its life processes, including synthesis of organic molecules from CO_2. Vent and seep animals either digest the bacteria directly or absorb organic molecules synthesized by their symbiotic partners. These vent and seep communities are rare today but were likely typical of Earth's environment 3000 mya.

Animals exhibit behavior of various kinds, such as attraction to light, avoidance of noxious chemicals, and sensing of dissolved gases and temperature. Such behavior is found in members of all five kingdoms, but animals have most elaborated this theme. Early in the history of the animal kingdom, more than 500 mya, nervous systems with brains evolved in several lineages. Organisms in other kingdoms lack nerve cells (neurons) and brains.

In form, the animals are the most diverse of all organisms. The tiniest animals are termed microbes. Smaller than many protoctists, these animals require a microscope to be seen. Many of these minute animal species make up the heterotrophic fraction of the plankton (Greek *planktos*, wandering); planktic animals—together with photosynthetic planktic species—constitute the base of freshwater and most marine food webs.

The largest animals today are whales, sea mammals in our own class (Mammalia) and phylum (Craniata, A-37). The members of most animal phyla inhabit shallow waters. Truly land-dwelling forms are found in only four phyla: chelicerates such as spiders (A-20), mandibulates (uniramians) such as insects (A-21), crustaceans such as sowbugs (A-21), and craniates such as reptiles, birds, and mammals (A-37). Species that live on land in the soil (for example, velvetworms) belong to several phyla, but they require constant moisture, they have not freed themselves from an aqueous environment. In fact, animals of most phyla are aquatic worms of one kind or another, except insects and others of phylum Mandibulata. Probably more than 99.9 percent of all the species of animals that have ever lived are extinct and are studied in paleontology rather than zoology.

Of all organisms, only the animals have succeeded in actively invading the atmosphere. Representatives of all five kingdoms (for example, spores of bacteria, fungi, and plants) spend significant fractions of their life cycles airborne in the atmosphere, but none in any kingdom spends its entire life in the air. Active flight evolved only in animals. Locomotion of animals through the air independently evolved several times but only in two phyla: (1) Mandibulata, class Insecta and (2) Craniata, classes Aves (birds), Mammalia (bats), and Reptilia (several extinct flying dinosaurs).

For many years and even now, some biologists assign animals to one of two large groups: the invertebrates, animals without backbones, and the vertebrates, animals with backbones. Animals, except members of our own phylum (Craniata), are invertebrates. Today, about 98 percent of all living animals are invertebrates. This invertebrate–vertebrate dichotomy amply accounts for our skewed perspective. Our pets, beasts of burden, and sources of food, leather, and bone—that is, terrestrial animals closest to our size and most familiar—are members of our own phylum. From a less human-centered point of view, traits other than lack of a backbone are better indicators of early evolutionary divergence. We prefer to describe these mostly marine animals by their unique traits rather than to collectively dismiss them as *in*vertebrates.

The animal phyla are described here in approximate order of increasing morphological complexity. Three phyla of animals, Placozoa (A-1) with its single genus, Myxospora (A-2), reduced symbiotrophs of fish, and Porifera, the better-known sponges (A-3), constitute subkingdom Parazoa ("alongside animals"); members of these phyla lack tissues organized into organs and most have an indeterminate form. The Rhombozoa and Orthonectida (A-8 and A-9) are animals whose evolution seems to have been independent of the true metazoa; rhombozoans and orthonectids do not fit the criteria of either Parazoa or Eumetazoa, the other animal subkingdom. The other 32 phyla constitute subkingdom Eumetazoa (true metazoans); most have tissues organized into organs and organ systems.

Broad overviews of organ systems that carry out circulation, respiration, digestion, support, and reproduction will be discussed in each phylum essay. Certain organisms have open circulation with blood circulating at least partially in body spaces rather than in veins, arteries, and capillaries. In other organisms, blood is confined to arteries, capillaries, and veins in what are referred to as closed circulatory systems. A circulatory system transports dissolved gases—oxygen and carbon dioxide—whereas an excretory system functions to rid an organism of toxic wastes, such as nitrogenous waste and salts. In regard to reproductive systems, some species are monoecious (one house, hermaphroditic), having both genders within one individual organism; other species are dioecious (two houses), with separate male and female organisms. Monoecious organisms may be either simultaneous hermaphrodites or sequential hermaphrodites—first male and then female or first female and then male.

The Eumetazoa comprises two branches: radially symmetrical and bilaterally symmetrical animals. The Radiata—radially symmetrical organisms—are Coelenterata (A-4), and Ctenophora (comb jellies, A-5) are biradially symmetrical organisms. Many species in these phyla are planktic. Comb jellies and cnidarians encounter a uniform aquatic environment on all sides; their bodies are radially symmetrical, both internally and externally. Other eumetazoan phyla have bilateral symmetry, at least at some time in their life histories. Echinoderms for example are often radially symmetrical as adults, but they are bilaterally symmetrical as larvae.

Characteristics of the body cavity including its embryonic origin allow most of the bilaterally symmetrical phyla to be assigned to one of three groups, but about eight phyla cannot yet be assigned because too little is known about the origin of their body cavities. Those that lack a body cavity between gut and outer body-wall musculature are Acoelomata (A-4 through A-8), although neither Phylum A-5 nor Phylum A-7 has outer body-wall musculature; those that have a body cavity called a pseudocoelom—not a coelom—are Pseudocoelomata (A-11 through

A-15); and those that develop a true coelom are Coelomata (A-20 through A-37). What is the difference between a pseudocoelom and a coelom? Gastrulation leads to the development of two, and eventually three, tissue layers in all animals more complex than the placozoans (A-1), sponges (A-3), Coelenterata (A-4), ctenophores (A-5), rhombozoans (A-8), and orthonectids (A-9). The three tissue layers are called the endoderm, mesoderm, and ectoderm (in order from the inside out) and are the masses of cells from which the organ systems of animals develop. In general, the intestine and other digestive organs develop from endoderm; the muscle, skeletal, and all other internal organ systems except the nervous system develop from mesoderm; and the nervous tissue and outer integument develop from the ectoderm. In the coelomates, the embryonic mesoderm opens to eventually form an internal body cavity lying between the digestive tract and outer body-wall musculature. This body cavity is called the coelom. A pseudocoelom is also an internal body cavity lying between the outer body-wall musculature and gut. Unlike the coelom, though, it is lined by loose cell masses rather than with mesoderm, and the pseudocoelom generally forms by persistence of the blastocoel, the embryonic cavity of the blastula.

Problematic phyla that do not easily fit these categories include nemertines (A-10). If one accepts that the nemertine proboscis cavity and blood vessels are homologues of the coelom, then phylum Nemertina may be assigned coelomate status and located beside sipunculans (A-23) on the phylogenetic tree.

For animals of eight phyla whose members are bilaterally symmetrical—particularly for kinorhynchs (A-15), loriciferans (A-18), and a newly proposed Cycliophora phylum—the nature of the body cavity is uncertain. Embryological studies are needed to determine the origin of the body cavity for members of these phyla before they can be placed in any of these groups. For priapulids (A-16), gastrotrichs (A-17), entoprocts (A-19), bryozoans or ectoprocts (A-29), brachiopods (A-30), and phoronids (A-31), relationships to other phyla in the animal phylogeny are uncertain; some uncertainties are discussed in the phyla essays. In establishing classification schemes, when we cannot place a species with any previously established phylum —Loricifera, for example—we create a new place for it. In creating a new phylum, we set that phylum forth as a hypothesis to be tested by studying relationships. Investigation of the origin of the body cavity is one mode of studying relationships between animals. Affinities of the new Cycliophora (*Symbion*) phylum to existing phyla Entoprocta and Bryozoa warrant testing.

Two groups of coelomate animals are distinguished according to the fate of the blastopore—the site of invagination of the blastula. In protostome ("first mouth") animals (A-20 through A-28), the blastopore is the site of the mouth of the adult. In deuterostomes (A-32 through A-37), the blastopore becomes the anus—the rear end of the intestine; the mouth forms as a secondary opening at the end opposite the anus. The deuterostome phyla are thought to have common ancestors more recent than their protostome ancestors. This protostome–deuterostome divergence occurred at least 540 mya, as judged from the presence of both protostomes and deuterostomes in the Lower (early) Cambrian fauna. Molecular biology shows the following phyla to be related to lophophorates including Loricefera, Bryozooa, and Phoronids. They are "Ecdysozoa," meaning they are recognized as molting marine animals. The relationship of lophophorates (A-29, A-30, and A-31), for example, to either deuterostome or protostome coelomates is not established.

Biologists agree that animals evolved from ancestral protoctists. Which protoctists, when, and in what sort of environments are questions that are still debated. Earl Hanson (Wesleyan University) amassed much information on the protoctist–animal connection and suggested that the question remains open. Patricia Wainright of Rutgers University (and colleagues) did the same for the fungus–animal connection. The Porifera (A-3), evolved from the choanomonads (Phylum Choanomastigota, Pr-36), as deduced from both molecular systematics (ribosomal nucleotide sequences) and details of fine structure of the cells. It is likely that the animal phyla other than poriferans, especially the eumetazoans, had different ancestors among the protoctists.

Box A-i: Larval transfer

Larval transfers result from hybridization events between distantly related taxa, whereby the acquired sets of genes encode two different body plans and life histories that evolve to function within the life history of the hybrid offspring. The distinct forms of the progenitor species manifest themselves in sequential expression of the genomes and associated morphology through the ontogeny of the offspring, first as larvae and then as adults (Williamson, 1992, 2003). The sequential transition from one form to another occurs through a metamorphic phase. Larvae undergo histolysis (disintegration of organic tissue) and regeneration of new tissues that develop into the morphologically distinct adult. Larval transfer therefore represents an exception to the biological species concept, defined as a population of organisms that successfully interbreed, and is reproductively isolated from other populations (Mayr, 1942). It is an important mode of rapid evolution in the animal kingdom.

In other kingdoms, exceptions such as interspecific mating and gene exchanges are common. Lateral gene transfer in bacterial conjugation or viral transduction is frequent, but gene transfer also occurs in eukaryotes. Plant hybridization and polyploidization (Box Pl-i) is the most important mode of speciation in angiosperms. Sperm penetration of alien eggs is likewise a means of interspecific gene transfer shaping animal evolution. Williamson suggests that during the Phanerozoic Eon, 542 mya, perhaps 10–50 separate interspecific hybridizations led to permanent partnerships that express more than a single parental genome (Figure A-i-1, *Luidia sarsi*).

Myriad larval morphologies have been described (Figure A-i-1). Many larval body plans occur as adult animals without larvae, such as Rotifers. Onychophorans (A-28), vermiform in shape, resemble caterpillars but do not metamorphose into moths or butterflies. Larvae of some brittle stars (class Ophiuroidea) resemble those of sea urchins (class Echinoidea); others undergo direct development from egg to adult. Some animals like decapod shrimp can express three or four sequential larval forms during their ontogeny. Williamson interprets multiple sequence metamorphosis as evolution from multiple permanent hybridizations. The bewildering dissimilarities between larvae and adults suggest that adult stages of the same animal were independently acquired as larval forms by members of separate higher taxa (families, classes, even phyla).

Larval transfer has probably occurred infrequently but may have been facilitated by an observable fact today: gametes of many invertebrate animals, freely released into water, fuse with gametes

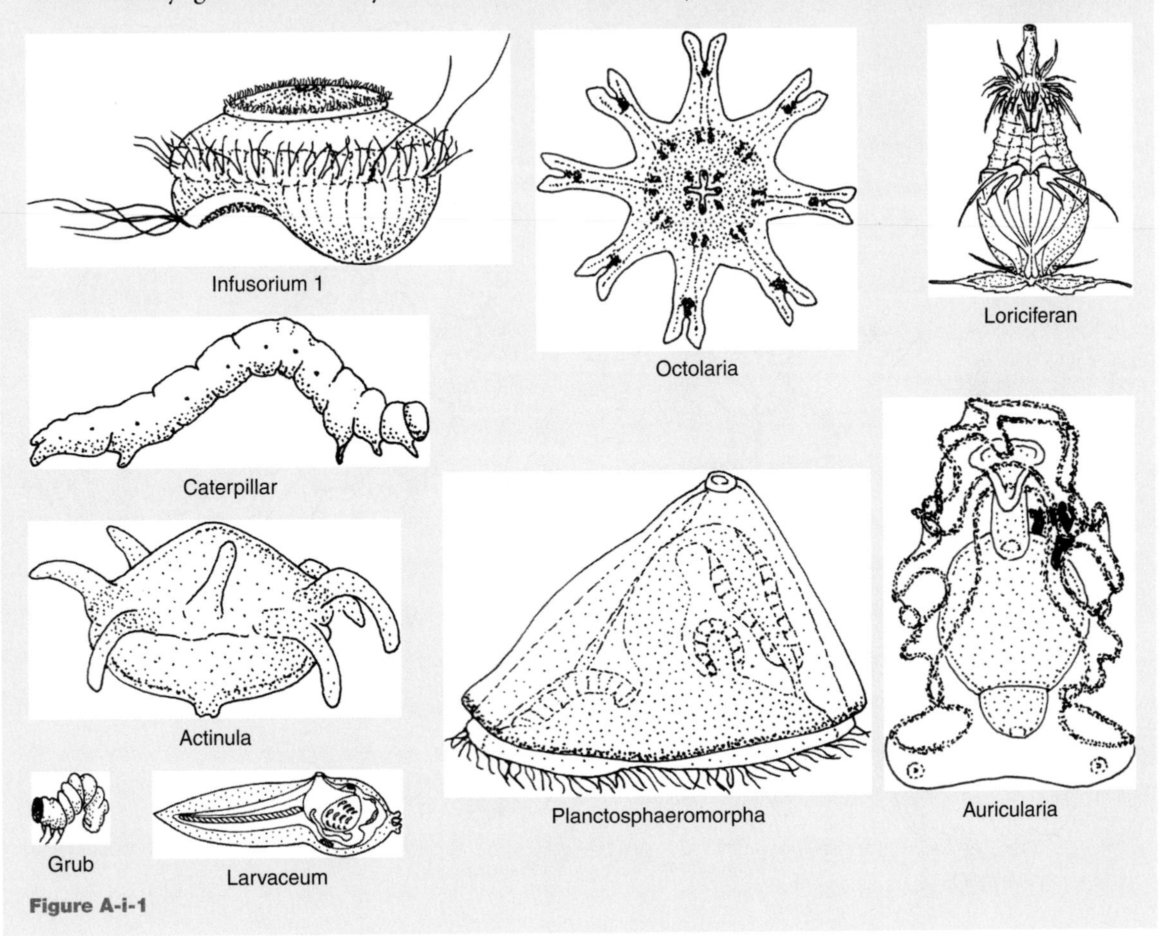

Figure A-i-1

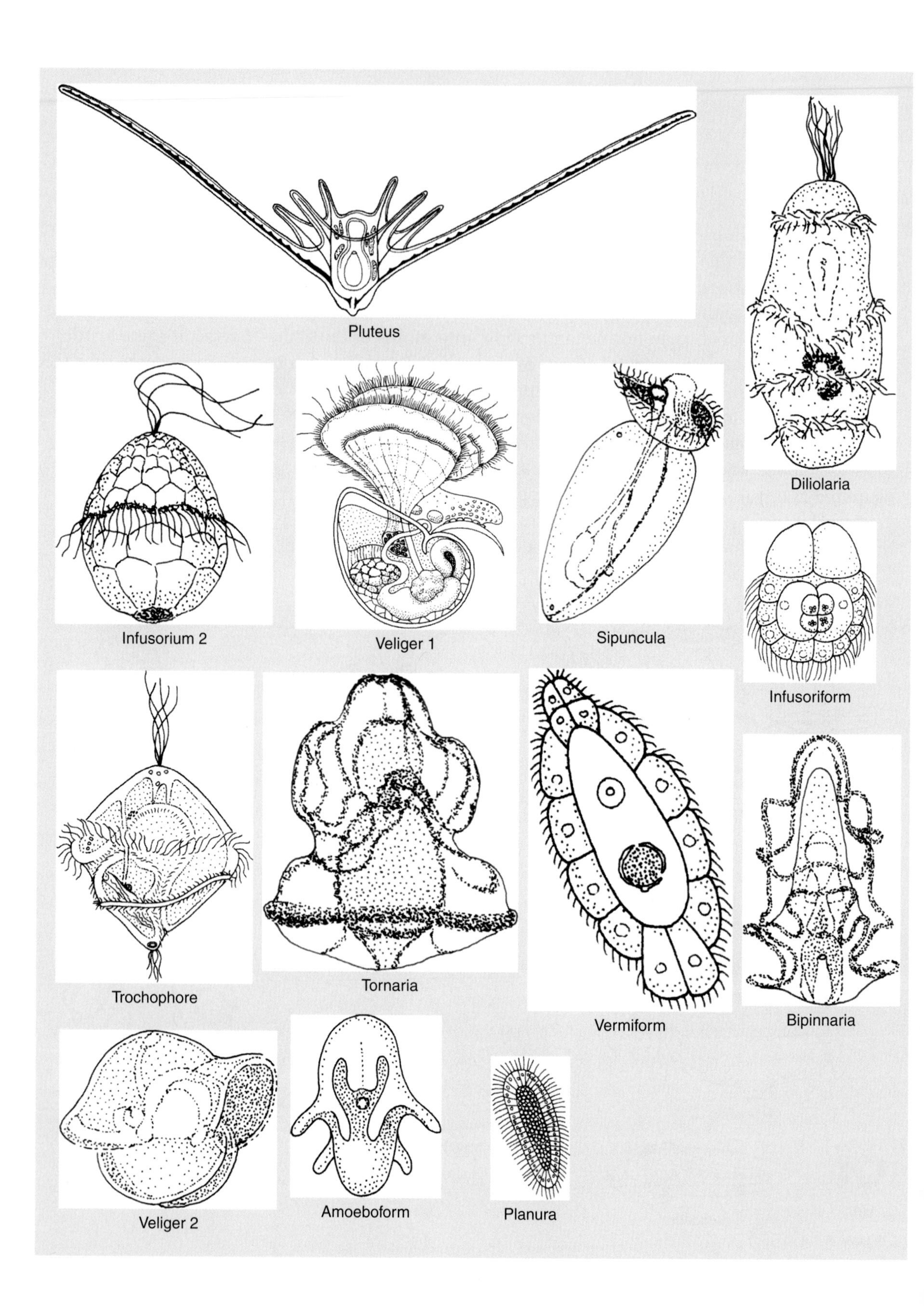
Pluteus
Diliolaria
Infusorium 2
Veliger 1
Sipuncula
Infusoriform
Trochophore
Tornaria
Vermiform
Bipinnaria
Veliger 2
Amoeboform
Planura

of other species. Such external breeding may also have promoted hybridization on land. Certain apterygote (wingless) insects, for example, reproduce when males lay spermatophores on the ground, later picked up by females. Amphibians deposit eggs in lakes, puddles, or moist soils, where males fertilize them by shedding sperm. The water frog, *Rana esculenta*, is a permanent hybrid of a cross between *R. lessonae* and *R. ridibunda*.

Williamson (1992) observed experimental gamete fusion between distantly related taxa as eggs of a urochordate sea squirt (A-35) were fertilized by sea urchin sperm (A-34), and a pluteus larva developed. Somatic cell hybrids are produced in the laboratory by use of sendaivirus to fuse membranes. Mouse, human, and other somatic cells in tissue culture may fuse naturally through close contact, even without viral induction, and then divide and reproduce in the fused state. Fusion produces viable offspring cells that contain genomic material from both parents, although often during replication, material from one genome is excluded.

The fossil record indicates that evolution is not always gradual. Neo-Darwinists teach that evolution occurs by gradual accumulation of mutation (variation) followed by natural selection for beneficial traits. However, periods of gradual change have been interrupted by bursts of rapid speciation ("punctuated equilibrium;" Eldredge and Gould, 1972). Symbiogenesis results in the acquisition of entire foreign genomes within just a few generations. Organelles such as mitochondria and plastids were acquired through symbiotic mergers with oxygen-respiring and blue-green photosynthetic bacteria, respectively (Margulis, 1993). Merger of lineages through larval transfer is likewise a genome acquisition event. A few such unique associations were favored by selection, as animals sequentially colonized new niches: open-ocean planktonic veliger larvae, for example, settle on reef surfaces as mussels or clams (Margulis and Sagan, 2002).

Williamson's theory of larval transfer does not contradict Darwinian evolution by natural selection. Rather, like conjugation, meiotic sex followed by fertilization, and symbiogenesis, larval transfer is a combinatorial mode of evolutionary change. Such anastomosing of evolutionary branches contributes variation to the animal kingdom (Williams and Vicker, 2007).

References

Eldredge, N., and S. J. Gould, "Punctuated equilibria: An alternative to phyletic gradualism." In: T. J. M. Schopf and J. M. Thomas, eds., *Models in paleobiology*. Freeman Cooper; San Francisco; pp. 82–115; 1972.

Margulis, L., *Symbiosis in cell evolution: Microbial communities in archean and proterozoic eons*. W.H. Freeman; San Diego; 1993.

Margulis, L. and D. Sagan, *Acquiring Genomes: A Theory of the Origins of Species*. Basic Books: New York; 2002.

Mayr, E., *Systematics and the origin of species*. Columbia University Press; New York; 1942.

Williamson, D. I., *Larvae and evolution: Toward a new zoology*. Chapman and Hall; New York, London; 1992.

Williamson, D. I., *The origins of larvae*. Kluwer Academic Publishers; Dordrecht, London, Boston; 2003.

Williamson, D. I., and S. E. Vickers, *The origins of larvae*. American Scientist; 95: 509–519; November 2007.

SUBKINGDOM (Division) PLACOZOA (no nerves or antero-posterior asymmetry)

A-1 Placozoa (Trichoplaxes)

Greek *plakos*, flat; *zoion*, animal

GENERA
Trichoplax

Soft and so small as to be barely visible to the naked eye, placozoans are among the least complex of animals. They do not contain sensory, digestive, excretory, or respiratory structures, nor do they possess organs of any type. *Trichoplax*, the only described genus, looks and behaves like a very large ameba; the grayish body continuously alters shape. Placozoans are, however, multicellular animals. They are between 0.2 and 1.0 mm in diameter, and are larger in cultures.

The entire surface of this animal is ciliated with cells containing a single undulipodium. Flat cells and intercellular shiny lipid droplets make up the dorsal epithelium. The ventral surface is composed of gland cells (without an undulipodium) and cylinder (columnar) cells. Between the dorsal and ventral cell layers, a contractile mesenchyme-like syncytium floats in a layer of fluid. *Trichoplax* has, indeed, an upper, or "dorsal," and lower, or "ventral," side, but it lacks both anterior–posterior (head–tail) and right–left symmetry. The animal moves along by means of undulipodia on its ventral side. If overturned, the animal can right itself to a dorsal–ventral orientation. Although placozoans have been traditionally considered benthic in their lifestyle, they have been collected commonly in the water column among the swimming plankton.

Reproduction in placozoans is primarily asexual. In one mode, large individuals divide by fission into two multicellular organisms. In another mode, round buds—which contain dorsal and ventral cells, and mesenchyme—form and detach. These are called "swarmers" and presumably give rise to new individuals. Swarmers may be the same as a solid spherical form previously observed in culture. Another type of spherical body has been observed in culture: a hollow form, the outer cell layer being made up of ventral-layer cells. Cut placozoans heal rapidly, and a complete animal regenerates from almost any fragmented part.

Sexual reproduction has been inferred from observations of what appear to be eggs and, conceivably, of sperm. Typical animal sperm with a single undulipodium have, however, never been seen, nor has meiosis or fertilization ever been observed. Putative eggs appear in the body, often during times of stress. Cleavage has been reported in these bodies through to the 64-cell stage but then ceases. The best evidence for sexual reproduction comes from molecular studies in which genetic "signatures" of outcrossing and inbreeding have been reported. No other information on placozoan life history currently exists.

Because they lack any form of digestive system, placozoans form an inverted cuplike space over organic debris and most likely ingest food by glandular secretion and intracellular absorption.

Trichoplax adhaerens, the only described species, was discovered in the seawater aquarium of the Graz Zoological Institute in Austria in 1883. For many years it was considered a larval form of Coelenterates and was largely ignored. Only when its identity as a distinct phylum was established, did it become more intensely studied. *Trichoplax* has always been found in seawater. It is now known to exist widely in warm waters around the world. Recent study has revealed that many genetically distinct "species" exist and as many as 11 have been identified (Figure A).

A placozoan cell has little DNA, of the order of 10^{10} daltons (molecular mass units), a quantity which is more like that of a bacterial nucleoid or the nucleus of some small protists than like that of the nuclei of animal cells. Compared with those of most animals, the chromosomes are very small, from 0.6 to 1.0 μm in length, the size of a bacterium. Nuclei of dorsal and ventral cells contain 12 chromosomes. Those of the syncytial mesenchyme have twice as much DNA as a typical cell; thus, either the mesenchyme layer has an extra set of chromosomes or the extra DNA is due to the presence of bacteria. *Trichoplax* is interpreted to be a diploid animal.

Molecular phylogenetic studies provide strong evidence that placozoans represent a distinct entity, not a reduced form of some other phylum, specifically the Coelenterata (A-4). It is still conjectured that they may represent a simplified form of an as yet unidentified ancestral metazoan. As research continues on the Placozoa, however, their relationship to other metazoans should become clearer. The Placozoa are truly multicellular organisms and provide us the simplest known model for the study of early metazoan evolution.

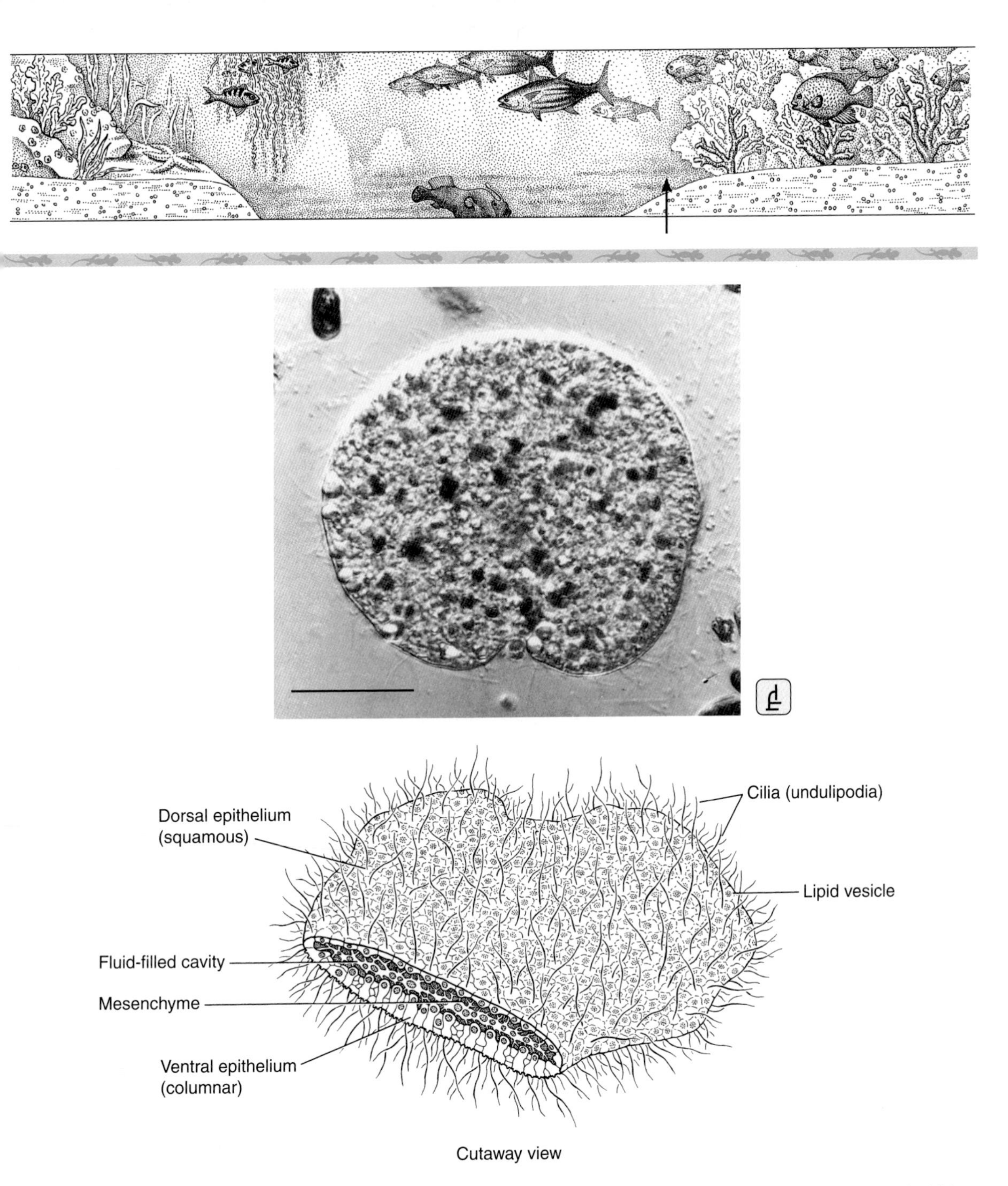

Figure A *Trichoplax adhaerens*, the simplest of all animals, found adhering to and crawling on the walls of marine aquaria. LM, bar = 0.1 mm. [Photograph courtesy of K. Grell; drawing of cutaway view by L. Meszoly; information from R. Miller.]

A-2 Myxospora
(Myxozoa)

Greek *myxa*, mucus; Latin *spora*, spore

GENERA

Tetracapsuloides
Myxobolus
Ceratomyxa
Enteromyxum
Myxidium
Sinuolinea
Sphaerospora
Henneguya
Chloromyxum
Parvicapsula
Kudoa

Myxozoans are microscopic, spore-forming symbiotrophs. Despite their multicellular nature, this group continued to be classified as Protozoa until molecular systematic approaches confirmed their placement among the Metazoa. Another major revision of the phylum was precipitated by the discovery that at least some members of the classes Myxosporea and Actinosporea are alternate life stages within the same life cycle. This caused suppression of the latter class, resulting in cumbersome nomenclature until life cycles are resolved. In 2000, the discovery that an obscure worm-like symbiotroph of bryozoans was a primitive myxozoan again caused revision of the phylum. The Myxozoa now comprise two classes: Myxosporea, the larger, more diverse group containing over 2000 species, and Malacosporea, with only four species.

Few life cycles are known from either group, but those described for the Myxosporea involve two hosts, with the symbiotroph alternating between actinospore and myxospore forms. Actinospores develop in marine and freshwater oligochaetes and polychaetes and marine sipunculids. Myxospores most commonly develop in fishes, however amphibians, reptiles and marine invertebrates are hosts for a number of species, and recently, myxozoan infections were reported from homeothermic vertebrates: birds, moles and shrews. Malacosporeans have been detected in fish and in bryozoans. In the bryozoan host the spore stages have a worm-like form (Figures A and B).

Many of the life cycles have been determined by matching the 18S ribosomal RNA gene sequences of actinospore and myxospore stages. Life cycles of several species have been established in the laboratory using cultures of aquatic annelids and appropriate fish hosts. While a two-host life cycle has been accepted as the model for the Myxosporea, direct transmission between fish has been demonstrated for some species. Life cycles of malacosporeans have not been completely resolved, although one species, *Tetracapsuloides bryosalmonae* infects salmonid fish and as well as bryozoans. It is considered likely that myxozoans evolved as necrotrophs of marine annelids (Myxosporea) or bryozoans (Malacosporea), with direct life cycles. The addition of teleost hosts would thus have afforded the necrotroph greater dispersal potential, as did further radiation into migratory birds.

Using the life cycle of *Myxobolus cerebralis* to describe a generalized model for this group, actinospores are released from the annelid hosts into the water column where they encounter their intermediate fish host. These spores anchor to the skin by injecting their polar filaments, then the infectious multinucleate sporoplasm penetrates the epidermis or mucous ducts. The symbiotroph divides asexually (presporogonic development) and begins migrating to its target tissue. Development of myxozoan plasmodium (sporogonic development) occurs in tissues (histozoic) or cavities of body organs (coelozoic). During spore formation, the multinucleate cell differentiates into capsulogenic, valvogenic and sporoplasmogenic cells, which will mature to form spores. Myxospores are typically 5–20 μm long, bilaterally symmetrical, have hard shells (valves), polar capsules with a coiled filament and an infective sporoplasm. Infection of the annelid (definitive) host occurs when myxospores released from infected fish are ingested. The myxospore valves open, releasing the sporoplasm(s) which penetrates the worm's intestinal epithelium. The sporoplasm undergoes mitotic nuclear and cellular division to form uninucleate and multinucleate cells. Some cells form binucleate stages that undergo meiotic division to produce eight zygotes in a pansporocyst. Sporogenic development results in eight actinospores. In contrast to myxospores, actinospores have triradial symmetry, a large number of infectious cells and usually inflate on contact with water, becoming up to 500 μm across.

That most myxozoans do not cause serious disease indicates that they have co-evolved with their teleost hosts. However, there are exceptions to this, and some diseases cause large losses in cultured and wild fish populations. Species infecting tissues are generally more pathogenic than those that favor hollow organs, such as urinary and gall bladders. In fresh water, *Myxobolus cerebralis* invades cartilage tissue in young trout, causing whirling disease. *Ceratomyxa shasta* has an affinity for intestinal tissues of salmonid fish, *Henneguya ictaluri* infects the gills of catfish, and *Tetracapsuloides bryosalmonae* causes a cascading immune response, proliferative kidney disease, in salmonids. In salt water, *Kudoa thyrsites* causes a condition known as soft flesh in a variety of marine fishes and species of *Enteromyxum* have caused losses in important Mediterranean fisheries.

Myxozoan taxonomy remains in a state of flux. Most recently, phylogenetic analysis of multiple protein-coding genes of a myxozoan symbiotroph bryozoans showed it had closest affinity with cnidarian groups Scyphozoa or Hydrozoa. This discovery may ultimately result in subsumation of the Myxozoa into the phylum Cnidaria, a hypothesis that has been proposed several times in the past because of the structural similarity between myxozoan polar capsules and cnidarian nematocysts.

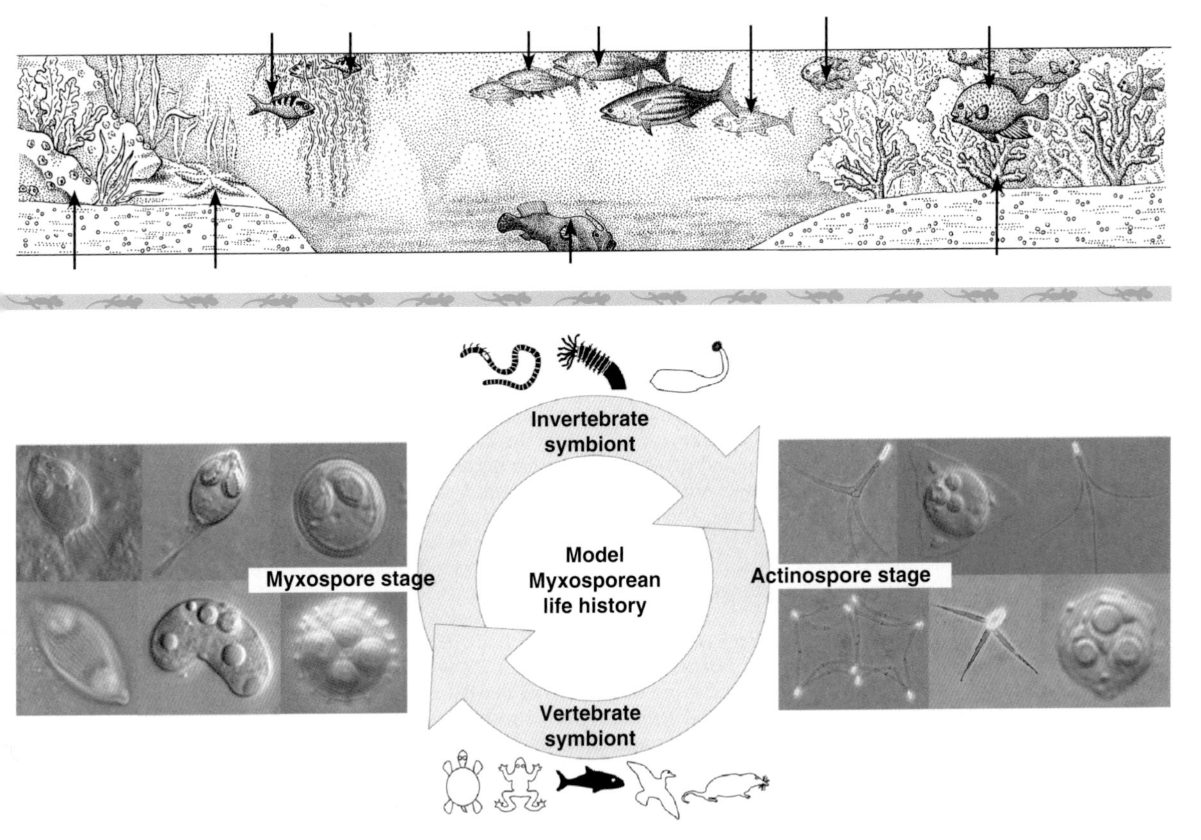

Figure A A model life history for the Myxosporea showing their invertebrate and vertebrate habitats, and examples of alternating myxospore and actinospore stages. Animal drawings in bold (oligochaetes, polychaetes, and fish) are those from which two-animal tissue habitat-life history stages have been described. The others, turtles, amphibians, birds, shrews, are those for which life histories have not been demonstrated. [Drawing and photographs courtesy of S. Atkinson.]

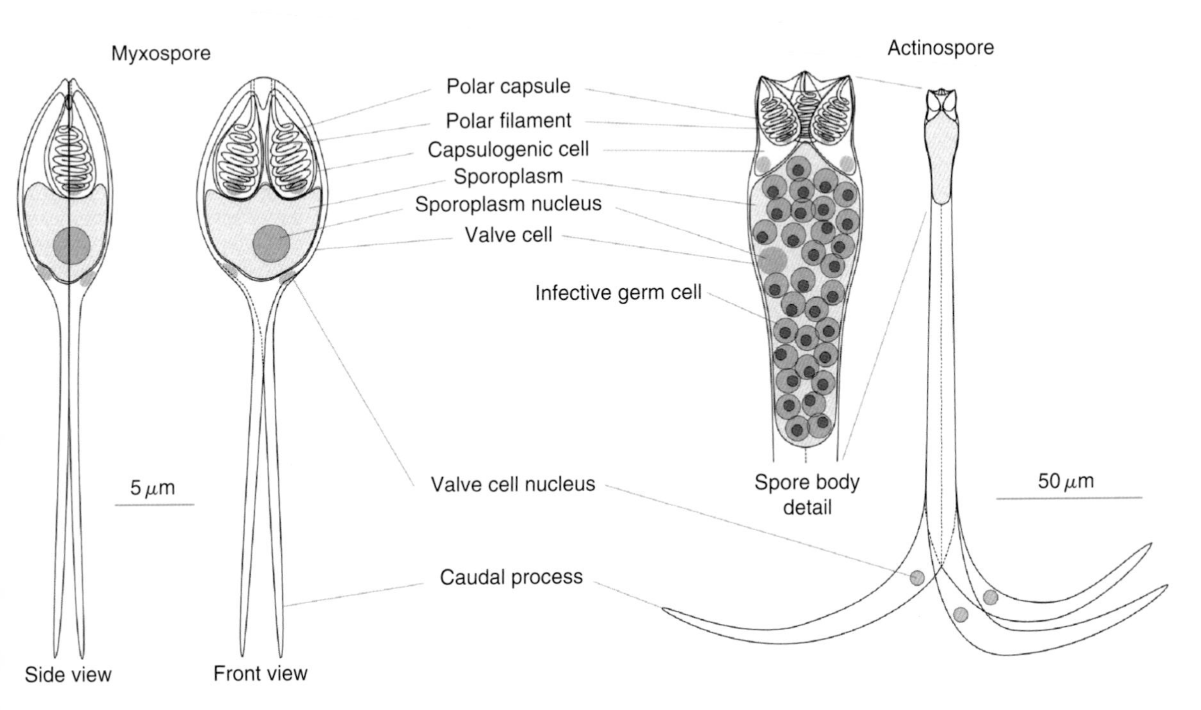

Figure B Characteristic structures of Myxosporea shown using *Henneguya* sp. as a representative myxospore and a triactinomyxon to represent the actinospore stage. [Drawing by S. Atkinson.]

SUBKINGDOM (Division) PARAZOA (nerve nets)

A-3 Porifera

(Sponges, poriferans)

Latin *porus*, pore; *ferre*, to bear

GENERA

Agelas
Asbestopluma
Euplectella
Gelliodes
Grantia
Halisarca
Hippospongia
Leucosolenia
Microciona
Scolymastra
Speciospongia
Spongia
Spongilla
Stromatospongia

Poriferans, the sponges, are named for their pores. Most sponges have thousands of pores into which water flows; thus, their body forms contribute in large part to current-induced water flow through their bodies. Because poriferans have no tissues or organs, like placozoans (A-1), they belong to the Parazoa. Like placozoans, most sponges lack anterior–posterior and left–right symmetry. About 5000 to 10,000 species of sponges are known; all are aquatic and only 150 species live in freshwater. Their fantastic forms—fans, cups, crusts, and tubes—range from a few millimeters wide to the two meters tall *Scolymastra joubini*, a barrel-like glass sponge of the Antarctic. Most classes encompass various body forms.

The classification of sponges is being revised because, like certain extinct sponges, recently described living sponges—chaeteiids, stromatoporids, and sphinctozoans—deposit calcium carbonate in patterns; eventually, the number of classes recognized in this phylum may increase or decrease. Currently, the sponges may be grouped into four classes: Calcarea, Demospongiae, Sclerospongiae, and Hexactinellida. The Calcarea, calcareous sponges such as *Leucosolenia*, comprise about 500 marine species bearing spicules of calcium carbonate. Sponges in other classes have siliceous spicules fashioned from opaline silica (SiO_2). The SiO_2 is extracted from seawater, concentrated, and laid down in delicate patterns inside sclerocytes—cells that secrete spicules. The Demospongiae, some 4000 marine and freshwater species of horny sponges (Figure A), are supported by spongin, fibers of protein related to the keratins of hair and nails. Whether demosponges contain silica spicules but not $CaCO_3$ depends on the species. The commercial bath sponges *Spongia* and *Hippospongia* are demosponges; *Hippospongia* has no spicules. Coralline sponges (Sclerospongiae, 15 marine species, sometimes included in Demospongiae or Calcarea, depending on the species) have a stony calcareous skeleton of $CaCO_3$, siliceous spicules, and spongin. Corallines are tropical reef sponges such as *Stromatospongia*. The Hexactinellida, 600 species of glass sponges, contain six-rayed siliceous spicules. Most of the elegant glass sponges (for example, *Euplectella*) dwell in the deep sea. Some biologists propose a new phylum for glass sponges because of their unique features, such as the syncytial outer net and syncytial choanoderm (undulipodiated collar cells) layers; syncytia are made up of nuclei not separated by cellular membranes.

Sponges are made of two cell layers; between the layers lies a cellular, gel layer called the mesoglea, which contains ameboid cells (amebocytes) and support spicules (skeletal needles) or spongin fibers (Figure B). Amebocytes are also called archaeocytes because they can differentiate into sperm, eggs, nutrient storage cells, spicule-secreting cells, spongin-secreting cells, and waste-eliminating cells. Although both inner and outer layers contain some specialized cells, the sponge body lacks the cell cohesiveness and coordination typical of tissues in other animals (A-4 through A-37 but not A-8 or A-9). As multicellular organisms, sponges have intercellular junctions (desmosomes and septate junctions), basement membrane–like structures, and matrix components (collagen and fibronectin). Nearly all sponges are sessile or encrusting, permanently attached. Water turbulence and the amount of available space tend to determine their shape and size. Many coating rocks and logs are so shapeless that they are not recognizable as animals.

Sponges lack mouths, intestines, muscle, nerves, and respiratory and circulatory organs. A nonnervous conducting system capable of halting the excurrent water flow in response to stimuli has been demonstrated, but it is not made up of nerves. Oxygen diffuses through the sponge body wall. Food is filtered from the copious quantities of water that flow through the sponge body. Plankton, such as dinomastigotes (Pr-5), makes up about 20 percent of their food; another 80 percent consists of detrital organic particles. Choanocytes (Figure B), cells with a collar of microvilli that surround the undulipodium are partly responsible for moving water (body form passively generates water flow in part) into the porocyte-lined pores and along incoming waterways called inhalant canals. Food particles either stick to the choanocyte collars or are directly engulfed by ameboid phagocytic cells lining the canals. When food is inside a phagocyte, these cells digest nutrients and allow other cells to absorb food. Wastes either diffuse out of the sponge directly through the body wall or flow out of the spongocoel (body cavity) through the excurrent opening.

Most sponges are hermaphrodites; mature organisms bear both eggs and sperm. In a few species, however, the genders are separate. Choanocytes develop into sperm. Eggs develop from either choanocytes or amebocytes. In most sponges, clouds of sperm released through the excurrent opening of one sponge enter another sponge with incoming water. Choanocytes capture the sperm and then transform into ameboid cells, which convey captive sperm to eggs in the mesoglea. Thus, fertilization is internal in most species. In synchrony with sperm release, the tube sponge *Agelas* releases a mass of eggs in strands of mucus to the reef to which it is attached. Zygotes develop into ciliated, free-swimming, multicellular, nonfeeding larvae. Some sponges with internal fertilization release their developing larvae into the water; others retain larvae for some time. Sponge larvae are of two main types: parenchymula or amphiblastula. Parenchymula larvae are solid and almost entirely ciliated. Amphiblastula larvae are hollow and ciliated at one end. The amphiblastula larvae turn inside out, bringing their formerly internal cilia outside. Although sponge larvae develop from blastulae, their embryonic development is not homologous to that of other animals. Eventually, sponge larvae metamorphose into adults.

Some sponges reproduce uniparentally. Fragments break off, are wafted away by water currents, and then grow to be individual sponges. Because of their capacity for regeneration, sponge cuttings are used to restock Florida's depleted sponge beds. Many freshwater and a few marine sponges also reproduce by means of propagules called gemmules. These overwintering

buds—composed of nutrient-laden ameboid cells surrounded by a tough outer layer of epithelial cells with spicules—disperse and, after dormancy, grow to form new sponges.

Many sponges harbor symbiotic bacteria and algae that provide food and probably oxygen, remove waste, and screen sunlight. Symbionts may be cyanobacteria (B-6) or red (Pr-33), green (Pr-28), or brown (Pr-17) algae. In some species, symbionts are transmitted from adult sponges to their offspring by adhering to gemmules. Because of their algal symbionts and their own pigments, sponges may be red, orange, yellow (most with carotenoids), green, blue, purple, brown, black or white. Some sponges luminesce.

Symbionts may be epibionts or may reside in the mesoglea. Many eubacterial and archaebacterial symbionts, are heterotrophs. In some demospongiae, the microbial symbionts can represent between 40 and 60 percent of the sponge volume. Some sponge symbionts are members of the deep branching planctomyces (B-11). They are called poribacteria and have only been found in sponges. The poribacteria (like their cousins *Gemmata*) have what appears to be a nucleoid membrane around their DNA (see Fig. D p. 95) region. Transmission electron micrographs show structures suggestive of nuclear pores. In some electron micrographs, the gerophore material seems more condensed (or just electron dense) than the nucleoid region seems in common laboratory bacteria like *E. coli.*

Endosymbiotic sponges have not been described, and most animals prefer other sources of food. Sponge spicules are evidently unsavory—a needle-sharp deterrent to predators, with the exception of snails and nudibranch molluscs (A-26), a few sea stars (A-34), and a few fish (A-37). Feather stars (A-34) perch on sponges to feed.

A carnivorous demosponge has been discovered in a nutrient-poor, deep, cold, sea cave. *Asbestopluma* passively snares crustaceans (A-22) on filaments covered with hook-shaped spicules. Once captured, thin new filaments engulf the prey as nutrients are transported along the strands to the rest of the sponge's body. External ingestion occurs by excretion of hydrolytic enzymes into the seawater. The entire process may take 8–10 days for larger prey, such as crustaceans 8 mm in length, to be thoroughly digested. This feeding habit—the capture and digestion of prey by sponges that lack any digestive cavity—appears to be a response of the sponge to oligotrophic life in the deep sea. Carnivorous deep sea sponges show that transitions from external "filter feeding" to a bodily reorganization for internal feeding in a coelenteron, stomach or other digestive cavity can occur. Such bodily transitions have been hypothesized for the early evolution of animals.

Figure A *Gelliodes digitalis*, one of the simpler sponges, shown live in its marine habitat. Water enters through its pores and exits through a single excurrent opening, called the osculum, on top. Bar = 10 cm. [Courtesy of W. Sacco.]

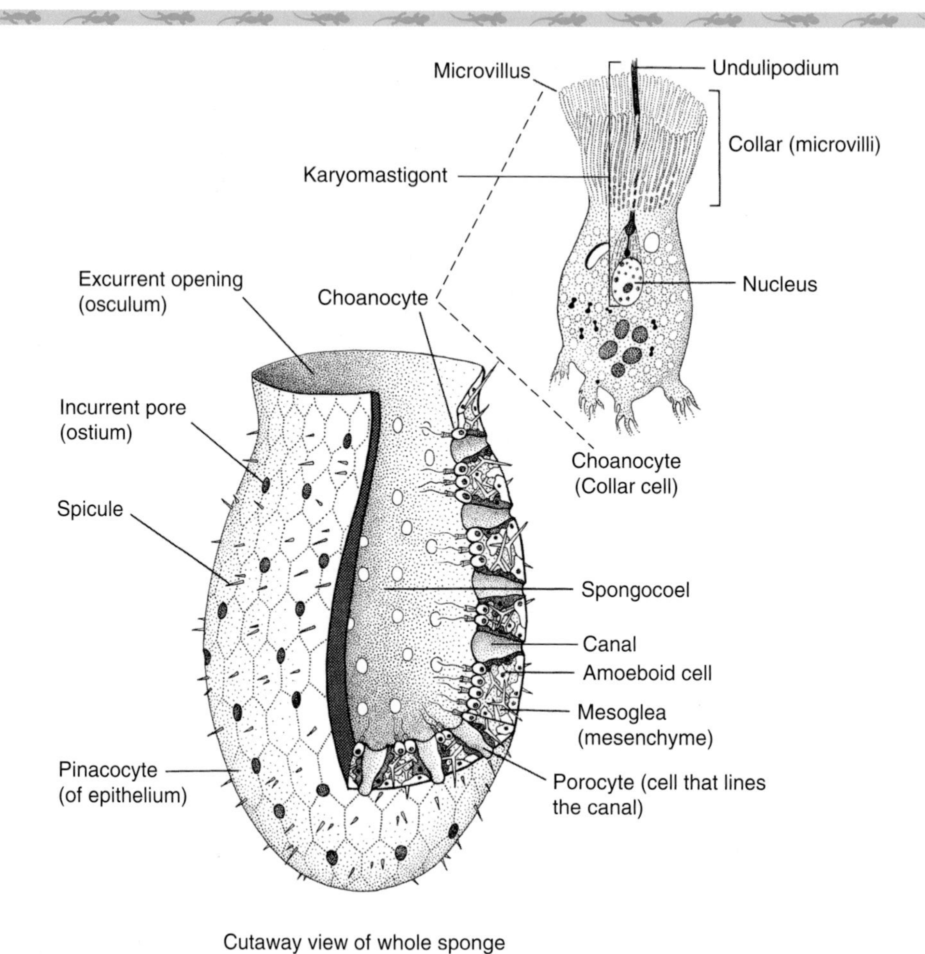

Figure B Cutaway diagram of simple sponge. The water current through the sponge is generated, in part, by two factors: (1) sponge shape (architecture) passively generates flow through incurrent pores (whose canals are lined by porocytes) and out at the large excurrent opening (the osculum) and (2) choanocytes (collar cells) actively generate additional water flow and capture food and sperm. [Drawing by L. Meszoly; information from S. Vogel and W. Hartman.]

The oldest sponges in the fossil record, the well-preserved Burgess shale sponges, date from the early Cambrian period, about 500 mya. Sponges seem to form a dead end, not ancestral to metazoans. Of all animals, poriferans are the easiest to relate directly to their protoctists ancestors—free-living colonial choanomastigotes (Pr-36) that are remarkably similar in structure to sponge choanocytes. Sponges may have been the earliest animal lineage to diverge from protoctists, as inferred from ribosomal RNA data.

Sponges harvested off the West Indies, Florida, Mexico, and the Philippines are used by lithographers, potters, silversmiths, and bathers. Oysters and other shellfish are sometimes destroyed by sponges that bore into them as the sponges excavate living space in their shells.

A-4 Coelenterata
(Cnidarians, hydras)

Greek *knide*, nettle; *koilos*, hollow; *enteron*, intestine

GENERA

Acropora
Alcyonium
Antipathes
Atolla
Aurelia
Branchioceranthus
Cassiopea
Chironex
Corallium
Craspedacusta
Cryptohydra
Cyanea
Dendronephthya
Ediacara
Haliclystus
Heliopora
Hydra
Metridium

(continued)

Sea anemones, jellyfish, hydras, and corals are among roughly 10,000 species of Coelenterates. These outwardly radially symmetrical invertebrates are the most morphologically complex members of subkingdom Parazoa. All coelenterates are aquatic and nearly all are marine. The five classes of coelenterates are the Anthozoa (most corals and sea anemones), Cubozoa (sea wasps and box jellies), Hydrozoa (hydras, hydroids, and hydromedusae), Scyphozoa (true jellyfish), and Staurozoa (stalked jellyfish). Coelenterates' tentacles and oral arms are replete with stinging cells called cnidoblasts, each containing an intracellular nematocyst, unique to this phylum.

Coelenterates have numerous life stages, with either the polyp or the medusa being the adult form (Figures A and B). Many species having a colonial polyp stage (that is, hydroids' siphonophores) have polymorphic individuals. Polyps such as *Hydra* and *Manania* have elongated bodies with one end having a mouth and the other end attaching to the substrate (Figures C and D). Most polyps are sedentary, but some glide, somersault, or employ tentacles as legs. Medusae (*Craspedacusta*, for example) usually swim free, the Frisbee-, umbrella-, or box-shaped bell pulsating, mouth downward, wit tentacles trailing (Figures A and B). The tentacles of medusae resemble the snaky locks of the mythical Medusa.

Coelenterate cells are assembled into tissues, in comparison with placozoans, which lack epithelial tissues. Between a cnidarian's outer layer of epidermis and its inner layer of gastrodermis lies an intermediate layer called the mesoglea. This mesogleal layer contains translucent secretions and often loose cells but is not organized as a tissue. The gastrodermis lines the gastrovascular cavity or stomach.

Coelenterates have nerve nets but have no central nervous system. The pacemaker of the nerve net maintains the swimming rhythm of medusae. Motion and light-sensitive cells on the edges of many medusae enable them to detect light and orient themselves. The nerve fibers of coelenterates are the only truly naked nerves in the animal kingdom—all others are covered by sheaths of insulating material such as myelin. Most coelenterate synapses (nerve junctions) are bidirectional. All other animals transmit impulses in only one direction.

The contractile system of polyps consists in part of a layer of epitheliomuscular cells; at the base of the epitheliomuscular cells are contractile fibers that run longitudinally and are anchored in the mesoglea. Coelenterates, in addition, have nutritive-muscular cells on the inside (below the epithelium) with contractile fibers that run circularly; nutritive-muscular cells contract, taking up and digesting food particles from the gastrovascular cavity. A medusa has muscle fibers in its swimming bell. Although they have no bones, polyps and medusae are stiffened by fluid pressure in the gut itself and by the mesoglea with its collagen matrix. Hydrocorals (Hydrozoa) and true corals (Anthozoa) secrete calcium carbonate exoskeletons within their soft polyps shelter. Medusae lack a calcium carbonate skeleton.

Most coelenterates are carnivores, preying on small animals, which are usually caught with the tentacles. Coelenterates sting when they contact active prey—crustaceans, worms, fish, comb jellies, diatoms, and other protoctists. Cnidoblasts triggered by touch, chemical stimulation, or both forcibly discharge harpoon-like barbs from their nematocysts along with venom (Figure E). A reef consisting of herbivorous soft coral (*Dendronephthya*), which lacks symbiotic algae and has poorly developed nematocysts, feeds almost exclusively on photoplankton, and some jellyfish are able to take up dissolved organic matter. Coelenterates digest food with their saclike gastrovascular cavity, or stomach, which opens through the mouth. The mouth squirts

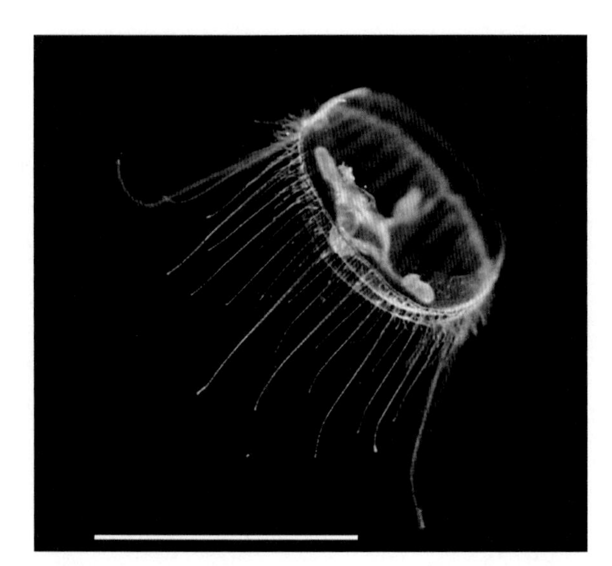

Figure A *Craspedacusta sowerbii*, the living medusa of a freshwater coelenterate. Contraction of the bell expels water, thereby propelling the medusa. Class Hydrozoa. Bar = 10 mm. [Courtesy of C. M. Flaten and C. F. Lytle.]

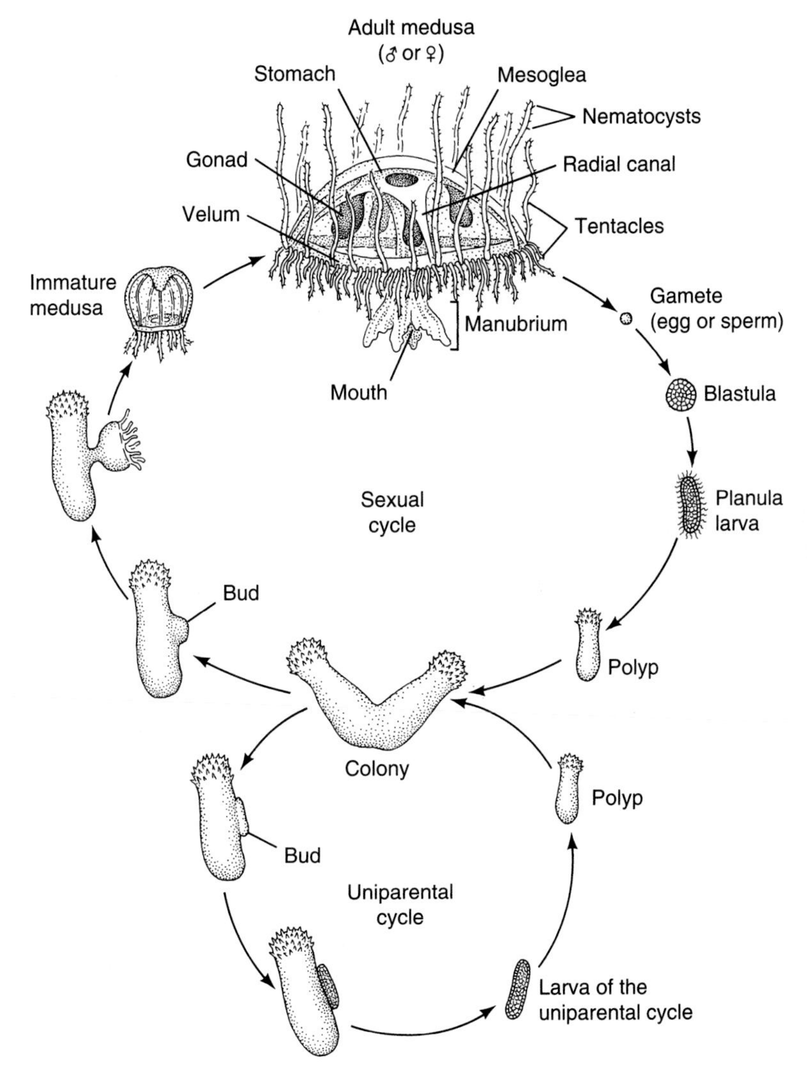

Figure B The life history of *Craspedacusta sowerbii*, a freshwater hydrozoan, and the anatomy of the adult medusa. The mouth of the medusa opens at the external end of the manubrium; the stomach is at the internal end. [Drawing by L. Meszoly; information from C. F. Lytle.]

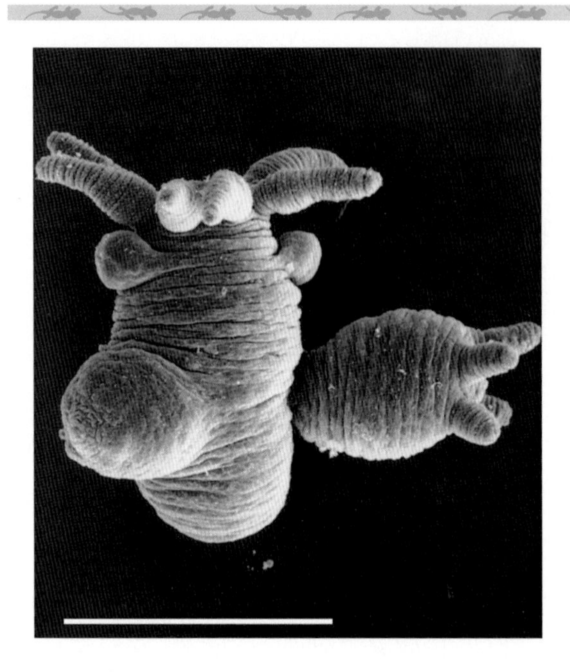

Figure C A sexually mature *Hydra viridis* (Ohio strain). The tentacles are at the top of the upright sessile form, two spermaries are located below the tentacles, a large swollen ovary is shown at the lower left in this picture, and bud is at the right. These green hydra are normally about 3 mm long when extended, but this one shrank by about 1 mm when it was prepared for photography. These green hydras harbor *Chlorella* (Pr-28) in their endoderm (gastroderm) cells. The photo symbionts are maternally inherited on the external surface of the egg after it is released from the ovary. SEM, bar = 1 mm. [Courtesy of Glyne Thorington.]

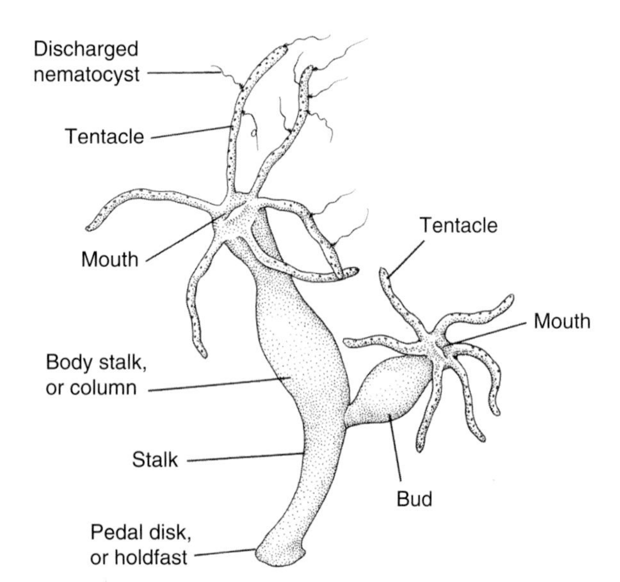

Figure D Overall view of a typical *Hydra*, between 0.5 and 3 cm long, depending on species. [Drawing by L. M. Reeves.]

to dissolved oxygen and carbon dioxide as well as nutrients from the stomach to the periphery of the medusa.

Within the cells of many corals, and a few medusae, algae are symbiotic partners. The algae sustain the animal partner with photosynthate and oxygen. Symbionts are not common in scyphozoans, with the notable exception of the upside down jellyfish, *Cassiopea*.

Anthozoans, as solitary or colonial marine polyps, include about 6500 species—sea anemones, sea pens, sea fans, sea pansies, stony (true or hard) corals, and soft corals. Anthozoans form polyps exclusively, never medusae. Some anthozoan species are hermaphroditic, whereas others have separate genders. Fertilized anthozoan eggs usually develop into planula larvae that settle, attach, and then metamorphose into polyps, cemented by secretions of their pedal disk. Some anemones are viviparous: offspring of sexually mature polyps are "born." Anemones may also reproduce uniparentally by growth into two, by budding, or by pedal laceration—splitting off part of the pedal disk. Without their photosymbionts, most anthozoans survive, but they grow more rapidly in sunlight and corals deposit limestone faster. Anthozoans possess a ciliated grove along one side of their mouths, essentially making their symmetry bilateral. Recent studies have shown that certain developmental genes in the anemone *Nematostella* are expressed in bilateral patterns, causing some to hypothesize that coelenterates may be ancestrally bilateral.

Cubozoa, or box jellies and sea wasps, bear one or a group of tentacles at each of the four corners of their translucent bells. *Chironex*, *Tripedalia*, and roughly 30 other cubozoan species are active swimmers in tropical and subtropical seas. Nematocysts of cubotoans usually cause nasty stings for humans, some are even fatal. Lenses, retinas, and light-sensing pigment spots make cubozoan eyes among the most complex among animals. The cubozoan planula gives rise to a polyp then metamorphoses into a single medusa, which reproduce sexually.

Hydrozoans, with about 3500 described species, include colonial hydroids and siphonophores such as the Portuguese man-of-war and fire corals, as well as the freshwater hydras and jellyfish. A velum—characteristic of most hydrozoan medusae—forms a rim around the umbrella margin. Several tiny (<1 mm) hydrozoans, including *Psammohydra*, live between sand grains through which they move using cilia. Hydras are named for the nine-headed dragon slain by Hercules in Greek myths. Hydrozoan life cycles are diverse; many have small or ephemeral medusae, or lack medusae altogether. Hydrozoan polyps usually reproduce by laterally budding offspring polyps to form polyp colonies and medusae, whereas hydrozoan medusae reproduce sexually by release of eggs and sperm from gonads along the radial canal or the manubrium (Figure B). Most hydrozoans produce either eggs or sperm, but some species are hermaphrodites. The zygote develops from a fertilized egg into a microscopic blastula and then into a free-swimming, ciliated, solid mouthless larva, the planula. Planula larvae metamorphose into polyps.

All 200 or so species of Scyphozoa are marine—most have a benthic polyp stage and all have free-swimming medusae. Medusae of Scyphozoa and Hydrozoa are frequently called jellyfish because their mesoglea is thick relative to that of other cnidarians. Scyphozoan medusae do not have velums. They produce zygotes that grow into planulae. Though some open-ocean species develop directly into medusae, most settle and grow into sessile polyps. At their oral ends, polyps produce distinctive juvenile medusae (ephyrae) serially in a process called strobilation.

Staurozoa or stalked medusae comprise roughly 50 species that live in temperate and polar seas. The adult form is sessile, usually living on red, brown, or green algae. It has eight clumps of tentacles and eight gonads. Adults produce gametes which are shed into the water. Fertilization produces creeping nonciliated planulae which grow into polyps. Polyps metamorphose. The tentacles are resorbed into the adult. As is the case with coelenterate groups, relatively a few life histories are known in detail.

Coral reefs—underwater limestone ridges in shallow tropical seas—usually form by combined secretions of several species of coelenterate and other carbonate-precipitating organisms such as chlorophytes (Pr-28) and rhodophytes (Pr-33). Soft corals predominate in Atlantic reefs; a soft coral does not lay down an exoskeleton, only internal spicules. Hard corals are more important in the Pacific. Below a depth of 60 m, corals tend not form reefs because the shortage of light limits photosynthesis by the algal symbionts. Most symbionts of anthozoans are dinomastigotes (Pr-5); for example, *Symbiodinium microadriaticum* (also called zooxanthellae) are yellow in color. Zooxanthellae inhabit coral polyp tissue in densities as high as 5 million/cm^2. Corals, live to a depth of 3000 m. Reefs provide a sea haven for protoctists, fish and other marine animals. Increasing carbon dioxide levels in the oceans are making them less basic. This pH change creates condition less favorable for skeletal production than the more alkaline one.

Coelenterates are eaten in Korea, Japan, and China. Jewelry has been carved from the internal limestone skeletons of the red coral *Corallium rubrum* since pre-Roman times, from black coral *Antipathes*, and from the blue coral *Heliopora*. Overcollecting of corals prompted the United States to forbid the importation of coral. Biocoral, a biomaterial derived from natural coral, is being used for jaw and face bone grafts in Europe and the United States; the porous structure of coral facilitates movement into the substitute bone graft by cells that form bone (osteoblasts). The biocoral graft is partly replaced by normal bone when the graft is resorbed.

Many of the newly discovered deep-sea medusae bioluminesce sparkling blue green. One species sheds its bioluminescent tentacles on attackers; the bell then pulses off into the wine dark sea and eventually regenerates tentacles. *Branchioceranthus*, a

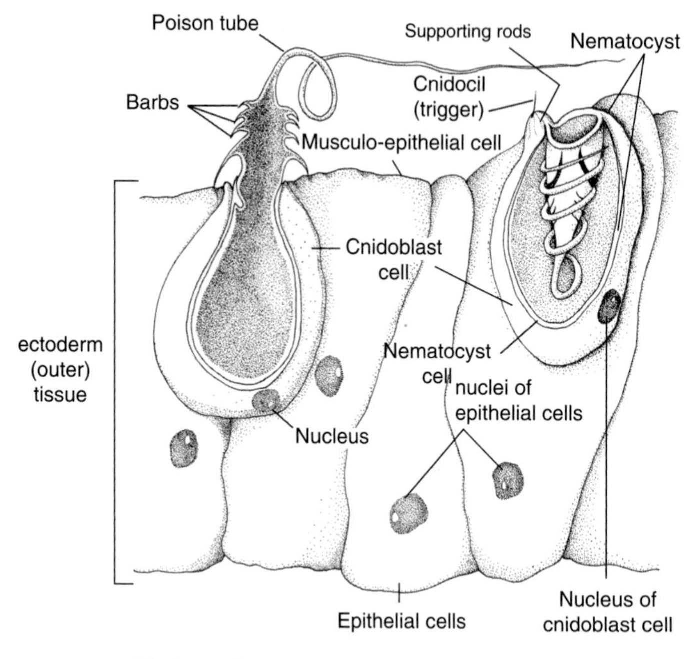

Figure E Discharged and undischarged nematocysts. Toxin is injected through the poison tube. The undischarged nematocyst is about 100 μm long. [Drawing by L. M. Reeves.]

hydrozoan polyp, may reach 2 m in length. This is a giant seafloor-deposit feeder. The sea blubber, or lion's mane, *Cyanea*, is the largest medusa—its bell is more than 3.6 m wide with tentacles more than 30 m long. The tiniest coelenterates, such as *Cryptohydra*, are numerous and diverse polyps and medusae smaller than 2 mm in their longest dimension.

Coelenterates may be among the oldest fossil animals. Although *Ediacaria*, and many others (mainly phosphate-embedded and sandstone imprinted pre-Phanerozoic fossils touted as "fauna") are claimed to be jellyfish and worms, it is doubtful that they are remains of animals at all. This enigmatic well-preserved, visible and varied fossil biota, has been found worldwide in rocks 600 million years in age. Ediacaran assemblages persisted until the base of the Cambrian Period 542 mya, perhaps some slightly beyond. Now reported from some two dozen localities they are entirely extinct. Nearly fifty distinct Ediacaran genera and species (e.g., *Dickinsonia, Kimberella, Mawsonites, Pteridium, Spriggina, Tribrachidum,* etc.) have been described. Their indeterminate growth, three-fold and larger symmetries, lazy beach-dwelling lifestyles and often quilt-like structures do not permit us to suggest any more about their taxonomic and evolutionary status than that it is likely many harbored photo- or chemoautotrophic endosymbionts and none were Prokarya. Perhaps they represent early macroscopic Eukarya, maybe undisturbed sunbathing protoctists in a prePhanerozoic world prior to the evolution of the familiar: the bioturbating, locomotory, predatory, rapacious and sensitive animals. Cambrian rocks 500 million years old include fossils interpreted as anthozoans, hydrozoans, and scyphozoans. Fossil corals of various groups are known from this time until the present.

Although it is clear that Coelenterates diverged relatively early in the history of animals, their precise position on the tree of Animalia remains unknown. Some molecular data indicate that sponges and ctenophores branched from protoctist ancestors prior to the divergence of coelenterates, placozoans, and the metazoans. Within Coelenterata, a primary divergence separated Anthozoa from Medusozoa. Within Medusozoa, Cubozoa and Scyphozoa appear to have a close relationship. These two taxa appear to be the sister group of Hydrozoa. Staurozoa may be the sister group of all other Medusozoa.

A-5 Ctenophora

(Comb jellies)

Greek *ctena*, comb and *pherein*, to bear

GENERA

Bathyctena
Beroë
Bolinopsis
Cestum
Coeloplana
Ctenoplana
Euchlora
Eurhamphea
Lampea (Gastrodes)
Mertensia
Mnemiopsis
Ocryopsis
Pleurobrachia
Thalassocalyce
Tjalfiella
Velamen

Sea walnuts, sea gooseberries, cat's eyes, Venus' girdle, and 90–150 more comb jellies are placed in the phylum Ctenophora. Typically, ctenes, or ciliary comb plates (Figure A), 2–5 mm long, are stacked in eight columns running from the animal's oral to its aboral end. Each ctene supports thousands of cilia with lengths as great as 2 mm—the longest cilia known—and their rapid, coordinated rowing away from the mouth results in continuous, smooth movement and reliable, mouth up orientation. With body lengths ranging from 0.4 cm to more than 1 m long, ctenophores are the largest animals propelled by cilia.

Ctenophores are found in every marine environment, from near shore to open ocean, and from the surface to mid-water to deep sea and comprise a considerable portion of the plankton's biomass. While weak swimmers, ctenophores are carried by currents, tides, and wind through the sea from the Antarctic to the Arctic, sometimes in great schools. Some ctenophores employ muscular flapping while escaping from predators. Others achieve pulsating or undulating motion through muscular contraction, and benthic ctenophores crawl over the seafloor.

The coherent packing of cilia in ctenes diffracts light and produces iridescence, giving rise to a rainbow of colors as the cilia beat. Otherwise, most ctenophores are translucent. Those living near the surface may be virtually transparent, whereas deep-living species may be highly pigmented. Tropical ctenophores may be tinted violet, rose, yellow, or brown by symbiotic algae, sometimes concentrated in spots, while deep-sea representatives may be purple or red. Many ctenophores are also bioluminescent, emitting a faint blue-green light when disturbed. The source of ctenophoran bioluminescence is a luciferin chromophore identical to that found in Coelenterates (A-4). One species of ctenophorans releases a red cloud that glows with a luminous blue.

Ctenophores are compact, carnivorous marine animals with biradial symmetry—like an American football—and a diploblastic body plan in which a mesoglea is sandwiched between two cell layers. The mesoglea has a jellylike consistency—accounting for ctenophores' common name, comb jellies—providing buoyancy, canals for distributing food and storing reserves, a home of ameboid cells, a nerve net, smooth muscle cells, and bands of muscle especially in tentacles, around the mouth and pharynx, and surrounding the aboral area. An outer epidermis (or ectoderm) is composed of two layers of epithelial cells; it surrounds the mesoglea and dips into the animal's apical mouth and pharynx or gullet. A glandular and ciliated gastrodermis (or endoderm) lines the pendulous stomach or central digestive cavity and a network of gastrovascular canals begins with eight meridional canals running beneath comb plates. Digestion is extracellular in the stomach, via the action of enzymes secreted by the gastrodermis. Dissolved and suspended food is distributed through canals, where intracellular digestion completes the process. The indigestible refuse is expelled through the mouth or through two anal pores.

Ctenophora is divided into two classes: Tentaculata and Nuda. Tentaculata is the larger class, consists of animals with two opposing tentacles that emerge about midway on the body (Figure A). Members of the smaller class, Nuda, lack tentacles entirely in both larvae and adults (Figure B). Typically, tentaculates travel mouth first. Orientation depends on the ability of an aboral sensory structure, a statocyst that controls ciliary beating. This saclike structure contains several hundred mineralized statoliths 5–10 μm in diameter and four mechano-responsive balancers consisting of 150–200 sensory hairs of fused cilia. Elevated bundles of long, thin, microtubule-filled processes originate from each side of the apical organ, run through parallel grooves in epithelial cells, and bridge the distance to opposing pairs of comb plates. Balancers may induce rapid or slow beats depending on the sign of the geotaxis signal, calcium ion concentration, and neural input.

Many tentaculate ctenophores have adhesive colloblast or lasso cells (Figure C) equipped with "heads" having sticky bodies that may also secrete a poisonous or anesthetic mucus. The "head" lies atop two filaments, one spiraled around the other absorbing the shock and tug of struggling prey. Colloblasts and, indeed, tentacles are continuously regenerated (Figure C).

Tentacles snare prey such as arrow worms (A-32), copepods (A-21, Crustacea), fish, eggs, or larvae. Creeping ctenophores extend their tentacles as rafts, picking up sea currents, while pelagic ctenophores, such as members of the genus *Bolinopsis* (Figure A), extend tentacles into sinuous nets as much as 100 times the length of the body. Prey is unloaded with a wipe of tentacles across the mouth or through the action of currents produced by cilia.

Members of the tentaculate order Cydippida (for example, *Pleurobrachia*), typically 3 cm or fewer, have long tentacles with branches (known as tentalia or tentilla), while members of the order Lobata (*Bolinopsis* and *Mnemiopsis*) are larger but have reduced tentacles that may remain within expandable oral lobes equipped with colloblasts on either side of the mouth (Figure A). Lobates also have a pair of ciliated flaps or ribbonlike projections called auricles (Figure A) that participate in feeding. Pelagic tentaculates of the order Cestida have ribbonlike bodies with comb plates running along the aboral surface that may reach 20 cm in height, while deep-sea species may have oral lobes stretching a meter across. Benthic members of the order Platyctenida are unusual in lacking ctenes as adults.

Members of the class Nuda, such as the thimble jelly, *Beroë gracilis* (Figure B), tend to be cylindrical or flat, but expand like a sac over and gulp down captured comb jellies and salps (soft urochordates, A-35). Ingestion occurs through the action of muscular lips lined with toothlike "macrocilia," fused standard but long undulipodia, extensions of epithelial cells (see Figure I-3) sensory receptors, and adhesive strips of epithelial cells equipped with Velcro-like, highly folded membranes. Macrocilia are attached by bundles of microfilaments to underlying smooth muscle. A continuous latticelike plexus of giant nerve fibers (6–8 μm in diameter) makes synaptic contact with longitudinal muscle and epithelial adhesive cells.

The efficiency of predatory behavior by comb jellies was strikingly illustrated by the consequences of the unintended introduction of the tentaculate *Mnemiopsis leidyi* into the northern Black Sea. Within a decade, and coupled to other anthropogenic effects, such as consequences of intense overfishing, *Mnemiopsis* brought about the collapse of the pelagic

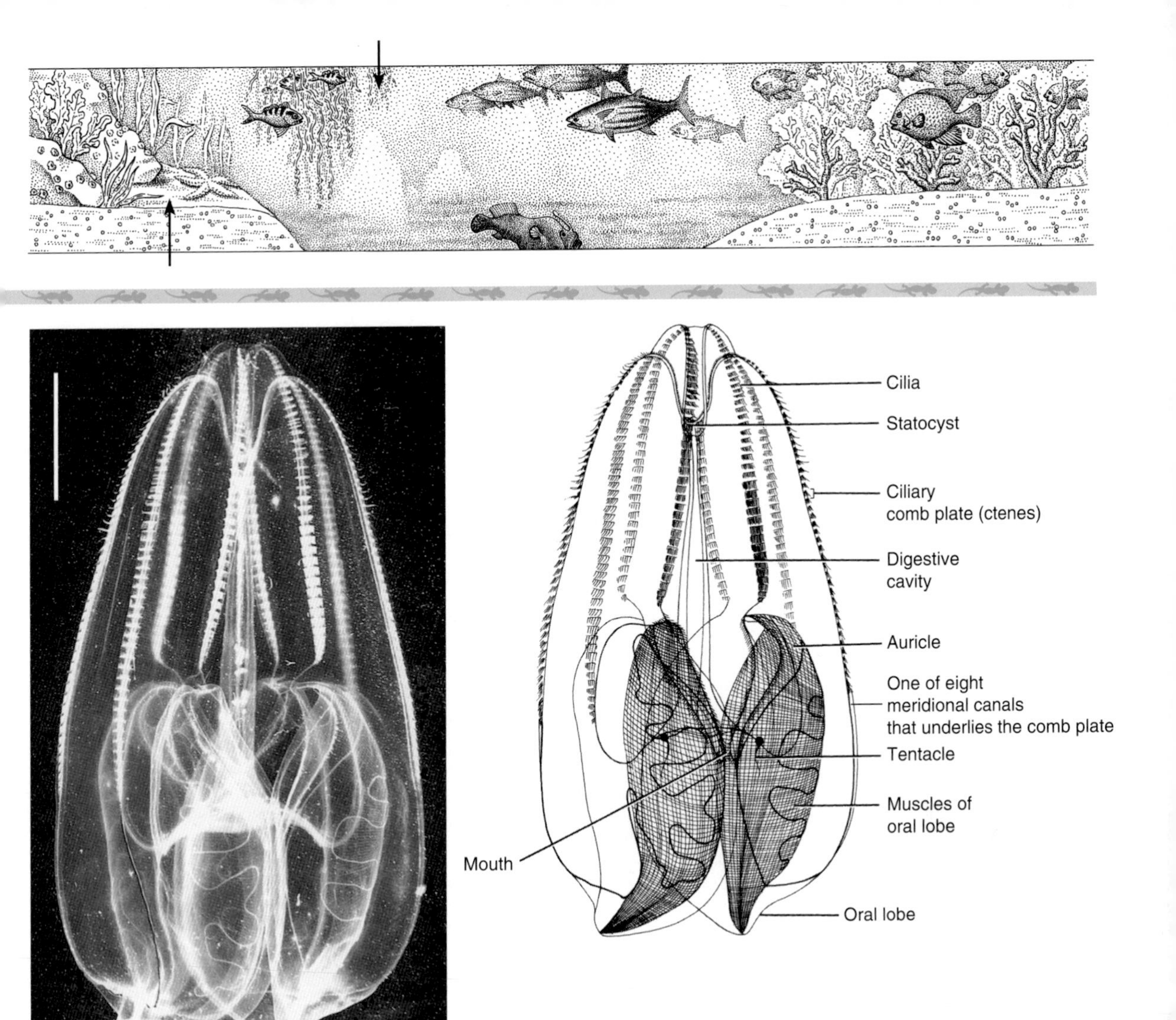

Figure A *Bolinopsis infundibulum*, a common northern comb jelly. A planktonic species of class Tentaculata, *Bolinopsis*, swims vertically with its mouth end forward at the bottom of this figure and enmeshes prey in mucus on its extended ciliated oral lobes. Tentacles are short in this adult but long in the young. Bar = 1 cm. [Photograph courtesy of M. S. Laverack; drawing by I. Atema; information from M. S. Laverack.]

ecosystem. Similar collapses now threaten the Caspian Sea and even the Mediterranean. Zooplankton, fish spawn, and larvae only began their recovery in the Black Sea after the appearance in 1999–2000 of the cannibalistic Nuda, *Beroë ovata*. Cannibalism may also account for the bioamplification of arsenic in *Beroë cucumis* which prey on *Bolinopsis mikad*.

Most ctenophores are hermaphrodites and mature early. Gonads develop beneath meridional canals, and eggs and sperm are released through the mouth into ocean habitat. The fertilized eggs of tentaculates typically develop into cydippid larvae bearing long feeding tentacles. At metamorphosis into an adult, members of the order Cydippida retain the tentacles, while members of the Lobata and Cestida replace them or lose their tentacles entirely. Larvae belonging to the class Nuda never form tentacles at all.

Ctenophores have extraordinary powers of regeneration as adults, but their development was long thought to be determinate, with the fate or prospective significance of early blastomeres determined in advance or as they were cleaved from the egg or other blastomeres. Recent experimental results, however, have exposed the role of intercellular interactions and influences on development suggesting that a degree of regulation moderates determinism during early development.

The discovery of fossils with spheroidal morphology and eight comb rows in the lower Cambrian suggests an ancient origin for ctenophores. Indeed, the parazoan ctenophores have been placed at the base of animal evolution as a sister group to metazoa. The presence of circular mitochondrial DNA (mtDNA) in ctenophores may confirm Ctenophora's position in the mainstream of metazoan evolution, and possibly its basal position. With the

Figure B A living comb jelly, *Beroë cucumis*, which lacks tentacles. A member of class Nuda, *Beroë* is common in plankton from Arctic to Antarctic seas. It engulfs prey with muscular lips visible here at the bottom of the animal. Bar = 1 cm. [Photograph courtesy of M. S. Laverack.]

exception of the Medusozoa among the Coelenterates, all metazoa have circular mtDNA. The large SOX family of transcription factors found in a ctenophore implies the early diversification of these determinants of differentiation began in the parazoan and metazoan lineages. Moreover, ctenophores exhibit the same telomere sequences found in vertebrates, indicative of extreme conservation of a motif. Ctenophore ectoderm and endoderm express genes associated with mesoderm-derived tissues in the bilateria. But belief in the strategic position of the ctenophore's mesoglea in the evolution of the metazoan mesoderm was shattered recently when the only known ctenophoran gene resembling a gene active in mesoderm development (a putative member of the Tlx family) was seen to be expressed in ctenophoran epidermis and possibly gastrodermis but not in the mesoglea.

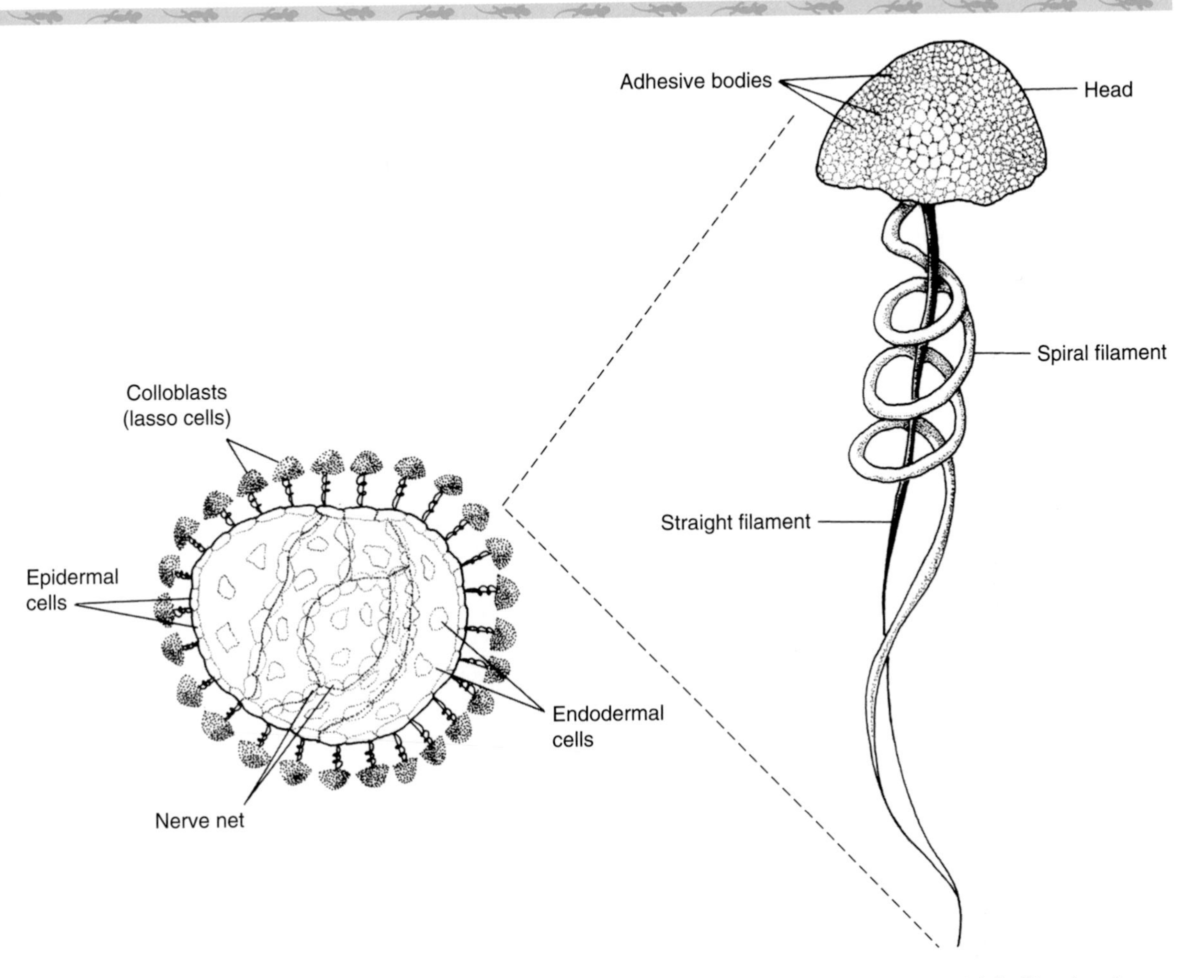

Figure C Cross section of a tentacle covered with colloblasts and a single colloblast (lasso cell) of a comb jelly. [Drawings by L. M. Reeves.]

A-6 Gnathostomulida

(Jaw worms)

Greek *gnathos*, jaw; *stoma*, mouth

GENERA

Austrognatharia
Gnathostomaria
Gnathostomula
Haplognathia
Mesognatharia
Nanognathia
Onychognathia
Problognathia
Pterognathia
Semaeognathia

Gnathostomulids are translucent free-living jaw worms that have unique, toothed jaws at their ventral mouths. These worms graze on fungi, bacteria, and protoctists among grains of sea sand. Like platyhelminths, jaw worms are bilaterally symmetrical, acoelomate eumetazoans. As triploblastic animals they have three body layers, the middle layer of mesodermally derived muscle is exterior to the digestive cavity and interior to the epidermis.

About 80 species in 20 genera of gnathostomulids have been described. They are from Maine, the Florida Keys, the Bahamas, the Caribbean, California, the Indo-Pacific, and the White Sea. Probably, more than a thousand species of these worms live today in shallow oceans down to a depth of several hundred meters throughout the world. Jaw worms adhere to sand particles or live on leaves of marine plants, such as the eel grass *Zostera*, turtle grass *Thallassia*, and marsh grass *Spartina* (Pl-12), and on thalli (leafy parts) of algae. Because they are recognizable only when living, the natural history of gnathostomulids remained unknown for a long time, until sophisticated techniques were devised to pull jaw worms off the surfaces on which they live. In California, jaw worms live near the roots of surf grass, *Phyllospadix*, in anoxic sand. Some inhabit sulfureta, communities in black, sulfide-rich fine-grained sediments. Often deep and underneath the white oxidized layer of marine sand bottoms, sulfureta smell like rotten eggs. The odor emanates from hydrogen sulfide, a gas produced by marine bacteria under anoxic conditions in absence of molecular oxygen. Gnathostomulids in sulfureta tolerate low levels of O_2 and high quantities of sulfide. In certain sediments, gnathostomulids may outnumber even nematodes (A-11). Population densities of more than 6000 gnathostomulids per liter of sediment have been reported.

Gnathostomulids are hermaphrodites. The single ovary produces large eggs that mature one at a time. Posterior to the ovary are paired testes (Figure A). Male organs in the different species vary. Sperm may be undulipodiated, mushroom shaped, or dwarful spheres. During copulation, one worm injects hundreds of sperm within a mucus ball beneath the skin of a second worm. A female genital pore is present in certain species. The penis of some species, like the *Problognathia* shown here, is stiffened with a stylet, facilitating hypodermic impregnation. Sperm in the prebursa, a mucus ball filled with sperm, migrate to a storage sac called a bursa, part of the female reproductive apparatus. They fertilize the most mature egg. Afterward, the fertilized egg ruptures the body wall and is released. Because development is from egg to adult with no larval stage, we say that development is direct. In at least some species, a nonsexual feeding stage alternates with a distinct nonfeeding sexual stage year that is required to complete the cycle to the mature adult.

The gnathostomulid body lacks an external cuticle; its average length is 1.5 mm, the size range is from 0.3 to 3.5 mm. A slight constriction separates the bristly head from the trunk in some; in others, an elongate rostrum forms the anterior end. Circulatory and respiratory organs, a coelom, and a skeleton are lacking, as in platyhelminths (A-7). Gas exchange occurs across the body wall of the jaw worm's minute body. The modest nervous system is made up of longitudinal nerve fibers and ganglia (frontal, buccal, caudal, and penile). In some species,

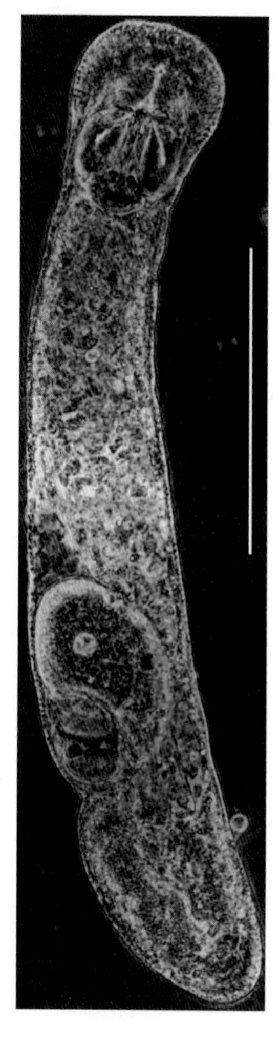

Figure A An adult *Problognathia minima*. It glides between sand grains in the intertidal zone and shallow waters off Bermuda. LM (phase contrast), bar = 0.1 mm. [Photograph courtesy of W. Sterrer, *Transactions of the American Microscopical Society* 93:357–367; 1975; drawing by L. Meszoly; information from W. Sterrer.]

the head bears tactile organs collectively called the sensorium: stiff bundles of sensory bristles and pits lined with cilia. On the ventral portion of the head a cuticular comblike based plate that can be protruded through the mouth and a pair of toothed lateral jaws comprise the feeding structures (Figure B). Contraction of the muscular pharynx snaps the jaws open and closed in about 0.25 seconds. Food particles are passed into the digestive cavity with its single opening. Undigested solid waste leaves through a temporary dorsal connection to the epidermis, a temporary anus or through the mouth (Figure A). Dissolved waste is disposed of protonephridia composed of ciliated excretory cells that open through pores to the exterior.

Gnathostomulids move by cilia (undulipodia). They nod their heads from side to side, swim, glide, and twist. Under the epidermis lie weak circular muscle fibers, with three or four paired longitudinal muscles underneath. The body-wall muscles

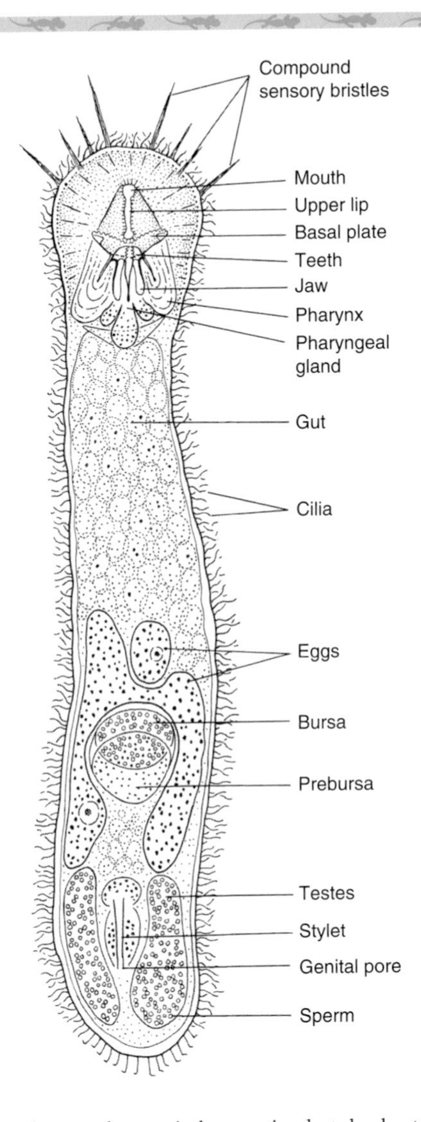

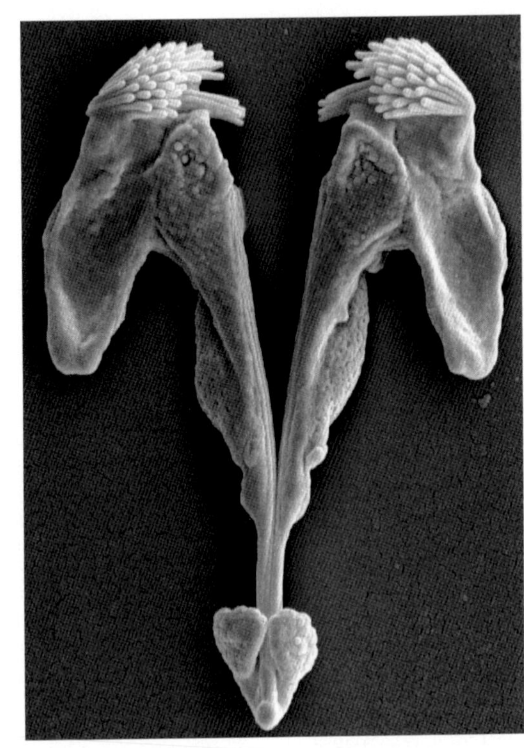

Figure B Jaws of *Haplognathia ruberrima* Sterrer, 1965 from Belize, in ventral view. Overall length about 20 μm. [Scanning electron micrograph by M. V. Sørensen, *Journal of Morphology* 253:315; 2002.]

do not take part in locomotion but do shorten the body. Polygonal cells of the external epithelium bear only one cilium each, and ciliary propulsion can reverse the animal's direction. These traits distinguish gnathostomulids from turbellarian flatworms (A-7), which gnathostomulids otherwise resemble. Anterior bristles and less flattened bodies of gnathostomulids distinguish them from the flatworms.

Gnathostomulids have been recognized as an independent phylum since 1969, and were initially grouped with Platyhelminthes (Phylum A-7) because of broad similarities in organization. However, the architecture of the cuticular mouth parts suggests a close relationship with rotifers (Phylum A-14), possibly within a taxon called Gnathifera (which also includes Micrognathozoa and Acanthocephala). Both jaw worms and flatworms are externally ciliated, lack a coelom, have a dead-end gut and lack an anus (at least a permanent anus). They have protonephridia and are hermaphroditic. Evidence against a close relationship between jaw worms and flatworms is that the sperm tails of gnathostomulids are typical undulipodia with the [9(2)+(2)] cross section or sperm with no tails. Sperm tails of flatworms different in their arrangement of microtubules. The jaw structure may relate gnathostomulids to rotifers (A-14); however, monociliated epithelium is found only in gastrotrichs (A-17) and gnathostomulids. Gnathostomulid adhesive cells are not homologous to the dual adhesive glands of platyhelminth or gastrotrichs.

Fossil conodonts were once thought to be remains of the tough parts of ancient gnathostomulids found in rocks that extend in age from about 540 million to 200 million years. However, the basal plates of modern gnathostomulids differ from those of conodonts in that gnathostomulid toothlike feeding structures are made of acellular rather than cellular organic material. Conodonts, are made of cells that, like bone, precipitate calcium phosphate. Since conodonts are now assigned a fossil phylogenetic position in the Craniata (A-37) gnathostomulid fossil history remains to be discovered.

A-7 Platyhelminthes
(Flatworms)

Greek *platys*, flat; *helmis*, worm

GENERA

Bothrioplana	*Echinococcus*	*Planaria*
Convoluta	*Fasciola*	*Procotyla*
Dipylidium	*Hymenolepis*	*Schistosoma*
Dugesia	*Opisthorchis*	*Taenia*

Platyhelminthes are free-living and symbiotrophic flatworms. The soft body of the flatworm is bilaterally symmetrical. Structures for capturing and consuming prey are localized in the anterior end except in many free-living flatworms, in which the mouth is anterior. Flatworm organs are composed of tissues and are organized into systems. Flatworms are the simplest metazoans to possess an embryonic intermediate tissue layer, the mesoderm. The platyhelminth worm, like the cnidarian, lacks an anus. The flatworm middle tissue layer, a loose mesoderm called parenchyma, never splits into a cavity (coelom) in which internal organs are suspended. Flatworms and other animals without a coelom are called acoelomates. Flatworms, having three tissue layers, are triploblastic, have spiral cleavage in their eggs, and yet are among the least complex of bilaterally symmetrical true metazoans.

Flatworm classification is constantly being reviewed and revised. The symbiotrophic forms, which undergo a transformation of the epidermis to a cuticle-like neodermis during development, are generally placed among two or three traditional classes within the group Neodermata. Two established classes contain the Trematoda, or flukes, and the single to multi-host Cestoda, the tapeworms. Some workers divide the Trematoda into the Digenea, or multi-host trematodes, which are internal necrotrophs, and Monogenea, which typically have a single host and are largely external necrotrophs. The free-living flatworms make up a number clusters or clades (using cladistic methodologies), the relationships of each remain unresolved. The most distinctly primitive group, the Acoela (Acoelomorpha), may not be a platyhelminth group at all.

There are about 20,000 species of flatworm altogether. Some species are richly colored. Others harbor symbiotic algae called zoochlorellae producing a green color. Most necrotrophic forms and those free-living forms that inhabit caves and underground water are colorless. Tapeworms are the largest platyhelminths; some reach a length of more than 30 m. The smallest are less than 1 mm in length.

Flatworms are masters of adaptation, exploiting an enormous variety of habitats. Some live in bat guano, others in the mantle fold of various Mollusca (Phylum A-26) where as symbionts they feed on the particles not consumed by the mollusc host. Members of many animal phyla, certainly an enormous number of vertebrates, play host to ubiquitous flatworm symbiotrophs. In sediments low in molecular oxygen, a few flatworms utilize energy by oxidizing hydrogen sulfide. A survey of the phylum reveals that flatworms tolerate an immense temperature range from minus 50 degrees to plus 47 degrees.

Some free-living flatworms are marine inhabitants; most are freshwater forms, and several dwell in moist soil. Soil flatworms are mainly tropical whereas aquatic forms are more abundant in temperate than in tropical waters. Free-living and non-neodermatid necrotrophic forms have several undulipodia per cell. By simultaneously sweeping ventral cilia through secreted mucus and generating muscular waves, free-living flatworms glide over surface films on water, plants and soil. On the ventral surface of free-living flatworms, duoglands (adhesive and releaser cells) secrete either adhesive that attaches the worm to its substrate or a substance that detaches it. Most free-living aquatic species are benthic, a few swim with undulations or loop along substrate like caterpillars, and some live among the plankton.

Free-living flatworms are detritus feeders, carnivores, and scavengers. They eat insects or crustaceans (A-21), tunicates (A-35), bivalve molluscs (A-26), other worms, bacteria, mastigotes (Phylum Pr-28), ciliates (Phylum Pr-8), and diatoms (Phylum Pr-18). Most free-living flat worms are marine; some inhabit the digestive tract of sipunculans (A-23) and echinoderms (A-34); a few are terrestrial in damp habitats or are freshwater species. Digestive systems of free-living flatworms range from a straight or branched gut to absence of a gut; food moves from the pharynx of acoel (lacking a gut) free-living flatworms into loose digestive cells. Some jab food by using a proboscis separate from the mouth. Others "vacuum out" soft parts of their prey by using a tubular, muscular pharynx, which may project through the mouth on the ventral side. Digestive enzymes secreted into the gut begin digestion; intestinal cells continue digestion by engulfing food in food vacuoles.

Necrotrophic flatworms undergo a phenomenon in which the epidermis is replaced by a new skin, the neodermis, during maturation. Thus, all cilia are lost and movement within the host is carried out by detaching and reattaching the sucker or variations of the sucker.

All flukes (digenean and monogenean trematodes) are internal or external necrotrophs, usually of vertebrates. The digenean trematodes have a life cycle that includes several types of larvae and sometimes an intermediate host or hosts. Trematode larvae include miracidium, sporocyst, redia, cercaria, and metacercaria. Schistosomiasis (bilharziasis), caused by several species of the blood fluke *Schistosoma*, is currently the second most prevalent infectious disease worldwide (malaria is first). Cercariae, which are distinctive swimming larvae with a tail and sucker, are carried by snails that spread schistosomiasis. Snails release cercariae; the cercariae swim, attach to and penetrate human skin between the fingers and toes, and then mature into adult worms and migrate to take up residence in the liver and other organs of the human host. The disease results from the human host's immune response to schistosome eggs deposited in host tissues by activation of lymphocytes and other immune cells; urinary tract and bowel blockage also can result. Monogenean trematodes typically have only one larval type, the onchomiracidium, which is released alive. Larvae move about the host or locate other hosts and attach. The larva matures, a neodermis replaces the ciliated epidermis, and the cycle repeats itself. Trematodes have one or two suckers; some trematodes feed through their oral suckers.

Tapeworms (cestodes) are exclusively internal necrotrophs that usually attach inside the gut of vertebrates by means of a specialized structure, the scolex. The scolex may contain exclusively suckers or a combination of suckers and other structures enabling a firm grasp of the host tissue. Tapeworms lack a gut. Microvilli (minute tissue projections) absorb nutrients (amino acids and sugars) from the hosts exploited by tapeworms. The typical tapeworm body plan consists of the scolex followed by repeated segments, each with reproductive organs; these sexually

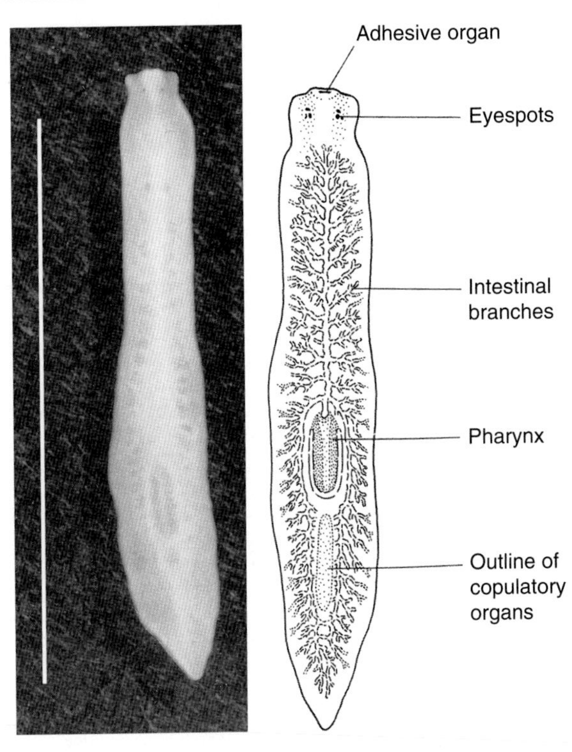

Figure A Dorsal view of gliding *Procotyla fluviatilis*, a live freshwater turbellarian flatworm from Great Falls, Virginia. Its protrusible pharynx connects to a branched intestine visible through its translucent body. Bar = 1 cm. [Photograph courtesy of R. Kenk; drawing by L. Meszoly; information from R. Kenk.]

reproducing segments bud from the tapeworm's neck. Many tapeworms, however, lack segments. Like flukes, most tapeworms have intricate life cycles with several distinctive larval types. Others, typically without segments, have simpler life cycles, and may represent progenetic forms of more typical segmented tapeworms.

The large surface area of free-living and symbiotrophic flatworms relative to their volume has physiological implications. Oxygen, carbon dioxide, and ammonia exchange across the body surface. Like cnidarians and ctenophores, flatworms are blood-, lung-, and heartless. In symbiotrophic flatworms, gases and nutrients diffuse into tissues of the flatworm from the host digestive system or from water. Free-living flatworms ingest food. Platyhelminth worms that have a gut discharge solid waste through their mouths because they lack an anus. Protonephridia are the organs of excretion and osmoregulation in flatworms, except in free-living flatworms that lack a gut. Protonephridia are extremely primitive excretory structures that are composed of ciliated cells, called "flame cells", that collect dissolved wastes. Protonephridia regulate water and ions by wafting liquid through ducts that exit to the outside through pores.

The simplest flatworm nervous system consists of light-sensitive pigment-cup eyespots (either single or in groups) connected to a cluster of nerve cells (brain) in the head and ventral, longitudinal nerve cords. The nervous system of flatworms ranges in complexity from this simple system to the more primitive nerve net of acoel free-living flat worms resembling that of cnidarians and ctenophores. Free-living flatworms detect chemicals, food, objects, and currents with sensory pits or tentacles on the sides of the head. When flatworms wander away from a scent source, they turn from side to side more frequently and so eventually home in on the source.

Triclad free-living flatworms and cestodes have prodigious powers of regeneration and reproduce sexually or asexually. Slices of *Dugesia*, a triclad, regenerate to form entire worms. Planarians (freshwater species of triclad free-living flatworms, an order of free-living flatworms characterized by a gut having three branches) and taeniid cestodes reproduce asexually as well as sexually. Almost all flatworm species are simultaneous hermaphrodites. Each individual flatworm bears ovaries and testes. Self-fertilization is rare in free-living flat worms and common in cestodes. In mating pairs of hermaphrodites a copulatory bursa receives sperm or, in some free-living flatworms, a hypodermic-like penis injects sperm through the body wall into the body of the mate. Some flatworm sperm have no tails; others have a [9(2)+1] arrangement or a [9(2)+0] arrangement (see Figure I-3). Ribbons of fertilized free-living flatworm eggs are laid in cocoons. Freshwater flat worms "glue" eggs to stones. Eggs of most free-living flat worms hatch into miniature adults. A few marine free-living flatworms develop ciliated larvae known as Muller's larvae. Necrotrophic flatworms frequently have complex reproductive cycles, with a succession of larval stages. *Schistosoma*, a fluke, is dioecious. *Bothrioplana*, a free-living flat worm, exhibits parthenogenesis—females asexually produce females.

The origin of this phylum is uncertain because flatworms fossilize poorly. Prevailing theory maintains that the bilaterally symmetrical, triploblastic, acoelomate pattern of early free-living flat worms was ancestral to the coelomates. Arguments that the Platyhelminthes represent secondarily reduced coelomates persist, however, and some data from molecular studies suggests they are spiralian protostomes; their coelom, anus, and multicellular excretory organs having been lost. Free-living flatworms logically preceded parasitic forms, but the relationship between platyhelminth worms and other metazoan phyla remains unclear.

A-8 Rhombozoa
(Rhombozoans)

Greek *rhombus*, a spinning top; *zoion*, animal

GENERA
Conocyema
Dicyema
Dicyemennea
Microcyema
Pseudicyema

The Rhombozoa—also called Dicyemida, and historically combined with the Orthonectida (A-9) as the Mesozoa—are small multicellular organisms that live in the nephridia of cephalopod Mollusca (A-26). Except for placozoans (A-1), rhombozoans are the least morphologically complex animals. The name of the former phylum Mesozoa indicates that they were at one time considered intermediate between the protoctists, which lack tissues, and the more complex metazoans, where millions of cells are integrated into tissues, organs, and organ systems. Rhombozoans are small wormlike bilaterally symmetrical animals. Although they are multicellular and contain metazoan cell junctions, their cells are arranged in one or two layers and morphologically cannot be considered diplo- or triploblastic animals. Moreover, the inner cell layer of rhombozoans does not correspond to endoderm. Rhombozoans lack body cavity and circulatory, respiratory, skeletal, muscular, nervous, excretory, and digestive systems.

The outer layer of rhombozoans (Figure A) consists of from 20 to 30 "jacket" cells, a constant number (eutelous) in studied species. Like Platyhelminthes (A-7), rhombozoans are ciliated. The ciliated cells form a jacket enclosing one or several long, cylindrical *axial cells*. Within the axial cell or cells are from 1 to about 100 cells called *axoblasts*, each containing a polyploid nucleus. The function of the large specialized axial cell is solely to surround the axoblast cells (also called agametes). The axoblast cells are reproductive cells that ultimately produce eggs and sperm.

Phylum Rhombozoa encompasses both dicyemids (*Dicyema*, *Pseudicyema*) and heterocyemids (*Conocyema*, *Microcyema*). Stages in the rhombozoan life history include rhombogen, nematogen, vermiform larvae, and *infusoriform larvae*. The nematogen of heterocyemids has a syncytial, unciliated outer cell layer, whereas the nematogen of dicyemids (Figure A) has a cellular outer layer. Rhombozoans have both uniparental and sexual generations.

Adult dicyemids (Figure B) range in size from fewer than 500 μm to more than 5000 μm in length in the nematogen stage (Figure A). About 65 species of rhombozoan have been described. All rhombozoans are symbiotrophs that live in the bodies of *benthic* (seafloor-dwelling) cephalopods mainly in temperate waters, especially cuttlefish and octopods. They probably evolved from more complex free-living ancestors. Because they frequently inhabit cephalopods that are widespread in shallow seas, rhombozoans are common. Dicyemids live in the kidneys of octopus and other benthic cephalopod molluscs (A-26) but not pelagic (open ocean–dwelling) squid. The dicyemid microhabitat is the interface between urine and mucus-covered epithelial kidney tissue of the mollusc host. The dicyemid attaches loosely by its anterior end to the kidney. The rest of the dicyemid body hangs free in urine. These minute symbiotrophs absorb nutrients directly from the urine of their host.

Octopus urine has organic solutes that may sustain the dicyemids. When cultured, dicyemids consume glucose in culture media; thus, they are aerobes. Dicyemids contribute to the acidification of urine, facilitating ammonia excretion by the mollusc; the dicyemid–cephalopod relation is thus a metabolic symbiosis.

Adult dicyemids have morphologically similar but functionally distinct reproductive phases called nematogen, the nonsexually reproductive stage, and rhombogen, the sexually reproductive stage. Rhombogens develop from nematogens and are sexual adults. All rhombozoans have a hermaphroditic gonad. The gonad is a single cell, rather than a multicellular organ. When a population of adult dicyemid rhombogens becomes dense, the rhombogens develop nonciliated, sexual *infusorigens* nested within the axial cells.

The life cycle has been described for several dicyemids. Agamout nematogen adults arise from the formation of a wormlike *vermiform larva*. The larva originates as an agamete that undergoes a series distinct and species-specific series of mitotic divisions. During development, specific cell lineages give rise to specific regions of the larva. Immature vermiform (wormlike) larvae are released into the urine of the young mollusc. Vermiform larvae attach to the kidney–urine interface and grow into nematogen adults. As long as the cephalopod is immature, new generations of nematogens are uniparentally produced this way. In older cephalopods and high densities of nematogens, rhombogens develop and there is either a mixture of nematogens and rhombogens or just rhombogens.

The sexual phase (rhombogen) involves the singe-celled hermaphroditic gonad, the infusorigen. Within the gonad, both eggs and ameboid sperm are produced, but do not emerge from the enclosing rhombogen. Self-fertilization takes place within the axial cells of the rhombogen parent. Oocytes do not complete the first meiotic nuclear division until after fertilization. The resulting zygotes develop into infusoriform larvae. These ciliated infusoriform larvae escape from their parent into the molluscan host urine and are shed into the sea, thus completing the life cycle. The sexually produced larva is the infusoriform. These larvae are the dispersal stage of the dicyemid. The spherical or top-shaped dicyemid infusoriform larva is about 40 μm long. It consists of 28 cells—each of the four interior cells contains another cell. Like sets of Chinese boxes and Paramyxa protoctists (Pr-30), one cell is packed inside the other. The larva grows by differentiation and enlargement of its existing cells rather than by mitotic cell division.

Newly hatched, free-swimming infusoriform larvae (Figures A and C) are soon weighted to the sea bottom by two cells filled with a high-density substance, magnesium inositol hexaphosphate. The larvae are acquired somehow by young bottom-dwelling cephalopods. (The dotted arrow in Figure A indicates a possible intermediate host.) The fate of larvae in

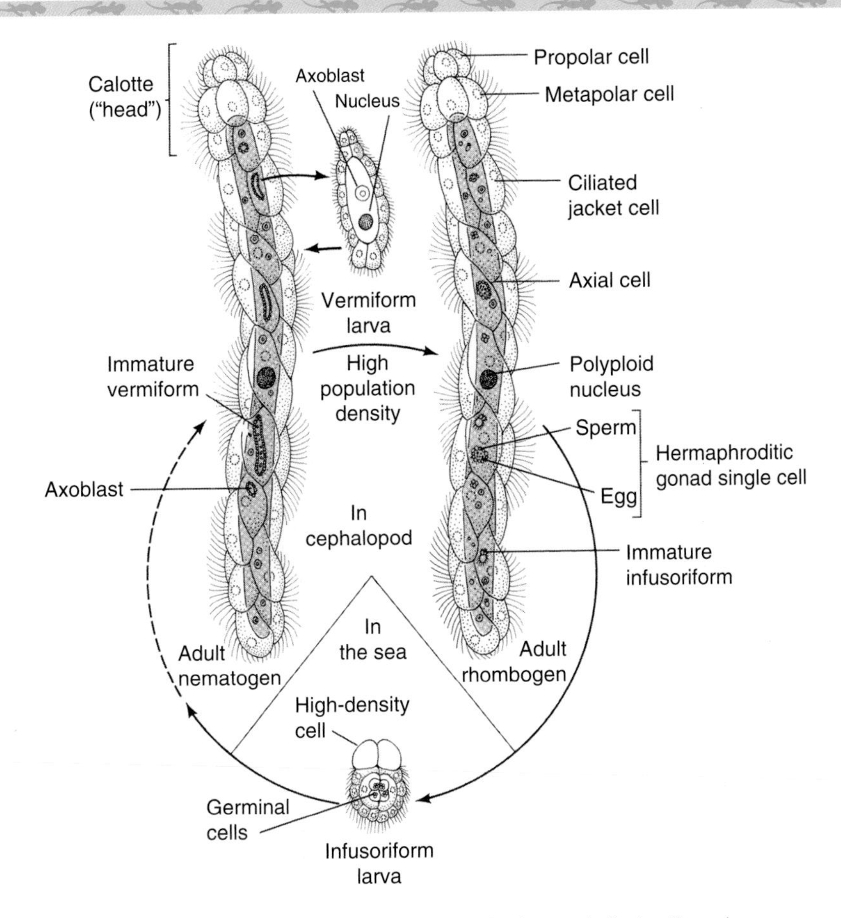

Figure A *Dicyema truncatum* life cycle. The dashed arrow indicates the unknown mode by which infusoriform larvae enter their cephalopod habitat. [Drawing by L. Meszoly; information from E. Lapan.]

the ocean is unknown; it is possible but not likely that intermediate hosts transfer rhombozoans to other cephalopods. Infusoriform larvae have been found in cephalopods younger than 3 weeks old. Larvae seem not to infect older cephalopods. The entry route of the larva into the cephalopod is obscure, although experimental infection in laboratory aquaria has been achieved. The larvae enter the kidneys of their host and attach lightly by their anterior cells. Larvae develop into the adult form at the interface between kidney epithelial cells and urine in the mollusc kidney. Infusoriform larvae develop into nematogens only, but not into rhombogens.

It has been theorized that rhombozoans are simplified Platyhelminthes (A-7). Symbiotic lifestyles often result in morphological simplification. Like most platyhelminths, rhombozoans are symbionts integrated at the metabolic level. Evidence that rhombozoans are not degenerate platyhelminths includes properties that are unique to rhombozoans and Orthonectida (A-9) as well: the cell-within-a-cell arrangement of rhombozoans

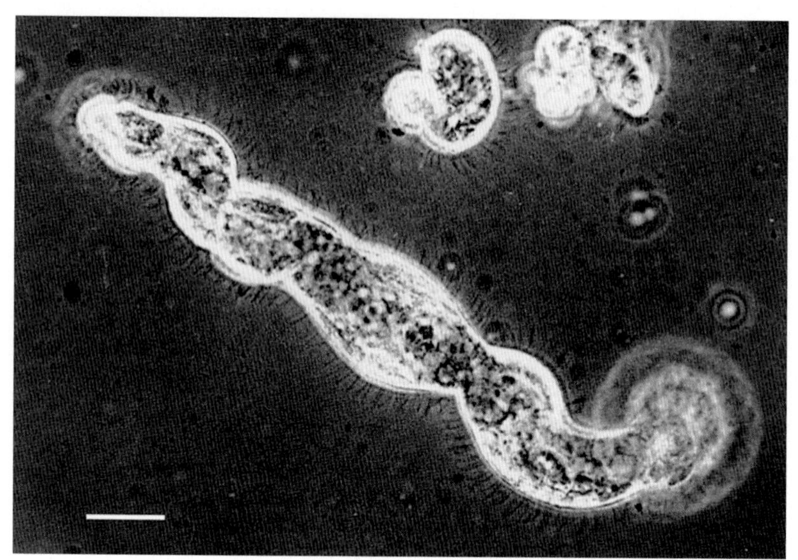

Figure B An extended adult *Dicyema truncatum*, with a small contracted one above. LM, bar = 10 μm. [Courtesy of H. Morowitz.]

Figure C *Dicyema truncatum* larva found in the kidneys of cephalopod molluscs. Free-swimming larvae disperse the dicyemids. LM, bar = 100 μm. [Courtesy of H. Morowitz.]

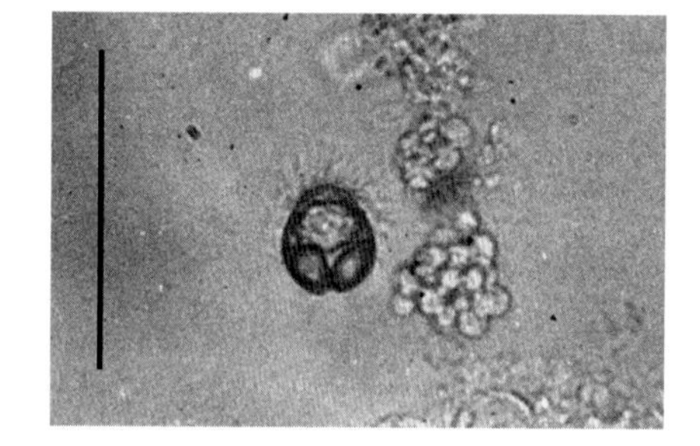

and in both rhombozoans and orthonectids the intracellular development of embryos; polyploid nucleus of the axoblast; and alternation of agamout and gamout generations. In other respects, though, what is known of rhombozoan biology suggests that rhombozoans and orthonectids have originated independently of each other and represent very early metazoans. Furthermore, the percentage of combined guanine and cytosine in the DNA of platyhelminths (35–50 percent) is considerably higher than that measured in dicyemids (*Dicyemennea* has 23 percent). Additionally, kinetid ultrastructure appears distinct from other metazoans. The issue is not resolved by recent studies on similar sequences of rDNA in *Dicyema* spp. so views on the origins of the Rhombozoa (and Orthonectida) continue to vary.

A-9 Orthonectida
(Orthonectids)

Greek *orthos*, straight; *nektos*, swimming

GENERA
Ciliocincta
Rhopalura
Stoecharthrum

The Orthonectida was once combined with the phylum Rhombozoa (A-8) in the Mesozoa, supposedly representing an intermediate stage between sponges or Porifera (A-3) and all other metazoans. The orthonectids, however, display many characteristics that are clearly unlike those of rhombozoans. Orthonectids are microscopic; none is larger than about 300 μm. Adults have one or two cell layers and are clearly multicellular but lack tissues and organs. Although there may be two layers of cells, there is no indication of true ectoderm or endoderm. Externally, the adult wormlike body consists of as many as 40 rings of jacket cells. For a given species, cell arrangement and number is rather constant. The jacket cells are multiciliated and when the animal is free living, it freely swims about. In adults of the genus *Intoshia*, thin bands of smooth muscle lie between outer jacket cells and an inner cell mass. The muscle forms a continuous band in immature individuals; it develops to more reduced, separate strips during maturity. The muscle extends the length of the animal; the muscle band uniquely connects with the jacket, cell kinetids cilia rootlets. Thus, unlike rhombozoans, locomotion and body shape involves both muscular and ciliary activity.

The Orthonectida are endosymbionts of marine animals. Approximately 50 species distributed among 10 genera have been described. Species have been found inside members of the Echinodermata (A-34), in Nemertina (A-10), free-living flatworms of the Platyhelminthes (A-7), in polychaete Annelida (A-22), and in bivalve Mollusca (A-26). Orthonectids reside in the gonads of their partners. Tissue fluids are probably the sole source of nutrition for orthonectids. In certain symbiotrophic relationships, the orthonectids castrate the animals in which they dwell unlike most rhombozoans.

Many species of orthonectids are dioecious. The gamont generation of orthonectids consists of tiny, ciliated free-swimming males (Figure A) and females (Figure B). The outer jacket cell layer encloses either eggs in the female (Figure C) or sperm with tails in the male (Figure D). A few species are hermaphroditic and self-fertilize. Orthonectids mate in the sea by bringing their genital pores, located about midway on the body, side by side. The male releases sperm, which bear undulipodia but lack acrosomes, into the female's pore. Fertilization is internal.

The fertilized eggs develop into ciliated larvae with two cell layers. The larvae, actively dispersed into the ocean through the female genital pore, somehow locate the potential symbiotic partner. Some orthonectid larvae shed their outer layer of cells when reaching the partner's gonad; the fate of the jacket cells is unclear. The inner cells undergo mitotic divisions to form a plasmodium (an ameboid multinucleated syncytium). Not all species follow this pattern. The "orthonectid plasmodium" may interact with or incorporate cells of the partner.

If the plasmodium develops, miniscule male and female worms develop within the syncytium. As the worms grow, they consume host-gonad tissue rendering the host infertile. Eventually the worms leave the plasmodium in the gonad to commence the free-living stage and gametic sexual reproduction.

Orthonectids with rhombozoans have been proposed as extremely simple relicts of early animal ("mesozoan") evolution. Some claim they are distant relatives, or possibly reduced examples, of Platyhelminthes (A-7). Because of the differences between orthonectids and platyhelminths (A-7), such a relationship is unlikely. The body plan, with its unique musculature, is unlike that of platyhelminths and the formation of adult worms from within a plasmodium is distinct from any of the near phyla including the Platyhelminthes. However, growing evidence, including the presence of a distinct muscle layer in some animals, structure of the kinetids, and analysis of rDNA sequences suggest that orthonectids in fact descend from a triploblastic ancestor, but probably not the Platyhelminthes.

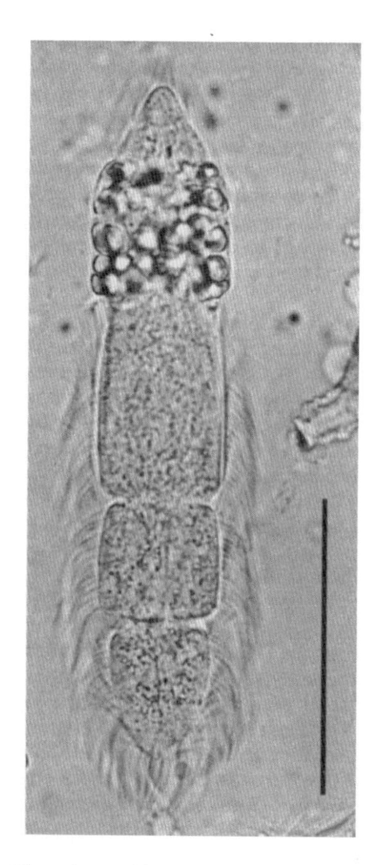

Figure A *Rhopalura ophiocomae* adult male showing ciliated outer jacket. This orthonectid feeds on tissue fluids of its echinoderm host, the brittle star *Amphipholis squamata* (A-34). LM, bar = 50 μm. [Courtesy of E. Kozloff.]

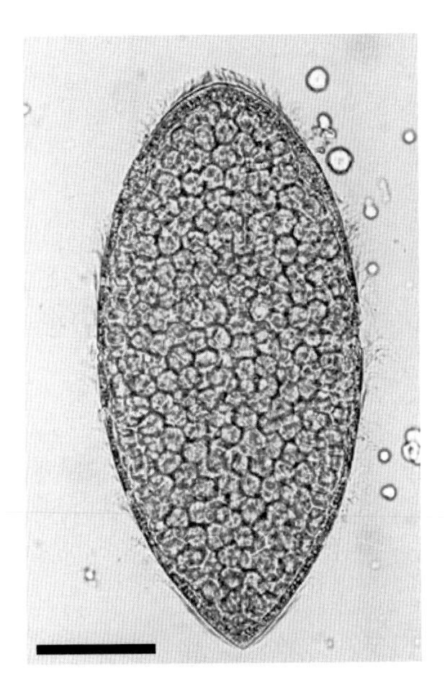

Figure B Ovoid type of mature female *Rhopalura ophiocomae* packed with oocytes. Near the anterior end, underlying the jacket cells, many small cells encircle the oocytes. LM, bar = 120 μm. [Courtesy of E. Kozloff.]

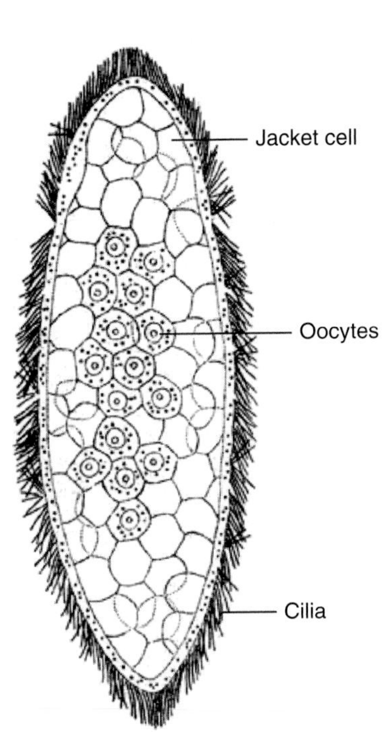

Figure C Mature female *Rhopalura ophiocomae*, as seen in optical section. The species has two types of females: elongate (left) and ovoid (right). Both types mate and then incubate fertilized eggs until larvae develop. LM, bar = 120 μm. [Drawing by E. Kozloff; courtesy of *Journal of Parasitology* 55:186; 1969; photograph courtesy of E. Kozloff.]

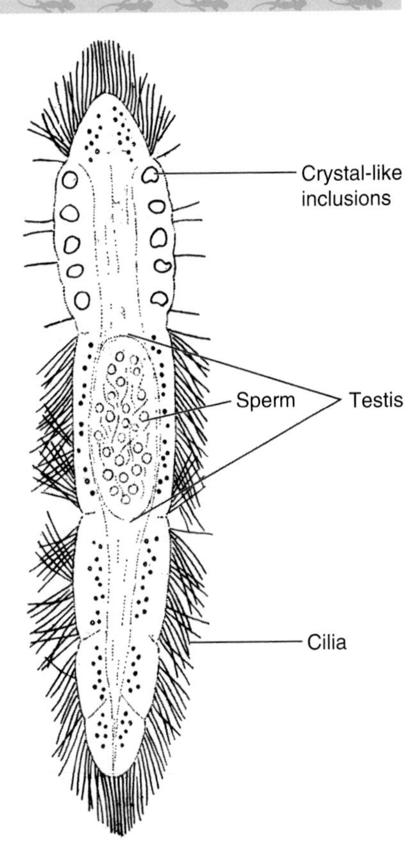

Figure D *Rhopalura ophiocomae.* An optical section brings a shallow slice of a living mature male into crisp focus and shows lipid inclusions, testis, and ciliated cells. In mature males, motile sperm with undulipodiated [9(2)+2] tails fill the testis. [Drawing by E. Kozloff; courtesy of *Journal of Parasitology* 55:172; 1969.]

A-10 Nemertina

(Ribbon worms, Nemertines, Nemertea, Rhynchocoela)

Greek *Nemertes*, a sea nymph

GENERA

Amphiporus
Carcinonemertes
Carinina
Cephalothrix
Cerebratulus
Emplectonema
Geonemertes
Gononemertes
Lineus
Malacobdella
Nectonemertes
Paranemertes
Prostoma
Tubulanus

Nemertina is a phylum consisting mostly of free-living worms found in marine, freshwater, and soil habitats. Their common name, ribbon worms, refers to their flat bodies and the brilliant color patterns of many species. A long, sensitive anterior proboscis that is separate from the digestive tract characterizes nemertines. This unusual organ branches in some species. The proboscis resides in a body cavity (rhynchocoel), from which the worm rapidly everts its proboscis as much as three times the length of its body. Muscular pressure on the fluid-filled proboscis chamber forces explosive eversion of the proboscis. So accurate is its aim that another common name for these worms is nemertine, based on the Greek term meaning the unerring one. Nemertines use the proboscis to explore the environment, to capture prey, to defend themselves, and to locomote. Annoyed nemertines release their proboscises, which they then regenerate.

Most ribbon worms live in the sea, in the intertidal marine sands, and in estuaries and are more abundant in temperate than in tropical oceans. *Carinina* is found in the abyss down to 4000 m. Some species are symbionts; *Gononemertes* lives in the branchial chambers of tunicates (urochordates, A-35). *Tubulanus* secretes a mucus-dwelling tube, whereas *Lineus* takes over empty burrows of *Chaetopterus* and other marine polychaete annelids (A-22). There are several freshwater forms; most are monotypic and several remain poorly known. *Prostoma* (Figure A) lives on aquatic plants in quiet freshwater along the Atlantic, Gulf, and Pacific coasts, in the US Midwest, and in Europe. Some nemertines are terrestrial in habit; *Geonemertes* inhabits moist soil of subtropical forests, between pandanus leaf bases, or can be arboreal. When introduced, *Geonemertes* thrives in greenhouse soils. Necrotrophic *Carcinonemertes* lives on crustaceans and feeds on the host's developing eggs as well as the host.

About 1150 nemertine species, placed among 250 genera, are known. Continued research using traditional morphology and, more recently, molecular data shows that the phylum is more complex than historical systematics indicates. Nemertines range from less than 0.5 mm to 30 m in length. *Lineus longissimus*, the iridescent bootlace worm, about 30 m long, is one of the longest invertebrates known. *Cerebratulus lacteus* (Figure B) can extend itself from about 1 to 10 m. *Emplectonema* is the only bioluminescent nemertine described. Most nemertines, especially the bottom dwellers, are pale, though a few are striped, speckled, or marbled multicolor ribbons.

Nemertines are abundant in the intertidal zone, although rarely seen; most are active at night, during low tide, burrowing, and feeding. They burrow by everting their proboscises into the mud; then they dilate the proboscis, forming an anchor. Nemertines pull themselves into their burrows by contracting body and proboscis muscles, pulling their bodies through the sediment.

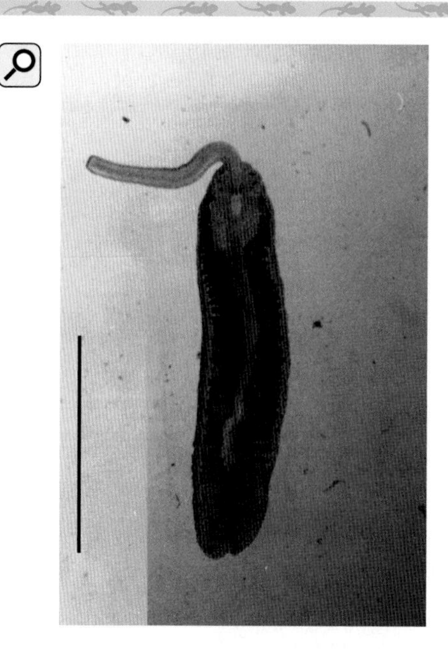

Figure A *Prostoma rubrum*, a live nemertine taken from Peck's Mill Pond in Connecticut. As a representative of the only freshwater nemertine genus, *Prostoma rubrum* has a proboscis armed with a stylet, unlike *Cerebratulus*, shown in Figure B. Bar = 1 cm. [Courtesy of J. Poluhowich.]

In tidal areas, nemertines have been observed to travel up to 10 m in search of prey. Like many other boneless animals, the nemertine's support is the incompressible liquid enclosed within its body wall. On tidal mud flats, nemertines can be found among algae, mussels, and tube-dwelling annelids. Nemertines sometimes creep out of seaweed placed in a dish of seawater.

Pelagic species (open-ocean dwellers), such as *Nectonemertes*, tend to be more leaf-shaped and have less well-developed muscles than do the benthic (seafloor-dwelling) nemertines. Pelagic nemertines float passively or swim with lateral undulations. Benthic nemertines crawl by muscular contractions, secreting a slime track. The smallest glide with their cilia against the resistance of their viscid mucus. Rarely, nemertines use their proboscises to attach themselves to an object and pull themselves forward.

Nemertines, unlike flatworms (A-7), have a blood vascular system through which contractile vessels and body muscle

contractions pump blood. Unidirectional valves and heart are lacking. Nemertine blood may be colorless, red, yellow, purple, or green, depending on the species. Some rhythmically take water into the vascularized foregut, presumably for gas exchange; most respire through the epidermis, like flatworms. The excretory system of these worms consists of protonephridia with flame cells that regulate ions, water, and possibly dissolved waste, which exits through lateral pores. Nemertina is the first phylum with openings at both ends of the digestive tract; solid waste leaves through the anus.

The nemertine nervous system resembles that of flatworms: a bilobed cerebral ganglion (brain) and longitudinal nerve cords with connecting nerves (Figure B). Their light-sensitive eyespots number from zero to as many as several hundreds. Functions suggested for the cephalic slits (those in the head) include chemotactic, auditory, excretory, respiratory, and endocrine. A cerebral organ opens into the cephalic slits. Papillae on the anterior end are sensory.

Most nemertines are predators, feeding on a wide variety of prey: annelids (A-22), crustaceans (A-21), flatworms (A-7), molluscs (A-26), roundworms (A-11), and even small fish (A-37). The mere presence of prey will elicit activity, attracting large numbers of nemertines to the surface. Most predacious nemertines have a stylet at the end of the proboscis. *Prostoma* (Figure A) has a venomous stylet with which the worm repeatedly stabs and paralyzes prey. The sticky proboscis wraps around prey and transfers captured prey to the mouth. Mouth and proboscis may share a common opening, depending on the species. Prey are sucked completely into the mouth; however, if the prey is too large to swallow, juices are sucked out of it instead. Cilia move food from the mouth along the foregut. Phagocytosis and extra- and intracellular digestion take place in the intestine, which has

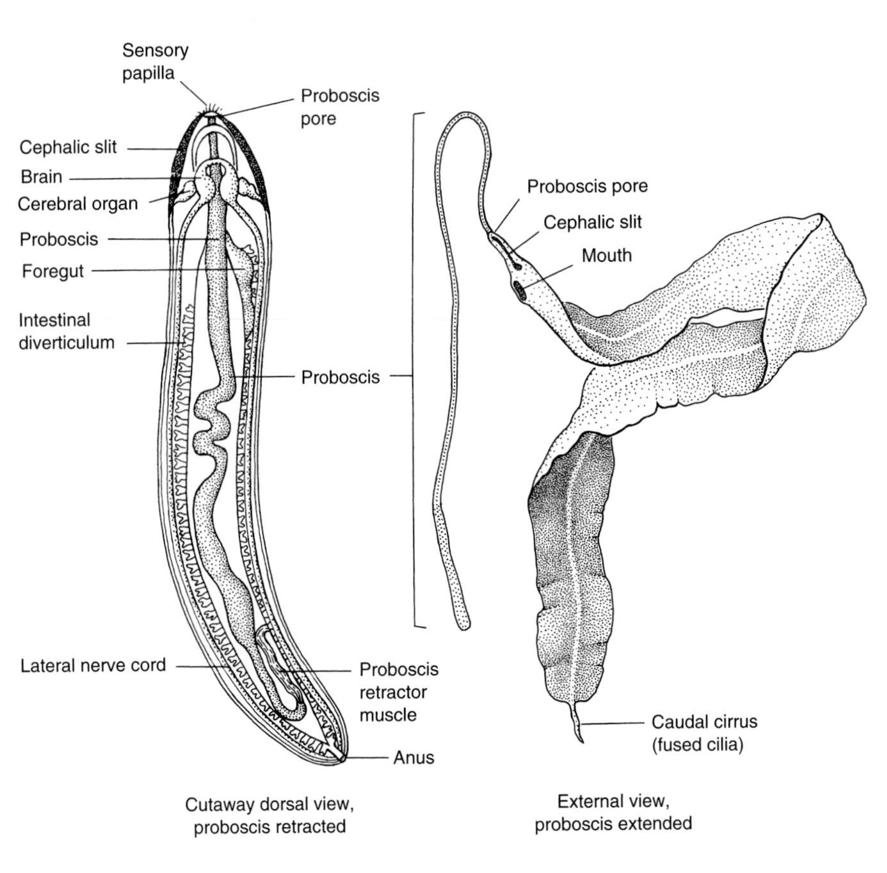

Figure B *Cerebratulus*, a marine nemertine, shown in dorsal view with proboscis retracted (cutaway view, left) and with unarmed proboscis extended (external view, right). This free-living ribbon worm swims small distances by undulating. [Drawing by L. Meszoly; information from P. Roe.]

numerous pouches (diverticula). At least one species of the terrestrial *Geonemertes* feeds on terrestrial molluscs *Malacobdella* is the unique filter-feeding nemertine, living commensally within the mantle cavity of clams (A-26). *Malacobdella* filters bacteria, algae, diatoms, and other protoctists from water within its host's mantle cavity through ciliated papillae in its foregut.

Nemertines' prodigious regeneration is a potential research model for tissue culture. The worms reproduce asexually by fragmentation—each fragment regenerates a complete worm. *Carcinonemertes* reproduces by parthenogenesis. Nemertines can also reproduce sexually. In most species, the sexes are separate. In sexual reproduction, numerous temporary gonads form during the breeding season in mesenchyme tissue between intestinal pouches. Each gonad opens to the outside through its own surface pore. Eggs are laid in gel strings. Fertilization typically takes place in the water but is internal in some species. The eggs develop either directly into adults or first into pilidium (free-swimming) larvae, which look like ciliated caps with earflaps and apical tuft. The juvenile nemertine forms around the stomach of the pilidium, and it can break free while the larva continues swimming. This is an example of "overlapping metamorphosis," the coexistence of larva and juvenile of the same individual. Comparable cases occur in polychaete annelids (A-22), echinoderms (A-34), and doliolid tunicates (A-35), and they are explicable in terms of larval transfer (Box A-i). Other species have Iwata larvae or Desor larvae, named for embryologists. Desor larvae are characteristic of *Lineus* and other heteronemertines. The Desor is ciliated, oval, postgastrula stage that stays within the egg membrane (unlike pilidium larvae); it lacks oral lobes and the apical tuft of pilidia. A few species are protandric hermaphrodites—each individual is first male and then becomes female. Members of the hermaphroditic terrestrial genus *Geonemertes* bear live young. In *Nectonemertes*, a genus of active swimmers, males clasp females with special attachment organs during mating. Knotted balls of about 30 *Cephalothrix* have been observed, perhaps mating, in breeding season beneath stones along the Yorkshire coast.

The fossil record of nemertines is sparse. The Cambrian *Amiskwia* was regarded as a nemertine fossil or as a chaetognath (A-32). Its phyletic position is obscure. Structures common to both nemertines and flatworms are parenchyma tissue that encloses organs; lack of body cavity, respiratory organs, and segmentation; ciliated epidermis that moves the animal along a mucus track; similar sensory and excretory organs; and multiple reproductive organs. Some flatworms, annelids, and molluscs also have anterior probosces but only those of nemertines are in fluid-filled, cell-lined cavities separate from the gut. Common features were thought to point to a close relationship between flatworms and nemertines despite their differences in food getting, digestive systems (the one-way nemertine gut is assumed more efficient than the dead-end flatworm gut) and oxygen–carbon dioxide exchange. However, comparison of ribosomal RNA sequences places *Cerebratulus*, the nemertine studied, near sipunculids, annelids, and molluscs (protostomous coelomates—the embryonic blastopore is the site of the adult mouth) and more distant from flatworms (acoelomates). These molecular data are supported by ultrastructural evidence that places nemertines as protostomous coelomates; the blood vascular system, gonadal sacs, and the rhynchocoel cavity are modified coelomic cavities, originating from mesoderm. Thus, it is now thought that the acoelomate attributes of nemertines may have evolved secondarily from a more typical coelomate ancestor. If this affiliation is accepted, nemertines will move to the protostome coelomate position in the phylogenetic tree.

A-11 Nematoda
(Nematodes, thread worms, round worms)

Greek *nema*, thread

GENERA
Ancyclostoma
Ascaris
Caenorhabditis
Dioctophyme
Diplolaimella
Dirofilaria
Dracunculus
Enterobius
Leptosomatum
Necator
Pelodera
Rhabdias
Trichinella
Trichodorus
Tricoma
Wuchereria

Nematodes (roundworms) are unsegmented pseudocoelomate mostly microscopic necrotrophs of plants and animals including humans. Their body cavity is a pseudocoel, defined as a space between embryonic endoderm and ectoderm; the pseudocoel lacks a peritoneum, the mesodermal lining of the coelomic body cavity. Nematodes are probably the most abundant animals living on Earth (Figure A). About 80,000 species of nematodes have been described in the scientific literature; researchers estimate that nearly 1 million living species exist. These worms range from only 0.1 mm (100 μm) to about 9 m in length. The female giant nematode *Dioctophyme renale* is 1 m in length; the male is only half as long. Free-living nematodes are slender and cylindrical, tapering at both ends, typically about 1 mm in length. Parasitic nematodes have a variety of shapes, many saclike; the longest is a 9 m long symbiotroph from a sperm whale.

Members of this phylum are grouped into two classes: Adenophorea and Secernentea. The Adenophorea lack phasmids and therefore are also called Aphasmida. *Trichinella* is a parasitic member of this class. Phasmids are sense organs, possibly chemoreceptors, found in the tail region particularly of "parasitic" (symbiotrophic) roundworms. Secernentea, also called Phasmida, do have phasmids. Many members of this class, in fact entire suborders, live in vertebrates, insects, or plants. Hookworms such as *Necator* and *Ancyclostoma*, some gapeworms, hairworms, stomach worms, lungworms, *Ascaris*, pinworms, and the filarial worms that cause filariasis (tropical diseases, elephantiasis, and river blindness) are all phasmidians.

Typically, nematodes are transparent, covered with a noncellular, patterned cuticle of collagen, a fibrous protein. Nematodes move with a characteristic flip of their bodies; unique oblique longitudinal muscles encircle their bodies, but nematodes lack circular muscles and so cannot extend and contract as segmented (annelid) worms do. Many parasitic nematode species look like microscopic dragons; they have well-developed teeth and are predaceous. They feed on amebas and other nematodes. Free-living nematodes devour rotifers and tardigrades. Necrotrophic species, such as the hookworms, have evolved specialized mouthparts with which they hook onto the intestinal wall or other tissues of their host.

Nematodes form three layers: ectoderm (ecto-, outside), mesoderm (meso-, middle), and endoderm (endo-, inside) during embryonic development and so are triploblastic. Ventral, dorsal, and lateral nerve cords are present, as is a nerve ring around the pharynx. The nematode digestive tract forms a tube complete with mouth and anus within the worm. The gut lacks cilia. The muscular pharynx pumps fluids into the gut. The pharyngeal pump counters internal hydrostatic pressure generated by contraction of the nematodes' longitudinal muscles and by its nonexpandable cuticle. This pumping mechanism differs from the circular muscle contractions that propel food through the human gut. Nematodes have no specialized organs for circulation, and soil waste is eliminated from the anus (in males, also used as an exit for gametes). The excretory system consists of terminal and lateral canals. Dissolved oxygen and carbon dioxide diffuse through the permeable body wall.

The sexes are separate in almost all species, the male being smaller than the female. *Caenorhabditis* is a simultaneous hermaphrodite. Reproduction is always sexual with internal fertilization. Males have copulatory structures called spicules near the posterior end. Single or paired ovaries connect to the outside through a midventral gonopore (genital pore) in females (Figure B). In males, a single testis produces ameboid sperm, which lack undulipodia.

Figure A *Rhabdias bufonis* (female), a nematode belonging to class Secernentea, necrotroph in the lung of the leopard frog, *Rana pipiens*. SEM bar = 1 mm. [Courtesy of R. W. Weise.]

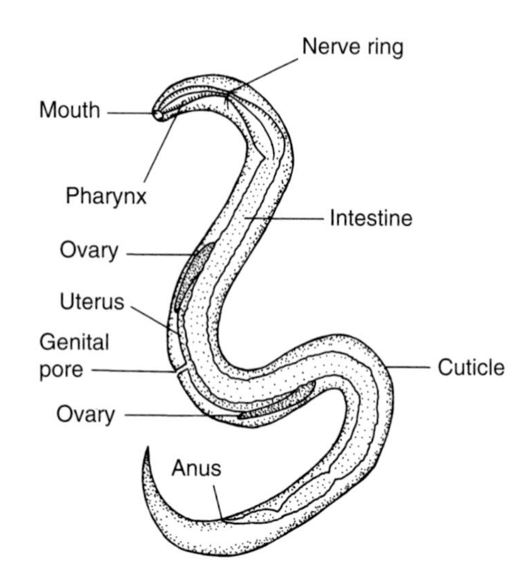

Figure B Diagram of a female nematode showing a well-muscled pharynx with which these worms pump liquid food into their digestive tracts. [Drawing by I. Atema; information from R. W. Weise.]

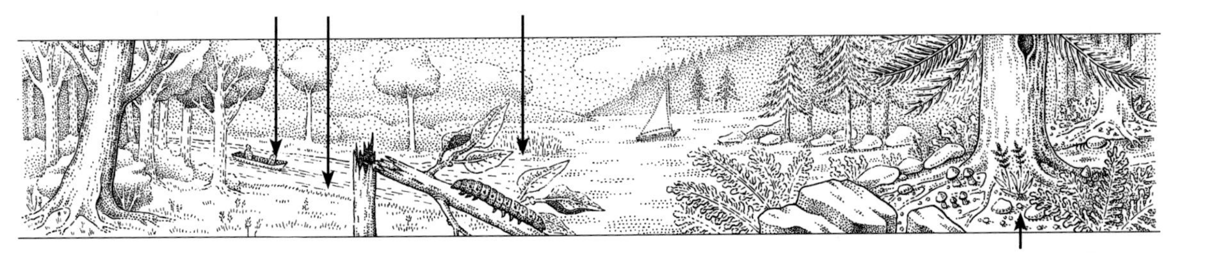

The reproductive capability of nematodes is prodigious: some females have been known to contain 27 million eggs and extrude a quarter million fertilized eggs a day. Nematodes lack a free-swimming larval stage—they hatch from the eggs as miniature adults. The life cycle consists of four stages, three juvenile stages and one adult stage. The young nematode molts its cuticle four times before the adult form.

About one of five humans harbor hookworms worldwide. Hookworms produce anticlotting molecules, proteases that dissolve antibodies released by the infected person, and antioxidant enzymes that neutralize oxidizing molecules secreted by the victim; these and at least half a dozen additional mechanisms enable hookworms to thwart the body's immunological and vascular protective mechanisms. Vaccines against hookworm are being tested.

Most plant symbiotrophic nematodes cause swellings of tissue in the host, mostly on the root system. Many produce resistant eggs that are well able to pause in their development indefinitely until harsh environmental conditions improve.

Because they infect human guts and domesticated plants and animals, some nematodes are being intensively studied. The heartworm, *Dirofilaria immitis*, infects cats and dogs. *Trichinella spiralis* is the infectious nematode responsible for causing trichinosis, acquired by eating infested pork or other meat that has not been cooked sufficiently. The minute juveniles of these worms are harbored as cysts in the striated muscles of wild pigs, cats, dogs, rats, and bears. If the flesh of an infested animal is eaten by another, the nematode cysts are digested, liberating the juvenile worms into the intestine of a new host. About 2 days after their release, the nematodes mature sexually, and male and female worms mate in the intestine. The male then dies. The females, about 4 mm long, then burrow into the muscles of the intestinal wall. Female worms are ovoviviparous in this species—they produce eggs that hatch within the mother's body and release hundreds of live juvenile worms, which enter the lymph and are carried to the bloodstream. From there, juveniles burrow into skeletal muscles and any other organs, where they coil up and become enclosed in cysts. Encysted larvae can remain dormant for months, even years, or the host may deposit calcium salts in the infected tissues, calcifying and killing the nematodes. When the skeletal muscle is eaten, they are passed to another host.

Although nematode-caused disease attracts public attention, most nematodes are not harmful. Free-living roundworms live almost everywhere; soil-, freshwater, and marine environments feeding primarily on bacteria. As many as a billion roundworms per acre have been counted in the top 2 cm of rich soil. Free-living nematodes aerate soil, consume detritus, and circulate mineral and organic components of soil and sea sediments.

Research on the free-living nematode *Caenorhabditis elegans* has revealed much of what we currently know about the expression of genes during development, genetic manipulation, and inheritance. *Caenorhabditis* is a good research animal; in *Caenorhabditis*, as in all nematodes, cell fates are permanently determined at the first cleavage of the zygote. The final fate of each of the embryonic cells of *Caenorhabditis* has been traced.

Nematodes are distinctive, placed in this phylum apart from other worms, although nematodes, rotifers (A-14), gastrotrichs (A-17), kinorhynchs (A-15), nematomorphs (A-12), and acanthocephalans (A-13) were formerly placed with aschelminthes or pseudocoelomates. Nematodes lack the circular body muscles and segmentation characteristic of annelid worms (A-22) and the eversible proboscis of ribbon worms (A-10), acanthocephalans, and certain flatworms (A-7). Lack of locomotory cilia sets nematodes apart from rotifers (A-14); nematodes do have nonmotile cilia in their sense organs. It seems likely that nematodes gave rise to no other phyla.

A-12 Nematomorpha

(Gordian worms, horsehair worms, nematomorphs)

Greek *nema*, thread; *morphe*, form

GENERA
Chordodes
Chordodiolus
Gordionus
Gordius
Nectonema
Parachordodes

Nematomorphs are commonly called horsehair worms. Their name stems from the once-held belief that these slender, cylindrical worms, observed in horse-watering troughs, spring from horsehairs. Adult nematomorphs coil and tangle with each other, so they are also known as gordian worms, after Gordius, king of Phrygia. Gordius tied a knot, declaring that whoever untied his intricate knot should rule Asia. Alexander the Great cut the Gordian knot with his sword and added Asia to his Greek Empire.

Although 240 species are known, nematomorphs are grouped into a very few genera. Gordian worms live all over the world in shallow oceans and lakes, temperate and tropical rivers, ditches, alpine streams, moist soil, and stock tanks. The only marine species in the continental United States is *Nectonema agile*, distinguished from other horsehair worms by a row of slate-colored bristles on each side of its gray-yellow or pale whitish body. Chances of observing *Nectonema* are highest in late summer, from July to October on moonless nights when the tide is receding. *Nectonema's* geographical distribution is poorly known. In the Gulf of Naples, Vineyard Sound (Massachusetts), Norway, and the East Indies, nematomorphs can be seen coiling and winding in shallow coastal waters. In fresh- and marine water, they make up only a small fraction of the plankton.

Nematomorph adults are free living, usually in fresh- or saltwater. All nematomorph species are endoparasitic (internal symbiotrophic symbiotrophs) for a part of their lives. Nematomorphs are rarely found in the human urethra or digestive system, and they do not seem to cause human disease. Hosts of nematomorphs include leeches (A-22); beetles, crickets, grasshoppers, and cockroaches, hermit crabs and spiders (A-20); and true crabs (A-21).

Nematomorphs are leathery, unsegmented invertebrates stiff as wire and generally brown, black, gray, or yellow in color. The head end is distinguishable by being a lighter color than the rest of the body (Figure A). A pair of caudal lobes posterior to the anus distinguishes the posterior from the anterior end. Nematomorphs range from 10 to 70 cm in length and from 0.5 to 2.5 mm in diameter, depending on the species. Body length also depends on the sex of these dioecious worms: females are longer than males. Polygonal or round thickenings ornament the hard, noncellular cuticle having fibrous layers—probably collagen—secreted by the epidermis. As they grow, nematomorphs molt the cuticle.

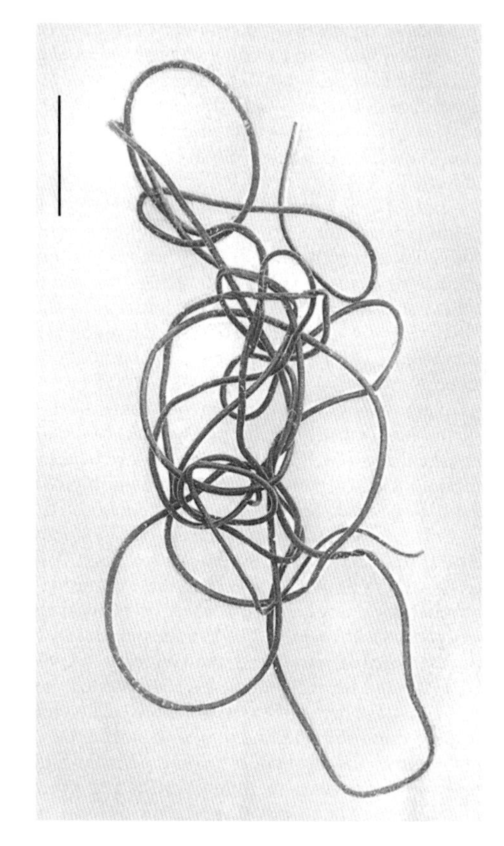

Figure A An adult female *Gordius villoti*, a horsehair worm. Bar = 1 cm. [Courtesy of Trustees of the British Museum (Natural History).]

Neither adults nor larvae ingest food; although the hind part of the digestive tube, the cloaca, is used in reproduction, the anterior part of the gut is degenerate. Because the posterior end of the nematomorph digestive tract receives gametes, it is called a cloaca; in other animals, both gametes and waste usually exit through the cloaca, as in male nematodes. During its symbiotrophic, larval phase, instead of ingesting food, a nematomorph absorbs nutrients across its body wall from its host animal. Respiratory, circulatory, and excretory organs are absent, as in many necrotrophs. Digestive, reproductive, and nervous systems are embedded in a matrix of collagen fibers packed with parenchyma. The nematomorph nervous system resembles that of kinorhynchs (A-15): a nerve ring encircles the pharynx, a single nerve cord runs down the ventral side, and some adults have eyespots composed of innervated sacs lying beneath transparent cuticle and backed by a pigment ring. Larvae lack eyes (Figure B) and have protrusible, spiny proboscises that resemble the proboscis of acanthocephalans (A-13). Like nematodes (A-11) and kinorhynchs (A-15), nematomorphs have only longitudinal muscles that permit whiplike swimming and serpentine coiling. Fluid in the pseudocoelom (body cavity) serves as a hydraulic skeleton.

Nematomorphs are dioecious. Eggs produced by ovaries or sperm produced in spermaries pass to the cloaca and then out through the anus. Adult male worms crawl or swim, especially in winter. In contrast, females are less active. The male wraps around the female, deposits sperm near her cloacal opening, and soon dies. The eggs are fertilized internally. The female drapes millions of fertilized eggs in gelatinous strings around aquatic plants. From 15 to 80 days later, the eggs hatch as tiny motile larvae. Nematomorphs lack an asexual reproductive mode.

The larvae of nematomorphs enter the body cavities of arthropods or leeches in a way that has not yet been observed. Larvae may be accidentally eaten or drunk by any of nematomorphs' host animals and may bore through a host's gut by using piercing mouthparts borne on their proboscises. Larvae of marine, soil-, and freshwater nematomorphs metamorphose within their hosts; then the sexually immature worms burst out of their hosts near or in water or during rain. How nematomorphs that spend part of their life cycle in water induce terrestrial hosts to seek water is a mystery. The exit of the larvae kills the host. The same larvae may pass through one or more hosts, the number of hosts depending on the species of nematomorph. If worms mature in autumn, they form cysts on waterside grasses and reenter water in spring. As a result, development from egg to adult worm may take a short time (2 months) or as long as 15 months.

The nematomorph body cavity is a pseudocoel, a body cavity that lacks a mesodermal lining. Nematomorphs probably did not evolve directly from any other pseudocoelomates—rotifers (A-14), kinorhynchs (A-15), acanthocephalans (A-13), or nematodes (A-11). Rather, each phylum of pseudocoelomates is thought to have evolved from acoelomates (lacking a body cavity between the gut and the outer body-wall musculature) at different times in several different ways. The pseudocoel is not a stage in the development of the true coelom; it developed independently.

Nematomorphs seem to be of no veterinary or medical importance.

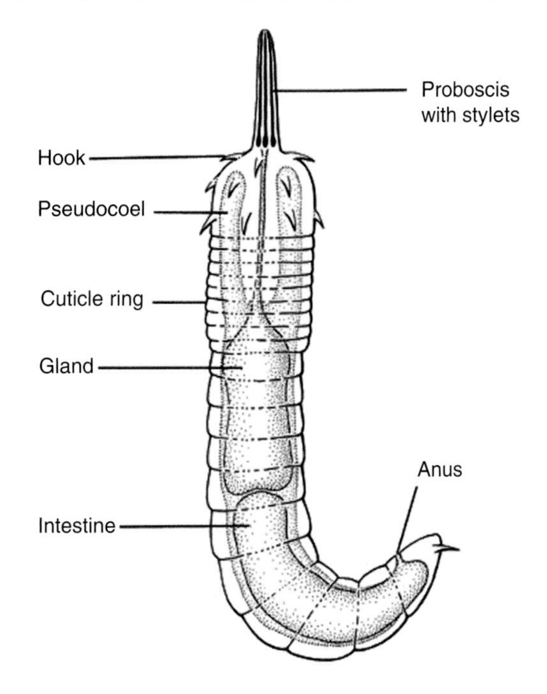

Figure B Parasitic larva of a gordian worm with its proboscis extended. The larva is about 250 μm long. [Drawing by L. Meszoly; information from L. Bush.]

A-13 Acanthocephala
(Thorny-headed worms)

Greek *akantha*, thron; *kephale*; head

GENERA

Acanthocephalus
Acanthogyrus
Centrorhynchus
Echinorhynchus
Gigantorhynchus
Leptorhynchoides
Macracanthorhynchus
Moniliformis
Neoechinorhynchus
Plagiorhynchus
Polymorphus

Acanthocephalans, also known as spiny-headed or thorny-headed worms, are necrotrophic worms that live as adults exclusively in the vertebrate's small intestine and exhibit an indirect life cycle, which utilizes an arthropod intermediate host. Acanthocephalans use their retractable and invaginable proboscis to attach to the intestine of their host. The proboscides, bearing hooks and spines, vary greatly among species in size and shape (Figure A). Spines or thorns may also occur on the body. Acanthocephalans also vary greatly in size and appearance. They may be less than 1 mm long to over 1 m and may be smooth or wrinkled. The phylum Acanthocephala comprises more than 1100 valid species. Bony fishes are the most exploited group of vertebrate Acanthocephalan hosts, followed by birds, mammals, amphibians, and reptiles.

The vertebrate hosts usually become infected by ingesting an infective larva, called a cystacanth, contained within an infected invertebrate intermediate host (Figure B). Within the intestine, the dioecious worms mature, mate, and upon patency ova containing the embryo known as an acanthor are passed in the host's feces. The intermediate host becomes infected by ingesting ova in fecally contaminated soil, food, water, or by feeding directly on the feces. The ovum hatches within the intestine of the arthropod intermediate host, releasing the acanthor that uses spines to penetrate the intestine. Within the body cavity of the arthropod, the worm develops into the infective cystacanth. Known intermediate hosts for those with terrestrial life cycles include insects (especially Coleoptera and Orthoptera), terrestrial isopods, or millipedes as intermediate hosts. Various decapods and other crustaceans serve as intermediate hosts for those with aquatic life cycles. No species of Acanthocephala has been demonstrated to require more than the arthropod intermediate host to develop infectivity to vertebrates. However, in some acanthocephalan life cycles, another vertebrate host is utilized between the arthropod intermediate host and the vertebrate definitive host. In such hosts, known as paratenic hosts, the worm penetrates the intestinal wall and localizes in the mesenteries or viscera, where it remains in the infective cystacanth stage. Although paratenic hosts have not been demonstrated to be physiologically required by the worms to attain maturity, they may be required to complete transfer of worms from intermediate hosts to the trophic level at which the final host feeds. For instance, insectivores such as shrews may be utilized by the representatives of the genus *Centrorhynchus* to facilitate the transfer of worms from an arthropod intermediate host to the birds of prey that serve as final host. Postcyclic transmission may also occur, which refers to the transfer of adult worms when one vertebrate host eats another. In some instances the presence of acanthocephalans may lead to modification in behavior exhibited by the intermediate host making it more susceptible to being preyed upon, thereby facilitating transmission. For instance, cockroaches infected with the acanthocephalan *Moniliformis moniliformis* move more slowly than do their uninfected counterparts.

Acanthocephalans are remarkably adapted to a symbiotrophic lifestyle in that they lack circulatory, respiratory, and digestive systems. Nutrients are absorbed directly across the body wall from the host's intestine. Respiration is facilitated by simple diffusion. Excretion is usually by diffusion or is rarely facilitated by ciliated flame cells called protonephridia, like those in flatworms, connected to excretory tubules. Like all pseudocoelomates, the body cavity lacks a peritoneum. The nervous system is simple, consisting of an anteriorly located mass of neural tissue known as the cerebral ganglion, from which nerves extend. A small genital ganglion is associated with the male reproductive system near the posterior end of the body. The innermost body wall contains longitudinal and circular muscles. When these muscles contract, hydrostatic pressure in fluid reservoirs called lemnisci are thought to cause the proboscis to evert. The lemnisci are continuous with a network of tubes running throughout the body wall. This lacunar system is thought to function in fluid transport, regulation of hydrostatic pressure leading to eversion and retraction of the proboscis and male's copulatory bursa, and possibly in the distribution of nutrients to muscles associated with the body wall.

The most prominent internal organs are those associated with the reproductive system. Eggs are produced by the female in the ovary, which fragments into ovarian balls. When gravid, the pseudocoel of the female is filled with ova, each constituting a shelled acanthor. A funnel called the uterine bell opens to the pseudocoel at one end and to the uterus at the other end. Males usually have a pair of testes from which sperm ducts lead to a penis. During copulation, the male injects seminal fluid from its seminal vesicle into the bursa. The Saefftigen's pouch, a muscular sac connected to the lacunar system, contracts causing hydrostatic pressure that everts the copulatory bursa. The shape of the copulatory bursa "holds" the female during copulation. Sperm are released through the penis which extends into the male's everted bursa, into the female's genital pore. Cement-gland secretions of the male, released through the penis, cap the female's posterior end preventing loss of sperm and possibly preventing subsequent mating by other males with the female. Sperm travels up the reproductive system of the female and into the pseudocoel, where fertilization occurs. Basic acanthocephalan anatomy is illustrated in Figure C.

Phylogenetic relationships of the Acanthocephala are unclear. Recent morphological and molecular studies suggest that, among the pseudocoelomates, acanthocephalans are most closely related to the rotifers. More specifically, recent molecular analysis using 18S rDNA led to the suggestion that Acanthocephala share a most-recent common ancestry with rotifers of the class Bdelloidea. Some have suggested that acanthocephalans are actually a "highly derived" group of rotifers, and have further suggested lumping the acanthocephalan and rotifers into a single phylum, Syndermata. Other analyses point toward monophyly of the acanthocephalan; most authorities argue in favor of retaining the phylum status for Acanthocephala.

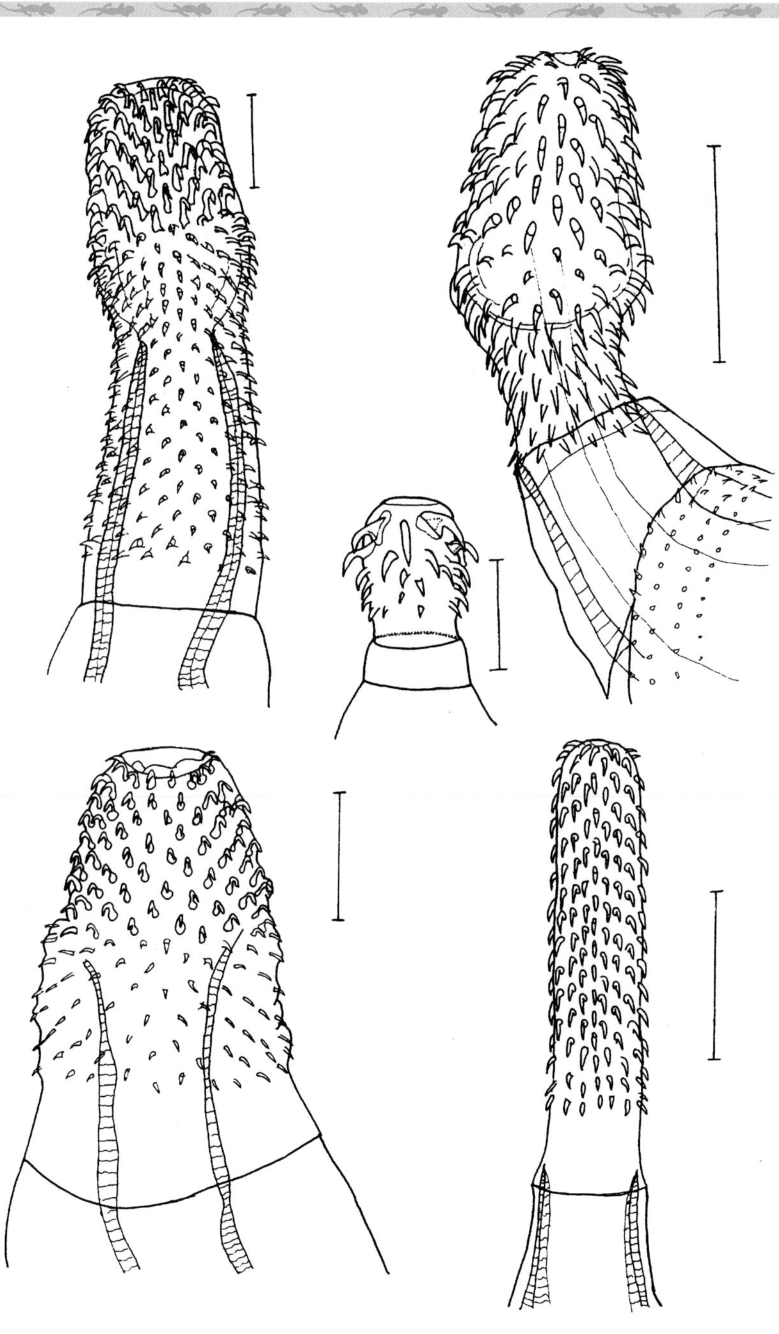

Figure A Proboscides of *Centrorhynchus robustus* from a northern spotted owl *Strix occidentalis*, upper left (redrawn from D. J. Richardson and B. B. Nickol, *Journal of Parasitology* 81:769; 1995) bar = 250 μm; *Polymorphus cucullatus* from a hooded merganser *Lophodytes cucullatus*, upper right [redrawn from H. J. Van Cleave and W. C. Starrett, *Transactions of the American Microscopical Society* 59:351; 1940], bar = 500 μm; *Oligacanthorhynchus tortuosa* from a Virginia opossum *Didelphis virginiana*, center [redrawn from D. J. Richardson, *Comparative Parasitology* 73:3] bar = 100 μm; *Mediorhynchus centurorum* from a red-bellied woodpecker *Centurus carolinus*, lower left [redrawn from B. B. Nickol, *Journal of Parasitology* 55:325; 1969], bar = 220 μm; *Plagiorhynchus cylindraceus* from an American robin *Turdus migratorius*, lower left [redrawn from G. D. Schmidt and O. W. Olsen, *Journal of Parasitology* 50:726; 1964], bar = 1 mm. [Drawings by D. J. Richardson.]

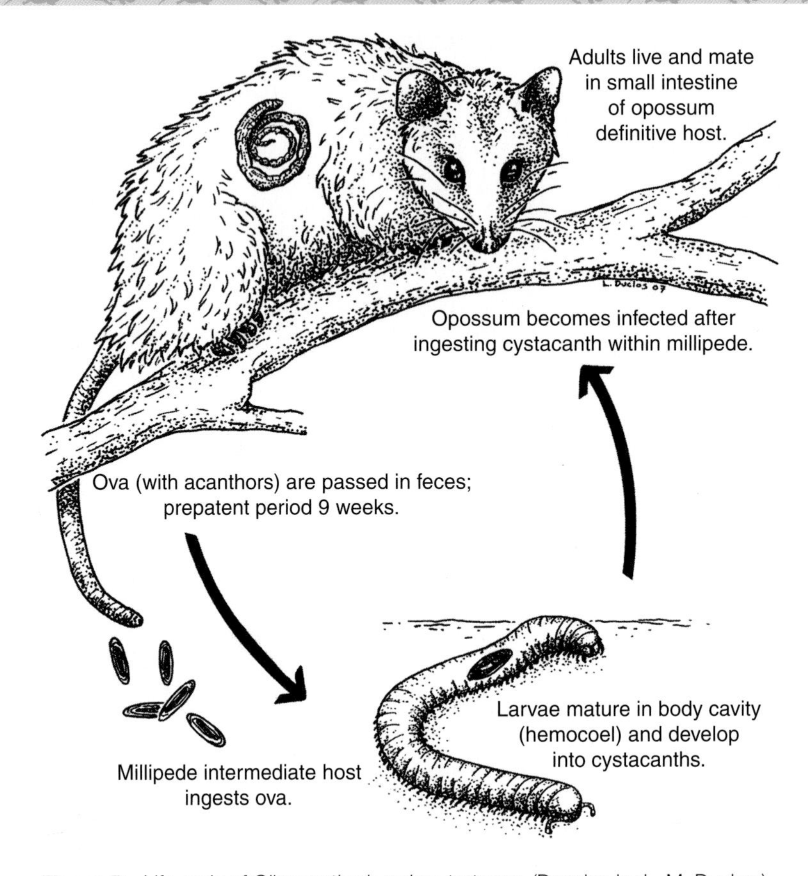

Figure B Life cycle of *Oligacanthorhynchus tortuosa*. (Drawing by L. M. Duclos.)

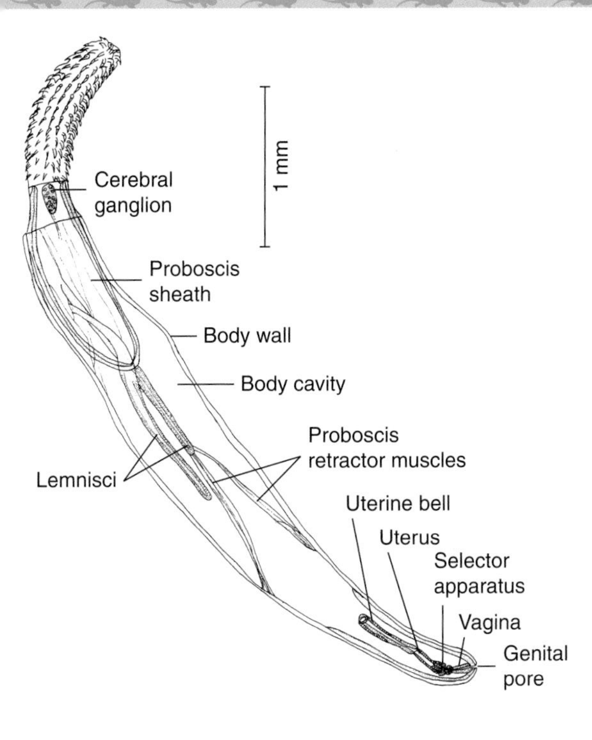

Figure C Young female *Leptorhynchoides thecatus* from the intestine of a large-mouth bass *Micropterus salmoides*, bar = 1 mm. (Drawing of male by L. Meszoly; female by D. J. Richardson; information from B. B. Nickol.)

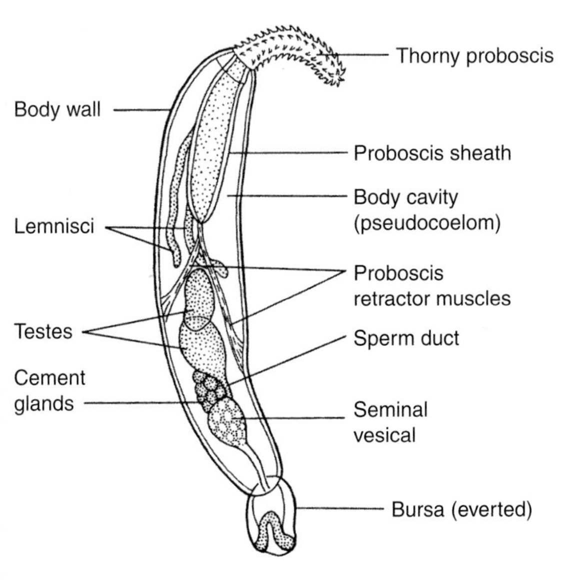

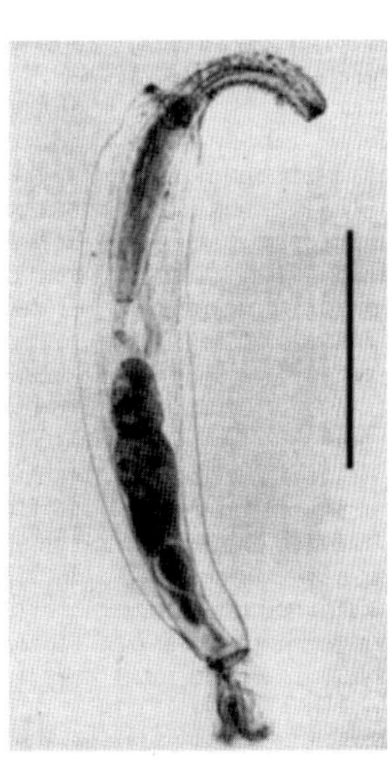

Figure D *Leptorhynchoides thecatus*, a young male symbiotrophic acanthocephalan from the intestine of a large-mouth black bass *Micropterus salmoides*. LM (worm fixed and stained), bar = 1 mm. [Photograph courtesy of S. C. Buckner; drawing by L. Meszoly; information from B. Nickol.)

A-14 Rotifera

(Rotifers)

Latin *rota*, wheel; *ferre*, to bear

GENERA
Albertia
Asplanchna
Brachionus
Chromogaster
Conochilus
Cupelopagis
Embata
Euchlanis
Floscularia
Notommata
Philodina
Proales
Seison
Stephanoceros
Synchaeta
Trochosphaera

The rotifer is a minute aquatic animal named for an optical illusion—waves of beating cilia on its head appear to be a rotating wheel. The cilia are dual-purpose: they propel the swimming rotifer and direct food currents to the mouth. Behind the anterior mouth lies a feeding apparatus unique to rotifers: rigid jaws called trophi are manipulated by and embedded in the mastax, a muscular pharyngeal region between the mouth and esophagus. Trophi suck, grab, or grind food.

About 2000 rotifer species have been reported; only about 50 species are marine, being benthic or pelagic. Some rotifers are epizoic (living on other animal species) or symbiotrophic. Many marine rotifers live interstitially (between sand grains) with loriciferans, tardigrades, kinorhynchs, and gastrotrichs. Most are free swimming, although *Conochilus* swims as a revolving colony of about 100 individual organisms anchored in a jelly sphere. *Stephanoceros* is sessile, permanently attached in a secreted jelly case. *Floscularia* molds pellets of mucus and feces to form an exquisite dwelling tube. Rotifers are the most abundant and cosmopolitan of the freshwater zooplankton. They inhabit bogs, sandy beaches, lakes, rivers, glacial muds, birdbaths, gutters, ditches, moss and lichen pads, rocks, and tree barks. Rotifers resist desiccation by secreting a protective envelope of gel around their bodies.

Rotifer animal hosts are exclusively invertebrate. *Seison* lives exclusively on the leptostracan crustacean *Nebalia. Seison* is toeless and attaches to its host with a posterior adhesive disk. It moves leechlike over its host's gills, where it is nourished by both eating its host's eggs and food in the ocean current. *Proales* lives on *Daphnia*, the common freshwater water flea (A-21), and in snail eggs (A-26), the heliozoan *Acanthocystis* (Pr-31), *Vaucheria* filament tips (Pr-16), cnidarians (A-4), and *Volvox* (Pr-28). Other rotifers are endosymbiotrophs of annelids and shell-less molluscs. The rotifer *Albertia* is a wormlike obligatory symbiotroph living in the coelom of annelids (A-22).

Free-living rotifers typically have eclectic diets, feeding on bacteria, suspended organic matter, protoctists, and other small animals including rotifers. A few sessile (permanently attached) rotifers are trappers; *Cupelopagis* traps protoctists by means of a retractable funnel. *Chromogaster* specializes in grabbing dinomastigotes (Pr-5) with its pincerlike mastax, drilling through the test (the firm covering) of the prey and sucking it dry.

The flexible body of some rotifers is covered with a nonchitinous layer that is not molted. In species of class Monogononta, the thick cuticle is called a lorica. In many species, the lorica can collapse and expand like a portable telescope. Rotifer shapes range from trumpetlike to spherical. Free-swimming rotifers, such as *Brachionus*, have spines that enhance flotation and swim in spirals propelled by their ciliated crowns. The head cilia may be fused into tentacles or platelike structures or they may be lacking altogether in some species. Other free-living species stick temporarily and sessile species attach permanently to substrates with secretions from the pedal cement glands on their contractile toes; moving like inchworms, free-living rotifers attach and reattach. On the posterior end, rotifers may have nonadhesive (lacking cement glands) spurs and as many as four pairs of adhesive toes. Rotifers are usually translucent. Some are colored green, orange, red, or brown from food in the gut, which can be seen through the translucent body wall. The largest rotifers reach 2.0 mm in length; the smallest are about 0.04 mm long (Figure A).

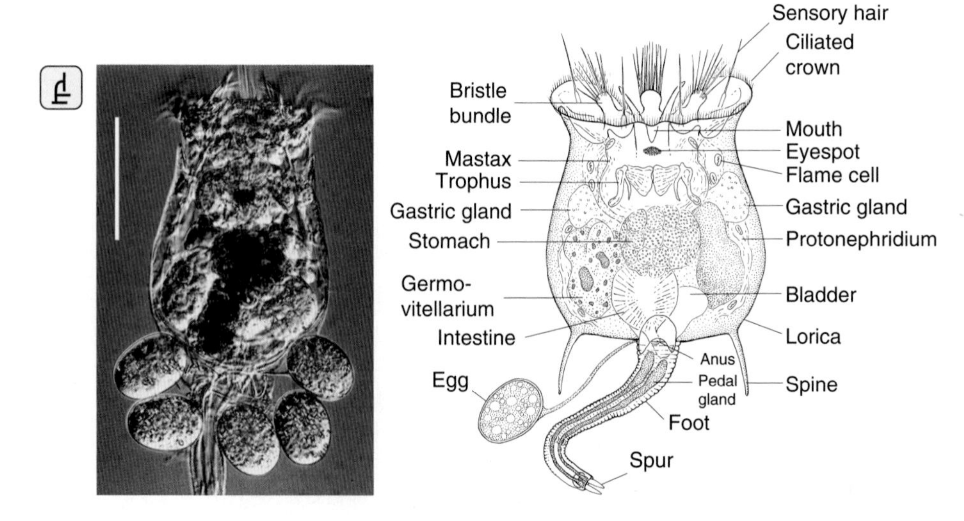

Figure A Living *Brachionus calyciflorus*, a freshwater female rotifer. Thin filaments attach the eggs to the female until they hatch. LM (interference phase contrast), bar = 0.1 mm. [Photograph courtesy of J. J. Gilbert; drawing by L. Meszoly; information from J. J. Gilbert.]

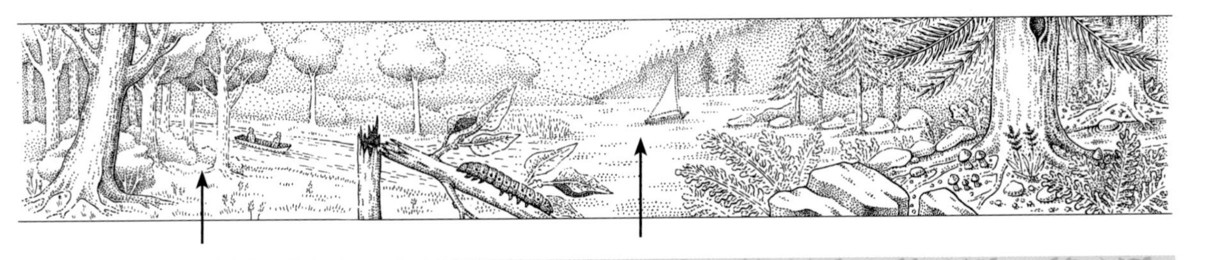

The crown of rotifers that feed on suspended food directs water to a food groove; the ciliated groove conveys particulate food to the mouth. Prey is ground by the trophi as it passes through the mastax into the ciliated stomach, into which gastric glands secrete digestive enzymes. Digestion is mostly extracellular. After passing through the ciliated intestine, solid waste leaves the anus. Dissolved wastes are collected within a pair of protonephridia, convoluted tubules that drain to a bladder, then to the cloaca or to nephridiopores, and out to the environment. Beating cilia of protonephridia maintain water flow and salt balance through the excretory system in freshwater rotifers; marine species lack protonephridia. Striated muscles move rotifer appendages and enable looping over the substrate. Rotifers have smooth muscle also, for more sustained contraction. Dorsal to the mastax is the small brain, from which nerve pairs extend to muscles and organs. Hair cells respond to mechanical and chemical sensation. *Asplanchna brightwelli* has photoreceptors containing membranes layered like cabbage leaves, but, unlike vertebrate photoreceptors, rotifer photoreceptors are not formed from cilia. The *Euchlanis* photoreceptor contains a lens. Rotifers respire through the body surface and lack a blood circulatory system.

Reproduction in rotifers differs among the three classes in this phylum: Seisonidea, Bdelloidea, and Monogononta. Most free-living rotifers are females, which reproduce parthenogenetically. Eggs mature in a germinarium (ovary). Embryos develop in a vitellarium from which they are released from their mothers. The combined site for the germinarium and vitellarium is a germovitellarium. In most, young are born live through the female's cloaca. All species in class Bdelloidea are free living. Adult bdelloid rotifers produce diploid eggs that develop into females; males are unknown and so, consequently, is sexual reproduction. Some, such as *Brachionus* (class Monogononta), carry their eggs on an external filament (Figure A). Monogonont rotifers reproduce either asexually by means of parthenogenesis or sexually, depending on environmental stimuli. Monogonont rotifers produce two kinds of eggs: diploid (containing two sets of chromosomes) and haploid (containing one set). Unfertilized diploid eggs hatch into female adults—40 or more generations may form per year. However, if ponds dry up, haploid eggs form by meiosis. These haploid eggs may be fertilized. If not fertilized, haploid eggs hatch into small, degenerate males, incapable of feeding but able to produce sperm that may fertilize haploid eggs. The males produce two distinct types of sperm: (1) ordinary sperm, which penetrate the female body wall and then fertilize the egg in the ovary, and (2) rod-shaped bodies thought to assist the regular sperm. Heavy-shelled resting, or winter, eggs, the products of monogonont rotifer fertilization, enter developmental arrest—cryptobiosis—until spring. Winter eggs hatch into females that produce female young asexually. A very few monogonont females produce both diploid eggs by means of mitosis and haploid eggs by meiosis. In class Seisonidea, the sexes are separate. *Seison*, a dioecious ectosymbiotroph, and others in this class reproduce exclusively sexually by either hypodermic copulation (males inject sperm into the females pseudocoelom) or cloacal copulation; from the cloaca, sperm move up to the ovary. All rotifers, including sessile species, bear free-swimming young.

Rotifers have no larvae. Females of *Trochosphaera* (class Monogonata) are spheroidal and have an equatorial band of cilia. They resemble the trochophore larvae of polychaete annelids (A-22), some molluscs (A-26) and several other phyla. It has been suggested that rotifers are descended from a "persistent" trochophores, which matured without metamorphosis, but under the larval transfer theory (Box A-i), rotifers were the adult source of trochophore larvae.

Membranes between somatic cells disappear in the epidermis of adult rotifers; such tissues are said to be syncytial. Like those of gastrotrichs, organs of rotifers of each species have a constant number of nuclei. Nuclear constancy and failure of cells to divide in the mature adult may account for the lack of healing and regeneration in rotifers.

For other animals in freshwater communities, rotifers are a major food source. Rotifers also aid in soil decomposition. Rotifer fossils are unknown. Bilateral, ciliated flatworms (A-7) were likely ancestral to rotifers because pharynx, cilia, and protonephridia are similar in these two phyla. Rotifers are informally grouped with nematodes (A-11), kinorhynchs (A-15), acanthocephalans (A-13), nematomorpha (A-12), and perhaps gastrotrichs (A-17) as pseudocoelomates. Although all pseudocoelomates lack a peritoneum-lined body cavity and circulatory and respiratory organs and have specialized pharyngeal regions, a one-way gut (except acanthocephalans, which lack a gut), and an external cuticle, it now appears that, at various times in the past, the pseudocoelomate phyla may have evolved independently from ancestors that were acoelomate.

A-15 Kinorhyncha
(Kinorhynchas)

Greek *kinein*, to move; *rynchos*, snout

GENERA
Campyloderes
Cateria
Centroderes
Echinoderes
Kinorhynchus
Pycnophyes
Semnoderes

Kinorhynchs are free-living marine animals generally 1 mm or smaller, somewhat larger than rotifers and gastrotrichs. A kinorhynch moves ahead by forcing body fluid into its head and thereby everting, anchoring with scalids—spines—on its head in sand grains or mud and then hauling itself forward as it retracts its head. Kinorhynchs do not swim. The name kinorhynch, meaning moveable snout, refers to the kinorhynch head. Microscopic neck plates cover the head when the kinorhynch inverts its protrusible snout. Some kinorhynch species retract both head and neck.

Kinorhynchs have been collected all over the world on seafloors, in estuaries, and on muddy marine beaches between the intertidal zone and to a depth of 5000 m as far north as Greenland, as far south as Antarctica, and in the Black Sea. About 150 species have been described. Most are colorless or yellow brown, perhaps colored by food (diatoms) in their gut. Kinorhynchs feed on bacteria, minute algae, diatoms (Pr-18), and organic debris. No kinorhynchs are known to be symbiotrophs (symbiotrophs), although some seem to be commensal with hydrozoans (A-4), bryozoans (A-29), and sponges (A-3). Kinorhynchs are food for shrimp, snails, and other bottom-feeding marine animals.

A cross section of the body is triangular in one of the two classes (Homalorhagida), oval in the other (Cyclorhagida). Segmented plates of tough cuticle armor the kinorhynch with one curved dorsal plate and two flat ventral plates on each trunk segment. Flexible cuticle lies between the plates. The eversible head is a single segment, as is the neck; there are 13 segments altogether. The dorsal plates and the forked hind end bear moveable, hollow spines and secretory tubules (adhesive tubes). Smaller spines cover the body (Figure A).

The kinorhynch body cavity—a pseudocoelom (false body cavity)—is not really a cavity; rather, it is filled with cellular material, as in rotifers (A-14), nematodes (A-11), acanthocephalans (A-13), nematomorphs (A-12), and loriciferans (A-18). Like nematodes, kinorhynchs lack circular body muscles. Fluid within the pseudocoelom circulates dissolved oxygen, carbon dioxide, and nutrients. This fluid also serves as a hydraulic

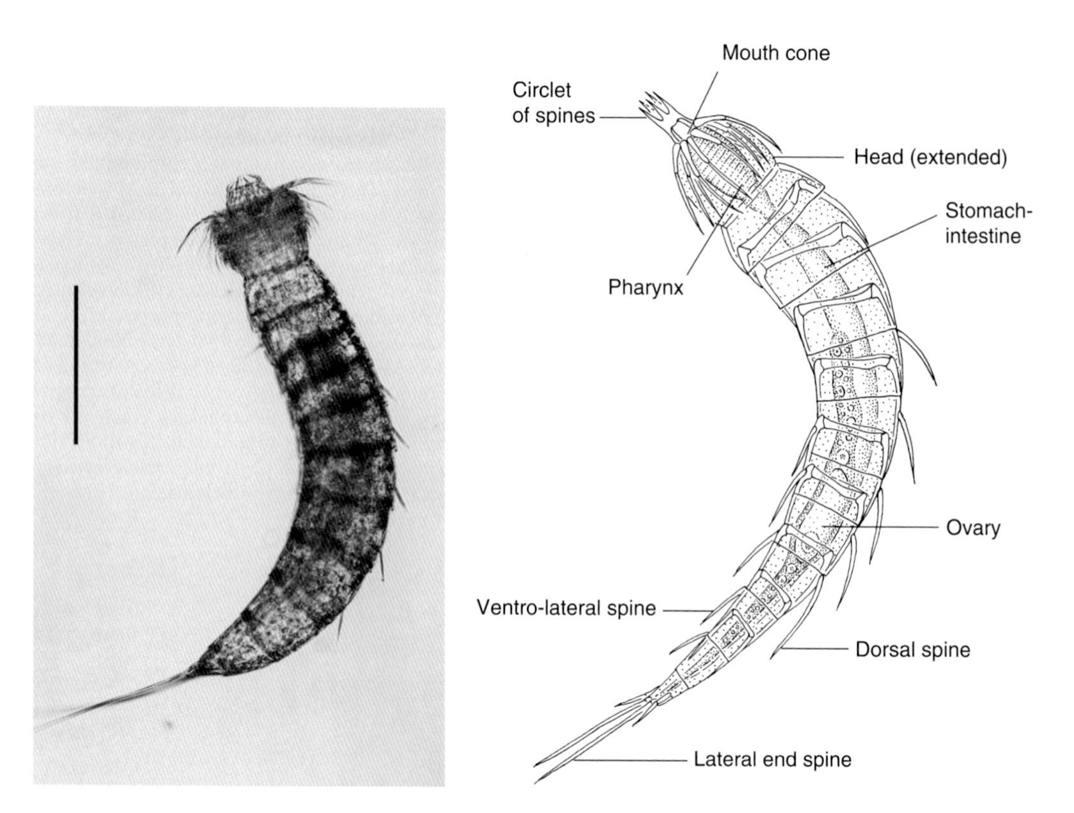

Figure A An adult kinorhynch, *Echinoderes kozloffi*, with its head extended. LM, bar = 0.1 mm. [Photograph courtesy of E. Kozloff; drawing of female *Echinoderes* courtesy of R. P. Higgins.]

skeleton, enabling kinorhynchs to burrow slowly through sediment. The segmented cuticle lacks free cilia. Kinorhynch muscles and nervous system are also segmented.

The oral styles are probably used to grasp microscopic algae. Then the muscular, cuticle-lined pharynx sucks food into the terminal mouth, from which the food moves to the foregut where extracellular digestion is believed to take place. At the posterior end, undigested matter passes out through the anus. Protonephridia, a pair of rudimentary blind-ended tubules with a single undulipodium, collect dissolved waste and discharge it through an excretory pore (sieve plate) in the eleventh segment. This constitutes the kinorhynch's water-balance and excretory system.

The kinorhynch nervous system is composed of a nerve ring circling the pharynx, a ventral double nerve cord and scattered clusters of ganglia ("brain"), and sensory bristles on the trunk. In some species, red pigmented light sensors called ocelli (singular: ocellum) lie behind the mouth cone.

Individual kinorhynchs are either male or female. No obvious external features distinguish the sexes from each other, though there is a minor sexual dimorphism in copulatory spines at the rear of males. Females, such as that illustrated in the adjoining drawing, have paired ovaries; males have paired testes. From these gonads, a gonoduct opens on the terminal segment. The ovary contains two types of nuclei: (1) germinal nuclei that form eggs and (2) nuclei that nourish the eggs. Males and females have not been observed copulating, but females have a seminal receptacle and fertilization is believed to be internal. Males of some species deposit a sperm packet called a spermatophore in females. Juveniles molt their cuticles at least six times as they develop into adults. Kinorhynchs lack a larval stage. A complete life cycle has not yet been observed for any kinorhynch species.

Because ancient kinorhynchs left no fossils, kinorhynch evolution is inferred by comparing living organisms. Kinorhynchs resemble priapulids (especially *Tubiluchus* larvae), rotifers, gastrotrichs, nematodes, and loriciferans. The cuticle of kinorhynchs and priapulids is chemically unique. The kinorhynch spiny cuticle, adhesive tubes, and (in some species) forked posterior end resemble those of gastrotrichs; however, the kinorhynch cuticle lacks external cilia in comparison with the gastrotrich ciliated epithelium. Kinorhynchs are likely descendants of free-living flatworms (A-7) and have been aligned with priapulids (A-16).

A-16 Priapulida
(Priapulids)

Latin *priapulus*, little penis

GENERA

Halicryptus	*Meiopriapulus*	*Priapulus*
Maccabeus	*Priapulopsis*	*Tubiluchus*

Priapulids are short, plump, exclusively marine worms. The priapulid proboscis terminates in a mouth and inverts as it retracts, so the proboscis (presoma) is called an introvert. Spiny papillae called scalids stud the introvert. The trunk consists of 30–100 superficial rings (bands of circular body-wall muscle) covered with spines and warts. *Tubiluchus corallicola* has a long, contractile, postanal appendage, called a tail, in some priapulids that may anchor or retract the trunk (Figures A–C). Species of the genus *Priapulus*, after which the phylum is named, have one or two retractable caudal appendages into which the body cavity extends. *Meiopriapulus*, *Halicryptus*, and *Maccabeus* lack tails. Seventeen species of priapulids have been described in this phylum. All are free living. About half are meiobenthic (seafloor dwellers less than 0.5 mm in length), and half, all cold-water forms, are macrobenthic (longer than 0.5 mm). The smallest priapulid is *Tubiluchus corallicola*, 0.05 cm long, and the largest is a new species of *Halicryptus*, 32 cm long. Body and tail length very much depend on the individual worms' state of contraction.

Priapulid worms have been collected from a variety of depths in the sea from waters as shallow as intertidal pools and as deep as the ocean abyss. *Tubiluchus corallicola* lives in coral sand, silt, and mud in warm shallow waters of the Caribbean, Bermuda, Cyprus, and Fiji. Cold-water priapulids live in the Arctic Ocean, off North America north of Massachusetts and California, in the Baltic and North Seas south to Belgium, around Patagonia, in the Antarctic, and in cold deep waters off Costa Rica. In an Alaskan bay, *Priapulus caudatus* larvae are abundant members of the macrofauna (organisms greater than 0.5 mm), as many as 85 adults per square meter; larvae have been found with densities of 58,000 larvae per square meter. Priapulids' spotty distribution in the seas may be an artifact of collecting.

Priapulids burrow by alternately anchoring their anterior and posterior ends; longitudinal muscles that line the body wall push the body through the sediment. *Maccabeus cirratus* and *M. tentaculatus* are filter feeders, utilizing hollow tentacles; they lack an eversible proboscis. *Maccabeus* builds a permanent tube that is open at both ends and buried in the seafloor. The tube of *Maccabeus* is flimsy, built of secreted material in which plant fragments are encased in a longitudinal pattern. Unlike some other burrowing worms, priapulids do not maintain water currents through their temporary burrows. Most priapulids lie with their mouths flush with the sea bottom, their bodies in sediment.

Smaller priapulid species, the meiofaunal *Meiopriapulus*, for example, feed on bacteria that coats sand grains. Larger priapulids are carnivorous; using the spines that line their mouths, they seize polychaete annelids (A-22) and other priapulids. Both the spine-studded proboscis (Figure B) and the terminal mouth roll inside and out again, passing prey completely into the muscular, toothed pharynx. Food is digested as it passes down the straight intestine, which is surrounded by longitudinal and circular muscle and leads to the rectum. Nutrients may be distributed by body cavity fluid. Solid waste is evacuated from the anus.

The cuticle-covered priapulid body lacks internal and external segmentation. The body wall contains longitudinal and circular muscles. *Priapulus caudatus* has red cells containing the pigment hemerythrin as does nemertine blood and amebocytes that circulate in the body fluid. Hemerythrin is an iron-containing protein that stores or carries oxygen; hemerythrin is also present in brachiopods and polychaete annelids. Priapulid respiration is not well understood. The tail of *Priapulus* may function in gas exchange or chemoreception, but its function has not yet been demonstrated. However, removal of its tail does not kill a priapulid (which regenerates the appendage), so other modes of gas exchange must exist.

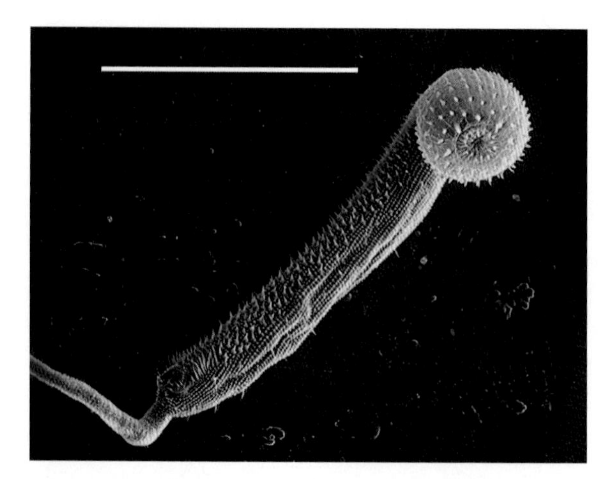

Figure A *Tubiluchus corallicola*, an adult priapulid taken from the surface layer of subtidal algal mats at Castle Harbor, Bermuda. SEM, bar = 0.5 mm. [Courtesy of C. B. Calloway; from *Marine Biology* 31:161–174; 1974.]

Figure B The presoma of *Tubiluchus corallicola*, showing the retractile proboscis everted. SEM, bar = 0.1 mm. [Courtesy of C. B. Calloway; from *Marine Biology* 31:161–174; 1974.]

A circumpharyngeal nerve ring connects to a single, ventral nerve cord, which runs down the body and from which peripheral nerves extend. Raised bumps (papillae or tubercules) on the body surface seem to be sensory. Tiny flower-shaped flosculi (composed of microvilli) of unknown function are scattered on the trunk. The gonads in both sexes are tubular (Figure C). The excretory system consists of a pair of protonephridia, waste-collecting, ciliated tubules. The protonephridia and gonads share ducts that open into nephridiopores, one on each side at the posterior end of the trunk. The body cavity may be a pseudocoel or a coelom.

Because any individual worm is either male or female, priapulids are dioecious ("two houses"). Whether females are similar to males or differ from them in external appearance depends on the species. Some priapulids, such as tube-dwelling *Maccabeus*, probably reproduce parthenogenetically, because only females have been found. Along the Scandinavian coast, spawning takes place in winter; eggs and sperm are shed into the sea. Macrobenthic priapulids such as *Priapulus* and *Halicryptus* are assumed to have external fertilization; meiobenthic priapulids such as *Meiopriapulus*, *Tubiluchus*, and *Maccabeus* are assumed to have internal fertilization. After external fertilization, eggs of most species develop into larvae that are smaller versions of the adults. *Meiopriapulus* is the only genus that lacks a larval stage. Of all the other species, *Tubiluchus* larvae have a longitudinally ridged cuticle, but other larvae have a lorica, a firm cuticle, of plates. Larvae adhere to mud and other sediment with adhesive from their hollow toes and are never free swimming. A series of larvae (morphologically indistinct stages) molt as they attain adult size. Adults also shed their partly chitinous cuticle. *Meiopriapulus fijiensis* females brood their young embryos and then release postembryonic young from the urogenital pore; *Meiopriapulus* lacks larvae—these juveniles develop directly into adulthood.

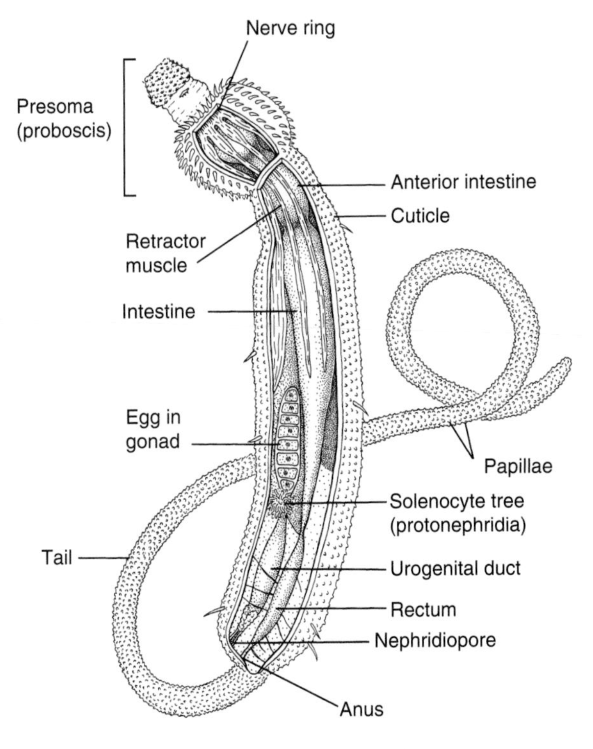

Figure C Morphology of an adult female *Tubiluchus corallicola*, a minute priapulid of the meiobenthos. [Drawing by L. Meszoly; information from C. B. Calloway.]

Middle Cambrian fossil priapulids abound in the Burgess shale, deposited about 500 mya. When polychaete annelids (A-22) with jaws evolved during the Ordovician, about 440–500 mya, polychaetes displaced priapulids from their role as abundant marine carnivores.

The relationship of priapulids to other marine animal phyla continues to be uncertain. Until the early 1900s, priapulids were placed with echiurans (A-24) and sipunculid worms (A-23) in phylum Gephyrea. In 1961, when the spacious priapulid body cavity was discovered to be seemingly lined with mesentery and thus considered a coelom, some researchers moved priapulids from pseudocoelomate to protostome coelomate status. Investigation of priapulid embryonic development is needed to clarify whether the priapulid body cavity is indeed a coelom, because other zoologists group priapulids with pseudocoelomate animals; that is, animals having a body cavity not lined with mesoderm. Priapulids resemble nematomorphs (A-12), nematodes (A-11), and juveniles of kinorhynchs (A-15) and loriciferans (A-18) in that they molt their cuticles from time to time. Also like nematomorphs, priapulids are dioecious and have a cylindrical unsegmented body without internal septa (cross walls). In 1980, newly revealed morphological features of priapulids, nematomorphs, kinorhynchs, and loriciferans led V. V. Malakhov, a Russian invertebrate zoologist, to propose that these four animals be included in a newly created phylum named Cephalorhyncha. Priapulids may also be related to rotifers (A-14) or to acanthocephalans (A-13).

A-17 Gastrotricha
(Gastrotrichs)

Greek *gaster*, stomach; *thrix*, hair

GENERA
Acanthodasys
Chaetonotus
Dactylopodola
Lepidodermella
Macrodasys
Tetranchyroderma
Turbanella
Urodasys

The phylum Gastrotricha derives its name from the cilia that cover the ventral side, the term from Greek meaning "hairy stomach." Elsewhere on the animal, bristles, spines, or scales ornament the surface, but not the underside. The miniscule, transparent body with lobed head ranges from less than 0.1 to 3.5 mm in length. Gastrotrichs make up the meiofauna, communities of animals that live among and within the spaces of particles, measuring about 0.040 mm, in marine and freshwater substrates. Many species also live close to and are associated with submerged vegetation. Gastrotrichs can attach to objects by means of secretion of adhesive materials from duo-glands, much like those found in Platyhelminthes (A-7). When not attached gastrotrichs can freely swim about.

About 400 species are known and these are distributed between two large orders. The phylum occurs in both marine and freshwater habitats. Species are known from the Arctic to southern South America. In marine habitats, gastrotrichs are truly meiobenthic and are plentiful in tidal and subtidal sands. Many marine species are known from coral reefs in tropical regions. Several marine species are Holarctic in their distribution. Freshwater species live chiefly among vegetation and are plentiful in quiet ponds, especially those with thick vegetation.

Gastrotrichs exchange oxygen and carbon dioxide by diffusion with the surrounding water. Circulatory and respiratory organs are absent. A thin unsegmented cuticle of lipoprotein and nonchitinous polysaccharide covers the entire body. Like rotifers (A-14), gastrotrichs do not shed their cuticles. Undulipodia direct water currents that bear organic debris, bacteria, algae, foraminiferans (Pr-3), and diatoms (Pr-18) to the mouth. Gastrotrichs have a complex pharyngeal foregut that includes a triradiate organization of underlying muscle and a monociliated epithelium. The muscular pharynx pumps food into the stomach—intestine. Food is digested intracellularly. Undigested material passes out through the anus. In those species in which gametes are also passed through the same opening, it is referred to as a cloaca.

Most gastrotrichs, especially freshwater species, have a pair of protonephridia with a midventral excretory pore. Marine species lack protonephridia. Gastrotrichs swim short distances, steering by means of contractions of longitudinal muscles. Muscles that encircle the body move the bristles and adhesive tubes (Figure A). Fused ganglia, forming the brain, surround the pharynx, with a pair of longitudinal and lateral nerve cords extending posteriorly. On the head and trunk are sensory bristles and tufts of sensory undulipodia. Some forms have reddish spots near the brain that may represent photoreceptors.

Many individual gastrotrichs are simultaneous hermaphrodites, producing both eggs and sperm, particularly marine gastrotrichs. Some species are sequential hermaphrodites, switching sexes during different periods. In *Dactylopodola* and *Urodasys*, sperm packets called spermatophores are transferred from individuals that behave as males to individuals behaving as females. The end of the sperm duct serves as a penis for sperm transfer. Fertilization is internal. Freshwater gastrotrich populations including *Lepidodermella* are entirely female; they reproduce parthenogenetically. The diploid eggs develop into females of the next generation in the absence of sperm. However, rodlike sperm are found in some *L. squammata* individuals that also contain eggs. Thus a hermaphroditic trait exists in the species suggesting former sexual reproduction. Female gastrotrichs lay from one to five large eggs in a lifetime, depositing eggs on algae, debris, or pebbles (Figure B). When in contact with water, an eggshell forms. Freshwater gastrotrichs produce two types of eggs. One type, usually laid at the end of growing season, typically falls in northern climates, may remain dormant for up to 2 years, a form of diapause, which may also require exposure to drying, freezing, or some other harsh environmental condition, before germinating. The second egg type is a thin-walled example that develops immediately after deposition. Gastrotrichs develop directly and hatch at almost adult size.

Because gastrotrichs scavenge dead bacteria and plankton from beaches and estuaries, they contribute to the entire ecology of the tidal zone. The high number and diversity of gastrotrichs in tidal substrates is still being discovered, and rich faunas are continuously observed. Gastrotrichs are themselves preyed upon by amebas (Pr-2), hydras (A-4), free-living flatworms (A-7), insect larvae and crustaceans (A-21), and annelids (A-23).

Adult gastrotrichs lack a body cavity, and various studies claim gastrotrichs are acoelomates. Others maintain that gastrotrichs are pseudocoelomate or even coelomate. Thus there is no general consensus on the condition of the body cavity or lack thereof. The gastrotrich body form, musculature, and protonephridia are like those of rotifers, which are pseudocoelomates. Gastrotrichs obviously lack the distinctive features of rotifers, the mastax and corona or crown. Gastrotrichs share with many other pseudocoelomates, an ornamented cuticle, tripartite muscular pharynx, eutely, adhesive glands, and a paucity of circular muscle. The monociliated epithelium in the gastrotrich pharynx is reminiscent of the Gnathostomulida (A-6) and other simple and presumed primitive multicellular phyla, thus implying a primitive position.

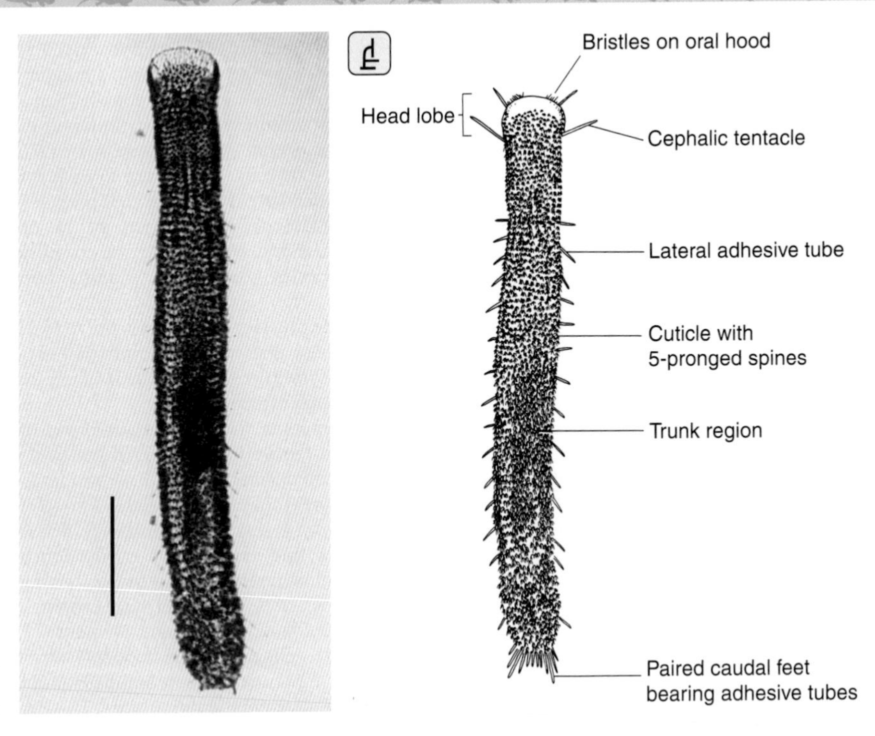

Figure A Living adult *Tetranchyroderma* from a New England beach. Adhesive tubes secrete glue that temporarily anchors it to sand in the intertidal zone. LM, bar = 0.1 mm. [Photograph courtesy of W. Hummon; drawing by I. Atema; information from W. Hummon.]

Figure B *Acanthodasys*, a marine gastrotrich and simultaneous hermaphrodite. After fertilized eggs are laid through a temporary opening in the body wall, the wall heals. LM, bar = 0.25 mm. [Photograph by David Scharf/ Peter Arnold, Inc.; information from W. Hummon.]

A-18 Loricifera
(Loriciferans)

Latin *loricus*, corset, girdle; *fero*, carry, bear

GENERA
Nanaloricus
Pliciloricus
Rugiloricus

A loriciferan is a microscopic marine animal with a mouth cone—a flexible anterior mouth tube—a head carrying club-shaped and clawlike spines, a neck (thorax), and a girdle of plates called a lorica that covers the abdomen. The head and neck can be inverted into the abdomen. Because the mouth is terminal, the spiny invertible head is, by definition, an introvert. The neck telescopes down to half its length when a loriciferan is disturbed. Heavily sculptured ventral lorica plates or longitudinal folds close over the retracted animal. Longitudinal lorica plates are sculptured and patterned with pores. The head and anterior neck are armed with spines called scalids, which fold together like the ribs of an umbrella. Adults are from about 100 to 400 μm in length, ranking in size with small rotifers.

The phylum was first described in 1983, and about 25 species are currently known. Preliminary studies suggest at least 100 species including many as yet undescribed. *Nanaloricus* (Latin *nana*, dwarf) has been found in shelly gravel dredged near the Atlantic shore in North Carolina and Florida, the Gulf of Mexico in the United States, and the Azores Islands. Other loriciferans have been collected from the Arctic Ocean, off Denmark, in the Mediterranean, and in the Coral Sea from depths of 10–480 m. The animals are thought to be a cosmopolitan part of the interstitial fauna—living between particles of coarse sand.

What loriciferans eat is still unknown. Adult loriciferans are sedentary and may be either ectosymbiotrophs (ectobionts), living attached to other animals, or attached to gravel grains. Larvae are probably free living and travel over the ocean floor by using two or three ventral locomotory spines (Figure A). The leaflike toes of *Nanaloricus* bear spines and rotate in ball-and-socket joints. These toes serve as paddles for swimming and enable larvae to change direction. Other species have straight toes. Glands that open into each hollow toe probably secrete adhesive. The mouth cone of the larva retracts into the introvert and, in turn, can be inverted into the neck (thorax). Unlike the adult mouth cone, the larval mouth cone lacks oral stylets in some species. A disturbed larva retracts its many-plated neck (thorax) like accordian bellows.

Researchers are undecided whether the body cavity of loriciferans is a pseudocoelom or a coelom—an understanding of the loriciferan body cavity during embryonic development is needed to resolve this question. The fluid in the body cavity serves both circulatory and respiratory functions.

The flexible head, called an introvert, everts, and has a terminal mouth with extrusible stylets (Figures B and C). The mouth opens into a buccal canal into which salivary glands open in turn. The buccal canal is somewhat folded within the mouth cone. The mouth tube surrounds the extruded buccal canal; both can be extruded through the mouth (Figure C). Mouth, buccal canal, and muscular pharyngeal bulb retract in some species. A short esophagus passes into a midgut, followed by a short rectum. Both anus and saclike gonads open at the posterior end. One pair of protonephridia constitutes the excretory system.

The loriciferan nervous system consists of a large brain within the introvert. Nerves run to each scalid. A large ventral ganglion innervates the thorax and probably continues into the abdomen. Flosculi, each a rosette of five microvilli or a single papilla, are found on the posterior dorsal side of the abdomen of some species; a sensory function has been proposed for flosculi.

Like the muscles of kinorhynchs (A-15), at least some of the large muscles that retract the introvert are cross-striated. The excretory system comprises a pair of protonephridia that may open through nephridiopores.

Differences in the most anterior rows of spines distinguish male from female loriciferans. In males, these spines are presumed to be olfactory. Males have two large dorsal testes (containing sperm) that fill the lorica. Although females have paired ovaries, only a single enormous egg develops. Fertilization is likely to be internal, judging by *Nanaloricus.* The female may have a seminal receptacle near the lorica hinge.

Each in a series of larval stages molts, and then secretes a fresh cuticle. Young loriciferans are named Higgins larvae, after Dr. Robert P. Higgins, expert zoologist, formerly at the National Museum of Natural History, Smithsonian Institution. From his knowledge of animal communities, Dr. Higgins had predicted the existence of organisms similar to loriciferans before their discovery.

Loriciferans are unique in their combination of characters. The mouth cone with stylets and cross-striated introvert muscles resembles those of kinorhynchs (A-15). Similarities between the flexible mouth cone with stylets of adult loriciferans and that of the tardigrade *Diphascon* (A-27) are examples of convergence. The unciliated cuticle of loriciferans resembles that of kinorhynchs and priapulids (A-16); the rotifer cuticle is intracellular (A-14). Larvae of loriciferans and of nematomorphs (A-12) bear hook-shaped ventral spines. If loriciferan glands prove to be adhesive, they may resemble the adhesive glands of kinorhynchs. Sensory organs (flosculi) are similar in loriciferans, priapulids (especially the larvae of *Tubiluchus*), and kinorhynchs. No fossil record for loriciferans has been reported. Loriciferans have so many characteristics in common with kinorhynchs, nematomorphs, and priapulids that some zoologists favor placing all four in one phylum. Until more details of loriciferan biology are known, their classification status will remain ambiguous.

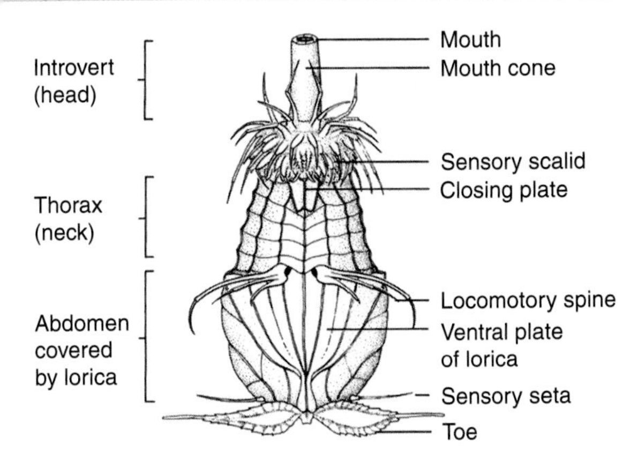

Figure A Ventral view of larva of *Nanaloricus mysticus*, a loriciferan. The toes are swimming appendages. [Drawing by L. Meszoly; information from R. P. Higgins.]

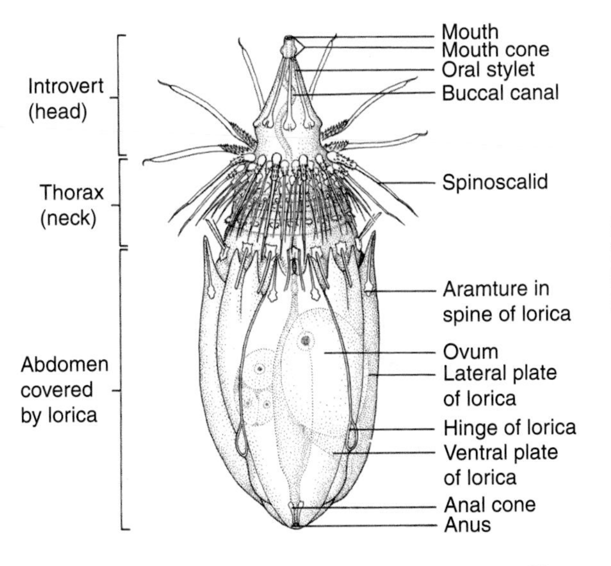

Figure B *Nanaloricus mysticus*, adult female loriciferan. The head and neck can be inverted into the abdomen. [Drawing by L. Meszoly; information from R. P. Higgins.]

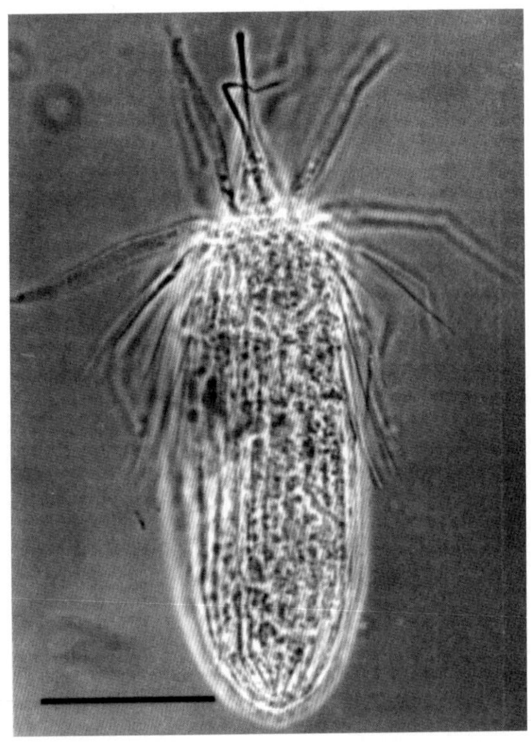

Figure C *Pliciloricus enigmaticus.* The mouth cone with its long mouth tube extended is visible centered on the head. LM, bar = 100 μm. [Courtesy of R. P. Higgins, Smithsonian Institution.]

A-19 Entoprocta
(Entoprocts)

Greek *entos*, inside; *proktos*, anus

GENERA
Barentsia
Loxosoma
Loxosomella
Pedicellina
Pedicellinopsis
Urnatella

Entoprocts are tiny, transparent animals, primarily marine. Most are sessile, living in colonies permanently or firmly attached by stalks, horizontal stolons, and basal disks to rocks, pilings, shells, algae, or other animals. A crescent of ciliated tentacles that contract and fold over the mouth and anus is located at the free end of the entoproct (Figures A and B). An individual entoproct has from 6 to 36 tentacles, depending on its age and species. A conspicuous muscle bulb at the stalk base permits the flicking movements of individual entoprocts (Figure B).The lightly tinted, cup-shaped body is called a calyx. Some entoprocts look and creep like hydrozoan cnidarians (A-4) or like those ectoprocts (A-29) that can creep; but, unlike either ectoprocts or cnidarians, entoprocts fold the tentacles of their cup-shaped bodies when disturbed. Entoprocts do not withdraw their tentacles into their bodies. The whole entoproct animal, tentacles, calyx, stalk, may extend to 10 mm.

A little over 150 species are known. Many form large colonies or "animal mats" on living and nonliving substrates. A few entoproct species are solitary, such as *Loxosomella davenporth*, one of the few that is mobile as an adult. *Loxosomella* somersaults with its tentacles over its basal disk. Marine entoprocts, making up nearly all species, are widely distributed along the seacoasts of every continent including Antarctica. A few species are found both in Antarctic and Arctic waters. For years only a single freshwater species was known, *Urnatella gracilis*, which has been reported from every continent except Antarctica. Recently a second species was described from rivers in central Thailand.

Using their tentacles and uncoordinated beats of the lateral cilia on the tentacles, entoprocts filter food suspended in the water (Figure B). Entoproct tentacles set up a feeding current that enters between the tentacles and exits at the top of the body. Diatoms (Pr-18), desmids (Pr-32), other plankton, and detritus are trapped in mucus on the extended tentacles, swept along the ciliated tracts into their mouths. Digestion occurs in a U-shaped, complete digestive tract. A recently discovered has in addition to tentacles, nematocyst-like structures that extend from the body.

The body cavity of entoprocts may be a pseudocoelom. Gelatinous material containing cells fills the entoproct body

Figure A A living laboratory culture of *Pedicellina australis*, with tentacles folded. A marine colonial entoproct, part of a fixed colony; from Falkland Islands. LM, bar = 1 mm. [Courtesy of P. H. Emschermann.]

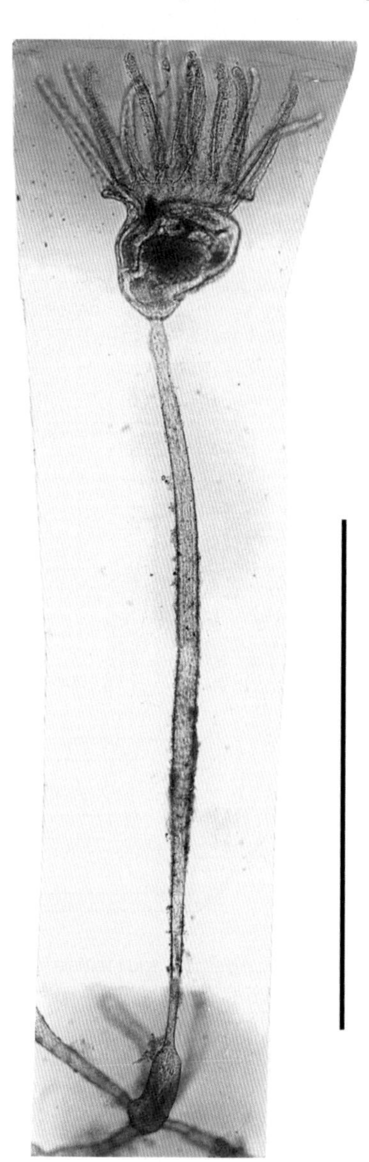

Figure B An individual entoproct, *Barentsia matsushimana*. Rows of cilia are visible on the extended tentacles. LM, bar = 1 mm. [Courtesy of P. H. Emschermann.]

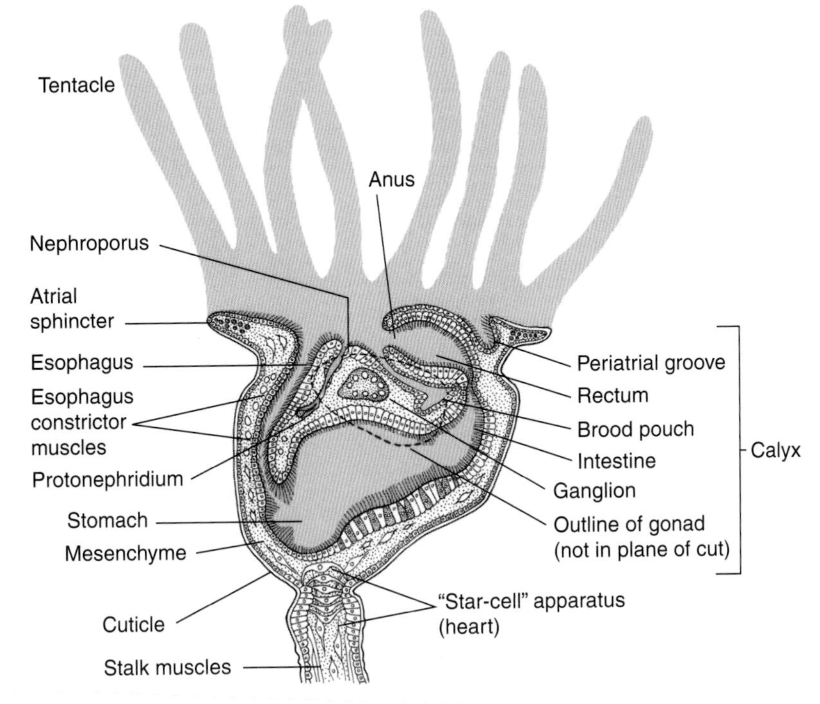

Figure C *Barentsia matsushimana*. A vertical section shows digestive, nervous, excretory, and muscle systems within the cup-shaped calyx. [Drawing by L. Meszoly; information from P. H. Emschermann.]

cavity and extends into tentacles. There is no heart or blood vascular system. The star cell apparatus contains contractile cells and may be involved with moving nutrients through the stolon to the growth areas. Uric acid and guanine are secreted into the stomach cavity and out of the body (Figure C). Additional dissolved waste is collected in ciliated flame cells (protonephridia) and discharged to the exterior through nephridiopores located on the calyx surface.

Entoprocts have a weak muscular system, but both striated and smooth fibers occur. The calyx contains a circular sphincter to control its size and there are small bundles of fibers along the stalk that cause the characteristic "nodding" behavior. Like the muscle system, the nervous system is simple consisting of a ganglion lying above the digestive system with nerves extending to the tentacles, stalk, and body. Specialized cells around the body are thought to be involved with chemical, light, or vibration detection.

Entoprocts reproduce both sexually and asexually. In the absence of sexual reproduction, colonies commonly give rise to new individuals by budding. Many of the sexual entoprocts are dioecious, whereas only a few are hermaphroditic. Both simultaneous and sequential hermaphroditism have been observed. Female gametes are extruded through a gonopore to a brood pouch. The method of fertilization is unknown but sperm with tails have been observed. Embryos develop into free-swimming larvae that resemble examples produced by the Ectoprocta (A-29). Following a brief swimming phase, the larvae settle to a substrate—secrete a bonding material—and metamorphose into adult entoprocts. Metamorphosis involves the rearrangement of organs and a general reorientation of the body.

The relationship of entoprocts with other sessile, tentacle-bearing invertebrates remains unclear. Arguments continue to support a union of the Entoprocta with the Ectoprocta, under the name Bryozoa. Much of the debate centers on the nature of the body cavity, or its absence, and developmental features. Molecular evidence suggests that developmentally, entoprocts are protostomes and produce spiralian larva. However, their affinity with other lophophorate phyla and reasonably well-studied, established protostomes (for example, Annelida, A-22) remains unclear.

A-20 Chelicerata

(Chelicerates)

Greek *cheli*, claw

GENERA

Acanthoscurria	*Ixodes*	*Lycosa*
Argiope	*Latrodectus*	*Nymphon*
Carcinoscorpinus	*Leiobunum*	*Psoroptes*
Chelifer	*Limulus*	*Scorpio*
Dermacentor	*Loxosceles*	*Trachypheus*

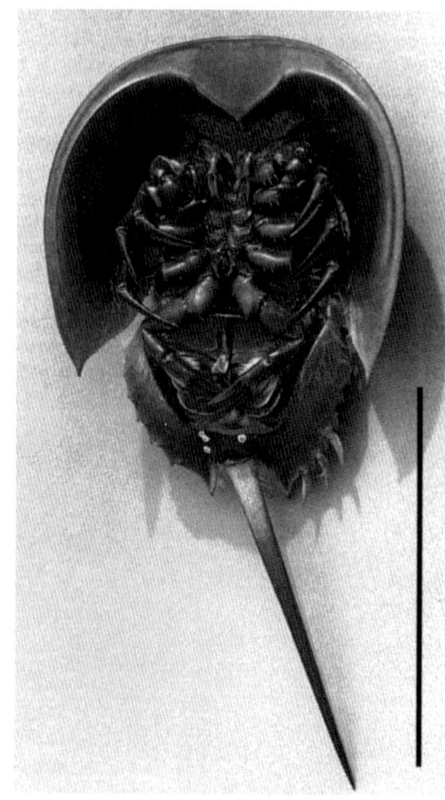

The chelicerates, which number about 93,000 species, include three classes: the horseshoe crabs (class Merostomata); sea spiders (class Pycnogonida); and spiders, scorpions, mites, ticks, chiggers, and harvestmen, also called daddy-longlegs (class Arachnida). Members of phylum Chelicerata have segmented bodies, jointed appendages, and a chitinous exoskeleton in common with other arthropods (crustaceans and insects, A-21). Chelicerates lack the antennae of crustaceans and insects; chelicerates also lack the mandibles (biting tips of the jaws for grinding and chewing food) formed from the distal part of appendages that are beside the mouth of crustaceans and insects. Chelicerae are the clawed, most anterior pair of jointed appendages characteristic of all chelicerates. Although some place chelicerates with insects, centipedes, crustaceans, and other joint-footed animals in one vast phylum Arthropoda, other zoologists, as we do, prefer to recognize chelicerates as a phylum, a unique and cohesive lineage. Chelicerates, mandibulates, and crustaceans' consistent distinguishing features imply that they are three independent evolutionary lineages.

The chelicerates are coelomate protostomes. The head and thorax of chelicerates are fused into a single unit called the cephalothorax or prosoma (Latin *pro*, forward; Greek *soma*, body). The prosoma bears a pair of feeding appendages called chelicerae, the most anterior appendages; posterior to the chelicerae is a pair of pedipalps, the first pair of walking legs and four or more additional pairs of walking legs. The abdomen, or opisthosoma (Greek *opisthen*, behind), is distinct from the cephalothorax. Abdominal appendages are variously used for gas exchange (for example, the book gills—named for their resemblance to pages of an open book—of the horseshoe crab and book lungs of some arachnids), for reproduction, or for silk extrusion, depending on the species.

Members of class Pycnogonida, sea spiders, include about a thousand species of marine chelicerates that resemble long-legged, slow, slender spiders. Because *Nymphon* and other sea spiders lack excretory and respiratory organs, the sea spiders are not true spiders. Sea spiders have been collected in habitats ranging from oceans 6800 m deep to shallow seas, from the Arctic to the Antarctic. Many sea spider young are commensals or symbiotrophs of medusae (A-4) and echinoderms (A-34). Most sea spider adults are free living and dull in color. Some deep-sea species are red. The bodies of sea spiders range from less than 1 to more than 10 cm in length, the largest with legs spanning almost 80 cm. The largest are deep-sea and polar species.

The sea spider cephalothorax has a muscular proboscis with a terminal mouth flanked by paired appendages: one pair of chelicerae; one pair of pedipalps; and one pair of ovigers—appendages often absent in females. Sea spiders clean their bodies with the ovigers and hold eggs on them. The nervous system consists of four eyes on a short projection above a brain in the cephalothorax, ventral nerve cords with ganglia, and sensory hairs. Four pairs of walking legs are usually clawed. Carnivory is the rule; through the terminal mouth on the proboscis, the pharynx pumps soft sponges (A-3), sea anemones and other cnidarians (A-4), and ectoprocts (A-29) into the extensive digestive tract. Compared with that of the horseshoe crab, the abdomen is very small relative to the cephalothorax; the digestive and reproductive systems fill the stumpy abdomen as well as much of the four to seven pairs of walking legs that attach to the cephalothorax. The sea spiders' large surface area may obviate the need for organs of excretion and gas exchange.

Reproduction in sea spiders is modified by the body form; because of the small cephalothorax, reproductive organs extend into the legs, where eggs are produced in the ovaries of females and sperm are produced in the testes of males. Gonads open through pores on the legs. Males gather fertilized eggs, produce adhesive that holds the amassed eggs on the ovigers, and shelter eggs until larvae hatch from them. Larvae become necrotrophic on colonial hydrozoans (A-4), stay on the ovigers of the male, or leave. After a series of molts, the larvae attain adult form. Fossil sea spiders are known from the Cambrian period.

Class Merostomata (merostomates) comprises three marine genera: the genus *Limulus* consisting of a single species, *L. polyphemus*—the horseshoe crab (Figure A); and two additional genera from the shallow seas of the Indo-Pacific Ocean—*Trachypheus* (*Tachypleus*) and *Carcinoscorpinus* (*Carcinoscorpius*).

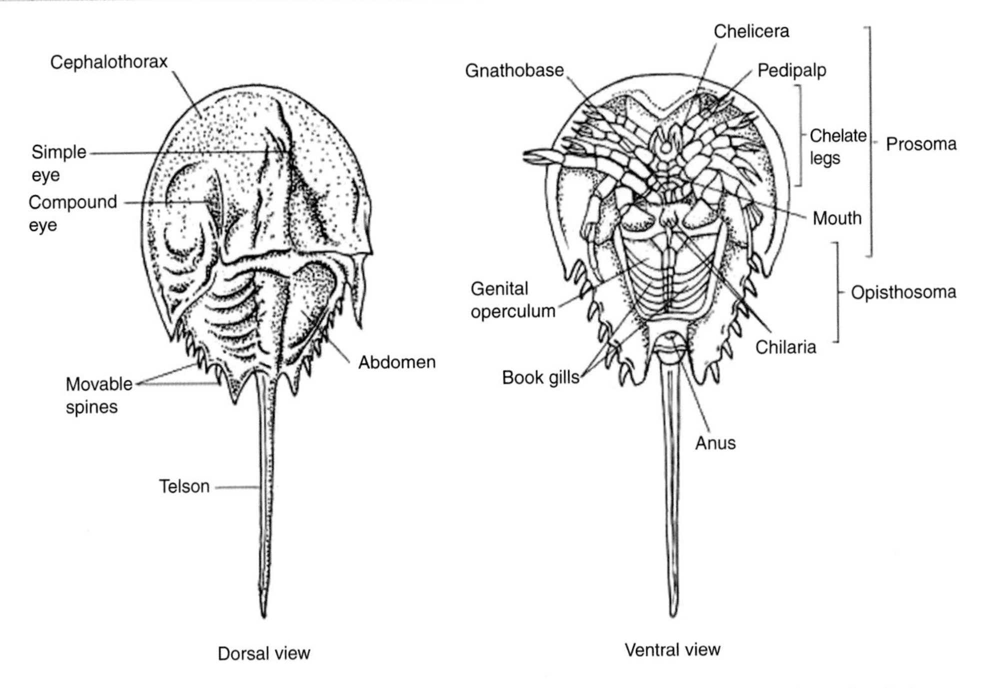

Figure A Beachcombers often encounter *Limulus polyphemus* along beaches from Nova Scotia to the Yucatan Peninsula in Mexico. In spring mating season, scores of the harmless horseshoe crabs become stranded as mature females come out of the shallows to lay eggs. The smaller male like the one in this photograph hitchhikes clasped on the female's abdomen and deposits sperm as the female drags him over the sandy nest. This adult male from the Florida Keys bears the clawed appendages that characterize chelicerates. Bar = 100 mm. [Photograph courtesy of C. N. Shuster, Jr.].

The horseshoe crab carapace is shaped like the iron shoe of a horse. *Limulus* lives in shallow waters of the Gulf of Mexico and the eastern seaboard from Nova Scotia to the Yucatan. *Limulus*, with its flexible, hornlike shell of chitin and protein, is not a true crab; true crabs are crustaceans and have a brittle, hard shell. *Limulus*, a familiar member of the shore community, reaches a length of 60 cm. The remaining three merostomate species are found in the Southeast Asian sea from India to Korea. Horseshoe crabs consume clams, other small animals, and plant material encountered in mud. The last of the horseshoe crabs five pairs of walking legs is specialized for paddling, for cleaning its gills, and for shoving mud as it burrows; these legs lack claws. The other four pairs of legs are clawed. *Limulus* uses not only its walking legs, but also its pedipalps for walking. The anteriormost appendages are a pair of small chelicerae that gather food, pass it to spines on the leg bases, which grind the food, and pass it to the mouth. The mouth is located between the pedipalps. *Limulus* regulates its internal ion concentration by means of excretory glands and tubules; a bladder passes urine through an excretory pore at the base of the last pair of legs.

The horseshoe crab brings its formidable tail spine into play when it rights its flattened, leathery body or shoves forward. Vast numbers of horseshoe crabs migrate to the shallows for nocturnal spawning. Females excavate a depression into which they lay about 300 eggs. Gonads open through genital pores that are under a flap on the abdomen. Fertilization is external; the smaller males clasp egg-laying females with modified pedipalps. The fertilized eggs hatch as swimming *trilobite* larvae, so called because they resemble the extinct trilobites, once common in Cambrian seas and now found as beautiful fossils in Utah, Nevada, Ohio, and worldwide. The name trilobite is applied to these larvae because the dorsal side of the cephalothorax has three longitudinal lobes. Horseshoe crab larvae molt more than a dozen times and attain maturity in about 3–10 years. Small horseshoe crabs can swim upside down. Larger crabs are bottom dwellers. Intertidal-dwelling horseshoe crabs tend to have a green tail because *Microcoleus* and *Spirulina* (B-5), *Thiocapsa* (B-3), and other photosynthetic bacteria adhere to it. If the crab happens to drag its tail over a suitable habitat, these bacteria can be seeded onto the substrate, leading to the formation of a microbial mat.

The *Limulus* body cavity is a hemocoel (blood-filled sinus in tissue); the coelom is very small. The *Limulus* circulatory pattern follows the general circulatory system of arthropods with variations in mode of gas exchange. A long tubular heart with ostia (perforations) pumps blood through an extensive arterial system of closed vessels into ventral sinuses from which blood flows into the book gills. The book gills contact seawater, underneath five pairs of gill flaps—abdominal appendages modified for gas exchange. Gill activity pumps blood back to the sinus around the heart. As the heart expands, blood is drawn in through ostia. *Limulus*' blue blood contains hemocyanin, a respiratory pigment, and amebocytes that facilitate clotting. A coagulating agent in *Limulus* blood is used to diagnose human bacterial infections. Bacteria produce endotoxins; when the horseshoe crab coagulating agent (a lysate) reacts with endotoxins, clotting of the horseshoe crab blood is triggered. The presence of fever-producing substances called pyrogens in intravenously administered medicines is also detected with the use of this same clotting reaction of the *Limulus* lysate. Another substance extracted from horseshoe crabs, lobster, shrimp, and crab shells is chitin, a polysaccharide that can be spun into fibers for surgical sutures and implantable drug containers because chitin does not elicit allergic reactions in humans.

The horseshoe crab nervous system includes a brain that encircles the esophagus, a ganglionated ventral nerve cord, a pair of compound eyes and a median simple eye, and a chemosensory organ (frontal organ). Merostomates are the only chelicerates with compound eyes. *Limulus* eyes probably sense movement but do not form images. Research studies of *Limulus* compound eyes have contributed to our knowledge of the physiology of vision.

Limulus has been virtually unchanged since it dwelt in shallow Silurian seas, about 425 mya. Along with horseshoe crabs, the Paleozoic seas were inhabited by another chelicerate group, eurypterids (sea scorpions) as well as trilobites, the arthropods of earlier lineages. Fossil merostomates, diverse and abundant in Paleozoic oceans, may be ancestors of class Arachnida.

By far the majority, more than 92,000 species, of chelicerates belong to the class Arachnida, which is the only air-breathing class of Chelicerata. Arachnids take their name from Arachne of Greek mythology, a girl turned into a spider for challenging the goddess Athena to a weaving competition. Spiders (*Lycosa*, the wolf spider; *Argiope*, the orb weaver spider; *Loxosceles*, the brown recluse spider; *Latrodectus*, the black widow spider) with 38,000 species are the second largest group of this class, which also contains mites (many, such as *Psoroptes*, or mange mite, are symbiotrophs), ticks (also symbiotrophs, such as *Ixodes*, which transmits Lyme disease), harvestmen, scorpions, pseudoscorpions, camel spiders (Solifugae), and a few other exotic minor groups. Many spiders engage in elaborate courtship. Like *Limulus*, the arachnid body consists of prosoma (cephalothorax) and abdomen. In spiders, the abdomen is usually distinct from the prosoma, linked by a narrow pedicle, whereas in mites and ticks, the abdomen is fused to the prosoma. Scorpion bodies have 18 segments: 6 segments in the prosoma, 7 segments of anterior abdomen, and 5 segments of posterior abdomen. Chelicerae, the most anterior appendage pair, tear food and function as fangs. The second appendages—pedipalps—kill prey, handle food, or have sensory or reproductive functions. The final four appendage pairs on the prosoma are walking legs. Some arachnid species have an abdomen with internal book lungs (similar to horseshoe crab book gills) that open to the environment by closeable orifices called spiracles. Instead of book lungs, tissues of some arachnids are linked to the outside by a network of branched tubing called tracheae; body movements pump air through the tracheae. Most spiders are terrestrial, and all are air breathers; a few freshwater spiders capture bubbles from which they breathe air. Arachnid circulatory system is open, with hemolymph pumped through tissue sinuses.

Spiders secrete silk proteins from glands within their abdomen; silk extruded through tiny orifices in abdominal appendages, and spinnerets at the posterior end of the spider's abdomen solidifies in air. The spider uses silk to build nests, to snare insects, to weave sacks for eggs and spiderlings, to wrap food gift packages during courtship, and to spin threads for sailing on the winds. Spider silk is very strong and has been used as uniform, tough threads in bombsights of airplanes. Most arachnids are carnivorous predators that dine on liquids. A spider bites prey and paralyzes it with poison from its chelicerae, predigests the prey's insides by extruding enzymes from the arachnid digestive tract into the prey, and pumps in the liquid meal with the muscular action of its stomach.

Scorpions are the only arachnid group that has abdominal stingers, producing toxic *venom*. The bite of a black widow spider injects toxic, but seldom deadly venom. Arachnids feed mostly on insects; many arachnids are predators of agricultural insect pests.

Ticks and mites number more than 50,000 species. Certain mite species inhabit the dust in our houses and the follicles of our eyelashes. These chelicerates live in myriad terrestrial, marine, and freshwater habitats. Some species are *vectors* of Lyme disease and encephalitis. Other mites and ticks are pests of domestic animals. Mites and ticks feed on vertebrate blood, invertebrates, fungi, and plants.

The chelicerate exoskeleton and its remarkable anatomical derivations, the small body size of many species, and the extraordinary diversity of form and habitat contribute to the success of chelicerates as a phylum. Scorpions—present already in the Silurian as aquatic forms and in the Carboniferous as definitely terrestrial ones—are traditionally considered the most primitive arachnids. An alternative phylogeny places them as derived arachnid group, probably relatives of harvestmen. The affiliations between chelicerates, crustaceans and insects (A-21) are unclear. All arthropods may have derived from annelid ancestry, but this theory is currently contested.

A-21 Mandibulata

(Mandibulates, mandibulate arthropods)

Latin *mandere*, to chew; *mandibulum*, a jaw

GENERA

Anopheles	*Balanus*	*Cancer*	*Coronula*
Armadillium	*Blatta*	*Cephalobaena*	*Daphnia*
Armillifera	*Bombyx*	*Chironomus*	*Drosophila*
Artemia	*Calanus*	*Clibanarius*	*Euphausia*
Apis	*Cambarus*	*Cryptocercus*	*Gigantocypris*

Of the more than 30 million species of animals estimated to exist (only about 10 million of which have been formally described and named), phylum Mandibulata claims the largest number. There are more than 1.2 million species of insects alone not to mention centipedes, millipedes, and less familiar species in this phylum. Uniramia (one branch) indicates unbranched appendages. Both unbranched (*uniramous*), as in insects, and branched (biramous) appendages, as in crustaceans, consist of a linear series of joints, as do all arthropod appendages; but both insects and crustaceans are considered mandibulates. The mandibulate body has three distinct parts: head (cephalum), *thorax*, and abdomen. As the mandibulate changes size or form, it secretes a new exoskeleton and sheds the outgrown armor. Although many mandibulates are dioecious, parthenogenesis is common in insects. Many species of mandibulates metamorphose: the fertilized egg hatches a larva that develops into a series of juvenile forms that differ considerably from the adult.

Features common to mandibulates and all other arthropods (A-19 through A-21) include a cuticle containing chitin and protein. The adult arthropod coelom is very small. The principle body cavity is a hemocoel—a cavity in which blood pumped by the tubular heart circulates; from the hemocoel, blood enters the heart through slit-shaped ostia characteristic of arthropod hearts. Food is moved through the digestive tract by muscle action, and nutrients are distributed by the blood. Most mandibulates are terrestrial arthropods with internal gas exchange; through thin-walled *tracheae*—minute tubes that penetrate their tissues—body activity moves the gases. Air enters tracheae through pores in the cuticle (spiracles), passing through hairs that filter out small particles. Many species can close their spiracles, retarding moisture loss. The mandibulate nervous system is typically arthropod—a dorsal brain, ventral nerve cord ganglionated in each segment, and compound or single-unit (simple; that is, nonimage forming) eyes or both. Jointed appendages and segmented bodies characterize all mandibulates and chelicerates as arthropods. An insect has a single pair of antennae; a crustacean has two pairs of antennae (Figures A and B), and a chelicerate lacks antennae. A comparison of body parts reveals that most mandibulates have head, thorax, and abdomen; the chelicerate has a cephalothorax and abdomen. Mandibulates have mandibles (Figure C)—mouthparts that crush food; chelicerates lack mandibles. Here phylum Mandibulata includes insects, myriapods, and Crustacea.

The three largest classes of mandibulate arthropods are Hexapoda (class Insecta is an alternative name), the largest class of arthropods by far, Crustacea (crabs, shrimp, lobsters etc), and Myriapoda, centipedes and millipedes (Figure D). The other two classes in this phylum are Symphyla and Pauropoda. Symphyla, garden centipedes and their relations, are soft-bodied (covered with soft chitinous exoskeleton) mandibulates with 1 pair of antennae, 12 pairs of jointed legs with claws, and a single pair of unjointed posterior appendages. Less familiar are members of class Pauropoda, such as Pauropus, tiny soft mandibulates distinguished by branched antennae. They are similar to millipedes (*Spirobolus*) but with only 11 or 12 segments and 9 or 10 leg pairs.

All member of class Hexapoda are insects, with three pairs of legs, three body sections, generally one or two wing pairs, and one pair of antennae. Insects are the only terrestrial animals living, besides birds and bats in own phylum Craniata, that evolved flight capability (Figure H), for which the earliest evidence comes from winged insects preserved as fossils. Flight by insects is possible because the combination of small body size and highly specialized striated muscles for rapid, strong contraction enabled the predator avoidance afforded by flight (insect flight evolved before the evolution of insect-hunting bats and humans wielding butterfly nets). Water loss increases during flight because of increased movement of air past the insect body; dehydration is slowed by the waxy cuticle, by closeable spiracles, and by excretion of solid waste as almost dry uric acid by the Malpighian tubules—the excretory organs located in the insect hemocoel. The vast array of muscle and exoskeletal designs in many species of flying insects coupled with a well-developed respiratory system and adequate storage of energy supply meet

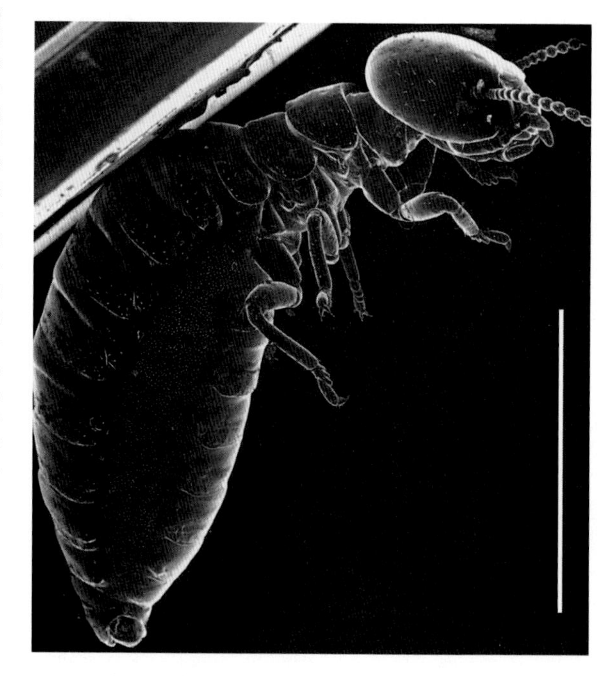

Figure A *Pterotermes occidentis*, the largest and most primitive dry-wood termite in North America. Its colonies are limited to the Sonoran Desert of southern Arizona, southeastern California, and Sonora, Mexico. The swollen abdomen of this pseudergate (worker) covers the large hindgut, which harbors millions of microorganisms responsible for the digestion of wood. SEM, bar = 0.5 mm.

S

Homarus	*Locusta*	*Periplaneta*	*Rothschildia*
Isaacsicalanus	*Melanopus*	*Polyxenus*	*Scolopendra*
Incisitermes	*Musca*	*Porocephalus*	*Spirobolus*
Kalotermes	*Pagurus*	*Pterotermes*	*Tenebrio*
Limenitis	*Pauropus*	*Raillietiella*	*Uca*
Linguatula	*Pediculus*	*Reighardia*	*Waddycephalus*

flight requirements of as many as 1000 wing beats per second. Finally, well-tuned visual and nervous controls for flight navigation make insect migration, mating on the wing, hovering, and food getting possible.

Insects, the dominant invertebrates in innumerable terrestrial and freshwater habitats, have evolved a few marine species. Only about 200 aquatic insects have been documented. Some insects are predators of plant pests, and many pollinate crop vegetables and fruit trees, which evolved simultaneously with their insect pollinators. Many insect species transmit disease associated with bacteria, fungi, or viruses. One such species is the mosquito *Anopheles*, which transmits the microbe that causes malaria. For many carnivorous animals, from ducklings to insect-eating bats, insects are crucial sources of protein. Carnivorous plants often live in nitrogen-poor soils and derive nitrogen from the protein of their insect prey bodies. Silkworms (larvae of moths) are the only entirely domesticated invertebrate animal.

Termites such as *Pterotermes occidentis* are social insects, like some bees, all ants, and some wasps. These termites live in colonies composed of a queen, king, workers, which may be female or male but always sterile, and soldiers. The division of labor among colony members is associated with marked differences in behavior and form. Workers protect and care for the queen and her offspring. The king mates with the queen. Certain workers become soldiers, which develop enlarged mandibles for defense of the colony and aggressive behavior patterns. The reproducing queen is much larger than other members of her colony, with the exception of the king; her role in the colony is reproduction.

All members of class Myriapoda (many feet) are centipedes or millipedes. Centipedes (order Chilopoda) comprise about 2800 species of predaceous, nocturnal carnivores; the centipede has a head and from 15 to 177 trunk segments, each with a pair of tiny jointed legs. The largest is *Scolopendra gigantea*, the giant African centipede, about 0.3 m in length. A few are harmless temperate species and most are tropical; they look like armored caterpillars. Centipedes secrete a neurotoxin that they inject from fangs (modified trunk appendages) into prey. The

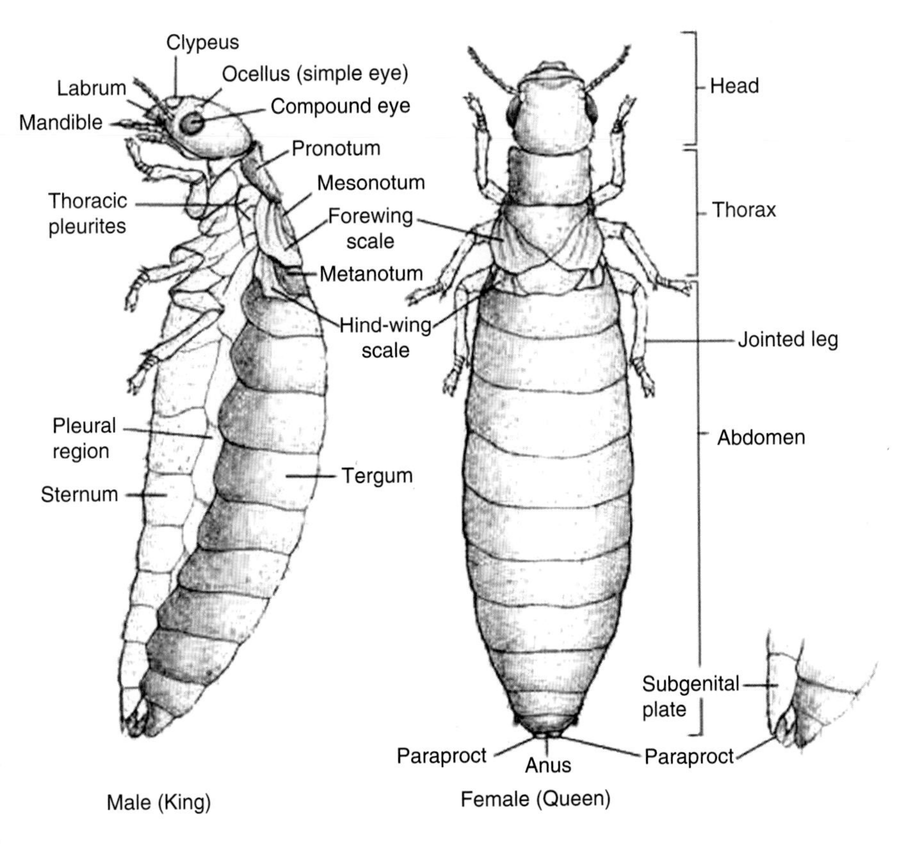

Figure B (Bottom right) Adult reproductive form of the termite *Pterotermes occidentis*. [Drawing by L. Meszoly; information from W. Nutting.]

A-21 Mandibulata
(continued)

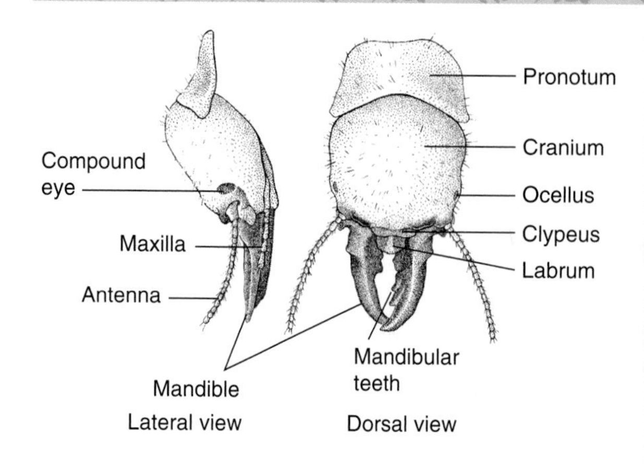

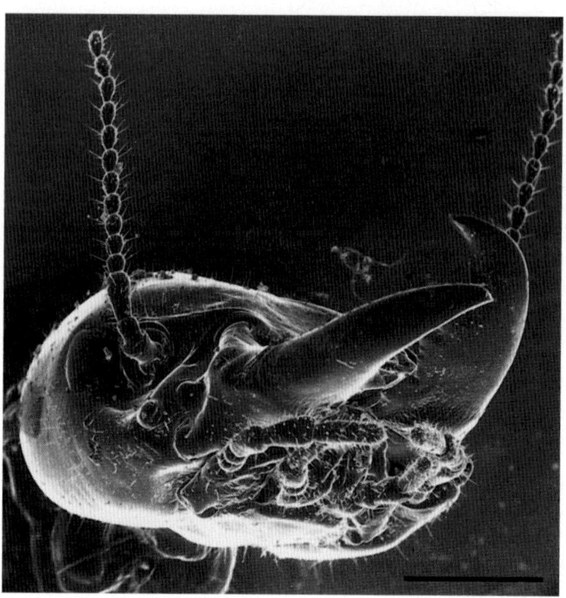

Figure C Head of soldier form of termite *Pterotermes occidentis*. The huge mandibles are mouthparts used to defend the colony against ants. The maxillae are mouthparts with which food is handled. SEM, bar = 0.5 mm. [Photograph courtesy of W. Ormerod; drawing by L. Meszoly; information from W. Nutting.]

Figure D The millipede *Polyxenus fasciculatus* repels attacks by sweeping its bristly tail tuft into ants and other predators. The bristles detach, tangling the ant's body hairs and thus incapacitating the ant. This millipede was discovered under the bark of a slash pine tree in Florida. SEM, bar = 1 mm. [Courtesy of *The New York Times*; photograph by Maria Eisner.]

millipede (order Diplopoda), such as *Spirobolus* and *Polyxenus*, has a head and from 20 to 200 trunk segments, each with two pairs of appendages and one pair of poison glands. Calcium carbonate like a calcified crustacean exoskeleton hardens the cuticle of the trunk. Upward of 11,000 millipede species have been described; those that are carnivores sometimes include hydrogen cyanide in their poison armamentarium and—because most millipedes are scavengers—their toxins serve to deter predators. Onychophorans (A-28), millipedes, and centipedes all employ adhesive defense, feed on fluids dissolved by the hunter's (centipede, millipede, or onychophoran) enzymes, have antennae, secrete cuticles of chitin, have hemocoels, and have tubular hearts. However, these traits may be due to convergent evolution, although analysis of ribosomal RNA suggests that onychophorans are specialized relatives of insects and other arthropods, evolved from a common ancestor.

The defining feature of class Crustacea is the possession of two pairs of antennae on the head, distinguishing crustaceans from all other arthropods. Many familiar animals belong to the Crustacea class—among them are crabs, lobsters, brine shrimp, and krill. The two pairs of jointed antennae of the crayfish *Cambarus* are typical crustacean features; the presence of antennae distinguishes crustaceans from chelicerates (A-20). Biramous (branched) appendages (Figure E) distinguish crustaceans from insects and other members of phylum Mandibulata with their unbranched appendages. Biramous appendages have two branches to each appendage (limb); the two branches connect to a single base, and each of the two branches consists of several jointed segments. The crustacean exoskeleton may be as thick as a lobster or as thin and flexible as the outer covering of a soft-shelled crab larva. We agree to place pentastomes, formerly phylum Pentastoma, as subclass Pentastomida within class Crustacea. Pentastomes, highly modified necrotrophs, are regarded by specialists to possibly be crustaceans, similar to branchiurans (marine crustaceans that infect fish skin). Crustacea number about 50,000 species.

As protostome coelomates, the crustacean body cavity is a coelom and, during embryological development of protostomes, the mouth originates at or near the blastopore, the opening from the outside of the blastula to its interior. The crustacean body plan resembles that of other arthropods in some respects (for example, nervous system), but most crustaceans have three body sections—head, *thorax*, and abdomen—as do insects; in comparison, chelicerates have two body sections (fused cephalothorax and abdomen). In some crustaceans, the head is fused with various thoracic segments to form a cephalothorax. The crustacean coelom is small and the main body cavity—a hemocoel, or cavity in which blood circulates—is part of the circulatory system, as in other arthropods. The crustacean mode of gas exchange, however, differs from that in other arthropods; gills—evident in crabs and lobsters—exchange gases with the water or air. A tube-shaped heart with ostia is the circulating pump for blood. In lobsters, one of the mouthparts is modified as a gill bailer, which the animal uses to generate a current of water through its gill chamber, enhancing gas exchange through the gills. Crustaceans that are semiterrestrial (for example, the hermit crab *Clibanarius*) inhabit the intertidal zone of the sea beach, a habitat that provides at least occasional moisture to their respiratory surfaces. Some crustaceans have blood vessels; others have no vessels and pump blood only to the hemocoel. A dorsal brain, ventral nerve cords with ganglia in each body segment, and sensory organs such as the compound eyes of *Daphnia* and statocysts constitute the nervous system. Krill chemically produce light in luminescent organs on their legs and eyestalks; their light displays may function in mating or in protecting them from predators, such as penguins, that are below them in the sea. All crustacean muscle is striated, typical of arthropods, and quicker to contract and relax than the smooth muscle that predominates in other invertebrates. The one-way digestive tract includes a mouth and stomach in which food is ground extensively; food then passes to a midgut sometimes associated with organs that facilitate digestion, such as the liver of a lobster. Crustaceans may be dioecious, often with sexual dimorphism that ranges from a difference in size between females and males to sexual dimorphism of appendages (males may have appendages used in mating) to the extreme sexual dimorphism of some isopods. Barnacles are hermaphrodites; the long penis of one barnacle fertilizes another nearby but permanently attached barnacle. A three-segmented, free-living larva called a nauplius is a characteristic crustacean larva. Some barnacles brood embryos to the nauplius larval stage. Some highly specialized crustacea are so bizarre in form that they can be classified only upon observation of their nauplius larvae. Diverse larvae having

Figure E Freshwater crayfish *Cambarus* sp., a component of the food web in a New Hampshire lake. The branched appendages are a distinguishing feature of this phylum. Crayfish are nocturnal lake-bed scavengers, feeding on aquatic worms and plant growth. In turn, loons, herons, black bass, and people prey on these crustaceans. Bar = 5 cm. [Photograph by K. V. Schwartz.]

forms as exotic as their names—*zoea*, megalops, phyllosoma, mysis—are present in decapod crustaceans, an order of class Malacostraca that includes crabs, some shrimp, hermit crabs, lobsters, and crayfish. Crustacean larvae are abundant in the ocean's plankton. Other species in class Malacostraca lack a larval stage; young hatch as juveniles. Some freshwater copepods and members of class Branchiopoda are parthenogenetically reproducing females. Dissolved nitrogenous wastes diffuse from the crustacean body through the gills. Freshwater crustaceans' green glands or antennal glands excrete water and resorb salts by way of nephridial canals in the anterior of the animal, collecting the resulting liquid in a bladder that discharges through an excretory pore near the antenna base. Chelicerate and crustacean excretory organs are similar; from the hemocoel, fluid is collected and eventually excreted as urine.

Crustaceans have evolved an enormous array of appendage diversity: gills, food collection, food shredding, defense, courtship display, walking, swimming, generating respiratory currents, brooding eggs, and clinging to seaweed. The crustacean head [two pairs of antennae, one pair of mandibles (food crushers), two pairs of maxillae (food handlers and current generators)], thorax (eight pairs), and abdomen (six pairs) all bear appendages. Females of one isopod species carry the male of the species on their antennae, the anteriormost appendages.

The covering of crustaceans provides a rigid skeleton that functions in locomotion, support, and protection; in general, the chitin–protein covering is similar in all arthropods, though less rigid in many species. Calcium carbonate is deposited in crustacean cuticle, strengthening the cuticle; insect cuticle lacks this calcification. Joints between sections of this armor are thinner cuticle, affording flexibility. The outermost layer of the cuticle is protein and often water-impermeable wax, rendering the soft body within somewhat resistant to water loss. At the same time, gas exchange across the body surface is severely reduced; oxygen and carbon dioxide diffuse from gills, which have great surface area and blood supply. Beneath the waxy layer lies a blue, red, yellow, or black pigmented chitin and protein layer underlaid by a calcified layer that hardens the cuticle. The epidermis, which secretes the cuticle, is the innermost layer. The crustacean must shed its rigid cuticle to accommodate growth. Enzymes partly break down the old exoskeleton, and new cuticle is secreted before the old cuticle splits. During ecdysis (escape), the crustacean not only slips out of its exoskeleton, but also jettisons the chitinous, internal linings of its fore- and hindgut.

Class Crustacea includes the major subclasses (largest and most diverse of the subclasses in number of species): Branchiopoda, Ostracoda, Copepoda, Cirripedia, and Malacostraca, as well as other subclasses including rare species or those in restricted habitats such as Remipedia, which inhabits underground saltwater caves in the tropics. In freshwater, including vernal pools, and saline habitats, crustaceans are the most prominent arthropods. Brine shrimp (*Artemia*), water fleas (*Daphnia*), fairy shrimp, tadpole shrimp, and others in class Branchiopoda mainly inhabit freshwater, although the brine shrimp withstand highly saline water. Subclass Branchiopoda also includes commensals (epibionts) that live on other crustaceans, herbivores, detritus feeders, and filter feeders. Ostracods (*Gigantocypris*) live mostly in marine and estuary habitats; a few species are abyssal or freshwater. Copepods (*Calanus*) are enormously abundant in diverse habitats: marine (from ocean surface to 5000 m deep), freshwater, estuarine, interstitial (between sand grains, belonging to the meiofauna); some species are necrotrophic or commensal (symbiotrophic). Many commensal copepods are color matched to their hosts—crinoid echinoderms (A-34). As the base of worldwide food webs that include commercially valuable fishes, copepods affect our own species. *Isaacsicalanus* is a newly described copepod from a hydrothermal vent. Subclass Cirripedia comprises all barnacles, such as *Balanus*. As adults, cirripeds cement themselves permanently to firm marine substrates. Research on this biomaterial reveals a strong new adhesive protein, promising for cementing artificial spare parts, such as artificial hip replacements, into humans. *Coronula diadema*, a huge planktonic barnacle, settles as an epibiont on *Megoptera*, the humpback whale. Many barnacles are sessile as adults; their calcium carbonate shelter houses the chitin-covered crustacean within, complete with biramous appendages.

Subclass Malacostraca comprises more than half of the 50,000 crustacean species: amphipods (sand fleas), isopods (sow bugs, pill bugs, woodlice), and decapods [lobsters (*Homarus*), crayfish, crabs (*Cancer*), some shrimp, and hermit crabs (*Pagurus*). Isopods dwell in caves, in all damp soil habitats, under logs and rocks, in the depths of the ocean, as necrotrophs of marine animals including whales, fish, molluscs, and other invertebrates, and as commensals with echinoderms, polychaetes, and bivalve molluscs. The krill (*Euphausia*) is one of the most ecologically critical malacostracan species; krill is the primary food for many fish marine species, seabirds, seals, filter-feeding whales, and other crustaceans worldwide. Exploitation of *Euphausia superba*, the krill of Antarctic seas, by humans for food may

directly reduce the base of marine food webs that are critical to life on this planet.

Members of the subclass Pentastomida (Greek *pente*, five; *stoma*, mouth), called tongue worms (confusing common name, see also Hemichordata A-33) because of their shape, have a flat, soft body covered by a nonchitinous cuticle. The relationship of tongue worms to other crustaceans is difficult to infer because the worms have been modified to live in damp passages of vertebrates. Although the body is not segmented internally, the exterior appears to be composed of about 90 rings, depending on the species. All 110 species are symbiotrophs of vertebrates. Two anterior pairs of leglike hooks hold the worms in place on their hosts. They live embedded in the lungs, nostrils, or nasal sinuses of mammals such as dogs, foxes, wolves, goats, and horses, as well as of reptiles and birds. Pentastomes are particularly prevalent in tropical and subtropical hosts.

Larval tongue worms are occasionally found in humans, who acquire them from other vertebrates, generally from domestic animals. Pentastome growth in humans is checked, however, because the nasal tissue responds by forming calcareous capsules that surround and isolate the worms. Tongue worms range in length from a few millimeters to more than 150 mm. The digestive tract is a straight tube. Excretory, respiratory, and circulatory organs are absent—common adaptations to necrotrophic life.

The sexes are separate in pentastomes, and the male is smaller than the female. In *Linguatula serrata* (Figures F and G, the male and the female are 2 and 13 cm long, respectively. Females have a genital pore near the anus, whereas males have a genital pore near the mouth. The eggs are minute, thick shelled, and without yolks. Fertilization is internal. The vagina of the female tongue worm is able to hold some half million fertilized eggs. As it fills with eggs, the vagina may stretch until it is 100 times its former size. The fertilized eggs pass through the genital pore onto plants with the nasal secretions and saliva of the host animal. The eggs must be eaten by herbivorous vertebrates to develop.

Tongue worms have three larval stages. The first stage develops within the eggs in a herbivorous intermediate host. Rabbits are the main intermediate hosts for *Linguatula*; fish are intermediate hosts for crocodile pentastomes. Second-stage larvae hatch from the eggs in the stomach of the herbivore; they swim and

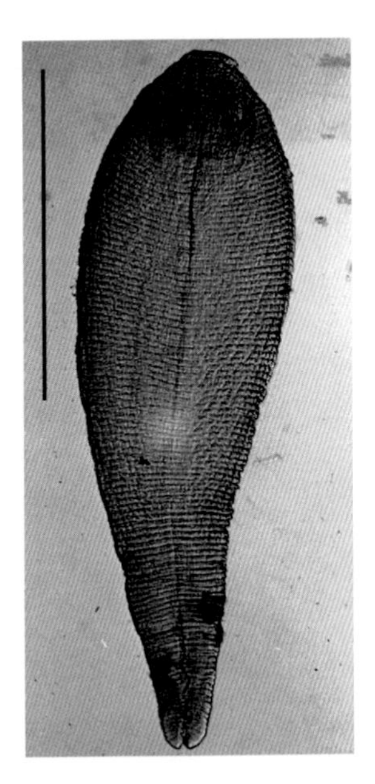

Figure F A living female tongue worm, *Linguatula serrata*, a pentastome that clings to tissues in the nostrils and forehead sinuses of dogs. Bar = 1 cm. [Courtesy of J. T. Self.]

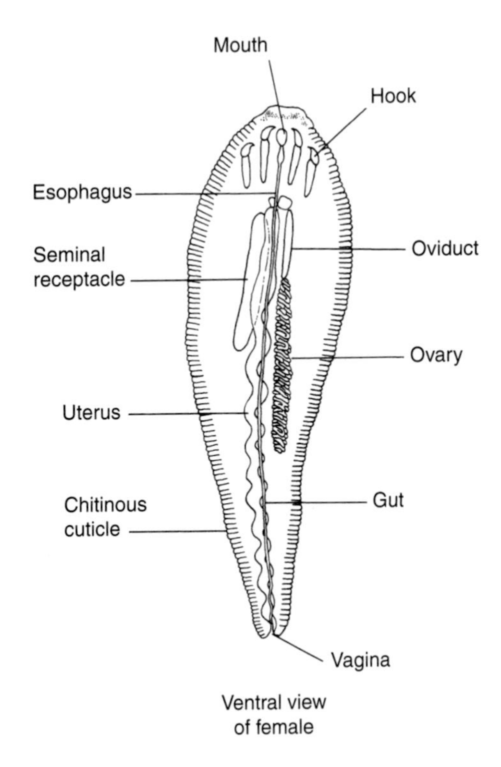

Figure G Ventral view of female tongue worm showing one of each pair of oviducts and seminal receptacles. [Drawing by R. Golder; information from J. T. Self and R. E. Haugerud.]

A-21 Mandibulata
(continued)

Figure H *Limenitis archippus* mating. These orange and brown viceroy butterflies, class Insecta, are found in central Canada and in the United States east of the Rockies. Insect wings are outfoldings of the body wall, supported by tracheae. A network of veins links the two wing pairs to the circulatory system. Bar = 7 cm. [Courtesy of P. Krombholz.]

bore with their mouths into the lungs or liver or into the linings of these host organs. In the tissues of the intermediate host, the larvae enter a third stage: they encapsulate to form cysts. It is possible that the cysts are digested when the herbivore host is eaten by a carnivore, and that the pentastome then moves with its hooks from the carnivore's stomach into its nasal or lung passages. However, there is evidence that the pentastome leaves its cyst while it is still in the carnivore's mouth. In either case, the adults embed themselves in the nasal sinuses or lung tissue, they mate, and the cycle begins again.

All appendages of most primitive arthropods are assumed to be similar with serial repetition of similar segments, resembling the bodies of extinct trilobites and of recently discovered remiped crustaceans. Affinities of crustaceans with the Chelicerata (A-20) are uncertain. Crustaceans may have an evolutionary origin independent from insects, centipedes, millipedes, and other terrestrial arthropods.

Chitin is present in all arthropods (A-19 through A-21), as well as in pogonophoran (A-25), annelid (A-22), brachiopod (A-30), and some echiuran (A-24) setae (bristles); in onychophoran cuticle (A-28), the radula (rasping organs) of molluscs (A-26), and cell walls of fungi and red algae (Pr-33); and in chytrid cell walls (Pr-35). Thus, chitin by itself is not a reliable indicator of evolutionary relationships.

Fossils of joint-footed animals are found in Ediacaran (Upper Proterozoic) rocks. *Trilobites*—common Cambrian (early Paleozoic) fossils having jointed legs and other arthropod features—are placed in their own phylum and are extinct. Trilobites were the earliest animals to have eyes. The trilobite also had jointed legs, one pair of antennae, and a carapace (hard covering) with head, thorax, and rear segment. (Trilobites are named for their three body lobes.) Insects arose in the Paleozoic era. At 300 mya, cockroaches appear in the fossil record. Arthropods—terrestrial mandibulates, crustaceans (almost all aquatic), and chelicerates—were traditionally thought to have evolved from annelid ancestors (A-22), but this view is currently contested. A number of data sets show phylogenetic grouping of arthropods with Nematoda (A-11). Phylogenetic relationships between arthropod phyla are not clear. Affinities of insects, centipedes, and millipedes to crustaceans and chelicerates are far from certain at present.

A-22 Annelida
(Annelid worms)

Latin *annellus*, little ring

GENERA

Aphrodite	*Limnodrilus*	*Nephthys*
Arenicola	*Lumbricus*	*Nereis*
Chaetopterus	*Macrobdella*	*Podarke*
Eunice	*Megascolides*	*Sabella*
Hirudo	*Myzostoma*	*Tubifex*

Annelid worms—polychaetes, earthworms (oligochaetes), and leeches (hirudineans)—are distinguished by linear series of external ringlike segments; the grooves between segments coincide with internal compartments, often separated by transverse sheets of tissue (septa), containing serially repeated nervous, muscle, and excretory systems. Anterior segments bear jaws, eyes, and cirri (singular cirrus, a slender appendage) in some species; the terminal segment may bear a cirrus (Figure A). Annelids have spacious, mesoderm-lined coeloms—except for leeches, in which tissue packs the coelom—and their coeloms are important in excretion, circulation, and reproduction. Chitinous lateral bristles called setae on each segment are used for locomotion or to anchor the annelid in substrate or burrow; leeches lack setae. Parapodia are unique to polychaetes; these thin, fleshy flaps protrude laterally from each body segment. Chitinous cuticle covers the entire body.

Annelids live in soil-, freshwater, and oceans including Antarctic seas. They may be striped or spotted, and pink, brown, or purple in color. Others are iridescent or luminescent. Some have colorful gills and cirri, modified parapodia. The endangered Australian earthworm *Megascolides*, which is 3 m in length, is the largest species; the smallest annelid is only 0.5 mm.

Active predation or scavenging is the feeding mode for most annelids. Many annelids burrow incessantly, turning over and exposing detritus and soil and aerating anaerobic muds and sands; these activities are known as bioturbation. Swimming annelids catch fish eggs or larvae. Filter-feeding marine annelids capture bacteria and feed selectively on sediment particles within tubes (which they build of mucus-cemented sand grains, calcium carbonate, protein and polysaccharide compounds, and other materials) buried in sand or mud. Some trap plankton on a mucus-covered, ciliated eversible proboscis. Others pop out of their tubes to seize prey. Certain species harvest algae growing on their tubes. The sea star *Luidia* (A-34) hosts the polychaete *Podarke* among its tube feet. Some carnivorous polychaetes have fangs with which they inject toxins into prey.

About 15,000 species of annelids are grouped into three classes: the Polychaeta, mostly marine and a few soil and freshwater bristle worms; the Oligochaeta, terrestrial and freshwater bristle worms; and the Hirudinea, or leeches. There are about 9000 species of polychaetes, including myzostomarians, a group of about 100 species of small polychaetes that live on or in echinoderms (A-34), 6000 species of oligochaetes, and 500 species of hirudinids.

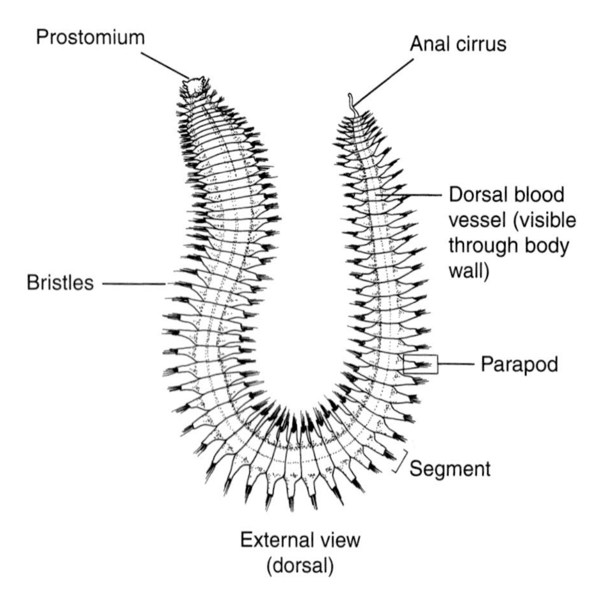

Figure A An adult *Nephthys incisa*, a polychaete (13 cm long) taken from mud under 100 feet of water off Gay Head, Vineyard Sound, Massachusetts. (Left) External dorsal view of adult *N. incisa*, showing thin parapodia that serve in locomotion and gas exchange in this polychaete. [Photograph courtesy of G. Moore; drawing by I. Atema; information from M. H. Pettibone.]

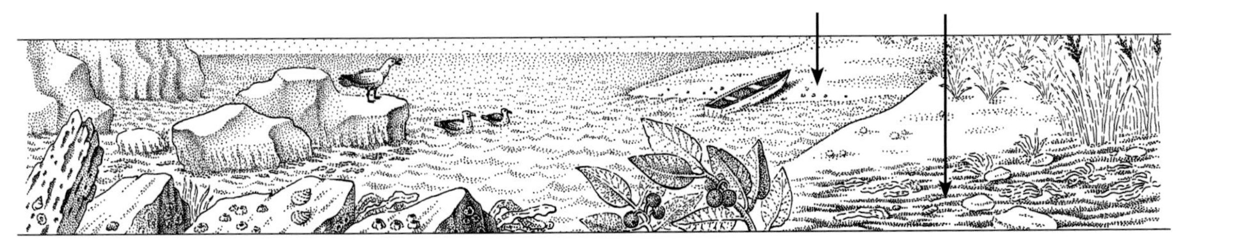

In most polychaetes (paddle-footed worms), a fleshy lobe (prostomium) projects over the ventral mouth and bears tentacles. Parapodia of all polychaetes with a few exceptions are stiffened by a bundle of chitinous bristles, enabling the parapodia to function as oars and levers. In a few polychaete species, chitinous rods called acicula support parapodia. Some tube-dwelling polychaetes leave their tubes, others do not; all of these tube builders are grouped as sedentary polychaetes. Free-living polychaete species are grouped as errant polychaetes. Marine polychaetes include *Aphrodite*, the hairy sea mice; lugworms (*Arenicola*), which burrow in sand and mudflats; sabellid and serpulid worms, whose tubes encrust shells, rocks, and algae, including peacock worms (*Sabella*), which construct mosaic tubes of sand or shell; economically important bait worms such as *Nereis*; and a few pelagic (ocean-dwelling) species. The concretelike tubes of some polychaetes foul ships. There are also some soil- and freshwater (in the Great Lakes and Lake Baikal) polychaete species. Oligochaetes include the earthworms and a few small freshwater, estuarine, and recently described deep-sea forms. The hirudinids, with anterior and posterior suckers, are popularly called leeches. Most leeches are free-living predators of frogs, turtles, fish, birds, and invertebrates of soil, foliage, algal thalli, fresh- and saltwater; a few symbiotrophically exploit vertebrates and invertebrates. Oligochaetes and leeches usually lack distinct eyes, tentacles, and parapodia including gills. Light-sensitive cells and sensory hairs in the earthworm epidermis that connect to the central nervous system alert earthworms to light and other environmental stimuli. Leeches lack setae. Annelids, except for the leeches, regenerate lost body parts from bristles to terminal body segments. Some polychaetes reproduce by budding.

Bloodletting by the freshwater leech *Hirudo medicinalis* is used to control swelling subsequent to the reattachment of severed fingers or transplanted tissue. Saliva of bloodsucking leeches contains the anticoagulant hirudin as well as an anesthetic. This leech harbors a symbiotic bacterium *Aeromonas hydrophila*; not only does the bacterium digest blood, but also produces an antibiotic that kills other bacteria.

Burrowing polychaetes turn over 1900 tons of seafloor sand per acre each year. Charles Darwin calculated that earthworms bring 18 tons of soil to the surface per acre each year. Suction by a muscular pharynx draws soil into the earthworm mouth. Food is passed through the esophagus by the peristaltic movements of digestive tract muscles to a crop, where the food is temporarily stored. The muscular gizzard of the earthworm grinds seeds, eggs, larvae, small animals, and plants ingested with soil. Annelids' longitudinal and circular body-wall muscles work against the coelomic fluid, which is, like all fluids, relatively incompressible; this system functions as a hydraulic skeleton. Thin peritoneal sheets called septa separate adjacent segments. As body-wall muscles contract, colorless coelomic fluid flows from segment to segment through openings in each septum. Food is pushed through the gut from mouth to anus by cilia or by peristaltic contractions of muscles that encircle the digestive tract. The aquatic annelid pumps water through its burrow with peristaltic body waves, cilia, and parapodia. The water current brings in food and dissolved oxygen and removes waste.

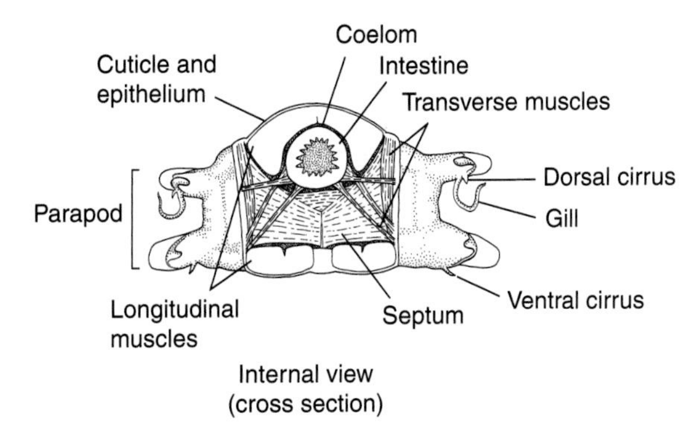

Figure B Cross section of one body segment of the polychaete *Nephthys incisa*. Contraction of the longitudinal muscles shortens the annelid, increasing the diameter of its body. [Drawing by L. Meszoly; information from M. Pettibone.]

Three iron-containing pigments (hemerythrin, hemoglobin, and the green pigment chlorocruorin) transport oxygen in blood vessels and coelomic fluid in most annelids, in corpuscles or in solution in blood. The annelid dorsal blood vessel is contractile with unidirectional valves, forcing blood through five aortic arches ("hearts") that act as pressure regulators and then into the ventral blood vessel. From the ventral vessel, blood moves to the digestive tube wall, the body wall, and the nephridia. Blood carried to the body wall exchanges oxygen and carbon dioxide through highly vascularized parapodia (in polychaetes)—sometimes modified into gills—and through the moist body wall, even through the protective cuticle (Figure B). The continuous coelom of the leech lacks septa; most circulatory functions in leeches are carried out by coelomic fluid that is transported within the contractile channels and sinuses of the coelom itself. The closed annelid circulatory system, with contractile heart, blood vessels, and capillaries, is quite unlike the circulation pattern in arthropods.

The annelid excretory system consists of nephridia; within most body segments, ciliated paired tubules—nephridia—draw in wastes from coelomic fluid and then discharge dissolved waste through external pores called nephridiopores. Additional waste moves into nephridia from the nephridial blood vessels. Gametes also exit through nephridiopores in many annelids and through the mouth in a few species. Nephridia also regulate the water content of the coelomic fluid. Castings of intertidal polychaetes are sand, cleaned of organic matter as it passed through the gut and out of the anus.

Segmental ganglia, bilobed cerebral ganglia (brain), and single or paired ventral nerve cords are the main components of the nervous system. Most polychaete annelids have eyes, some with retinas and lenses. Chemoreceptors, touch receptors, vibration receptors, and statocysts (balance organs) concentrated at the head end link to the ventral nerve.

Breeding polychaetes swarm by the millions, their hormones triggered by phases of the moon, the tides, or changes in temperature. Polychaetes are usually dioecious, oligochaetes are usually monoecious but cross-fertilize, and leeches are monoecious. Polychaete gametes arise from the coelom walls in a number of body segments; polychaetes lack permanent gonads. Nephridia usually discharge gametes through nephridiopores in addition to urine, and polychaete fertilization is external. Many polychaete adults brood their young; in some species, the male protects and aerates the eggs. Development of most polychaete annelids includes a free-swimming, ciliated, planktonic larva, the trochophore, which is also often a feeding larva (Figure C). Metamorphosis from the trochophore is to a segmented form that may be a swimming larva, called a nectochaete, or a benthic juvenile. The head of the segmented form develops at the apex of the larva, while segments bud off at anal end. These two parts link up to give a wriggling worm, with its own nervous system, protruding from the ciliated larva. This is another example of "overlapping metamorphosis," in which the larva and the succeeding phase coexist. Comparable cases occur in nemertine worms (A-10), echinoderms (A-34) and doliolid tunicates (A-35), and they are explicable in terms of larval transfer (Box A-i). In some sedentary polychaete species, individuals are budded off or are transformed into epitokes, which are sexually mature, swimming, gamete-bearing individuals. Epitokes of *Eunice viridis*—the palolo—swarm on the sea surface in the South Pacific, in coincidence with lunar cycles. Male epitokes, stimulated by a pheromone from female epitokes, shed sperm. Responding to sperm, females release eggs. Samoans and other South Pacific people gather palolo epitokes to eat.

In contrast with polychaetes, leeches and oligochaetes are usually hermaphroditic (monoecious); each copulates with another individual. Sperm and eggs are produced in ovaries and testes, rather than in the peritoneal coelom lining. Oligochetes transfer sperm from one worm to its partner, which stores sperm in seminal receptacles until egg laying. The oligochaete lays eggs and releases stored sperm into a secreted mucus band; embryos develop in a secreted cocoon until a juvenile worm escapes. Some leeches attach to their partners with suckers and forcibly drive spermatophores, packets of sperm, into their mates' bodies; fertilization is internal. Leeches and oligochaetes incubate eggs in a cocoon—an adaptation to terrestrial life. They hatch as miniature adults, without a free-living larval stage.

Those free-living polychaetes that evolved in the Cambrian seas are the most ancient annelid group. From polychaetes, oligochaetes evolved. Leeches are of oligochaete ancestry. Fossil polychaetes have been found in rocks of the late Proterozoic eon that contain the Ediacaran fossil assemblages (about 700 million years old) and are well preserved in the middle Cambrian rocks of the Burgess shale (500 million years old) of western Canada. Terrestrial oligochaetes probably did not evolve before the Cretaceous, when angiosperms, which contributed humus

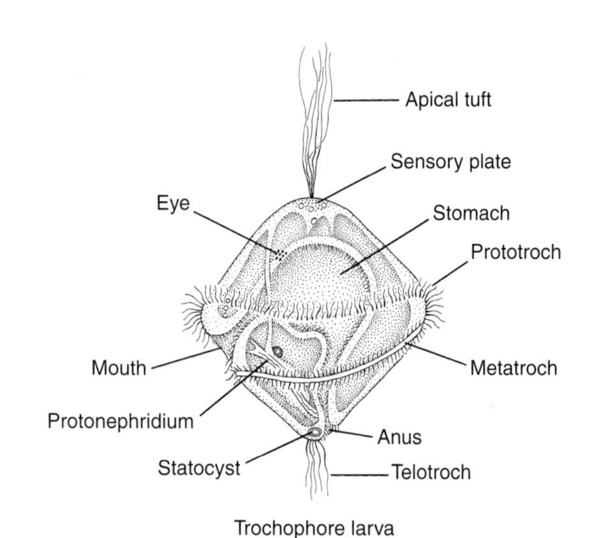

Figure C Annelid trochophore, free-swimming larva that is the dispersal form of marine polychaete annelids. Ciliary bands—telotroch, metatroch, and prototroch—are distinctive features of trochophore larvae. [Drawing by L. Meszoly; information from M. Pettibone.]

in which earthworms live, arose. Annelids are possible ancestors of sipunculans (A-24) and echiurans (A-25)—at least some species of these phyla have trochophore larvae, produce gametes from peritoneal tissue rather than ovaries and testes, and have similar body-wall anatomy, nervous systems, excretory systems, and patterns of gamete production. Immunological data support annelids as closer relatives of sipunculans than are molluscs; however, the evidence from paleontology, biochemistry, and embryology supports molluscs as closer relatives of sipunculans than are annelids. Data from ribosomal RNA indicate that pogonophorans (A-25) may have evolved from annelids. Molluscs (A-26), which also have trochophore larvae, may have evolved from annelid ancestors or from sipunculans; alternatively, the trochophore larvae may have arisen independently, an example of convergent evolution. Unlike wormlike nemertines (A-10), nematodes (A-11), nematomorphs (A-12), and acanthocephalans (A-13), annelids have linearly segmented bodies, coeloms, a ventral nerve cord, and distinct eyes (in polychaetes). Onychophorans (A-28) and annelids may have a common ancestor. The phylogenetic relationship between annelids and arthropods is currently controversial.

A-23 Sipuncula

(Sipunculans, sipunculids, peanut worms)

Latin *siphunculus*, little pipe

GENERA
Aspidosiphon
Golfingia
Lithacrosiphon
Nephasoma
Onchnesoma
Phascolion
Phascolopsis
Siphonosoma
Sipunculus
Themiste

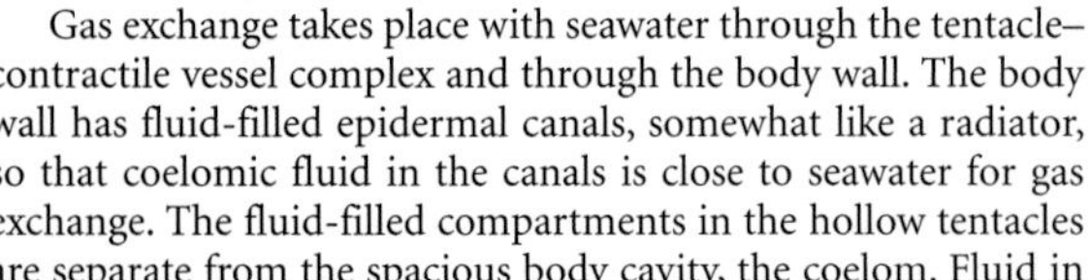

Sipunculans are crevice-dwelling and burrowing sea animals. The wormlike, unsegmented body contains an introvert that when contracted constitutes an invagination. The introvert ranges in length from one-third to many times longer than the length of the trunk portion of the body. Minute spines or papillae stud the distal portion of the introvert in some species. Ciliated bushy or fingerlike mucous-covered tentacles encircle the terminal mouth (Figure A) in one group. In the large group, tentacles are arranged into a crescent above the mouth. The introvert is everted by hydraulic pressure exerted by contraction of body muscles on the coelomic fluid. It is retracted by muscles within the coelom. Sipunculans can adjust their body length telescopically and when fully contracted resemble a peanut, hence the common name peanut worms.

There are about 150 species of sipunculans distributed among several families. The relationships of higher taxa within the phylum remain unclear. Typically, the trunk is from 15 to 30 mm long. Extremes range from the smallest species, which is about 2 mm long to the longest species reaching 0.5 m in length.

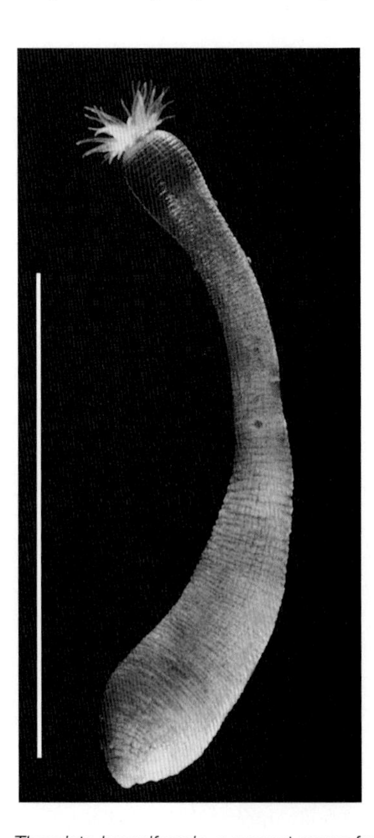

Figure A *Themiste lageniformis*, a peanut worm from the Indian River, Fort Pierce, Florida, with introvert and tentacles extended. Bar = 1 mm. [Courtesy of W. Davenport.]

Their rubbery, cuticle-covered body walls are iridescent pearl gray, yellow, or dark brown. The cuticle is mostly collagen; the underlying epidermis contains no chitin and lacks undulipodia and setae.

Gas exchange takes place with seawater through the tentacle–contractile vessel complex and through the body wall. The body wall has fluid-filled epidermal canals, somewhat like a radiator, so that coelomic fluid in the canals is close to seawater for gas exchange. The fluid-filled compartments in the hollow tentacles are separate from the spacious body cavity, the coelom. Fluid in the tentacles is stored in a reservoir called the contractile vessel attached to the esophagus; the tentacle cavity and the contractile vessel communicate. In some sipunculans, such as *Themiste* sp., the oxygen-carrying respiratory pigment hemerythrin is dissolved in the coelomic fluid; together with myohemerythrin in retractor muscles, this can be considered a closed vascular system. Heart and blood vessels are not present, but oxygen and nutrients are distributed by the coelomic fluid. Metanephridia collect dissolved wastes from the coelom. Ammonia is a sipunculan's main nitrogenous waste. In some species, urn cells form on the peritoneum. These ciliated groups of cells are of two types, fixed or free. Urn cells gather waste from the coelomic fluid, then leave the worm through the nephridia or leave the waste on the body wall. The sipunculan nervous system consists of a bilobed brain (cerebral ganglion) located above the esophagus and connected to an unsegmented ventral nerve cord. Body and tentacle surfaces bear protrusible ciliated cells, presumably sensory. Sipuncilans also have pigmented, ciliated photoreceptor cells and likely chemoreceptors near the mouth.

Some sipunculans feed on detritus in mud; others on diatoms, other protoctists, and animal larvae. Sipunculans inhabiting coral scrape algal films with hooks on their introverts. A few filter feeders trap food from seawater on mucous-covered tentacles; cilia move the food toward the mouth. Dissolved organic matter in seawater may supply as much as 10 percent of sipunculan nutrition. Selective deposit-feeding sipunculans remove nutrients from ingested mud. A ciliated groove moves food through a U-shaped, spiral digestive tract; feces are removed at the anus, which is located at the base of the introvert (Figure B). The U-shaped digestive system allows the animal to remain in its burrow without reversing its position.

Sipunculans live in burrows, which they create by digging. The introvert is thrust into soft sediment by rapid contraction of the body-wall muscles. The introvert is then swollen to form an anchor in the substrate. Retractor muscles in the body pull the body forward, and the process is repeated until the animal is lodged in its new mucous-lined burrow. Some species live inside large *Spheciospongia* sp. sponges (A-3) or among mangrove roots, eel grass, or reef-forming coral. Others live beneath rocks or in annelid tubes. In general, sipunculans are inhabitants of the lower tidal zone and the greatest abundance of species live in warm mid-latitude waters. A few common species are circumtropical. The highest density of sipunculans recorded is about 4000 per square meter in Indian River lagoon, Florida. Several species, however, live at extreme depths (7000 m) and still others are known from polar seas. Recent research suggests that

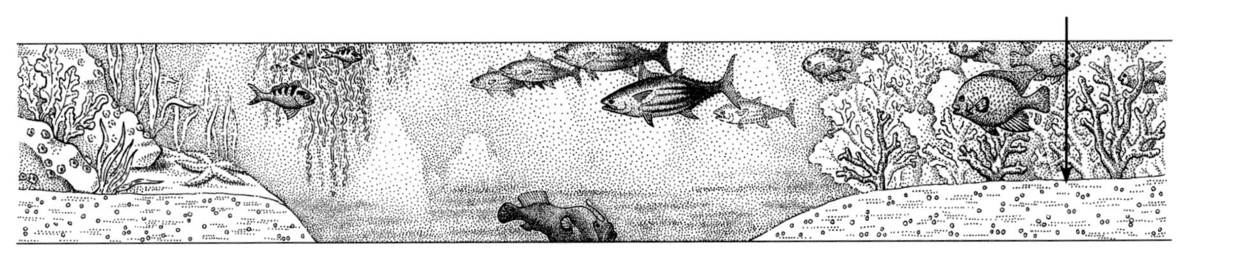

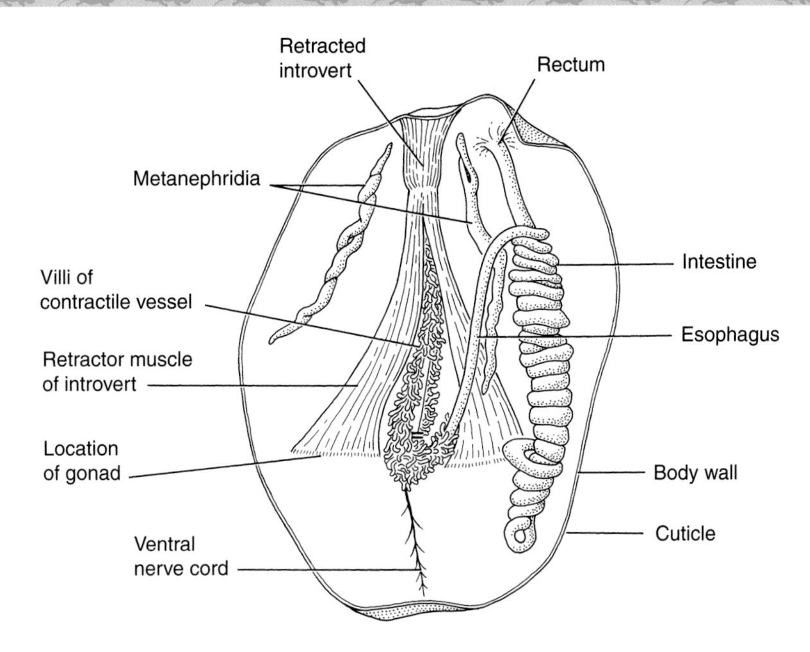

Figure B A cutaway view of a *Themiste lageniformis* with its introvert retracted. Contractile vessels push fluid into the hollow tentacles, causing them to extend. Contraction of the introvert retractor muscles forces fluid back into the contractile sacs. [Drawing by L. Meszoly; information from M. E. Rice.]

sipunculans are as common in deep seas as in tidal areas. Sipunculans, like annelids (A-23), are important in bioturbation; by turning over large amounts of sediment, they enable oxidation of sedimentary material. Sipunculans are also important eroders of coral reefs; their ability to burrow into the calcareous reef skeleton weakens the entire structure.

Sexes are generally separate in sipunculans, but males and females cannot be distinguished externally. Inconspicuous and temporary gonads form seasonally at the base of the introvert retractor muscles. *Nephasoma minutum* is the only monoecious species known; it fertilizes itself. Of the dioecious species, females tend to be more abundant than males. Mature females shed gametes into the coelom, where the oocytes accumulate yolk and become ova. Males release sperm into the coelom from which it is disbursed to the sea through nephridiopores. The presence of sperm induces the female to likewise discharge ova through the nephridiopores. Fertilization takes place in seawater. Development includes three different patterns. In some sipunculans, development is direct without a larval stage, in most species a trochophore (Annelida) larva is produced, which is free swimming for a short period, but eventually settles to metamorphose into an adult worm. The third type includes the trochophore stage but it is succeeded by the pelagosphera larva that remains in the plankton for extended periods of time during which it is distributed by oceanic currents. Asexual reproduction has been observed in *Aspidosiphon* and *Sipunculus* spp. Buds are formed on the posterior trunk, containing a new ganglionic mass. The fragmented bud then grows into new individual. Sipunculans are able to regenerate the nervous system in severed individuals.

Sipunculans are traditionally thought to be most closely related to annelids. They share a similar larva, metanephridia, simple and temporary gonads, and demonstrate identical coelom (schizocoely) and mouth (protostomous) formation. However, they lack any trace of segmentation, even in early development. Recent molecular, morphological, paleontological, and embryological studies point to a closer evolutionary relationship with the Mollusca (A-26).

A-24 Echiura
(Spoon worms)

Greek *echis*, snake; Latin *-ura*, tailed

GENERA

Bonellia	*Listriolobus*	*Thalassema*
Echiurus	*Ochaetostoma*	*Urechis*

Echiurans are a group of about 150 species of soft-bodied, benthic marine invertebrates that live in burrows made in marine sediments from the intertidal region to the abyss.

Echiurans have a varied classification history. The first specimens were described in 1766 and classified within a defunct group, the Gephyrea. They were thought to be a bridge group linking annelid (segmented) worms and holothurians (sea cucumbers). Later, the echiurans were moved into the group with the annelid worms, and then in 1940, they received phylum status. Recent evidence from DNA sequence studies leads many scientists now to believe that they are indeed members of the marine polychaete annelid worm group.

The echiuran body has two distinct regions: the trunk and the prostomium (sometimes referred to as a proboscis) (Figure A). The trunk is oval or elongated and may be a few millimeters to over 20 cm long. The mouth is a small opening at the anterior end and the anus opens at the posterior terminus. The body wall may be smooth and translucent or thick and covered with papillae. Longitudinal muscle bands may be present and visible. Typically, two hooked setae are present just behind the mouth on the ventral surface. These structures straddle the ventral midline and are used in burrowing and anchorage within the burrow. A few species have a ring of setae surrounding the anus at the posterior end. In the bonellid echiurans and a few others, the trunk is covered by a velvety green pigment called bonellin. The trunk remains within the burrow at all times. Echiurans pass waves of contractions over their trunk to pump water through the burrow. The water is necessary for respiration and, in some species, for feeding.

The prostomium is a feeding structure attached to the trunk dorsal to the opening of the mouth. The shape may be thin and narrow with a rounded or blunt anterior end. In other species, the anterior end of the proboscis may be forked as is the case in the bonellid echiurans.

The complex musculature of the prostomium allows it to extend out the burrow aperture onto the sediment surface ventral side upward to pick up organically enriched sediment particles (Figure B). Cilia on the ventral surface carry the sediment particles to the mouth where they are consumed (Figures C and D). In large echiurans, the prostomium may extend up to a meter from each burrow aperture. The spoonlike nature of the proboscis when used in feeding is the origin of the common name, spoon worm, used for this group of organisms.

In the genus *Urechis*, the prostomium is a short stubby tag of tissue surrounding the dorsal side of the mouth. These echiurans do not feed on surface sediments. Instead, they secrete a very fine mucous net within the burrow and trap food particles that are suspended in the water that the animals pump through the burrow. When the net becomes clogged with food, the animal simply eats the net with its nutritious load and then forms a new net. One famous echiuran of this type is *Urechis caupo*, which lives in burrows as deep as 50 cm in muddy estuaries of the central California coast. It is commonly known as the "fat innkeeper worm" because it harbors several symbiotic animals (polychaete worms, small clams, crabs, and a fish) within its burrow. Presumably, these symbionts benefit from the continuous flow of water and the protection afforded by the burrow.

The outside of the echiuran body wall is often covered by numerous papillae. Beneath the epidermis lie three muscle layers. On the outside is a layer of circular muscle that, when contracted, reduces the diameter and lengthens the trunk. Next is the longitudinal muscle layer. In some species, these muscles may be present in bundles that are visible from the outside as stripes. Contraction of the longitudinal muscles shortens the length of the trunk and increases its diameter. An oblique muscle layer is innermost. Collectively the body-wall musculature allows an echiuran to advance or back up, rotate or even turn around in its burrow. Turning around allows them to feed on sediment surrounding all burrow apertures.

The digestive tract includes the pharynx, gizzard, a long midgut region, and short hindgut. Nutrient absorption occurs in the midgut. A small collateral tube called the "siphon" runs alongside the midgut for most of its length. No functional role has been demonstrated for this structure. The hindgut bears two thin-walled sacs called anal vesicles that may extend in the coelom (body cavity) at least half way toward the anterior end of the trunk. The sacs may be simple elongated pouches or multiply branched. Small ciliated funnels may be present on the coelomic surface of the sacs. Seawater is intermittently drawn in through the anus to fill these sacs. After a short interval, the water is released back out through the anus. This action is thought to be a means by which the animal oxygenates the internal organs.

A single ventral nerve cord lies within the body cavity on the ventral midline. At the anterior end, the nerve extends into the prostomium and forms a loop.

Most echiurans have a closed circulatory system. Blood vessels in the prostomium join in the body cavity. One branch, the ventral vessel, follows the ventral nerve cord to the posterior end. Another branch, the dorsal vessel, extends to the foregut–midgut junction and then goes on to join the ventral vessel near the anterior end. Echiurans in the genus *Urechis* have an open circulatory system without blood vessels. Blood cells in these forms are free within the body cavity.

Echiurans are dioecious (separate sexes) and most are not sexually dimorphic. Eggs and sperm begin their formation in a small indistinct gonad usually located on the ventral blood vessel at its posterior end. Early in gametogenesis, the gametes are shed into the body cavity where they complete their differentiation. Differentiated gametes are collected into gamete storage organs frequently referred to as nephridia (excretory organs). Technically these organs are gonoducts (reproductive organs) because an excretory function for them has not been demonstrated. Usually one or two pair, but occasionally three or more pairs, of gonoducts are present near the anterior-ventral region of the body cavity. Eggs and sperm are spawned to the outside through the gonopores and become fertilized by mixing in the seawater. Development produces a trochophore larva similar to the kind produced by the polychaete annelids, the group to which the echiurans are now considered to be related. Late stage

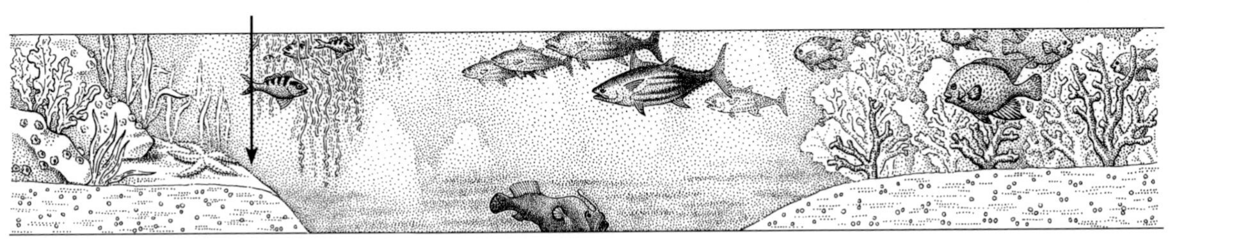

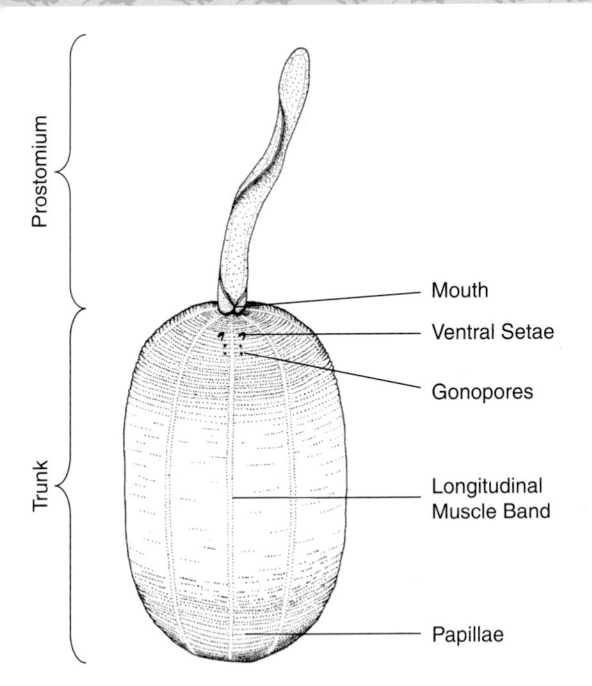

Figure A External anatomy of *Listriolobus pelodes*. (Ventral view). [Drawing by J. Pilger.]

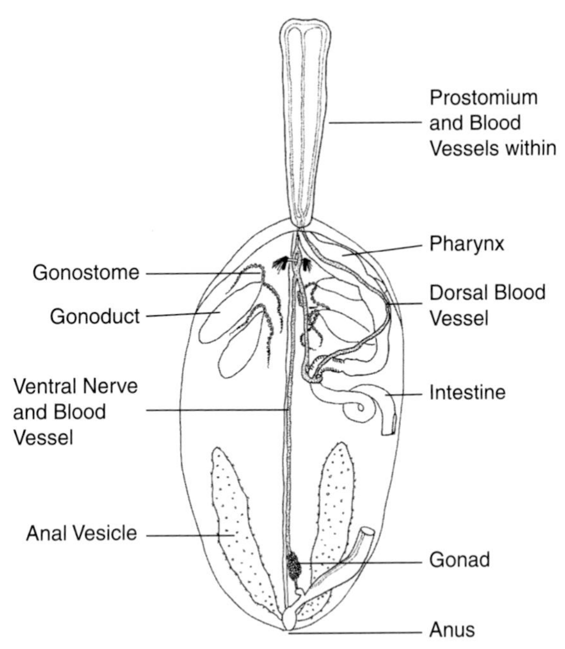

Figure B Internal anatomy of *Listriolobus pelodes*. (Dorsal view) Note that midgut is shown cut out for clarity. [Drawing by J. Pilger.]

larvae gradually change into juveniles and settle to the sea bottom where they form a burrow and take up residence.

Echiurans in the bonellid group are unique and famous for their sexual dimorphism and mechanism of sex determination. Females have the typical echiuran body plan and a trunk that may be 3–5 cm long. The forked prostomium may extend over a meter from the burrow during feeding. In stark contrast to the females, males are only about a millimeter long and they lack a distinct prostomium, trunk, and digestive tract. Further, they are not free living but reside within the single, unpaired gonoduct of the female. The sole role of these males is to produce sperm and fertilize the eggs as they pass out of the female's body during spawning. Bonellid larvae are sexually indifferent (neither male nor female). If one settles on sediment that is not fed upon by a female, it will become a female. If one settles on sediment on which a female has been feeding, it will be picked up by her while feeding and carried on the prostomium to the trunk. There it avoids being eaten at the mouth and enters the single, unpaired gonopore and gonoduct. There it will develop the degenerate morphology of a male. The powerful sex determining substance produced by the female has not been characterized.

Echiurans are believed to have been well established by the Silurian period (408–438 mya). The evidence supporting this includes fossil burrows and feeding traces on sediment similar to those made by contemporary echiurans (Figure D). Actual fossils of an echiuran have not been found since soft-bodied organisms commonly do not form fossils.

As noted earlier, echiurans are now considered by many to be within the group of marine polychaete worms. Morphological evidence supporting this hypothesis includes the structure of the body-wall musculature, the presence of chitinous setae, the method of cell division in early development, and the formation of a trochophore larva. However, the body of polychaetes and all other annelids is segmented but the echiuran body is not. Recent DNA sequence data, evidence of segmental organization in the ventral nerve cord, and the reconsideration of the distinction between segmentation and metamerism (repetition of body structures along the anterior–posterior axis) have shifted the weight of evidence toward inclusion of these interesting animals with the polychaete worms.

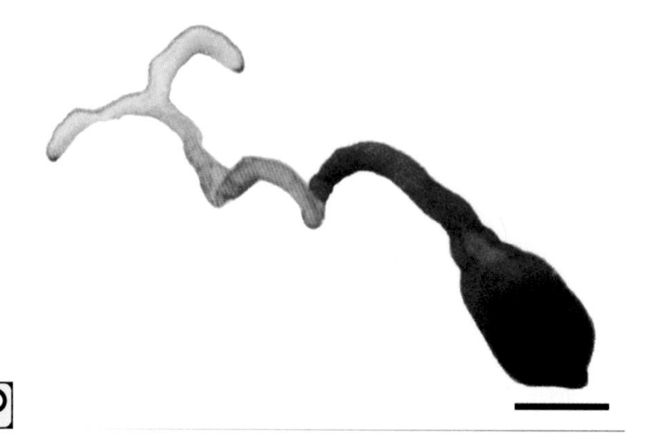

Figure C External view of a female bonellid echiuran from Belize. Note the forked prostomium and characteristic velvet-green trunk. Bar = 1 cm. [Photograph courtesy of John Pilger.]

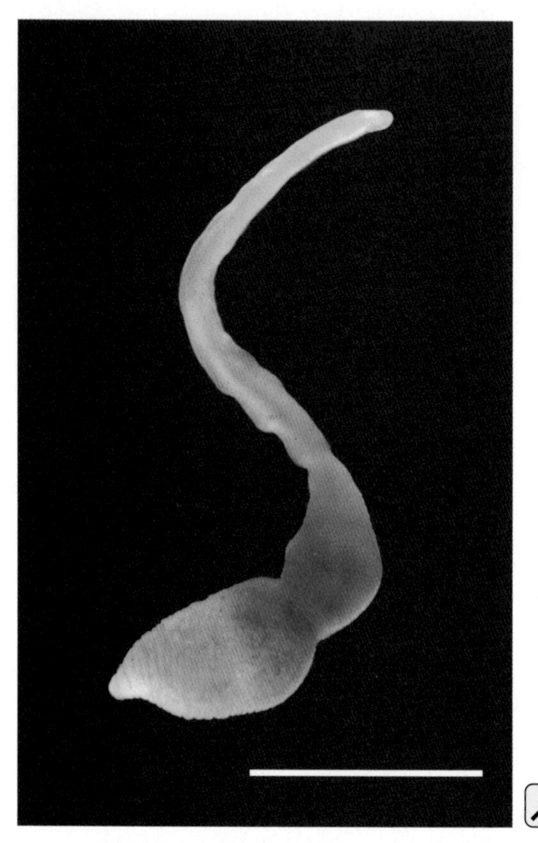

Figure D External view of *Thalassema hartmani* from Florida. The pink trunk has a wave of contraction passing over it and the ribbon-like prostomium is extended. Bar = 1 cm. [Photograph courtesy of John Pilger.]

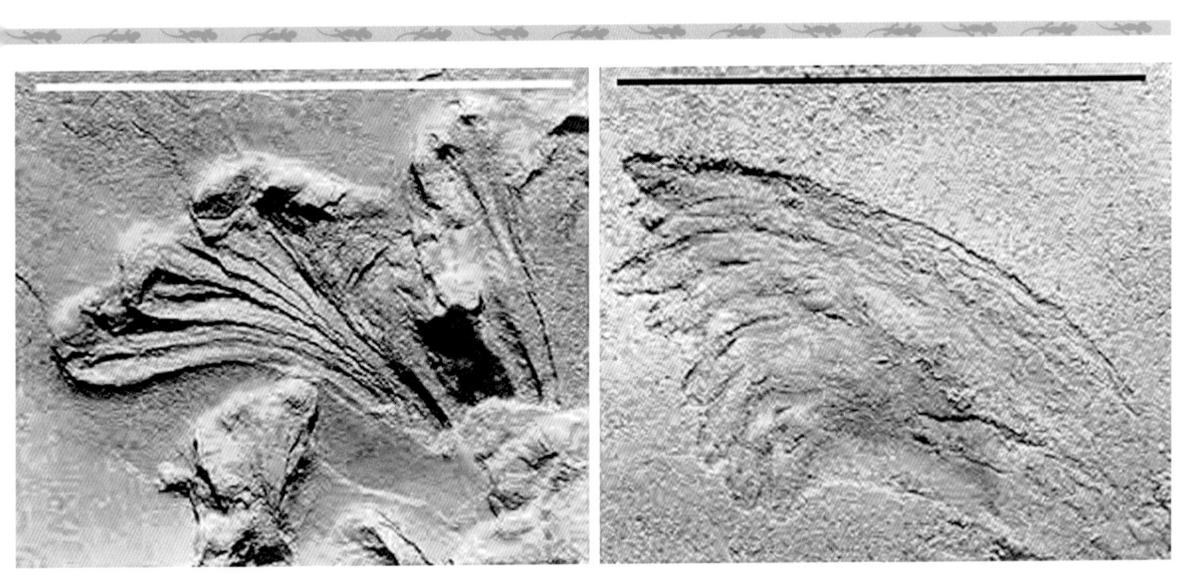

Figure E Trace fossil in Silurian sandstone from southern Ontario, Canada (left) and (right) recent feeding trace made by proboscis of *Listriolobus pellodes.* Bar = 6 cm for both. [Photographs courtesy of Michael J. Risk.]

A-25 Pogonophora

(Beard worms, pogonophorans, tube worms)

Greek *pogon,* beard; *pherein,* to bear

GENERA

*Escarpia**
Heptabrachia
*Lamellibrachia**
Lamellisabella
*Oasisia**
Oligobrachia
Polybrachia
*Ridgeia**
*Riftia**
Siboglinum
Spirobrachia
*Tevnia**
Zenkevitchiana

Pogonophorans, the beard worms, are sessile benthic marine worms that live in fixed upright chitin tubes that they secrete on sediments, shell, or decaying wood on the ocean floor. This is the only phylum of free-living animals in which all lack digestive tracts as adults. The phylum contains about 120 species in two classes, Class Perviata (perviates) and Class vestimentifera (Obturata vestimentiferans). Meredith Jones of the Smithsonian Institution assigns the six vestimentiferan genera, marked with asterisks in the list of examples above, to a proposed separate Vestimentifera phylum that includes all 10 or so vestimentiferan genera, leaving the remaining pogonophorans in Phylum Pogonophora.

Pogonophorans are probably distributed throughout the world in cold (2°–4°C), deep ocean waters, in shallow Arctic and Antarctic seas, as well as in hot (10°–15°C) submarine vents (hydrothermal vents), where the water contains hydrogen sulfide and methane. Pogonophorans of these deep, dark communities do not feed—they derive nutrients and energy from symbiotic chemoautotrophic bacteria (Phylum B-2). Although hydrogen sulfide is usually toxic to animals, pogonophorans in this hot vent habitat use their red extracellular hemoglobin to carry both hydrogen sulfide—bound so that it cannot absorb the hemoglobin site that binds oxygen—and oxygen from their tentacles to the site of bacterial oxidation. Hydrogen sulfide thus reversibly bound is not toxic to the worm.

Beard worms are collected along continental slopes, usually in seas deeper than 100 m and, more rarely, as shallow as 20 m. The first beard worms, in Class Perviata, were dredged up off the coast of Indonesia in 1900; members of Class Vestimentifera were not described until 1969. The greatest diversity of pogonophorans has been found in the western Pacific, as far south as New Zealand and Indonesia. Along the Atlantic coast, beard worms have been brought up from near Nova Scotia down to Florida, the Gulf of Mexico, the Caribbean, and Brazil. In the eastern Atlantic, pogonophorans are reported from Norway to the Bay of Biscay.

Long and skinny, most pogonophorans in Class Perviata are from 10 to 40 cm long and less than 1 mm wide. These beard worms move up and down inside their tough tubes, which are usually open at both ends (Figure A). Tubes are often banded yellow or brown. Each tube is anchored upright in soft sediment. Perviate bodies have three sections: a short forepart, a long trunk, and a short rear part (Figure B). The forepart includes a cephalic lobe containing the brain and bearing the beardlike tentacles that give the phylum its name. A given species may have a single ciliated spiral tentacle or thousands of them (see Figure A). Also in the forepart and posterior to the cephalic lobe is a glandular region that secretes the chitinous, sometimes flexible, sometimes stiff, tube. The long trunk bears papillae (bumps), cilia, and belts of toothed, chitinous bristles called setae (see Figure B). Circular and longitudinal striated muscles make up the body wall. The hind region, called the opisthosoma (Greek *opisthen,* hind; *soma,* body), is composed of 5 to 25 short segments with setae. Because the perviate body is longer than its tube, the opisthosoma facilitates burrowing; the extended narrow opisthosoma protrudes from the bottom end

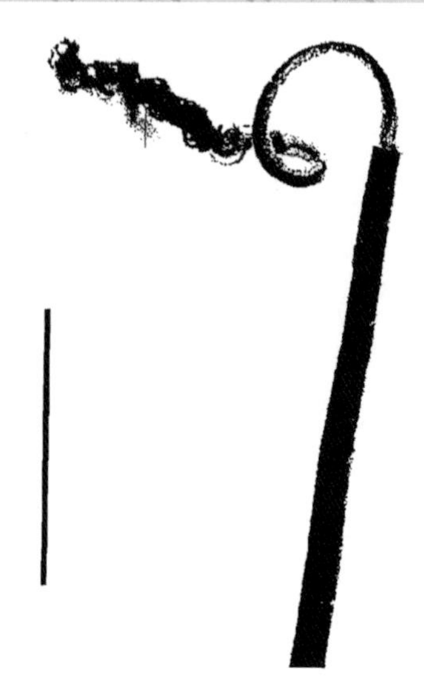

Figure A Front end of body and tentacle crown of *Oligobrachia ivanovi,* a perviate, partly dissected out of its tube. Bar = 1 cm.[Courtesy of A. J. Southward.]

of the tube and is pushed into the sediment. As blood flows in, the opisthosoma expands and the setae protrude, anchoring the posterior end, which allows the upper body to be pulled further into the substrate, anchoring the perviate. Because most perviate pogonophorans are collected by dredging the ocean floor, it took a generation to discover that all have an opisthosoma—this missing region is easily lost during dredging.

Vestimentiferans (Figure C), which include about a dozen species, are heavy-bodied pogonophorans from 2.5 to 4.0 cm wide and as much as 2 m long; they are much more plump than most other pogonophorans. The vestimentiferan body has four regions. At the top of the first region is the branchial plume—a crown of tentacles fused into sheets that is supported by a structure called the obturaculum. The obturaculum plugs the tube when the worm draws in. The second region bears body-wall folds that form an external collar called the vestimentum; inside are a heart, a brain, genital apertures, and glands that secrete tube material. Excretory pores open on the vestimentum. The third and longest region is a trunk that contains the gonad and a heavily vascularized trophosome, brown spongy tissue packed full of symbiotic chemolithoautotrophic bacteria from which these tube worms obtain energy and food. Sulfide and methane oxidized by the bacteria diffuse into the animal across the vestimentiferan's tentacles and are delivered by dissolved hemoglobin (not in cells) in both blood vessels and in coelomic fluid to the trophosome. There the reversibly bound sulfide and methane are available to the bacteria. The fourth and posterior region, called the opisthosoma, has setae that correspond to internal segments. Vestimentiferans have two compartments

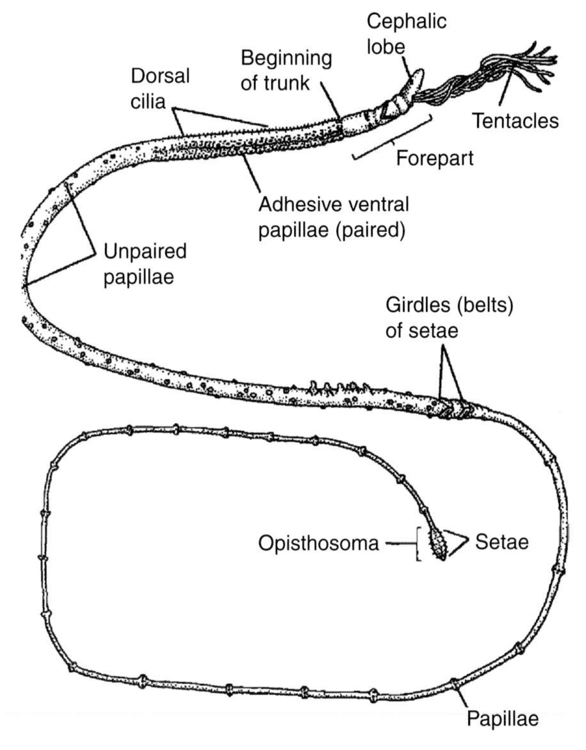

Figure B Diagrammatic and shortened view of pogonophoran removed from tube. This thin beard worm, belonging to Class Perviata, has a segmented hind region—the opisthosoma—that bears chitinous setae. [Drawing by L. Meszoly.]

Figure C *Riftia pachyptila,* vestimentiferans in their flexible tubes. Taken at a 2500 m depth off the Galapagos Islands, this is the first photograph of a live colony in situ. Bar = 25 cm.[Courtesy of J. Edmond; information from M. Jones.]

in each segment of the opisthosoma (in comparison with one compartment per segment in perviate pogonophorans). Unlike the perviate opisthosoma, the vestimentiferan opisthosoma secretes partitions that wall off the tube base, so the bottom end of the tube is closed. Perviates lack both the vestimentum and obturaculum of vestimentiferans.

The coelom of pogonophorans extends into all three body segments in perviates and into all four body segments in vestimentiferans, including the tentacles. The trunk coelom of pogonophorans contains gonads and the trophosome; the trunk coelom is unseptate, and the trunk is unsegmented. The opisthosoma coelom of all pogonophorans is segmented by muscular septa (walls) that divide it into separate compartments that correspond to the posterior region's external segmentation. The nervous system includes ventral ganglionated nerve cords including giant axons that may innervate muscles by which pogonophorans retract into their tubes.

Pogonophora adults lack a mouth, digestive tract, and anus. Gathering evidence for its mode of nutrition is a challenge, since many live in the abyss. Thin, small perviates feed by extention of their tentacles from the tube. They gather organic detritus and plankton. Suspended food particles are trapped on tentacles—cilia drive seawater through a funnel formed by the tentacles. The worms take up nutrients in solution: amino acids, glucose, and fatty acids by phagocytosis, pinocytosis and absorption through ciliary (undulipodial) membranes. Absorptive feeding by active uptake of dissolved organic matter is the rule—at least for small pogonophorans—even when pogonophorans do not extend their tentacles beyond the tube.

Studies of perviates have shown that absorption of dissolved organic matter does not suffice to meet the metabolic requirements of all species; so additional feeding modes are probably present. But, for the considerably larger vestimentiferans, the trophosome—loaded with as many as 1000 million chemolithoautotrophic bacteria per gram of trophosome tissue—is the source of nutrition. The bacteria oxidize methane (CH_4) or sulfide (H_2S) to synthesize organic molecules from carbon dioxide.

The chemolithoautotrophic bacteria generate ATP by oxidaion of vent gases. The ATP is used to "fix" CO_2 into organic compounds of the body as the source of both energy and food. Juveniles incorporate the "correct" free-living marine bacteria and transport them through a transient duct to the trophosome; a duct that functions in this manner has been described in *Lamellibrachia, Riftia, Ridgeia,* and *Oasisia* juveniles. Once the sulfide-oxi-dizing bacteria are in the larval gut, epithelial cells phagocytose but do not digest the bacteria to establish the symbioses.

A closed blood vascular system with dorsal and ventral vessels extends into each tentacle. A distinct heart and nearby paired excretory organs are located in the anterior region. The respiratory surface of the microvillus-covered tentacles is multiplied by pinnules (protrusions of the surface), major sites of gas exchange for beard worms. Their respiratory pigment in blood and coelomic fluid is red hemoglobin that carries H_2S as well as oxygen.

The genders are usually separate and externally indistinguishable. Hermaphroditic species are known. In the trunk coelom are two cylindrical gonads. Perviate pogonophore males package sperm into packets (spermatophores). Sperm are found in bundles of several hundred—not in spermatophores—in *Riftia pachyptila* and other vestimentiferans. Sperm morphology may be related to the mode of sperm transfer from male to female, suggesting the possibility of direct transfer of sperm or internal fertilization or both. Sperm are released in clouds into the water. Ciliated embryos have been taken from inside the tubes of females of at least some species, indicating that they brood their embryos. Larvae leave the tube, disperse, settle, and metamorphose into the adult form.

Studies of gastrulation in *Siboglinum* provide evidence that pogonophorans are protostome coelomates related to annelids (Phylum A-22): the mouth forms near the site of the blastopore and the body cavity is a coelom. Early embryos of two species of hydrothermal-vent pogonophorans—*Riftia* and *Ridgeia,* both vestimentiferans—are ciliated trochophore larvae. These annelid-like larvae have mouth-to-anus complete digestive tracts; later in development, the gut is lost.

A fossil *Hyolithellus* from the Lower Cambrian rocks in North America, Greenland, and northern Europe has been assigned *to* Pogonophora. It has been suggested that pogonophorans derived from a annelid-like marine worms during the early Cambrian.

Some researchers include pogonophorans in Phylum Annelida (Phylum A-22). Similarities that suggest that pogonophorans descended from annelid ancestors are as follows: muscular septa compartmentalizing the coelom; chitinous setae; hemoglobin dissolved in body fluid rather than in blood cells; trochophore larvae; and ribosomal RNA. Other specialists consider pogonophorans to be deuterostome coelomates. Therefore, they group Phylum Pogonophora—including vestimentiferans and perviates—with other deuterostome (the anus forming at the blastopore) coelomates: hemichordates (Phylum A-33), echinoderms (Phylum A-34), and chordates (Phyla A-35 through A-37). We reserve judgment regarding classification of vestimentiferans as a separate phylum from perviate pogonophorans until more is known. Whether vestimentiferans and perviates evolved convergently independently or share a common ancestor is not clear.

A-26 Mollusca
(Molluscs)

Latin *molluscus*, soft

GENERA
Aplysia
Arca
Architeuthis
Argonauta
Busycon
Chaetoderma
Charonia
Conus
Crepidula
Cryptochiton
Dentalium
Doris
Dreissena
Elysia
Haliotis
Helix
Littorina
Loligo
Mercenaria
Murex
(continued)

The concept Mollusca brings together much information about animals that look radically different from one another—snails, slugs, mussels, clams, oysters, octopuses, squids, and others. The diversity and disparity of this phylum has resulted in the recognition of eight named classes.

Estimates of the number of species alive today range from 50,000 to 130,000. Most of the shells found on modern beaches belong to molluscs, and molluscs are probably the most abundant invertebrate animals in today's oceans. Estimates of the number of known fossil species have little meaning because many new species are found annually. As an example, it was recently estimated that at least 50 percent of the fossil molluscs in the Paleozoic and Mesozoic rocks of Australia have yet to be documented.

Living molluscs range in size from microscopic snails and clams to 18 m (60-foot) long squids. Molluscs live in almost every marine and freshwater environment, and some snails and slugs live on land. In the oceans, molluscs range from the intertidal zone to the deepest ocean basins, and they may be bottom-dwelling, swimming, or floating organisms. Some squids actually glide through the air for short distances, behaving like the so-called flying fish. A few snails are ectosymbiotrophs on various invertebrates.

The word Mollusca is derived from Latin and refers to the soft body inside a shell of most species. The concept Mollusca is unified by anatomical similarities, by embryological similarities, and by evidence from fossils of the evolutionary history of the species placed in the phylum. All of this information indicates a common ancestry for the species placed in the phylum Mollusca.

The great majority of molluscs are free-living animals that have a multilayered calcareous *shell* or *conch* on their backs. Except in snails, the shell is ordinarily bilaterally symmetrical. This exoskeleton is in one piece, *univalved*, in the classes Gastropoda, Cephalopoda, Monoplacophora, and Scaphopoda. *Bivalved* shells are found in the class Pelecypoda (Bivalvia); they consist of two pieces that are on the right and left sides of the animal and at the top are held together by an elastic ligament. The class Rostroconchia has a *pseudobivalved* shell; this shell is univalved in the larval stage and bivalved in the adult stage. The class Polyplacophora has an eight-part *multivalved* shell. The class Aplacophora lacks a shell and the body is covered by calcareous *spicules*. Various living molluscs, such as slugs, octopuses, squids, some snails, and a few clams, that evolved from shelled ancestors have lost their shell, reduced it in size, or internalized the shell (Figure A).

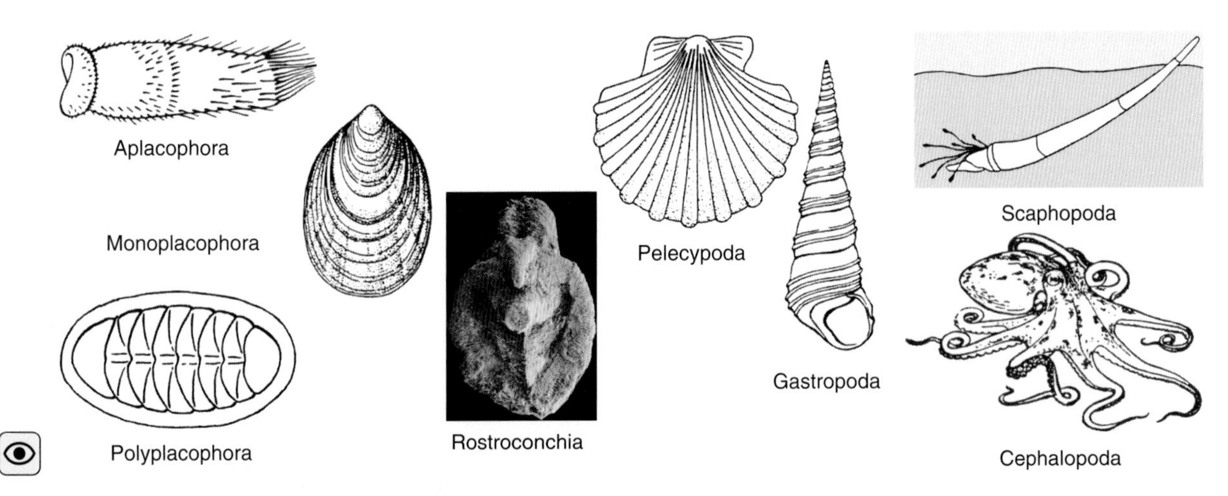

Figure A Figures of one member of each class of molluscs. The posterior view of the rostroconch shows the univalved larval shell above the bivalved adult shell.

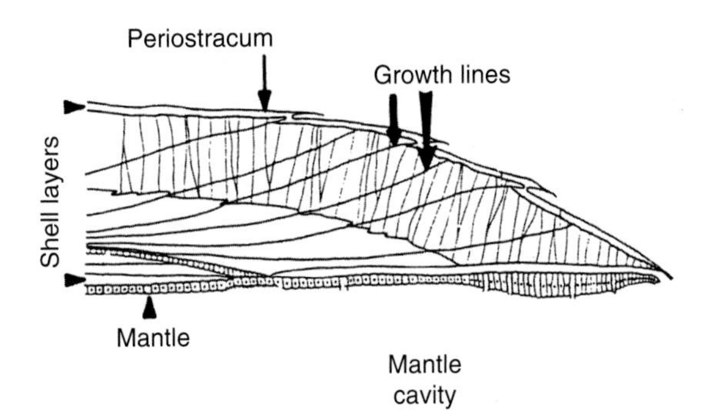

Figure B Schematic drawing of the growing edge of a pelecypod valve.

The shell provides support for the soft organs including the organs of digestion, respiration, excretion, and reproduction (*visceral mass*), the muscular *foot*, and serves as the hard surface for muscle insertions. Around all of the soft parts is a space called the *mantle cavity*, which is open to the outside. This cavity is a passageway for incoming feeds and respiratory currents, and an exit for the discharge of sex cells and waste products. The outer wall of the mantle cavity is a thin flap of tissue called the *mantle* that secretes the shell. The outermost layer of the shell is organic and is called the *periostracum*. The shell is marked by *growth lines* formed at the leading edge as the shell grows and enlarges (Figure B).

As with other organisms that have internal or external hard parts (skeletons), molluscs are readily fossilized and they have an extensive fossil record. The molluscan conch, its impression in the rocks, or the internal mold filling of the shell, is what is ordinarily available for paleontologists to study.

In the fossil record, molluscs predate the appearance of trilobites. Molluscs are already present in earliest Cambrian time (about 542 mya). Early and Middle Cambrian molluscs include many unfamiliar and extinct species, but by Late Cambrian time (501 mya) six of the eight named classes of molluscs occur in the fossil record [Gastropoda, Cephalopoda, Monoplacophora, Polyplacophora, Rostroconchia, and Pelecypoda (Bivalvia)]. By Late Ordovician time (461 mya), all classes of molluscs having a shell occur in the fossil record and at least 5000 species of molluscs have been documented in these oldest Paleozoic rocks (Figure C).

Cenozoic	Quaternary	
	Tertiary	65 mya
Mesozoic	Cretaceous	
	Jurassic	
	Triassic	251 mya
Paleozoic	Permian	
	Pennsylvanian	
	Mississippian	
	Devonian	
	Silurian	443 mya
	Ordovician	
	Cambrian	542 mya

Figure C Simplified Geologic time scale.

The only class lacking a fossil record is the spiculose and shell-less Aplacophora. For a variety of reasons, the multivalved Polyplacophora and the univalved Scaphopoda have poor and spotty fossil records. The oldest known polyplacophorans are found in Upper Cambrian rocks (490 mya), and the oldest known scaphopods occur in Middle Ordovician rocks (465 mya). All known aplacophorans, scaphopods, and polyplacophorans occur only in marine environments.

The other five classes of molluscs have extensive fossil records beginning in Cambrian time.

A-26 Mollusca
(continued)

Mya
Mytilus
Nautilus
Neomenia
Neopilina
Pecten
Sepia
Teredo
Tridacna
Vema

The univalved *monoplacophorans* (no common name) are probably the ultimate ancestors of pseudobivalved, bivalved, and the other univalved classes of molluscs. Monoplacophorans occur in the oldest Cambrian rocks (542 mya) and they are still found in modern oceans. Today, they occur in relatively deep to very deep cold marine waters below the depths to which light penetrates. However, their Cambrian and Ordovician ancestors are found in rocks that were deposited in shallow marine water in tropical to temperate climate regimes. The species diversity and morphological disparity of living monoplacophorans is but a fraction of what is known from their Paleozoic fossil record. All known monoplacophorans occur only in marine environments.

Gastropods (snails and slugs) usually have asymmetrically coiled shells. In the number of species, gastropods are the most diverse molluscs in modern oceans; they have diversified into all freshwater environments and are the only class of molluscs that also has terrestrial species. Various marine and terrestrial species have internalized the shell or lost it entirely. Their fossil record begins in Early Cambrian time (542 mya). Their first major diversification and adaptive radiation was at the beginning of Ordovician time (488 mya). The esthetic appeal of gastropod shells is highly prized by collectors and artists.

Cephalopods (*Nautilus*, ammonites, squids, octopuses, and cuttlefish) mostly are represented in modern seas by species that lack a shell or have an internal shell remnant. However, from their earliest fossil record in Late Cambrian time (490 mya) through Mesozoic time (65 mya), most cephalopods had external shells. In modern seas, only a few species of the pearly or chambered *Nautilus* have an external shell. A special structure in the external shell, called the *siphuncle*, allows gas or seawater to be added or removed from the chambers of the shell. This helps the animal move up and down the water column, much as a submarine does. As is the case with gastropods, the first major diversification and adaptive radiation of cephalopods was at the beginning of Ordovician time (488 mya). All known cephalopods occur only in marine environments. Along with pelecypods, cephalopods are a major source of human food (Figure D).

Rostroconchs is the only extinct class of molluscs. They are known only from Paleozoic rocks, Early Cambrian (542 mya) through Late Permian (255 mya). Rostroconchs are known only from rocks deposited in marine environments, and they are the likely ancestors of pelecypods (bivalves) in Early Cambrian time (540 mya) and scaphopods (460 mya) in Middle Ordovician time. Some of the unfamiliar shell forms of rostroconchs are shown in Figure E.

Pelecypods (bivalves, clams, scallops, mussels, cockles) have been a highly successful and diverse group of marine molluscs since early Middle Ordovician time (470 mya). They first occur in Early Cambrian time (540 mya) as microscopic species. Pelecypods invaded freshwater environments in Middle Devonian time (390 mya). Many types of pelecypods serve as human food.

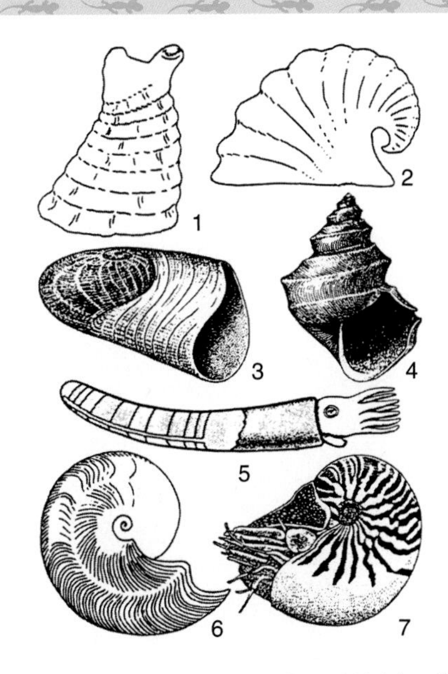

Figure D D.1 and D.2: *Yochelcionella* and *Helcionella* are small Middle Cambrian *monoplacophorans* from Australia, ranging in size from about 4 to 7mm.D.3 and D.4: *Dyeria* and *Lophospira* are Ordovician *gastropods* from Ohio and New York, about 30mm in size. D.5 and D.7: Fossil and extant *cephalopods*. D.5: Reconstruction of the curved Upper Silurian genus *Glossoceras* from Sweden; part of the shell has been cut away to show the chambers and siphuncle, about 25mm in size. D.6: The coiled Middle Cretaceous ammonite *Falciferella* from England, about 25mm in size. D.7: The living coiled shelled genus *Nautilus* from New Caledonia, about 150mm in size.

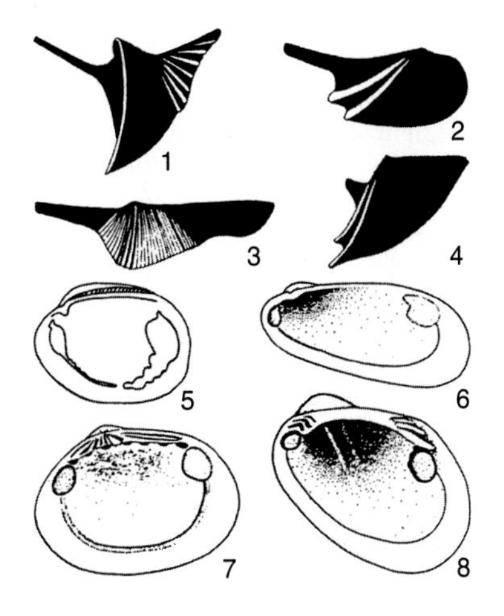

Figure E E.1–E.4: Silhouettes of the exteriors of the right valves of *rostroconchs*. E.1: *Hippocardia* from Lower Mississippian rocks of Ireland, about 50mm in size. E.2: *Technophoris* from Upper Ordovician rocks of Ohio, about 25mm in size. E.3: *Conocardium* from Mississippian rocks of England, about 35mm in size. E.4: *Bigalea* from Middle Devonian rocks of Michigan, about 12mm in size. E.5–E.8: Cambrian and Ordovician *pelecypods* showing variation in shape and internal features. E.5: *Fordilla* from Lower Cambrian rocks in New York, about 5mm in size. E.6: *Pholadomorpha* from Upper Ordovician rocks of Ohio, about 60mm in size. E.7: *Cycloconcha* from Upper Ordovician rocks in Ohio, about 15mm in size.E.8: *Cyrtodonta* from Middle Ordovician rocks in Kentucky, about 40mm in size.

A-27 Tardigrada

(Water bears, tardigrades)

(Tardigrades, water bears, moss piglets)

GENERA

Archechiniscus	*Echiniscus*	*Milnesium*
Batillipes	*Halobiotus*	*Pseudechiniscus*
Coronarctus	*Hypsibius*	*Tetrakentron*
Echiniscoides	*Macrobiotus*	*Thermozodium*

Tardigrades, or water bears, are one of the lesser-known phyla of invertebrate animals. They were first seen and reported in the 1770s. Their lumbering bearlike gait caused a German pastor, Johann August Ephraim Goeze, to call them *kleiner WasserBär*, little "water bear" giving them their descriptive common name, and an Italian monk (Lazzaro Spallanzani), to call them *il Tardigrado* (*tardi*, slow; *grado*, walker), from which the phylum Tardigrada and common name tardigrade are derived.

A tardigrade is roughly cylindrical with bilateral symmetry. There are four pairs of short stubby lobopodial limbs (that is, soft and without joints) that terminate with claws, toes, or adhesive disks. Their body length varies from about 50 μm (0.05 mm), after hatching from the egg, to a maximum of 1250 μm (1.25 mm). Most adults fall within the range 200–500 μm (0.2–0.5 mm). Tardigrades have five distinct segments including the head, three trunk segments each with a pair of legs and the caudal segment with the terminal fourth pair of legs. They have a fluid-filled body cavity (hemocoel) that functions in circulation and respiration, and a nervous system consisting of a dorsal lobed brain and ventral nerve cord with fused paired ganglia. The cuticle is composed of chitin, protein, and lipids. It also lines the fore- and hindgut and is molted several times throughout life. The digestive system consists of a complex bucco-pharyngeal apparatus consisting of a buccal tube, armed with stylets, triradiate and muscular pharynx, followed by the esophagus, midgut, and hindgut. The bucco-pharyngeal apparatus and hindgut are ejected and re-formed during molting.

Within the classification of life, the phylum Tardigrada is closely associated with the Onychophora and Arthropoda forming the Panarthropoda. In growth, tardigrades shed their cuticle—ecdysis—and are grouped with other similarly molting invertebrate animals under the title Ecdysozoa. There are two classes within the phylum Tardigrada; the Heterotardigrada, distinguished by cephalic sensory cirri and including most of the marine species and terrestrial forms with "plated" cuticles; and the Eutardigrada, lacking cephalic sensory cirri are mostly limno-terrestrial, "naked" forms.

Tardigrades live in marine, freshwater, and terrestrial environments. More than 1000 species of tardigrade have been described from the polar regions to the tropics; cryoconite holes on glaciers to hot springs; mountain springs to the intertidal zones to abyssal depths. All tardigrades require a film of water around their body to permit active life and prevent desiccation. The terrestrial environments are most studied, accounting for about 83 percent of the species; the remaining 17 percent are marine. Limno-terrestrial habitats include mosses (Figures A and B), liverworts, lichens, algae, forest litter, and soils. Truly aquatic environments can include lakes, streams, rivers, ephemeral pools, and miniature pools created by plants (for example, bromeliads or teasel *Dipsacus* spp.). Tardigrades in these environments live along side rotifers, nematodes, protoctists, and other meiofauna. Marine species are found in interstitial habitats of the littoral zone (for example, *Batillipes* spp.); benthic on algae [for example, *Styraconyx sargassi* on *Sargasssum* spp.], plates of barnacles (for example, *Echiniscoides* spp. (Figure C)) or other substrates and including a commensal, *Pleocola limnoriae* living on the isopod *Limnoria lignorum*, and an ectosymbiotroph

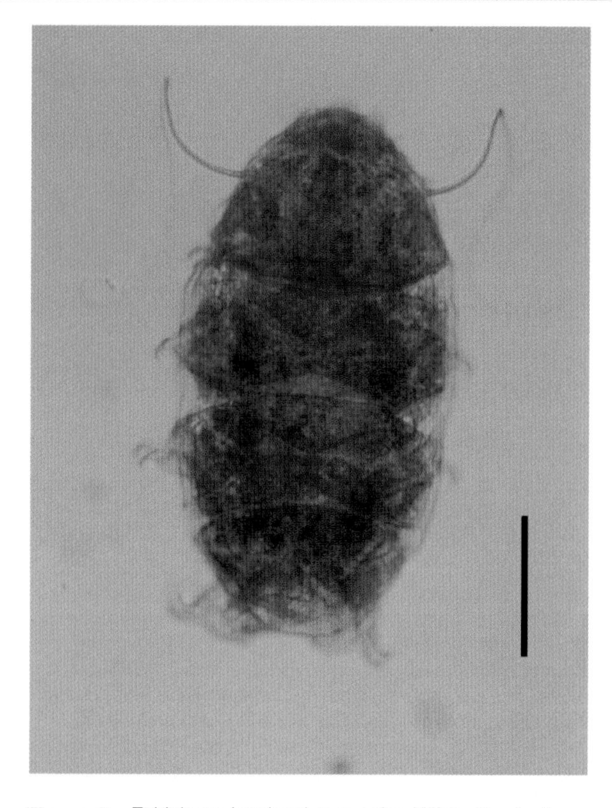

Figure A *Echiniscus jenningsi*; true color, Köhler illumination photomicroscopy. Terrestrial species found in moss from Signy Island, South Orkney Islands. Bar = 50 μm. [Image courtesy of S. McInnes, BAS.]

Tetrakentron synaptae living on the holothurian *Leptosynapta galliennei*; and deep-sea benthic habitats such as manganese nodules (for example, *Angursa capsula*). Marine habitats are relatively stable environments but the limno-terrestrial environment can experiences great changes in light, humidity, and temperature on a seasonal or even diurnal timescale, which tardigrades must be able to survive.

Tardigrades feed on plant and animal cells by piercing the cell with their stylets and then sucking out the contents via the musculature of the pharyngeal bulb. Some graze on the bacteria that cover the surfaces of their habitat, whereas others are able to ingest whole animals such as rotifers, nematodes, and other tardigrades (Figure D).

In reproduction, tardigrades are usually dioecious (that is, have males and females), which group includes all marine species. In the limno-terrestrial group, both parthenogenesis (favoring pioneer species) and hermaphroditism have been reported. A single dorsal saclike gonad is present in all tardigrades.

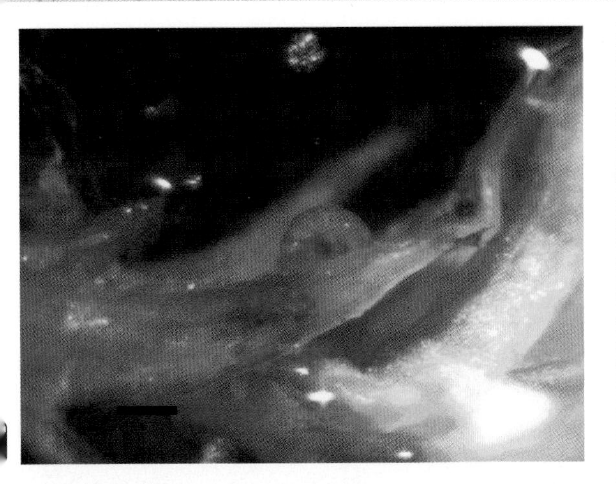

Figure B *Echiniscus* sp. on moss leaf; true color, dissecting microscope. Bar = 100 μm. [Image courtesy of S. McInnes, BAS.]

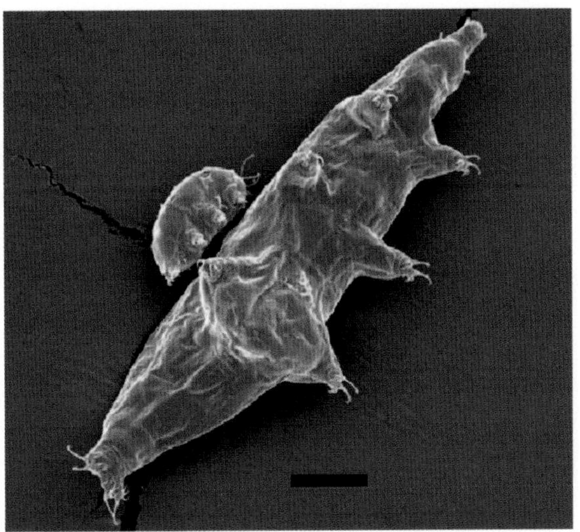

Figure D *Milnesium antarcticum* and *Echiniscus* sp.; false color, SEM image. A terrestrial species, *Milnesium* spp. are carnivorous and feed on other tardigrades and rotifers. This specimen came from Alexander Island, Antarctica. Bar = 50 μm. [Image courtesy of Plymouth University Electron Microscope Unit.]

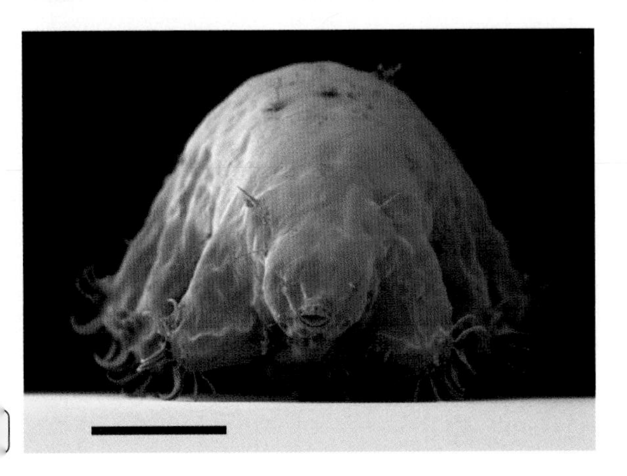

Figure C *Echiniscoides sigismundi*; false color, SEM image. Marine species found in algal holdfasts in the intertidal zone, Marion Island, Prince Edward Islands. Bar = 50 μm. [Image courtesy of S. McInnes, BAS.]

Differences between the two classes of tardigrade can be seen in the reproductive anatomy. In the eutardigrades paired male sperm ducts or a single female oviduct open into a cloaca. In the heterotardigrades, the sperm ducts and oviduct open into a ventrally situated preanal gonopore. Mating in gonochoristic species can be between a single female and one or more males. For limno-terrestrial eutardigrades fertilization usually occurs inside the body, while some marine heterotardigrades have seminal receptacles, and fertilization is external. Eggs can be grouped into four types: smooth eggs deposited directly into the substrate (marine heterotardigrades), sculptured eggs deposited in the substrate [terrestrial and freshwater eutardigrades], smooth eggs deposited in the cast cuticle (terrestrial heterotardigrades, marine eutardigrades, some freshwater and terrestrial eutardigrades), and smooth eggs deposited in the cast cuticle which is then carried (this limited parental care has been observed in a few freshwater eutardigrades Figures E-G). Seasonal (summer and winter) sculptured egg types are seen in some species. The time from egg deposition to hatching is variable between species and due to environmental factors, but usually ranges from 5 to 40 days. After embryonic development, the young tardigrade hatches by using stylets, hind legs, or hydrostatic pressure. Immature eutardigrades appear similar to adults, whereas immature heterotardigrades are noticeably different. Both take two to four instars (molts) to become sexually adult.

Marine environments are relatively constant without rapid fluctuations between temperature and humidity, and so marine tardigrades have not had to evolve a physiological mechanism to withstand oscillations of environmental extremes. The habitats occupied by the limno-terrestrial and littoral tardigrades do fluctuate and therefore these tardigrades have evolved the typical characteristics of cryptobiosis (*crypto*, hidden; *biosis*, life). In cryptobiosis metabolism, growth, reproduction, and senescence are reduced or temporarily cease, giving the organism a greater resistance to environmental extremes such as drought, cold, heat, chemicals, or ionizing radiation. Cryptobiosis in tardigrades can be divided into four main environmental factors

Figure E Egg of *Dactylobiotus* sp.; false color, SEM image. Terrestrial species found in lake sediments from Boeckella Lake, Antarctica. [Image courtesy of J. Gibson, Tasmanian Aquaculture and Fisheries Institute, Hobart, and Plymouth University Electron Microscope Unit.]

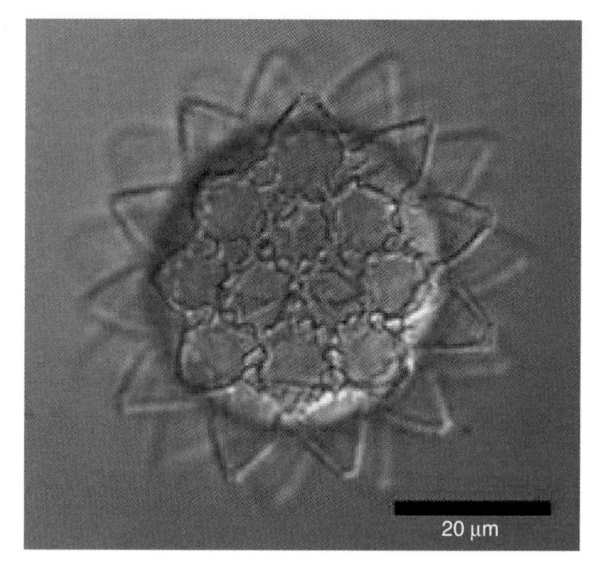

Figure G Egg of a *Macrobiotus* sp.; false colour, Differential Interference Contrast microscopy. Terrestrial species found in moss from Botswana, Africa. [Image courtesy of S. McInnes, BAS.]

Figure F Egg of a *Macrobiotus furciger*; false colour, SEM image. Terrestrial species found in lake sediments from Boeckella Lake, Antarctica. [Image courtesy of J. Gibson, Tasmanian Aquaculture and Fisheries Institute, Hobart, and Plymouth University Electron Microscope Unit.]

enforcing the shutdown of metabolism; anhydrobiosis (dehydration), cryobiosis (reduction in temperature), anoxybiosis (reduction in oxygen concentration), and osmobiosis (elevated solution concentration).

In drying conditions, some tardigrades form "tuns" (contracting into a barrel shape), which is associated with the production of trehalose, glycerol, and other membrane protectants. Cold is essentially another form of desiccation, so the membrane protectants of trehalose and glycerol can also act as freeze protectants. Anoxybiosis caused by rapid decline in oxygen concentration results in extended, turgid, and immobile animals that can survive from a few hours to five days depending on species. Slower lowering of oxygen concentrations allows some animals to form cysts, which can survive for over a year but are less resistant to other environmental extremes (for example, high temperatures). Osmobiosis is experienced by tardigrades in intertidal and euyhaline environments. However, most limno-terrestrial tardigrades form tuns in saline conditions.

Tardigrades have been described as "overengineered." In the resistant stage of a "tun," limno-terrestrial tardigrades have shown resistance to temperatures close to absolute zero and up to 151°C. They can also resist high levels of ionizing radiation and ultraviolet light, methyl bromide fumigation, vacuum and up to 6000 atmospheres of pressure.

The ability of tardigrades to resist unfavorable environmental conditions particularly allied with parthenogenetic reproduction makes the tardigrades very potent primary colonizers. If a food source is available and conditions favorable for a period of time, then the tardigrades can survive. They have been found on nunataks—isolated rocky out crops—in Antarctica that are set in a 500–1500 m deep ice sheet and separated from their nearest neighbors by 20–150 km. New islands emerging from the sea are rapidly colonized, with tardigrades arriving among the earliest. Artificial structures are similarly very rapidly colonized.

The tardigrade ability to survive in a desiccated state not only helps with primary colonization, but also offers survival over longer periods. It is a popular myth that tardigrades can survive in the desiccated form for over 100 years. Recent studies have demonstrated that the period is closer to 7 years for herbarium and 8 years for samples kept at −20°C. This is still an impressive ability for any animal to survive adverse conditions.

A-28 Onychophora

(Velvet worms, onychophorans, peripatuses)

Greek *onyx*, claw; *pherein*, to bear

GENERA

Cephalofovea
Epiperipatus
Macroperipatus
Mesoperipatus
Ooperipatus
Peripatoides
Peripatopsis
Peripatus
Speleoperipatus
Symperipatus

Onychophorans are commonly known as velvet worms; their cuticles are studded with minute bumps that feel like velvet. Velvet worms crawl like caterpillars, raising each pair of legs in waves, along with a wave of body contraction (Figure A). Paired claws at the tip of each little foot are the basis of their phylum name, Onychophora. Velvet worms usually walk on walking pads; on rough, hard substrates, they extend their claws. Onychophorans are just 14–200 mm long (*Macroperipatus*) and may be mistaken for arthropods (A-19 through A-21) or annelid worms (A-23). Females in one species may be 50 percent longer and weigh twice as much as males. The onychophoran bodies may be iridescent green, blue black, orange, red, or whitish, although most are brown. The onychophoran walks at a rate of less than 1 cm/s with its 14–43 pairs of unjointed, hollow legs. Like the muscles of annelids, its circular, longitudinal, and diagonal body-wall muscles are smooth (nonstriated). These muscles work against the hydraulic skeleton to move the velvet worm. Vascular (hemal) channels that encircle the velvet worm body, like wire in a vacuum cleaner hose, are unique to velvet worms. Hydrostatic pressure maintains the firmness of the legs as leg muscles bend and shorten the limbs; a valve at each leg base enables each leg to be extended independently by altering the pressure in the hemocoel, the main body cavity. Velvet worms are coelomates, but their coeloms are vestigial, having been reduced to gonoducts and tiny sacs surrounding the nephridia.

Velvet worms are terrestrial. The thin chitinous cuticle of the onychophorans permits water loss—to resist desiccation they require high-humidity habitats such as forest litter, the underside of logs and rocks, bromeliads, and the tunnels of termites. Velvet worms are believed to be unable to make tunnels themselves even in soft substrate but can reduce the diameter and increase the length of their soft bodies to fit into small spaces. During rain or at night, onychophorans venture forth to hunt or mate, avoiding drying by sunlight. Some hunt partly exposed from their burrow entrances. Certain species hunt in tree foliage. *Speleoperipatus* in Jamaica and *Peripatopsis alba* in South Africa inhabit caves.

About 10 genera and 100 species have been described. Onychophoran species fall into two natural groups: a northern group in tropical India, the Himalayas, West Equatorial Africa (*Mesoperipatus*), tropical America (*Epiperipatus*) as far north as Mexico, the West Indies (but not Cuba), and Malaya; and a southern group in New Guinea, temperate Australia (*Symperipatus*), New Zealand (*Peripatoides* and *Ooperipatus*), Tasmania, Madagascar, South Africa (*Peripatopsis*), and Chile. Recently, this dichotomy has been explained as being linked to Jurassic–Pliocene biogeography. These two onychophoran groups diverged some 200 mya, when a vast desert separated the northern and southern groups on the ancient southern continental mass called Gondwana. Contemporary distribution of onychophorans corresponds to the modern separated continental remnants of Gondwana. In fact, the fossil record of onychophorans, which traces from the middle Cambrian period, has been used to reassemble the historical pattern of drifting continents.

The onychophoran ventral nervous system consists of two eyes—one at the base of each antenna, a brain, and longitudinal ventral cords having transverse connections but lacking the ganglia (bundles of nerve cell bodies) found in arthropods and

Figure A *Speleoperipatus speleus*, a blind and unpigmented onychophoran, or velvet worm, taken from a cave in Jamaica. This troglodyte (cave-dwelling) species lacks eyes; other nontroglodytic onychophoran species have eyes. Bar = 1 cm. [Courtesy of R. Norton.]

annelids. Onychophoran eyes have a chitinous lens and retina and are used to direct viscous adhesive at prey and predators. Velvet worms avoid light between 470 and 600 nm; this photonegative behavior may protect them from desiccation. Cave-adapted *Speleoperipatus* and *Peripatopsis alba* are eyeless and lack body pigment. Sensory bristles on body papillae and sensory antennae orient the velvet worm to touch and perhaps to water vapor.

All onychophoran species for which the diet is known are carnivores, attacking and eating isopods and spiders (A-20), crickets and termites (A-21), and molluscs (A-26). When hunting or disturbed, velvet worms squirt secretions from adhesive glands (Figure B), modified nephridia that open through perforations in oral papillae beside the mouth. This spray congeals into bitter, elastic, sticky white threads, entangling and immobilizing their prey. An onychophoran holds prey to its mouth by sucking, then slices off bits with bladelike jaws, and liquifies tissues inside the prey with saliva secreted from salivary glands behind the jaws. While waiting for prey to be liquified, the hunter consumes much of the proteinaceous threads and then sucks in its liquid diet. Food passes through the mouth, pharynx, and esophagus; the internal organs are suspended by mesenteries within the body cavity. The midgut secretes a tubular peritrophic membrane, which encloses the food, and deposits uric acid on the inner side of the peritrophic membrane. Food is digested (whatever was not predigested in the body of the prey) and absorbed across the peritrophic membrane in the intestine, whereas waste and uric acid remain within the membrane and both are expelled, still in the membrane, from the gut through the terminal anus by the onychophorans swallowing air. Nephridia collect dissolved waste of unknown composition, which is discharged by contractile bladders at the leg bases. Onychophorans decrease water loss by nocturnal hunting, by occupying humid habitats, and by resting in pairs in body contact. Young velvet worms have been observed riding on their mothers' back.

Like the ostia in an arthropod heart, the muscular tubular dorsal heart has slits in each body segment. Pumping of the heart and body movements circulate colorless blood in the body cavity, which is a hemocoel, and in subcutaneous hemal channels, circulation is open. The hemal channels afford body firmness, which is essential for movement. Oxygen enters over the whole body through the thin cuticle and is transported from hundreds of spiracles (surface pores) to internal organs by tracheae, fine canals that carry oxygen directly to tissues. The spiracles of onychophorans cannot be closed to control water loss, unlike the spiracles of insects. Some species may also respire with thin-walled, fluid-filled, contractile, eversible sacs on their legs. These sacs function roughly as minilungs.

The sexes are separate, and each velvet worm has a pair of testes or ovaries. Females are usually larger than males. Larger females bear more offspring. In some species, fertilization is external; in others, internal. The *Peripatopsis capensis* male deposits spermatophores, packets of sperm, on the sides or back of the female. Beneath each spermatophore, the cuticle of the female erodes; eventually sperm probably travel through her hemocoel to her ovaries. In other species, the male impregnates the female by injecting sperm through her body wall. The male of *Cephalofovea tomahmontis* moves spermatophores into his cephalic pit, perhaps bending to transfer spermatophores from his genital pore to the pit on his head; the female of that species assists in the mating by pressing the male head to her cephalic pit with her legs. After mating, the blood amebocytes of the female may break down the skin of the female's cephalic pit, providing a path for sperm to travel

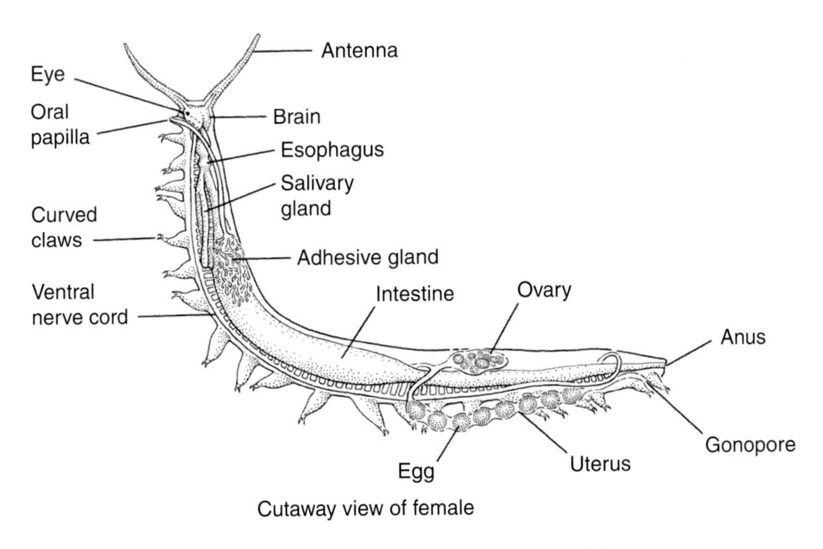

Figure B Cutaway drawing of a female velvet worm. The paired claws are extended when the velvet worm grips or climbs. [Drawing by L. Meszoly; information from R. H. Arnett.]

through her hemocoel fluid to the eggs in her reproductive tract. In any case, she stores sperm in her uterus until they are needed to fertilize eggs. The oviduct is the site of fertilization in species for which it is known. A given female may be impregnated once in her lifetime and may reproduce more than once. Development time is from 6 to 17 months for eggs and from 11 to 13 months for viviparous velvet worms; development is direct. Many tropical species are viviparous; the female nourishes her tiny embryos internally by means of a true placenta attached to the uterine wall. In viviparous velvet worms, material taken from the maternal hemocoel is stored in the placenta and released into the embryo cavity. The viviparous female bears half-inch long live young through the gonopore located between the last or penultimate leg pair. Placental development is an example of convergent evolution of placental mammals and onychophorans. In oviparous species such as *Ooperipatus*, shelled fertile eggs are laid that develop outside the female's body. In species that are ovoviviparous, shelled eggs are retained in the female until the embryos hatch. Ovoviviparous onychophorans are found in temperate areas where food supply and climate are less stable. Females in a population of *Epiperipatus imthurni* in Trinidad are parthenogenetic.

Onychophorans may link mandibulate arthropods (A-21) and annelids (A-22). Another view is that onychophorans are a sister group of annelids and that the onychophoran lineage originated from segmented ancestors. Under the larval transfer theory (Box A-i), a former onychophoran has been submitted as the adult source of caterpillar larvae of winged insects of the orders Lepidoptera, Mecoptera, and some Hymenoptera (A-21). Burgess shales from British Columbia contain a middle Cambrian fossil now considered to be an onychophoran, *Aysheaia pedunculata* Walcott. Newly discovered fossilized mandibles of onychophoran-like animals in Yunnan, China, strengthen the evidence that these fossils are onychophorans.

All three groups—arthropods, annelids, and onychophorans—have chitinous cuticles secreted by the epidermis, although the cuticles of many arthropods are nearly impervious. The velvety skin of onychophorans covered with minute papillae is unique to velvet worms. Developmental patterns are similar in arthropods and onychophorans. Both have tracheae tubes that facilitate gas exchange between tissues and the environment by openings through the body wall (spiracles). Jaws of arthropods and of onychophorans are derived from the differentiation of appendages. The onychophoran has only a single pair of jaws; among the arthropods, the crustacean has more than a single pair of jaws (mandibles and maxillae), chelicerates lack mandibles, and insects have mandibles and maxillae. An open (lacking capillaries) circulatory system with a hemocoel and a tubular, dorsal heart with ostia is similar in arthropods and onychophorans. Velvet worm adhesive secretions, used in defense and to capture prey, resemble those of millipedes and centipedes (A-21). Jaw muscles are striated in onychophorans and arthropods; onychophoran muscle other than jaw muscle is smooth.

Figure C Onychophoran with young; US quarter (25-cent piece) for scale is approximately 2 cm in diameter. (Credit: M. Chapman.)

Onychophorans also resemble annelids. Members of both these phyla molt their flexible, thin cuticles in patches; however, the onychophoran cuticle is unsegmented except for the antennae. Both onychophorans and annelids have segmentally arranged, paired nephridia that open at the base of each leg (in each annelid segment); onychophorans lack the Malpighian tubules, the principle excretory organs, of insects, centipedes, and millipedes. Eyes, when present in annelids, are ocelli, like those of onychophorans; thus onychophoran eyes resemble those of annelids more than compound insect eyes. Both annelids and onychophorans have ciliated reproductive tubules. Their chief internal organs, suspended in mesentery, are arranged similarly; onychophorans lack the internal septa of annelids. The fine details of sperm structure of annelids and onychophorans are more similar to each other than to those of arthropods. The skeleton of annelids and onychophorans (and centipedes) is hydraulic. Appendages are unjointed in onychophorans and annelids. A comparison of the mechanics of locomotion in polychaete annelids and onychophorans suggests that polychaetes are probably not direct ancestors of onychophorans, although onychophorans probably arose from annelids. This view is supported by the evidence that onychophoran legs lack joints and internal stiffening bars and are quite different from the appendages of polychaete annelids.

Analyses of mitochondrial ribosomal RNA sequences suggest that velvet worms are highly modified chelicerates or are related to chelicerates, crustaceans, and insects. This interpretation contradicts the other current major hypotheses of onychophoran evolutionary affiliations based on morphological, behavioral, and physiological evidence. On the basis of the present data of reproductive features, biogeography, and phylogeny, onychophorans may be placed between the polychaete annelids and the arthropods or as ancestors to their larvae.

A-29 Bryozoa

(Ectoprocta, bryozoans, ectoprocts, sea mats, moss animals)

Greek *bryon*, moss; *zoion*, animal
Greek *ektos*, outside; *proktos*, anus

GENERA
Alcyonidium
Bugula
Cristatella
Fredericella
Lichenopora
Membranipora
Pectinatella
Plumatella
Selenaria
Stomatopora
Tubulipora

Bryozoans—also called ectoprocts, sea mats, or moss animals—are aquatic, primarily sessile filter feeders. Like brachiopods (A-30) and phoronids (A-31), bryozoans have a tentaculate lophophore surrounding the mouth, filled with fluid and externally ciliated. The lophophore is protruded for food capturing and gas exchange. Bryozoans are modular, or colonial, organisms; each individual colony being built of a few (two or three) to many (perhaps a million or more) functionally independent units termed zooids. Colonies may form encrusting sheets, nodules or erect, branching tree forms; most are calcified to a degree, and may be bushy and flexible, or rigid, while uncalcified taxa may be rubbery or gelatinous, or lax and shrubby.

There are presently 5700 described marine species, and at least another 1000 undescribed, and about 50 species are limited to freshwaters. Marine bryozoans occur in the intertidal zone, encrusting rock, shell and seaweed, and achieve maximum diversity and density on shallow hard grounds offshore; they remain common on slope habitats worldwide, and range to beyond 5000 m depth. Freshwater bryozoans encrust stones, wood, and live plant substrates in most freshwater habitats; their colonies are solid, spongy mounds, repent branching chains, or soft gelatinous masses. Bryozoan zooids are usually less than 1 mm long; colonies may be individually small, fewer than 50 zooids and just a few millimeters in diameter, while others may exceed 0.5 m in extent. The outer body wall of the zooid, ectocyst, is underlain by an epidermis, and in a majority of species is strengthened by an internal calcite skeleton. The body cavity is lined by peritoneum, is fluid filled, and contains the polypide, comprising the lophophore, enclosed within an introverted tentacle sheath, a U-shaped gut and associated musculature. The most prominent of these is a powerful retractor muscle, allowing rapid withdrawal of the lophophore, while others close the aperture of the zooid. A network of mesenchyme, the funiculus, within each zooid extends to communication pores between neighboring zooids and faciliatates distribution of nutrients and metabolites throughout the colony.

Members of this phylum are grouped into three classes. Stenolaemata is an ancient class including one living order, Cyclostomata, comprising calcified marine species, with tubular zooids and robust colonies, such as *Tubulipora* (Figure A). Zooids are typically monomorphic, and the terminal aperture, through which the lophophore is protruded, is closed by a sphincter muscle. Class Gymnolaemata consists of two orders: the Ctenostomata (Figure B), in which the zooids are uncalcified, and the aperture is closed by a sphincter muscle, and the Cheilostomata (Figures C and D) in which zooids are more or less calcified and the aperture is closed by a chitinized, hinged flap of body wall, the operculum. Most extant marine bryozoans, and a few freshwater species, are grouped in this class. Ctenostomate zooids show limited polymorphism; variously shaped kenozooids (empty chambers lacking polypides) are modified as attachment structures, or in series from cylindrical stolons. Many cheilostomate species show striking zooid polymorphism. The autozooid, with feeding lophophore, is the basic unit of the colony, and kenozooids of varying size and shape are modified as spines and stolons, and as structural units that contribute to the architecture of the colony.

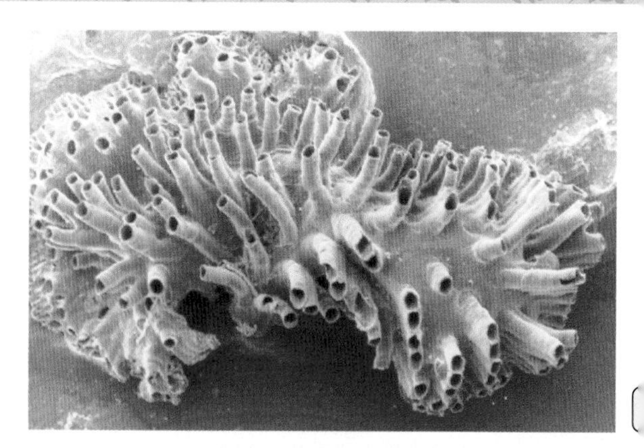

Figure A A colony of the cyclostomate *Tubulipora liliacea*, with characteristic slender, tubular zooids. [Courtesy of P. J. Hayward.]

Sexual dimorphism is seen in many species, with distinct male and female zooids interspersed with autozooids, and most brood chambers, or ovicells, consist of single or multiple polymorphic units. Avicularia and vibracula are highly modified polymorphs in which the zooid operculum forms a mandible or seta; these serve defensive functions, repelling micropredators and discouraging settlement of invertebrate larvae, and dispersing silt and debris. Class Phylactolaemata comprise most freshwater ectoprocts, such as *Plumatella* (Figure E) and *Pectinatella magnifica* (Figure F) are uncalcified. In many species the lophophore is U-shaped, but no species displays polymorphism. Colony form is limited in this class, from simple repeat, branching chains to spongy or gelatinous mounds; most are sessile, but a *Cristatella* colony can creep with its muscular shared foot. Phylactolaemates produce asexual buds called statoblasts, that develop in summer on the funiculus as balls of cells enclosed within paired chitinous valves, usually disk shaped and often with a marginal float and hooked spines. In the fall, statoblasts are released as the colony disintegrates and may be dispersed by wind and water or attached to the breasts (perhaps feet) of waterfowl. Statoblasts survive desiccation and winter cold, and in spring, the valves open and a new zooid grows from the enclosed cells to found a new colony. Statoblasts are unique to phylactolaemates; formation of this resting overwintering stage is ecologically comparable to tardigrade (A-27) resting eggs, rotifer (A-14) resting eggs, and freshwater sponge gemmules (A-3).

Bryozoan zooids have a U-shaped digestive system within the body cavity. Both the mouth and the anus face upward toward the surface; the mouth opens within the lophophore and the anus opens outside the lophophore and just beneath it. A water hydraulic system extends from the body cavity into the ring of hollow tentacles; body-wall muscles contract against the incompressible fluid in the body cavity, forcing the tentacles to protrude. Cilia (and possibly touch receptors) stud the tentacles on the lophophore. With a coordinated beat, cilia waft currents bearing fine particles toward the mouth; selected food particles pass through an esophagus and a stomach, which leads

Figure B An erect, bushy colony of the ctenostomate *Bowerbankia citrina*. Uncalcified, bottle-shaped zooids are spiraled around slender branching stolons. [Courtesy of P. J. Hayward.]

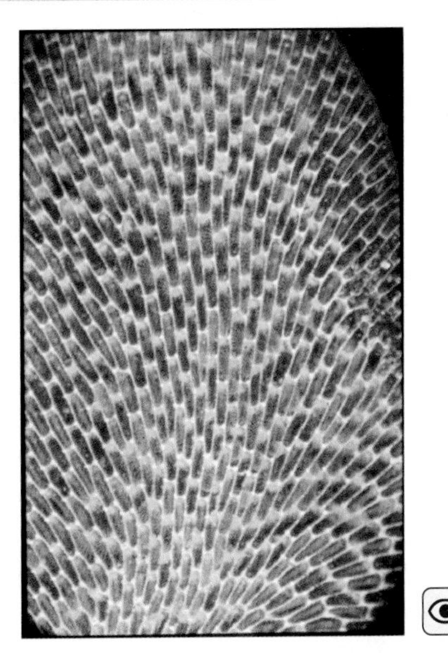

Figure C A unilaminar colony of the cheilostomate *Membranipora membranacea* encrusting a kelp blade. The polypides appear as white streaks through the membranous frontal walls of the zooids, many obscured by protruding lophophores. [Courtesy of P. J. Hayward.]

downward to a large cavity called the cecum. Food is digested by intracellular modes in the large stomach. The intestine leads back up to the anus, which discharges solid waste. Gas exchange takes place directly from all soft tissues of the zooid to water; adults lack respiratory, circulatory, and excretory organs.

Colony growth in both marine and freshwater bryozoans occurs through asexual budding of zooids. A colony is founded when a sexually produced larva attaches to a substrate and metamorphoses into a single founding zooid, or ancestrula. This immediately commences budding and the colony develops as a single genetic entity, comprising a few to many contiguous zooids; colony form may be indeterminate, as in the sheet-encrusting *Membranipora* (Figure C), or have an ordered structure, as in the fan-shaped or spiraled branching colonies of *Bugula* species. All bryozoan species are hermaphroditic; zooids are most often protandric hermaphrodites, a given zooid forming sperm and then eggs in sequence, but in many species separate male and female zooids are found in a single colony. Gonads lack ducts; sperm discharged into the body cavity leave through pores in the lophophore and are entrained in the ciliary currents of female zooids, and fertilization occurs within the body cavity of the female. Embryos may develop within the body cavity, the polypide degenerating as the embryo increases in size, or within a brood chamber, or ovicell. Fertilized eggs pass from the female body cavity through a pore (coelomopore) at the base of the lophophore and into the associated ovicell. In *Bugula* ovicells are single modified zooids, each attached to the maternal zooid; in other species the ovicell may be a part of the female zooid, or a compound polymorphic structure. Cleavage is radial, and in most marine bryozoans the egg develops as a nonfeeding coronate larva that settles within hours of release. In a minority of species, the larva is a type termed a cyphonautes, which is enclosed within a bivalved chitinous shell, has a functional gut, and a planktonic existence of several weeks. Coronate larvae have a girdle or crown of cilia used for swimming, an anterior tuft of long cilia, and a posterior adhesive sac. At first, larvae are positively phototactic and swim toward light, but soon become negatively phototactic and tend to settle in shade. Settlement is induced by biological stimuli, microbial films, and many species settle on only a narrow range of substrata; in particular, epiphytic species may be restricted to just one or two algal species. At settlement, the adhesive sac everts and secretions fasten the larva to the substratum; the larva undergoes

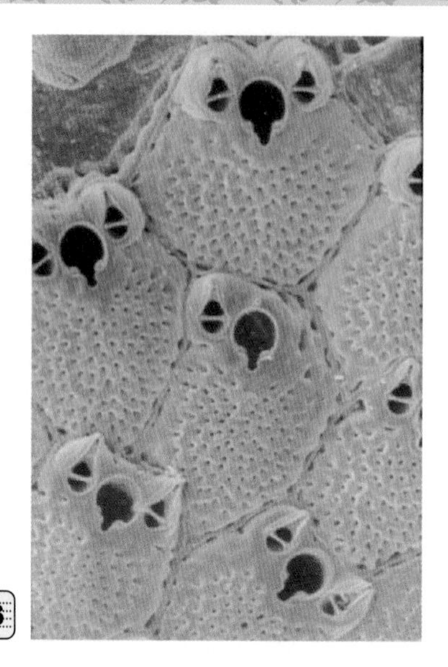

Figure D Zooids of the cheilostomate *Schizoporella magnifica*, cleaned of tissue to show the calcite skeletons. Each sinuate zooid orifice is flanked on one or both sides by a small adventitious avicularium. [Courtesy of P. J. Hayward.]

Figure E A single living zooid of *Plumatella casmiana*, showing the retractile horseshoe-shaped collar, the lophophore, from which ciliated tentacles originate. LM, bar = 0.5 mm. [Courtesy of T. S. Wood.]

Figure F This colony of freshwater Bryozoa, *Pectinatella magnifica*, extends 16 or more feet on a sunken tree trunk in Puffers Pond, Amherst, MA. Although typical sizes are about 1 foot, the massive colony closest to the camera is approximately 20″ in diameter. [Photo courtesy of James MacAllister.]

a profound metamorphosis and a single zooid, the ancestrula, develops. Because of this thorough remodeling of larval tissue, we cannot know if the adult body cavity is a true coelom.

Marine bryozoans are often pioneer species in epifaunal fouling communities, and may be abundant on ships' hulls, in docks and marinas, and on offshore installations; the most prevalent are invasive species with worldwide distributions but of unknown origin. Coral reef communities are rich in bryozoan species, although none contributes significantly to reef biomass, but temperate shell gravels often include a significant proportion of bryozoan skeletal remains as well as a substantial living biomass. In freshwaters, some phylactolaemate species may be so abundant as to cause severe fouling and blockage of water circulation systems.

Bryozoans are not known from the Cambrian, but the five orders of Stenolaemata flourished from the beginning of the Ordovician period and thousands of fossil species have been described. The class Gymnolaemata first appears in Jurassic sediments, with a sparse fauna of ctenostomates and cheilostomates, and the latter commenced a massive radiation which is probaly at its peak today.

Ectoprocts and entoprocts (A-19) were formerly classified together as subphyla of the Bryozoa. A case for reuniting the two is still occasionally made, but they have fundamentally different body plans, and each is now assigned a separate phylum; the tentacular crown of the entoprocts is not a lophophore. It surrounds both anus and mouth and is not retractable. Furthermore, entoprocts are pseudocoelomate, while ectoprocts possess a body cavity, although its homology cannot be determined with certainty.

The phylogenetic relationships of the three lophophore-bearing phyla—Phoronida, Brachiopoda, and Ectoprocta—are still discussed. All three were considered to display mixtures of deuterostome and protostome characteristics, but most recent concensus considers them to belong to the Protostomia. Molecular, morphological, and developmental evidence now links the Brachiopoda and the Phoronida, but unites the Bryozoa with Annelida and Mollusca, in the Lophotrochozoa. Phylogenetic debate has been clouded in the past by the apparently anomalous Phylactolaemata, considered as bryozoans sharing characteristic especially with phoronids. Perhaps unsurprisingly, recent molecular genetic research suggests that phylactolaemates show greater similarity to phoronids than to gymnolaemate bryozoans.

A-30 Brachiopoda
(Lampshells, brachiopods)

Latin *brachium*, arm; Greek *pous*, foot

GENERA
Argyrotheca
Dallinella
Discinisca
Glottidia
Hemithiris
Lacazella
Lingula
Megathyris
Neocrania
Notosaria
Terebratula

Brachiopods have two apposed hard shells (valves), superficially similar to clams and other bivalved molluscs (A-26). However, brachiopod valves are each bilaterally symmetrical; the ventral valve is often larger while the dorsal valve is smaller, whereas the shells of bivalved molluscs are similar in size and arranged on the right and left sides of the body. Brachiopods are called lampshells because the most common type of brachiopods resembles ancient oil lamps ("Aladdin's lamps") on-end, in life position (Figure A). Adult brachiopods living today range in size from roughly 1 mm to nearly 100 mm in shell length. The two valves are secreted by dorsal and ventral folds of epithelial tissue called mantles, which also enclose the soft body of the brachiopod, and are covered by an organic layer of periostracum (Figure B). Brachiopod mantles are not thought to be homologous with molluscan mantles. Various organs of the brachiopod body occupy the posterior part of the space between the two convex valves, whereas the anterior part is occupied by a large mantle cavity (Figure B) containing the ciliated tentacle-bearing lophophore (Figure C). The lophophore contains coelomic extensions continuous with the main body coelom, and functions in gas exchange, food gathering, and gamete release. The name Brachiopoda derives from the paired "arms" of the lophophore, which were thought, erroneously, to help the brachiopod locomote, like a foot in a bivalved mollusc.

Brachiopods living today are either attached to a hard substrate by a stalklike structure called a pedicle (for example, *Notosaria*, *Terebratula*), are pediculate and free living and burrow in soft substrates (for example, *Lingula*, *Glottidia*), or live with the ventral valve cemented directly to a hard substrate (for example, *Neocrania*). Brachiopods are always solitary, never colonial, but often occur in gregarious clusters of individuals (Figure A). They are all fully marine, live from intertidal habitats to more than 4000 m depth, and are relatively intolerant of all but normal marine salinity. They are quite patchy in their distribution and are more abundant in subtidal to deeper water habitats. They are cosmopolitan in their biogeographical distribution, but, unlike many other marine invertebrate animals today, they are distinguished by a prominently antitropical diversity gradient, where their peak diversity occurs in the temperate to polar regions, rather than in the tropics.

Because most brachiopods are attached to a hard substrate by a pedicle, they are sessile, not mobile. Once they settle out of the water column as larvae, they metamorphose into adults and cannot move from the place where they settle. Most live epifaunally and can be vulnerable to potential predators. Interestingly, incidents of boring predation on extant brachiopods are rare, and common nonboring predators (sea stars and crabs) rarely attack brachiopods, perhaps because the food rewards inside are meager.

Brachiopods are bilaterian coelomate animals, currently thought to be protostomes and members of the lophotrochozoan clade, together with annelids and molluscs. Phoronids (A-31) possess a lophophore, shared by virtue of close common ancestry with brachiopods. Both also share extensive nuclear and mitochondrial gene sequences. Bryozoans (ectoprocts; A-29) are also lophophorates, although they appear to be more distantly related to both brachiopods and phoronids within the Lophotrochozoa, on the basis of both anatomical and molecular evidence.

Three groups (subphyla) within Brachiopoda are currently recognized, each named after their most ancient living representative: Rhynchonelliformea, Craniiformea, and Linguliformea.

Figure A Three living rhynchonelliform brachiopods, *Terebratulina retusa*, dredged from a depth of about 20 m in Crinan Loch, Scotland. Bristlelike setae project from antero-lateral margins of valves. Animals attached to hard substrate by pedicle (obscured). Bar = 1 cm. [Courtesy of A. Williams.]

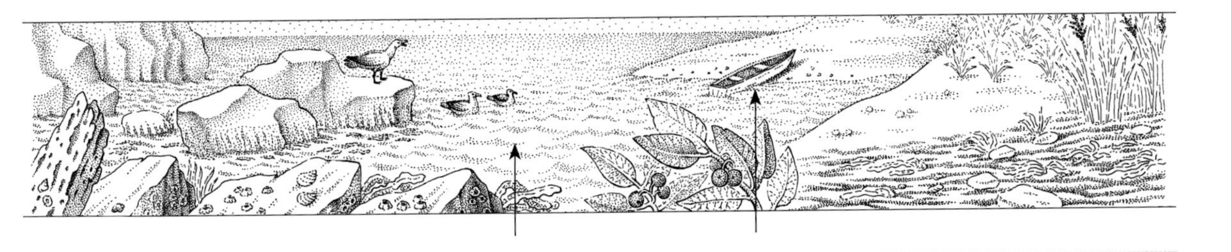

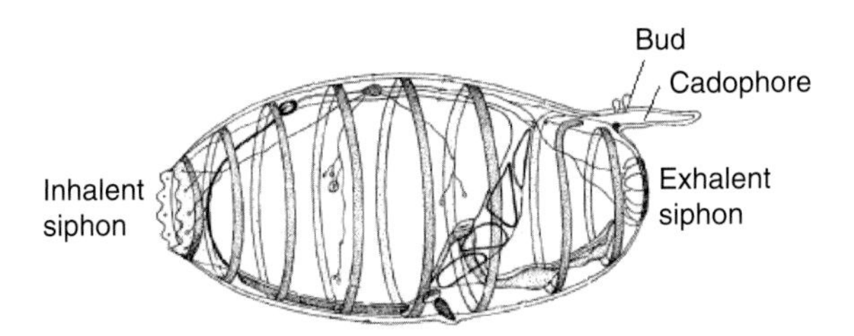

Figure D Asexual phase of the doliolid *Doliolum rarum* (about 50 mm). [Redrawn by D. I. Williamson, after Brien.]

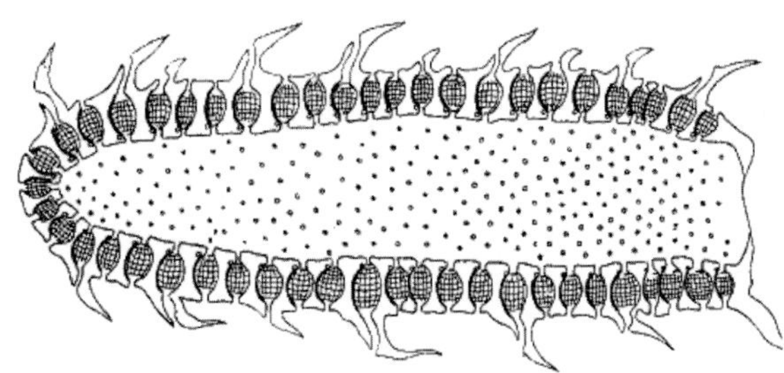

Figure F Colony of *Pyrosoma* (about 500 mm). [Redrawn by D. I. Williamson, after Brien.]

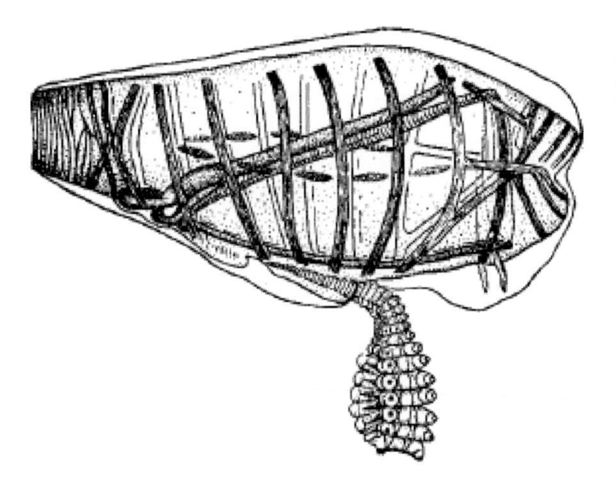

Figure E Asexual phase of the salp *Cyclosalpa pinnata* (about 60 mm). [Redrawn by D. I. Williamson, after Brien.]

niobium, and iron. Vanadium is deposited in the tunic, where concentrations may reach 105–106 times that of the surrounding seawater. The brain (cerebral ganglion) is at the upper end of the body. When a solitary ascidian is disturbed, it contracts the muscle bands in the body wall and closes both siphons. The contractions around the trapped internal water cause the tunic to become turgid, which forces the tunic spicules to project outward, possibly deterring potential predators. *Halocynthia roretzi* is cultivated for food in Korea and Japan. *Halocynthia pyriformis* (Figure B) is found in shallow water along the Atlantic coast of North America from Maine northward. *Lissoclinum*, a coral reef ascidian, hosts a photosynthetic bacterium *Prochloron* (chloroxybacteria, B-7) in the wall of its combined excretory and reproductive canal.

Although they are simultaneous hermaphrodites, ascidians seldom discharge eggs and sperm together. Many colonial species retain eggs in the oviducts until they hatch. The egg hatches as a tadpole larva, with no mouth or alimentary system (Figure C). A dorsal nerve cord (neural tube) and notochord extend almost to the tip of the muscular tail. There is a ganglion or brain at the anterior end of the nerve cord and a pigmented light receptor or ocellus. The larva has an anterior attachment organ and usually settles within a few days. The juvenile ascidian starts to form within the tadpole at hatching. The juvenile does not adopt the orientation of the larva, and most larval tissues and organs are discarded at metamorphosis. The larval and future adult nervous systems exist side by side or one above the other, before the larval system disintegrates. The ascidian ganglion is a new structure, and the tadpole's brain is discarded.

Class Thaliacea (pelagic tunicates) are divided into three orders: Doliolida (doliolids; for example, *Doliolum*), Salpida (salps; for example, *Salpa*) and Pyrosomida (pyrosomes; for example, *Pyrosoma*) (Figures D through F). A band of perforations occurs on each side of the pharynx of a doliolid; a salp has a single pharyngeal bar on each side, equivalent to one perforation; each pyrosome has numerous pharyngeal slits. Both doliolids and salps have transverse muscle bands, continuous in doliolids but discontinued ventrally in salps, and these muscles are used to produce feeding and swimming currents. They propel themselves by closing the inhalent siphon and contracting the muscles to expel water from the exhalent siphon. They can reach speeds of up to 50 body lengths per second. Both doliolids and salps have sexually and asexually reproducing generations. In doliolids, both generations are barrel-shaped. In the asexually reproducing phase (Figure D), a long series of buds migrate to a dorsal projection, the cadophore, before release as solitary individuals without cadophores, which will mature as sexually reproducing forms. Eggs of the sexual form hatch as tadpole larvae (discussed later). In the barrel-shaped asexual phase in salps (Figure E), ventral buds form a long chain of individuals which will eventually separate and become sexually reproducing forms, pointed at each end. Pyrosomes (Figure F) are always colonial, and a colony can be several meters long. The zooids, which have no transverse muscle bands, occupy a common cylindrical test, open at one end. The inhalent apertures are external, and the exhalent apertures discharge into the common lumen.

Salps and pyrosomes have no larvae, but most doliolids hatch as tadpoles, similar to those of ascidians but without attachment organs (Figure G). A complete juvenile grows at the anterior end of the doliolid tadpole, with only a narrow connection between the larval and juvenile bodies, and it breaks free

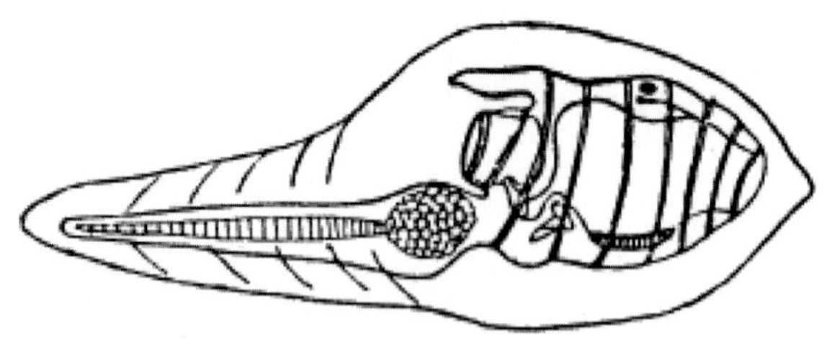

Figure G Late larva of *Doliolum* (about 0.5 mm), showing developing juvenile at anterior end. [From Borradaile *et al.*, 1935.]

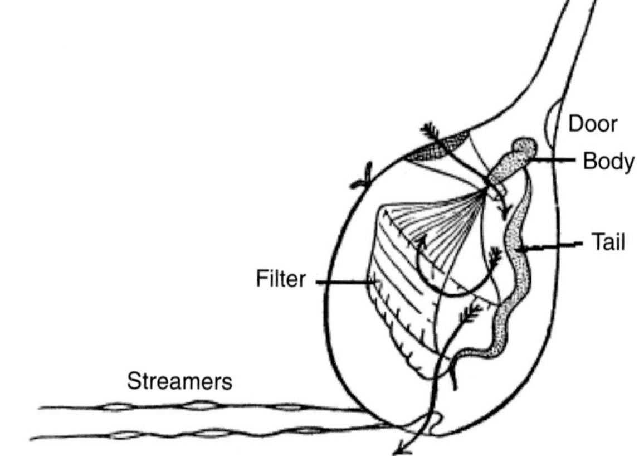

Figure H *Oikopleura dioica* in its house, showing currents (animal about 5 mm). [From Borradaile *et al.*, 1935.]

while the tadpole is still swimming. This is another example of "overlapping metamorphosis," coexistence of larva and juvenile of the same individual. Comparable cases occur in nemertines (A-10), polychaete annelids (A-22), and echinoderms (A-34). Overlapping metamorphosis is explicable in terms of larval transfer (Box A-i).

Class Larvacea or Appendicularia (for example, *Oikopleura*: Figure H) consists of tadpoles that do not metamorphose. They are planktonic and solitary and bear a general resemblance to the tadpole larvae of ascidians and doliolids. Each has an alimentary canal, with an endostyle and a pair of pharyngeal slits, and mature specimens have ovaries and testes. Each appendicularian secretes a gelatinous "house," which encloses the body and a filtration apparatus. Old houses are discarded and new ones secreted several times a day. Two theories have been proposed to explain the similarity of larvaceans to tunicate larvae. One claims that larvaceans are "persistent larvae," descended from a doliolid tadpole that matured in the larval state. The other theory claims that larvaceans are not closely related to tunicates, but they were the source of tunicate tadpole larvae: an example of larval transfer (Box A-i). The exceptionally small genome of *Oikopleura* (15,000 genes), with a very few junk genes, fits with the larval transfer explanation. Under this theory, the Larvacea are the only members of the Urochordata, and the Tunicata are a separate phylum, some of which have acquired urochordate larvae.

A-36 Cephalochordata
(Amphioxus, lancelets, Acrania)

Greek *cephalo*, head; Latin *chroda*, cord

GENERA
Branchiostoma (Amphioxus)
Epigonichthys

Cephalochordates, like larvaceans, tunicate tadpoles (A-35) and amphibian tadpoles (A-37), are acraniate chordates; that is, chordates that lack a skull. Cephalochordates range from about 5 to 15 cm in length and live on shallow sandy seafloors. A few of the 23 species in the two genera that make up this invertebrate phylum inhabit estuaries. *Branchiostoma* has a double row of gonads (Figure A); *Epigonichthys* has gonads on its right side only. Both are fishlike but scaleless and without bones and cartilage. Because cephalochordates are lance shaped, they are also called lancelets. All three defining features of chordates notochord, hollow nerve cord dorsal to the notochord, and pharyngeal gill slits persist in adult cephalochordates. The lancelet notochord persists throughout its life, like that of the larvacean urochordates (A-35). The gill slits that open in the sides of the pharynx also persist in adult cephalochordates, like those in urochordates. (In terrestrial vertebrates, such as our own species, gill slits appear as transitory embryonic structures.) Like other coelomates (A-20 through A-37), including urochordates (A-35) and craniate chordates (A-37), the cephalochordate has a coelom.

Lancelets swim to escape predators or to move to a new feeding locale. The stiff notochord flexes when the lancelet swims by contracting the longitudinal muscles in its tail; the notochord itself cannot shorten or lengthen. These clearly segmented muscles can be seen through the translucent skin of the tail. The notochord stiffens the lancelet's body, just as the vertebral column stiffens the body of a swimming fish. Fin rays in the lancelets dorsal fin may provide additional stiffening. Lancelets lack bony vertebrae. Feeding lancelets shove down into sand, turn, and emerge with their heads protruding above the sand. Twelve tiny tentacles called oral cirri (buccal tentacles) at the top of the lancelet's head screen out large particles from seawater and pass small plankton and organic particles through the pharynx to the mouth and the digestive system. Cilia in the pharynx generate water flow through the pharyngeal gill slits. This mode, ciliary filter feeding, is similar to the way in which some ascidian and thaliacean urochordates (A-35) waft water through the pharynx. Seawater that has passed through the gill slits continues to flow into the atrium—a chamber around the pharynx—and out to sea through the atriopore, which is an outlet midway along the lancelet's body. As in urochordates, an endostyle secretes a sheet of mucus that coats the gill slits. Food caught in the mucus is wrapped into a mucus-food string, which cilia pass to the intestine. Extracellular digestion and phagocytosis facilitate digestion and absorption of nutrients. A liver (hepatic cecum) extends from the intestine. Excretion is by paired nephridia, like those in annelid worms (A-22) and craniate chordates (A-37). The pharyngeal gill slits are vascularized and serve as gas exchangers as well as food gatherers. After being oxygenated in the gill slits, colorless blood is pumped by contractile blood vessels to the rest of the body in a pattern similar to the closed circulatory pattern of fish.

The dorsal nervous system of amphioxus contrasts with the solid ventral nerve cords found in nonchordate animals with well-developed nervous systems, such as arthropods (A-19 through A-21). Nerves from muscles and body wall connect to

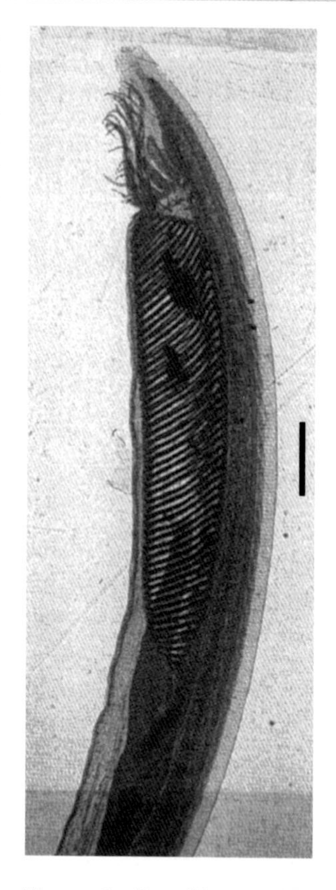

Figure A *Branchiostoma*. This best-known cephalochordate lives with its head projecting out of the sandy bottom of a warm, shallow sea. This lancelet (amphioxus) resembles the larvae of ascidian tunicates (A-36) and has segmented swimming muscles with nerves in addition to notochord, dorsal hollow nerve cord, and gill slits. Oral cirri on the head (top) sweep in phytoplankton on the water current entering by ciliary action, strain the water through gill slits visible on the pharynx into the atrium that leads to the atriopore posterior to the gills. Rays in the dorsal fin (right) are visible, as is the finger-shaped hepatic cecum behind the gills. (Tail not included in this image.) Bar = 0.5 cm. [Courtesy of Wards Natural Science Establishment, Rochester, NY.]

the hollow dorsal nerve cord of lancelets. Chemosensory cells and touch receptors are located precisely where food-bearing water enters the little body at the anterior end. Pigmented light receptors called ocelli are found along the nerve cord. Cephalochordates lack a cerebral ganglion (brain), unlike ascidian, larvacean, and thaliacean urochordates.

In breeding season, lancelets emerge from the sand to breed. Lancelets are dioecious, unlike the urochordates, which are usually

monoecious. Each sex forms gonads beneath the forepart of the intestine. Sperm from males and eggs from females exit through the atriopore into the ocean. Fertilization takes place externally. The parents return to the bottom at dawn, wriggle headfirst into the sea sand, and turn to lie head upward, partly buried. Lancelets have no larvae, but the ammocoete larvae of lampreys (A-37) resemble lancelets, one of many examples of larvae with an adult counterpart in another taxon (Larval transfer, Box A-i).

Ribosomal RNA comparisons suggest that cephalochordates are the closest relatives of vertebrates, confirming the notochord, pharyngeal gill slits, and hollow dorsal nerve cord that we have in common.

Although most animals consumed by humans are vertebrates, molluscs, and crustacea, the cephalochordates are eaten in China.

A-37 Craniata

Greek *kranion*, brain

GENERA

Alligator
Ambystoma
Amia
Anas
Anser
Aptenodytes
Apteryx
Archilochus
Ardea
Balaenoptera
Bos
Bradypus
Bubalus
Bubo
Bufo
Buteo
Camelus
Canis
Casuarius
Cavia
(continued)

Members of this phylum, our own, are the most familiar of all the animals. Craniates include about 45,000 species, including most animals of direct economic importance, except molluscs and arthropods. The major living groups of craniates are Agnatha (jawless fishes), Chondrichthyes (sharks and other cartilaginous fishes), Osteichthyes (bony fishes), Amphibia [frogs, salamanders, caecilians (Figure A)], Reptilia, (turtles, lizards, snakes, alligators), Aves (birds), and Mammalia (mammals).

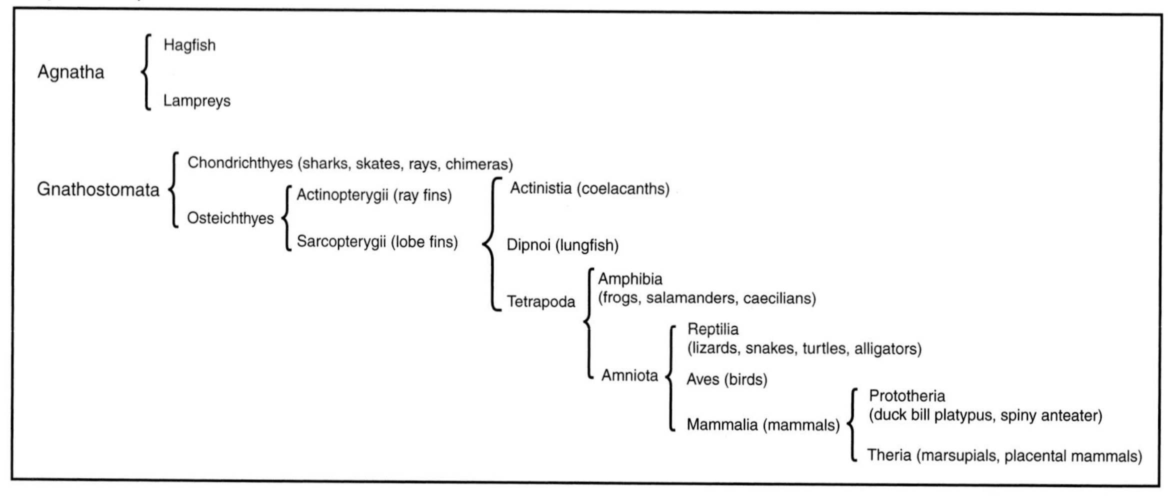

Illustration by M. Coombs

All craniates have a brain that lies within a cranium (braincase; Figure B), which distinguishes members of this phylum from the acraniate chordates—urochordates (A-35) and cephalochordates (A-36). Craniates are grouped by many as the subphylum Vertebrata in the phylum Chordata, together with the subphyla Urochordata and Cephalochordata. The presence of three defining chordate characteristics in craniates suggests that this phylum and the other chordates have common ancestry. In craniates, the first chordate characteristic—the dorsal, single, hollow (fluid-filled) nerve cord—becomes the brain and spinal cord during embryogenesis. The second characteristic defining chordates—a rod called the notochord—forms dorsal to the gut in the early craniate embryo. This slender flexible cylinder of cells containing a gelatinous matrix and sheathed in fibrous tissue extends the length of the body and persists throughout life in lampreys, hagfish, and some other groups. In all other craniates, the embryonic notochord is largely replaced by the bony or cartilaginous vertebral column—the backbone—during later development. The third chordate feature is the presence of gill slits in the pharynx at some stage of the life history, as in urochordates and cephalochordates (A-35 and A-36). Gill slits reveal the water-dwelling ancestry of the phylum. These slits are present only in the embryo and larva of tetrapod craniates with a few exceptions, such as the axolotl, a Mexican amphibian that retains gills throughout life. Gill slits persist in adults of all fish (nontetrapods). In tetrapods, gill slits grow shut and the gill skeleton is transformed into other structures such as the stapes, hyoid apparatus, and laryngeal cartilages, so gill slits are absent as such in the adult.

Craniates, like all chordates, are bilaterally symmetrical animals that develop from three embryonic germ layers: ectoderm, mesoderm, and endoderm. A well-developed coelom lined by a tissue layer called the peritoneum arises from the embryonic mesoderm. Thin membranes called mesenteries suspend the internal organs in this coelom. The bodies of craniates are partly segmented. The backbone is a series of vertebrae (Figure B) associated with nerves and muscles that are replicated in a series mirroring the segmented muscles present in cephalochordates. All craniates have a digestive tract complete with mouth and anus (Figure C). Although craniates reproduce sexually, a few species also reproduce parthenogenetically. These species consist of uniparental females in which no fertilization of the egg is required for development of the offspring. In the vast majority of species, male and female are separate individuals. Sexual reproduction requires the fertilization of a comparatively large egg by a much smaller, undulipodiated sperm.

Craniates grouped as Agnatha lack jaws and paired appendages. All the other craniates are the Gnathostomata, which have jaws—facilitating food getting and defense—and paired appendages. Extant Agnatha include, scavenging hagfish, and parasitic lampreys (*Petromyzon*), which have gill slits and a round mouth like a suction cup with horny teeth on a protrusible

Figure A *Ambystoma tigrinum*, the tiger salamander, a member of the class Amphibia and family Ambystomatidae, photographed in Nebraska. *A. tigrinum* is one of the most widespread salamander species in North America and may grow to be more than 20 cm long. The adults are black or dark brown with yellow spots. Bar = 10 cm. [Courtesy of S. J. Echternacht.]

tongue. Some authors group hagfish and lampreys together as the Cyclostomata (circular mouths), whereas others suggest that hagfish are the most basal craniates. Hagfish and lampreys have a cartilaginous skeleton and little or no trace of vertebrae. The notochord remains well developed throughout life. The cranium is a rigid box of cartilage that protects the brain. Hagfish and lampreys have smooth, scaleless skin and lack the paired fins of cartilaginous fish (such as *Squalus*) and bony fish. The oldest fossil remains of undoubtedly vertebrate animals are jawless fishes. Many of the ancient jawless fishes were heavily armored, filter-feeding bottom dwellers. The term ostracoderm is sometimes used to refer to the armored groups. Fossil agnathans greatly expand the known diversity of jawless fishes. Some apparently had paired pectoral fins and may have been more closely related to gnathostomes (jawed vertebrates) than other groups.

Extinct groups of gnathostome fishes include the armored placoderms and the multispined acanthodians. Neither group has any living descendants, though the acanthodians may be distant relatives of the Osteichthyes.

The Chondrichthyes (cartilaginous fish) and Osteichthyes (bony fish) are both gnathostomes. They have paired fins and jaws. Many species of sharks and bony fish are currently endangered by commercial exploitation and destruction of their near-shore nursery areas.

Members of Chondrichthyes lack bones; their skeletons are made instead of a softer but more flexible material called cartilage. Marine cartilaginous fish include sharks, skates, stingrays, and chimeras. Their scales, called placoid scales and composed of a dentine plate covered by an enamel-like substance, make their skin surface rough.

All other fish besides cartilaginous fish belong to the Osteichthyes, the bony fish. The Osteichthyes consists of two main groups: Actinopterygii (ray fins) and Sarcopterygii (lobe fins). The Actinopterygii includes sturgeons, gars, salmon, tuna, bass, and many other salt- and freshwater fish. Scales may be thick as in sturgeons or gars, or thin, cycloid or ctenoid, according to whether the outer edge of the scale is rounded (Greek *kyklos*, a circle) or toothed (Greek *ktenos*, comb). The Sarcopterygii includes the living coelacanth, *Latimeria* and the lungfish with three living genera that breathe with both lungs and gills. Some lungfish live in African and South American freshwater lakes with low dissolved oxygen. When a lake dries up, these lungfish secrete a cocoon, dig into the mud, and gulp air into their exposed mouths. The Sarcopterygii also includes the extinct ancestors of tetrapods.

The remaining groups of gnathostomes, Amphibia, Reptilia, Aves, and Mammalia, are all tetrapods (Greek *tetra*, four; *podos*, foot), having four limbs (except when lost, as in snakes). Modern tetrapods are the most visible animals in terrestrial habitats today.

Members of the Amphibia, about 200 described species, including salamanders (*Salamandra*), frogs (*Rana, Xenopus*), and toads (*Bufo*), lack scales and respire through their moist, flexible, scaleless skin and across the moist mouth lining or through relatively small lungs. Some salamander species lack lungs and thus respire entirely across their moist skins. A few others retain externally visible gills throughout their lives. Unlike fish, which have a single atrium and ventricle in the heart, from which blood passes to the gills and then to the body, amphibians have two atria and a single ventricle. One heart circuit carries

Chordeiles	*Cynocephalus*	*Gadus*	*Llama*
Chrysemys	*Dendrohyrax*	*Gallus*	*Manus*
Colaptes	*Elephas*	*Gavia*	*Megaceryle*
Colius	*Equus*	*Geococcyx*	*Meleagris*
Columba	*Felis*	*Gorilla*	*Mesocricetus*
Corvus	*Fregata*	*Homo*	*Micropterus*
Crotalus	*Fulica*	*Latimeria*	*Mus*
Cygnus	*Fundulus*	*Lepus*	*Myotis*

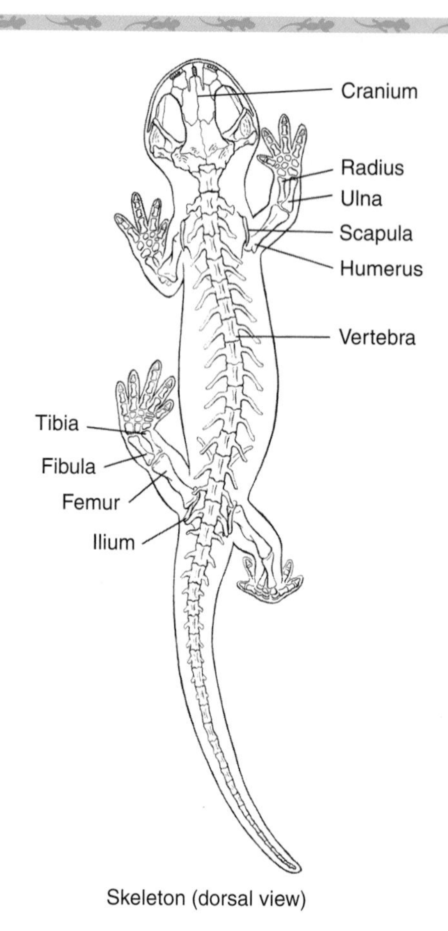

Figure B Skeleton of a generalized salamander, dorsal view. Bony vertebrae enclose the dorsal hollow nerve cord. The cranium (skull) encloses the brain. [Drawing by L. Meszoly; information from R. Estes.]

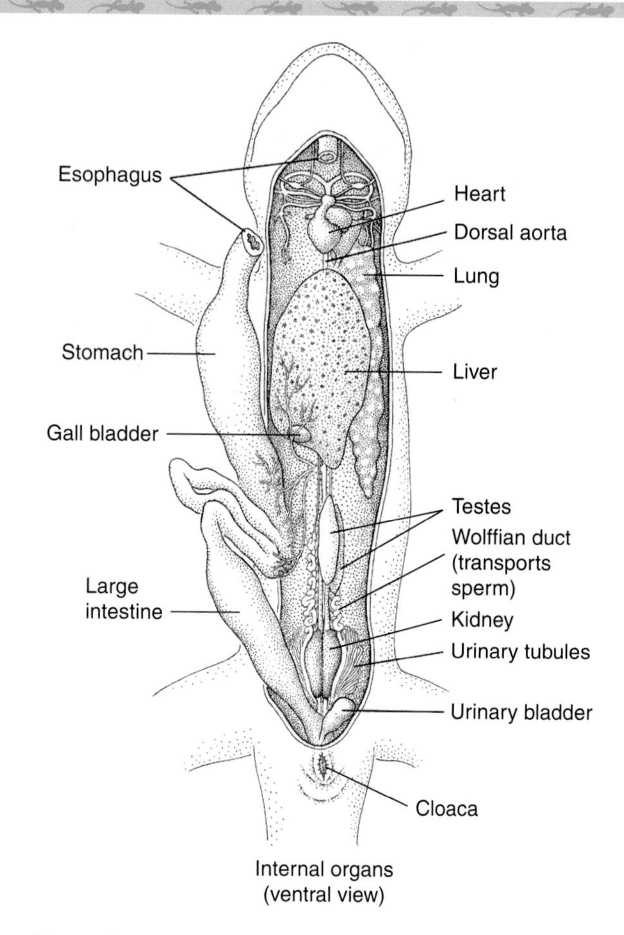

Figure C Internal, ventral view of a generalized male salamander. [Drawing by L. Meszoly; information from R. Estes.]

deoxygenated blood to the lungs, whereas the other carries oxygenated blood to body tissues. The circuits are not completely separate, however, because of the single ventricle. Many but not all amphibians lay eggs and release sperm in water, where fertilization takes place. Most amphibians spend at least the earlier part of their lives as aquatic larval, juvenile forms called tadpoles. A few tropical frogs lack larvae—their embryos develop directly into miniature adults carried on the parent's body or in the minipond formed by a bromeliad. Bullfrogs and leopard frogs live much of their adult lives in freshwater. The Mexican axolotl retains its gilled larval form even as a sexually mature, aquatic adult.

Reptiles, birds, and mammals are called amniotes. They develop from an internally fertilized egg more adapted to life on land, with a fluid-filled compartment called the amnion, in which the embryo can develop in its own small sea. Amniotes thus lack aquatic tadpoles. Members of Reptilia have dry skin covered with scales—affording protection from desiccation and predation. The skin layer (epidermis) from which the scales of reptiles develop differs from the dermal layer from which scales of fish develop. Turtles (*Chrysemys*), lizards, snakes (*Crotalus*), crocodiles, and alligators together total about 5000 species. The popular Mesozoic era dinosaurs belonged to this class. Living reptiles are ectothermic (regulating body temperature via the external environment). Some species are adapted to a wide range of internal body temperatures; others regulate their temperature with behavioral adaptations such as basking in the sun or taking shelter from direct sun by moving to cooler, moister microhabitats. Reptiles breathe by using lungs, although their scaly skin is slightly permeable to gas. The reptilian heart has two atria, and

Numida	*Petromyzon*	*Rangifer*	*Struthio*
Opisthocomus	*Phascolarctus*	*Rhea*	*Sus*
Ornithorhynchus	*Phasianus*	*Salamandra*	*Tachyglossus*
Orycteropus	*Phoca*	*Sorex*	*Tinamus*
Oryctolagus	*Podiceps*	*Sphenodon*	*Trichechus*
Ovis	*Pongo*	*Squalus*	*Vulpes*
Pan	*Puffinus*	*Sterna*	*Xenopus*
Pavo	*Raja*		

Figure D *Cygnus olor*, the mute swan, the swan common in parks and occasionally established in the wild. Swans are members of the class Aves (Latin *avis*, bird). All birds are aviators, except flightless species such as ostriches and the extinct dodo. Bar = 100 cm. [Photo by W. Ormerod.]

the ventricles are not completely separated (permitting partial mixing of oxygenated and deoxygenated blood), except in alligators and crocodiles, which have a complete septum between the left and the right ventricles. Most reptiles are well adapted to terrestrial life. Several sea turtle and snake species are endangered—overharvested for meat, shell, eggs, and skins. Baby sea turtles hatch from eggs laid on land but are disoriented by artificial night-lights along the ocean, which they must reach to feed and breed. Floridians and other coastal residents are urged to turn down outdoor night-lights to allow turtle hatchlings to reach the ocean.

Feathered reptiles—that is, birds—have traditionally been placed in their own class Aves (Figure D). However, most paleontologists agree that they are surviving dinosaurs descended from small to moderate-sized bipedal carnivorous members of the saurischian dinosaur group. Nearly 9000 living bird species—among them, *Phasianus*, *Podiceps*, *Puffinus*, and *Rhea*—are recognized. To understand avian structure and behavior, it is best to think of birds as feathered dinosaurs. Aves have land-adapted eggs with porous calcium carbonate shells. The forelimbs of many but not all (*Struthio*) bird species are modified for flight as wings, and their bones are hollow. Their scaly skin is studded with feathers; bird feathers and scales (look at the leg of a bird) are modified reptilian scales. The saying "scarce as hen's teeth" is based on biological fact: birds today lack teeth. Birds have a four-chambered heart with two atria and two ventricles, and—like mammals—they regulate their internal temperature metabolically. Animals with this ability are called endotherms. *Gavia*, the loon, has a lower body temperature than that of most other bird species.

Class Mammalia contains about 4500 living species, including *Homo sapiens*, *Bos*, *Felis*, *Phoca*, *Rangifer*, and *Sus*. As mammals, we nourish our young with milk, the nutritious and immunoprotective secretions of the mammary glands of the mother. Like birds (Aves), all mammals are endotherms—internal regulators of body temperature. Some species, such as our own, allow only a small range of temperature variation. Others—opossums and hibernators such as woodchucks and ground squirrels—evolved a much broader range of body temperatures. The hair that covers the skin of many mammals at some stage of life is one of several physical and behavioral temperature-control features. The mammalian heart has four chambers. Mammals have complete double circulation: the oxygen-rich blood in the arteries does not mix with the oxygen-depleted blood in the veins. Parental care, although not absent in other classes of vertebrates and even some invertebrates, is well developed in many mammals. Mammals have complex and differentiated teeth.

There are nearly 20 orders of mammals in 2 main groups: Prototheria and Theria. Prototheria includes the egg-laying mammals of Australia, New Guinea, and Tasmania. The spiny anteater, *Tachyglossus aculeatus*, and the duck-billed platypus, *Ornithorhynchus anatinus*—both prototherians—have lower body temperatures than do most other mammals (28.3°C compared with 38°C). The cloaca of prototherians, like that of birds and reptiles, is a common chamber for digestive waste, excretory products, and eggs or sperm. Both the spiny anteater and duck-billed platypus also have horny beaks or bills but lack true teeth. They lay shelled and yolk-rich cherry-sized eggs. All egg-laying mammals nourish their young with milk from primitive mammary glands after hatching.

Theria, mammals that do not lay eggs but retain the embryos inside the female until they are born live, includes all other mammals—Metatheria (marsupials such as opposums and *Phascolarctus*, the koala) and Eutheria (placental mammals—*Balaenoptera*, *Cavia*, *Pan*, *Vulpes*). The young of kangaroos and other metatherians are extremely immature at birth after a brief sojourn in the mother's uterus. Their relatively well-developed forelimbs permit metatherian newborns to crawl into an exterior pouch in which young suckle milk while attached to the mammary glands as they continue development. The metatherian female has a cloaca, two vaginas, and a double (Y-shaped) uterus. The eutherian female has a single vagina. Eutherian young undergo considerable development inside the mother's uterus, where they are nourished inside her body by the transfer of nutrients and by gas exchange through the placenta before she gives birth. Eutherian orders include Insectivora (hedgehogs, shrews, moles), Primates (lemurs, tarsiers, monkeys, apes, humans, chimpanzees, gorillas), Hyracoidea (hyraxes), Chiroptera (bats), Dermoptera (flying lemurs), Rodentia (porcupines, mice, squirrels, chipmunks, capybaras), Carnivora (dogs, cats, bears, otters, seals, sea lions), Scandentia (tree shrews), Xenarthra (sloths, armadillos, anteaters), Pholidota (pangolins), Lagomorpha (hares, rabbits), Cetacea (whales, dolphins), Tubulidentata (aardvark), Proboscidea (elephants), Sirenia (sea cows, manatees), Perissodactyla (horses, tapirs, rhinoceroses), and Artiodactyla (pigs, camels, llamas, deer, cattle, bison) (Box A-ii).

BOX A-ii: Karyotypic fissioning—Speciation in mammals

A karyotype is a representation of the entire species-specific set of mitotic metaphase chromosomes arranged in homologous pairs. The usual convention is an arranged photograph of the chromosome set, numbered from large to small (Todd, Figure A-ii-1). Polymorphic karyotypes in eukaryotes have resulted from diverse chromosomal rearrangements in many lineages, for example in pigs, where a given population often is heterogeneous for karyotype. Some pigs have 36, others 37, and still others 38 chromosomes. Karyotypes of plants have been shaped by polyploidy (Box Pl-i). Duplications are among the most common sorts of structural mutations in animals and give rise to gene families such as the tubulins or the hemoglobins. Chromosomal changes, for example, translocations, fusions, and fissions, are driving forces in evolution.

Chromosomal fission and fusion are processes that lead to altered karyotypes. Typically, larger metacentric chromosomes divide at the centromere to form two smaller acrocentrics. Pericentric inversions often return acrocentric chromosomes to the metacentric state. Thus, the entire genome is rearranged, but little if any DNA content is lost during these events (Kolnicki, Figure A-ii-3).

Neil Todd (1975) has proposed that fission of such whole chromosome sets is a process that underlies chromosomal evolution and speciation in mammalian taxa. Todd predicts that speciation events known from molecular clock studies and the fossil record should overlie patterns of chromosomal diversification by karyotypic fissioning. Derived taxa generally have higher numbers of smaller chromosomes compared to fewer and larger chromosomes in their ancestors.

Karyotypic sequences observed in carnivores, artiodactyls, and primate groups (old world monkeys and apes) are best explained as the result of fission. Ancestral taxa usually have large, metacentric chromosomes, whereas derived species have more numerous, smaller, often acrocentric chromosomes. Stained chromatin preparations yield banding patterns as do DNA *in situ* hybridization assays, both demonstrate homologies.

Several explanations are proposed for "division at the centromere," for example: (1) centromeric cleavage, resulting in two functional halves attached to the separated chromosomal segments, (2) preduplication of centromeric domains or activation of preexisting centromeres, and (3) "neocentromere formation," that is, epigenetic generation of new centromeric domains where none existed before. The stability and function of the new karyotype depends on the retention of genetic material, preservation of homology and meiotic symmetry, protection of broken chromosome ends, and the viability of the centromere on each new chromosome. Centromeric drive due to a bias during female meiosis, whereby higher numbers

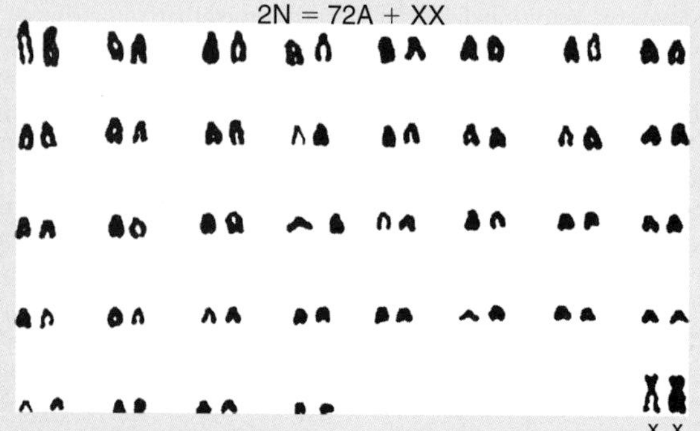

Figure A-ii-1 Karyotype of *Atelocynus microtis*, showing 36 pairs of autosomes (AA) plus one pair of sex chromosomes (XX) in this female individual. (Courtesy of N. Todd.)

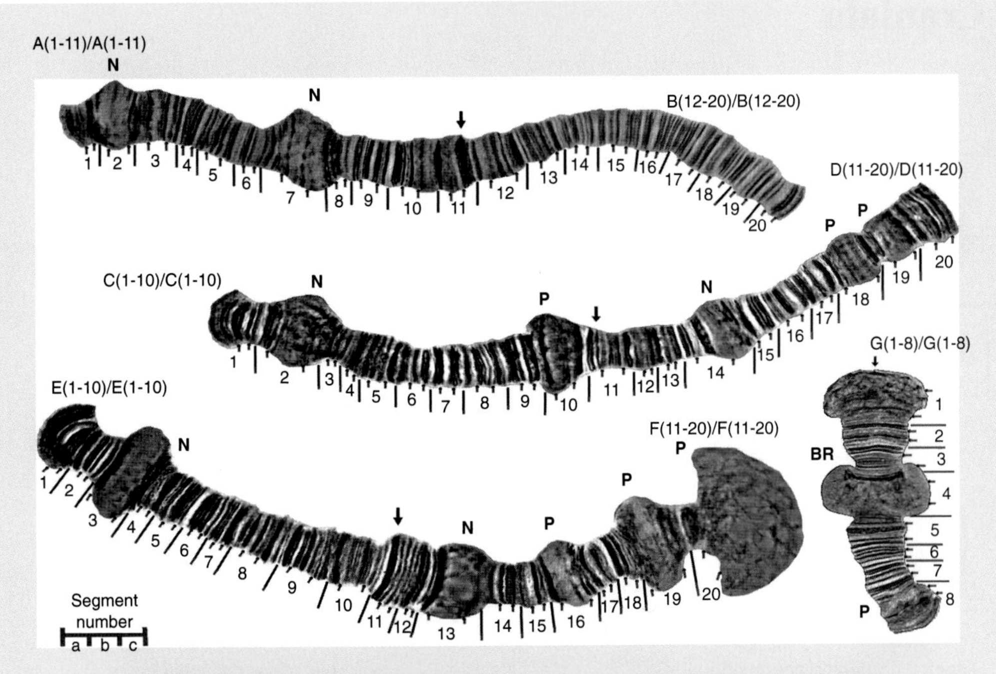

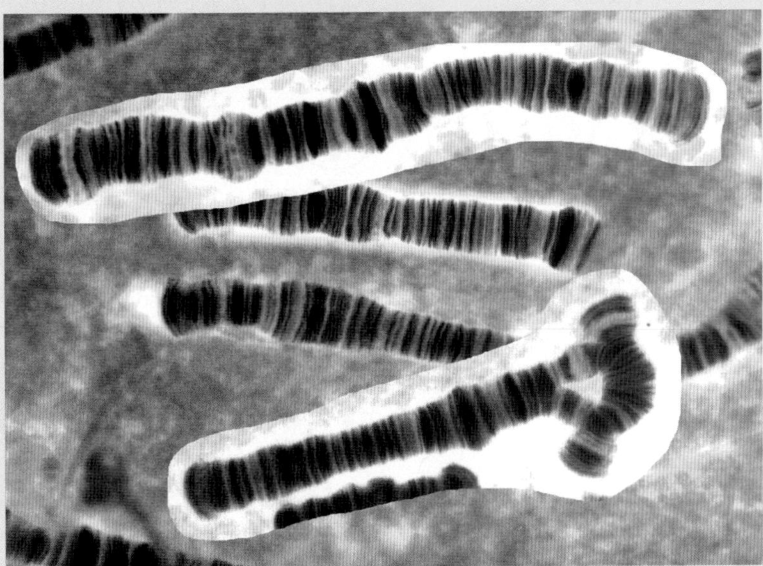

Figure A-ii-2 *Axau sp.* Phylum Mandibulata (A-21) a dipteran (fly) larva. The immature (maggot) form of the midge in its salivary glands has replicated its chromosomal DNA many times to form these thick polytene chromosomes. With this dramatic amplification of the DNA quantity the bands of nuclear genetic material can be stained to be seen, identified, counted and studied. Although the reason for differential production of total chromosomal DNA is unknown it is possibly related to the high quantity of protein needed to wrap the red maggot in silk in its burrow hole beneath the water level of the Connecticut River where it lies dormant all winter. In A-ii (top) the linear order of the genes (represented by stained bands) can easily be traced whereas in A-ii-2 bottom the triangular "loop" represents a chromosomal inversion related to reduced fertility between the populations in the northern section of old Glacial Lake Hitchcock and the southern section near the present location of the Pioneer Valley in Massachusetts to Rocky Hill Connecticut. The Glacial Lake was separated into two parts for so long that the original one species of flies began the process of evolutionary divergence into two new descendant species: north and south. The Glacial Lakes persisted for 8,000 years, long enough to separate into two populations in the incipient new species (northern without the inversion) and southern (with the inversion). The great Lake formed over 20,000 years ago and lasted until 12,000 years when the dam broke at Rocky Hill. The water drained and flowed into Long Island Sound and thus formed to the Connecticut River that separates the states of Vermont on the east and of New Hampshire on the west. The superb preservation of these events in the chromosomes of the live flies correlated with the cyclical glacial varves permits us to observe arthropod evolution in action.

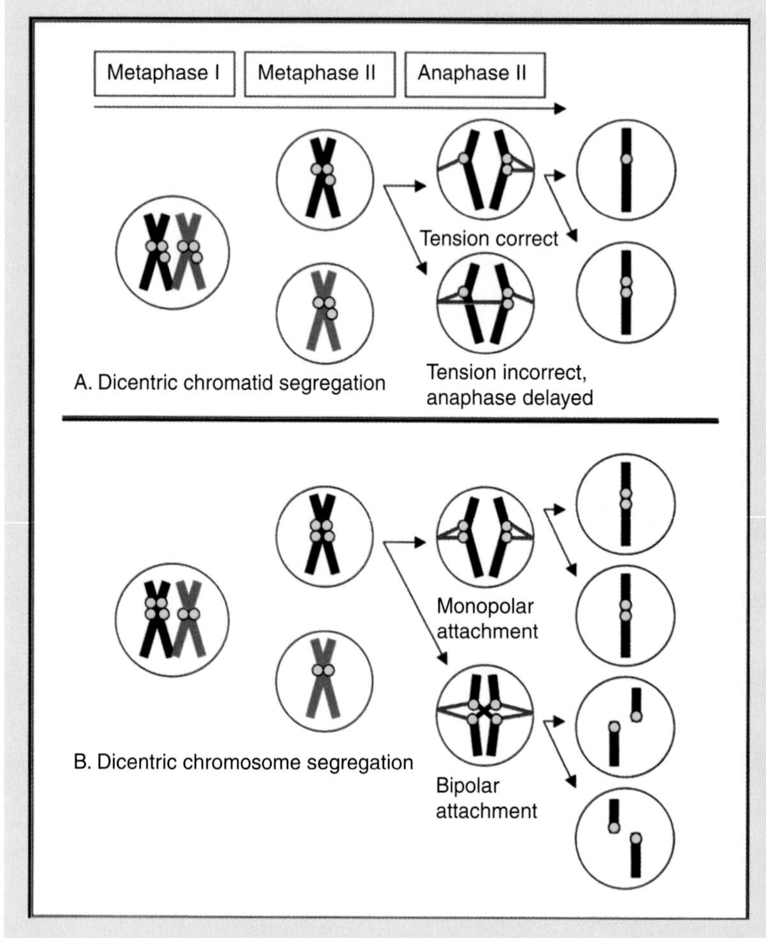

Figure A-ii-3 Karyotypic fission (= neocentromere formation as in Figure Aii-3B) is a process of change in mammalian chromosome organization, total number and size of chromosomes but does not affect the total quantity of DNA (genes in the genome) or their relative proportions to each other. For the entire karyotype of a mammal to fission so that he leaves fertile descendants in the infissioned original population the synthesis of a set of new centromeres occurs at once in development of his sperm. The newly fissioned mammal's cells have about twice the number of half-sized chromosomes. In herding mammals such a fissioned mammal can mate with his sisters and daughters to begin the spread of the new karyotype through the population. This way in which this process established itself and strongly influences speciation in mammals such as lemurs and other old world primates is discussed by Kolnicki, 2000 and 2010.

of centromeres are captured by the egg versus polar body, may favor multiple centromeric domains on chromosomes (that is, increased probabilities of fission during cell division). The new karyotype in a single individual must spread through the population to fixation; in alpha male-dominated small populations, the fissioned karyotype may prevail within a few generations.

Fission that results from the production of new centromeric domains with new kinetochores was proposed by Kolnicki (2000). Centromeric duplication can occur from DNA tandem replication. Fissioned chromosomes may form dicentric chromatids. During meiotic reduction division, the tension on dicentrics paired with unfissioned homologues is sufficient to bypass cell checkpoints such that viable gametes are produced. Gametogenesis and reproduction from fissioned chromosomes that pair with unfissioned homologues, as in the aforementioned example with pigs, has been documented. (Figure A-ii-3)

Karyotypic fissioning played a major role in the diversification of mammalian species: sequences from ancestral to descendant karyotypes within families of lemurs, artiodactyle (boars to domestic pigs) and carnivores (wolves to dogs) illustrate this. While further details are needed, the evolutionary significance of this process in mammals is beyond dispute.

References

Coghlan, A., EE. Eichler, SG. Oliver, AH. Paterson, and L. Stein, "Chromosome evolution in eukaryotes: A multi-kingdom perspective." *Trends in genetics* 21(12):673–682; 2005.

Giusto, JP., and L. Margulis, "Karyotypic fission theory and the evolution of old world monkeys and apes." *BioSystems* 13:267–302; 1981.

Godfrey, L. R., and J. Masters, "Kinetochore reproduction theory may explain rapid chromosomes evolution." *Proceedings of the National Academy of Sciences USA* 97(18):9821–9823; 2000.

Kolnicki, RL., "Kinetochore reproduction in animal evolution: Cell biological explanation of karyotypic fission theory." *Proceedings of the National Academy of Science USA* 97(17):9493–9497; 2000.

Kolnicki, R. L., and I. Rodriguez, "Centromere/Kinetochore fission in lemurs." In: L. Margulis, C. Asikainen, and W. Krumbein, eds., *Chimeras and Consciousness: Evolution of the Sensory Self.* Chelsea Green; White River Junction, VT; 2010.

Palestis, B. G., A. Burt, R. N. Jones, and R. Trivers, "B chromosomes are more frequent in mammals with acrocentric karyotypes: Support for the theory of centromeric drive." *Proceedings of the Royal Society of London B Series Supplement* 271:S22–S24; 2004.

Perry, J., HR. Slater, and KH. Andy Choo, "Centric fission—simple and complex mechanisms." *Chromosome Research* 12:627–640; 2004.

Todd, N. B., "Karyotypic fissioning and canid phylogeny." *Journal of Theoretical Biology* 26:445–480; 1970.

Bibliography: Animalia

General

Adrianov, A. V., and V. V. Malakhov, "The phylogeny and classification of the phylum Cephalorhyncha." *Zoosystematica Rossica* 3(2):181–201; 1995.

Barnes, R. D., *The invertebrates.* Blackwell; Oxford, UK; 1995.

Bayer, F. M., and H. B. Owre, *The free-living lower invertebrates.* Macmillan; New York; 1968.

Baldauf, S. L., and J. D. Palmer, "Animals and fungi are each other's closest relatives: Congruent evidence from multiple proteins." *Proceedings of the National Academy of Sciences, USA* 90:11558–11562; 1993.

Brasier, M., *Darwin's lost world. The hidden history of animal life.* Oxford University Press, Oxford UK; 2009.

Brusca, R. C., and G. J. Brusca, *Invertebrates.* Sinauer; Sunderland, MA; 1990.

Burton, M., ed., *New Larousse encyclopedia of animal life.* Bonanza Books; New York; 1980. Revised edition of Larousse encyclopedia of animal life.

Cloud, P. E., "Pre-metazoan evolution and the origins of the metazoa." In: Drake, E. T., ed. *Evolution and environment.* Yale University Press; New Haven, CT; 1968.

Conway Morris, S., J. D. George, R. Gibson, and H. M. Platt, eds., *The origins and relationships of lower invertebrates.* Systematics Association, Special Volume 28; Clarendon Press; Oxford; 1985.

Dorit, R. L., W. F. Walker, and R. D. Barnes, *Zoology.* Saunders; Philadelphia; 1991.

Erwin, D. H., *The great Paleozoic crisis: Life and death in the Permian.* Columbia University Press; New York; 1993.

Erwin, D. H., "The mother of mass extinctions," *Scientific American* 275(1):72–78 July 1996. Two hundred fifty million years before the present, mass end–Permian extinction led to expansion of mobile marine animals and decrease in immobile forms.

Gilbert, L. E., and P. H. Raven, *Co-evolution of animals and plants.* University of Texas Press; Austin, TX; 1980.

Glaessner, M. F., *The dawn of animal life*. Cambridge University Press; Cambridge, UK; 1984.
Gosner, K. L., *Guide to identification of marine and estuarine invertebrates: From Cape Hatteras to the Bay of Fundy*. Wiley (Interscience); New York; 1974. Books on Demand, Ann Arbor, MI.
Gould, S., *Wonderful life*. Norton; New York; 1989.
Grassé, P.-P., ed., *Larousse encyclopedia of the animal world*. Larousse; New York; 1975.
Grassé, P.-P., ed., *Traité de zoologie: Anatomie, systématique, biologie*, 17 vols. Masson; Paris; 1948. continuing.
Hanson, E. D., *Origin and early evolution of animals*. Wesleyan University Press; Middletown, CT; 1977. Books on Demand; Ann Arbor, MI.
Higgins, R. P., and H. Thiel, eds., *Introduction to the study of meiofauna*. Smithsonian Institution Press; Washington, DC; 1988.
House, M. R., ed., *Origin of major invertebrate groups*. Academic Press; London; 1979.
Hyman, L. H., *The Invertebrates*, 6 vols. McGraw-Hill; New York; 1940–1967. Vol. 1, Protozoa through Ctenophora; 1940. Vol. 2, Platyhelminthes and Rhynchocoela; 1951. Vol. 3, Acanthocephala, Aschelminthes, and Entoprocta; 1951. Vol. 4, Echinodermata; 1955. Vol. 5, Smaller coelomate groups; 1959. Vol. 6, Mollusca, Part 1; 1967.
Kinchin-Simonetta, A. M., and S. Conway Morris, *The early evolution of metazoa and the significance of problematic taxa*. Cambridge University Press; Cambridge, UK; 1991.
Kozloff, E. N., *Invertebrates*. Saunders; Philadelphia; 1989.
McMenamin, M. A. S., "The emergence of animals." *Scientific American* 256(4):94–102; April 1987.
McMenamin, M. A. S., and D. L. S. McMenamin, *The emergence of animals: The Cambrian breakthrough*. Columbia University Press; New York; 1990.
Meglitsch, P. A., and F. R. Schram, *Invertebrate zoology*, 3d ed. Oxford University Press; New York; 1991.
Nichols, D., J. A. L. Cooke, and D. Whitely, *Oxford book of invertebrates*. Oxford University Press; New York and London; 1971.
Nybakken, J., *Diversity of the invertebrates: A lab manual*. W. C. Brown; Dubuque, IA; 1996.
Nybakken, J., *Marine biology: An ecological approach*, 4th ed. Benjamin Cummings; Redwood City, CA; 1997.
Nybakken, J., and J. L. McClintock, *Diversity of the invertebrates: A laboratory manual*. Gulf of Mexico version. W. C. Brown; Dubuque, IA; 1997.
Parker, S. P., ed., *Synopsis and classification of living organisms*, 2 vols. McGraw-Hill; New York; 1982.
Pechenik, J. A., *Biology of the invertebrates*, 3d ed. W. C. Brown; Dubuque, IA; 1996.
Pennak, R. W., "*Ecological affinities and origins of freeliving acoelomate freshwater invertebrates.*" In: E. C. Dougherty, ed. The lower metazoa: Comparative biology and phylogeny. University of California Press; Berkeley, CA; pp. 435–451; 1963.
Pennak, R. W., *Freshwater invertebrates of the United States*, 3d ed. Wiley (Interscience); New York; 1989.
Ruppert, E. E., and R. D. Barnes, *Invertebrate zoology*, 6th ed. Saunders; Philadelphia; 1994.
Ruppert, E. E., and R. S. Fox, *Seashore animals of the southeast United States*. University of South Carolina Press; Columbia, SC; 1988.
Sterrer, W., *Marine fauna and flora of Bermuda: A systematic guide to the identification of marine organisms*. Wiley (Interscience); New York; 1986.
Thorpe, J. H., and A. P. Covich, eds., *Ecology and classification of North American freshwater invertebrates*. Academic Press; New York; 1991.
Vogel, S., *Life in moving fluids: The physical biology of flow*, 2d ed. Princeton University Press; Princeton, NJ; 1994.

Whittington, H. B., *The Burgess shale*. Yale University Press; New Haven, CT, and London; 1985.
Williamson, D. I., *Larvae and evolution: Toward a new zoology*. Chapman and Hall; London; 1994.
Williamson, D. I., and S. E. Vichers, "The origins of larvae". *American Scientist*; 95:509–517; 2007.
Willmer, P., *Invertebrate relationships*. Cambridge University Press; New York; 1990.

A-1 Placozoa

Dellaporta, S. L., A. Xu, S. Sagasser, W. Jakob, M. A. Moreno, and L. Buss, "Mitochondrial genome of *Trichoplax adhaerens* supports Placozoa as the basal lower metazoan phylum." *Proceedings of the National Academy of Sciences, USA* 103:8751–8756; 2006.
Ender, A., and B. Schierwater, "Placozoa are not derived cnidarians: Evidence from molecular morphology." *Molecular Biology and Evolution* 20:130–134; 2003.
Miller, D. J., and E. E. Ball, "Animal evolution: The enigmatic phylum Placozoa revisited." *Current Biology* 15:R26–R28; 2005.
Pearse, V.B., and O. Voigt. "Field biology of placozoans (*Trichoplax*): Distribution, diversity, biotic interactions." *Integrative and Comparative Biology*; In Press.
Signorovitch, A. Y., S. L. Dellaporta, and L. W. Buss, "Molecular signatures for sex in the Placozoa." *Proceedings of the National Academy of Sciences, USA* 102:15518–15522; 2005.
Voigt, O., A. G. Collins, V. B. Pearse, A. Ender, H. Hadrys, and B. Schierwater, "Placozoa—no longer a phylum of one." *Current Biology* 14:R994–R995; 2004.

A-2 Myxospora

Canning, E. U., and B. Okamura, "Biodiversity and evolution of the Myxozoa." *Advances in Parasitology* 56:44–131; 2004.
Fiala, I., "The phylogeny of Myxosporea (Myxozoa) based on small subunit ribosomal RNA gene analysis." *International Journal for Parasitology* 36:1521–1534; 2006.
Jiménez-Guri, E., H. Philippe, B. Okamura, and P. W. H. Holland, "*Buddenbrockia* is a cnidarian worm." *Science* 317:116–118; 2007.
Kent, M. L., K. B. Andree, J. L. Bartholomew, M. El-Matbouli, S. S. Desser, R. H. Devlin, S. W. Feist, R. P. Hedrick, R. W. Hoffman, J. Khattra, S. L. Hallett, R. J. G. Lester, M. Longshaw, O. Palenzuela, M. E. Siddall, and C. Xiao, "Recent Advances in Our Knowledge of the Myxozoa." *The Journal of Eukaryotic Microbiology* 48:395–413; 2001.
Lom, J., and I. Dyková, "Myxozoan genera: definition and notes on taxonomy, life cyce terminology and pathogenic species." *Folia Parasitologica* 53:1–36; 2006.

A-3 Porifera

Fieseler, L., M. Horm, M. Wagner, and U. Hentschel, "Discovery of the novel candidate phylum Poribacteria in marine sponges." *Applied and Environmental Microbiology* 70:3724–3732; 2004.
Imhoff J. F., and R. Stohr, Sponge-Associated bacteria: General overview and special aspects of bacteria associated with *Halichondria panacea*. pp. 35–57. In W.E.G Muller ed. SPoonges Spinger. 2003.
Vacelet, J., "Etude en microscopie electronique de l'association entre bacteries et spongaires du genre *Verongia*." *J. Microsc. Biol. Cell* 23:271–288; 1975.
Vacelet, J., and C. Donadey, "Electron microscopy study of the association between some sponges and bacteria." *Journal of Experimental Marine Biology and Ecology* 30:301–314; 1977.
Vacelet, J., and E. Duport, "Prey capture and digestion in the carnivorous sponge." *Asbestopluma hypogea* (Porifera: Demospongiae) *Zoomorphology* 123:179–190; 2004.

A-4 Cnidaria

Argo, V. N., "Mechanics of a turnover: Bell contractions propel jellyfish." *Natural History* 74:26–29; August–September 1965.

Brown, B. E., and J. C. Ogden, "Coral bleaching." *Scientific American* 268:64–70; January 1993.

Gao, T. J., "Morphological and biomechanical differences in healing in segmental tibia defects implanted with Biocoral [bone substitute] or tricalcium phosphate cylinders." *Biomaterials* 18:219–223; 1997.

Gould, S. J., "A most ingenious paradox." *Natural History* 93:20–29; December 1984.

Gowell, E. T., *Sea jellies: Rainbows in the sea*. Franklin Watts; New York; 1993.

Grange, K., and W. Goldberg, "Fjords down under." *Natural History* 102:60–69; March 1993.

Jacobs, W., "Floaters of the sea." *Natural History* 71:22–27; August–September 1962.

Kramp, P. L., "Synopsis of the medusae of the world." *Journal of the Marine Biological Association of the United Kingdom* 40:1–469; 1961.

Lane, C. E., "The Portuguese man-of-war," *Scientific American* 202:158–168 March 1960. Physalia.

Muscatine, L., and H. M. Lenhoff, eds. *Coelenterate biology*. Academic Press; New York; 1974.

Rees, W. J., ed., *The Cnidaria and their evolution*. Academic Press; New York; 1966.

Roux, F. X., D. Brasnu, B. Loty, B. George, and G. Guillemin, "Madreporic coral: A new bone graft substitute for cranial surgery." *Journal of Neurosurgery* 69:510–513; 1988.

Shick, J. M., *A functional biology of sea anemones*. Chapman and Hall; New York; 1991.

A-5 Ctenophora

Hardy, A., *Great waters*. Harper & Row; New York; 1967.

Matsumoto, G. I., and G. R. Harbison, "*In situ* observations of foraging, feeding, and escape behavior in three orders of oceanic ctenophores: Lobata, Cestida, and Beroida." *Marine Biology* 117:279–287; 1993.

Purcell, J. E., and J. H. Cowan, Jr., "Predation by the scyphomedusan *Chrysaora quinquecirrha* on *Mnemiopsis leidyi* ctenophores." *Marine Ecology Progress Series* 129:63–70; 1995.

Robison, B. H., "Light in the oceans midwaters," *Scientific American* 273:60–64 July 1995. Exploring the lives of bioluminescent sea animals by using ROVs (remotely operated vehicles) and submersibles.

Russell-Hunter, W. D., *A biology of lower invertebrates*. Macmillan; New York; 1968.

Tamm, S. L., and S. Tamm, "Reversible epithelial adhesion closes the mouth of Beroë, a carnivorous marine jelly," *Biological Bulletin* 181:463–473 1991. How a pink, blimp-shaped ctenophore feeds.

Vogel, S., "Natures pumps." *American Scientist* 82:464–471; September–October 1991.

A-6 Gnathostomulida

Briggs, D. E. G., "Conodonts: A major extinct group added to the vertebrates." *Science* 256:1285–1286; 1992.

Fenchel, T. M., and R. Riedl, "The sulfide system: A new biotic community underneath the oxidized layer of marine sand bottoms." *Marine Biology* 7:255–268; 1969.

Gould, S. J., "Natures great era of experiments." *Natural History* 92:12–21; July 1983.

Lammert, V., "Gnathostomulida." In: F. W. Harrison and E. E. Ruppert, eds., *Microscopic anatomy of invertebrates, Vol. 4: Aschelminthes*. Wiley-Liss; New York; pp. 19–39; 1991.

Sterrer, W., M. Mainitz, and R. M. Rieger, "Gnathostomulida: Enigmatic as ever." In: S. Conway Morris, D. George, R. Gibson, and H. M. Platt, eds., *The origins and relationships in lower invertebrates*. Oxford University Press; Oxford; pp. 181–199; 1986.

A-7 Platyhelminthes

Balavoine, G., "Are platyhelminthes coelomates without a coelom? An argument based on the evolution of Hox genes." *American Zoologist* 38:843–858; 1998.

Cheng, T. C., *Parasitology*, 2d ed. Academic Press; New York; 1986.

Dawes, B., *The Trematoda*. Cambridge University Press; Cambridge, UK; 1946.

Erasmus, D. A., *The biology of trematodes*, Crane, Russak; New York; 1974. Arnold; London; 1972.

Ferguson, M. A., T. H. Cribb, and L. R. Smales, "Life-cycle and biology of *Sychnocotyle kholo* n.g., n. sp. (Trematoda: Aspidogastrea) in *Emydura macquarii* (Pleurodira: Chelidae) from southern Queensland, Australia." *Systematic Parasitology* 43:41–48; 1999.

Jennings, J. B., "Physiological adaptations to entosymbiosis in three species of graffillid rhabdocoels." *Hydrobiologia* 84:147–153; 1981.

Poddubnaya, L. G., J. S. Mackiewicz, and B. I. Kuperman, "Ultrastructure of *Archigetes sieboldi* (Cestoda: Caryophyllidea): relationship between progenesis, development and evolution." *Folia Parasitologica* 50:275–292; 2003.

Reuter, M., O. I. Raikova, U. Jondelius, M. K. S. Gustafsson, A. G. Maule, and D. W. Halton, "Organisation of the nervous system in the Acoela: an immunocytochemical study." *Tissue Cell* 33:119–128; 2001.

Roberts, L.S. and Janovy, Jr. "Foundations of Parasitology." 7th Edition.

Smyth, J. D., *The physiology of cestodes.* W. H. Freeman and Company; New York; 1969.

Smyth, J. D., and D. W. Halton, *The physiology of trematodes,* 2d ed. Cambridge University Press; New York; 1983.

A-8 Rhombozoa

Lapan, E. A., and H. Morowitz, "The Mesozoa." *Scientific American* 227:94–101; December 1972.

McConnaughey, B. H., "The Mesozoa." In: M. Florkin and B. T. Scheer, eds., *Chemical zoology*, Vol. 2. Academic Press; New York; pp. 557–570; 1968.

Noble, E. R., and G. A. Noble, *Parasitology*, 5th ed. Lea & Febiger; Philadelphia; 1982.

A-9 Orthonectida

Hanelt, B., D. Van Schyndel, C. M. Adema, L. A. Lewis, and E. S. Loker, "The phylogenetic position of *Rhopalura ophiocomae* (Orthonectida) based on 18s ribosomal DNA sequence analysis." *Molecular Biology and Evolution* 13:1187–1191; 1996.

Slyusarev, G. S., "The fine structure of the muscle system in the female of the orthonectid *Intoshia variabili* (Orthonectida)." *Acta Zoologica* 84:107–111; 2003a.

Slyusarev, G. S., "Muscle formation in the sexual generation of *Intoshia variabili* (Orthonectida)," *Parazitologiia* 37:216–220 2003b. (In Russian).

Slyusarev, G. S., and M. Ferraguti, "Sperm structure of *Rhopalura littoralis* (Orthonectida)." *Invertebrate Biology* 121:91–94; 2002.

Slyusarev, G. S., and R. Kristensen, "Fine structure of the ciliated cells and ciliary rootlets of *Intochia variabili* (Orthonectida)." *Zoomorphology* 122:33–39.

A-10 Nemertina

Gerlach, J., "The behaviour and captive maintenance of the terrestrial nemertine (*Geonemertes pelaensis*)." *Journal of Zoology* 246:233–2237; 1998.

Gibson, R., *Nemerteans*. Hutchinson University Library; London; 1972.

Gibson, R., "Nemertean genera and species of the world: An annotated checklist of the original names and description citations, synonyms, current taxonomic status, habitats and recorded zoogeographic distribution." *Journal Natural History* 29:271–561; 1995.

Theil, M., "Nemertines as predators on tidal flats—high noon at low tide." *Hydrobiologia* 365:241–250; 1997.

Thollesson, M., and J. L. Norenburg, "Ribbon worm relationships: A phylogeny of the phylum Nemertea." *Proceedings Royal Society London* B 270:407–415; 2003.

A-11 Nematoda

Bird, A. F., and J. Bird, *Structure of nematodes*, 2d ed. Academic Press; New York; 1991.

Cheng, T. C., *Parasitology*, 2d ed. Academic Press; New York; 1986.

Goodey, J. B., *Soil and freshwater nematodes*. Wiley; New York; 1963.

Hope, W. D., ed., *Nematodes: Structure, development, classification, and phylogeny*. Smithsonian Institution Press; Washington, DC; 1994.

Hotez, P. J., and D. I. Pritchard, "Hookworm infection," *Scientific American* 272:68–74 June 1995. Biology of hookworm infection and vaccine development.

Lee, D. L., and H. J. Atkinson, *The physiology of the nematodes*, 2d ed. Columbia University Press; New York; 1977.

Maio, J. J., "Predatory fungi," *Scientific American* 199:67–72 July 1958. Nematodes trapped by predatory fungi.

A-12 Nematomorpha

Cheng, T. C., *Parasitology*, 2d ed. Academic Press; New York; 1986.

Croll, N. A., *Ecology of Parasites*. Harvard University Press; Cambridge, MA; 1966.

Noble, E. R., and G. A. Noble, *Parasitology*, 5th ed. Lea & Febiger; Philadelphia; 1982.

Poinar, G. O., "Nematoda and Nematomorpha,". ch. 9 In: J. H. Thorpe and A. P. Covich, eds., *Ecology and classification of North American freshwater invertebrates*. Academic Press; New York; pp. 273–282; 1991.

A-13 Acanthocephala

Baer, J. G., *Animal Parasites*. World University Library; London; 1971. McGraw-Hill; New York; 1971.

Conway Morris, S., and D. W. T. Crompton, "The origins and evolution of Acanthocephala." *Biological Reviews* 57:85–113; 1982.

Crompton, D. W. T., *Parasitic worms*. Taylor and Francis; Bristol, PA; 1980.

Moore, J., "*Parasites* that change the behavior of their host." *Scientific American* 250:108–115; May 1984.

Nicholas, W. L., "The biology of Acanthocephala." In: B. Dawes, ed., *Advances in parasitology 5*. Academic Press; New York; pp. 205–206; 1967.

Noble, E. R., and G. A. Noble, *Parasitology*, 5th ed. Lea & Febiger; Philadelphia; 1982.

Olsen, O. W., *Animal Parasites: Their life cycles and ecology*. Dover; New York; 1986.

A-14 Rotifera

Barron, G., "Jekyll-Hyde mushrooms." *Natural History* 101(3):47–52; March 1992.

Donner, J., *Rotifers*. Stuttgart; 1956. Reprint, Frederick Warne; London; 1965.

Eddy, S., and A. C. Hodson, *Taxonomic keys to the common animals of the north central states*, 4th ed. Burgess; Minneapolis; 1982.

Edmondson, W. T., H. B. Ward, and G. C. Whipple, eds., *Freshwater biology*, 2d ed. Wiley; New York; 1959; pp. 420–494.

Nogrady, T., R. L. Wallace, and T. W. Snell, *Biology, ecology, and systematics*, Vol. 1. SPB Academic Publishing; The Hague; 1993.

Pennak, R. W., "Ecological affinities and origins of freeliving acoelomate freshwater invertebrates." In: E. C. Dougherty, ed. *The lower metazoa: Comparative biology and phylogeny*. University of California Press; Berkeley, CA; pp. 435–451; 1963.

A-15 Kinorhyncha

Dougherty, E. C., ed., *The lower metazoa: Comparative biology and phylogeny*. University of California Press; Berkeley, CA; 1963.

Higgins, R. P., *A historical overview of kinorhynch research*. In N. C. Hullings, ed., Proceedings of the first international conference on meiofauna. Smithsonian Contributions to Zoology 76:25–31; 1971.

Higgins, R. P., and H. Thiel, eds., *Introduction to the study of meiofauna*. Smithsonian Institution Press; Washington, DC; 1988.

Morell, V., "Life on a grain of sand." *Discover* 16:78–86; April 1995.

Russell-Hunter, W. D., *A biology of lower invertebrates*. Macmillan; New York; 1968.

A-16 Priapulida

Adrianov, A. V., and V. V. Malakhov, "The phylogeny and classification of the phylum Cephalorhyncha." *Zoosystematica Rossica* 3(2):181–201; 1995.

Hammond, R. A., "The burrowing of *Priapulis caudatus*." *Journal of Zoology* 162:469–480; 1970.

Higgins, R. P., V. Storch, and T. C. Shirley, "Scanning and transmission electron microscopical observations on the larvae of *Priapulus caudatus* (Priapulida)." *Acta Zoologica (Stockholm)* 74(4):301–319; 1993.

Morse, P., "*Meiopriapulus fijiensis*, n.gen., n. sp.: An interstitial priapulid from coarse sand in Fiji." *Transactions of the American Microscopical Society* 100:239–252; 1981.

Por, F. D., and H. J. Bromley, "Morphology and anatomy of *Maccabeus tentaculatus*." *Journal of Zoology* 173:173–197; 1974.

Shirley, T. C., "Ecology of *Priapulus caudatus* Lamarck, 1816 (Priapulida) in an Alaskan subarctic ecosystem." *Bulletin of Marine Science* 47:149–158; 1990.

A-17 Gastrotricha

Brunson, R. B., "Aspects of the natural history and ecology of the gastrotrichs." In: E. C., Dougherty, ed. *The lower metazoa: Comparative biology and phylogeny*. University of California Press; Berkeley, CA; pp. 473–478; 1963.

D'Hondt, J.-L., "Gastrotricha." *Oceanography and Marine Biology: An Annual Review* 9:141–192; 1971.

Hummon, W. D., "Biogeography of sand beach Gastrotricha from the northeastern United States." *Biological Bulletin* 141:390; 1971.

Ruppert, E. E., "Comparative ultrastructure of the gastrotrich pharynx and the evolution of myoepithelial foreguts in aschelminthes." *Zoomorphology* 99:181–220; 1982.

Warwick, N., and M. A. Todaro, "Observations on Gastrotricha from a sandy beach in southeastern Australia, with a description of *Halichaetonotus australis* sp. nov. (Gastrotricha, Chaetonotida)." *New Zealand Journal Marine Freshwater Research* 39:973–980; 2005.

A-18 Loricifera

Anonymous, "New phylum found." *Bioscience* 34:321; 1984.

Higgins, R. P., and R. M. Kristensen, "New loricifera from southeastern United States coastal waters." *Smithsonian Contributions to Zoology* 438:1–70; 1986.

Higgins, R. P., and H. Thiel, eds., *Introduction to the study of meiofauna*. Smithsonian Institution Press; Washington, DC; 1988.

Kristensen, R. M., "Loricifera, a new phylum with Aschelminthes characters from the meiobenthos." *Zeitschrift fur zoologische Systematik und Evolutionsforschung* 21:163–180; 1983.

Kristensen, R. M., "Loricifera: A general biological and phylogenetic overview." *Verhandlungen der Deutschen Zoologischen Gesellschaft* 84:231–246; 1991.

A-19 Entoprocta

Emschermann, P., "On Antarctic Entoprocta: Nematocyst-like organs in loxosomatid, adaptive developmental strategies, host specificity, and bipolar occurrence of species." *Biological Bulletin* 184:153–185; 1993.

Funch, P., and R. M. Kristensen, "Cycliophora is a new phylum with affinities to Entoprocta and Ectoprocta." *Nature* 378:711–714; 1995.

Mackey, L. Y., B. Winnepenninckx, R. De Wachter, T. Backeljau, P. Emschermann, and J. R. Garey, "18S rDNA suggests that Entoprocta are protostomes, unrelated to Ectoprocta." *Journal of Molecular Evolution* 42:552–559; 1996.

Nielsen, C., "The phylogenetic position of Entoprocta, Ectoprocta, Phoronida, and Brachiopoda." *Integrative and Comparative Biology* 42:685–691; 2002.

Wood, T., "*Loxosomatoides sirindhornae*, new species, a freshwater kamptozoan from Thailand." *Hydrobiologia* 544:27–31; 2005.

A-20 Chelicerata

Cracraft, J., and M. J. Donoghue, eds., *Assembling the tree of life*. Oxford University Press; New York; 2004.

Foelix, R., *Biology of spiders*, 2d ed. Oxford University Press; 1996.

Gupta, A. P., *Arthropod phylogeny*. Van Nostrand Reinhold; New York; 1979.

Kaston, B. J., *How to know the spiders*, 3d ed. W. C. Brown; Dubuque, IA; 1978.

King, P. E., *Pycnogonids*. Hutchinson University Library; London; 1973.

Polis, G. A., "The unkindest sting of all." *Natural History* 98(7):3439; July 1989.

G. A., Polish, ed.,*The biology of scorpions*. Stanford University Press; Stanford, CA; 1990.

Snodgrass, R. E., *A textbook of arthropod anatomy*. Cornell University Press; Ithaca, NY; 1952.

Ubick, D., P. Paquin, P. E. Cushing, and V. Roth, *Spiders of North America: An identification manual*. American Arachnological Society; 2005.

Walter, D. E., and H. C. Proctor, *Mites: Ecology, evolution and behavior*. University of New South Wales Press; Sydney, and CABI, Wallingford; 1999.

A-21 Mandibulata

Batra, L. R., and S. W. T. Batra, "The fungus gardens of insects." *Scientific American* 217:112–120; November 1967.

Borrer, D. J., and R. E. White, *A field guide to the insects of America north of Mexico*. Houghton Mifflin; Boston; 1970.

Cottam, C. A., and H. S. Zim, *Insects: A guide to familiar American species*. Golden; New York; 1987.

Cracraft, J., and M. J. Donoghue., eds, *Assembling the tree of life*. Oxford University Press; New York; 2004.

Dillon, E. S., and L. S. Dillon, *A manual of common beetles of eastern North America*, 2 vols. Dover; New York; 1972.

Grimaldi, D., and M. S. Engel, *Evolution of the insects.* Cambridge University Press; New York; 2005.

Gupta, A. P., *Arthropod phylogeny.* Van Nostrand Reinhold; New York; 1979.

Marshall, S. A., *Insects: Their natural history and diversity: With a photographic guide to insects of easter North America.* Firefly Books; 2006.

D. E., Bliss, ed. *The biology of Crustacea*, Vols 1–9. Academic Press; New York; 1982–1987.

Cracaft, J., and M. J. Donoghue, eds. *Assembling the tree of life.* Oxford University Press; New York; 2004.

Emerson, M. J., and F. R. Schram, "The origin of crustacean biramous appendages and the evolution of Arthropoda." *Science* 250:667–669; 1990.

Fleminger, A., "Description and phylogeny of *Isaacsicalanus paucisetus*, n. gen., n. sp. (Copepoda: Calanoida: Spinocalanidae) from an east Pacific hydrothermal vent site (21 degrees N)." *Proceedings of the Biological Society of Washington* 96:605–622; 1983.

Gould, S. J., "Of tongue worms, velvet worms, and water bears." *Natural History* 104:6–15; January 1995.

Gupta, A. P., *Arthropod phylogeny.* Van Nostrand Reinhold; New York; 1979.

Haugerud, R. E., "Evolution in the Pentastomids." *Parasitology Today* 5:126–132; 1989.

Nichols, D., J. Cooke, and D. Whiteley, *Oxford book of invertebrates.* Oxford University Press; New York; 1971.

Schram, F. R., *Crustacea.* Oxford University Press; New York; 1986.

Self, J. T., "Biological relationships of the Pentastomida." *Experimental Parasitology* 24:63–119; 1969.

A-22 Annelida

Brinkhurst, R. O., "Evolution in the Annelida." *Canadian Journal of Zoology* 60(5):1043–1059; 1982.

Dales, R. O., *Annelids*, 2d ed. Hutchinson University Library; London; 1967.

Darwin, C. R., *The formation of vegetable mould through the action of worms with observations on their habits. 1881. Reprinted as Darwin on earthworms The formation of vegetable mould through the action of worms.* Bookworm Publications; Russelville, AR; 1976.

Edwards, C. A., and J. R. Lofty, *Biology of earthworms*, 2d ed. Chapman and Hall; London; 1972.

Eernisse, D. J., J. S. Albert, and F. E. Anderson, "Annelida and Arthropoda are not sister taxa: A phylogenetic analysis of spiralian metazoan morphology." *Systematic Biology* 41:305–330; 1992.

Laverack, M. S., *The physiology of earthworms.* Macmillan; New York; 1963.

Wells, G. P., "Worm autobiographies," *Scientific American* 200:132–142; 1959. Annelid behavior patterns.

A-23 Sipuncula

Cutler, E. B., *The Sipuncula: Their systematics, biology and evolution.* Cornell University Press; Ithaca, NY; 1994.

Rice, M. E., J. Piraino, and H. F. Reichardt, "Observations on the ecology and reproduction of the sipunculan *Phascolion cryptus* in the Indian River lagoon." *Florida Scientist* 46:382–396; 1983.

Scheltema, A. H., "Aplacophora as progenetic aculiferans and the coelomate origin of molluscs as the sister taxon of Sipuncula." *Biological Bulletin* 184:57–78; 1993.

Schulze, A., E. B. Cutler, and G. Giribet, "Reconstructing the phylogeny of the Sipuncula." *Hydrobiologia* 535–536:277–296; 2005.

Rajulu, G. S., and N. Krishnan, "Occurrence of asexual budding in Sipuncula." *Nature* 223:186–187; 1969.

A-24 Echiura

Kohn, A., and M. Rice, "Biology of Sipuncula and Echiura." *Bioscience* 21:583–584; 1971.

MacGinitie, G. E., and N. MacGinitie, *Natural history of marine animals*, 2d ed. McGraw-Hill; New York; 1968.

Risk, M. J., "Silurian echiuroids: Possible feeding traces in the Thorold sandstone." *Science* 180:1285–1287; 1973.

Stephen, A. C., and S. J. Edmonds, *The phyla Sipuncula and Echiura*. British Museum (Natural History); London; 1972.

Wolcott, T. G., "Inhaling without ribs: The problem of suction in soft-bodied invertebrates." *Biological Bulletin* 160:189–197; 1981.

A-25 Pogonophora

George, J. D., and E. C. Southward, "A comparative study of the setae of Pogonophora and polychaetous Annelida." *Journal of the Marine Biological Association of the United Kingdom* 53:403–424; 1973.

Gould, S. J., "Microcosmos," *Natural History* 105(3) March 1996, 21, 23, 66, 68.

Ivanov, A. V., *Pogonophora*. Consultants Bureau; New York; 1963.

Jones, M.L., "The Vestimentifera: Their biology, systematics and evolutionary patterns" (Biology and Ecology Symposium, Paris, 4–7 November, 1985, Proceedings), L. Laubier, ed. *Oceanologica Acta*. Special Volume 8:69–82; 1988.

Nørevang, A., ed., *The phylogeny and systematic position of Pogonophora* (special issue of *Zeitschrift für Zoologische Systematik und Evolutionsforschung*). Verlag Paul Parey; Hamburg and Berlin; 1975.

Southward, A. J., and E. C. Southward, "Pogonophora." In: T. J. Pandian and F. J. Vernberg, eds., *Animal energetics*, Vol. 2. Academic Press; New York; pp. 201–228; 1987.

Webb, M., "*Lamellibrachia barhami*, gen. nov., sp. nov. (Pogonophora), from the northeast Pacific." *Bulletin of Marine Science* 19:18–47; 1969.

A-26 Mollusca

Abbott, R. T., *American seashells: The marine Mollusca of the Atlantic and Pacific coasts of North America*, 2d ed. Van Nostrand Reinhold; New York; 1974.

Brooks, W. K., *The oyster*. Johns Hopkins University Press; Baltimore; 1996.

Lane, F. W., *Kingdom of the octopus: The life history of the Cephalopoda*. Jarrolds; London; 1960. Sheridan House; New York; 1960.

Morris, P. A., *A field guide to the shells of the Atlantic and Gulf coasts and the West Indies*, 3d ed. Houghton Mifflin; Boston; 1973.

Morton, J. E., *Molluscs*, 4th ed. Hutchinson University Library; London; 1967.

Page, H. M., C. R. Fisher, and J. J. Childress, "Role of filter-feeding in the nutritional biology of a deep-sea mussel with methanotrophic symbionts." *Marine Biology* 104:251–257; 1990.

Scheltema, A. H., "Aplacophora as progenetic aculiferans and the coelomate origin of molluscs as the sister taxon of sipuncula." *Biological Bulletin* 184:57–78; 1993.

Solem, A., *The shell makers*. Wiley (Interscience); New York; 1974.

Vogel, S., "Flow-assisted mantle cavity refilling in jetting squid." *Biological Bulletin* 172:61–68; 1987.

K. M., Wilbur, ed. *The Mollusca*, 12 vols. Academic Press; New York; 1983–1988.

Yonge, C. M., *Oysters*, 2d ed. Collins; London; 1966.

A-27 Tardigrada

Guidetti, R., and R. Bertolani, "Tardigrade taxonomy: An updated check list of the taxa and a list of characters for their identification." *Zootaxa* 845:1–46; 2005.

Kinchin, I. M., *The biology of Tardigrades.* pp. 186. Portland Press; London; 1994.

Nelson, D. R., and S. J. McInnes, "Tardigrades." In: S. D. Rundle, A. L. Robertson, and J. M. Schmid-Araya, eds., *Freshwater meiofauna: Biology and ecology buckhuys*, Leiden, 177–215; 2002.

Ramazzotti, G., and W. Maucci, "Il Phylum Tardigrada. III edizione riveduta e aggiornata." *Memorie dell'Istituto Italiano di Idrobiologia* 41:1–1011; 1983.

A-28 Onychophora

Boudreaux, H. B., *Arthropod phylogeny with special reference to insects.* Wiley (Interscience); New York; 1979.

Ghiselin, M. T., "A movable feaster." *Natural History* 94(9):54–61; September 1985.

Gould, S. J., "Of tongue worms, velvet worms, and water bears." *Natural History* 104(1):6–15; January 1995.

Manton, S. M., *The Arthropoda: Habitats, functional morphology, and evolution.* Oxford University Press (Clarendon Press); Oxford, UK; 1977.

Monge-Najera, J., "Phylogeny, biogeography and reproductive trends in the Onychophora." *Zoological Journal of the Linnean Society* 114:21–60; 1995.

Monge-Najera, J., "Jurassic–Pleiocene biogeography: Testing a model with velvet worm (Onychophora) vicariance." *Revista de Biologia Tropical* 44(1):159–175; 1996.

Peck, S. T., "A review of the New World Onychophora with the description of a new cavernicolous genus and species from Jamaica." *Psyche* 82:341–358; 1975.

Ross, H. H., *Textbook of entomology.* [Reprinted Krieger; Melbourne, FL; 1991.]., 3d ed. Wiley; New York; 1965.

Snodgrass, R. E., "Evolution of the Annelida, Onychophora, and Arthropoda." *Smithsonian Miscellaneous Collection* 97:1–159; 1938.

A-29 Bryozoa

Boardman, R. S. Cheetham, A. H. Oliver, W. A. Jr. eds., *Animal colonies: Development and function through time.* Dowden, Hutchinson Ross; Stroudsburg, PA; 1973.

Larwood, G. P., *Living and fossil Bryozoa.* Academic Press; New York; 1973.

Larwood, G. P., and B. R. Rosen, *Biology and systematics of colonial organisms.* Academic Press; New York; 1979.

Nielsen, C., *Animal evolution: Interrelationships of the living phyla.* Oxford University Press; New York; 1995.

Pennak, R. W., *Freshwater invertebrates of the United States*, 3d ed. Wiley (Interscience); New York; 1989.

Ryland, J. S., *Bryozoans.* Hutchinson University Library; London; 1970.

Woollacott, R. M., Zimmer, R.L. eds., *Biology of bryozoans.* Academic Press; New York; 1977.

A-30 Brachiopoda

Erwin, D. H., "The mother of mass extinctions." *Scientific American* 275:72–78; July 1996.

Gould, S. J., and C. B. Calloway, "Clams and brachiopods: Ships that pass in the night." *Paleobiology* 6:383–396; 1980.

Jorgensen, C. B., *The biology of suspension feeding.* Pergamon Press; New York; 1966.

LaBarbara, M., "Water flow patterns in and around three species of articulate brachiopods." *Journal of Experimental Marine Biology and Ecology* 55:185–206; 1981.

Richardson, J. R., "Brachiopods." *Scientific American* 255:100–106; September 1986.

Rudwick, M. J. S., *Living and fossil brachiopods.* Hutchinson University Library; London; 1970.

Russell-Hunter, W. D., *Biology of higher invertebrates.* Macmillan; New York; 1969.

Williams, A., "The calcareous shell of the Brachiopoda and its importance to their classification." *Biological Reviews of the Cambridge Philosophical Society* 31:243–287; 1956.

Williams, A., *et al.*, "Brachiopoda, Part H (2 vols.)," In: R. C., Moore, ed. *Treatise on invertebrate paleontology.* Geological Society of America; Boulder, CO; 1965. University of Kansas Press; Lawrence, KS; 1965.

A-31 Phoronida

Kozloff, E. N., *Seashore life of the northern Pacific Coast: An illustrated guide to the common organisms of northern California, Oregon, Washington, and British Columbia.* University of Washington Press; Seattle; 1983.

MacGinitie, G. E., and N. MacGinitie, *Natural history of marine animals,* 2d ed. McGraw-Hill; New York; 1968.

Zimmer, R. L., "Morphological and developmental affinities of the lophophorates." In: G. P., Larwood, ed. *Living and fossil Bryozoa.* Academic Press; New York; pp. 593–599; 1973.

A-32 Chaetognatha

Alvari–o, A., "Chaetognaths." In: H., Barnes, ed. *Oceanography and marine biology: Annual review 3.* Allen and Unwin; London; pp. 115–194; 1965.

Bieri, R., D. Bonilla, and F. Arcos, "Function of the teeth and vestibular organ in the Chaetognatha as indicated by scanning electron microscope and other observation." *Proceedings of the Biological Society of Washington* 96:110–114; 1983.

Darwin, C., "Observations on the structure and propagation of the genus *Sagitta.*" *Annals and Magazine of Natural History* 13 (Series 1, No. 81):1–6 and Plate 1; January 1844.

Eakin, R. M., and J. A. Westfall, "Fine structure of the eye of a chaetognath." *Journal of Cell Biology* 21:115–132; 1964.

Ghirardelli, E., "Some aspects of the biology of the chaetognaths." *Advances in Marine Biology* 6:271–375; 1968.

Grant, G. C., "Investigations of inner continental shelf waters off lower Chesapeake Bay 4: Descriptions of the Chaetognatha and a key to their identification." *Science* 4:107–119; 1963.

A-33 Hemichordata

Barrington, E. J. W., *The Biology of Hemichordata and Protochordata.* W. H. Freeman and Company; New York; 1965.

Berrill, N. J., *The origin of vertebrates.* Oxford University Press; New York; 1955.

Harrison, F. W., and E. E. Ruppert, eds., *Microscopic anatomy of invertebrates, vol. 15. Hemichordata, Chaetognatha, and the invertebrate chordates.* Wiley-Liss; New York; 1996.

Spengel, E. J. W., "*Planctosphaera pelagica.*" *Scientific Results of the Michael Sars North Atlantic Expedition* 1910 5(5); 1932.

Wada, H., and N. Sato, "Details of the evolutionary history from invertebrates to vertebrates, as deduced from the sequences of 18S rDNA." *Proceedings of the National Academy of Sciences USA* 91:1801–1804; 1994.

A-34 Echinodermata

Binyon, G., *Physiology of Echinoderms.* Pergamon Press; New York; 1972.

Borradaile, L. A., F. A. Potts, L. E. S. Eastham, and J. T. Saunder, *The invertebrata.* Cambridge University Press; New York; 1935.

Clark, A. M., and M. E. Downey, *Starfishes of the Atlantic.* Chapman and Hall; London; 1992.

Hendler, G., J. E. Miller, D. L. Pawson, and P. M. Kier. *Sea stars, sea urchins, and allies: Echinoderms of Florida and the Caribbean.* Smithsonian Institution Press; Washington, DC, and London; 1995. Natural history and identification of echinoderms of the Bahamas, Florida, and Caribbean.

MacGinitie, G. E., and N. MacGinitie, *Natural history of marine animals*, 2d ed. McGraw-Hill; New York; 1968.

Nichols, D., *Echinoderms*, 4th ed. Hutchinson University Library; London; 1969.

Nichols, D., *The uniqueness of echinoderms (Oxford/Carolina Biology Reader).* Oxford University Press; New York and London; 1975.

A-35 Urochordata

Borradaile, L. A., F. A. Potts, L. E. S. Eastham, and J. T. Saunders, *The invertebrata.* Cambridge University Press; New York, 1935.

Brien, P., "Embrachement des Tuniciers." In: Grassé, P. P., ed. *Traité de Zoologie, Anatomie, Systématique, Biologie, Tome XI, Échinodermes, Stomocordés, Procordés.* Masson, Paris; pp. 553–894; 1948.

Deibel, D., and G. Paffenhofer, "Cinematographic analysis of the feeding mechanism of the pelagic tunicate *Doliolum nationalis.*" *Bulletin of Marine Science* 43:404–412; 1988.

Lambert, G., "Ultrastructural aspects of spicule formation in the solitary ascidian *Herdmania momus* (Urochordata, Ascidiacea)." *Acta Zoologica (Stockholm)* 73:237–245; 1992.

Lambert, G., and C. C. Lambert, "Spicule formation in the solitary ascidian *Herdmania momus.*" *Journal of Morphology* 192:145–159; 1987.

Millar, R. H., "The biology of ascidians." *Advances in Marine Biology* 9:1–100; 1971.

Pearse, V., J. Pearse, M. Buchsbaum, and R. Buchsbaum, "Invertebrate chordates: Tunicates and lancelets."*Living invertebrates.* In: Blackwell Scientific; Pacific Grove, CA; 1987; pp. 737–752.

Plough, H. H., *Sea squirts of the Atlantic continental shelf from Maine to Texas.* Books on Demand; Ann Arbor, MI. Johns Hopkins University Press; Baltimore; 1978.

A-36 Cephalochordata

Lytle, C., and J. E. Wodsedalek, *General Zoology Laboratory Guide*, 11th ed. W.C. Brown; Dubuque, IA; 1991.

Pearse, V., J. Pearse, M. Buchsbaum, and R. Buchsbaum, "Invertebrate chordates: Tunicates and lancelets."*Living invertebrates.* Blackwell Scientific; Pacific Grove, CA; 1987; pp. 737–752.

A-37 Craniata

Alexander, R. M., *The chordates.* Cambridge University Press; London; 1975.

Benton, M. J., *Vertebrate palaeontology*, 3d ed. Blackwell Publishing; Malden, MA; 2005.

Duellman, W. E., and L. Trueb, *Biology of amphibians.* Johns Hopkins University Press; Baltimore; 1994.

Liem, K. F., W. E. Bemis, W. F. Walker, Jr., and L. Grande, *Functional anatomy of the vertebrates: An evolutionary perspective*, 3d ed. Harcourt College Publishers; Fort Worth, TX; 2001.

Pough, F. H., C. M. Janis, and J. B. Heiser, *Vertebrate life*, 7th ed. Pearson Prentice Hall; Upper Saddle River, NJ; 2005.

Young, J. Z., *The life of vertebrates*, 3d ed. Oxford University Press; New York; 1981.

CHAPTER FOUR

KINGDOM FUNGI

Schizophyllum commune. [Courtesy of W. Ormerod.]

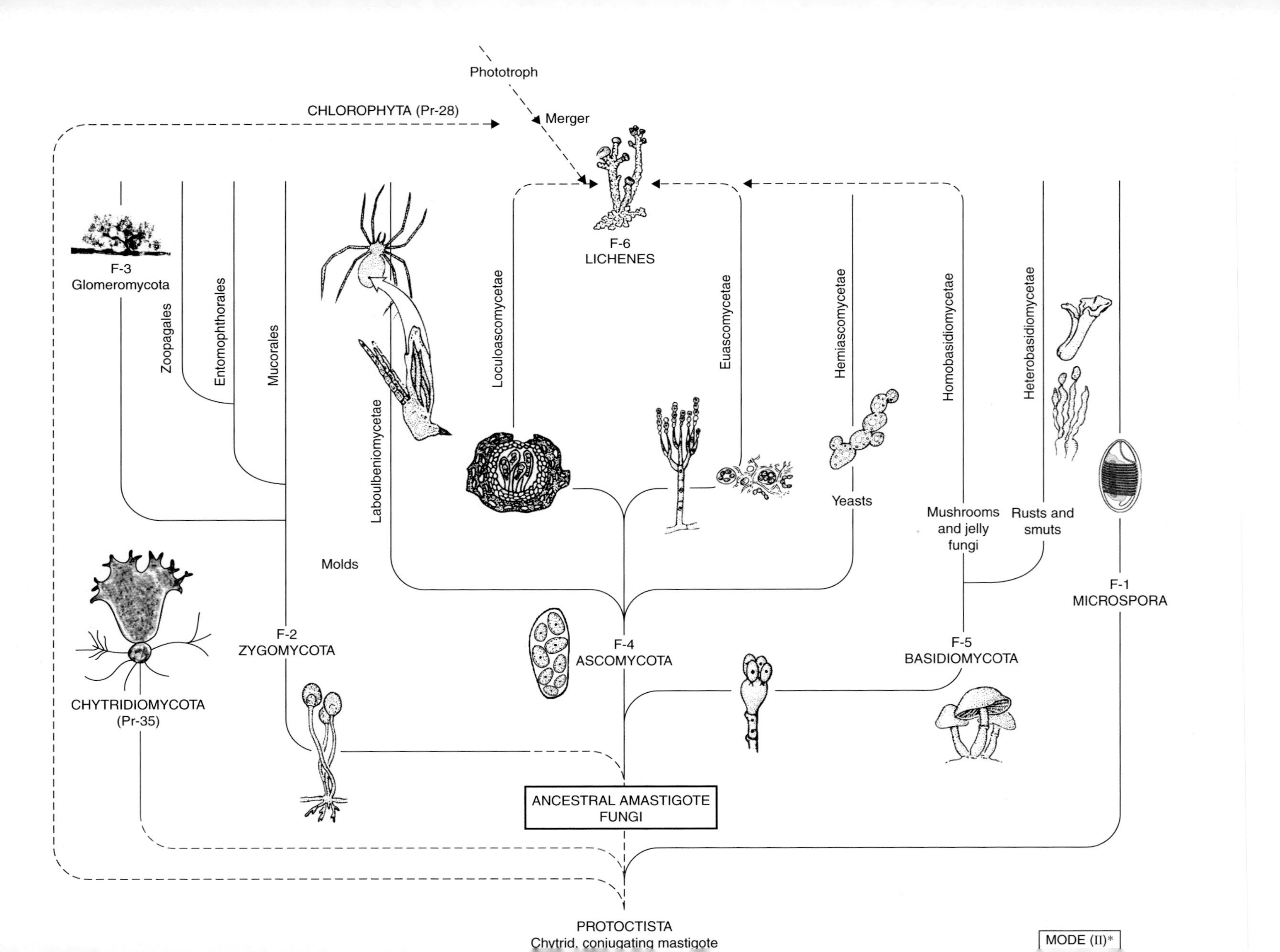
Phototroph
CHLOROPHYTA (Pr-28)
Merger
F-6
LICHENES
F-3
Glomeromycota
Zoopagales
Entomophthorales
Mucorales
Laboulbeniomycetae
Loculoascomycetae
Euascomycetae
Hemiascomycetae
Homobasidiomycetae
Heterobasidiomycetae
Yeasts
Mushrooms
and jelly
fungi
Rusts and
smuts
Molds
F-1
MICROSPORA
F-2
ZYGOMYCOTA
F-4
ASCOMYCOTA
F-5
BASIDIOMYCOTA
CHYTRIDIOMYCOTA
(Pr-35)
ANCESTRAL AMASTIGOTE
FUNGI
PROTOCTISTA
Chytrid, conjugating mastigote
MODE (II)*

KINGDOM FUNGI

Latin *fungus*, probably from Greek *sp(h)ongos*, sponge

MODE II*

Haploid or dikaryotic organisms with zygotic meiosis. Fossil record extends from 450 mya to present. Conjugating (hypha- or cell-fusing) haploid (monokaryotic) or dikaryotic osmotrophs that develop from resistant nonmotile fungal spores. Cells, including spores, have chitinous walls. Spores are produced by mitosis or by zygotic meiosis. Undulipodia (kinetosomes and axonemes) lacking at all stages. Equivalent to Kingdom Mycota or Eufungi. Fossil record extends from the lower Paleozoic era (450 million years ago) to the present.

Kingdom Fungi, as defined in this book, is limited to eukaryotes that form chitinous, resistant propagules (fungal spores) and chitinous cell walls and that lack undulipodia (that is, are amastigote or immotile) at all stages of their life cycle. Of the 1.5 million species of fungi estimated to exist, about 60,000 have been described; most are terrestrial, although a few truly marine species are known. Because fungi often differ only in subtle characteristics such as the details of structure, pigments, and complex organic compounds, it is likely that many have not yet been recognized as distinct species.

Fossil fungi date from the Ordovician period, 450–500 mya. The ancestry of fungi is not well understood. Absorptive heterotrophy (derivation of nutrients from the digestion of living or dead tissue) associated with multicellularity, syncytia (many nuclei per cell), and propagules (the fungal way of life) have evolved many times and in many protoctist groups: slime molds (Pr-2 and Pr-23) and slime nets (Pr-19), oomycotes (Pr-21), hyphochytrids (Pr-14), and even such ciliates as *Sorogena* (Pr-6). True fungi may have descended from conjugating protoctists and thus share an ancestor with the rhodophytes (Pr-33) or the gamophytes (Pr-32) through the zygomycotes (F-2); alternatively, chytrids (Pr-35) may be the protoctists from which fungi evolved. Molecular systematics supports our classification: only chytrids show a close relationship with fungi as the kingdom is presented here. Some classification systems incorrectly place funguslike microbes that include a motile stage (such as plasmodiophorans, oomycotes, and hyphochytrids) in the fungi kingdom. Both molecular evolutionary and morphological studies show that funguslike microbes with motile cells—that is, cells that have undulipodia—are protoctists (Pr-14, 20, 21 and 37).

The ascomycotes and basidiomycotes are more closely related to each other than to zygomycotes and probably descended from a common ancestor. Organisms that lack sexual stages were traditionally grouped together

* See page 119.

as Deuteromycota. Because deuteromycotes clearly descended from either the ascomycotes or the basidiomycotes by loss of sexual stages, we have simply returned them to their relatives and have now abandoned the artificial taxon "Deuteromycota." Lichenes (F-6) no doubt evolved by association, mostly of Ascomycota with either Cyanobacteria (B-6) or Chlorophyta (Pr-28) or both. We elevate lichens to phylum status.

Fungi were traditionally aligned with plants, and some classification schemes formerly considered the fungi to be a subkingdom of kingdom Plantae. However, fungi are clearly more closely related to animals than to plants, considering that chitin is the main component of both fungal cell walls and the arthropod exoskeleton (A-20 through A-21). In comparison, plant cell walls instead contain cellulose, a polysaccharide similar to chitin. At any rate, fungi differ from animals and plants in life cycle, mode of nutrition, pattern of development, and many other ways. We thus support here the many mycologists who feel that fungi constitute their own kingdom.

Fungi lack an embryonic stage and develop directly from spores. Often borne in sporangia (singular: *sporangium*), spores may be of mitotic or meiotic origin. The spores germinate into hyphae or in the case of yeasts, into single growing cells. The hyphae (singular: *hypha*) that grow from fungal spores are slender tubes divided into cells by cross walls called septa (singular: *septum*). The nuclei increase by mitosis as hyphae grow. Each such cell may contain more than one nucleus. The septa seldom separate the cells completely. Thus, cytoplasm, including nuclei, mitochondria, and other inclusions, can flow more or less freely through the hyphae. The hyphae of some fungi have no septa at all. The hyphae of an individual fungus are collectively called a mycelium (plural: *mycelia*), which is the feeding and growing form of most fungi. As part of an absorptive organism, the fungal mycelium has a morphology that is beautifully fitted to its ecological role: great surface area and growth at the periphery of the tips of hyphae. The yeasts, in comparison, remain as single cells and do not form mycelia.

Spores produced in the absence of any sort of sexual fusion are called conidia (singular: *conidium*). These propagules can form on the hyphae of zygomycotes or ascomycotes (Figure 4.1). The hyphal structures bearing the conidia are called conidiophores. Most conidia and many sporangia are dispersed by the wind and can endure conditions of heat, cold, and desiccation unfavorable to the growth of fungi. Under favorable conditions, the conidia and sporangiospores (spores carried in the sporangia) grow into hyphae and form mycelia. This is the general reproductive pattern, although there are many variations. New individual mycelia also result from fragmentation of mycelia. Yeasts reproduce asexually by budding. Nearly all fungi, even those that have sexual stages, form spores directly—that

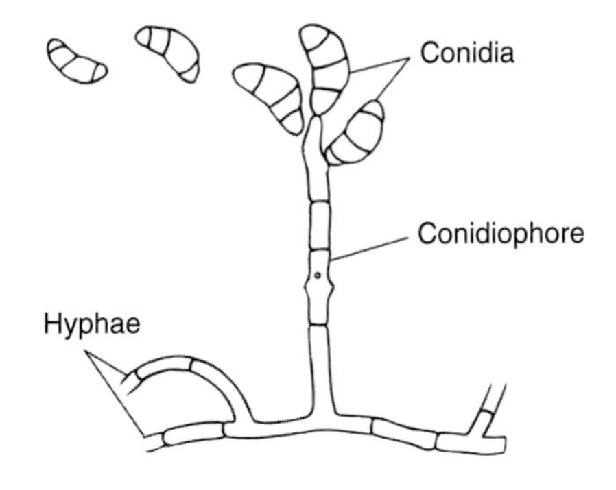

Figure 4-1 Conidiophore and multicellular conidia of a typical ascomycote *Curvularia lunata.* [Adapted from W. B. Kendrick and J. W. Carmichael. "Hyphomycetes," in G. C. Ainsworth, F. K. Sparrow, and A. S. Sussman (eds.), *The Fungi: An Advanced Treatise, vol. IVA: A Taxonomic Review with Keys: Asomycetes and Fungi Imperfecti.* New York: Academic Press (1973).]

is, asexually. In fact, most reproduce by mitotically produced (asexual) spores more often than they do sexually.

From time to time, most fungi form a sexual stage, including reproductive structures called gametangia. All basidiomycotes form spores in basidia, zygomycotes form spores in sporangia, and ascomycotes form spores in asci. Fungi bear spores that are the products of meiosis—basidiospores, zygosporangia, or ascospores. The reproductive structures of basidiomycotes are masses of tightly packed hyphae called basidiomata. The reproductive structures of ascomycotes are called ascomata; they sometimes originate from a single cell and are derived from a three-dimensional proliferation of cells. Such spore-bearing structures are commonly noticed as molds, jelly fungi, and mushrooms. The largest and most complex fungi are the large mushrooms and shelf fungi, all of which are products of sexual mergers. Some of these recognizable structures arise from underground mycelia that are kilometers in length and meters in diameter. Many other fungi are microscopic—for example, the unicellular yeasts.

Sexual reproduction in fungi is by conjugation, in which hyphae of complementary mating types come together and fuse (see F-2). One parent fungus can have hyphae of two different mating types. Conjugation always consists of at least two processes: hyphal (cytoplasmic fusion, or cytogamy) and nuclear fusion (karyogamy). In fungi, after cell or hyphal fusion, the nuclei, which are always haploid, do not immediately fuse. This situation is analogous to the delayed fusion of animal egg and sperm nuclei. In fungi, instead of immediate nuclear fusion, each parental nucleus grows and divides within a common cytoplasm, often for long periods of time. The offspring nuclei remain in pairs, one nucleus descended from each mating type. A hypha containing paired haploid nuclei, whether or not they have been shown to come from separate mating types, is called dikaryotic. A mycelium of such hyphae is called a dikaryon. If the nuclei of each pair differ in genotype, the mycelium is called a heterokaryon. If the paired nuclei are from the hyphae of a single mating type (that is, have the same genotype), the mycelium is called a homokaryon. If the hyphae contain only single, unpaired, nuclei, the mycelium is called a monokaryon. The paired nuclei eventually fuse to form diploid zygotes. Diploidy is transient; the zygotes immediately undergo meiosis, which results in the formation of haploid spores and thus reestablishes the haploid state.

All fungi form some sort of spores. In all cases, fungal spores are haploid (although some are dikaryotic) and are capable of germinating into haploid hyphae or yeasts. In most fungi, hyphae of complementary mating types fuse later in the life cycle, and the dikaryotic or sexual stage follows, in which nuclei are transiently diploid. Organisms that lack sexual stages produce spores only by mitosis. The nuclei of hyphae in species that never develop sexual stages seem to be permanently haploid.

Nearly all fungi are aerobes, and all of them are heterotrophs that characteristically absorb their food. They excrete powerful enzymes that break food down into smaller molecules outside the fungus; dissolved nutrients are then transported into the fungus through the fungal membrane.

Fungi are tenacious, resisting severe desiccation and other environmental challenges. Their cell walls, composed of the nitrogenous polysaccharide chitin, are hard and stiff and resist water loss. Some grow in acid; others survive in nitrogen-poor environments such as bogs. Fungal strategies for survival include the production of such complex organic compounds as penicillin derivatives that block the formation of bacterial cell walls and amanita alkaloids that induce hallucinations or even death in mammals. The penicillin derivatives are antibiotics inhibiting bacterial growth; the amanita alkaloids deter feeding. Fungi are the most resilient of the eukaryotes, although by no means invulnerable. For example, red squirrels of the New England forest nip off mushrooms and then carry them up in trees, where the mushrooms dry in the sun, to be eaten later or stored.

Many fungi cause diseases, especially in animals, including humans, and plants. However, many more form constructive, intimate associations with plants. The roots of nearly all healthy tracheophytes (vascular plants; Pl-4 through Pl-12) have symbiotic relations with fungi. Fungi inhabiting (endomycorrhizal; see Glomeromycota (F-3)) and coating (ectomycorrhizal) the roots of grasses and trees are especially critical to plants growing in nutrient-poor soils; the fungi function as root extensions. Such mycorrhizal associations enhance the transport of soil nutrients, such as phosphates, nitrates, copper, zinc, and manganese. Most orchid seeds, for example, require specific fungal partners to germinate. (The first plant organ to emerge from the seed is the rootlet, or radicle.) The fungi that transport nutrients through soil may have helped prevent the earlier plants from succumbing to desiccation and direct sunlight as plants made the transition from water onto land. Plant–fungus relations became truly terrestrial. In any case, a strong association between most plants and some fungi has persisted for at least 400 million years.

Many fungi yield products useful to humans. Some fungi are sources of citric acid and pharmaceuticals. The fungal molds *Rhizopus nigricans* and *Curvularia lunata* carry out fermentations in the manufacture of cortisone, hydrocortisone, and prednisone—steroids used to control inflammation. Molds and yeasts are used in the production of cheese, beer, wine, and soy sauce.

Penicillin, which is an antibiotic produced naturally by the mold *Penicillium chrysogenum*, has been joined by several thousand additional antibiotics. An antibiotic is any low-molecular-weight compound—generally excreted by a living organism—that specifically inhibits the growth of other microbes when present in low concentration. Microbiological production of pharmaceuticals from molds also includes vitamins, interferons (which prevent rather than cure viral infections), and steroid hormones. Microorganisms carry out syntheses that produce penicillin, cephalosporins from the marine mold *Cephalosporium*, erythromycin, and streptomycin (see F-4). Antibiotics have been found that interfere with nearly every phase of the bacterial life cycle. Penicillins (several have been discovered) and cephalosporins interfere with the construction of the bacterial cell wall. Bleomycins and anthracyclines, isolated from *Streptomyces verticillus* (Actinobacteria, B-12), interfere with DNA replication; rifamycins block the transcription of DNA into mRNA; erythromycin interrupts the ribosomal synthesis of protein. In the fungal cell that normally contains the antibiotic, it is thought that the antibiotic may inhibit growth of competing organisms. This ability was amplified many-thousand fold by treating *Penicillium* with radiation and chemicals, which caused mutations that resulted in increased production of penicillin.

The genetic information to make penicillin is present in the unaltered genome of the mold. To improve antibiotics such as penicillins and cephalosporins, after the starting molecule has been produced by fermentation, a second step—substitution of a part of the original antibiotic molecule—is accomplished synthetically. Such substitutions can broaden the spectrum of organisms against which the antibiotic acts and reduce its toxicity.

Although many substances have been discovered that counter bacterial infections, fewer have been discovered that destroy invading fungi in, for example, the human body. One such substance is the antibiotic amphotericin B, produced by *Streptomyces* and *Streptoverticillium* (B-12), which interrupts the function of the cell membrane in fungi.

Fungi add flavor, color, protein content, and preservative qualities when used in the production of beverages and food. In the conversion of cabbage, green olives, and cucumbers into sauerkraut, cured olives, and pickles, the microbial actions of lactic acid bacteria are followed by fermentation by the yeasts *Saccharomyces* and *Torulopsis*. The mold *Rhizopus*, acting on soybeans, produces tempeh, or Japanese soybean cakes, and *Aspergillus* produces the Japanese product natto. Soy sauce results from the action of *Aspergillus oryzae* on wheat and soybeans, in addition to fermentation by the yeast *S. rouxii* and the bacterium *Pediococcus* (B-13). Strains of microbes are selected that either increase or decrease flavoring compounds and that change the ability of the microbe to ferment particular carbohydrates; for example, a strain of yeast that ferments dextrin contributes to one facet of beer production.

Yeasts that grow on the surface of Limburger cheese provide aroma; Camembert and Brie owe their ripening to *Penicillium* as well as yeasts. *Penicillium roquefortii* mold develops flavor in Stilton, Danish blue, Gorgonzola, and Roquefort cheeses. Swiss cheese holes originate with bubbles of carbon dioxide generated by yeasts. In baking, brewing, and wine making, yeast fermentation has been used since long before this one-celled fungus was discovered.

Box F-i: Mycosomes

Mycosomes are fungal propagules associated with plastids in plant and algal cells (Atsatt, 2003). Plastid-rich plant tissue, when externally sterilized, macerated, and allowed to senesce, will result in the germination of fungi along one of three developmental pathways: wall-less cells (protoplasts), yeast cells, or membrane-bounded spherules containing plastids. Grown in this way, most are identified as yeasts (chiefly *Aureobasidium pullulans*, but also *Candida* spp.), although other fungal species have also been described. Mycosomes are quite widespread in plants, occurring in green and nongreen angiosperms (respectively, *Castilleja* and *Cuscuta* spp.), conifers (*Sequoia*), *Ginkgo*, *Ephedra*, *Cycas*, and *Psilotum*. Algae from which mycosomes have been isolated include *Trebouxia* (lichen symbiont), *Coleochaete*, and other chlorophytes, *Spirogyra* and *Chlamydomonas*. The apparent ubiquity of mycosomes in plants and algae raises the possibility that they are very ancient, perhaps even ancestral to yeasts. The physical association of mycosomes with plastids involves complex membranous associations: mycosomes are observed inside senescing plastids but also may themselves incorporate plastids.

Although plants and algae have distant evolutionary origins, their biochemistry is strikingly similar in some ways. Fungi assimilate nitrogen and sulfate along similar pathways with both cyanobacteria and plants, for example. The resistant coat of pollen grains and fungal spores (sporopollenin) is found only in green plants and fungi (Gooday, 1981), and its precursors are synthesized in plastids (Mizelle *et al.*, 1989). Plant-defensive compounds, coumarins, are synthesized on chloroplast membranes (Goodwin and Mercer, 1983) and are also produced by fungi (Murray *et al.*, 1982). Both plants and fungi synthesize rubber (*cis*-1,4-polyisoprene) that can comprise up to 6 percent of the dry weight of some basidiomycetes (Kurihara *et al.*, 1962, 1963, 1964). Fungi pathogenic in plants secrete hormones of plant senescence: ethylene and abscisic acid, as well as gibberellins and cytokinins that alter plant morphology and redirect nutrients to sites of fungal infection (Angra-Sharma and Mandahar, 1993; Goodman *et al.*, 1986; Lopez-Carbonell *et al.*, 1998).

Plant–fungal symbioses are ubiquitous in nature. Forest ecosystems are dominated by broadleaf trees above ground. Carbon dioxide is "fixed," that is converted to cell nutrients, in photosynthesis. Below ground, however, mycorrhizal fungi consume dead plant material and respire, releasing carbon dioxide back to the atmosphere as they make groundwater nutrients available to the plants. Examples may be found of practically every possible level of symbiotic integration through study of plant–fungal associations, from opportunistic/necrotrophic interactions like potato blight (genus *Phytophthora*, Pr-21), through intracellular/extramembrane systems like vesicular-arbuscular mycorrhizae, to fully integrated, organellar symbioses like mycosomes within plastids. Mycosomes represent perhaps the deepest level of integration that fungi have evolved with plants. Their occurrence not only within plant cells but within the organelles of photosynthesis suggests that the origin of the mycosome symbiosis was very ancient, probably involving fungal ancestors (Chytridiomycota, Pr-35) incorporating plastids or cyanobacteria. Plants evolved vascular systems and colonized the land, obviating the need for motility in their proto-fungal symbiotic partners. Unlike chytrids, therefore, no true fungus possesses an undulipodium.

A relict of the ancient plastid-incorporation event that led to mycosomes may still be observed today in the fungus *Geosiphon pyriforme*, which incorporates *Nostoc* cyanobacteria and lives off the photosynthate (Kluge *et al.*, 1991; Mollenhauer and Kluge, 1994). The interconnected evolutionary pathways of plants and fungi continue to converge in far more pervasive ways than crop-focussed humans appreciate.

Atsatt, P. R., “Fungus propagules in plastids: The mycosome hypothesis. *Int Microbiol* 6:17–26; 2003.

References

Angra-Sharma, R., and C. L. Mandahar, "Involvement of carbohydrates and cytokinins in pathogenicity of *Helminthosporium-Carbonum*." *Mycopathologia* 121:91–99; 1993.

Gooday, G. W., "Biogenesis of sporopollenin in fungal spore walls." In: G. Turian and H. R. Hohl, eds., *The fungal spore: Morphogenetic controls*. Academic Press; New York; p. 486; 1981.

Goodman, R. N., Z. Kiraly, and K. R. Wood, *The biochemistry and physiology of plant disease*. pp. 249–286. University of Missouri Press; Columbia, MO; 1986.

Goodwin, T. W., and E. I. Mercer. *Introduction to plant biochemistry*. Pergamon Press; Great Britain; 1983.

Kluge, M., D. Mollenhauer, and R. Mollenhauer, "Photosynthetic carbon assimilation in *Geosiphon pyriforme*, an endosymbiotic association of fungus and cyanobacterium." *Planta*. 185:311–315; 1991.

Kurihara, K., Y. Kato, Y. Takei, and S. Shichiji, "Production of rubber by microorganisms." *Reports of the Fermentation Research Institute (Japan)*. 22:61–68; 1962.

Kurihara, K., Y. Kato, Y. Takei, and S. Shichiji, "Production of rubber by microorganisms." *Reports of the Fermentation Research Institute (Japan)*. 24:25–30; 1963.

Kurihara, K., Y. Kato, Y. Takei, and S. Shichiji, "Production of rubber by microorganisms." *Reports of the Fermentation Research Institute (Japan)*. 25:33–38; 1964.

Lopez-Carbonell, M., A. Moret, and M. Nadal, "Changes in cell ultrastructure and zeatin riboside concentrations in *Hedera helix*, *Pelargonium zonale*, *Prunus avium*, and *Rubus ulmifolius* leaves infected by fungi." *Plant Disease* 82:914–918; 1998.

Mizelle, M. B., R. Sethi, M. E. Ashton, and W. E. Jensen, "Development of the pollen grain and tapetum of wheat (*Triticum aestivum*) in untreated plants and plants treated with chemical hybridizing agent RH 0007." *Sexual Plant Reproduction*. 2:231–253; 1989.

Mollenhauer, D., and M. Kluge, "Geosiphon pyriforme." *Endocytobiosis and Cell Research* 10:29–34; 1994.

Murray, R. D. H., J. Mendez, and S. A. Brown, *The natural coumarins*. John Wiley and Sons; New York; 1982.

F-1 Microspora

GENERA
Encephalitozoon
Glugea
Ichthyosporidium
Nosema
Vairimorpha

Microsporans, heterotrophic amitochondriate eukaryotic microbes symbiotic in animals were classified in the kingdom Animalia as protozoan symbiotrophs since they were discovered. More recently microsporans were classified with other amitochondriates (such as archaemoebids, metamonads and parabasalids in T. Cavalier-Smith's "Kingdom Archezoa" of "early-diverging" eukaryotes or with Phylum Apicomplexa (Phylum Pr-7) as "sporozoan parasites" in either Kingdom Protoctista or Protist. However, microsporans differ from apicomplexans and other protoctist symbiotrophs in many ways. We applaud the molecular evidence that indicates microsporans (unlike the early-diverging eukaryotes of Phylum Pr-1, page 130ff) to be directly related to members of Kingdom Fungi. As free-living fungi that reduced to become intracellular single-cell tumor-forming highly specialized symbiotrophs we infer they lost their mitochondria. They began to use ATP supplied by the highly oxygenated mitochondriate animal tissue in which they reside. Microsporans, with small ribosomes reminiscent of those of prokaryotes, probably became streamlined as their increased capacity for intracellular animal protein synthesis evolved.

The 1200 known species of microsporans form a unique coiled organelle called the 'polar tube' or 'polar filament'. Because many different kinds of single-celled fungi are formally known as "yeast" we might now say that microsporans are better classified as intracellular yeasts. They are obligate symbiotrophs of animals—the greatest number of species from arthropods and fish. They have a great reproductive capacity. A thick-walled chitinous spore contains the conspicuous polar filament and infective nucleated cytoplasm "the sporoplasm". When penetrating animal tissue, the sporoplasm emerges from tis spore, squeezes through a narrow hollow tube derived from the polar filament and forces itself into the arthropod or fish. Microsporans independently of us evolved the injection needle.

Glugea stephani, illustrated here, is a microsporan that lives in the tissues of the starry flounder (Phylum A-37, Craniata). *Ichthyosporidium* also grows on fish. Microsporans tend to form large single-cell tumors in the tissue of the animals within which they dwell. They are seen in huge numbers in these animal tumor cells, many apposed to the animal mitochondria, suggesting that the oxygen for both amitochondriate intracellular fungus and animal is derived from the blood or hemolymph. The animal chromatin becomes severely polytene and the tissue cell full or ribosomes that clearly are inside the microsporan. These symbionts are integrated on the level of gene product: microsporan ribosomes inside tissue cells synthesize specific nucleoproteins and nucleic acids of animals. The least known microsporans are so well integrated into animal tissue that they cause no harm. Some better-known microsporans, like certain *Glugea* species are severe necrotrophs. Members of the genus *Encephalitozoon* live in warm-blooded vertebrates. *Nosema* has caused devastating damage to the silk industry, because members of the genus are agents of a disease of silkworm larvae, pebrine.

Multiple fission via mitosis generates large numbers of offspring. Uniparental cells committed to develop into spores divide by sporogony from a parent cell known as the sporont. One sporont divides into several sporoblasts that each then matures into a spore. The number of sporoblasts, that depend on the number of preceding mitotic cell divisions, is characteristic of the genus. Sporoblasts mature into spores. Life histories of different microsporan species vary considerably. Some have uniparental life histories in a single species of fish, arthropod or other animal whereas complex life histories that require transfer from one to multiple other animals, by for example, biting insects, have been described. Situated deep in muscle, intestinal or salivary gland tissue, live microsporans are difficult to study. The mature spores contain the anchoring disc, extrusion apparatus, and infective sporoplasm. Sexual life cycles, levels of ploidy and genetic organization in microsporans are not well documented. Microsporan cells are either uni- or binucleate per cell. Inside fish *Glugea* sp. may develop a multinucleate plasmodium. Typical of fungi, in the total absence of any undulipodiated sperm (cilia or other undulipodia) at any stage, nuclei of the plasmodium may fuse by twos. After the fusion of two nuclei, apparently haploid gamete nuclei, diploid zygotes are claimed to develop and become sporonts as mentioned above. Inside them the diploid nuclei undergo the zygotic meiosis characteristic of members of the Fungi kingdom. The meiotic products develop into cells that form a new the 'polar filaments' and thus are called "filamented spores".

Since the mid-1970's the incidence of microsporan infections has risen. At least fourteen species of microsporans have been reported to cause diseases in humans that affect digestive, urinary, respiratory or nervous system.

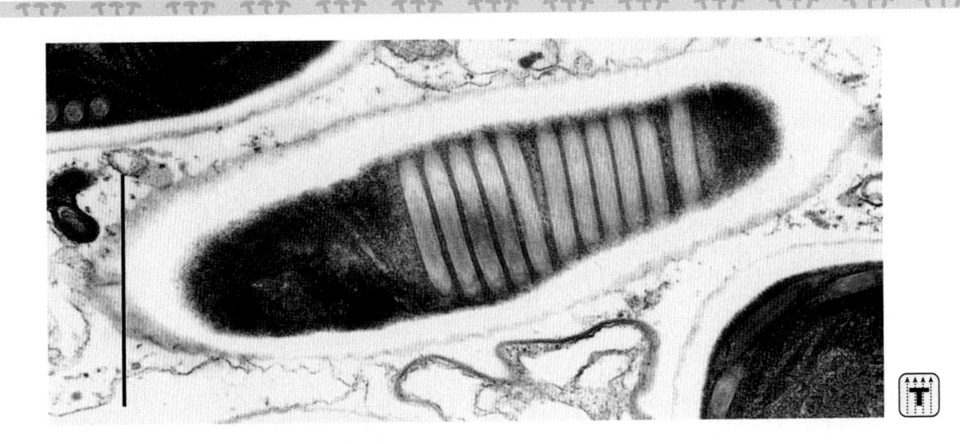

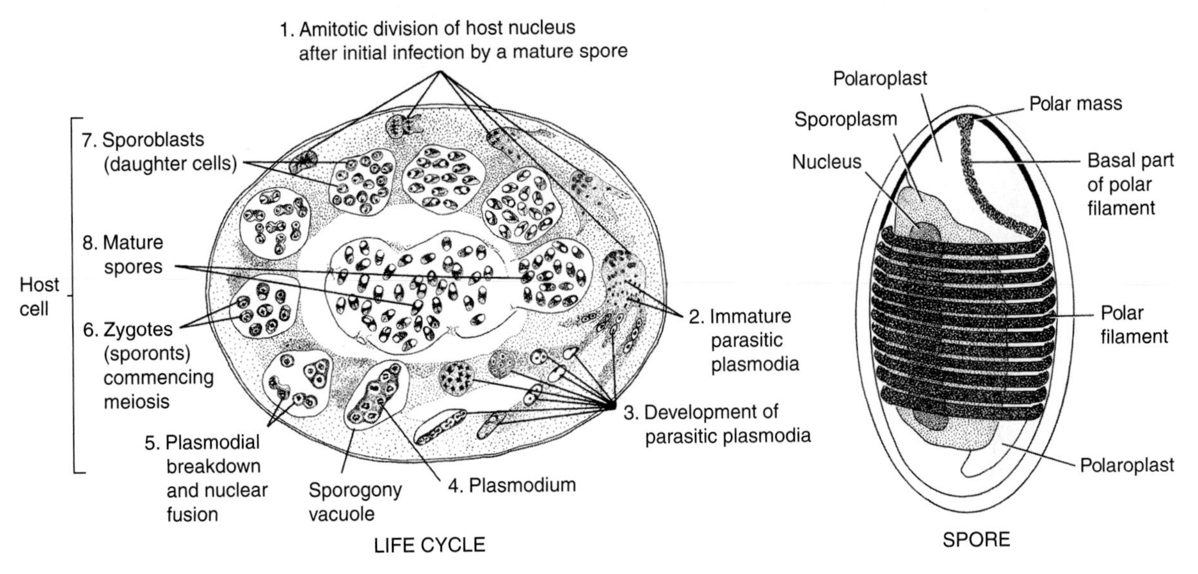

Figure A *Glugea stephani*, a microsporan necrotroph in the starry flounder, *Platiclothus stellatus* (A-37). Ultrastructure of a mature spore. TEM, bar = 1 μm. [Photograph courtesy of H. M. Jensen and S. R. Wellings, *Journal of Protozoology* 19:297–305 (1972); drawings by R. Golder.]

F-2 Zygomycota
(zygomycetes)

Greek *zygon*, pair; *mykes*, fungus

GENERA
Basidiobolus
Blakeslea
Chaetocladium
Cochlonema
Conidiobolus
Endocochlus
Endogone
Glomeromycota
Kickxella
Mortierella
Mucor
Phycomyces
Pilobolus
Rhizopus
Stylopage

The hyphae of most members of this phylum are internally continuous—they lack cross walls (septa), except for those between reproductive structures and the rest of the mycelium. About 700 species of zygomycetous fungi live on land throughout the world. Although not numerous, they are common and important fungi. Many are saprobic, feeding osmotrophically—absorbing nutrients—on stored or decaying plant material, which may be your food. Some are highly specialized, living on animals, plants, protoctists, or even each other. A few zygomycetes can attack mushrooms, for example, as can a few ascomycetes (F-4).

Two modes of reproduction characterize zygomycotes: (1) direct development, by mitosis, of sporangiospores in sporangia and (2) formation of sexual zygosporangia (sometimes misleadingly called "zygospores") in which, after karyogamy, meiosis occurs to form haploid propagules. During reproduction by direct sporulation, haploid sporangiospores develop inside sporangia, which arise at the top of specialized sporangiophores (Figures A through C). In (2) special hyphae of compatible mating types are called gametangia (often labeled "plus" and "minus" because they look exactly the same); they are attracted to each other by hormones and grow until they touch. The ends of the hyphae swell, the walls between them dissolve, and the two sources of cytoplasm merge (cytogamy). Many haploid nuclei from each mating type enter the fused structure, which then develops into a thick-walled zygosporangium. Zygomycete matings more resemble orgies than simple couplings, because many nuclei fuse pairwise (karyogamy), giving multiple diploid nuclei simultaneously. Eventually each diploid nucleus undergoes meiosis during germination. Zygosporangia behave like spores, surviving unfavorable conditions, then germinating into haploid mycelia or directly into stalked sporangia. Clearly, in the fungi, the mammal-centered terms "mating" (in this case, conjugation) and "sexual reproduction" (in reference to the number of offspring following conjugation) conceal, rather than reveal, knowledge of sexual practices (Figures D and E).

Taxonomists have recently revised the classification of this group, recognizing four subphyla: Mucoromycotina, Entomophthoromycotina, Zoopagomycotina, and Kickxellomycotina. The order Glomales has now been recognized as a separate phylum, Glomeromycota. We describe the behavior of the four remaining groups here.

The Mucorales (members of the Mucoromycotina) are saprobic organisms; they secrete extracellular digestive enzymes and absorb dead organic matter. They form mitosporangia and conjugate to develop zygosporangia. Many mitosporangia, each containing one or many spores, have delicate walls that break easily to release their contents. Members of this order include the common black bread mold *Rhizopus stolonifer* and the well-known genera *Mucor*, *Phycomyces*, and *Pilobolus*. The sensitivity of some members of the genus *Phycomyces* to light is truly astounding—only a few photons are required to initiate the development of the sporangium. The photoreceptor, which absorbs blue light, may be the common B vitamin riboflavin. *Pilobolus* (the "hat thrower") grows in horse dung and shoots its black mitosporangia away from the dung to a distance of 2 m.

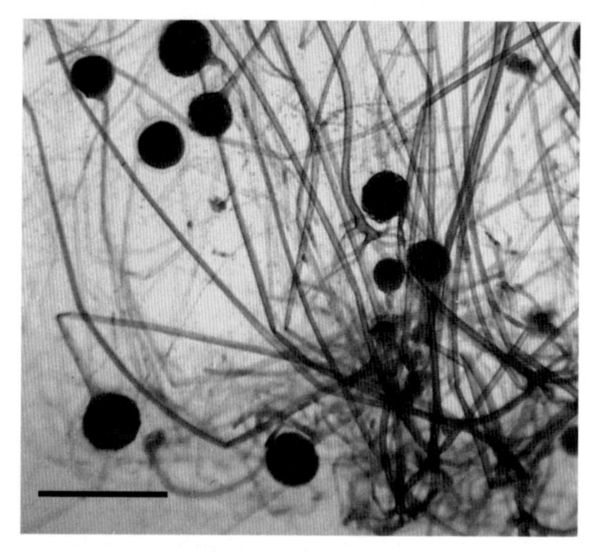

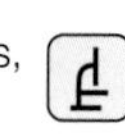

Figure A *Rhizopus stolonifer* hyphae, sporangiophores, and sporangia. LM, bar = 100 μm. [Courtesy of W. Ormerod.]

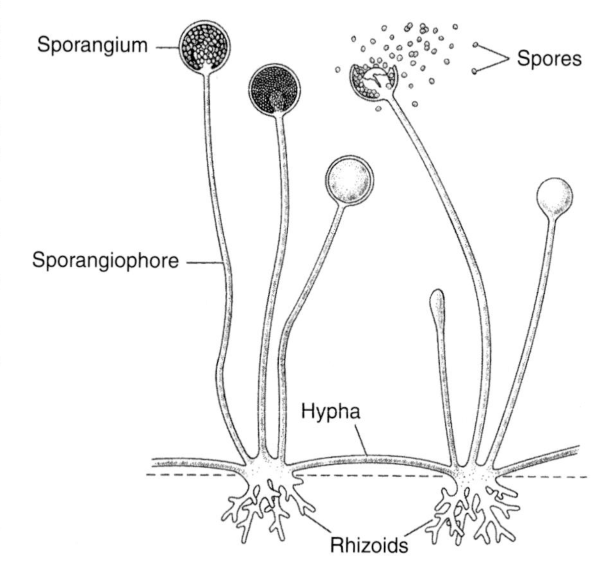

Figure B *Rhizopus* sp., black bread mold. Rhizoids anchor the fungus to the substrate. [Drawing by R. Golder.]

Most of the Entomophthorales (members of the Entomophthoromycotina) parasitize insects. *Entomophthora* attacks houseflies, reproducing directly after killing its host by forming tiny mitosporangia that, like those of *Pilobolus*, are forcibly

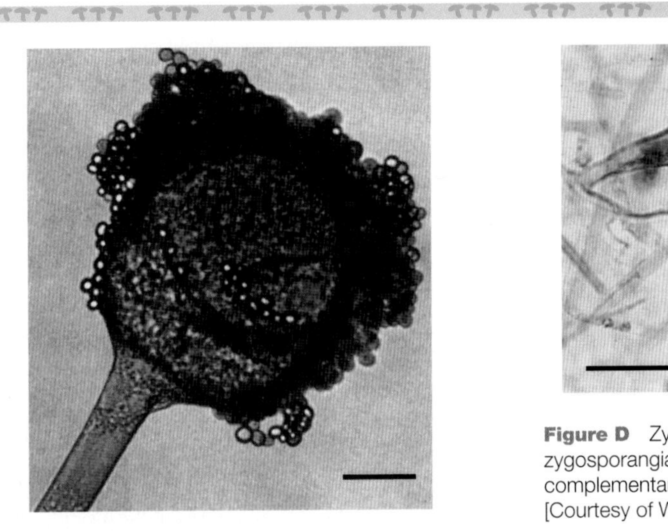

Figure C Broken *Rhizopus sporangium* showing sporangiophore bearing its propagules, spores produced by mitotic cell division. LM, bar = 10 μm. [Courtesy of G. Cope. Reproduced by permission of Elementary Science Study of Education Development Corporation, Inc., Newton, MA.]

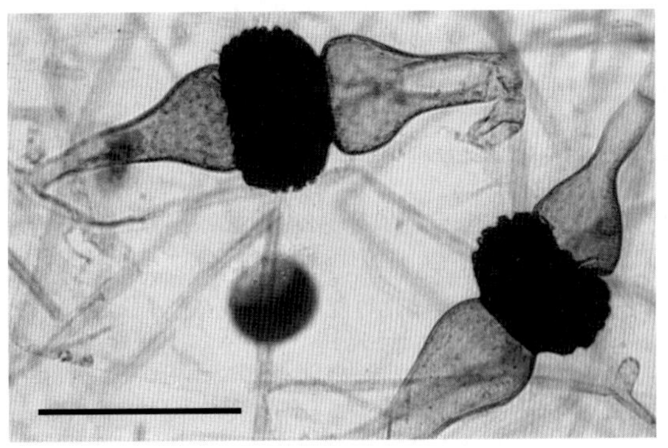

Figure D Zygomycote sexuality: black *Rhizopus* zygosporangia formed by conjugating hyphae of complementary mating types. LM, bar = 100 μm. [Courtesy of W. Ormerod.]

discharged. *Basidiobolus* is perhaps the best-known member of the Entomophthorales. It can be isolated worldwide from dung of frogs, reptiles, and bats but sometimes produces infections in humans (mostly in Africa, South America, and Asia).

The Zoopagales (members of the Zoopagomycotina) comprise about 65 species in 10 genera, all of which are symbiotrophic on amebas and other protoctists, on nematodes (A-11), and on other small animals. Most of them produce hyphae that penetrate the surface of the animal, coil inside it, and debilitate their victim. From the remains of the host emerge club-shaped gametangia that fuse and form zygosporangia. Among the genera are *Cochlonema* (*C. verrucosum*, a necrotroph of amebas), *Endocochlus*, and *Stylopage*.

The Kickxellales (members of the Kickxellomycotina) are distinctive in many ways: their hyphae are divided up by cross walls; they produce some of the most bizarre and complex asexual reproductive structures of any fungi, and they grow on such substrates as bat guano and the dung of rats in Death Valley.

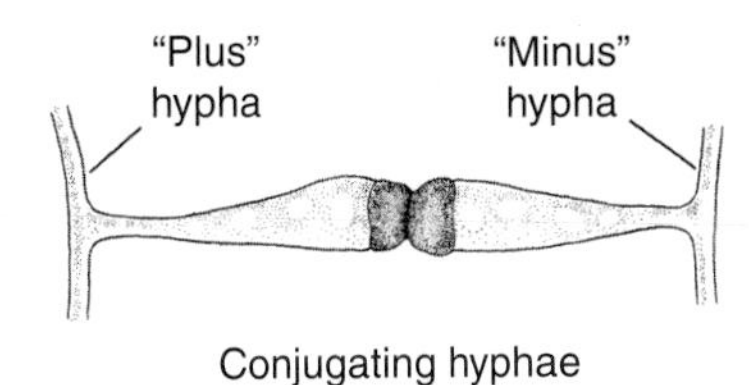

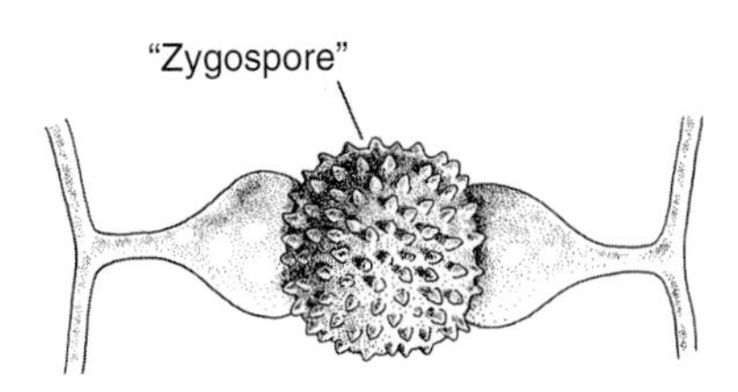

Figure E Conjugation in *Rhizopus*. [Drawing by R. Golder.]

F-3 Glomeromycota

Latin *glomus*, "ball of yarn"; *mykes*, fungus

GENERA

Acaulospora
Archaespora
Diversispora
Enthrophospora
Gigaspora
Glomus
Pacispora
Paraglomus
Sclerocystis
Scutellospora

The arbuscular mycorrhizal (AM) fungi [also called endomycorrhizal symbionts or vesicular-arbuscular (V-A) mycorrhizal fungi] and their closest nonmycorrhizal relative, *Geosiphon pyriforme*, are removed from Zygomycota and placed into the Glomeromycota, a new phylum (Redecker *et al.*, 2000; Schüßler, 2001). Formerly, the AM fungi were placed in a separate order, the Glomales, within the Zygomycota, but molecular evidence indicates that they are a distinct lineage (Schüßler *et al.*, 2001) that probably shares common ancestry with Ascomycota and Basidiomycota (Schüßler, 2001). The Glomeromycota are probably among the oldest terrestrial fungi. Their fossil record dates back to the Ordovician (460 mya; Pawlowska, 2005).

The AM fungi form symbioses with most land plant species, especially dicot angiosperms. Mycorrhizae (Figure A), usually stubby short fungal-root tissue, rather than roots alone are generally the organs of mineral nutrient uptake (Schüßler, 2001) (Figure A). AM fungi colonize plant roots and gain carbon from photosynthate. Phosphate, a highly insoluble ion in most soils, is transferred from the fungus to the plant (Helgason and Fitter, 2005). Consequently, AM fungi are important in terrestrial ecosystems and sustainable land management areas (Pawlowska, 2005). Members of more than 80 percent of extant vascular plant form these symbioses. Nonvascular plants such as liverworts and hornworts also form symbioses with AM fungi (Schüßler, 2001). The prevalence of mycorrhizal associations both in extant ecosystems and preserved with roots in Devonian fossils suggests that plants participated in fungal metabolic associations as soon as their ancestors colonized the land, more than 400 mya (Pirozynski and Malloch, 1975).

A close living relative of AM fungi is *Geosiphon pyriforme*. This fungus forms a unique endocytobiotic symbiosis with the filamentous, heterocystous, nitrogen-fixing *Nostoc* cyanobacteria that together grows into a green bladder a few centimeters long. This plantlike riverbank holobiont probably resembles ancient ancestral AM fungal-autotrophic symbioses. The *Geosiphon–Nostoc* association, hypothetically, is a living relic of symbioses that formed during the earliest stages of land colonization (Redecker *et al.*, 2000; Schüßler, 2001).

The AM fungi apparently are obligately symbiotic organisms that either lost or never evolved meiotic sexuality. Taken together with their long evolutionary history, the absence of sexual-reproductive structures suggests that Glomeromycota represent an ancient lineage of potential general interest to evolutionary biologists (Pawlowska, 2005).

The AM fungi influence plant biodiversity and help to control nematodes (A-11). Coiled mycelia of the AM fungus *Arthrobotrys oligospora* snare nematodes in a manner reminiscent of predatory constrictor snakes, for example (Bordallo *et al.*, 2002). Plants in association with AM mycorrhizae outcompete fungal pathogens in the rhizosphere. Mobilization of phosphate by V-A fungi promotes plant growth in acidic, nutrient-poor, and polluted soils. Thus, the Glomeromycota profoundly influence life on land in many ways (Schüßler, 2001).

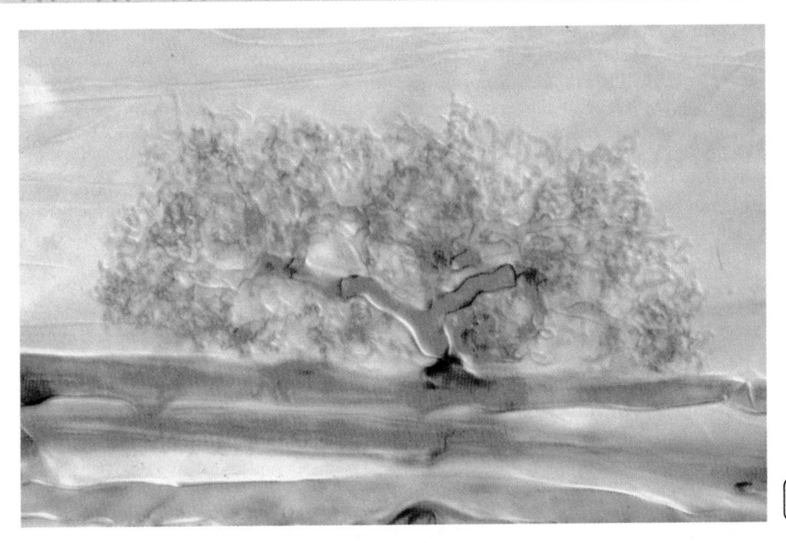

Figure A *Glomus mosseae,* a glomeromycotous fungus that forms mycorrhizae with onion root, *Allium cepa*. AMycorrhiza are symbiotic associations of fungi and roots, these are cruciae to food plant nutrition. [Credit: A. Schüßler.]

F-4 Ascomycota
(Ascomycotes, ascomycetes)

Greek *askos*, bladder; *mykes*, fungus

GENERA (d = deuteromycote)

Alternaria **(d)**	***Claviceps***	***Geotrichum*** **(d)**
Amorphomyces	***Clypeoseptoria*** **(d)**	***Histoplasma*** **(d)**
Aspergillus **(d)**	***Cryptococcus*** **(d)**	***Lichina***
Candida (Monilia)	***Cryptosporium*** **(d)**	***Morchella***
Cephalosporium	***Curvularia***	***Mycosphaerella***
Ceratocystis	***Elsinoe***	***Neurospora***
Chaetomium	***Fusarium*** **(d)**	

Familiar as baking and brewing yeasts, blue-green and black molds, cup fungi, morels, truffles and lichens, the ascomycotes are a large, diverse, and economically important group of fungi. About 33,000 species are known, and several times that many remain to be discovered and described. About 14,000 are the heterotrophic components of lichens (F-6), which are fungal symbioses with photoautotrophs [green algae (Pr-28) or cyanobacteria (B-6)]. The ascomycote symbionts make up 90–95 percent of the mass of the lichen.

Most ascomycetes form hyphae; others (many yeasts) are often exclusively unicellular. Ascomycotes are distinguished from other fungi by possession of the ascus (plural: *asci*), a microscopic reproductive structure (a tubular, spore-shooting meiosporangium) inside which haploid ascospores are produced. Two basic kinds of asci exist: (1) unitunicate, with a homogeneous wall and pressure-sensitive apical spore discharge mechanism such as an operculum or a ringlike sphincter; and (2) bitunicate, with a double wall—a thin inelastic outer wall and a thick inner wall that absorbs water, rupturing the outer wall and expanding upward, carrying spores with it.

The hyphae of ascomycotes are long, slender, branched tubes, and the mycelium that forms is a cottony mass. The hyphae are divided up into compartments by cross walls (septa) with simple pores, although a small spherical body sits on either side in case it is necessary to plug the pore in damage control. We have seen that Basidiomycotes often have an extended mycelial dikaryon. Ascomycotes have a much more restricted dikaryophase, which is initiated by fusion of compatible monokaryotic (haploid) hyphae only within the developing ascoma. In many species, the packed asci and their associated hyphae are so numerous that their organized mass forms a visible ascoma (plural: *ascomata*; formerly "ascocarp"). Ascomata are multicellular structures that act as platforms from which ascospores are launched. In "discomycetes" (cup fungi, morels, and the fungal component of many lichens), large numbers of asci form in an exposed hymenium on the surface of an apothecial ascoma (Figure A).

An ascus is produced when two hyphae of complementary mating types conjugate (cytogamy), as illustrated in Figure A. In the ascus, the partners' nuclei fuse (karyogamy), mingling the parental sets of chromosomes temporarily in one zygote nucleus. This transient diploid nucleus undergoes division by meiosis, producing four new haploid nuclei with the same number, but a different combination, of chromosomes as in the nuclei of the parent fungi. These newly formed nuclei then undergo another mitotic division to produce eight haploid nuclei, and some cytoplasm and a protective spore wall condense around each nucleus. Each of these eight cells is an ascospore. Ascospores are the kind of propagules that distinguish members of this phylum. Nestled in the ascus like peas in a pod, ascospores are released when mature and may be borne long distances by wind, water, or animals. If ascospores land in an appropriate nutrient-rich place, they germinate and send out hyphae of their own.

The formation of ascospores by the fusion of sexually different hyphae in what is called the teleomorph is not the sole means of ascomycote propagation. The production of propagules without sexual fusion is very widespread in ascomycotes, and the resultant asexual spores are called conidia. Specialized ascomycote hyphae may develop a succession of (blastic) conidia from conidiogenous cells by modifications of budding, or they may segment into huge numbers of (thallic) conidia, which, dispersed by wind, water, or animals (often insects), germinate elsewhere. In fact, large numbers of ascomycotes have lost all sexual processes and reproduce only by mitotic production of spores; these ascomycotes are the anamorphic taxa described later.

Dead or living plant and animal material nourishes ascomycotes; they secrete digestive enzymes into their immediate environment and absorb the dissolved nutrients thus formed. Ascomycotes play an essential ecological role by attacking and digesting resistant plant and animal molecules such as cellulose, lignin, keratin, and collagen. Valuable biological building blocks—compounds of carbon, nitrogen, and phosphorus, among others—locked in such macromolecules are thus recycled. The lichen symbioses are atypical in being photosynthesizers, rather than absorptive heterotrophs.

The ergot fungus, *Claviceps* (Figure B), causes a disease of rye flowers, and the resultant sclerotia are poisonous to humans and domesticated animals. Yet some alkaloids extracted from ergots are used to treat migraine and staunch uterine hemorrhages and are so valued that *Claviceps* is now artificially inoculated onto its host. Like most obligate symbiotrophs for which the nutrient or genetic contribution of the plant is unknown, *Claviceps* cannot grow in pure or axenic culture. Other ascomycotous pathogens have nearly eradicated such trees as the American chestnut and American elm. However, like the basidiomycotes (F-5), still other ascomycotes, such as truffles, form healthy mycorrhizal (fungus–root) associations with trees, shrubs, and other vascular plants.

There are three subphyla of Ascomycota:

(1) *Taphrinomycotina*, a small group including the agent of peach leaf curl, some yeasts with percurrent conidiogenesis, and even *Pneumocystis*, the agent of a form of pneumonia common in AIDS patients, which does not look fungal, having lost many fungal characteristics, but still has strong affinities with fungi at the molecular level;
(2) *Saccharomycotina*, about 1000 taxa of yeasts, including the baking and brewing varieties, but also *Candida albicans* that causes a disease of humans called Candidiasis (thrush is one form of this ailment);
(3) *Pezizomycotina*, all other ascomycetes, now divided up into ten classes.

The Taphrinomycetes are structurally among the most simple ascomycetes. They have short hyphae and form restricted mycelia on their hosts; the unitunicate asci are formed directly, rather than on ascogenous (ascus-forming) hyphae that grow from the conjugated cells; and they lack ascomata, just forming a layer of exposed asci on the surface of the host leaf. Their ascospores commonly bud while still in the ascus.

The Saccharomycotina are represented here by *Saccharomyces cerevisiae* (Figure C). Many yeasts do not grow hyphae or mycelia and have reverted to a single-celled way of life that might seem to

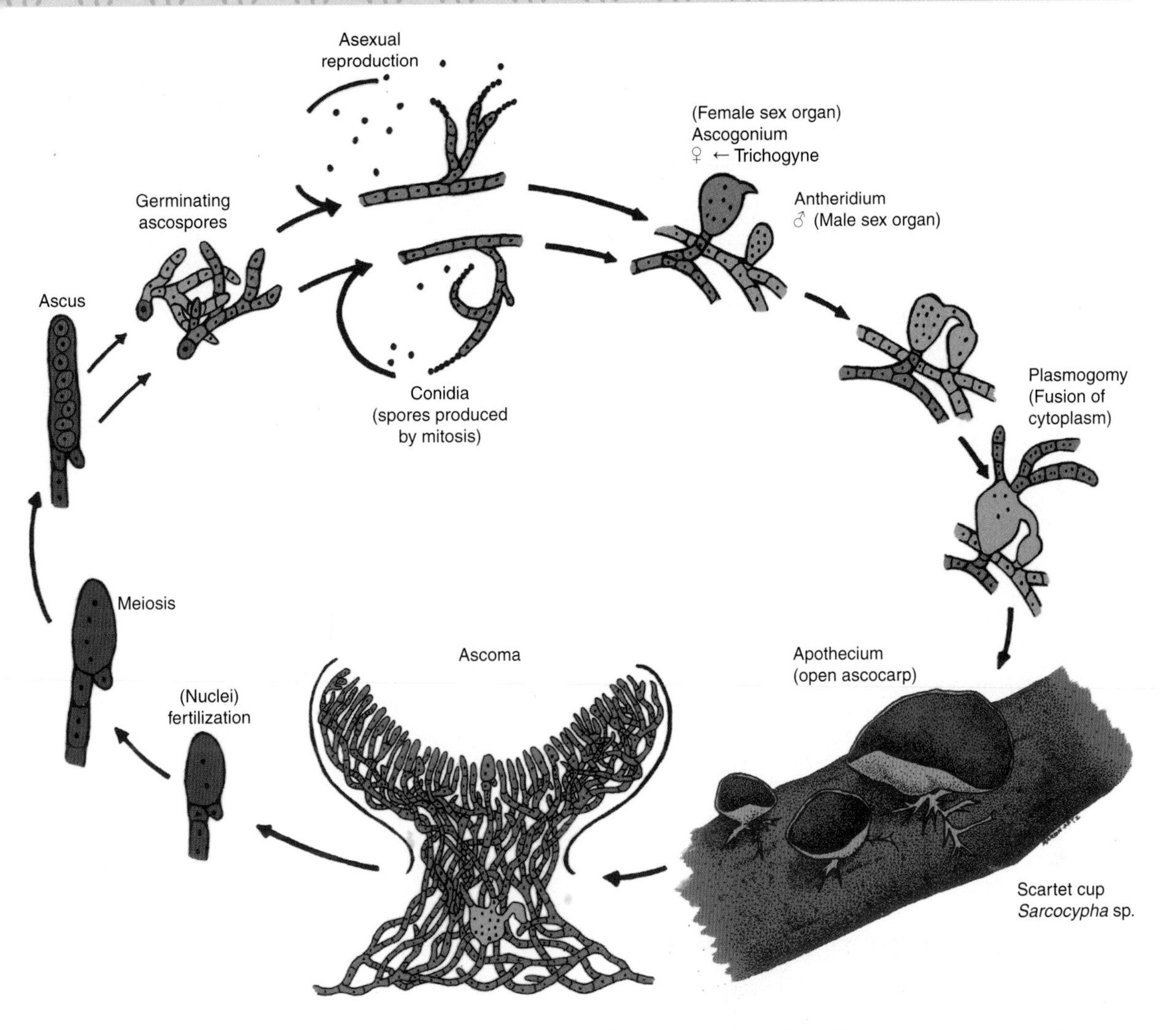

Figure A The ascus, produced as a result of cytogamy, is a specialized ascomycote cell in which meiosis takes place, giving rise at maturity to eight cells, ascospores, in a linear arrangement. In most ascomycotes, the mature ascus bursts to release a cloud of spores; spores may travel as far as 30 cm. Multiple asci form within an apothecium or within larger reproductive structures in the morels and cup fungi. [Illustration by Sheila Manion-Artz.]

place them with the protoctists. Yeasts grow by mitosis; after karyokinesis, the new offspring nucleus is injected, by microtubule spindle elongation, into the bud. The bud enlarges to the size of the parent, and cytokinesis produces two approximately equal offspring cells. Because many yeasts form asci, however, their resemblance to protoctists is superficial. Conjugation is by direct fusion of haploid yeast cells to form a diploid zygote, which undergoes meiosis and forms a meiosporangium (ascus). No ascoma is produced. The yeast ascospore arrangement is often tetrahedral (Figure D). Typical of yeast cells, the ascospores germinate, on release from the ascus, by budding (Figure E).

Yeasts ferment sugars such as glucose and sucrose to ethyl alcohol; this ability is utilized in the making of wine and beer. In the presence of gaseous oxygen, yeasts oxidize sugars to carbon dioxide, seen as gas bubbles in bread making. Brewer's and baker's yeasts have been cultivated for thousands of years. Now

F-4 Ascomycota
(continued)

Penicillium (d)	*Sordaria*
Pneumocystis	*Talaromyces* (d)
Rhizoctonia	*Torulopsis*
Rhizomyces	*Trichophyton* (d)
Saccharomyces	*Tuber*
Sarcoscypha	*Verticillium* (d)

Figure B Grain infected by *Claviceps purpurea*. Airborne spores of the fungus *C. purpurea* infect flowers of cereal grains, replacing normal seeds with a purple-black mass. The plant disease that we call ergot results. Diseased grain or flour, if consumed, produces ergotism in humans and livestock. Temporary insanity, painful involuntary muscle contractions, gangrene, and death result. Bar = 52 cm. [Courtesy of G. Bean.]

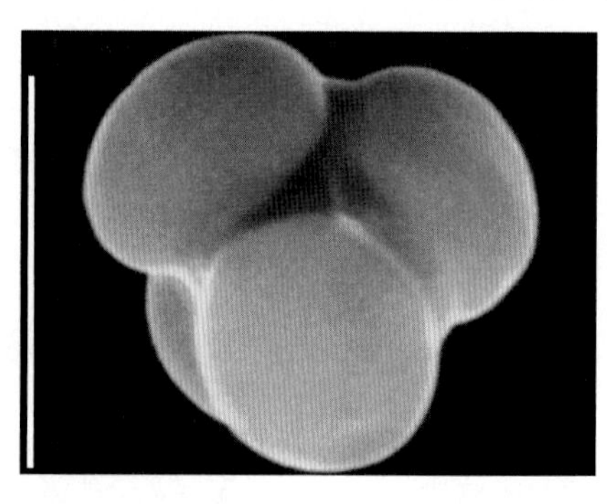

Figure D Tetrad of yeast ascospores formed after fertilization. Sexual reproduction in *Saccharomyces*. Cells of complementary mating types have fused and undergone meiosis. SEM, bar = 10 μm. [Courtesy of L. Bulla.]

Figure E Budding yeast cells after a day's growth. Cells reproduce by asymmetric mitotic cell division. LM, bar = 10 μm. [Courtesy of P. B. Moens.]

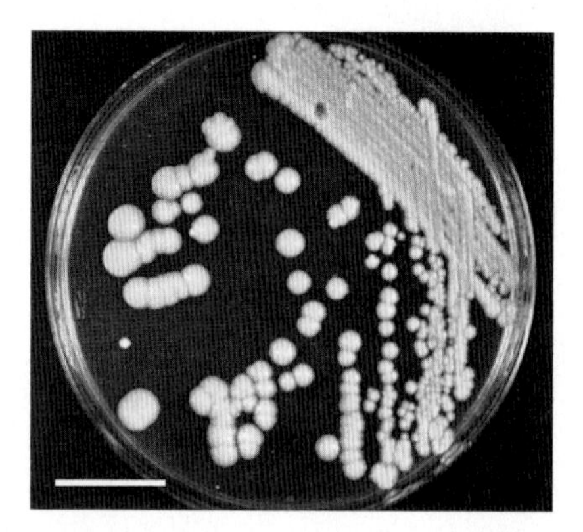

Figure C *Saccharomyces cerevisiae*. Yeast colonies on nutrient agar in petri dish. Bar = 1 cm. [Courtesy of P. B. Moens.]

yeasts are being modified by genetic engineering; they are used especially in the construction of artificial chromosomes. Their chromosomal centromeres can be placed on foreign DNA and used to propagate it.

The Pezizomycotina are the largest and best-known subphylum, incorporating almost all ascomycetes except yeasts and the leaf curl fungi. The asci generally develop from hyphae, which, in most cases, are part of an ascoma. Dikaryotic ascogenous hyphae grow through monokaryotic sterile tissue to the hymenium. The asci can be (1) unitunicate (the inner and outer layers of the ascus wall are more or less rigid and do not separate when spores are ejected), (2) bitunicate (the inner wall is elastic and expands beyond the outer wall when spores are released), or (3) prototunicate (the wall of the ascus dissolves at maturity and the ascospores are not forcibly expelled).

We will mention seven of the ten classes.

Class Pezizomycetes encompasses what have long been known as the operculate discomycetes—all ascomycetes producing apothecial ascomata containing asci with apical lids or opercula (Figure A). All ascomycetes that establish ectomycorrhizal symbioses with plants belong to this group. Many other members of the group are saprotrophic.

Class Dothideomycetes produces stromatic ascomata containing bitunicate asci. The genus *Mycosphaerella* has about 500 species, including one that forms part of the tissue of a brown seaweed (Pr-17) called *Ascophyllum*. Others feed on decaying soil organics, including rotting corn. The genus *Elsinoe* includes many pathogenic species and obligate symbiotrophs that cause diseases of citrus, raspberry, and avocado, among others. There are also some lichen-forming fungi, such as the Arthopyreniaceae, in this class.

Class Eurotiomycetes contains many fungi with prototunicate asci in small, closed ascomata, including the order Eurotiales, some of which are teleomorphs (sexual phases) of many important molds such as *Penicillium* and *Aspergillus*. The closely related Onygenales include many fungi that specialize in metabolizing keratin, allowing them and their asexual phases to grow on hoofs, on horns, and even on our skin, hair, and nails inciting diseases called tinens. There are over 40 species of such dermatophytic molds (anamorphs) placed in three genera. *Epidermophyton* has 2 species, *Microsporum* has 17 species, and *Trichophyton* has 24 species and varieties. The group also contains many lichen-forming fungi with bitunicate asci, including *Verrucaria*, which grows as a black band on rocks above high tide along the east and west coasts of North America.

Class Sordariomycetes includes most nonlichenized ascomycetes that produce perithecial ascomata—inside which thin-walled, unitunicate, inoperculate asci are produced and that have a narrow opening (ostiole) through which the ascospores are expelled. Perithecia can be single, as in *Neurospora*, or grouped in compound ascomata with many fertile cavities, as in *Xylaria*, which grows on wood. The genus *Neurospora* is widely used in genetic research. In each ascus, the four products of meiosis divide once by mitosis to form eight cells that remain fixed in a row in the order in which they were formed (Figure A). Each ascospore in an ascus can be picked up in that order and grown to determine its genetic constitution. The information thus obtained reveals the behavior of chromosomes during a single meiosis and the position of the genes on the chromosomes.

Class Leotiomycetes includes the majority of nonlichenized, fungi-producing apothecial ascomata containing unitunicate, inoperculate asci. These encompass the whitish powdery mildews (order Erysiphales) that attack the leaves of a wide range of plants, the tar spot fungi (Rhytismatales) that parasitize maple leaves, and the Leotiales (most other inoperculate discomycetes such as the teleomorphs of the *Monilia* that causes soft rot of stone fruits and the *Botrytis* that causes gray mold of strawberries) and many saprotrophs.

Class Lecanoromycetes is the largest group of lichenized ascomycetes, most of which produce apothecial ascomata, in which the asci are usually bitunicate with atypical kinds of apical dehiscence. Most common lichens with a variety of different thallus organization (foliose, crustose, fruticose) belong to this class—*Cladonia*, *Lecanora*, *Ramalina*, *Usnea*, *Lobaria*, *Peltigera*, *Physcia*, and so on.

Class Laboulbeniomycetes consists of all minute ectosymbiotrophs of insects. These organisms are highly host-specific—some will grow on only one sex of the host species or only one body part, such as the legs or the wings. Their ascospores, produced in bitunicate asci, germinate directly and develop eventually into reproductive structures between 0.1 and 1.0 mm in diameter. The ascospore first forms a septum delimiting two cells. The upper cell differentiates into the male reproductive organ, with several vial-shaped cells that produce male gametes called spermatia; the lower cell becomes the female reproductive organ, which is fertilized by the spermatia. The number of cells is fixed for each species. Genera include *Rhizomyces* and *Amorphomyces*.

The ascomycetous anamorphs are fungi that lack organs for sexual reproduction. Like the ascomycotes and basidiomycotes, they develop from spores, or conidia, into mycelia whose hyphae are divided by septa.

Although ascomycetous anamorphs lack meiotic sexuality, some of these fungi do exhibit a parasexual cycle. As documented in the fungal genetics laboratory, they form recombinant mycelia having different inherited traits by fusion of hyphae from two genetically marked, distinct organisms. From these recombinant mycelia, by processes not understood, new true-breeding haploid offspring appear and persist. The parasexual process does not require specialized mycelia or ascomata.

The term "anamorphic holomorphs" refers to those ascomycotes or basidiomycotes that have lost their potential to differentiate asci or basidia, but still reproduce asexually. It is a challenge to relate each anamorph to its sexual relative, as they have no morphological similarity, but molecular data now supply the missing information. Once these species have been reclassified, we are burdened with organisms having two valid names (for example, *Penicillium*, the well-known green mold, and *Talaromyces*, one of its poorly known sexual ascomycetous stages).

There are about 20,000 species of anamorphs, including some of great economic and medical importance (for example, athlete's foot fungus). Anamorphs are divided into three major groups distinguishable morphologically and functionally: hyphomycetes, coelomycetes, and mycelia sterilia.

The almost 10,000 coelomycetes reproduce by conidia borne on short, closely packed conidiophores that form a hymenial layer covered by part of the host plant. Acervular conidiomata can be subcuticular (covered only by the host cuticle), intraepidermal (arising within the cells of the epidermis), subepidermal, or developing beneath several layers of host cells. Acervular conidiomata can often be seen as flat, disk-shaped cushions on host plants. *Cryptosporium lunasporum* produces crescent-shaped conidia. Form-taxa are convenient groupings of structurally similar, although probably unrelated, organisms. As evolutionary information becomes available, form-taxa are replaced with standard or "phylogenetic" taxa.

The hyphomycetes comprise more than 10,000 anamorphic (asexually reproducing) species. In this group are many pathogenic and other yeasts that form neither asci nor basidia. But most hyphomycetes reproduce by means of conidia that develop in various ways at the tips of specialized hyphae. *Penicillium* (Figures F through H) belongs to this group.

The mycelia sterilia include those asexual fungi that lack any specialized reproductive structures; the mycelia simply grow without visible differentiation into spores. Of the two dozen or so genera belonging to this conglomerate group, the best known is *Rhizoctonia*, a common soil fungus that causes damping off and root rot of plants, including cultivated ones of economic importance. The basidiomycote *Pellicularia filamentosa* has *Rhizoctonia solani* as its anamorph.

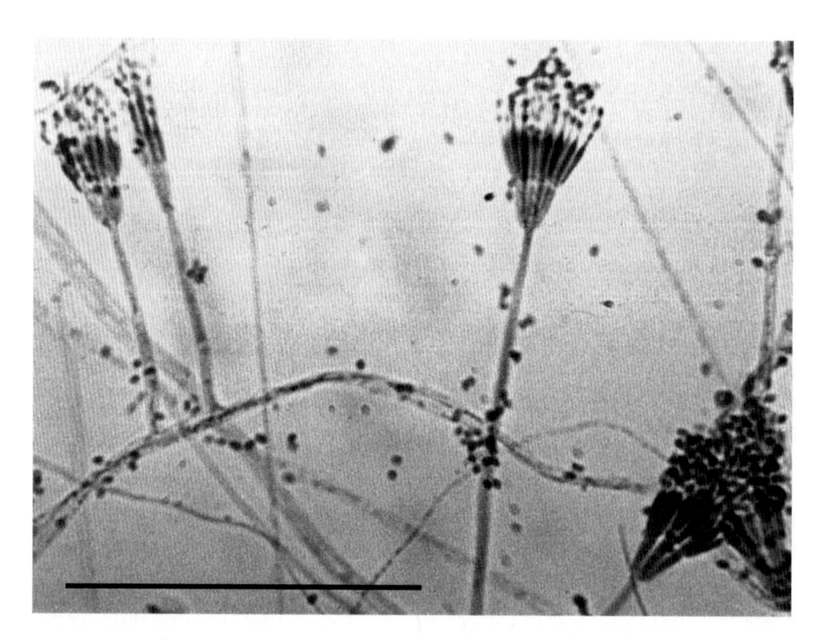

Figure F Hyphae of a *Penicillium* species with several conidiophores bearing conidia (spores) at their tips. LM, bar = 0.1 mm.

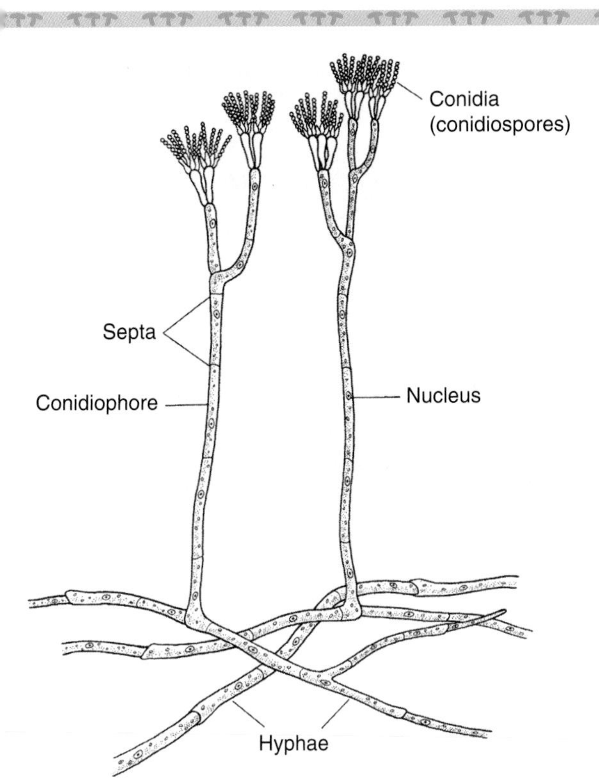

Figure G *Penicillium* sp. The antibiotic penicillin is a natural metabolic product of this mold. [Drawing by R. Golder.]

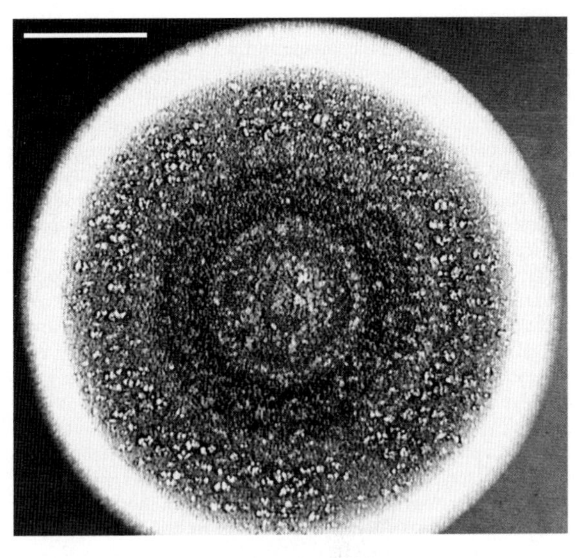

Figure H Colony of *Penicillium* derived from a single conidium growing on nutrient agar in a petri dish. Pigmented conidia form from the center, in the older parts of the colony; only unpigmented newer hyphae, as yet lacking conidia, are at the outer edge. Bar = 1 cm. [Courtesy of W. Ormerod.]

F-5 Basidiomycota
(Basidiomycotes, basidiomycetes)

Greek *basidion*, small base; *mykes*, fungus

GENERA

Agaricus	*Cronartium*	*Gymnosporangium*	*Psilocybe*
Amanita	*Cryptococcus*	*Lepiota*	*Puccinia*
Boletus	*Cyanthus*	*Lycoperdan*	*Rhizoctonia*
Calvatia	*Exiclia*	*Pellicularia*	*Schizophyll*[illegible]
Cantharellus	*Fomes*	*Phallus*	*Tilletia*
Clavaria	*Geastrum*	*Polyporus*	*Ustilago*

Basidiomycotes include the smuts, rusts, jelly fungi, mushrooms, chanterelles, shelf fungi, puffballs, and stinkhorns. The basidium (plural: *basidia*), a club-shaped microscopic reproductive structure from which their name is derived, distinguishes basidiomycotes from fungi in other phyla. The basidioma (plural: *basidiomata*; formerly "basidiocarp"), or mushroom, is the spore-producing body (Figure A). Each basidium usually bears four haploid spores, called basidiospores (Figure B), produced by meiosis. There are about 29,500 species of basidiomycotes. All are heterotrophs, many on agricultural crops and forest trees; many are symbiotrophs, forming important symbioses, called ectomycorrhizae ("outside fungus-roots"), with most forest trees and shrubs. Ectomycorrhizal fungi create a sheath of hyphae—a fungal mantle—around the outside of root. This connects to a network of hyphae (the Hartig net) that grows inside the root, between and around the cells of the root cortex, without penetrating them, and is the interface for exchange of nutrients. Fungal hyphae emanating from the mantle replace root hairs in many plants, as they can penetrate much further into the soil in search of phosphorus, nitrogen, and other inorganic nutrients. These they translocate and transfer to their plant partners via the Hartig net, in exchange for photosynthesis-derived carbohydrates, which move in the opposite direction. Others of this phylum, such as the saprobic cultivated button mushroom *Agaricus* and the plant symbiotrophic Mexican delicacy *Ustilago*, are popular in the human diet.

Figure A Basidiomata of *Boletus chrysenteron*, a yellow bolete mushroom of New England deciduous forest. Bar = 5 cm. [Courtesy of W. Ormerod.]

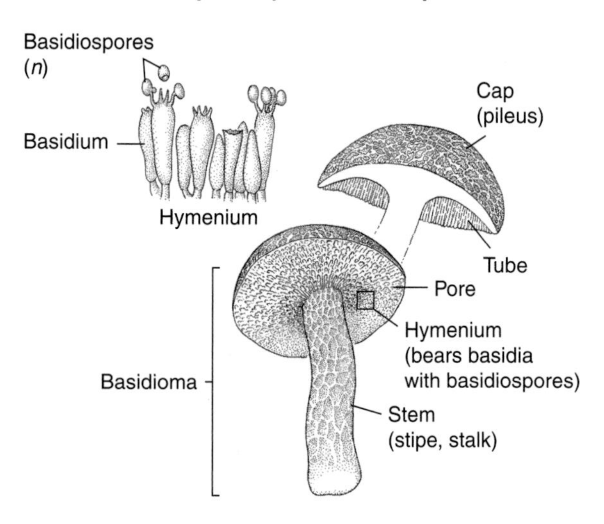

Figure B Basidioma, reproductive structure, of *Boletus chrysenteron*. The basidioma is composed of tightly packed hyphae. Sexual reproduction takes place in basidia; spores form on the basidia, which open in the tubes. [Drawing by L. Meszoly.]

Most basidiomycotes display structures from three stages of development: monokaryotic mycelia, dikaryotic mycelia, and dikaryotic basidiomata (but note the important exceptions below). A basidiospore germinates and grows into a septate mycelium that has one nucleus per compartment. When compatible monokaryotic mycelia meet, hyphal tips conjugate (Figures C and D). From the fused hyphae, a dikaryotic mycelium develops. Dikaryosis, the state of having a pair of compatible haploid nuclei in each hyphal segment, is common in this phylum. A dikaryotic mycelium grows by the simultaneous division of the two nuclei in a hyphal segment, a process that is facilitated by the development of a clamp connection, as follows. A backwardly directed side branch emerges from the hypha. This allows the two nuclei of the dikaryon to divide simultaneously—one in the main hypha, one in the branch—in such a way that after the branch fuses with the hypha again and two new septa are laid down (Figure C), the cells on either side contain a compatible pair of nuclei (Figure D—the clamp connection does not vanish after the nuclei have migrated but remains as a bump on the hypha, as shown in the fourth picture in Figure C).

By differentiation and growth, the dikaryotic, heterokaryotic mycelium in the largest subphylum, the Agaricomycotina, eventually forms the reproductive structure, the basidioma. In old literature, these basidiomata are called "fruiting bodies" or "fruits" because they were first well described by botanists. The basidioma reproduces by releasing huge numbers of meiotically produced propagules, the basidiospores. In basidiomata of the agarics and boletes, basidia form all over the surface of the gills (Figure E) or line the tubes (Figure B). After meiosis takes place within the basidium, the haploid basidiospores are produced at the ends of short projections from the basidia called sterigmata. After the spores have been actively expelled, the basidioma disintegrates, although the dikaryotic mycelium will usually persist, especially in ectomycorrhizal species. Thus, the life cycle of

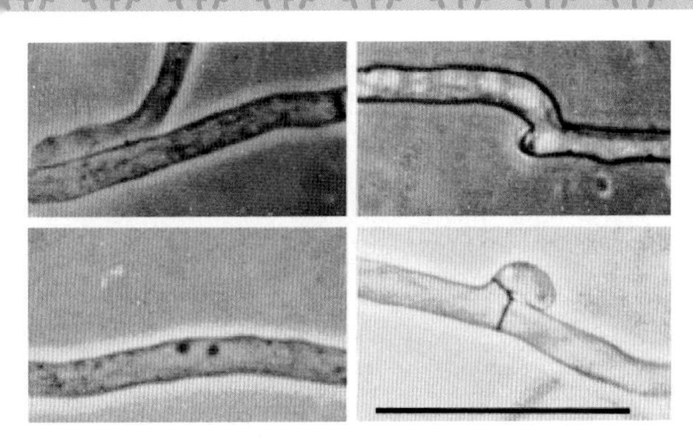

Figure C Conjugation of basidiomycote hyphae of complementary mating types. Sexual reproduction: (left top) approach of hyphae; (right top) incipient fusion; (left bottom) fused hyphae, two nuclei apparent; (right bottom) clamp connection. The species is *Schizophyllum commune*, shown in Figure E. LM, bar = 40 μm. [Photo by W. Ormerod.]

Figure E Underside of *Schizophyllum commune*, showing the gills. The white double lines of the gills bear the basidia in rows. Bar = 100 μm. [Photo by W. Ormerod.]

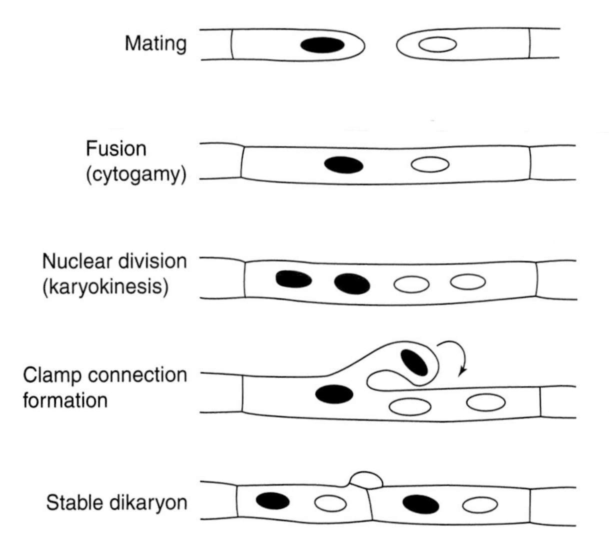

Figure D Conjugation. In the basidiomycete clamp connection, two forming cells of a dikaryotic hypha join laterally, ensuring that each cell contains two dissimilar nuclei.

basidiomycotes comes full circle: basidiospores, monokaryotic mycelia, dikaryotic mycelia, spore-producing basidiomata.

Three subphyla of basidiomycotes are recognized. The first two, numerically smaller, groups do not produce basidiomata because they are specialized plant symbiotrophs, living inside their hosts.

About a third of the basidiomycotes (8000 species so far described) belong to subphylum Pucciniomycotina, which includes 7000 rust fungi, order Pucciniales, necrotrophic on a wide range of plants. Rust fungi do not develop basidiomata. Their spore-bearing organs typically look like orange, rusty, or blackish eruptions on leaves and stems (and sometimes on fruits). Rust fungi produce thick-walled propagules called teliospores—diploid resting spores in which karyogamy and meiosis take place. From the teliospore, a septate basidium emerges and develops basidiospores. Some—for example, rust fungi that cause cereal crop diseases—have complex life cycles involving two host plants and up to five different kinds of spores, which are linked to seasonal conditions and the developmental biology of their host plants.

The second subphylum, Ustilaginomycotina, includes the smut fungi (Figure F), comprising about 1500 species of plant symbiotrophs. They usually have a haploid, saprobic yeast phase arising from germinating basidiospores, and a dikaryotic, hyphal, necrotrophic phase, at the end of which the teliospores are produced (Figure G). Like the rust fungi, they do not produce basidiomata. The teliospores often act as dispersal agents and later, after karyogamy and meiosis, give rise to basidiospores. Almost all smut fungi attack angiosperms, and most are found on grasses and sedges. Members of the class Ustilaginomycetes produce teliospores and comprise the order Urocystales, with nonseptate basidia, and the order Ustilaginales, with septate basidia. The numerically smaller Exobasidiomycetes include nonteliosporic plant symbiotrophs. The septa of this group have unusual pores enclosed by membrane caps on each side, and their wall carbohydrates mainly contain glucose and no xylose. Smut fungi sporulate in or on specific organs of the host plants—bulbs, stems, leaves, flowers,

F-5 Basidiomycota
(continued)

Figure F *Ustilago maydis*, corn smut. This fungus can be injurious to corn plants when it infests a field heavily. Although infestation is detrimental to the production of field corn, the smut itself has economic importance. The spore masses produced by this fungus are harvested, sold in Mexican marketplaces as cuitlacoche, and fried as a delicacy. In U.S. food shops, corn smut is sold under the name "corn mushrooms." Having been bred for low susceptibility to the smut, corn is now being bred for high susceptibility to produce corn smut as a gourmet food crop. The ear is approximately life size. [Courtesy of R. F. Evert.]

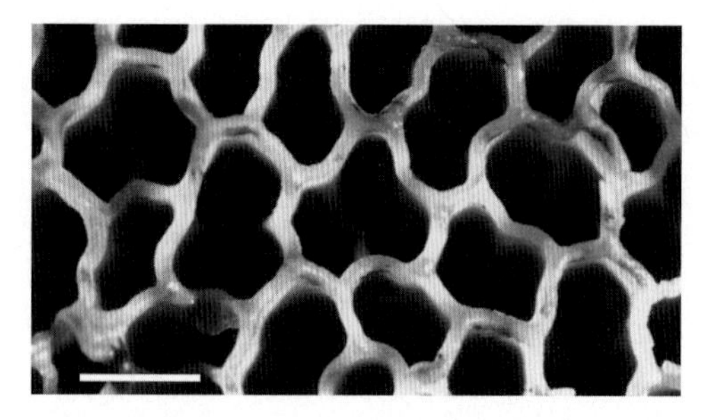

Figure G Pores of *Boletus chrysenteron*. Haploid basidiospores are produced within tubes (into which the pores lead) by meiosis. Spores exit through pores. Bar = 1 mm. [Photo by W. Ormerod.]

anthers, ovaries, or seeds—liberating characteristic masses of dark teliospores.

The third subphylum is the largest, the Agaricomycotina, with 20,000 species. These are divided among three classes, the Tremellomycetes, the Dacrymycetes, and the Agaricomycetes, all of which typically produce basidiomata.

The class Tremellomycetes include the largest group of jelly fungi, those with cruciately septate basidia, which constitute the order Tremellales, many of which are mycosymbiotrophs. Also placed here are a small group called the Filobasidiales, which include *Filobasidiella neoformans* and its *Cryptococcus neoformans* yeastlike anamorph (the causal agent of the human disease, Cryptococcosis).

The class Dacrymycetes are a small group of jelly fungi growing on dead wood (10 genera, 70 species), with "tuning fork" basidia, and very unusual basidiospores that become phragmoseptate (multiseptate).

The class Agaricomycetes are by far the largest group, as they include all the gill-bearing mushrooms (agarics; at least 10,000 species) and the tube-bearing boletes (over 1000 species), as well as the ear fungi, the shelf fungi, the resupinate basidiomycetes (basidiomata spread out on the surface of wood), the chanterelles, and the so-called gasteromycetes, including the stinkhorns, the puffballs, and earthstars.

The Agaricomycetes are divided into three groups:

(1) subclass *Agaricomycetidae* incorporating the orders Agaricales (10,000 mostly gilled mushrooms, about half of which are ectomycorrhizal and half saprobic) and Boletales (1000 mushrooms with fleshy tubes);
(2) subclass *Phallomycetidae*, which includes the orders Geastrales (earthstars), Gomphales (chanterelle-like, and club- or coral-like), Hysterangiales (hypogeous "false" truffles), and Phallales (stinkhorns, with a variety of amazing shapes);
(3) another ten orders, which are not yet grouped in any subclass categories. They are the *Auriculariales* (ear fungi), *Cantharellales* (chanterelles, plus coral-like, toothed, and resupinate forms), *Corticiales* (resupinates), *Gloeophyllales* (shelf fungi), *Hymenochaetales* (resupinate, chanterelle-like, agariclike, toothed, branched, and shelf fungi), *Polyporales* (shelf fungi, mostly saprobic), *Russulales* (about 1700 mostly ectomycorrhizal gilled fungi, mostly Russulas and Lactarii, plus poroid, toothed, shelflike and sequestrate [closed or gastroid] forms), *Sebacinales* (resupinate jelly fungi), *Thelephorales* (earth fans, all ectomycorrhizal), and *Trechisporales* (resupinates). The diversity of form displayed by several of these groups will not make the construction of a practical working classification easy.

F-6 Lichenes

GENERA

Calicium	*Lobaria*
Cladonia	*Ochrolechia*
Collema	*Parmelia*
Coniocybe	*Umbilicaria*
Herpothallon	*Usnea*
Lepraria	*Verrucaria*
Lichenothrix	

At least one-quarter of all described fungi, between 12,000 and 20,000 species, can establish symbiotic associations with algae to form lichens. Some 14,000 lichen species have so far been formally described. At least 40 genera of photosynthetic partners have been found. To the unaided eye, the lichen fungus partner and the isolated alga are entirely different from their associated counterparts (Figure A). The fungus dominates the morphology of the lichen and varies so conspicuously that, invariably, less attention is paid to the algal partner. Some lichens superficially resemble a plant such as a moss. For many years, lichens were called "pioneer plants" because they often grow on bare rock or sterile soil and are among the first organisms to cover burnt-out or newly exposed volcanic regions.

Although many lichens are pioneers, they are not plants. All lichens, upon analysis, are revealed to be symbiotic partnerships between a fungus, almost always an ascomycote (the mycobiont), and a photosynthetic organism, most often one of the Chlorophyta (Pr-28, usually *Trebouxia* or *Pseudotrebouxia*) or a cyanobacterium (B-6) such as *Nostoc* (the photobiont or phycobiont); but one xanthophyte (Pr-16) has been reported. Although we informally maintained lichens as a group because of their common features and easy recognition in nature, they have now formally been incorporated into the fungi, as the fungus essentially provides the morphology of the thallus (remember that the alga represents only 5–10 percent of the thallus).

Lichen fungi are not found without their algal partners in nature, but some lichen algae are found on their own. In the laboratory, few lichen partners have been separated and grown by themselves. For more than a century, biologists have realized that lichens evolved through of many independent symbiotic associations. Such associations probably began in response to attack by the fungus on a photosynthesizing protoctist. The attack led to a permanent truce, and the two partners persisted in a unique form. Lichens synthesize compounds—such as the lichenic acids and pigments—that are absent in the individual algae and fungi when these are grown alone.

Because so many different fungi have evolved symbiotic relations with members of the same algal and cyanobacterial genera (for example, *Trebouxia* or *Nostoc*), most lichenologists have insisted that they be classified with the ascomycote or basidiomycote groups to which the fungus belongs. Like the fungi that become lichens, various heterotrophic protoctists and animals, including at least half a dozen flatworms (A-7) and molluscs (A-26), have independently acquired some sort of photosynthetic partner. Far less diversity exists among the photobionts than among their associated heterotrophs. By analogy with the taxonomic convention for photosynthetic animals and protoctists, therefore, when we refer to 14,000 species of lichen-forming fungi, we are actually citing the number of distinct fungal species, the great majority of which are ascomycetes (there are a few basidiomycetes).

The lichens can thus be usefully classified, depending on whether the fungal partner is an ascomycete (as most are), a basidiomycote, or an anamorph (grouped with ascomycetes in our classification). Lichens composed of ascomycetes include *Lichina*, *Collema*, and *Cladonia*. Common in most of North America, *Cladonia* species are eaten by many animals (and, in emergencies, by people). They are at the base of the Arctic food webs that include caribou (reindeer) and Arctic people. *Lepraria* and *Lichenothrix*, woodland lichens, are lichen-forming fungi in which no sexual structures are known; they are lichenized anamorphs, which we group here with ascomycetes.

Naturalists group lichens according to their external appearance: crustose (low and crusty, closely attached to the substrate), foliose (leafy, with parts free from the substrate), or fruticose (finely branched or bushy; Figure A). The thallus—crusty, leafy, or bushy—is the growing part of the organism (Figure B).

Figure A *Cladonia cristatella*, the British soldier lichen of New England woodlands. Bar = 1 cm. [Courtesy of J. G. Schaadt.]

Figure B *Cladonia cristatella*, ascospore (mycobiont) and *Trebouxia* sp. (phycobiont) comprise the tissue. [Drawing by E. Hoffman.]

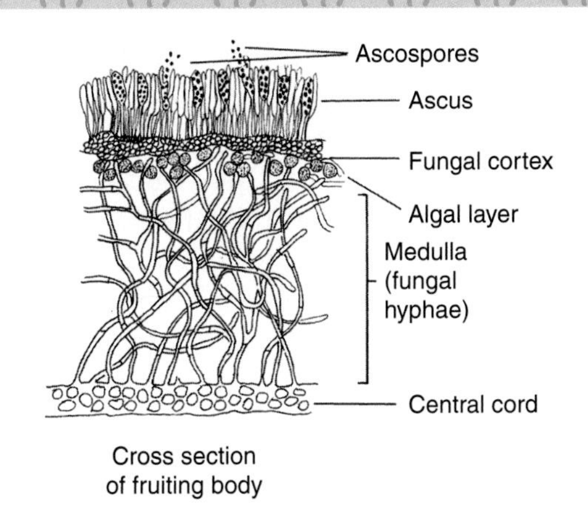

Figure C Enlarged view of a cross section of ascospore tissue. Some lichen fungi form asci. These sexual structures can generate spores that germinate to produce fungal hyphae. [Drawing by E. Hoffman.]

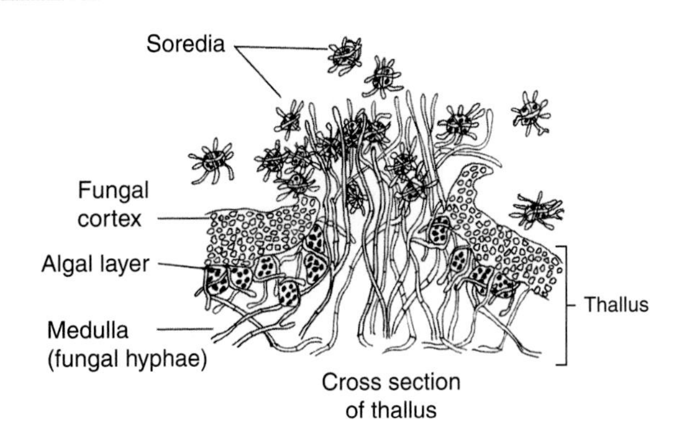

Figure D Enlarged view of a cross section of *Cladonia* thallus. Lichens can also reproduce by means of soredia, which are made up of algae (or cyanobacteria) and fungal hyphae. Dispersal of these soredia establishes new populations of lichens. [Drawing by E. Hoffman.]

Lichen reproduction is fungal. The sexual structure in which meiosis takes place and propagules form usually has a distinct appearance; lichens composed of ascomycotes typically produce open apothecial ascomata (Figure C). Sexual conjugation is followed by the formation of ascospores by meiosis. These ascospores germinate, but only in the presence of the proper algae, and produce fungal hyphae associated from the beginning with the algae, usually *Trebouxia* (Pr-28). Perhaps the most common form of lichen propagation is by somatic propagules in the absence of any sexual process. Lichens release soredia, small fragments consisting of at least one algal cell surrounded by fungal hyphae (Figure D). Soredia are easily dispersed by air currents; in a suitable environment, they develop into new lichens.

Lichens are most abundant in the Antarctic, the Arctic tundra, high mountains, the tropics, and northern old-growth forests; they commonly grow on tree barks or rocks. In the supratidal zone of rocky coasts, lichens become highly diversified in the sea-spray zone. *Verrucaria serpuloides* is a permanently submerged marine lichen. Lichens are well known for their resistance to desiccation; less well known is that they must have alternating dry and wet periods, because continuous drought or continuous dampness kills them, except for marine lichens. Desiccation is accompanied by cessation of photosynthesis; this hiatus may enable the lichen to survive otherwise harmful intense sunlight or extremes of heat or cold. The slow rate of growth of lichens is legendary. Many studies of lichens on gravestones and other dated monuments indicate that lichens grow from just 0.1 to 10.0 mm a year.

Lichens increase in total size slowly, but they metabolize quickly. Once metabolism begins, carbon dioxide is fixed into organic matter by the photobionts (the photosynthetic partner) within minutes. The photobiont transfers this photosynthate rapidly as sugar, sugar alcohol, or nitrogenous compounds (such as amino acids) to the fungal partner. Especially in lichens having nitrogen-fixing cyanobacteria, rapid metabolism—chemical production and transfer of nitrogen compounds—can be detected.

Lichens symbiotic with nitrogen-fixing cyanobacteria and perhaps other bacterial nitrogen fixers provide northern old-growth forests with much of their nitrogen. *Lobaria*, one of about 130 lichens recently found as an epiphyte in northern tree canopies, fixes as much as 75 percent of the nitrogen required by this Douglas fir environment. Rain and fog wash soluble nitrogenous compounds from the lichens to the forest floor, where the mycorrhizal fungi associated with the tree roots absorb them. Some tree species even send out roots from their branches into canopy lichens, thereby taking in fixed nitrogen directly.

Lichens are very important initiators of biological succession. By slowly wearing away and dissolving the minerals that compose the rocks on which some of them grow, lichens prepare the surfaces for the germination of seeds and the formation of rooted plant communities. Lichens thus accelerate weathering and initiate the formation of soils. Despite their hardiness, lichens are very sensitive to certain airborne materials—for example, the sulfur dioxide and volatile metal compounds that are released when coal is burned. Thus, the presence of lichens and the state of their health are used to indicate levels of air pollution.

Bibliography: Fungi

General

Aharonowitz, Y., and G. Cohen, "The microbiological production of pharmaceuticals." *Scientific American* 245(3):140–152; September 1981.

Ahmadjian, V., and M. E. Hale, eds., *The lichens.* Academic Press; New York; 1974.

Ahmadjian, V., and S. Paracer, *Symbiosis: An introduction to biological associations.* University Press of New England; Hanover, NH, and London; 1986.

Ainsworth, G. C., *Introduction to the history of medical and veterinary mycology.* Cambridge University Press; New York; 1987.

Ainsworth, G. C., *Ainsworth and Bisby's dictionary of the fungi, including the lichens*, 8th ed. Oxford University Press; New York; 1996.

G. C. Ainsworth, and A. S. Sussman, eds., *The fungi*, 4 vols. Academic Press; New York; 1965–1973.

Bonner, J. T., "The growth of mushrooms," *Scientific American* 194(5):97; May 1956; 98, 100, 102, 104, 106.

Brightman, F. H., *Oxford book of flowerless plants: Ferns, fungi, mosses and liverworts, lichens and seaweeds.* Oxford University Press; New York; 1966.

Emerson, R., "Molds and men," *Scientific American* 186(1):28–32; January 1952; Potato blight, ergot, wheat rust, athlete's foot, and citric acid from mold fermentation.

Gould, S. J., "A humongous fungus among us," *Natural History* 101(7):10–14; July 1992; 16, 18; *Armillaria bulbosa* covers 2.5 miles square—the largest organism?

Kendrick, W. B., *The fifth kingdom*, 2d ed. Focus Information Group, Inc; Newburyport, MA; and Mycologue Publications; Waterloo, Ontario; 1992.

Kosikowski, F. V., "Cheese," *Scientific American* 252(5):88–99; May 1985.

Large, E. C., *The advance of the fungi.* Dover; New York; 1962.

"Microbes for hire," *Science '85* 6(6):30–46; July/August 1985. A series of articles including: T. Monmaney, "Yeast at work," 30–36; D. Morgan and T. Monmaney, "The bug catalog," 37–41; and P. Preuss, "Industry in ferment," 42–46.

Moore-Landecker, E., *Fundamentals of the fungi*, 3d ed. Prentice-Hall; Englewood Cliffs, NJ; 1990.

Phillips, R., *Mushrooms of North America.* Little, Brown; Boston, MA; 1991.

Pirozynski, K. A., and D. L. Hawksworth, eds., *Coevolution of fungi with plants and animals.* Academic Press; San Diego, CA; 1988.

Rose, A. H., "The microbiological production of food and drink," *Scientific American* 245(3): 126–134, 136, 138; September 1981.

Wainwright, P. O., G. Hinkle, M. L. Sogin, and S. K. Stickel, "The monophyletic origins of the metazoa: An unexpected evolutionary link with fungi." *Science*, 260:340–343; 1993.

F-1 Microspora

Canning, E. U., "Phylum Microspora." In: L. Margulis, J. O. Corliss, M. Melkonian, and D. J. Chapman, eds., *Handbook of protoctista.* Jones and Bartlett; Boston, MA; 1990.

Dyer, P. S., "Evolutionary Biology: Microsporidia Sex–A Missing Link to Fungi." *Current Biology.* 18:1012–1014;2008.

Sprague, V., "Annotated list of species of Microsporidia." In: L. A. Bulla and T. C. Cheng, eds., *Comparative pathobiology. Vol. 2: Systematics of the microsporidia.* Plenum Press; New York; 1977.

F-3 Glomeromycota

Bordallo, J. J., L. V. Lopez-Llorca, H. B. Jansson, J. Salinas, L. Persmark, and L. Asensio, "Colonization of plant roots by egg-parasitic and nematode-trapping fungi." *New Phytologist* 154(2):491–499; 2002.

Helgason, T., and A. Fitter, "The ecology and evolution of the arbuscular mycorrhizal fungi." *Mycologist* 19(3):96–101; 2005.

Pawlowska, T. E., "Genetic processes in arbuscular mycorrhizal fungi." *FEMS Microbiology Letters* 251:185–192; 2005.

Pirozynski, K. A., and D. W. Malloch, "The origin of land plants: A matter of mycotrophism." *Biosystems.* 6:153–164; 1975.

Redecker, D., J. B. Morton, and T. D. Bruns, "Ancestral lineages of arbuscular mycorrhizal fungi (glomales)." *Molecular Phylogenetics and Evolution.* 14:276–284; 2000.

Schüßler, A., D. Shwarzott, and C. Walker, "A new fungal phylum, the Glomeromycota: Phylogeny and evolution." *Mycological Research.* 105:1413–1421; 2001.

Thorunn, H., and A. Fitter, "The ecology and evolution of the arbuscular mycorrhizal fungi." *Mycologist.* 19:96–101; 2005.

F-6 Lichenes

Brodo, I., S. D. Sharnoff, and S. Sharnoff, *Lichens of North America.* Yale University Press; New Haven, CT; 2001.

CHAPTER FIVE

KINGDOM PLANTAE

Pl-8B *Ceratozamia purpussi* cone. [Photograph by K. V. Schwartz.]

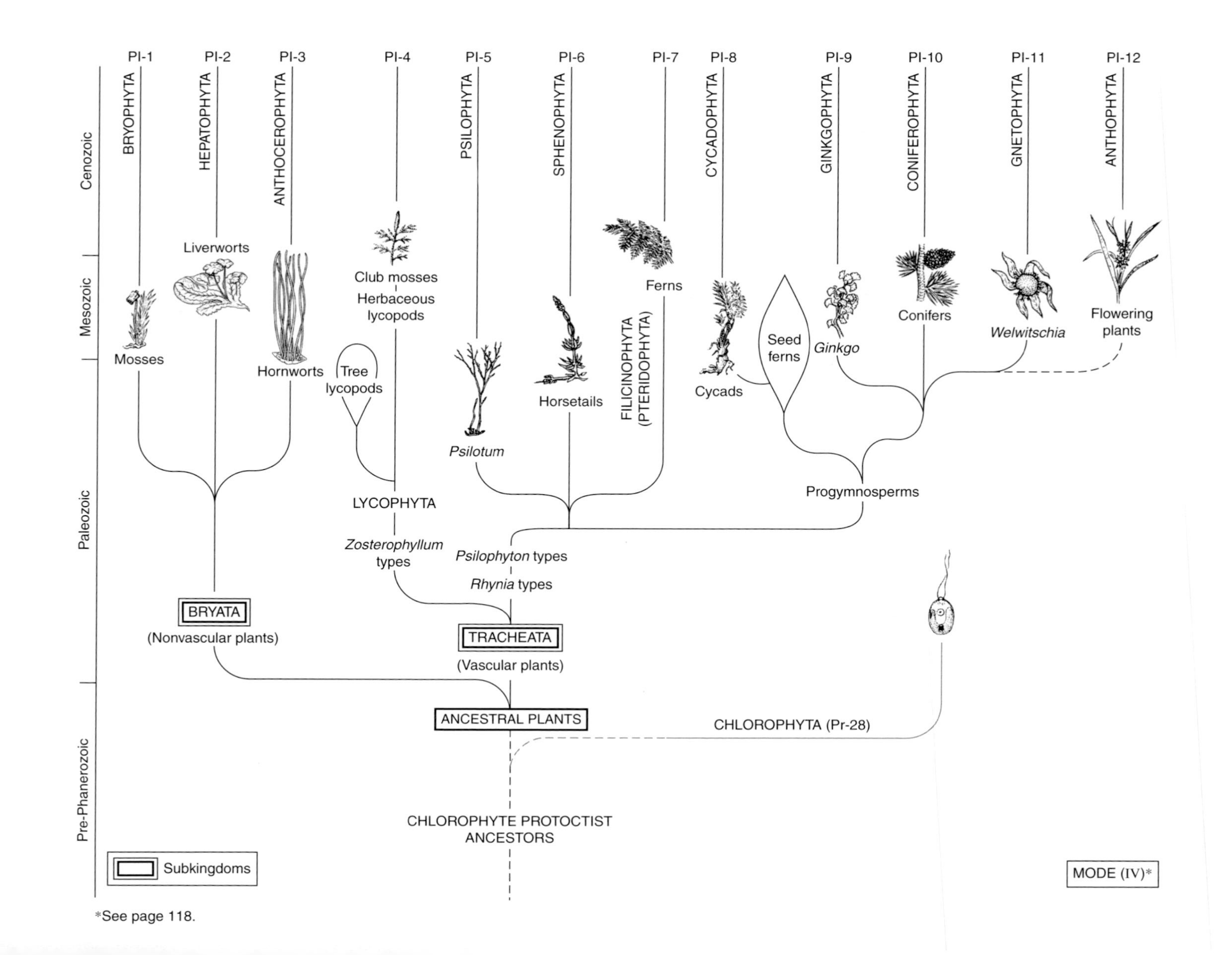

PI-1
BRYOPHYTA
PI-2
HEPATOPHYTA
PI-3
ANTHOCEROPHYTA
PI-4
PI-5
PSILOPHYTA
PI-6
SPHENOPHYTA
PI-7
PI-8
CYCADOPHYTA
PI-9
GINKGOPHYTA
PI-10
CONIFEROPHYTA
PI-11
GNETOPHYTA
PI-12
ANTHOPHYTA
Cenozoic
Mesozoic
Paleozoic
Pre-Phanerozoic
Liverworts
Mosses
Hornworts
Club mosses
Herbaceous lycopods
Tree lycopods
Ferns
FILICINOPHYTA (PTERIDOPHYTA)
Horsetails
Psilotum
Cycads
Seed ferns
Ginkgo
Conifers
Welwitschia
Flowering plants
LYCOPHYTA
Zosterophyllum types
Psilophyton types
Rhynia types
Progymnosperms
BRYATA
(Nonvascular plants)
TRACHEATA
(Vascular plants)
ANCESTRAL PLANTS
CHLOROPHYTA (Pr-28)
CHLOROPHYTE PROTOCTIST ANCESTORS
Subkingdoms
MODE (IV)*

*See page 118.

KINGDOM PLANTAE

Latin *planta*, plant

MODE IV*

Haplo-diploid organisms with zygotic meiosis. Fossil record extends from 450 mya to present that takes place in the adult diploid. These haploids produce gametes by mitosis. Fertilization by sperm (cytogamy and karyogamy) or pollen nucleus (karyogamy) leads to diploid embryo retained by the female haploid organism during early development. Fossil record extends from the lower Paleozoic era (450 million years ago) to the present.

Members of the plant kingdom develop from embryos—multicellular structures enclosed in maternal tissue (Figure Pl-1). Because all plants form embryos, they are all multicellular. Furthermore, because embryos are the products of the sexual fusion of cells, all plants potentially (although not always in reality) have a sexual stage in their life cycle. In the sexual stage, the male cell (sperm nucleus, haploid) fertilizes the female egg (embryo sac nucleus, haploid). Many plants grow and reproduce in ways that bypass the two-parent sexual fusion—all must have evolved from ancestors that formed embryos by sexual cell fusion. One example of asexual reproduction is the strawberry plant; plantlets form on extensions called runners extending from the parent plant. A second example is the asexual reproduction of little green balls of cells called gemmae (Latin gems or buds) by a parent moss or liverwort plant. Evolution of the embryo, protected by maternal tissue from drying and other environmental hazards, was a major factor in the spread of plants from oceans to dry land. Development in green algal (chlorophyte) ancestors of intimate symbioses with fungi may have been another factor in transitions from aquatic to terrestrial life, facilitating uptake of minerals and water by the plant. All plants are composed of eukaryotic cells, many having green plastids (Figure I-1). We distinguish plants from all other organisms by their life cycles rather than by their capacity for photosynthesis, because some plants (beech drops, *Epifagus*, for example) are entirely without photosynthesis throughout their lives. Photosynthesis by plants requires enzymes within membrane-bounded plastids. All plants that photosynthesize produce oxygen. (In comparison, in photosynthetic prokaryote species, enzymes are bound as chromatophores to cell membranes, not packaged separately. Prokaryote patterns of anaerobic and aerobic photosynthesis include formation of end products such as sulfur, sulfate, and oxygen.)

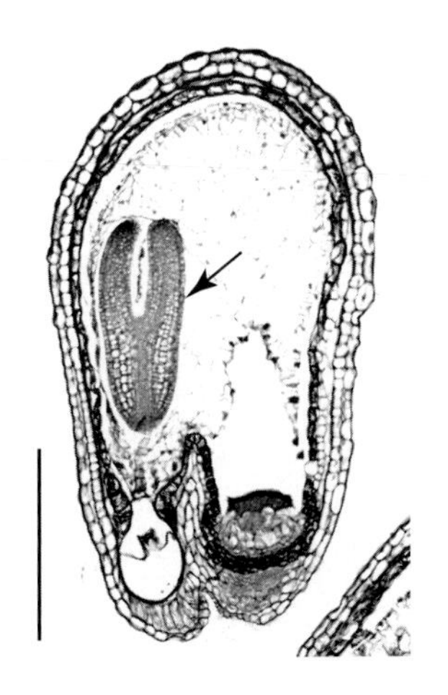

Figure Pl-1 Embryo (arrow) of shepherd's purse, *Capsella bursa-pastoris*. The two horseshoe-shaped cotyledons of the embryo have developed within the seed. Stored food (E, endosperm) surrounds and nourishes the body of the young sporophyte plant. LM, bar = 300 μm. [Photo by of W. Ormerod.]

Plants are adapted primarily for life on land, although many dwell in water during part of their life history. Plants are the organisms most responsible on land and in shallow marine environments for transforming solar energy, water, and carbon dioxide into photosynthate: food, fiber, coal, oil, wood, and other forms of stored energy. (In the open ocean, the protoctist plankton are the primary producers.) Houseplants, trees, and crop plants are members of the plant kingdom. Although most plants are multicellular, green, photosynthesizing organisms, a few genera such as dodder (*Cuscuta*) and Indian pipe (*Monotropa*) lost green pigment in the course of evolution and became sapro- or symbiotrophic. Photosynthetic organisms that were

* See page 119.

once classified as members of the plant kingdom on the basis of color and sedentary habit are no longer considered plants, because they lack embryos and other minimal criteria for plant classification. Cyanobacteria ("blue-green algae", B-6), green algae (Pr-28), all other photosynthetic protoctists (for example, Pr-15 through 18, Pr-27, Pr-32, and Pr-33), and lichens (F-6; fungi with bacteria or protoctist symbionts) are now placed with their relatives in the bacteria, protoctista, or fungi kingdom. Photosynthesis by plants sustains the rest of the biota not only by converting solar energy into food, but also by absorbing carbon dioxide and producing oxygen.

Some half million species of plants have been described. Because new species are found each year, especially in the tropics, probably another half million plants await discovery. Furthermore, this estimate is probably low; many plants resemble each other in form and will be distinguishable as separate species only by chemical analysis.

Two great groups—the nonvascular plants (informally called bryophytes, also called Bryata, Pl-1 through Pl-3) and the vascular plants (Tracheata, Pl-4 through Pl-12)—constitute the plant kingdom. We refer to the 12 "phyla" of the plant kingdom, but "division" is the term used by some botanists instead of "phylum." Tracheata, the familiar woody and herbaceous plants, are distinguished by vascular systems—lignified conducting tissues called xylem and phloem. Primary vascular tissues consist of cells derived from apical meristems (undifferentiated cells that give rise to new cells) and their derivatives. An example is a primary vascular bundle of xylem and phloem. Lignin is a complex macromolecule that stiffens the plant, impregnates xylem, and strengthens the wood of woody trees and shrubs. Under the bark of woody plants is a layer of cells (cambium) that generate new xylem and phloem throughout the life of a plant. This so-called secondary growth increases the diameters of tree trunks and shrub stems. Within the ring of cambium lies relatively undifferentiated tissue called pith. Herbaceous plants are nonwoody plants—for example, dandelions, ferns, and moss. Xylem cells transport water and ions from the roots through the plant. Phloem cells transport photosynthate—products of photosynthesis—throughout the plant body. Aboveground structures consisting of a shoot (the central upright axis) comprising a stem with branches and leaves and underground structures consisting of roots are unique to tracheophytes. The parts of the plant below ground that are anatomically similar to the stem are called a rhizome (Pl-5, Figure A), whereas those that differ from stems anatomically (in pattern of vascular tissue, for example) are roots, anchoring the plant and taking up water with dissolved minerals. A taproot is a single, large root (in carrot or cycad, for example) that may store nutrients (sugar beet) and water. True leaves and roots contain vascular tissues.

A true leaf consists of photosynthetic tissue covered by a cuticle (a waxy, water-resistant layer on the external surface) pierced by stomata—openings through which gases pass in and out of the leaf blade (Figure Pl-2). Vascular tissue is the plumbing system of the leaf, continuous with stem (and, eventually, root) through the leaf stalk, called the petiole. In comparison, mosses and many other nonvascular plants have leafy structures; a leafy structure lacks a vein of vascular tissue, is only one or a few cells thick, and may even lack cuticle and stomata. Mosses, liverworts, and some vascular plants also lack true roots and instead may have rhizoids, root-hair–like structures that lack vascular tissue veins. The site at which leaves and branches join the stem is called a node, evident in horsetails (Pl-6); internodes are stem regions between nodes. Branches that subdivide into two smaller branches are said to be dichotomous, as in psilophytes (Pl-5). Vascular plants may be grouped into seed-bearing (Pl-8 through Pl-12) and non-seed-bearing vascular plants (Pl-4 through Pl-7). The vast majority of plants living today are tracheophytes belonging to phylum Anthophyta, the flowering plants (Pl-12). The current number of gymnosperms (Greek *gymnos*, naked; *sperma*, seed) is about 720 species in 65 genera compared with approximately 240,000 species of flowering plants.

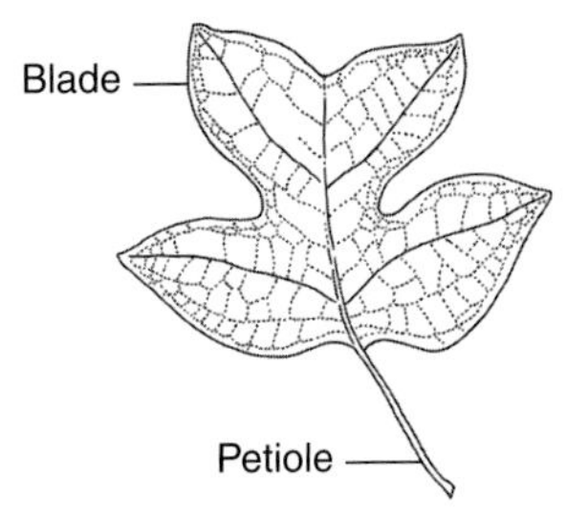

Figure Pl-2 Leaf of tulip tree, *Liriodendron tulipifera*. The net of veins is typical of a dicot megaphyll. [Drawing by L. Meszoly.]

Seed-bearing plants develop with exposed (naked) or enclosed seeds. A seed is formed by maturation of the ovule ("little egg") after fertilization; the ovule contains the female gametophyte with its egg cell, both surrounded by integuments. Minimally, the integuments surrounding the ovule form a seed coat. Gymnosperms (Pl-8 through Pl-11) develop seeds in cones, in comparison with protective, seed-enclosing fruits produced by flowering plants, or angiosperms (Pl-12). A cone is a reproductive structure that consists of a number of modified leaves clustered at the end of a stem; club mosses, horsetails, and gymnosperms bear cones. Cones are simple or compound. Scales (modified leaves) of a simple cone—a male pine cone, for example—bear sporangia and attach directly to the cone's central axis. In a compound cone—a female pine cone, for example—sporangia attach indirectly by a sterile bract to the cone's axis. Gymnosperms—cycads, ginkgos, conifers, and gnetophytes—produce pollen cones (male) and seed cones (ovule-bearing, female), whereas flowering plants produce pollen and ovules in the flower and, eventually, seed within fruit (mature, ripe ovules).

Four phyla of vascular plants do not produce seeds: club mosses (Pl-4), whisk ferns (Pl-5), horsetails (Pl-6), and ferns (Pl-7). All four of these phyla reproduce with spores. Spore dispersal in liverworts, hornworts, and horsetails is aided by elaters (Greek, "driver"), elongated structures that form within sporangia. The sporangium is an organ in which cells undergo meiosis and produce (haploid) spores. A spore (Pl-3, Figure B) is a reproductive cell capable of developing into a mature plant without fusing with another cell; in comparison, an egg or sperm fuses with its complementary reproductive cell to produce a new plant. A spore usually consists of a single cell and is produced in a sporangium. In club mosses and horsetails, sporangia are borne on modified leaves clustered in cones called strobili (Pl-6, Figure B). Club mosses and ferns also reproduce asexually by means of plantlets. These new plantlets are shed from the parent plant and, unlike moss gemmae, are diploid. Non-seed-bearing vascular plants have evolved a wide array of leaf structures. Leaves of Phyla Pl-7 through Pl-12 are megaphylls (Figure Pl-2)—that is, comparatively large leaves with a web of veins or parallel veins and a gap above the junction of leaf with stem. These veins connect through several strands of vascular tissue to the vascular tissues of the stem. Leaves may be simple, as is a tulip leaf, with undivided blade, or compound, as is a walnut leaf, with a leaf blade composed of many leaflets. The club moss leaf, called a microphyll, has only a single vascular strand and lacks the leaf gap characteristic of megaphylls. Horsetail leaves are very small and scalelike; psilophytes (Pl-5) lack leaves altogether.

We call the three phyla of nonvascular plants (Pl-1 through Pl-3) Bryata—mosses, liverworts, and hornworts. Opinion varies regarding *Takakia*, considered a moss by some, including us, and a separate phylum (or division) by others. All nonvascular plants have a thallus—a plant body without true leaves, stem, or roots (Pl-2, Figure B). Lacking true roots and vascular systems, they obtain moisture and nutrients from the environment by diffusion directly into their tissues. Within their bodies, diffusion, capillary action, and cytoplasmic streaming conduct fluids. Mosses also have conducting cells called leptoids and hydroids, but these are nonlignified. Hydroids are elongated cells that lack living cytoplasm at maturity. Their thin end walls are very permeable to solutes and water. Nutrient-conducting leptoid cells surround the hydroids in some mosses. Delicate uni- or multicellular filaments (rhizoids) anchor nonvascular plants to soil, rock, or tree bark (Pl-1, Figure C). Most of the Bryata flourish in moisture-saturated habitats such as acidic bogs. Given these similarities among the nonvascular plants, significant differences, such as presence or absence of cuticle, remain such that the three nonvascular plant phyla may have diverged independently of one another from green algal ancestors.

All plant cells at all stages harbor plastids, usually many. Minimal plastids are 1 mm in diameter, membrane-bounded, colorless organelles (as in roots, colorless sprouts, and symbiotrophic plants). Exposure to sunlight may transform colorless plastids into the chlorophyll-containing green form called chloroplasts. Fully developed plastids are so similar to those of green algae that biologists agree that these chlorophytes (Pr-28) were ancestral to plants. Other support for this hypothesis is that green algae and plants have similar cell-to-cell connections called plasmodesmata (singular: *plasmodesma*). Some chlorophytes, such as *Klebsormidium*, even have cellulosic walls and patterns of mitotic cell division identical with those of plants. In these green algae, as in plants, a cell wall structure called a cell plate (phragmoplast) develops perpendicularly to the mitotic spindle and separates the two daughter cells at the completion of mitosis.

Plants, like all other extant organisms, have aquatic ancestors, with land plants having evolved from only a small group of green algae. Plantlike fossils first appear in rocks of the Silurian period (430–408 mya) as rootless, leafless, but upright seaweed-like organisms. The earliest plants, for which the fossil record is abundant, were ancestral tracheophytes of two major types, represented during the Devonian period by the extinct *Zosterophyllum* and *Rhynia* (see phylogeny). Because nonvascular plants lack vascular tissue, they are presumed to have evolved before the appearance of vascular tissue—before tracheophytes. In apparent contradiction to this time sequence, though, the earliest bryophyte fossil found so far is only 350 million years old, which is later than the first tracheophytes in the fossil record. Better fossilization of lignified tissues of vascular plants compared with nonlignified tissue of nonvascular plants may explain this discrepancy.

The *Zosterophyllum* types gave rise to or share a common ancestor with lycopods (Pl-4), which have a fossil record as definite lycopods extending 400 million years into geological history. This tracheophyte group speciated extensively and included tree lycopods at the end of the Paleozoic era but is now reduced to a few genera of club mosses and their kin. Ancestral groups for psilophytes (Pl-5) and horsetails (Pl-6) are unknown at present.

Lycophytes and psilophytes have each been put forward as extant representatives of the first split in early lineages of vascular land plants. Chloroplast DNA studies tend to confirm the geological evidence that lycopods are more closely related to nonvascular plants (Pl-1 through Pl-3), whereas psilophytes are more closely related to vascular plants (Pl-4 through Pl-12) other than club mosses. The chloroplast gene order in modern lycopods is shared with that of *Marchantia*, a liverwort (Pl-2). Although psilophytes seem ancestral ("primitive")—they lack roots and have shoot protrusions that are probably branchlets rather than being homologous to leaves—their chloroplast DNA resembles that of ferns (Pl-7), gymnosperms (Pl-8 through Pl-11), and angiosperms (Pl-12). Psilophytes probably evolved directly from *Rhynia* types.

The *Rhynia* types of extinct tracheophytes were the ancestors of all the vascular land plants except club mosses. Many groups, such as the extinct phylum of seed ferns (Cycadofilicales or pteridosperms) and the phylum of horsetails (Sphenophyta, Pl-6), were far larger and more important in the past than they are now. We do not know if the ancestors of psilophytes (Pl-5) have any modern representatives other than *Psilotum* still living, although relationships among extant groups are now being sought with elegant molecular methods of inquiry. The *Psilophyton* types were ancestral to progymnosperms.

The details of seed, flower, fruit, and endosperm origins are under investigation, but we do know that these evolutionary innovations of flowering plants (Pl-12) changed the living world forever. Endosperm is a tissue, unique to flowering plants, that is neither sporophytic nor gametophytic. Endosperm develops from the union of sperm with polar nuclei of the central cell (female). Stored nutrients in endosperm are digested by the embryo. The remarkable innovation of the seed evolved by at least 360 mya (in the late Devonian) and more than once. At least one lineage of seed plants—progymnosperms, which had seeds but no flowers or fruit—gave rise to the great Mesozoic forests of cone-bearing plants: cycads, ginkgos, conifers, and other gymnosperms. In what may be the most primitive animal pollination system, cycad cones (Pl-8) produce odors and heat that attract pollinating insects. The anthophytes—the angiosperms, or flowering plants (Pl-12)—by their production of nectar (sugary liquid produced in flowers that serves to attract and reward pollinating animals), flowers, and fruit created an environment in which we and so many other animals could thrive. Flowering plants are an enormous group and are relatively young, having appeared on the scene only about 130 mya, the newest plant phylum. Like the cycads, flowering plants were considered by some biologists to have descended from seed ferns. Now there is considerable evidence that flowering plants evolved from an *Ephedra*-like gnetophyte (Pl-11); double fertilization occurs in both.

If it seems that there are far fewer plant groups than animal groups, it is partly because plant and animal taxa are defined by morphological criteria, and the diversity of internal and external anatomy is more extensive in animals than in plants. The differences between many plants are subtle, often involving chemical distinctions. Plants produce many chemical compounds that are secondary metabolites used in their defenses against fungi, animals, and other plants and are therefore only indirectly required for survival and reproduction. Secondary compounds include toxins, psychoactive compounds such as the marijuana alkaloids, and respiratory poisons such as cyanide—all of which deter predators from eating the plant. For example, black walnut trees leak compounds into the soil that prevent plants of other species from growing nearby. These poisons and other secondary metabolites, even gaseous compounds, are important in determining the distribution, growth rate, and abundance of plants in natural communities. Many of these compounds directly affect survival: the diterpenes, gibberellic acids, and—in reproduction, pollen and seed dispersal—flavonoids. Some activate genes, resulting, for example, in nodulation (swellings on roots) for symbiotrophy of *Rhizobium* in legumes and recognition of fungi in the establishment of mycorrhizae. Thousands of secondary metabolites are known, and many are used by the pharmaceutical industry as starting materials for the manufacture of drugs.

All plants develop from embryos, young diploid, multicellular organisms supported by sterile or nonreproductive tissue; in conifers, gnetophytes, and flowering plants (Pl-10 through Pl-12), the cotyledon (embryonic seed leaf) provides nutrients to the young embryo (Figure Pl-1). The cotyledon of monocots absorbs food, whereas cotyledons of many dicots store food. Unlike animal embryos, the plant embryo is not a blastula (Figure A-1). Unlike fungi, in which cells are either haploid (monokaryotic) or diploid (dikaryotic), except for the transient diploidy of the zygote during sexual reproduction, plants alternate haploid and diploid generations in their life cycle. Haploid plants (n) are called gametophytes; diploid plants ($2n$) are called sporophytes.

The life cycles of all nonvascular plants (Pl-1 through Pl-3) are dominated by the conspicuous green gametophyte (haploid), exemplified by a green mat of moss. To follow the life cycle of the moss (Figure Pl-3), in broad outline an example of the general plant life cycle, we begin with sexual reproduction. As its name implies, the gametophyte produces gametes. The male reproductive organ—the antheridium—produces sperm having a pair of forward-pointing undulipodia (Figure I-2 and Pl-1, Figure D). The female reproductive organ—the archegonium—produces an egg cell. Sperm of mosses are dispersed from the male mature gametophyte by splashing raindrops. In a magnified view of the archegonium, sperm can be seen swimming toward the egg. Sperm of cycads, ginkgos, gnetophytes, conifers, and flowering plants are carried to the egg in a pollen tube, formed after germination of a pollen grain. The tube nucleus of the mature pollen grain directs growth of the tube to the ovule. In flowering plants, the tube grows through stigma and style (Pl-12, Figure B). The pollen tube provides a moist environment for sperm; thus its evolution eliminated the requirement for environmental water during fertilization. There—within the archegonium and still on the female gametophyte—fertilization of the egg by the sperm takes place. Fertilization restores diploidy (the $2n$ condition) and initiates development of the zygote into the embryo. The embryo develops into the diploid sporophyte that emerges from the archegonium. The young sporophyte derives nutrients from the female gametophyte, on which it permanently

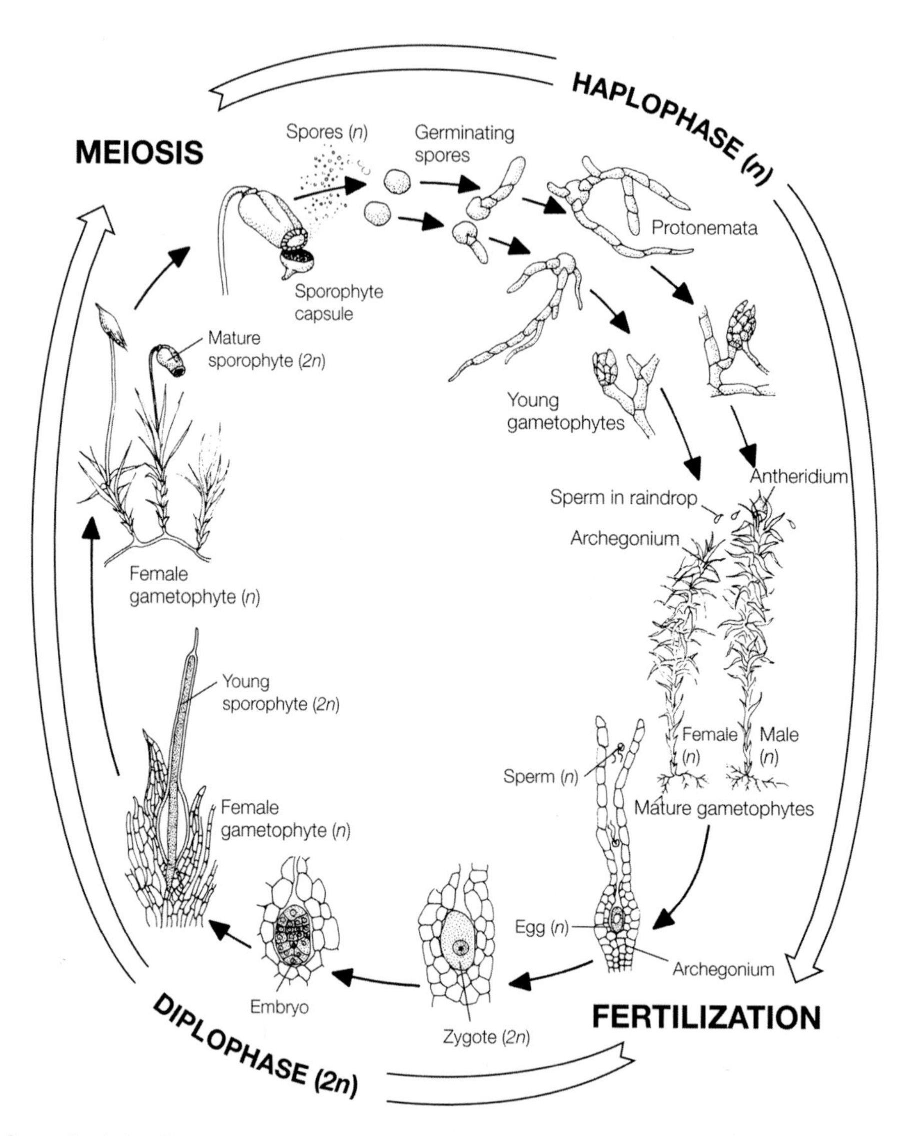

Figure PI-3 Generalized plant life cycle. In the life cycle of a moss, both the gamete-forming haploid phase and the spore-forming diploid phase are conspicuous. This moss, *Polytrichum*, is described in phylum PI-1. [Drawing by K. Delisle.]

perches. Early in its growth, the moss sporophyte becomes green and photosynthesizes. As its name implies, the mature sporophyte produces haploid spores by meiosis within the capsule. The nonvascular plant sporophyte is small and often brown by the time it releases spores, like the stalked spore capsule of the moss shown in the life cycle. Spores that land in a favorable site germinate, beginning the (haploid) gametophyte generation and developing a strand of photosynthetic cells (protonema) that resembles a green algal filament. This young gametophyte forms a little outgrowth of cells called a bud, which grows into a mature gametophyte. The sporophyte of nonvascular plants is generally dependent on the gametophyte. In the vascular plants, on the other hand, the sporophyte is greener, larger, and more conspicuous than the gametophyte. For example, the oak tree is a sporophyte. The sporophyte generation dominates the life cycle in the most recently evolved extant phyla [horsetails (Pl-6), ferns (Pl-7), and the seed bearers (Pl-8 through Pl-12)], and the gametophyte is reduced in size. In flowering plants, both male and female gametophytes (instead of living as separate plants like moss gametophytes do) are only small groups of cells entirely dependent on the sporophyte. The oak female gametophyte, as is the case for all flowers, is hidden within the flower; the oak male gametophyte is hidden within a grain of pollen. Extant seed plants, which produce nonmotile sperm in combination with pollen tubes, constitute one end of a series that extends from mosses, liverworts, hornworts, club mosses, horsetails, psilophytes, and ferns—all having swimming sperm—through ginkgos and cycads, which have pollen tubes as well as motile sperm, to conifers, gnetophytes, and flowering plants, with pollen tubes as conduits for nonmotile male gametes. Plant life histories are as elegantly diverse as their forms and colors.

SUBKINGDOM BRYATA

Pl-1 Bryophyta

(Mosses)

Greek *bryon*, moss; *phyton*, plant

GENERA

Andreaea
Bryum
Buxbaumia
Fontinalis
Funaria
Geothallus
Hypnum
Physcomitrella
Physcomitrium
Polytrichum
Sphagnum
Takakia
Tetraphis
Tortula

The nonvascular plants informally called bryophytes are the liverworts, hornworts, *Takakia*, and mosses, but only the mosses and *Takakia* are now included in the phylum Bryophyta. The presence of conducting cells and the lack of elaters—coils that facilitate spore dispersal—distinguish mosses from liverworts and hornworts. Mosses have leafy gametophytes with a stalk (stem) and multicellular rhizoids but lack the true leaves, stems, and roots present in vascular plants. The rhizoids, which are rootlike multicellular filaments, anchor the moss gametophyte to the substrate.

The moss life cycle alternates gametophyte and sporophyte—as is universal in plants (Figure Pl-2). Fusing in pairs, the egg and sperm form a zygote. Embedded in the female, the embryo develops into a (diploid) sporophyte. Spores (haploid), produced by the sporophyte, germinate and give rise to the next generation of young gametophytes.

Mosses are low-growing plants that flourish in moist habitats including freshwater. Some grow on oceanside rocks, but none are marine. Most of the nearly 10,000 moss species live in moist tropical environments, although mosses are also more conspicuous than other nonvascular plants in temperate North America. The three classes of this phylum are Sphagnopsida (peat mosses), Andreaeopsida (granite mosses), and with the majority of species, Bryopsida, or true mosses. *Takakia* (Figure A), a mosslike low-growing plant, is regarded by some workers as a moss similar to Andreaeopsida, by others as deserving separate phylum status. Many mosses are well adapted to withstand desiccation and survive as quiescent spores or dry gametophytes throughout the dry season. Several moss species are found in warm deserts, and mosses (for example, *Andreaea*) dominate the cold deserts of the Arctic tundra. Since its discovery in 1951, *Takakia* has been found in a great arc around the Pacific from Borneo to the Himalayas (Nepal, Sikkim, and China), Japan, the Aleutian Islands, the Alaskan panhandle, and British Columbia.

The conspicuous and familiar generation of mosses, including *Takakia*, is the green, leafy (not a true leaf) gametophyte—a haploid organism. Instead of the rootlike rhizoid of mosses, *Takakia* arises from an underground, branching rhizome. The leaves of mosses are one-cell thick with a midrib in some species, arrayed in a spiral around the stem. *Takakia*, in comparison, has phyllidia—solid, cylindrical appendages three to five cells thick that arise singly or in twos, threes, or fours in an irregular spiral from an erect shoot of the gametophyte (Figure B). Phyllidia are unique to *Takakia*. Gametophytes and sporophytes of mosses, including *Takakia*, lack the lignified xylem and phloem of vascular plants. Conducting tissues of mosses are water-conducting hydroids and (rarely) leptoids. All mosses have stomata—minute openings in the epidermis through which gases move

Figure A *Takakia* habit: female gametophyte plants bearing sporophytes with tapered capsules. Male plants nestled among female plants. [Photograph by permission of A. S. Heilman, D. K. Smith, and K. D. McFarland. From D. K. Smith and P. G. Davison, *Journal of the Hattori Botanical Laboratory*, 73:263–271 (1993).]

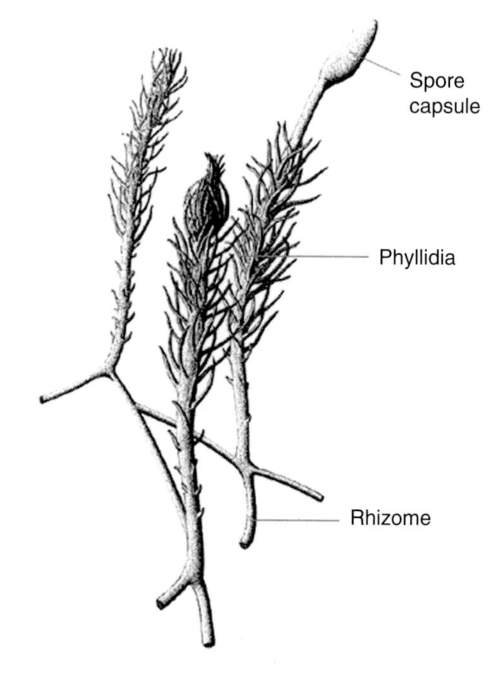

Figure B *Takakia* gametophyte upright axis with rhizome, phyllidia, and spore capsule. [Drawing by C. Lyons.]

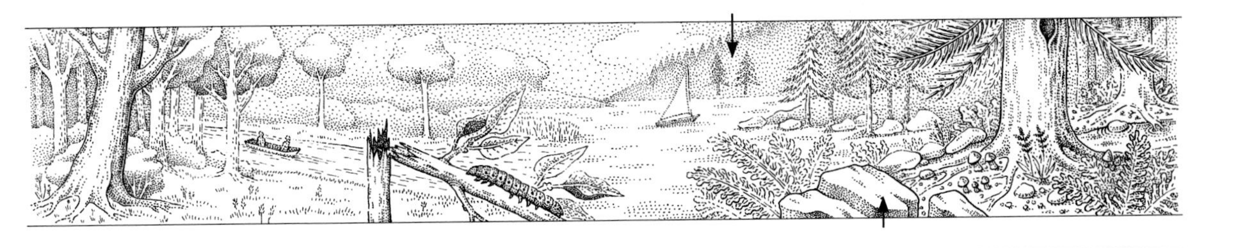

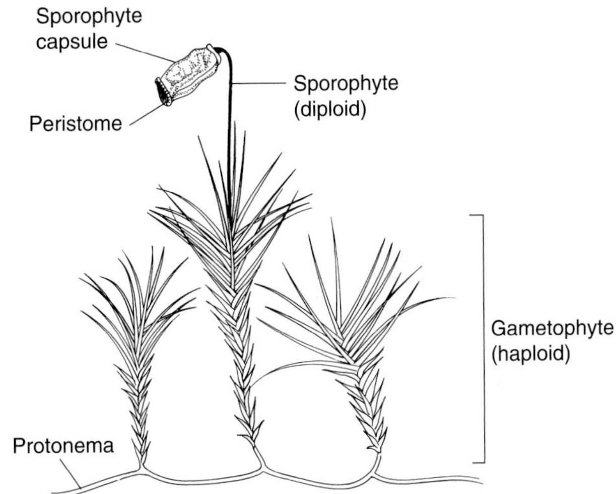

Figure C *Polytrichum juniperinum*, a common ground cover in the mixed coniferous and deciduous forest of New England. Leafy gametophyte (haploid) with mature sporophyte capsule (diploid). Bar = 3 cm. [Photograph courtesy of J. G. Schaadt; drawing by L. Meszoly.]

(like those of hornworts but of a different form from pores of liverworts)—on their gametophytes, and some have stomata on their sporophytes. *Takakia* lacks stomata. The stoma of some mosses is surrounded by a single bagel-shaped cell, in comparison with the paired dog bone or kidney–shaped guard cells surrounding stomata of vascular plants. Like hornworts, some mosses have a cuticle on their leaves; *Takakia* lacks a cuticle. *Polytrichum*, a common moss in temperate woodlands (Figure C), typifies moss species in which the entire plant is either male or female. *Takakia* gametophytes are male, female, or sterile, resembling one another in size. In some other mosses, two sexes, with their reproductive organs, are on the same plant. Fragmentation of the stem or leaf or production of minute green spheres called gemmae also can give rise to new individual mosses, which is also true of liverworts. Each individual plant produced asexually bears the haploid genetic information of its parent plant. Asexual reproduction is not known in *Takakia*.

The moss gametophyte produces gametes in multicellular gametangia. These reproductive organs are either archegonia, which produce eggs, or antheridia, which produce sperm. After antheridia are produced, the shoot tip of *Takakia* may resume growth, unlike other mosses. Moss sperm have two forward-directed undulipodia, as do the gametes and swimming cells of liverworts and hornworts (Figure D). Sperm of *Polytrichum* and other mosses are dispersed from a splash cup: the head of the antheridium collects rain into which sperm are released. Water droplets splash out along with the sperm, ejecting them long distances and enabling them to reach archegonia. Insects also disperse sperm-loaded water. At closer range, chemotaxis directs sperm in the water droplet to swim toward the moss egg.

Fertilization initiates development of the sporophyte embryo, which is retained in the archegonium (Figure Pl-3). The diploid sporophyte obtains nutrients from the female gametophyte, to which it attaches permanently by a foot. Later, the moss sporophyte becomes nutritionally independent and green; it photosynthesizes at least early in its growth, like hornwort and unlike liverwort sporophytes. Mature sporophytes of mosses release haploid spores in a sequence of events that includes the formation of spores by meiosis, drying and browning of the sporangium (capsule), opening of the capsule lid (operculum), and wind dispersal of spores (Figure E). In mosses of the class Bryopsida, tissues beneath the operculum split into a circlet of teeth called the peristome. As they dry, the teeth curl, releasing spores. The other moss classes lack peristomata. A single capsule may release as many as 50 million spores. Spore germination initiates the haploid gametophyte generation, growing the protonema, a strand of photosynthetic cells that resembles a green algal filament (Figure Pl-3). Protonemata form budlike structures that give rise to the leafy upright gametophyte. Some spores give rise to male gametophytes, other spores to female gametophytes. In mosses with male and female reproductive organs in the same individual plant, one spore produces both. The sporophyte development of *Takakia* suggests evidence for classifying *Takakia* as a moss. *Takakia* sporophytes are erect, about 2 mm tall, developing like a moss with a tapered capsule atop the gametophyte on a slender stalk (Figure F). The sporangium eventually breaks open, releasing ripe spores (Figure G).

The spongy *Sphagnum* (peat moss) grows in acid bogs of northern Eurasia and in North America and contributes to the natural development of new soils. In intense sunlight, sphagnum

Pl-1 Bryophyta
(Mosses)
(continued)

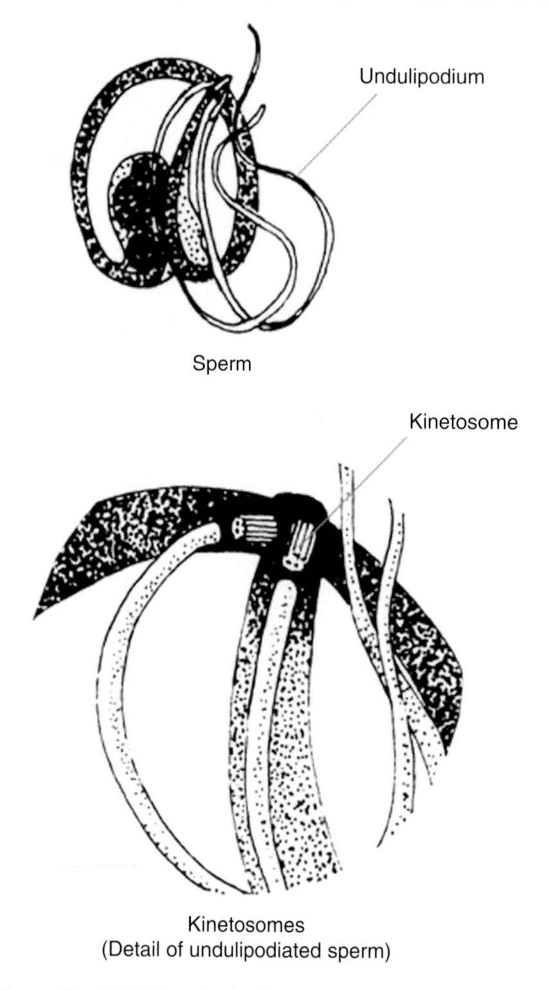

Figure D *Polytrichum juniperinum* sperm. Moss sperm swim with undulipodia. [Drawing by K. Delisle.]

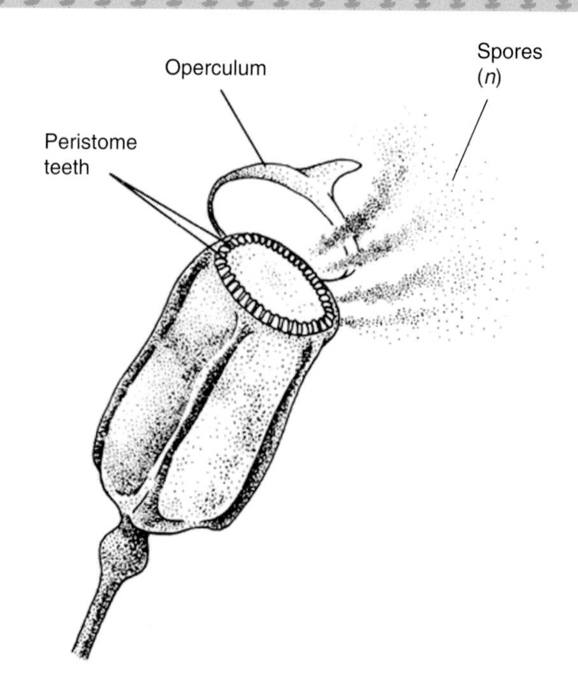

Figure E *Polytrichum juniperinum* capsule, the sporangium. Spores, produced by meiosis, develop into an adult without fusing with another cell. [Drawing by L. Meszoly.]

develops red pigmentation, anthocyanin. In North Carolina and Virginia, the Great Dismal Swamp is a peat bog that originally encompassed 2200 square miles, now reduced to 750 square miles. Gardeners and florists use sphagnum to increase the water-holding capacity of soil. Compressed, decayed sphagnum becomes carbonized in the absence of air and is dug, dried, and burned as peat fuel. Sphagnum has proved an effective wound dressing, particularly in Europe, where it was utilized between 1880 and World War I; sphagnum absorbs many times its weight in moisture and contains iodine.

Chloroplasts and pigments of mosses resemble those of chlorophytes (Pr-28)—they contain chlorophylls ***a*** and ***b*** and carotenoids such as beta-carotene. Moss cells store starch as food reserve within chloroplasts, further evidence that mosses evolved from ancestral green algae. Mosses have a fossil record dating from the late Paleozoic, about 395 mya, but probably were never the dominant land plant form. They do not seem to be the ancestors of vascular plants (Pl-4 through Pl-12) or of hornworts or liverworts.

Evidence put forward by scientists who place *Takakia* in its own phylum includes the presence and patterns of development of phyllidia, growth by multiple meristematic cells (cells that continue to divide throughout the life of the organism), and the branching rhizome. In common with mosses, *Takakia* has stalked photosynthesizing archegonia and spiral leaf (recall that neither *Takakia* phyllidia nor moss appendages are true leaves) array. In common with some liverworts, *Takakia* has underground rhizomes and pitted hydroids (conducting vessels). In common with some lycopod sporophytes, *Takakia* has a multicellular meristem. In common with hornworts, *Takakia* has fewer plastids in mitotically active cells than in mitotically inactive cells. As finer details of development and physiology of the tiny, fascinating *Takakia* become better known, the puzzling relationship between *Takakia* and other nonvascular plants may be clarified.

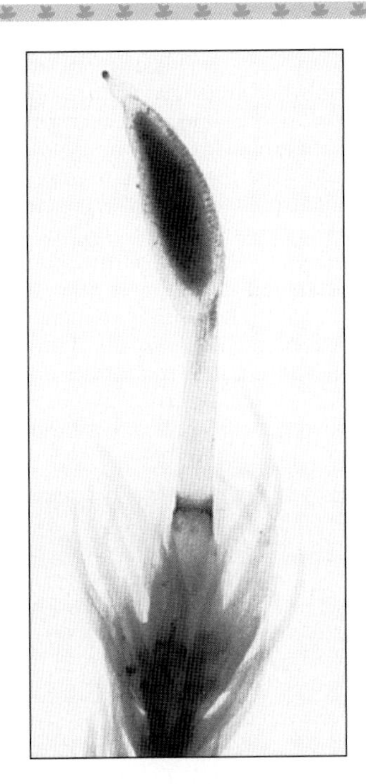

Figure F *Takakia ceratophylla* sporophyte perched on gametophyte. [Photograph by permission of A. S. Heilman, D. K. Smith, and K. D. McFarland.]

Figure G *Takakia* capsule breaking open and releasing mature spores. Bar = 0.1 mm. [Photograph by permission of A. S. Heilman, D. K. Smith, and K. D. McFarland. From D. K. Smith and P. G. Davison, *Journal of the Hattori Botanical Laboratory*, 73:263–271 (1993).]

Pl-2 Hepatophyta
(Liverworts)

Greek *hepat*, liver; *phyton*, plant

GENERA
Calypogeia
Carrpos
Conocephalum
Fossombronia
Frullania
Haplomitrium
Hymenophytum
Lejeunea
Lepidozia
Mannia
Marchantia
Marsupella
Monoclea
Neohodgsonia
Pellia
Plectocolea
Porella
Riccardia
Riccia
Ricciocarpus
Riella
Scapania
Sphaerocarpos
Targionia

Hepatophyta are commonly called *liverworts* — a term derived from the liver-shaped outline of their gametophyte. Liverworts are the simplest of all extant plants. They thrive in moist habitats and are less well known than the mosses (Pl-1). Like that of mosses, the gametophyte of liverworts lacks xylem and phloem, and therefore true leaves, stems, and roots are absent. However, both gametophyte form and a less complex sporophyte distinguish liverworts from mosses. The liverwort gametophyte, called a thallus, takes on one of two forms—either the ribbon-shaped or lobed form of "thallose" liverworts (Figure A) or the leafy shoot system of "leafy" liverworts (Figure B). Both forms usually are flattened and grow prostrate on substrates.

The gametophyte bears stalked reproductive organs (archegonia and antheridia) on the upper surface, and fine hairlike unicellular rhizoids project from the lower surface. All liverwort thalli lack the mucous-filled cavity present in hornworts (Pl-3). The thallose liverwort *Marchantia* grows on stream banks, among mosses on rocks, and in wet ashes after fires. Although all liverworts lack a cuticle (the waxy, water-resistant layer present in mosses and hornworts), *Marchantia* is one of numerous thallose liverworts characterized by a thallus with internally differentiated tissues, which exchange gases through barrel-shaped pores that open into air chambers within the thallus. Liverwort pores differ in form from the stomata of vascular plants. Liverwort sporophytes lack air pores. Liverwort rhizoids are single-celled in comparison with moss rhizoids, which are always multicellular. Rhizoids both anchor the thallus and help move water and dissolved minerals via capillary action. Many thallose liverworts have a midrib, a thickened region that runs down the center of each thallus lobe. The thallus lacks vascular tissue, although some species have specialized tissue to aid in conduction. In height, liverworts seldom exceed 5 cm.

Liverworts reproduce sexually (with gametes) and asexually (by spores, fragmentation, and gemmae), in broad outline like mosses. The liverwort egg is produced in the archegonium of the gametophyte (Figure C) by mitotic division. On a separate thallus (male gametophyte), antheridia produce motile, biundulipodiated sperm. Sperm transported by raindrops fertilize the egg. The liverwort embryo develops a sporophyte from the resulting diploid zygotes. The liverwort sporophyte is permanently attached by a minute stalk to the female gametophyte. The sporophyte consists of a capsule (sporangium), seta (stalk), and foot.

Meiotic cell division takes place at the sporophyte tip, leading to the production of haploid spores. After the capsule opens, spores are discharged by elaters, helical coils that twist as they dry and then snap suddenly, releasing spores. Hornworts and horsetails have cells similar to elaters, but mosses do not. Wind, animals, and water aid in spore dispersal. A spore germinates directly into a young thallus or, in a few genera, a filament of cells precedes

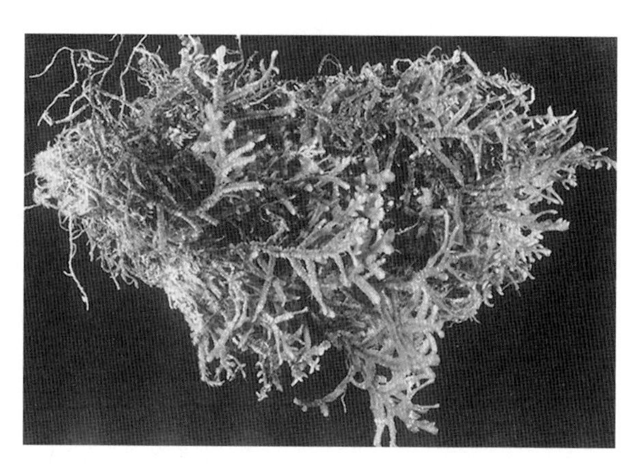

Figure B *Porella*, a leafy liverwort collected in northern California. Two rows of minute "leaves"—not visible—grow along the stem. [Photograph by L. Graham.]

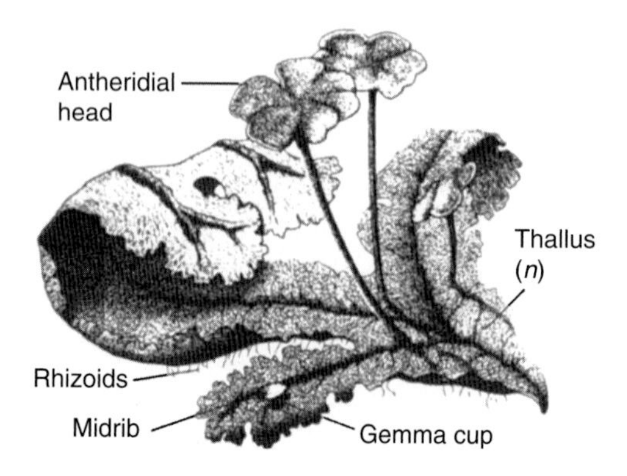

Figure A A common genus of liverwort, *Marchantia*. The gametophyte thallus with lobed, stalked reproductive structures bears antheridia on antheridial heads. Rhizoids differentiate on the lower surface of the thallus. [Photograph by K. V. Schwartz; drawing by C. Lyons.]

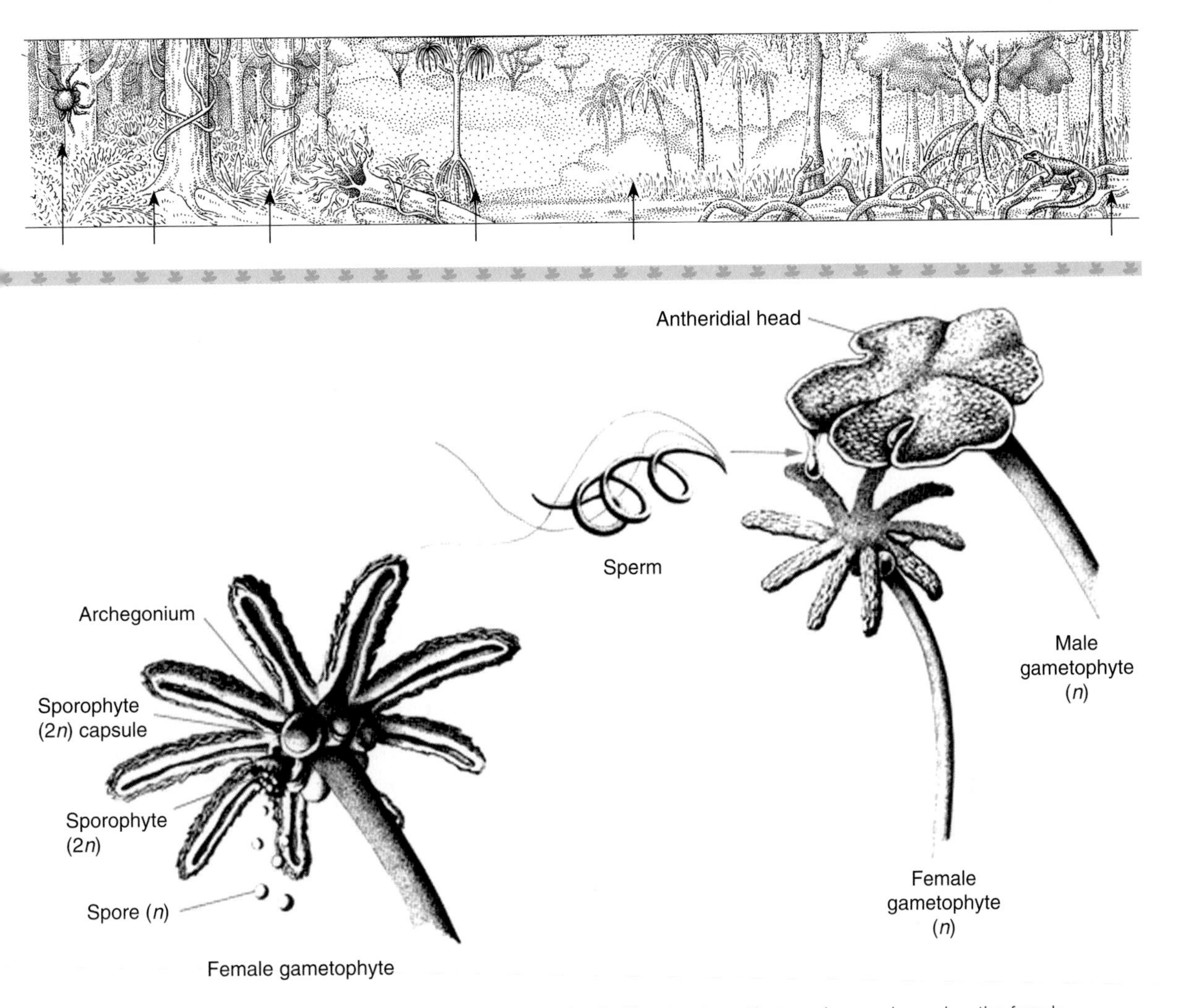

Figure C *Marchantia* habit. The green thallus bears lobed umbrella-like structures that produce archegonia—the female reproductive organs—which produce eggs. Antheridia—the male organs, which contain sperm—differentiate on a separate thallus on the upper surface of stalked disks. A raindrop carries sperm from the male to the female, egg-bearing gametophyte. [Drawing by C. Lyons.]

the thallus. This haploid gametophyte differentiates gametangia and the life cycle begins again. The haploid-dominated life cycle characterizes all mosses, liverworts, and hornworts.

In asexual reproduction by gemmae, liverworts reproduce haploid organisms that are genetically identical with the parent plant. In some thallose liverworts, small cup-shaped organs called cupules (Latin "little cups"; Figure A) form on the upper surface of the thallus. Within the cupules, little green spheres called gemmae grow. When gemmae are dispersed by raindrops to suitable damp soil, they grow into new haploid liverworts.

Most of the 6000 liverwort species live in tropical regions throughout the world, on rock, shaded trees, fallen logs, and soil. Liverworts are often found in waterfalls and other rapidly running freshwater and as epiphytes, organisms that grow on other organisms but are not symbiotrophic. A number of species are known in Antarctica, where they may survive harsh environmental conditions by production of "antifreeze."

During the Middle Ages, liverworts were believed to be useful in treating liver ailments. At that time, plants that looked like an organ were used to treat medical conditions affecting that organ. Liverworts are not currently credited with therapeutic value and are not eaten. Their value lies in their function as pioneer plants in burned areas and other inhospitable habitats.

Combined morphological and molecular evidence indicates that liverworts likely evolved from green algal ancestors but independently of either hornworts or mosses. Another way of saying this is that the three groups of nonvascular plants appear to be paraphyletic. Like hornworts and mosses, liverworts gave rise to no other plant lineages.

Pl-3 Anthocerophyta
(Hornworts)

Greek *anthos*, flower; *keros*, wax; *phyta*, plant

GENERA
Anthoceros
Dendroceros
Folioceros
Megaceros
Notothylas

Anthocerophyta, called horned liverworts or, simply, hornworts, derive their common name from the horn-shaped, elongate sporophyte that is embedded in the gametophyte by a foot. About 100 species live worldwide in temperate and tropical regions, on tree trunks, cliffs, and streamside water-splashed banks.

Members of this phylum superficially resemble mosses and liverworts—the gametophyte is a green dorsoventrally flattened thallus. Within the hornwort thallus is a mucus-filled cavity. A nitrogen-fixing cyanobacterium, *Nostoc* (B-6), lives inside the hornwort *Anthoceros*. Nitrogen-fixing organisms incorporate nitrogen from air into inorganic nitrogen-containing compounds available to plants. Thus, it is no surprise that hornworts are among plants that pioneer sterile substrate such as bare rock. Rhizoids connect the thallus to the substrate. Like liverworts (Pl-2), hornworts absorb moisture and inorganic ions across the flat thallus. Hornworts lack vascular tissue, stems, leaves, and roots. Some hornworts produce gemmae, small vegetative balls of cells that ultimately produce new thalli. Certain species of hornworts have both male and female sexual organs on the same thallus; others are unisexual. Sexual reproduction with swimming sperm is similar to the process in mosses (Figure Pl-3).

Sporophytes of hornworts bear stomata and are covered with cuticle, like moss sporophytes but unlike liverwort sporophytes. Hornwort sporophytes are green and photosynthesize (Figure A); like young sporophytes of mosses, they are nutritionally independent of the gametophyte. Unlike those of mosses and liverworts, which stop growing when they reach the height—as much as 4 cm—characteristic of each genus, the hornwort sporophyte keeps growing from a meristem at its base located between its foot and the sporangium. Meristem is a tissue consisting of undifferentiated cells that give rise to new plant cells. The mature sporangium eventually splits from tip to base, ejecting spores that are often multicellular (Figure B). Most hornworts resemble liverworts in the way that spores are discharged. Packed among the spores in the sporophyte are elongate, helical, sterile hygroscopic (moisture-absorbing) cells called elaters that, when exposed to air, dry and expand to help disperse the spores. These spores initiate young gametophytes directly, without forming protonemata.

Each hornwort cell has a single large chloroplast and each chloroplast contains a pyrenoid, a feature unique to this phylum of the plant kingdom. Pyrenoids, which are common among algae such as *Spirogyra* (Pr-32), are morphologically defined regions of the chloroplast that are associated with photosynthate (sugars and starches) and probably function in food storage.

Hornworts, mosses, and liverworts probably evolved independently of one another. Fossilized hornwort spores from the Cretaceous (144 million to 66 million years before the present) are younger than the oldest moss fossils, which date from the early Carboniferous. So far, no fossil that links mosses, liverworts, and hornworts has been recognized. The origin of hornworts cannot be deduced by examining the fossil record; therefore, workers must test hypotheses of ancestry based on clues in living hornworts.

Figure A *Anthoceros*. This hornwort commonly grows on damp soil. Female and male reproductive organs (not visible) are embedded in the rosettelike thallus. Bar 51 mm. [Photograph courtesy of E. Kozloff. From *Plants and Animals of the Pacific Northwest* (University of Washington Press, Seattle, WA, 1978), plate 33.]

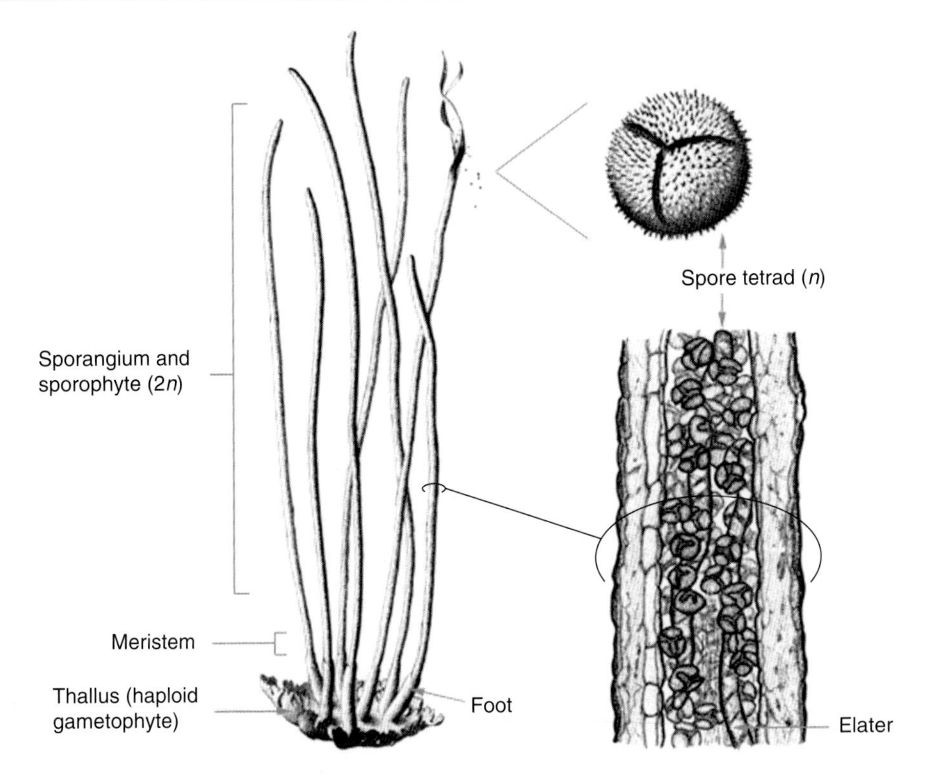

Figure B *Anthoceros*, a gametophyte with horn-shaped sporangia. Left: The mature sporangia split in two, releasing spores. Upper right: tetrad of haploid spores. Lower right: longitudinal section of segment of sporophyte. [Drawings by C. Lyons.]

SUBKINGDOM TRACHEATA

Pl-4 Lycophyta
(Club mosses, lycophytes, lycopods)

Greek *lykos*, wolf; *phyton*, plant

GENERA
Isoetes
Lycopodium
Phylloglossum
Selaginella
Stylites

Lycophytes—club mosses, spike mosses, and quillworts—are relicts of a glorious 400-million-year-old past. The current name of the phylum, Lycophyta, is a contraction of the earlier phylum name, Lycopodophyta. The derivation of Lycopodophyta is from the Greek *lykos*, wolf, and *podus*, foot—based on a resemblance between the pattern of a wolf foot and the branching form of lycophytes.

Both treelike and herbaceous lycopods are found in the fossil record. Only 10–15 genera comprising perhaps 1000 species are still living; many more that lived in the Devonian period are extinct. All living genera are herbaceous. However, *Isoetes*, *Selaginella*, and *Stylites* share characteristics [all are heterosporous (having two kinds of spores) and have ligules (projections from the modified leaves that bear sporangia)] with woody, ancient lycopods. The treelike lycopods—woody (fibrous) lepidodendrids—grew to heights of 40 m; they dominated the swampy Carboniferous coal forests long before the evolution of flowering trees until they died out some 280 mya. Giant lycopods are depicted in their Carboniferous community in the coal forest diorama at the Milwaukee Public Museum.

Lycophytes are evergreen vascular plants that bear neither seeds nor flowers. Most of the tropical species are epiphytes, depending on hosts for support. *Lycopodium* and *Selaginella* are two genera in temperate regions. *Lycopodium*—common club moss—consists of 200 species and is the most familiar lycophyte in the United States. A species of *Lycopodium* is used in winter decorations as a miniature conifer and is called ground pine or ground cedar by some and club moss by others. But these names are misleading—these plants are related neither to pines and cedar (Pl-10) nor to mosses (Pl-1).

The other well-known genus, *Selaginella* (spike moss), comprises about 700 species and flourishes in moist habitats such as Olympic National Park in Washington state. Paradoxically, the resurrection plant (*Selaginella lepidophylla*) is native to dry regions of Mexico and the southwestern United States. A curious feature of the resurrection plant is that it revives upon contact with water even after having been dry and dormant for months. Repeated cycles of desiccation and revival lead to no apparent loss of vigor.

Like all other plants, lycophytes alternate haploid and diploid generations. In lycophytes, the sporophyte (diploid) is more conspicuous than the gametophyte (haploid), as in other vascular plants. This is in contrast to the nonvascular plants (Pl-1 through Pl-3), in which the gametophyte is the more conspicuous form. The *Lycopodium* sporophyte consists of short, upright, branched stems with leaves attached, and creeping, branching rhizomes (underground stems) that lack leaves. Sparse adventitious roots attach to the rhizome.

The glossy leaves of *Lycopodium* are arranged in spirals or whorls, usually held close to the branches. The leaves characteristic of lycophytes and unique to them are called microphylls. These leaves probably evolved as outgrowths of the main photosynthetic axis of the plant. Eventually the outgrowths differentiated to form leaves with a single cylinder of vascular tissue that conducts water and nutrients. In contrast, the leaves called megaphylls have multiple strands of vascular tissue and probably originated by a different mechanism. Megaphylls are characteristic of ferns and seed plants (Pl-7 through Pl-12). Some microphylls are fertile; they bear sporangia. In some species, fertile microphylls—called sporophylls—and sterile microphylls (leaves) are interspersed; both are photosynthetic. The glossy leaves referred to earlier are sterile microphylls. In other species, such as *Lycopodium obscurum*, the fertile microphylls are nonphotosynthetic, scalelike structures grouped into cones (strobili; Figure A). These cones form at the tips of top branches; cones are the "clubs" for which club moss is named (Figure B).

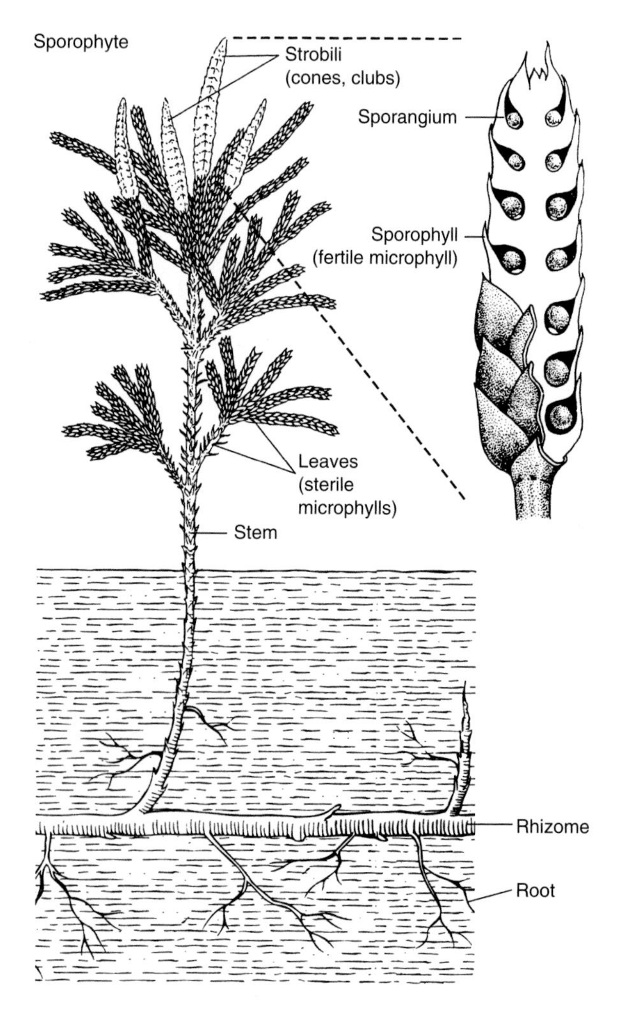

Figure A The club moss *Lycopodium obscurum* (shown here is a sporophyte) is widespread in the central and northeastern United States, in wooded areas under maples, pines, and oaks. The inset exposes the sporangia. Meiosis occurring in cells within the sporangia produces spores. [Drawing by R. Golder.]

Figure B The club moss *Lycopodium obscurum* (shown here is a sporophyte) is widespread in the central and northeastern United States, in wooded areas under maples, pines, and oaks. Bar = 6 cm. [Photograph by W. Ormerod.]

Some lycophytes, such as *Lycopodium*, are homosporous, producing only one kind of (haploid) spore. Others of the phylum—*Selaginella* and *Phylloglossum*, for example—are heterosporous, forming two kinds of haploid spores on different sporophylls of the same plant: megaspores and microspores. Spores, growing by mitosis, germinate into haploid gametophyte plants that produce haploid gametes (eggs or sperm) by mitosis. Megaspores germinate into female gametophyte plants, forming archegonia containing eggs. Microspores germinate into male gametophytes, which produce sperm in male reproductive organs (antheridia). Or the microspore may simply release sperm, as in *Selaginella.* After the parent plant sheds both microspores and megaspores, the sperm swim to and fertilize eggs close by. The young sporophyte eventually sprouts root, stem, and microphylls. In homosporous lycophytes, the spores germinate into gametophytes that produce antheridia as well as archegonia on the same gametophyte. The gametophytes of homosporous lycophytes may be white subterranean tissue harboring symbiotic, mycorrhizal fungi in their tissues, or they may be green and photosynthetic, living on the soil surface. These tiny gametophytes live inconspicuously for years. In all cases, fertilization of the egg by sperm requires at least a thin film of water so that the bi-undulipodiated sperm can swim into the nearby archegonium and fertilize the egg. As the resulting zygote develops into a green sporophyte, it may remain attached to the gametophyte on which it is nutritionally dependent, completing the life cycle.

Some club mosses—*Lycopodium lucidulum* and *L. selago*, for example—also reproduce by means of plantlets. Plantlets grow at the bases of the upper leaves. These small plants are produced asexually, are shed, and begin new diploid plants on their own. In comparison, mosses produce gemmae asexually, but gemmae are haploid.

Smooth-surfaced club moss spores—called lycopodium powder—have been used to coat pills and condoms. Ignited spores generated the flash for early photography and "pink lights," a type of fireworks.

Pl-5 Psilophyta
(Psilophytes, whisk fern)

GENERA
Psilotum
Tmesipteris

Greek *psilo*, bare, smooth; *phyton*, plant

The psilophytes, *Psilotum* and *Tmesipteris* (pronounced meziptéris), are unique among vascular seedless plants. They constitute the only phylum of vascular plants that—like the nonvascular liverworts, hornworts, and mosses—lack both roots and leaves. The dichotomously branched green stem has vascular tissue and alternate, minute outgrowths. These outgrowths—scalelike in *Psilotum* and leaflike in *Tmesipteris*—lack vascular tissue and are considered branchlets rather than microphylls or true leaves. *Psilotum*'s distinctive three-part synangia (fused sporangia) produce spores and are supported in the axil (crotch between stem and scale) by the scalelike outgrowths. A rhizome from which rhizoids arise anchors the psilophyte sporophyte.

Psilotum and *Tmesipteris* are the only two living genera in this phylum. A plant buff can recognize both species of *Psilotum* (*P. nudum* and *P. complanatum*) in the subtropics and can maintain them in the temperate zone in a greenhouse. *Psilotum nudum*, the whisk fern (Figure A), grows in the Florida woods. In Hawaii, *P. nudum*, known locally as moa, perches on tree trunks (in bits of soil), rock crevices, and soil. Interested naturalists can see *Tmesipteris* in Australia, New Zealand, and other South Pacific islands, growing as an epiphyte. A more likely opportunity to view *Tmesipteris* is in a world-class botanical garden such as the Royal Botanic Gardens, Kew, in London.

A casual glance at these herbaceous, leafless plants evokes images of a landscape rich in bacteria and protoctists some 400 mya. Then, in the late Silurian and early Devonian periods, Earth was barren except for early simple, rootless, leafless, seedless, flowerless plants (along with bacteria and protoctists). Are the living psilophytes direct descendants of *Rhynia*, one of the first land plants? How do we know about *Rhynia*? In the quarry of the Scottish town of Rhynie, black, smooth silica rocks have been known since the nineteenth century. Geologists tell us that these rocks, called cherts, probably precipitated in freshwater on the shores of an ancient lake. When cherts are cut and polished for microscopic study, some preserve ancient material so well that a multimillion-year-old covering of epidermal cells on the plants can still be distinguished. Fossil rhyniophytes have a leafless, dichotomously branching stem arising from a rhizome (an underground stem) with rhizoids like that of extant psilophytes. These beautifully preserved plant fossils, like the living psilophytes (Figure B), have vascular tissue in their stems, are cuticle covered, and have stomata. However, *Rhynia* sporangia were borne singly at the tips of the stems rather than in the axils of outgrowths as in present-day psilophytes. No intermediate fossils have been found that link modern psilophytes to ancient *Rhynia*; it is uncertain whether modern psilophytes are direct descendants of rhyniophytes.

Consistent with the idea of a direct relationship between the ancient *Rhynia* and modern *Psilotum* is the spectacular preservation of endomycorrhizae (fungus within the root) in 400-million-year-old fossils of the Rhynie chert. In the rhizoids of these rootless rhyniophytes, one can see spherical fungal reproductive structures (Figure C) that are remains of the ancient plant–fungus partnership. The spherical structure in this rhizome is interpreted as a fungal sporangium of an *Endogone*-like zygomycote (F-2). Rhizoids of the living *Psilotum* sporophyte also harbor mycorrhizal fungal hyphae that increase the flow of nitrate, phosphate, and organic compounds from soil to the nonphotosynthetic plant cells.

Chloroplast DNA comparisons suggest that psilophytes' closest relatives are nonlycophyte vascular plants such as ferns (Pl-7). And some botanists contend that today's psilophytes evolved directly from true ferns by simplification and loss of structure (rather than directly descending from ancient *Rhynia*).

[Photographs by W. Ormerod.]

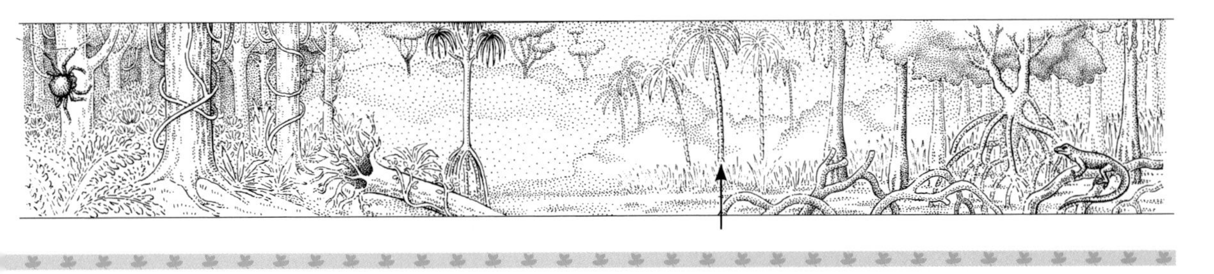

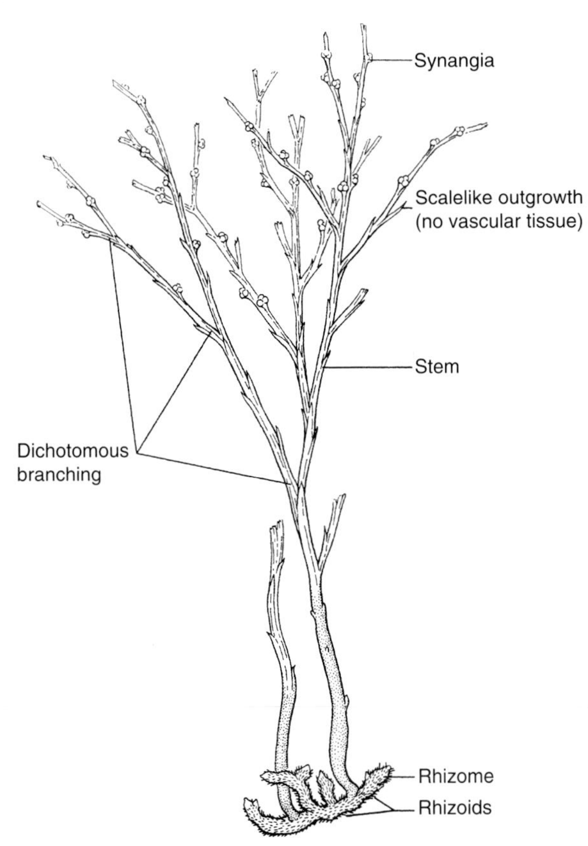

Figure A *Psilotum nudum*, whisk fern, showing dichotomous branching, scalelike outgrowths, and synangia. This specimen (in the photograph, see facing page), from ancestors in the Florida bush, has spent its life in a Boston greenhouse. [Drawing by L. Meszoly.]

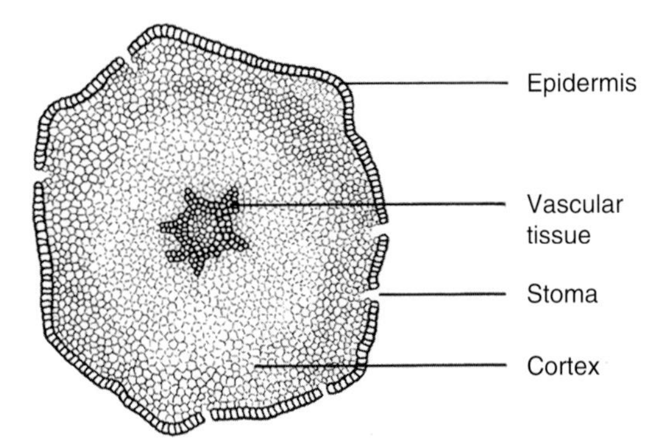

Figure B *Psilotum nudum* stem cross section showing vascular tissue. [Drawing by L. Meszoly.]

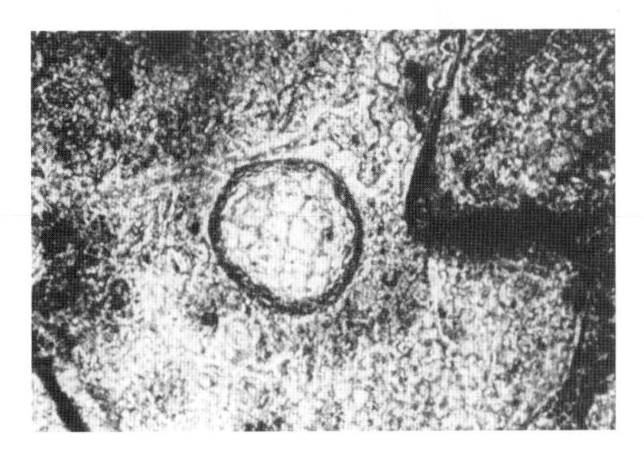

Figure C Fossil *Rhynia* tissue section of rhizome, showing 400-million-year-old plant–fungus relationship. From Rhynie chert. [Photograph by L. Read.]

Psilophytes, as well as most ferns, horsetails (Pl-6), and some club mosses (Pl-4), have a single type of spore. Plants in these four phyla also have similar life cycles. The final word is not in regarding relationships of ancient rhyniophytes to the modern psilophytes, *Tmesipteris* and *Psilotum.*

Botanists search for clues that point to closest relative(s) of modern psilophytes by chemical comparisons—modern ferns and living psilophytes both produce secondary compounds. These biochemicals are not absolutely necessary for plant development but often play a crucial role in plant development and ecology. However, secondary compounds of modern ferns differ distinctly from those of *Psilotum.* This chemical evidence—in contrast to chloroplast DNA evidence—fails to support a strong evolutionary relation between the psilophytes and the ferns.

Additional similarities between modern psilophytes and some modern ferns are subterranean gametophytes and endophytic fungi both in gametophytes and in rhizomes.

Within each of the three chambers of the yellow-brown synangia on the sporophyte (Figure A), one kind of haploid spore (homospore) is produced by meiotic cell divisions. Mature spores are released into the air, germinate in soil, and produce a bisexual haploid gametophyte called a prothallus (Figure D). Careful inspection reveals that the prothallus has fuzzy threads toward its center. These threads are endomycorrhizae.

Examination of the prothallus reveals two types of external sex organs (Figure E). The male sex organs, called antheridia, are microscopic bumps ringed with a layer of surface

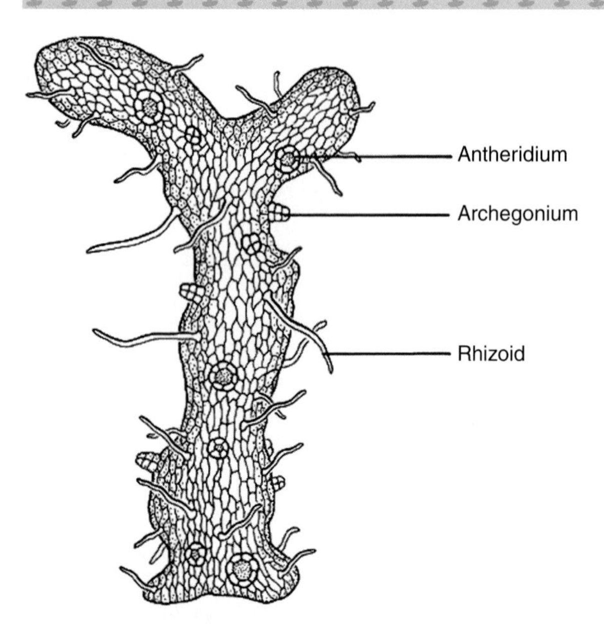

Figure D A prothallus. This subterranean, bisexual, independent gametophyte of *Psilotum nudum* bears antheridia and archegonia, the reproductive organs. [Drawing by L. Meszoly.]

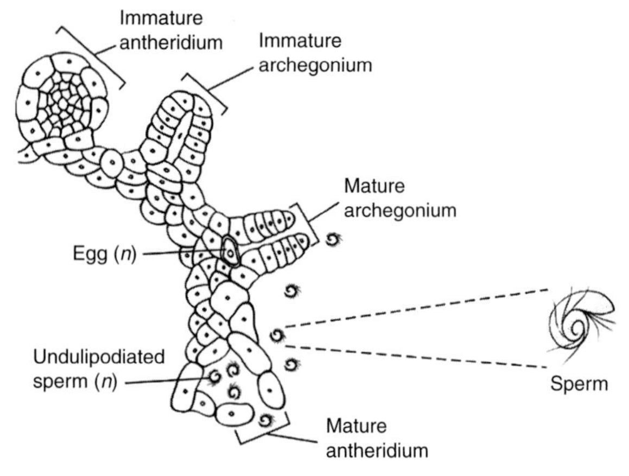

Figure E *Psilotum nudum* prothallus cross section. Mature antheridia release spirally coiled, undulipodiated, sperm (*n*) that swim to mature archegonia. Each archegonium contains an egg (*n*), which is fertilized by a sperm.

cells. A few cells away, on the same gametophyte (prothallus), are smaller female sex organs, archegonia. Each archegonium is composed of several ranks of cells with an opening that forms between them when the middle layer breaks down. The bisexual prothallus produces several archegonia, each with a mitotically produced egg at the base of the opening, as well as antheridia. Curled sperm with many undulipodia form inside the antheridia by mitosis. The sperms' undulipodia have the [9(2)+2] organization of microtubules that reveals the protoctist ancestry of these plants. Sperm fertilize eggs within the archegonia. Because the sperm that are released into the soil must swim, moisture must be present for fertilization to occur. The resulting zygote develops into the multicellular diploid sporophyte embryo characteristic of all plants. At first, the young sporophyte is nourished through a foot anchored in the gametophyte. Later, the sporophyte takes up an independent, photosynthetic life aboveground.

Hawaiian men once used *Psilotum* spores as powder to prevent groin irritation from loin cloths. By boiling the moa plant, Hawaiians made laxative tea and a medicine to treat thrush (a yeast infection).

Tmesipteris and *Psilotum* fire the imagination—we envisage a past reign of dichotomously branched land plants that early in their phylogenic history already had established symbioses with members of the kingdom Fungi. Some scientists hypothesize that this mycorrhizal association with fungi was a prerequisite to the coming ashore of all land plants.

Pl-6 Sphenophyta
(Sphenophytes, Equisetophyta, horsetails)

GENERE
Equisetum

Greek *sphen*, wedge; *phyton*, plant

Sphenophytes include common "horsetails" and are easily recognized by their jointed hollow stems, whorled branches and distinctly ridged surfaces. Indeed, the surfaces of some are very stiff and highly abrasive due to the presence of silica within epidermal cells of their stems, earning them the name "scouring rushes."

Like lycophytes (Pl-4), ferns (Pl-7), and probably psilophytes (Pl-5), sphenophytes are relicts of a far more glorious past. Because of their distinctive morphology, they are easily recognized in the fossil record and are especially prominent in tropical swamp "coal floras" of the Carboniferous Period, 354–290 mya. Some horsetails were woody treelike plants reaching 0.5 m in diameter and about 15 m in height. Others, called "sphenophylls" (hence the phylum name). were small herbs with whorls of simple wedge-shaped leaves. Today all 15 species belong to the single genus *Equisetum*; they thrive along roadsides, stream banks, and other mostly disturbed places in moist woods.

Horsetails, like other free-sporulating plants, have a life cycle consisting of two separate plant bodies. Most conspicuous is the sporophyte (diploid) generation. In some species, the sporophyte is dimorphic (that is, produces two kinds of aboveground shoots: one pale, bearing sporangia; and the other green and photosynthetic). In all, sporangia are borne in groups on umbrella-like structures termed *sporangiophores,* with as many as 50 sporangiophores grouped together into a conelike strobilus (Figures A and B). Meiosis takes place within the sporangia, resulting in haploid spores of a single size (the plant is homosporous). The outer wall of each spore differentiates four specialized coiled bands called elaters. Elaters uncoil as they dry out and thus help disperse the spores. If a spore settles in a sufficiently moist

Figure A *Equisetum arvense* fertile shoot, bearing a strobilus. This horsetail is common in wasteland and on silica-rich soils. Bar = 3 cm. [Photograph courtesy of J. G. Schaadt; drawing by I. Atema.]

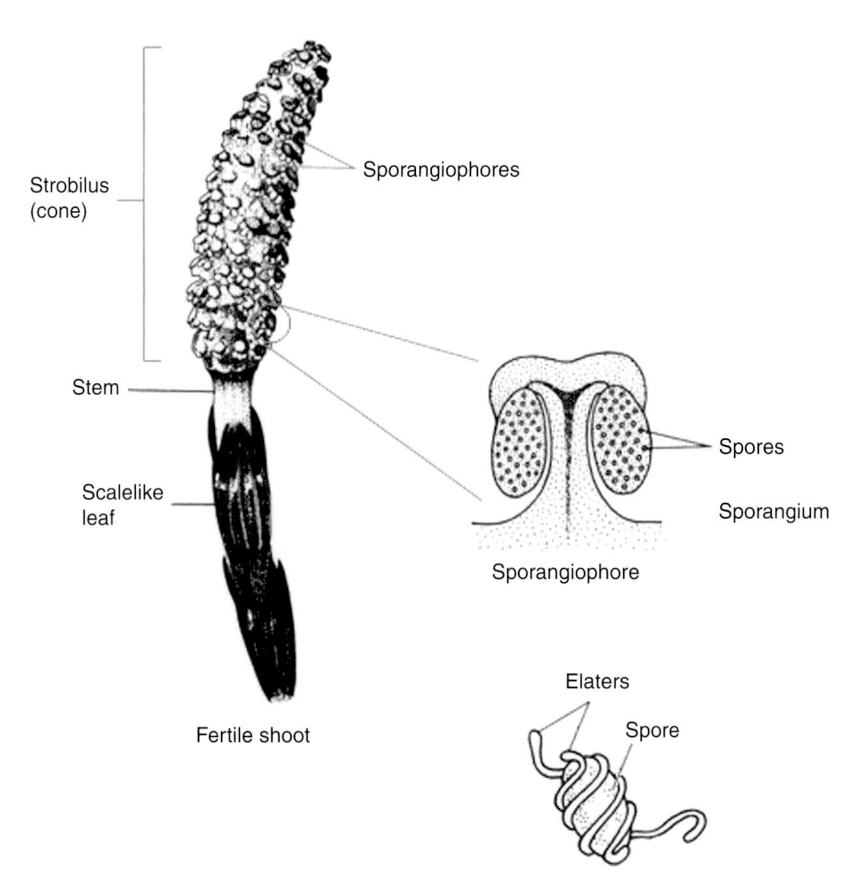

Figure B *Equisetum arvense* strobilus [drawing by I. Atema], sporangiophore vertical section [drawing by R. Golder], and spore [drawing by I. Atema].

place, it germinates to form a very small, free-living, green and photosynthesizing gametophyte (haploid) plant.

Gametophytes have many lobes of tissue emerging from small rootlike rhizoids that anchor the plant to the soil. Upper lobes of the gametophyte produce antheridia (multicellular sacklike organs that mature sperm) and archegonia (flask-shaped multicellular organs that produce eggs). The sperm bear many undulipodia and swim through a film of water to find archegonia. Several sperm, even from different plants, can fertilize the eggs on the same gametophyte. The resulting zygotes then develop into independent diploid sporophytes as the parent gametophyte dies. Horsetails also propagate vegetatively from underground stems (rhizomes), forming deeply rooted and extensive clones that are often very hard to control or eradicate (Figure C).

In addition to their use in washing pots, Native Americans, English, Tuscans, and Romans once consumed horsetails. Some contemporary references list them as edible. However, horsetails are known to be poisonous to livestock, especially cattle and horses. The toxicity is due to production of the enzyme thiaminase, which destroys the vitamin thiamine.

Figure C *Equisetum hiemale*, common even in urban areas. Jointed stems with conspicuous nodes are evident. Bar = 15 cm. [Courtesy of W. Ormerod.]

Box Pl-i: Plant hybridization and polyploidy

The zoocentric "species concept" has never applied well to plants. An animal species is a taxon whose members produce fertile offspring only through mating conspecifics. However, plants of different species, and even different genera (*Raphanobrassica*, for example, a hybrid of *Brassica*, cabbage, and *Raphanus*, radish), interbreed and give rise to fertile hybrids. Because plants are more tolerant than animals of major alterations to genomic structure (for example, polyploidy, aneuploidy, translocations, duplications), new species more easily arise through chromosomal hereditary changes. Agriculture exploits the genetic plasticity of plants to generate many important crop species: wheat, cotton, and tobacco are all hybrid polyploid crops. They evolved through efforts by ancient farmers to incorporate desirable characters (higher yield; large, tasty fruit or seeds; hardiness; simultaneous ripening; pest resistance) from diverse wild progenitor species. The initial hybrid from an interspecies cross is often infertile because interspecific chromosomes are not completely homologous at prophase I of meiosis. If the hybrids become polyploid (double both maternal and paternal chromosome sets), pairing between homologues is again enabled. The tetraploid, hexaploid, or octaploid descendants are then often fertile. They may possess desired traits from two, three, or more progenitor species. Polyploidy renders plants larger and more vigorous and often generates attractive multiplication of body parts, as in polyploid carnations or the florist's rose with numerous petals. Hybridization and polyploidy are major modes of saltatory evolution in plants.

Hybridization is encouraged by factors that promote plant outcrossing (panmixis): indiscriminate pollinators, proximity, timing, and human agency. Polyploidy, which may restore fertility to hybrids, arises in at least three ways. Diploid egg or sperm cells form through spontaneous or drug-induced nondisjunction of the entire chromosome set at meiosis. Fertilization between diploid cells generates a tetraploid. Fertilization of a diploid egg by a wild-type (haploid) sperm cell generates triploids. The cells of the "crown" tissue where stem meets root are naturally tetraploid in many plants. Grasses, ferns, and other herbaceous plants store enough nutrients in the root to mitotically regenerate grazed shoots and leaves from tetraploid crown cells. The tetraploid flowers of such a grass will then produce diploid pollen and ovules; a tetraploid fern will produce diploid spores. Such *autopolyploids* incorporate chromosome sets from only the parent species. If the polyploidy event occurs in a sterile allospecific hybrid, as in the crops, then the resulting *allopolyploid* will possess complete diploid chromosome sets from more than one parent species.

Meiotic fertility is actually undesirable in certain crops. Wild, bat-pollinated bananas are full of seeds and have little pulp. Growers selected for a triploid hybrid: *Musa X paradisiaca*, which is a hybrid between a pulpy-fruited form of diploid *M. acuminata* (AA) and tetraploid *M. balbisiana* (BBBB). Triploid supermarket bananas (AAB or ABB; $3n = 33$), although sterile due to their meiotic chromosomal asymmetry, are seedless, tasty and pulpy.

In variable habitats such as old-growth forest, many outcrossing wild plants live closely among related species with similar flowers. This affords ample opportunity to hybridize. Many crop ancestor species grow in close proximity to farmer's fields. Half of all angiosperm species (approximately 125,000) are estimated to be of hybrid origin. Like symbiogenesis (Box Pr-i), hybridization and polyploidy afford plants the ability to acquire and combine entire new genomes within one or just a few generations.

Hybridization may also be disadvantageous. Plants under antihybridization selective pressures may evolve the self-fertilizing habit (apomixis), altered timing of male versus female sexual maturity (protandry/protogyny), or floral shapes and colors designed to attract specific pollinators. Such pollinators usually coevolve closely with their target flower. Specialization helps to conserve pollen, whereas the pollinator outcompetes nonspecialists. The well-studied coevolution of figs with wasps is a good example: the anatomy of the compound fig ovary (syconium) attracts and admits only correct wasp species.

Species are categorized by selection pressures under which they evolved. ***K***-selected or "prudent" species tend to be specialized and few in number. Only one individual monopolizes the carrying capacity (***K***) of a habitat. Broadleaf trees—oaks and beeches for example—that shade out undergrowth in climax temperate forests are examples of ***K***-selected plants. A side effect of ***K***-selection is that only a few offspring of a given individual survive to maturity. An oak may produce thousands of acorns, but the vast majority of seedlings are shaded out by the parent. Ecological ***K***-selection, common in habitats with variable selection pressures, dominates climax temperate forests and tropical and subtropical rainforests.

By contrast, "weedy" species are ***r***-selected, nonspecialized and highly fecund. They maximize reproductive rate (***r***) and quickly colonize a new habitat. Examples include self-pollinating or parthenocarpic plants such as dandelions, ragweeds, and most grasses. Ecological ***r***-selection occurs in habitats with strong but unchanging selective pressures, like the American Great Plains.

Oenothera, evening primrose, is a plant whose breeding system has evolved in the last 1.5 million years under ***r***-selection. All *Oenothera* species have outcrossing floral characters: large fragrant flowers, nutritious pollen, and a nectar reservoir. But like other plants under antihybridization forms of selection, *Oenothera* has recently evolved floral timing that favors inbreeding. The anthers shed pollen a full day before the flower opens; by the time pollinators arrive all ovules are already fertilized.

Oenothera evolved as a ***K***-selected plant in the variable but richly populated "rain shadow" zone of the Sierra Madre mountains. In the course of its move eastward across the continent, however, *Oenothera* encountered ***r***-selection and evolved the self-pollination characters mentioned earlier. Ancestral *Oenothera* species are outcrossing and display the familiar seven pairs of chromosomes at meiotic prophase I. Derived species, by contrast, have all chromosomes involved in a translocation ring. The chromosome ring suppresses Mendelian independent assortment. It is a result of anti-hybridization selection, which helps to preserve ***r***-selected characters in unchanging habitats such as the prairie. Mitochondria and plastids have coevolved with the ***r***-selected nucleus in most *Oenothera* species, so that forced hybrids, generated in the greenhouse by surgical emasculation and hand-pollination, are chlorotic, weak, and relatively infertile. Through its evolution of the full 14-chromosome translocation ring, a meiotic configuration no animal could survive, *Oenothera* has recently become "weedy" in a quintessentially plantlike way.

Pl-7 Filicinophyta (Pterophyta, Pterodatina, Pteridophyta, ferns)

Latin *felix*, fern; Greek *phyton*, plant; Greek *pteridion*, little wing, feather

GENERA
Adiantum
Anemia
Asplenium
Azolla
Botrychium
Cyathea
Dennstaedtia
Dicksonia
Dryopteris
Hymenophyllum
Marattia
Marsilea
Matteucia
Osmunda
Platyzoma
Polypodium
Polystichum
Pteridium
Pteris
Salvinia

Ferns are seedless vascular plants that, like bryophytes, psilophytes, lycophytes, and sphenophytes (Pl-1 through Pl-6), reproduce and disperse by means of spores. But, unlike plants in these other phyla, fern sporophytes have megaphylls (Greek, "large leaf"), formerly called fronds in reference to ferns, which consist of the blade (expanded leaf part) and leaf stalk (stipe) that attaches to the rhizome (Figure A). The megaphyll is a relatively large leaf with a web of veins, in comparison with the single-veined microphylls of lycophytes. The fern megaphyll is usually compound, divided into leaflets called pinnae. Fern megaphylls may be fertile—bearing sporangia on the undersurface of modified leaves (as in *Polypodium*, Figures A and B) or on specialized stalks that emanate from the rhizome (as in *Osmunda*; Figure C)—or sterile (nonreproductive). Fern sporangia tend to develop in clusters called sori (singular: *sorus*). In certain species, sori are bare. In many species, sori are covered with the indusium, a tissue that shrivels and folds back to expose the ripe sporangia. Many sporangia have an annulus, a strip of cells having a thin-walled outer surface. When mature sporangia dehydrate, annulus cells contract along their outer surface, ripping open the sporangium. When the water-surface tension in the annulus wall breaks, air penetrates the cell wall and the annulus snaps back, expelling the spores.

Most ferns form only one kind of spore and thus are homosporous; a few are heterosporous, producing both small and large spores. The sporophyte commences sexual reproduction with meiosis, producing haploid spores in sporangia. Spores store nutrients for future use, including some similar to proteins found in angiosperm seeds. Spores also accumulate the hormone abscisic acid, which may bring about dormancy in partly dehydrated spores. The waxy wall secreted by a spore prevents deadly dehydration. Water and usually light stimulate germination of a wind-borne spore. The young gametophyte that develops from the spore grows as a green photosynthesizing filament (protonema) toward the light. Blue light from sunlight switches development of the filament to lateral growth, forming a flat, heart-shaped gametophyte, called the prothallus. On the lower surface of the prothallus are numerous rhizoids that anchor it to the substrate and, in some ferns, form species-specific symbioses with fungi.

Depending on the species, fern gametophytes bear only antheridia (containing sperm) or only archegonia (containing eggs) or both (that is, they are bisexual, or hermaphroditic). Homosporous ferns usually produce bisexual gametophytes. In heterosporous ferns, the smaller male spores form thin gametophytes that develop only antheridia, whereas the larger female spores form rounded gametophytes that develop only archegonia. In some cases, environmental factors such as crowding can induce changes in the proportion of gametophytes with both sorts of sexual organs; a hormone secreted by gametophytes that stimulates antheridial development is responsible.

A film of water around each gametophyte (female and male) is required for fertilization. Water enters the antheridium and pops open the antheridium cap, enabling sperm to swim free. From several to thousands of motile undulipodia facilitate a sperm to move through the neck of the archegonium toward the egg. As the neck cells of the archegonium swell by water uptake, they part, creating a canal. Water dissolves the mucus secreted by the neck cells. These processes probably release sperm-attracting molecules as by-products of cell respiration. Swimming fern sperm enter the archegonia, attracted by chemicals such as malic acid. A sperm (haploid) fertilizes an egg (haploid), beginning

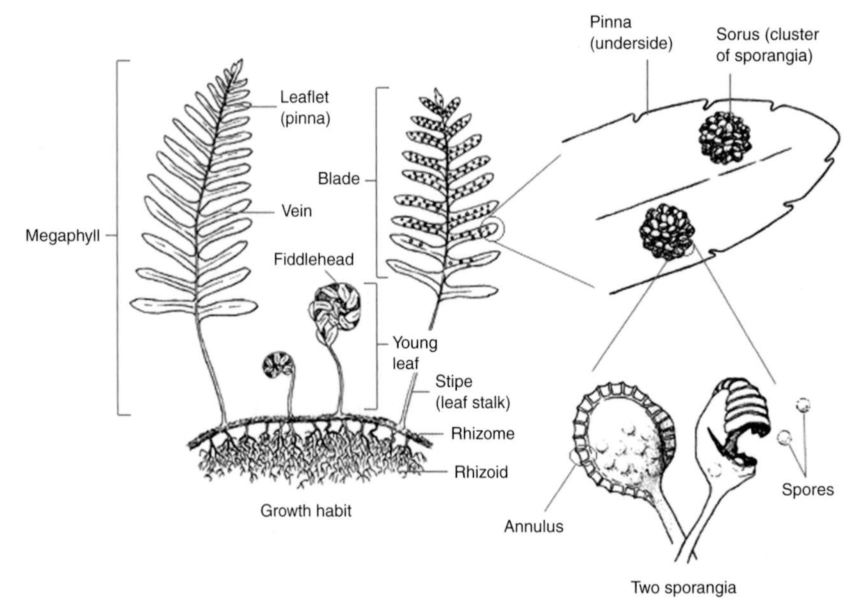

Figure A Growth habit and reproductive structures of the sporophyte polypody fern. The name "polypody" is derived from Greek *poly* (many) and *pous* (foot), alluding to the branching rhizoids. [Drawing by R. Golder.]

Figure B *Polypodium virginianum*, the rock polypody, showing clusters of sporangia on the underside of a fertile megaphyll (leaf). *Polypodium* is extensively distributed in North American and Eurasian woods. Bar = 10 cm. [Photograph by K. V. Schwartz.]

Figure C *Osmunda cinnamomea*, the cinnamon fern, a species widespread in moist, shady areas, especially along the edges of ponds and streams. The sterile (nonreproductive) lateral leaves are easily distinguished from the upright fertile leaves. Bar = 50 cm. [Photograph by K. V. Schwartz.]

the diploid (sporophyte) generation. The embryo sporophyte may retain its connection to the gametophyte (prothallus), but soon organ development begins; the first leaf and rhizome grow out from under the gametophyte. This sporophyte grows into the familiar independent fern plant as the gametophyte dies, after having provided physical support, nutrients, and possibly hormones. Its rhizomes (subterranean or creeping stems) send out aerial shoots (young leaves) along their lengths or near their tips. Adventitious roots originate on the rhizome, anchoring the rhizome and absorbing moisture and nutrients.

The developmental pathway is flexible. Environmental conditions can sometimes induce development of either gametophyte or sporophyte from cells other than spore and zygote. For example, ferns sometimes reproduce asexually—cells of the megaphyll tips divide by mitosis and form new diploid plants that fall to the ground. These vegetatively produced offspring are genetically identical with their parent, like club moss plantlets.

Fern spores and leaf impressions first appeared in the Devonian period (408–360 mya). Ferns abound in the fossil record from the Carboniferous through the present. Of plants in phyla that do not form seeds, cones, or flowers, ferns are the most diverse. Their sperm swim to the egg, limiting ferns to habitats that are at least occasionally moist. About 12,000 living species are known, two-thirds of them in tropical regions. The genus *Polypodium* comprises nearly 1200 species, mostly tropical. Many tropical rain forest ferns grow high in the canopy, where they are watered by mist and rain and obtain nutrients from dust and decomposing organisms that land among their leaves. A few species live north of the Arctic Circle; like many ferns of cold areas, they produce new leaves from their rhizomes each growing season. Species are distinguished by spore morphology and details of their life histories. For example, although heterospory—production of two differing spore types—is uncommon in ferns, water ferns such as *Marsilea*, *Salvinia*, and *Azolla*, as well as in *Platyzoma*, native to northern Australia, are heterosporous. Fern species are also distinguished by the nature of their sporangia. Like other seedless vascular plants, some fern species, such as *Ophioglossum* and *Marattia*, form eusporangia. Multicellular in origin, the eusporangium has a wall consisting of several cell layers. Other fern species (water ferns and *Polypodium*, for example) form leptosporangia, which develop from single cells and have single-cell-thick walls. Among the smallest ferns is the aquatic *Azolla*. These floating ferns harbor, in cavities on the undersides of their leaves, the nitrogen-fixing cyanobacterium *Anabaena* (B-6). The symbiotic complex provides nitrogen to rice paddies. *Azolla* leaves are typically 1 cm long, whereas the leaves of tree ferns may be 500 times as large. The largest ferns are thick-trunked tree ferns, with leaves as long as 5 m, stems 30 cm in diameter, and heights more than 25 m. "Tree" is a misnomer, though; tree ferns lack the bark and fibrous woody tissue of trees. When moist, warm Carboniferous forest inhabitated what is now temperate North America and Europe, tree ferns flourished.

Fern fiddleheads—coiled, young sporophytes—are a delightful spring vegetable (Figure A). Ostrich fern, *Matteucia struthiopteris*, is commonly eaten and commercially grown in the United States. Ferns of all sizes provide texture and shape in landscaping. Thatch (from megaphylls), emergency food (starch from rhizomes and tree-fern pith), tea (from leaves), dye, and medicines are all products derived from ferns. Medicine that expels parasitic worms is prepared from the rhizome and root of *Polypodium aureum* in Puerto Rico. Hawaiians used fluff of tree-fern fiddleheads to stuff pillows.

Pl-8 Cycadophyta
(Cycads)

Greek *kykos*, a palm; *phyton*, plant

GENERA
Bowenia
Ceratozamia
Chigua
Cycas
Dioon
Encephalartos
Lepidozamia
Marcrozamia
Microcycas
Stangeria
Zamia

Some cycads are small shrubs, such as *Zamia*, which is about 0.3 m in height, whereas others, such as *Cycas* and *Microcycas*, are palmlike trees more than 18 m tall. Members of the genus *Cycas* are sometimes called sago palms. Although they resemble palms, these cycads are not true palms, which flower and fruit, and belong to Anthophyta (Pl-12). Cycads bear seeds (a structure formed after fertilization by maturation of the ovule in seed plants). Because their seeds are naked, cycads are classified as gymnosperms, along with conifers (Pl-10), ginkgos (Pl-9), and gnetophytes (Pl-11). As in other gymnosperm phyla, cycads lack flowers, and their seeds are exposed on female cones, instead of being enclosed in a fruit. In *Cycas* and *Dioon*, the petioles (leaf stalks) of shed leaves cover the trunk (the principal axis of the cycad, also called the stem). Shiny palmlike or fernlike leaves cluster at the apex of the stem. Like all vascular plants, cycads have megaphylls, each attached by a petiole to the stem, or trunk (Figure A). The leaves of some cycad species are subdivided into pinnules. Cycads have coralloid roots—named for their coral-like appearance, which is unique to this phylum—that provide nitrogen to the cycad.

Like other vascular seed plants, cycads are heterosporous. Among all vascular seed-bearing plants, sexual reproduction in cycads is extraordinarily unusual owing to properties of the male gamete. Cycad sperm are motile, like sperm of ginkgo; sperm of conifers and gnetophytes lack undulipodia and are not motile. Cycad sperm are conveyed to the cycad egg in a pollen tube, as are sperm of conifers, ginkgo, and flowering plants (Pl-12). The combination of motile sperm and pollen tubes is characteristic of cycads and ginkgos and unique to these phyla; it is believed to be an evolutionary link between, on the one hand, ferns (Pl-7) and mosses (Pl-1), which have swimming sperm, and on the other hand extant seed plants (Pl-8 through Pl-12), with pollen tubes.

Separate male cycad plants bear male cones (Figure B). Cycad and conifer cones are analogous; they probably evolved independently rather than having derived from a common ancestor. Paleobotanical investigations were once believed to indicate that cycads were among the closest living relatives of flowering plants, related through their common ancestors, the seed ferns, now extinct. However, seed ferns and living cycads are no longer believed to be direct ancestors of flowering plants.

About 185 living species of cycads are grouped into 11 genera, all living in the tropics and subtropics. All cycads are listed as endangered species—vulnerable to habitat destruction and overcollection for gardens, clinging precariously to life in rain forests, deserts, grasslands, and even mangrove swamps. At Foster Garden (Honolulu, Hawaii) and Fairchild Garden (Coral Gables, Florida), we may observe these fine plants, whose ancestors shared the early Earth with dinosaurs. In temperate zones, cycads are occasionally grown in greenhouses. *Zamia*, the only genus native to the continental United States (Georgia and Florida), is also found in the West Indies, Mexico, Central America, and northern South America. *Zamia* can be seen in Everglades National Park.

Cycads tend to have unbranched trunks, underground or aboveground, with pith but little wood. The layer of cambial cells, the source of new woody tissue, divides sluggishly throughout

the life of the cycad. As a result of limited cambial growth, cycads form little wood. Cycads have unique contractile trunks (stems) and roots that provide protection against adverse environments. When subject to drought or fire, subterranean stems of cycads contract as much as 30 percent in length, drawing the plant down into the protection of the soil. The contraction is due to the collapse of cells in the cortex—tissue of a root or stem bounded externally by epidermis and internally by vascular tissue—and pith, which reduces the root and stem in length.

Most cycads have taproots, some as long as 12 m, which reach deep into sand. In addition, the coralloid roots grow on or even above the soil surface (Figure A). All cycads harbor nitrogen-fixing symbiotic cyanobacteria, generally *Anabaena* or *Nostoc* (B-6), in these coralloid roots (Figure C). The cyanobacteria lie as a layer of single green cells just under the surface of cycad roots; when free living, they live as filaments of cells. The fixation of atmospheric nitrogen by the symbiotic bacteria probably permits cycads to populate areas where soils are depleted of nitrates. Other

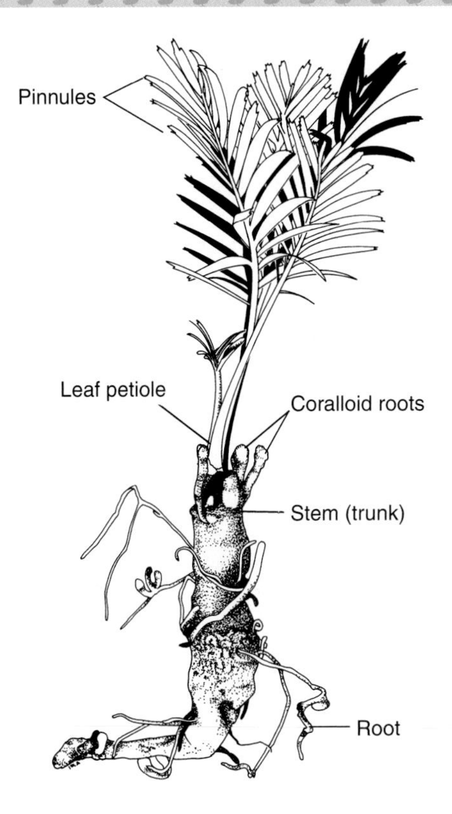

Figure A *Macrozamia communis*, a very young sporophyte tree from sandy soil near Melbourne, Australia. Bar = 10 cm. [Photograph courtesy of C. P. Nathaniels and I. A. Staff; drawing by I. Atema.]

Figure B A male cone of *Ceratozamia purpusii*, a cycad native to Mexico. Bar = 50 cm. [Photograph by K. V. Schwartz.]

nitrogen-fixing bacteria, *Pseudomonas radicicola* (B-3) and *Azotobacter* (B-3), also are associated with coralloid roots.

Cycads bear their reproductive structures in cones that, in some species, are brilliant orange or velvety brown. Cycad sporophytes are either male or female (dioecious)—different individual mature plants bear reproductive structures of only one sex, either male, pollen-producing cones called microsporangiate cones (Figure B) or female, seed-producing cones called megasporangiate cones. Female cones tend to be shorter and plumper than male cones of the same species. Male cones develop microsporangia on the lower surface of microsporophylls. The microsporophylls are packed into male cones. The female cones have megasporangia, borne on the surface of megasporophylls—modified leaves (cone scales); the megasporophylls are either packed into female cones or more loosely arranged in a leafy crown called a pseudocone. The cycad gametophytes are greatly reduced in size to only a multicellular structure in the ovule in the female and to pollen in the male. Within the ovule (the structure containing the gametophyte with its egg cell), meiosis results in megaspores that produce a haploid female gametophyte which produces haploid egg cells. Within the microsporangium, meiosis results in haploid microspores that produce pollen grains—the immature male gametophyte.

Ceratozamia mature cones of both sexes give off musty odors; these odors probably attract insects. Beetles, particularly weevils, lay their eggs in the male cycad cones. For example, the weevil *Tranes lyterioides* (A-21) is associated with *Macrozamia communis.* Both pollen and beetle larvae mature inside male cycad cones. The beetles feed on tissues of the cone but not on pollen. When adult beetles exit from the male cone, they chew through the microsporophylls and are dusted with pollen. (Some beetles remain behind, laying eggs and feeding on the male cone, continuing the beetle life cycle.) Beetles, as well as wind, transport pollen from male to female cones.

Heat, odor, and possibly sugar and amino acid–containing fluids produced by mature ovules attract insects to cycad cones at pollination time. The adult, pollen-dusted beetles enter female cones through cracks between the scalelike megasporophylls. A pollen grain enters through a canal called a micropyle and is pulled into the ovule by a pollination droplet (also called micropylar fluid) secreted by the female gametophyte. The pollen (immature male gametophyte) germinates in the ovule, growing a haustorial pollen tube. Pollen that forms a tube that penetrates and absorbs is called haustorial pollen. The haustorial pollen tube transports the motile sperm to the neighborhood of the eggs, so water is not required to convey the sperm to the eggs, as it is for mosses and ferns. The mature male gametophyte consists of the germinated pollen grain, which produces sperm and a pollen tube. In a fluid-filled fertilization chamber inside the ovule beside the eggs, the pollen tube releases two large sperm, nearly 0.5 mm in diameter—the largest in the plant kingdom—each with some 40,000 undulipodia. The sperm swim in the fertilization chamber before fertilization; the entire process, from pollination to fertilization, takes as long as 5 months in cycads. (In comparison, angiosperm fertilization is often completed in a few hours.) Cumbersome, lengthy fertilization, like the unusual combination of motile sperm and pollen tubes, is unique to cycads and ginkgos. Fertilization results in a diploid embryo. Even when more than one egg produced by one female gametophyte is fertilized, generally only a single embryo survives. The embryonic sporophyte matures within a seed coat, nourished by the surrounding female gametophyte. The outer layer of the cycad seed coat is fleshy, is brightly colored, and contains starch; starch functions as a food reward for animals, from parrots to elephants. Birds, mammals, and water disperse cycad seed.

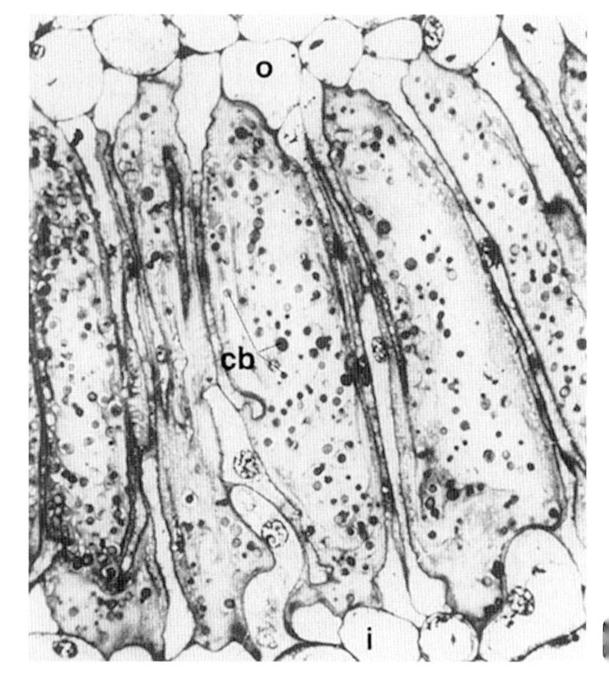

Figure C Transverse section through a coralloid root of *Macrozamia communis*, showing the inner (i), outer (o) and cyanobacterial (cb) layers of the cortex. LM, bar = 10 μm.

Similarities in the pollination of cycads in Australia and Africa by beetles suggest that this beetle–plant relation originated when Australia and Africa were still joined, in the Paleozoic era before the breakup of Gondwana. Pollinating beetles are attracted to a mature female cycad by an attractant, probably a scent, in what may be the most primitive animal pollination system. Thus, cycads provide clues to the origins of insect pollination.

Starch from cycad stem pitch or seed kernels is a potential source of industrial alcohol by fermentation. The starchy seed inner kernel (female gametophyte) is roasted or made into food starch. The outer fleshy layer of the seed also is eaten, as is oil from this layer. Uncooked seeds are toxic. In certain species, the toxin is confined to the kernel. Edible starch leached from subterranean stems and roots of *Zamia* once was cooked as the staple of Florida's Seminoles. Kaffirs of Mozambique and Bantus of South Africa ate fermented stem pith of *Encephalartos*, called Kaffir bread. Cycad starch prepared from stem pith and seeds has furnished flour and bread to people wherever cycads grow from Asia to the Americas, Australia, and Africa. These cultures developed techniques, including drying, fermentation, and leaching with water followed by cooking, that serve to remove a toxic glycoside called cycasin (macrozamin) in cycad starch that they prepared for consumption. The glycoside is a neurotoxin; failure to detoxify cycad causes illness and even death. Cycad toxicity may be a causal factor of the enigmatic neurodegeneration called Lytico-Bodig in Guam. Livestock that feed on raw leaves develop a paralysis called zamia staggers.

Cycad leaves are used as thatch; they are dried for rituals and window displays. Gum exuding from cuts in cones, stems, and leaves is used as adhesive. Gum, prounded cycad seeds, and crushed buds may be used to dress ulcers, boils, and wounds. Cycads are extensively collected from the wild for interior and exterior landscaping.

Pl-9 Ginkgophyta

Japanese *ginkyo*, silver apricot; Greek *phyton*, plant

GENERA
Ginkgo

The ginkgo tree, *Ginkgo biloba*, is the only genus and species of the phylum Ginkgophyta, a phylum of vascular seed plants. In number of species (one), this is the smallest plant phylum. Features that characterize the ginkgo include leaf veins that each branch into two smaller veins, active cambium (cells that produce wood), and fleshy, exposed ovules. Ginkgos, like all other gymnosperms—conifers (Pl-10), gnetophytes (Pl-11), and cycads (Pl-8)—are naked-seed plants; their ovules are enclosed only in the integument (outer layer of the ovule), giving rise to seeds not enclosed by fruit (a ripe, mature ovary—protective, seed-enclosing tissue). Like ferns (Pl-7), ginkgos have megaphylls, roots, motile sperm, and two spore types (heterospory). In contrast to the small, but independent, fern gametophyte, that of ginkgo is microscopic and totally dependent on the sporophyte, as is the case with cycads. The ginkgo sporophyte is a tree. As in all woody trees, the trunk, branches, and root of ginkgo increase in girth by division of cambium cells. Each growing season, a cylinder of cambium adds new xylem cells to the inside and phloem to the outside. These vascular tissues, primarily living xylem, are called sapwood; they conduct water and dissolved minerals from the soil to the leaves. As the ginkgo ages, older xylem dies and becomes heartwood, a fibrous support tissue. Tissues external to the cambium, collectively called bark, surround the trunk and branches. Additional features found in ginkgo but not in phyla P1-1 through Pl-7 include anemophilous (windborne) pollen, sperm-conveying pollen tubes, haustorial pollination, ovules, and seeds. Among gymnosperms, haustorial pollination—present only in cycads and ginkgo—is a transition between motile sperm (as in seedless plants such as ferns) and the pollen tubes combined with nonmotile sperm (as in conifers, gnetophytes, and flowering plants).

Ginkgo biloba is the only living descendant of a great group of plants known from the fossil record to have been more extensive during the age of dinosaurs. The Ginkgo family originated in the Permian period of the Paleozoic era along with the cycads and conifers. Petrified stumps of these great Mesozoic era trees still stand on the northwest coast of North America. The ginkgo tree is native to warm, temperate forests of China. On steep slopes in southeastern China, a few semiwild ginkgo populations can still be found. *Ginkgo biloba* has been cultivated on temple and garden grounds in Asia for centuries. Photographs of living ginkgo trees are usually encumbered with house and telephone wires because ginkgos, resistant to pollution, have been widely planted in urban settings (Figure A).

Common names of the living *Ginkgo* are ginkgo and maidenhair tree, the latter derived from the resemblance between the bilobed leaves of ginkgo and the leaves of the maidenhair fern (*Adiantum*, P1-7). Ginkgo leaves with a bilobed, notched outline grow at tips of long shoots (branches) and seedlings. Leaves with a differing, fan-shaped outline are borne close to the stems on spur shoots (Figure B) that are shorter than long shoots, giving ginkgos a characteristic silhouette recognizable even from a distance.

Ginkgo is dioecious—female and male sexes are on separate plants—and heterosporous. On male trees in hanging inflorescences, haploid microspores develop in microsporangia that are grouped into microstrobili (cones; Figure C). On female trees, haploid megaspores develop in megasporangia. The megasporangium, megaspore, and its protective integument constitute a single structure called the ovule (Figure B). Cells within the mega- and microsporangia give rise by meiosis to megaspores and microspores, respectively.

The haploid megaspore develops into a haploid female gametophyte. This gametophyte develops within an ovule composed of tissue of the parent sporophyte (female tree), on which it depends entirely. Within the ovule, the female gametophyte develops a gamete—the egg.

In the microsporangium, each microspore develops into an immature male gametophyte, a grain of pollen. Wind transports pollen to female trees, where some pollen will land on exposed ovules. The female gametophyte secretes a liquid pollination droplet near the ovule. A pollination droplet pulls each pollen grain through a canal (micropyle) in the integument into a pollen chamber. There the haustorial pollen grain germinates, forming a much-branched pollen tube. Two sperm in each pollen grain develop only after the branched pollen tube

Figure A *Ginkgo biloba*, the ginkgo tree, in an urban setting in China. Bar = 5 m. [Photograph Courtesy of Peter Del Tredici.]

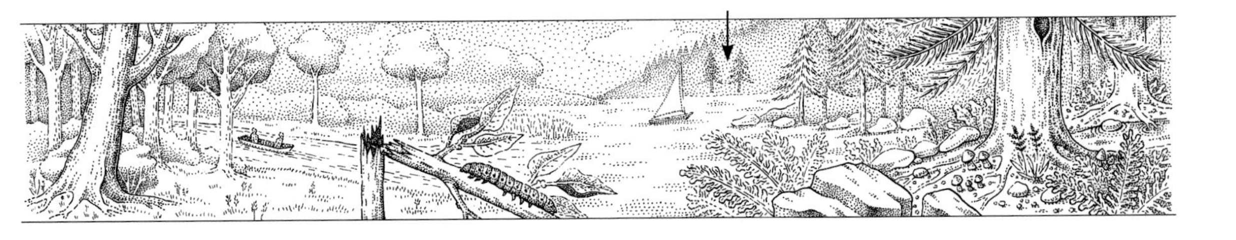

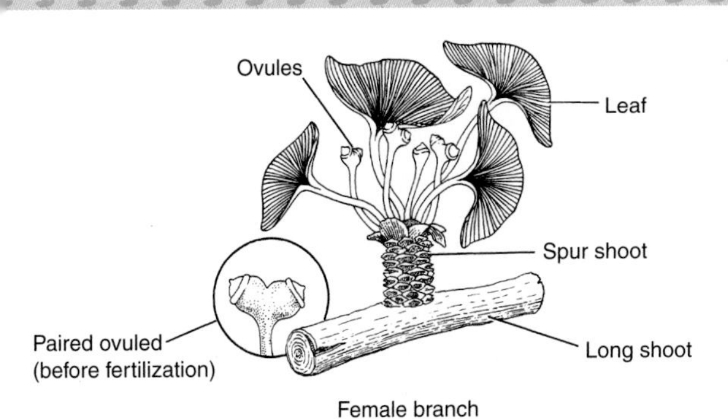

Figure B *Ginkgo biloba* female branch, showing immature ovules. [Drawing by R. Golder.]

Figure D *Ginkgo biloba* fleshy seeds (mature ovules) and fan-shaped leaves with distinctive venation—on a branch. Bar = 5 cm. [Photo by W. Ormerod.]

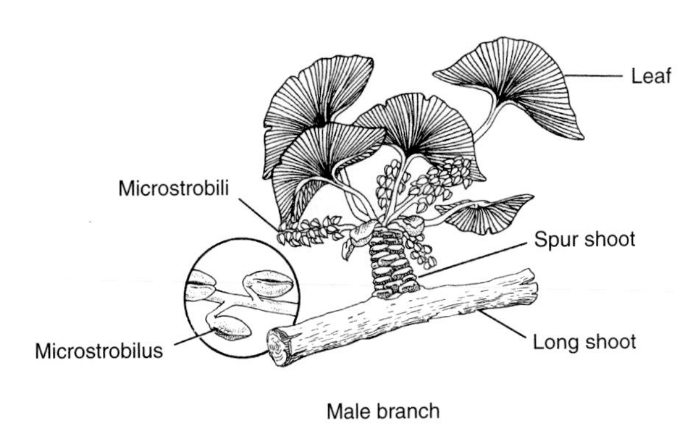

Figure C *Ginkgo biloba* male branch, showing a microstrobilus comprising many microsporangia.

has entered the micropyle. After about 4 months of growth nourished by the ovule, the pollen tube releases motile, helically coiled sperm, each having hundreds of undulipodia; one sperm fertilizes the egg. Because the ginkgo pollen tube delivers sperm within a liquid-filled pollen chamber, ginkgo fertilization needs no environmental water (as do ferns, for example). Embryonic development occurs before the seed has been shed from the parent tree. If pollination occurs in early April, fertilization takes place in September, embryo development is completed during spring, and seed germinates the following May. The embryo and the female gametophyte that nourishes it develop within a true but naked seed. This means that unlike seeds of flowering plants (Pl-12), ginkgo seeds are not enclosed in a fruit. Instead the outer skin (integument) of the ginkgo ovule develops a fleshy seed covering (seed coat) after fertilization (Figure D). A single female tree in one productive day can drop thousands of highly odoriferous seeds, which were originally dispersed by now-extinct Mesozoic animals. In contemporary Asia, tree squirrels feed on *Ginkgo* seeds, as do several members of the Carnivora family, perhaps attracted by their strong smell. Because these animals swallow their food rather then grind it, they may act as effective dispersal agents for the seed, which if it ends up in a favorable site, germinates and develops into a sporophyte seedling, completing the life history.

Ginkgo's major use today is ornamental—understandably, city dwellers prefer to cultivate odor-free male trees. The leaves of ginkgo turn golden before they are shed in autumn. *Ginkgo* is a traditional source of food in China and Japan—the outer fleshy seed covering (mistakenly called fruit) is discarded and the inner seed kernel is roasted. *Ginkgo* leaf extracts have traditionally been used in Asian medicine to increase mental alertness and ameliorate asthma, allergies, and heart ailments. Recently, the ginkgo leaf's medicinally active compounds, called ginkgolides, have been synthesized in the laboratory, and their therapeutic properties are being actively investigated.

Pl-10 Coniferophyta (Conifers)

Latin *conus*, cone; *ferre*, to bear; Greek *phyton*, plant

GENERA
Abies
Araucaria
Cedrus
Cryptomeria
Cupressus
Juniperus
Larix
Metasequoia
Picea
Pinus
Podocarpus
Pseudotsuga
Sequoia
Sequoiadendron
Taxodium
Taxus
Thuja
Tsuga

Most of these cone-bearing gymnosperms are trees, although some are shrubs and creeping, prostrate conifers. As in all seed plants (Pl-8 through Pl-12), the heterosporous, megaphyll-bearing conifer sporophyte is the conspicuous generation, whereas the gametophyte is smaller and nutritionally dependent on the sporophyte. Among the naked-seed plants (gymnosperms), conifers, with 6–8 families, 650 living species, are the most familiar. Conifers are grouped into some 65–70 genera, including *Pinus* (pine; Figure A), *Taxus* (yew), *Abies* (fir), *Pseudotsuga* (Douglas fir), *Picea* (spruce), and *Larix* (larch). Two genera in phylum Coniferophyta are the largest living plants: *Sequoiadendron giganteum*, the giant sequoias growing to 100 m high and 8 m wide in the Sierra Nevada mountains of northern California; and *Sequoia*, the redwoods of the California and Oregon coasts that exceed 115 m in height and are the tallest trees on the planet. *Araucaria*, the curly-branched monkey puzzle tree, originated in the southern hemisphere, where it covers vast tracts of mountainous terrain; it now also thrives in the congenial northern climate of California. Monkey puzzle tree is named for its twisted branches, said to deter climbing animals.

Conifers dominate many northern temperate forests; they are also common in the tropics and southern temperate forests. In mountainous habitats or in the far north, associations of symbiotic fungi called mycorrhizae of conifers are especially characteristic. Contemporary conifers with a variety of mycorrhizal partners ring the cold northern temperate zone; this cold desert is arid in the sense that water frozen into ice is unavailable to plants for much of the year. The fungus sheaths but does not penetrate the conifer roots, forming ectomycorrhizal associations. By enhancing water and nutrient uptake from soil, thus promoting growth of the conifer, ectomycorrhizae make the trees more tolerant of drought.

Most conifers have leaves that are needle shaped, except *Podocarpus*, which has flat, narrow, strap-shaped needles. All conifer leaves are simple—undivided into leaflets (Figure B), unlike compound leaves, such as those of walnut, which are divided into leaflets. Conifer needles are arranged in a fascicle, a bundle comprising one to eight needles with abbreviated leaves at their bases (Figure C). Needles are borne on a spur shoot, borne in turn on a long shoot. A heavy, transparent wax cuticle retards water loss from needles yet allows light to enter cells in the interior for photosynthesis. Ducts, through which resin flows, penetrate the compact interior cells of the needles. Resin is a viscous, yellow brown, organic substance secreted by conifers, thought to protect wounds in the conifer from infection. Resin ducts and cuticle are particularly conspicuous in pines that inhabit arid places. Members of most genera shed their needles gradually—remaining evergreen and photosynthesizing even in winter. Some genera are deciduous, that is, shed their needles; *Larix* (larch) and *Taxodium* (bald cypress) needles turn gold and are shed each autumn. Beneath the bark, cambium cells divide throughout the life of the plant, producing wood.

Most conifers are monoecious—female and male reproductive structures, the cones, are borne on the same plant. Resin ducts are present in the cones. Cones contain cone scales that bear sporangia, the spore-producing structures. There are two

Figure A *Pinus rigida*, a pitch pine on a sandy hillside in the northeastern United States. Bar = 5 m. [Photograph by K. V. Schwartz.]

Figure B *Pinus rigida* branch, showing bundles of needle-shaped leaves and a mature female cone. Bar = 10 cm. [Photograph by K. V. Schwartz.]

types of sporangia and spores: male spores in microsporangia in small, male cones; and female spores in megasporangia in the larger, more familiar, female cones (Figure C). In the megasporangium, meiosis produces a megaspore cell that divides several times to produce a female gametophyte (haploid), which forms two haploid eggs. Similarly, in the microsporangium, meiosis produces microspores that develop into the immature male gametophyte, the pollen grain. Pollen, generally yellow and dustlike, is carried passively by wind from the odorless male cone. Some pollen may land on a female cone: this process is called pollination. A drop of sugary liquid secreted by the ovary dries, pulling pollen grains through an opening called the micropyle

toward the ovule. When pollen reaches the ovule, the male gametophyte completes maturation, producing two immotile sperm and one pollen tube. One of the two sperm produced by each pollen grain degenerates. The pollen tube conveys the other sperm to the vicinity of the female gametophyte. Because conifer sperm are not motile and are transported by pollen tubes, fertilization is not dependent on environmental water. After pollination, fertilization may be delayed; in the genus *Pinus*, fusion of the male and female nuclei may be delayed for a year or more after pollination while the female gametophyte forms eggs. Eventually the haploid egg and a sperm nucleus fuse, and the resulting diploid zygote develops into an embryo, the diploid sporophyte. (Even when more than a single egg is fertilized, only one embryo develops in each ovule.)

Conifer seeds are naked and borne on compound cones. Embryos are embedded in the nutrient-providing female gametophyte, enveloped by an integument that becomes the seed coat. Each scale of a female cone carries two ovules; these together with a sterile projection called a bract make up a seed-scale complex (Figure C). Male cones, in comparison, are simple, bearing the microsporangia directly on modified leaves called sporophylls. No fruit is produced, because the embryo is not covered by an ovary wall, such as develops into fruit in flowering plants (Pl-12). Two conifer embryos are visible as a pair of raised areas on the underside of each female seed scale. The embryos are dormant, young sporophytes. Conifer seeds have multiple cotyledons, or seed leaves (usually eight) and, like all seeds, contain stored nutrients (gametophyte tissue). The winged, wind-dispersed seeds usually separate from female cones at maturity. If the seed reaches a suitable location, it germinates, and the seedling sporophyte commences to grow. In some pines, ripe seeds fail to separate from the cones until fire scorches the parent tree. Because their seeds germinate only after having been subjected to extreme heat, these pines repopulate forests after fires.

We value conifers immensely, as sources of lumber and paper pulp and as horticultural trees and shrubs. Turpentine, pitch, tar, amber, rosin, and resin are products of conifer metabolism. Vanillin—a fragrant compound made by vanilla orchid seedpods and used to flavor perfumes and foods—can also be synthesized from conifers. Old English crossbows were made of the wood of yew trees. A drug used to combat ovarian cancer, Taxol, is derived from the Pacific yew, *Taxus brevifolia*, and has been partially synthesized. Taxol is also produced by cell culture of *T. media.* Thousands of conifers are cut each year in Europe and the Americas as Christmas trees. North Americans, Italians, Russians, and others eat edible seeds (pine nuts) of several pines. *Pinus sabiniana* (Digger pine), *P. coulteri* (Coulter pine), *P. lambertiana* (sugar pine), and several species of pinyon pine (*P. edulis*) in the United States are sources of protein-rich pine nuts.

Conifers likely descended from the progymnosperms, which may have given rise independently to the various phyla of gymnosperms (Pl-8 through Pl-11). (See plant phylogeny.) Conifers themselves gave rise to no other plant phyla.

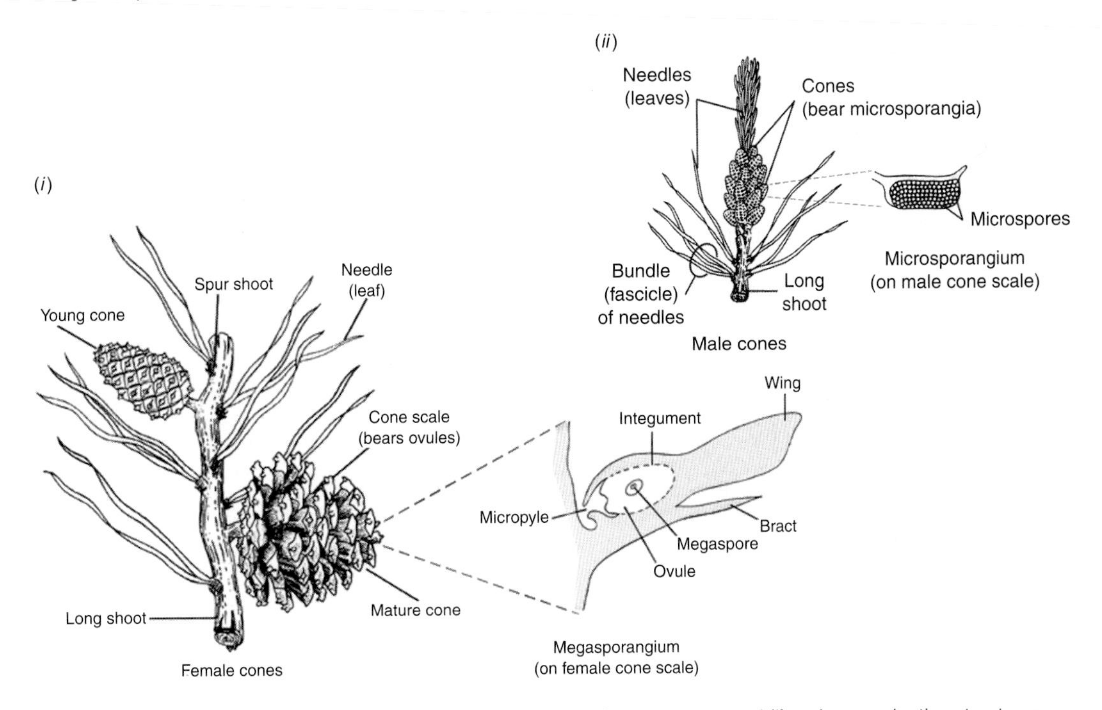

Figure C Reproductive structures of *Pinus rigida*: (*i*) female reproductive structures and (*ii*) male reproductive structures. [Drawings by R. Golder.]

Pl-11 Gnetophyta
(Gnetophytes)

GENERA
Ephedra
Gnetum
Welwitschia

Latin *gnetum* from Moluccan Malay *ganemu*, a gnetophyte species found on the island of Ternate; Greek *phyton*, plant

Gnetophytes living today number about 70 species in three vastly different genera: *Welwitschia*, *Gnetum*, and *Ephedra*. Gnetophytes are distinct from other gymnosperms in the anatomy of their long water-conducting tubes (vessels), which are composed of linking vessel element cells rather than meshing tracheids. Such vessels form when cells join end to end in the xylem tissue and are common in flowering plants. Most gnetophytes produce naked (exposed) seeds, like the other gymnosperms—ginkgo, cycads, and conifers. *Gnetum leyboldii*, unlike most gymnosperms, produces seed enclosed in a juicy, fruit-mimicking layer. Motile sperm are absent from gnetophytes, as in conifers.

Like other seed plants, gnetophytes are heterosporous. Reproduction of gnetophytes resembles conifer reproduction in several ways. Microsporangiate (male) cones and megasporangiate (female) cones produce micro- and megaspores, respectively. Both the female and male gnetophyte cones are compound, like female (but not male) conifer cones. Gnetophyte cones and leaves lack the resin ducts present in conifer cones and leaves.

Most gnetophytes are dioecious, bearing male and female cones on different plants. *Welwitschia* cones (strobili) are borne on small branches that arise from the outer rim of the stem (Figure A). The male cone contains sterile ovules (female parts), suggesting that functioning male and female reproductive structures once resided on the same plant, as occurs in conifers and many flowering trees (Pl-12; Figure B). *Ephedra*'s and *Welwitschia*'s male and female reproductive structures are borne on bracts or cone scales in minute cones that superficially resemble cones of the conifer hemlock (Figure B). In *Gnetum* and *Ephedra*, the cones are attached along the stem at the nodes.

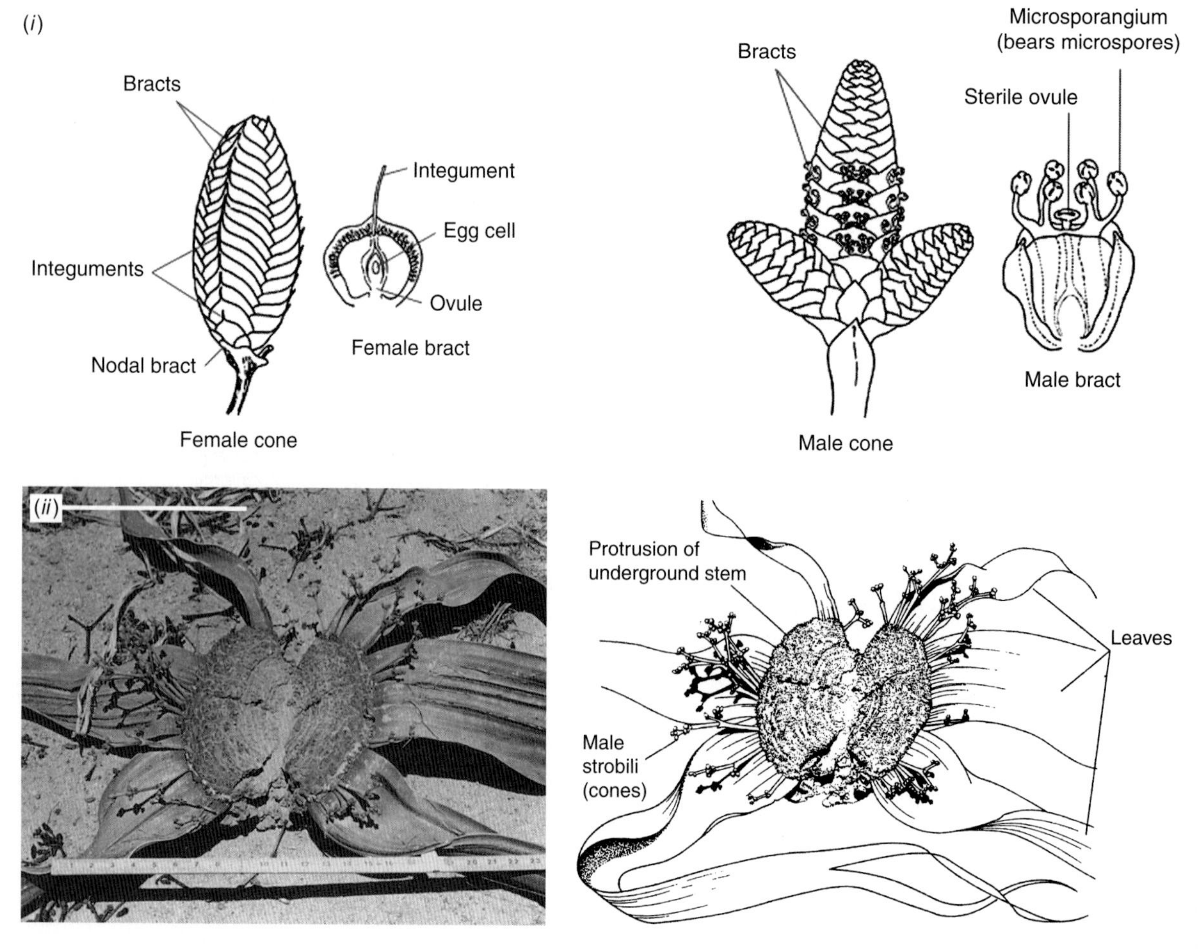

Figure A (*i*) Cones of *Welwitschia mirabilis*. [Drawings by R. Golder.] (*ii*) *Welwitschia mirabilis* (male), growing in the desert in southwestern Africa. Bar = 25 cm. [Photograph courtesy of E. S. Barghoorn; drawing by I. Atema.]

In the microsporangia of male cones, meiosis produces microspores that mature into pollen. After the pollen has been shed from the male cone, it germinates, producing immotile sperm. In the megasporangia of female cones, meiosis results in megaspores that produce eggs. One *Ephedra* sperm (haploid) fertilizes an egg cell (haploid), producing a diploid embryo. A second sperm nucleus fuses with a female gametophyte nucleus, producing another embryo. After this process—called double fertilization—both resulting zygotes initiate development. An integument—outer layer of the ovule—covers *Ephedra*, *Gnetum*, and *Welwitschia* seeds, but no gnetophyte embryo is enclosed in a fruit or nourished by endosperm. The gnetophyte seed resulting from the fertilization of the egg stores its food reserves in a pair of cotyledons, which are the embryonic leaves. *Gnetum gnemon* also undergoes double fertilization although it is unresolved whether double fertilization occurs in other species of *Gnetum*. *Gnetum gnemon* (and perhaps *Welwitschia*) does not form egg cells; instead, as a product of meiosis, it forms egg nuclei that are free in the female gametophyte. Each ovule forms a pollination droplet that retracts, pulling pollen into the ovule. In *Gnetum*, each of two nuclei in a binucleate sperm from one pollen tube fertilizes a separate female (haploid) nucleus; two separate fertilizations form two (diploid) embryos. Ultimately, just one of the two embryos reaches maturity in each seed. Subsequent to fertilization, several female *Gnetum* nuclei fuse, developing polyploid, embryo-nourishing tissue that is the functional equivalent of endosperm but of different origin. Double fertilization in *Gnetum* and *Ephedra*—whether it takes place in *Welwitschia* is unknown—results in two zygotes; in comparison, double fertilization in many flowering plants results in the production of one zygote and polyploid endosperm. Researchers suggest that the gnetophyte pattern of rudimentary double fertilization with formation of two embryos evolved in a common ancestor of gnetophytes and anthophytes. After divergence of gnetophytes from their common ancestry, the flowering plants may have modified one of the two zygotes resulting from double fertilization to form triploid endosperm, unique to flowering plants.

Gnetum survives in the angiosperm-dominated tropical rain forest even though individual plants are widely spaced, as are individual plants of angiosperm species. In several ways, dioecious *Gnetum* plants resemble many angiosperms: insect pollination and seed dispersal by fruit-eating birds. At least some species of *Gnetum*, *Ephedra*, and *Welwitschia* produce nectar, which enhances the efficiency of pollination. Insects pollinate *Gnetum*; wind and insects probably pollinate the other gnetophytes. *Gnetum leyboldii* seed is covered by a fleshy integument layer called a pseudofruit. Its sweet flesh entices the chestnut-mandible toucan to feed on the pseudofruit. Later the toucan regurgitates the seed, sometimes dispersing *Gnetum* seed.

Welwitschia was discovered in 1859 by an Austrian, Friedrich Welwitsch, in arid southwestern Africa. This genus of extraordinary plants contains but one extant species, *Welwitschia mirabilis*. Some *Welwitschia* plants may be 2000 years old. Each plant grows two strap-shaped leaves as much as several meters long. These leaves grow at their attachment points during the entire life of the plant and tatter with age, lying on the sand. The leaves are attached to the outer margin of a woody, top-shaped stem that protrudes a meter or more aboveground. Depending on the plant's age, the stem may be as much as a meter wide. Below ground, the stem tapers to a root that branches after a meter or so. Morning fog forms perhaps 100 days per year in *Welwitschia*'s coastal Namib Desert habitat, equivalent to about 50 mm of water per year. Although water absorption by the leaves on foggy mornings has been proposed as the main method of water uptake, the taproot may take up water as well.

The genus *Gnetum* comprises about 30 species of woody, large-leafed trees, shrubs, and lianas (woody vines) that are native to the tropical deserts, rain forests, and mountains of Asia, Africa, and Central and South America. *Gnetum gnemon* is a small tree that grows on islands in the Pacific. About 7 m high, it has glossy leaves and scarlet seeds. This tree of the forest is also cultivated for its edible seeds. Wood is produced by old stems of *Gnetum* vines as well as by *Gnetum* trees.

Ephedra, called joint fir, is native to southwestern deserts and uplands of North America, to the Mediterranean, and across the Himalayas to Mongolia. The 40 or so *Ephedra* species are evergreen perennial shrubs as tall as 3 m. Greenish, slender, jointed branches are longitudinally ribbed. Scalelike leaves sheath the nodes, where two stem joints meet. Reduction in leaf surface is probably a response to water loss in semiarid regions. At first glance, *Ephedra* resembles joint-stemmed horsetails (Pl-6), but *Ephedra*'s shrubby habit and cones (Figure B) distinguish it from horsetails.

The oldest gnetophyte fossil dates from the Triassic period, which began 245 mya (Figure I-4). Gnetophytes are believed not to have given rise to any other plant lineage. Fossil pollen evidence dating back 280 million years suggests that *Welwitschia* originated more than 300 mya from progymnosperm-derived cone-bearing plants that are common ancestors not only of gnetophytes but also of modern conifers.

Double fertilization, which was once considered exclusive to angiosperms, either has evolved independently twice or else *Ephedra* and *Gnetum* are closely related to angiosperms through an extinct common ancestor. Molecular evidence suggests that gnetophytes are the phylum closest to flowering plants; data from ribosomal RNA are supported by ribulose-1,5-bisphosphate carboxylase (*rbcL*) gene sequence information. Other similarities between gnetophytes and flowering plants (especially

Pl-11 Gnetophyta
(Gnetophytes)

(continued)

Figure B Reproductive structures of *Ephedra trifurca*, the long-leaf ephedra. This desert shrub ranges from Texas to Baja California, Mexico. Left: Pollen-producing, microsporangiate cones—male cones. Each pollen grain produces a single pollen tube and two sperm. Right: Megasporangiate cones—female cones. A glistening pollination droplet at the tip of each 0.5-cm-long cone draws pollen into the ovule. Upon fertilization, two sperm fertilize a binucleate egg cell, producing two embryos. Ultimately, one embryo will mature and become a seed. [Courtesy of K. Niklas, Cornell University.]

dicots) include vessels in the wood, seeds with two cotyledons, and leaves with netlike veins (in *Gnetum*). Vessels are present in groups of plants only distantly related to each other—gnetophytes, angiosperms (Pl-12), several fern species (Pl-7), *Equisetum* (Pl-6), and some *Selaginella* (Pl-4). Vessels and pseudofruit exemplify convergent evolution—that is, the independent development of similar structures by distantly related organisms.

Native Americans and settlers brewed teas from *Ephedra*'s yellow male cones and dried stems, giving rise to its common names Mormon tea, Brigham tea, and Mexican tea. *Ephedra* and *Gnetum* seeds can be roasted and eaten. *Gnetum* seeds are also edible raw. Native Americans ground *Ephedra* seeds into meal. Some *Ephedra* species native to Asia and the Mediterranean are harvested for the alkaloids ephedrine and pseudoephedrine. These compounds have diverse therapeutic uses: they dilate bronchioles and are thus used to treat nasal congestion, hay fever, emphysema, and colds; as vasoconstrictors, they staunch nosebleeds and alleviate hypotension during anesthesia. In Chinese medicine, *Ephedra* has been used for at least 5000 years as a remedy for colds, malaria, headaches, and cough. Although extracts from the North American *Ephedra* species were used in folk treatment of venereal disease (leading to the name "whorehouse tea"), this therapeutic use is unproved. Gum from *Gnetum nodiflorum* is used as a medicine to reduce swelling caused by muscle damage. On islands of Southeast Asia, young leaves of *Gnetum* are eaten as a cooked vegetable.

Pl-12 Anthophyta
(Angiospermophyta, Magnoliophyta, flowering plants)

Greek *anthos*, flower; *phyton*, plant

GENERA (m = monocot; d = dicot)

Acer (d)
Agave (m)
Allium (m)
Artocarpus (d)
Aster (d)
Atropa (d)
Avena (m)
Beta (d)
Brassica (d)
Camellia (d)
Capsella (d)
Chondrodendron (d)
Chrysanthemum (d)
Cinchona (d)
Cocos (m)
Coffea (d)
Colchicum (m)
Cucurbita (d)
Cuscuta (d)
Datura (d)
Daucus (d)

Flowering plants, the angiosperms, are superstars of diversity and abundance. More than 230,000 angiosperm species are grouped into about 350 families. If we had more botanist explorers to identify them, the number of described angiosperm species would probably be closer to a million. The flower (Figure A), the reproductive organ common to all flowering plants, reveals their common ancestry. The gametophyte is barely visible within the flower; so, in this phylum, the sporophyte is the more familiar generation. Flowers and fruits uniquely distinguish this phylum (Figure B). Flowering plants have female reproductive structures (ovules enclosed in a carpel) as well as male reproductive structures (stamens), which may be on the same flowers and plants or on different ones. Fertilized eggs in the ovules become seeds, and fruits develop around the seeds as flowering plants mature. In corn, for example, the tassel is the male reproductive structure and the corn ear is the female structure. Fertilized eggs become corn kernels and, when mature, corn seed. Familiar broad-leaved trees, shrubs, garden plants, crops, and wildflowers that produce flowers and fruits are members of this phylum. Flowering plants, like gnetophytes, ginkgo, cycads, and conifers (Pl-8 through Pl-11), are seed plants.

The detail of this vast diversity of plants is unknowable by a single person. How, then, are flowering plants organized? The major subdivisions established by Antoine-Laurent de Jussieu (1748–1836) that divide plants lacking seed leaves, such as mosses, and those bearing one (monocot) or two (dicot) seed leaves are still valid. Monocotyledones and Dicotyledones are subphyla within phylum Anthophyta. A seed leaf, also called a cotyledon, is a leaflike structure of the seed-plant embryo; the seed leaf often contains stored food (in dicots), absorbs food (in monocots), and provides nutrients used during seed germination (Figure Pl-1).

Among the ca. 65,000 monocot species are bananas (*Musa*), cattails (*Typha*), coconut palm (*Cocos*), grasses such as maize (*Zea*), and crocus (*Colchicum*). Monocots are easily distinguished from dicots: in addition to the defining characteristic of one-seed leaf (cotyledon), monocots display a complex array of primary vascular bundles in their stems, their leaf veins run in parallel through the leaf, and their petals and other flower parts often grow in threes. Primary growth increases plant length by means of growth at tips of shoots and roots. Monocots lack secondary cambium, the tissue that secondarily increases stem girth. Most monocots are herbaceous rather than woody shrubs or woody trees: for example, the banana plant is herbaceous; it lacks secondary (woody) growth and thus is not a tree. Palms, monocots that appear to have woody trunks, have trunks stiffened by strands of fiber rather than the woody growth characteristic of dicots.

By far the largest group of flowering plants, the dicots comprise some 170,000 species. Dicots include roses (*Rosa*), sunflowers (*Helianthus*), maples (*Acer*), pumpkins (*Cucurbita*), grapes (*Vitis*), peas (*Pisum*), tulip tree (*Liriodendron*; Figure C), and asters (*Aster*; Figure D). Dicot characteristics include two-seed leaves (two cotyledons), branching leaf veins, primary vascular bundles in a ring within the stem, flower parts in fours or fives, and the presence of secondary (woody) growth.

Our major food plants are just a few species of flowering plants. Our cereals—rice, maize (corn), wheat, oats, barley, millet, rye—are all seeds of monocots. We consume the monocot seeds of maize and coconut (coconut meat) and press edible oils from them. Sugarcane (a monocot) is crushed to release its stem juices; sugar is crystallized from the resulting liquid. Coffee, soybeans, potatoes, tomatoes, beans, lentils, buckwheat, and apples are dicots. Cotton is fiber surrounding the cotton seeds. Linen and hemp are fiber from plant stems. Most pharmaceuticals are of plant origin. Digitalis, a medicine that slows the heartbeat, is extracted from leaves of the foxglove, dicot *Digitalis.* Atropine—an antisecretory drug—is extracted from belladonna, *Atropa belladonna.* Bark of the cinchona tree *Cinchona calisaya* is the source of quinine, an antimalarial. Morphine and codeine, prescribed to relieve pain, are extracted from *Papaver somniferum*, the opium poppy.

Figure A *Liriodendron tulipifera* (tulip tree) flower. Bar = 10 cm. [Photograph by K. V. Schwartz.]

Both female and male flower structures evolved from the leaf as a modification of a whorl of leaves into a shoot specialized for reproduction. Clues for this line of reasoning come from plant development: leaf and flower development share many similarities. The female flower structure, the carpel (composed of stigma, style, and ovary; Figure B) began as a folded leaf blade. The ovules of the earliest angiosperms probably formed in rows along the inner surface of the carpels. One or more carpels make up the ovary, enclosing the maturing seeds in a protective layer. Several carpels probably fused in the evolution of the more complex flowers; tomato—a fruit that grows from a single flower that has fused carpels—is an example. The stamen, the male flower part (consisting of a stalk, or filament, bearing an anther), also is a modified leaf. The leaf shoots that form the various floral parts—sepals, petals, carpels, stamens—shortened and fused; so the whorled arrangement of the evolved leaves is undetectable in most species.

The flower bears both sporophylls (modified leaves that bear sporangia) and gametophytes. Female and male reproductive

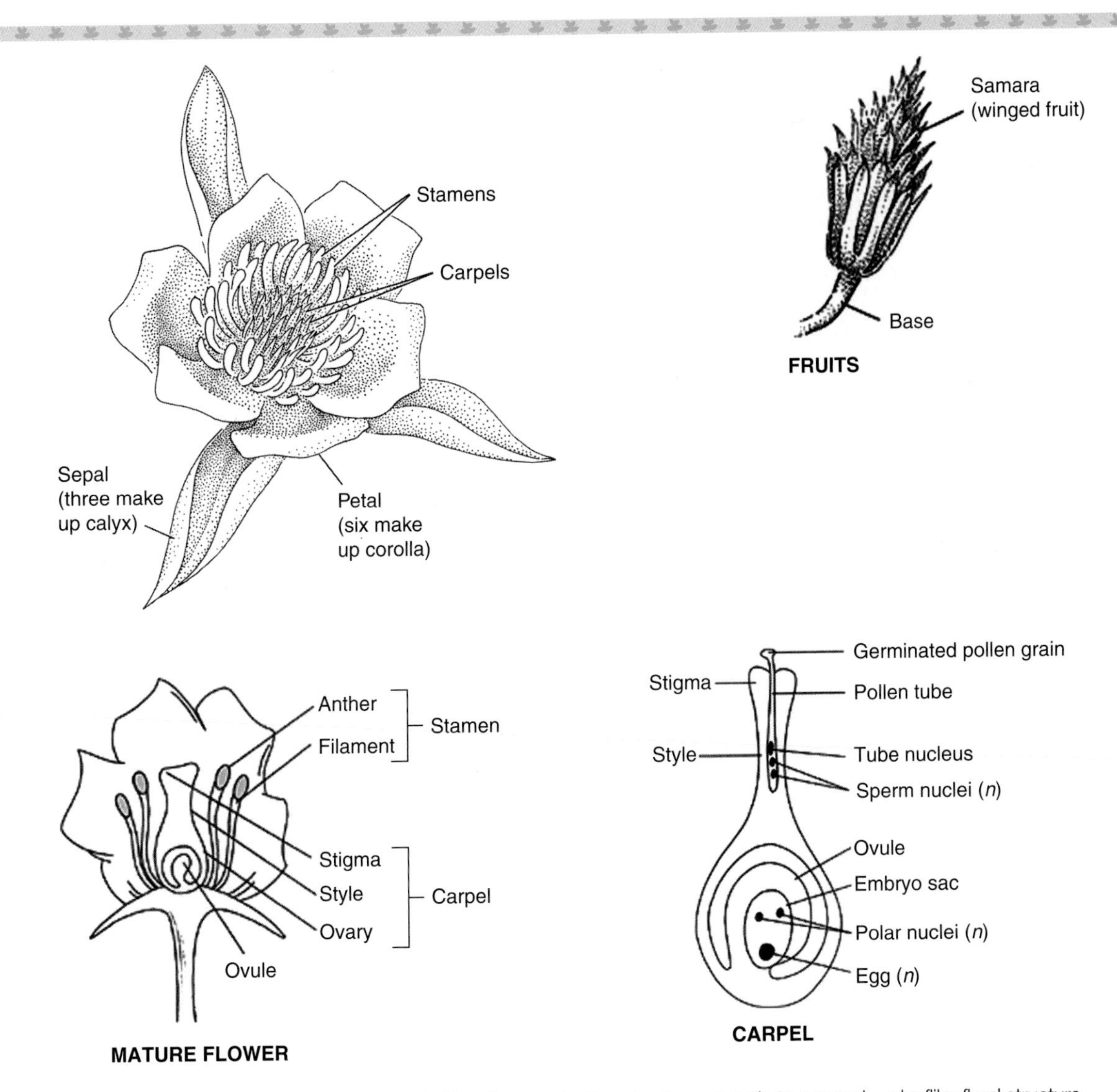

Figure B *Liriodendron tulipifera*, flower and fruit. Female reproductive structures constitute a carpel—a leaflike floral structure including stigma, style, and ovary. Within the female gametophyte, which is called the embryo sac, are two polar nuclei and one egg nucleus. Male reproductive structures constitute the stamen—composed of anthers and the filaments that carry them. Within the male gametophyte—the germinated pollen grain—is a tube nucleus and a pair of sperm nuclei. Only a single carpel and four stamens are illustrated; however, many flowers, including that of the tulip tree, have multiple carpels and stamens. [Drawings by L. Meszoly.]

parts can be present in the same flower, as in magnolia and apple, in different flowers of the same plant, as in corn, or in different plants, as in date palm. In corn, tassels bear male flowers, whereas female reproductive parts are borne within the young ear. Megaspores (female) form by meiosis in the ovule; microspores (male) form by meiosis in the anther. From these haploid spores, gametophytes germinate; a gametophyte consists of fewer than a dozen specialized cells hidden inside the flower and is physically dependent on the sporophyte plant. The haploid gametophytes develop haploid gametes, sperm in the male and eggs in the female. The archegonium and antheridium of mosses and other nonvascular plants have thus been replaced

Pl-12 Anthophyta

(Angiospermophyta, Magnoliophyta, flowering plants)

(continued)

Digitalis (d)
Echinocactus (d)
Elodea (m)
Epifagus (d)
Eucalyptus (d)
Fagus (d)
Glycine (d)
Haplopappus (d)
Helianthus (d)
Hevea (d)
Hordeum (m)
Ipomoea (d)
Kalanchoë (d)
Lemna (m)
Lilium (m)
Liriodendron (d)
Magnolia (d)
Manihot (d)
Monotropa (d)
Musa (m)
Oreodoxa (m)
Oryza (m)
Papaver (d)
Phaseolus (d)

Figure C *Liriodendron tulipifera*, the tulip tree, in summer in Illinois. Bar = 5 m. [Courtesy of the Arnold Arboretum, Harvard University.]

in angiosperms as in all other seed plants—cycads, ginkgos, conifers, and gnetophytes (Pl-8 through Pl-11)—with male and female gametophytes greatly reduced in size.

Male flower parts of a typical angiosperm consist of a stalk carrying anthers, which form microsporangia called pollen sacs. Microsporogenesis takes place in the microsporangia, generating haploid microspores, which develop into pollen grains, the male gametophytes. A pollen grain consists of a tough outer coat enclosing two or three haploid nuclei within a larger tube cell. When the mature anther releases the pollen, wind or animals carry the pollen grains to the female part of the same flower or of another flower, a necessary precondition for fertilization.

The female flower part, the carpel, consists of the stigma, style, and ovary. Inside the ovary are the ovules, the female sporophyte flower tissue in which megasporogenesis produces megaspores. In the ovule, the megasporocyte (a diploid cell) divides meiotically to form four haploid megaspores. Three of these haploid megaspores usually degenerate and the other one divides mitotically to form several haploid nuclei. One nucleus becomes the egg, several others are short-lived, and two nuclei form a binucleate central cell ($n + n$), separate from the egg cell. Growth and development of the nuclei and associated cytoplasm leads to the mature female gametophyte (megagametophyte), also called the embryo sac. The egg—the female gamete—lies within the embryo sac along with the central cell. Production of female gametes is complex and yet remarkably uniform in thousands of flowering plants.

If the pollen grain reaches a receptive female reproductive part (the stigma), it germinates and produces a pollen tube. The germinated pollen grain is the mature male gametophyte (microgametophyte), consisting of three haploid nuclei—two sperm nuclei and one tube nucleus—with surrounding cytoplasm all in a single cell. Directed by the tube nucleus, the pollen tube grows down through the style into the embryo sac, usually conveying two sperm nuclei. One (haploid) sperm fertilizes the (haploid) egg to form the zygote (diploid); the second sperm nucleus (haploid) usually fertilizes two female nuclei (of the binucleate central cell) of the embryo sac, forming a cell—often triploid—that becomes the endosperm. This is "double fertilization," which also occurs only in *Ephedra* and *Gnetum* (gnetophytes), indicating that gnetophytes and flowering plants may have a common ancestor. The zygote sporophyte embryo is surrounded by the triploid endosperm tissue; endosperm is destined never to reproduce but to grow as the nutritive support tissue of the seed and, in some species, the seedling. Maternal diploid sporophyte tissue (integument) of the preceding generation forms a protective seed coat around both embryo and endosperm. As zygote and endosperm develop into a mature seed, the ovary matures into fruit. Reserve nutrients are mobilized from the endosperm and stored in the cotyledons that emerge from the embryo as it grows within the seed coat; many but not all plant embryos do this—for example, the dicot peas. Many dicots and some monocots use all or most of the endosperm before dormancy—arrested growth. In dicots, such as the peanut, the embryo nestles between two cotyledons, the kernels of the peanut. Many dicots develop nutrient-storing, fleshy cotyledons that themselves provide food for both the developing embryo and the young seedling. A plump acorn is such a dicot seed; if you dig up an oak seedling, you will see the acorn. In monocots, such as wheat, the embryo is tucked toward one end of the single cotyledon. Peas absorb most of the stored nutrients before the pea seed becomes dormant, well before the seedling grows forth. Corn finally absorbs the food reserves of the endosperm at germination. Endosperm is present only in angiosperms, whereas gymnosperm seeds are nourished by stored nutrients provided by the female gametophyte. Other seed plants—conifers, gnetophytes, cycads, and ginkgos—also have cotyledons. In comparison with the embryo of flowering plants, the conifer embryo absorbs food stored not in the cotyledons or endosperm but in the female gametophyte tissues within the conifer seed.

Eventually, the fruit sheds its seed. Humans consume the fruit but not the seed (such as an apple), only the seed (such as a lima bean), or the fruit including the seed (such as green beans). Seeds are dispersed by ejection (jewelweed), wind (milkweeds), water (coconut), and animals (apples). Birds consume

Pisum (d)
Primula (d)
Prunus (d)
Quercus (d)
Rauwolfia (d)
Rosa (d)
Saccharum (m)
Solanum (d)
Spartina (m)
Spiradela (m)
Theobroma (d)
Tradescantia (m)
Trifolium (d)
Triticum (m)
Tulipa (m)
Typha (m)
Viburnum (d)
Vitis (d)
Wolffia (m)
Yucca (m)
Zea (m)
Zostera (m)

Figure D *Aster novae-angliae*. The New England aster blooms purple in meadows and along roadsides. This wildflower belongs to Compositae, second largest of the plant families. Bar = 1.5 cm. [Photograph by K. V. Schwartz.]

wild cherries, later depositing the seeds with their feces sometimes distant from the parent cherry tree. The cherry seed itself is both bitter and toxic, discouraging animals that eat the cherry fruit from breaking the seed itself up and digesting it. Eventually, the seed may sprout and develop into a seedling, the diploid sporophyte.

These fertile machinations led to one of the greatest of all evolutionary innovations—the angiosperm seed (Figure Pl-1). Seeds are commonly genetically programmed to remain dormant until conditions are favorable for resumption of growth. Seed coats are prepared for many contingencies, depending on the species; many contain chemical inhibitors that maintain dormancy until the seed passes through an animal's digestive tract or the inhibitor is otherwise removed. Some seed coats require burning or scarring (cutting or abrasion of the seed coat by rain and weathering) to germinate; others must be chilled or frozen; still others must be exposed to certain wavelengths of light. Chilling followed by warming and light signals spring. Environmental cues signal suitable conditions for germination of the seed, involving hormone production and mobilization of food reserves, and subsequent growth of the young sporophyte.

Strawberry plants and a number of other flowering plants also reproduce asexually by sending out runners on which new tiny plants grow, by rhizomes (potato), and from unfertilized (autodiploid) seed (dandelion). These mechanisms are somewhat analogous to asexual reproduction by fragmentation or production of gemmae and plantlets in mosses, liverworts, club mosses, and ferns.

We find the earliest angiosperm fossils—angiosperm pollen grains—in the early Cretaceous period, near the end of the Mesozoic about 140 mya. At that time, flowering plants became the dominant plant phylum worldwide, although they undoubtedly evolved earlier in the Permian. The earliest well-preserved flowering plant fossil (120 million years old) is the extinct *Koonwarra* collected in southeastern Australia. Some early fossil angiosperms resemble large-flowered magnolias; others, like the herbaceous Koonwarra, are smaller. The presence of double fertilization, net-veined leaves, seeds with two cotyledons, and vessels (water-conducting cells) in the wood in both gnetophytes and angiosperms suggests that they may share a common ancestor (see plant phylogeny). *Welwitschia*—a gnetophyte—has strobili that contain both male (stamens) and female (ovules, although the ovules are aborted) structures, somewhat like flowers that contain both male and female structures. Seventy-five million years ago, great Mesozoic forests of gymnosperms—seed ferns, conifers, cycads, and ginkgos—were displaced by flowering plants. By the opening of the Cenozoic era, 65 mya, many modern families and even genera (for example, *Viburnum*) of angiosperms appear in the fossil record. Flowers reveal certain evolutionary trends with time. Older, earlier-evolving flowers, such as those of tulip tree (Figures A and B), have many parts, indefinite in number, and tend to be radially symmetrical. Younger, later-evolving flowers have fewer parts, definite in number, and tend to be bilaterally symmetrical (such as orchids).

Angiosperm dominance in the plant kingdom is related, in part, to characteristics that afforded angiosperms greater success in completing their life cycle than that of gymnosperms. Such a characteristic is the coevolution of plants with land animals, especially with insects (A-21) and terrestrial chordates (classes Reptilia, Aves, and Mammalia of phylum A-37). Flower color and the animal species that transfer pollen for that flower evolved together, with the result that color often advertises to a certain animal pollinator. Yellow crocuses, for example, are attractive to insect pollinators, whereas white night-blooming cactus is more visible in low light and therefore attracts nocturnal moths and bats. Wind pollination of conifers is much less efficient than animal pollination; animal pollination permits production of a smaller quantity of pollen. In angiosperms, fertilization rapidly follows pollination, in contrast to the delay between pollination and fertilization in conifers. Flowers are pollinated and seeds enclosed in fruit are dispersed by beetles, butterflies, moths, mice, bats, birds, and hundreds of other animal species. Chipmunks stash beechnuts and acorns for later consumption; forgotten seeds may sprout, dispersing the tree seeds. The flowers of the mango tree are pollinated by bats, blueberry blossoms and sunflowers by insects, columbine by hummingbirds, and sage and alfalfa by honey bees. Water-dispersed seeds, such as coconut, tend to be buoyant. These angiosperm innovations—efficient pollination and seed dispersal, reduced water loss resulting from seasonal shedding of leaves (also characteristic of the gymnosperm larches and bald cypress), short life cycle compared with conifers, tough seed coat, chemical diversity, and others—led to the dominance of angiosperms in the tropical and temperate regions of Earth.

Bibliography: Plantae

Literature

Attsatt, P. R., "Fungi and the origin of land plants." In: L. Margulis and R. Fester, eds., *Symbiosis as a source of evolutionary innovation.* MIT Press; Cambridge, MA; 1991.

Balick, M. J., and P. S. Cox, *Plants, people, and culture: The science of ethnobotany.* Scientific American Library, W. H. Freeman and Company; New York; 1996.

Bazzaz, F. A., and E. D. Fajer, "Plant life in a CO_2-rich world." *Scientific American* 266:68–74; January 1992.

Britton, N. L., and A. Brown, *An illustrated flora of the northern United States and Canada*, 3 vols.. Dover; New York; 1970.

Carpenter, P. L., and T. D. Walker, *Plants in the landscape*, 2d ed. W. H. Freeman and Company; New York; 1989.

Chaloner, W. G., and P. MacDonald, *Plants invade the land.* Royal Scottish Museum; Edinburgh; 1980.

Cox, P. A., and M. J. Balick, "The ethnobotanical approach to drug discovery." *Scientific American* 270:82–87; June 1994.

Fernald, M. L., *Gray's manual of botany: A handbook of the flowering plants and ferns of the central and northeast United States and adjacent Canada.* Dioscorides (Timber Press); Portland, OR; 1987.

Friedman, W. E., "The evolutionary history of the seed plant male gametophyte." *Trends in Ecology and Evolution* 8:15–20; 1993.

Gifford, E. M., and A. S. Foster, *Morphology and evolution of vascular plants*, 3d ed. W. H. Freeman and Company; New York; 1989.

Goulding, M., "Flooded forests of the Amazon," *Scientific American* 266:114–120 March 1993. Plant and animal biodiversity and adaptations.

Govindjee, and W. C. Coleman, "How plants make oxygen," *Scientific American* 262:50–58 February 1990. Green plant photosynthesis compared with bacterial.

Graham, L. E., *Origin of land plants.* Wiley; New York; 1993.

Handel, S. N., and A. J. Beattie, "Seed dispersal by ants." *Scientific American* 263:76–83; August 1990.

Harlan, J. R., "Plants and animals that nourish man." *Scientific American* 235:88–97; September 1976.

May, R. M., "How many species inhabit the earth?." *Scientific American* 267:42–48; October 1992.

May, R. M. "Origin and relationships of the major plant groups." *Annals of the Missouri Botanical Garden* 81; 1994. Special edition.

Pirozynski, K., "Gall, Flowers, Fruits, and Fungi." In L. Margulis and R. Fester, eds., *Symbiosis as a source of evolutionary innovation.* MIT Press; Cambridge, MA; 1989.

Raven, P. H., R. F. Evert, and S. Eichhorn, *Biology of plants*, 6th ed. Worth; New York; 1998.

Rosenthal, G. A., "Chemical defenses of higher plants." *Scientific American* 254:94–99; January 1986.

Scagel, R. F., R. J. Bandoni, G. E. Rouse, W. B. Schofield, J. R. Stein, and T. M. C. Taylor, *Plants: An evolutionary survey.* Wadsworth; Belmont, CA; 1984.

Scientific American 235; September 1976. (Special issue: Food and Agriculture.)

Scientific American 271; October 1994. (Special issue: Life in the Universe.)

Stebbins, G. C., Jr., *Variation and evolution in plants.* Books on Demand; Ann Arbor, MI; 1950.

Stix, G., "Back to roots," *Scientific American* 268:142–143 January 1993. Pharmaeuticals derived from plants.

Taylor, T. N., and E. L. Taylor, *Biology and evolution of fossil plants.* Prentice-Hall; Englewood Cliffs, NJ; 1993.

Van De Graaff, K. M., S. R. Rushforth, and J. L. Crawley, *A photographic atlas for the botany laboratory*. Morton; Englewood, CO; 1995.

Pl-1 Bryophyta

Conard, H. S., and P. L. Redfearn, Jr., *How to know the mosses and liverworts*, 2d ed. Brown; Dubuque, IA; 1980.

Crandall-Stotler, B., "Morphogenesis, developmental anatomy and bryophyte phylogenetics: Contraindications of monophyly." *Journal of Bryology* 14:1–23; 1986.

Mishler, B. D., L. A. Lewis, M. A. Buchheim, K. S. Renzaglia, D. J. Garbary, C. F. Delwiche, F. W. Zechman, T. S. Kantz, and R. L. Chapman, "Phylogenetic relationships of the green algae and bryophytes." *Annals of the Missouri Botanical Garden* 81:451–483; 1994.

Richardson, D. H. S., *The biology of mosses*. Wiley; New York; 1981.

Schofield, W. B., *Introduction to bryology*. Macmillan; New York; 1985.

Smith, D. K., and P. G. Davison, "Antheridia and sporophytes in *Takakia ceratophylla* (Mitt.) Grolle: Evidence for reclassification among the mosses." *Journal of the Hattori Botanical Laboratory* 73:263–271; 1993.

Pl-2 Hepatophyta

Schofield, W. B., *Introduction to bryology*. Macmillan; New York; 1985.

Schuster, R. M., *The Hepaticeae and Anthocerotae of North America east of the hundredth meridian*, 6 vols. Columbia University Press; New York; 1966–1992.

Pl-3 Anthocerophyta

Paolillo, D. J., Jr., "The swimming sperm of land plants." *BioScience* 31:367–373; 1981.

Schofield, W. B., *Introduction to bryology*. Macmillan; New York; 1985.

Pl-4 Lycophyta

Cobb, B., *A field guide to ferns*, 2d ed. Houghton Mifflin; Boston; 1984.

DiMichele, W. A., and J. E. Skog, "The Lycopsida: A symposium." *Annals of the Missouri Botanical Garden* 79:447–449; 1992.

Hoshizaki, B. J., *Fern growers' manual*. Knopf; New York; 1975.

Mickel, J. T., *How to know the ferns and fern allies*. Brown; Dubuque, IA; 1979.

Smith, G. M., *Cryptogamic botany. Vol. 2. Bryophytes and pteridophytes*, 2d ed. McGraw-Hill; New York; 1955.

Wherry, E. T., *The fern guide*. Doubleday; Garden City, NY; 1961.

Pl-5 Psilophyta

Banks, H. P., *Evolution and plants of the past*. Wadsworth; Belmont, CA; 1970.

Bierhorst, D. W., *Morphology of vascular plants*. Macmillan; New York; 1971.

Cooper-Driver, G., "Chemical evidence for separating the Psilotaceae from the Filicales." *Science* 198:1260–1262; 1977.

Hoshizaki, B. J., *Fern growers' manual*. Knopf; New York; 1975.

Raven, P., R. Evert, and S. Eichhorn, *Biology of plants*, 6th ed. Worth; New York; 1998.

Sporne, K. R., *Morphology of pteridophytes*, 4th ed. Hutchinson University Library; London; 1975.

Pl-6 Sphenophyta

Cobb, B., *A field guide to ferns*, 2d ed. Houghton Mifflin; Boston; 1984.

Hoshizaki, B. J., *Fern growers' manual*. Knopf; New York; 1975.

Smith, G. M., *Cryptogamic botany. Vol. 2. Bryophytes and pteridophytes*, 2d ed. McGraw-Hill; New York; 1955.
Wherry, E. T., *The fern guide*. Doubleday; Garden City, NY; 1961.

Pl-7 Filicinophyta

Brooklyn Botanical Garden, *Handbook on ferns, a special printing of plants and gardens 25(1)*. Brooklyn Botanical Garden; Brooklyn, NY; 1969.
Camus, J. M., A. C. Jermy, and B. A. Thomas, *A world of ferns*. Natural History Museum Publications; London; 1991.
Cobb, B., *A field guide to ferns*, 2d ed. Houghton Mifflin; Boston; 1984.
Foster, G., *Ferns to know and grow*. Timber Press; Portland, OR; 1993.
Hoshizaki, B. J., *Fern growers' manual*. Knopf; New York; 1975.
Lellinger, D. B., *A field manual of the ferns and fern-allies of the United States and Canada*. Smithsonian Institution Press; Washington, DC; 1985.
Pryer, K. M., A. R. Smith, and J. E. Skog, "Phylogenetic relationships of extant ferns based on evidence from morphology and the rbcL sequences." *American Fern Journal* 85:205–282; 1995.
Sporne, K. R., *Morphology of pteridophytes*, 4th ed. Hutchinson University Library; London; 1975.
Tryon, R. M., and A. F. Tryon, *Ferns and allied plants with special reference to tropical America*. Springer-Verlag; New York; 1982.

Pl-8 Cycadophyta

Bold, H. C., C. J. Alexopoulos, and T. Delevoryas, *Morphology of plants and fungi*, 5th ed. Harper College; New York; 1990.
Chamberlain, C. J., *Gymnosperms: Structure and evolution*. Dover; New York; 1966. (Reprint of 1935 edition.)
Dallimore, W., and A. B. Jackson, *A handbook of Coniferae and Ginkgoaceae*. 4th ed., rev. S. G. Harrison. St. Martin's; New York; 1966.
Jones, D. L., *Cycads of the world*. Smithsonian Institution Press; Washington, DC; 1993.
Norstog, K., "Cycads and the origin of insect pollination." *American Scientist* 75:270–279; May 1987.
Stevenson, D., ed., "Biology, structure, and systematics of the Cycadales." *Memoirs of the New York Botanical Garden*, Vol. 57. Symposium Cycad '87, Banyuls-sur-Mer, France, April 17–22. New York Botanical Gardens; New York; 1987.

Pl-9 Ginkgophyta

Chamberlain, C. J., *Gymnosperms: Structure and evolution*. (Reprint of 1935 edition.). Dover; New York; 1966.
Dallimore, W., and A. B. Jackson, *A handbook of Coniferae and Ginkgoaceae*, 4th ed., rev. S. G. Harrison. St. Martin's; New York; 1966.
Dirr, M. A., *Manual of woody landscape plants*, 4th ed. Stipes; Champaign, IL; 1990.
Gifford, E. M., and A. S. Foster, *Morphology and evolution of vascular plants*, 3rd ed. W.H. Freeman and Company; New York; 1988.
Hori, T., ed., "*Ginkgo biloba*—a global treasure—from biology to medicine." *Journal of Plant Research*. Springer-Verlag; Tokyo; 1997. Special issue.
Sargent, C. S., . (Reprint of 1922 edition; Houghton Mifflin; Boston.).*Manual of the trees of North America*, Vol. 1. Dover; New York; 1961.

Pl-10 Coniferophyta

Chamberlain, C. J., *Gymnosperms: Structure and evolution*. (Reprint of 1935 edition.). Dover; New York; 1966.

Dallimore, W., and A. B. Jackson, *A handbook of Coniferae and Ginkgoaceae,* 4th ed., rev. S. G. Harrison. St. Martin's; New York; 1966.
Denison, W. D., "Life in tall trees," *Scientific American* 228:74–80 June 1973. . High forest canopy community of old-growth Douglas fir: Animals, lichens, and nitrogen pathways.
Hartzell, H., *The yew tree.* Hulogosi Communications; Eugene, OR; 1991.
Nemecek, S., "Rescuing an endangered tree," *Scientific American* 274:22 March 1996. A Florida conifer, *Torreya taxifolia.*
Nicolaou, K. C., R. K. Guy, and P. Potier, "Taxoids: New weapon against cancer." *Scientific American* 274:94–98; June 1996.
Rushforth, K. D., *Conifers.* Facts on File; New York; 1987.

Pl-11 Gnetophyta

Beck, C. B., ed., *Origin and evolution of gymnosperms.* Columbia University Press; New York; 1988.
Bold, H. C., C. J. Alexopoulos, and T. Delevoryas, *Morphology of plants and fungi*, 4th ed. Harper and Row; New York; 1980.
Bornman, C. H., *Welwitschia: Paradox of a parched paradise.* Struik; Cape Town, South Africa; 1978.
Carmichael, J. S., and W. E. Friedman, "Double fertilization in *Gnetum gnemon* (Gnetaceae): Its bearing on the evolution of sexual reproduction within the Gnetales and the anthophyte clade." *American Journal of Botany* 83:767–780; 1996.
Carmichael, J. S., and W. E. Friedman, "Double fertilization in *Gnetum gnemon*: The relationship between the cell cycle and sexual reproduction." *Plant Cell* 7:1975–1988; 1995.
Chamberlain, C. J., *Gymnosperms: Structure and evolution.* (Reprint of 1935 edition.). Dover; New York; 1966.
International Journal of Plant Sciences 157(6); November 1996. (Supplement: *Biology and Evolution of the Gnetales: A Symposium.*)
Stewart, L., *Guide to palms and cycads.* Angus and Robertson; Sydney; 1994.

P1-12 Anthophyta

Bray, F., "Agriculture for developing nations," *Scientific American* 271:30–37 July 1994. Rice polyculture compared with monoculture.
Brockman, C. F., *Trees of North America.* Golden; New York; 1968.
Doyle, J. A., M. J. Donoghue, and E. A. Zimmer, "Integration of morphological and ribosomal RNA data on the origin of the angiosperms." *Annals of the Missouri Botanical Garden* 81: 419–450; 1994.
Friedman, W. E., "Double fertilization in *Ephedra,* a nonflowering plant: Its bearing on the origin of angiosperms." *Science* 247:951–954; 1990.
Friedman, W. E., "Evidence of a pre-angiosperm origin of endosperm: Implications for the evolution of flowering plants." *Science* 255: 336–339; 1992.
Friis, E. M., W. G. Chaloner, and P. R. Crane, *Origin of angiosperms and their biological consequences.* Cambridge University Press; Cambridge, UK; 1987.
Gleason, H. A., and A. Cronquist, *Manual of vascular plants of northeastern United States and adjacent Canada*, 2d ed. New York Botanical Garden; Bronx, NY; 1991.
Perry, D. R., "The canopy of the tropical rain forest." *Scientific American* 251:138–147; November 1984.
Rickett, H. W, ed. *Wildflowers of the United States*, 6 vols. New York Botanical Garden; Bronx, NY; 1970–1990.

Robacker, D. C., B. J. D. Meeuse, and E. H. Erickson, "Floral aroma." *BioScience* 38:390–396; 1988.

Swaminathan, M. S., "Rice," *Scientific American* 250:80–93; 1984. Compare with Bray, F., "Agriculture for developing nations." *Scientific American* 271:30–37; 1994.

Symonds, G. W., *Shrub identification book*. Morrow; New York; 1973.

Symonds, G. W., *Tree identification book*. Morrow; New York; 1973.

GENERAL GLOSSARY

A form (pl. **A forms**) In foraminifera with alternation of generations, the sexual organism, the gamont. In dinomastigotes (Pr-5), arrangement of thecal plates

AAA pathway Biosynthetic metabolic pathway forming the amino acid lysine. This pathway is characteristic of some protoctists and fungi and is entirely different from the diaminopimelic acid pathway of lysine biosynthesis, found in bacteria, other protoctists, and plants; see *Diaminopimelic acid pathway*

A-axis (pl. **A-axes**) Morphological term describing foraminiferal shells; the A-axis is the shortest axis of the hexagonal pattern. The C-axis is the longest axis; see *C-axis*

aboral Away from the oral opening

abyssal Ocean depths of 4,000 meters or more and the organisms that live at those depths

acanthopod (pl. **acanthopods**) Fine tapering pseudopod of *Acanthamoeba* (Pr-2)

acanthopodia See *acanthopod*

acanthopodium See *acanthopod*

acellular slime mold (pl. **acellular slime molds**) - Plasmodial slime molds (Pr-23); myxomycotes (that is, myxomycetes and protostelids)

acentric chromosome (pl. **acentric chromosomes**) Chromosome that lacks a centromere

acentric mitosis Mitosis that occurs in the absence of centrioles, centriolar plaques, kinetosomes, or other microtubule-organizing centers at the poles of the cell

acetabularian life history Stages in the development of chlorophytes belonging to the genus *Acetabularia* (Pr-28) and their class relatives

acrasin Chemical attractant secreted by dictyostelid amebas (Pr-2) that signals aggregation into a multicellular structure (for example, cyclic adenosine 3′,5′-monophosphate (cAMP) or glorin, a dipeptide)

acritarch (pl. **acritarchs**) Hollow, organic unidentified microfossil; may be spherical, ellipsoidal, or polygonal, smooth or granulated, or with spinose projections. Proterozoic to Recent. Probably eggs or protoctist cysts, but lacks sufficient morphological detail to be classified

acrobase (pl. **acrobases**) Groove or surface marking that extends anteriorly from the sulcus onto the epicone of unarmored dinomastigotes (Pr-5)

acrocentric chromosome (pl. **acrocentric chromosomes**) Chromosome with a terminal or nearly terminal centromere (kinetochore)

acronematic See *acroneme*

acroneme (pl. **acronemes**) Smooth undulipodium with a fine fibril at the distal end

actines Main branches of an ebridian skeleton, arising from the rhabde (longitudinal rod)

actinopod (pl. **actinopods**) Spine, or thin cell process, characteristic of heliozoans, acantharians, phaeodarians, and polycystines; underlain by microtubules. Used in feeding, locomotion, etc. Informal name of organism in phylum Actinopoda (Pr-31)

adelphoparasite (pl. **adelphoparasites**) Parasite (necrotroph) closely related to its host; shares family or lower taxon with its host; see *alloparasite*

adenosine triphosphate A primary energy carrier molecule for cell metabolism and motility

adhesive disk (pl. **adhesive disks**) Cup-shaped attachment of some protists; for example, in mastigotes and some spirotrichous ciliates (Pr-6), a thigmotactic cup-shaped organelle at the aboral end of the protist used for attachment to its substratum (usually the surface of a host organism). Organelle of the diplomonad *Giardia* (phylum Archaeprotista, Pr-1) that attaches *Giardia* to animal epithelium. Supported by complex cytoskeleton and delimited by a ridge, the lateral crest, it is composed of tubulin and a 30-kDa protein, giardin

adhesorium (pl. **adhesoria**) Adhesive organelle (for example, of plasmodiophorans, Pr-20)

adoral Toward the oral opening

aerobe Organism active and capable of completing its life cycle only in the presence of gaseous oxygen, O_2

aerophyte (pl. **aerophytes**) Air dweller; refers to algae and plants

aerophytic See *aerophyte*

aerotolerant Anaerobic, but not inhibited by low concentrations of O_2

aethalia One of three types of sporophores in myxomycotes (Pr-23); large syncitial sporocarp of certain plasmodial slime molds; see *plasmodiocarp, sporangium*

aflagellate (pl. **aflagellates**) Amastigote; protoctist cell lacking undulipodia, either for the entire life cycle or for the stages of the life cycle; nonmotile cell; also, a specific morphological stage in the life cycle of Trypanosomatidae (Pr-11), which is rounded, lacks external undulipodia, but has a prominent kinetoplast and a short, internal undulipodium

agamogony Series of nuclear or cell divisions producing individuals that are neither gametes nor capable of forming gametes

agamont Reproducing organism at a stage in its life cycle during which it lacks gametes or other sexual structures (for example, foraminifera, Pr-3, schizonts of apicomplexans, Pr-7)

agar Type of phycocolloid; sulfated carbohydrate composed of β1,3-linked D-galactose and α1,4-linked anhydro-l-galactose extractable from cell walls and intercellular spaces of the rhodophytes *Gelidium* and *Gracilaria* (Pr-33). Resistance to digestion and transparency make agar (to which nutrients are added) an ideal matrix upon which to grow microbes

agglutinated test (pl. **agglutinated tests**) Glued test; covering or shell produced by protoctist from sediment particles including tests of other organisms, usually with organic lining (for example, foraminifera)

aggregation center (pl. **aggregation centers**) Structure of dictyostelid cellular slime molds formed by the coming together of hundreds of amebas

agnotobiotic culture (pl. **agnotobiotic cultures**) Mixed culture; pertaining to a heterogeneous culture in which the microbiota is unidentified; see *gnotobiotic*

agnotoxenic Pertaining to a culture medium contaminated by one or more unknown organisms; see *monoxenic, polyxenic*

akinete (pl. **akinetes**) Type of propagule; nonmotile single or few-celled structure formed by thickening of the cell wall of a growing cell; capable of passive propagation and germination, usually of cyanobacteria (B-6) or algae (for example, conjugating green algae, xanthophytes, Pr-32, Pr-16)

akont (pl. **akonts**) See *aflagellate*

algae Photoautotrophic protoctists; all oxygenic phototrophs exclusive of cyanobacteria, chloroxybacteria, and plants; ecological term for aquatic oxygenic phototroph

alginate Salt of alginic acid, produced in walls of phaeophytes; a polysaccharide with β-1,4-linked D-mannuronic acid and β-1,4-linked L-guluronic acid in varying ratios

algivore Mode of nutrition; referring to organisms that feed on algae (Table 2)

algivorous See *algivore*
algivory See *algivore*
algology Science of the study of algae
allelochemic Ecological term referring to chemical substances (secondary metabolites) which, when released into the surroundings of organisms, influence the behavior or development of other individuals of different species; see *antibiotic, pheromone, secondary metabolite*
allelochemicals See *allelochemic*
allometric transformation (pl. **allometric transformations**) Growth in three dimensions that can be described by simple rules (for example, in protostelids)
alloparasite (pl. **alloparasites**) Parasite (necrotroph) not closely related to its host; that is, does not share family or lower taxon with its host; see *adelphoparasite*
allophycocyanin Type of phycobiliprotein; water-soluble extract is blue; found in cyanobacteria, rhodophytes, and cryptomonads; see *phycocyanin, phycoerythrin*
allosteric Describing the change in some enzymes whereby a small molecule combines with a site on the protein (other than the active site) resulting in a change in catalytic activity via a change in protein conformation
allozyme (pl. **allozymes**) Alternative enzyme forms encoded by different alleles at the same genetic locus
aloricate Lacking a lorica; see *lorica*
alpha aminoadipic acid pathway See *AAA pathway*
alpha spore (pl. **alpha spores**) Spore of rhodophytes, typically diploid, released from a carposporangium
alpine Characteristic or descriptive of the mountainous regions lying between timber and snow line
alternation of generations Description of life cycles of plants and protoctists usually in which haploid (1N, gametophyte) generation alternates with diploid (2N, sporophyte) generation; the haploid and diploid organisms may be morphologically identical or extremely different. Also refers to alternation of morphological types in a given life cycle even when there is no change in ploidy (for example, hydroid–medusoid transformation in coelenterate animals, A-4)
alveolus (pl. **alveoli**) Small cavity or pit (for example, bubblelike cytoplasmic compartments filled with either fluid or gas, and often forming a soap bubblelike frothy layer around certain spumellarian radiolaria (Pr-31); cavities in the cortex of ciliates (Pr-6) or valves of diatoms (Pr-18); pellicular alveoli enclose the thecal plates in the armored dinomastigotes (Pr-5))
amastigote (pl. **amastigotes**) See *aflagellate*
ameba (pl. **amebas**) Member of the phylum Rhizopoda; also refers to stages in the life cycles of other organisms that move by means of pseudopods; descriptive term for habit, that is, movement by pseudopod formation and protoplasmic streaming (for example, in dictyostelids, Pr-2)
ameboid See *ameba*
amoebomastigote (pl. **amoebomastigotes**) Amebas that undergo transformation to mastigote stage. Informal name of members of the class Amebomastigota, Pr-22
amicronucleate Lacking micronuclei (for example, in ciliates)
amitosis Cell divisions of eukaryotes that lack chromosome changes typical of mitosis
amitotic See *amitosis*
amoeba (pl. **amoebas**) See *ameba*
amoeboid See *ameba*
amphibious The ability to live both on land and in water
amphiesma Outer, peripheral complex of dinomastigotes, consisting of a cell membrane, a single layer of (amphiesmal) vesicles, trichocysts, and sometimes a pellicle

amphiesmal vesicles Vesicles directly under the amphiesma of dinomastigotes, thought to be responsible for test production

ampulla (pl. **ampullae**) Accessory branch systems, usually congested in appearance (for example, in rhodophytes, ciliates Pr-33, -6)

amylopectin Storage polysaccharide of algae composed of β-1,4 glucoside linkages, with β-1,6-linked side chains. Can be detected by its red-purple staining when treated with iodide–potassium iodine (Lugol's) solution

anabiosis Reviving; restoring to active metabolism and growth from a deathlike or suspended condition; resuscitation from dry or frozen state; see *cryptobiosis*

anadromous Organisms that normally live in a marine environment, but mate in freshwater (for example, salmon)

anaerobe Organism active and capable of completing its life cycle in the absence of gaseous O_2

anaerobic See *anaerobe*

analogous Of macromolecules, structures or behaviors that have evolved convergently; similar in function but different in evolutionary origin

anano Planktonic protists in the 2–20 μm size range that lack plastids

anaphase Stage in mitosis in which chromatids separate (segregate) at their kinetochores and move toward opposite poles; see *mitosis*

anastomose The process of linking branches, filaments, or tubes by fusion to form networks

anastomosis See *anastomose*

anchoring disk (pl. **anchoring disks**) Structure of microsporan (F-1) spores that develops from a vesicle and into which the base of the polar tube is inserted (for example, *Nosemoides vivieri*)

aneuploid Possessing a number of chromosomes that is not an exact multiple of the typical haploid set for the species; bearing translocations or other chromosome abnormalities; see *euploid*

animal (pl. **animals**) Multicellular, diploid organism that develops from a blastula, product of fertilization of eggs and sperm, generally heterotrophic

anisofilar Description of a filamentous structure that, in the everted state, is composed of stretches of markedly unequal width along its length (for example, microsporan (F-1) polar tube, myxosporan polar filaments (A-2)); see *isofilar*

anisogametes Gametes of a given species that differ in size or form; see *isogametes*

anisogametous See *anisogametes*

anisogamontous See *anisogamonts*

anisogamonts Gamonts of a given species that differ in size or form

anisogamous Pairing of gametes that differ in size or form (anisogametes); see *isogamy*

anisogamy See *anisogamous*

anisokont A cell with undulipodia (or other motile organelles) unequal in length or unlike in movement or form (for example, usually one with short mastigonemes and the other smooth, lacking mastigonemes); see *isokont*

anisoplanogametes Motile gametes (swarmers) of different sizes

anlage Primordium; the first recognizable part of a developing organ in an embryo or organelle in a cell (for example, immature ciliate macronucleus or suctorian tentacle, Pr-6)

anlagen See *anlage*

annular Structure or part resembling a ring (for example, the central area on the valve face of some centric diatoms); general term for a girdle or an equatorial belt, band, or groove

annulus See *annular*

anoxia Oxygen deficiency; lack of gaseous oxygen

anoxic See *anoxia*

anoxic layer Layer of water or air in which molecular oxygen (O_2) is absent

antapical Opposite to or on the other side from the apex or tip

anterior flagellum (pl. **anterior flagella**) Forwardly directed undulipodium

anterior undulipodium (pl. **anterior undulipodia**) See *anterior flagellum*

antheridium (pl. **antheridia**) Male sex organ; sperm-producing gametangium

antherozoid Male motile gamete; undulipodiated sperm

antibiotic Substance produced and released typically by bacteria or fungi that injures or prevents growth of organisms belonging to a different species. Kind of allelochemical

antibody Protein produced by vertebrate blood cells capable of defending an animal against a specific foreign substance (antigen)

antigen Foreign material that, upon introduction into a vertebrate animal, stimulates antibody production

antigenic determinant Part of a molecule recognizable by an antibody (for example, part of an amino acid residue or a few amino acids residues in a protein); epitope; the portion of a macromolecule to which an antibody binds

antigenic variation Change in the surface antigen-type expressed. A process that enables necrotrophs to evade the host's immune response (for example, in trypanosomes)

antiplectic Synchronous waves of movement of adjacent undulipodia resulting from a tight viscous–mechanical coupling (hydrodynamic linkage). Symplectic metachronal waves refer to those which pass over the field of undulipodia in the same direction as the effective stroke of the beat. Antiplectic metachronal waves are those which pass in a direction more or less opposite to the effective stroke

APC Complex of openings at the apex of a structure or organism (for example, in dinomastigotes, Pr-5)

aperture (pl. **apertures**) Opening; for example, the major openings to the exterior through which the cytoplasm extends in foraminiferal tests (Pr-3); usually larger than the pores

apex Pertaining to top, tip

aphanoplasmodium (pl. **aphanoplasmodia**) One of three types of plasmodium formed by myxomycotes (Pr-23). Intermediate between phaneroplasmodia and protoplasmodia in size and complexity. Formed by most members of the subclass Stemonitomycetidae; see *phaneroplasmodium*, *protoplasmodium*

aphotic zone Region of water where light does not penetrate

aphytal Ecological term for nonphotic, benthic zones in aquatic environments

apical See *apex*

apical complex Structure at the apex of members of the phylum Apicomplexa (Pr-7) generally consisting of two apical conoidal rings, a conoid, and a polar ring, to which subpellicular microtubules and electron-dense membrane-bound organelles composed of rhoptries and micronemes are attached. The name of the phylum Apicomplexa is derived from this structure, which facilitates attachment and penetration of the protoctist to its host cell

apical conoidal ring (pl. **apical conoidal rings**) Cone-shaped structure at the anterior end of an apicomplexan cell that is part of the apical complex (Pr-7)

apical depression (pl. **apical depressions**) Depression in the anterior portion of a cell (for example, the epicone of a dinomastigote, Pr-5)

apical growth Growth at the tip or apex

apical pore complex See *APC*

aplanospore (pl. **aplanospores**) Nonundulipodiated spore; a nonsexual, nonmotile propagule

aplastidic nanoplankton See *anano*

aplerotic Descriptive of oogenesis in oomycote oospores in which the oospore clearly does not fill the oogonial cavity; see *plerotic*

apoagamy Development of organism without fusion of gametes; development of a diploid phase from a haploid phase without fertilization in organisms with sexual ancestors

apochlorotic The lack of photosynthetic pigments in organisms or cells that once contained them or whose ancestors once contained them; usually refers to algae

apogamic See *apoagamy*

apogamous See *apoagamy*

apogamy See *apoagamy*

apomeiosis Nuclear division without meiosis in a cell that generally divides by meiosis

apomictic Altered meiosis or fertilization such that mixis is bypassed (for example, parthenogenesis); condition of being formerly sexual; see *mixis*

apomixis See *apomictic*

apophysis Swollen region

aposeme Seme identified as an altered form of an earlier seme; see *seme*

aposymbiotic Condition of lacking symbionts in formerly symbiotic organisms

appressorium Swollen hyphal tip used to penetrate other organisms (usually plants) by means of attachment, building of turgor pressure, and growth of a thin hyphal peg into the host organism

approsorium Specialized cell structure that functions in the penetration of a host cell wall and presumably in the uptake of nutrients in the host. In plasmodiophorans (Pr-20), it arises at the end of a short germ tube that is formed by an encysted zoospore

aragonite A mineral, such as calcite, composed of calcium carbonate ($CaCO_3$), but differing from calcite in having orthorhombic crystallization, greater density, and less distinct cleavage

aragonitic See *aragonite*

arbusculate Having the form of a bush or tree (for example, some protoctists, fungi in mycorrhizal associations, F-3)

archegoniate Female sex organ; the egg- or oogonium-producing gametangium characteristic of some plants and protoctists

archegonium (pl. **archegonia**) See *archegoniate*

archeoplasmic spheres Proteinaceous structures visible as dots by light microscopy and resolvable by electron microscopy as spherical organelles from which spindle microtubules seem to emerge. Associated with the rostrum of hypermastigotes (Pr-1)

archeopyle Opening or rupture commonly observed in cysts of dinomastigotes (Pr-5) and their microfossils; its position is of taxonomic significance

arenaceous test (pl. **arenaceous tests**) Test or outer covering composed of sand grains bound together by organic cement (for example, of testate amebas, Pr-2)

areola (pl. **areolae**) Small area between or about structures (for example, around a vesicle), especially a colored ring; the regularly repeated perforation through the siliceous layer of a diatom frustule; spore-connection scar (hilum) at their points of contact with each other (for example, in acrasids, Pr-2); see *hilum*

argentophilic Part of cell or tissue that stains black with silver stains

argillaceous Claylike in texture or structure

armored plate (pl. **armored plates**) Latitudinal series of articulated plates that make up the dinomastigote theca

articulate Segmented or jointed in appearance, bearing joined segments

asexual reproduction Increase in number of individuals in the absence of conjugation, fertilization, or any other sexual process

assemblage Group of relatively homogeneous organisms; a group of fossils that occurs at the same stratigraphic level

assimilatory hair (pl. **assimilatory hairs**) Filaments or rows of cells capable of assimilating nutrient materials from hosts or the environment (for example, in the phaeophyte order Cutleriales, Pr-17)

aster (pl. **asters**) Stellate, polar, paired structures of animal eggs and other mitotically dividing cells; conspicuous but ephemeral star-shaped microtubule-organizing centers usually at distal ends of the mitotic spindle

astral microtubule (pl. **astral microtubules**) Microtubules that arise from a microtubule-organizing center (aster) at the spindle poles

astropyle (pl. **astropylae**) Main opening of the central capsule of phaeodarian actinopods, Pr-31; usually accompanied by two or more secondary openings, the parapylae

athecate Lacking a theca or covering; see *theca*

ATP See *adenosine triphosphate*

atractophore (pl. **atractophores**) Fibrillar rod-like structure arising from the kinetosome that is or serves the role of centriole or centrocone in the formation of the mitotic spindle (for example, in some trichomonad mastigotes, foraminiferans, and radiolarian actinopods, Pr-1, -3, -31)

attenuate (**v.**) Becoming noticeably reduced (for example, in diameter, narrowing to a point; in size, quantity, strength, force, or severity); especially of light intensity

A-tubule (pl. **A-tubules**) Tubule (subfiber) of doublet of axoneme of undulipodium; the innermost tubule of axoneme doublets to which dynein arms are attached or the microtubule comprising the wall of the complete tubule of centriolar triplets is called the A-tubule

Aufwuchs community Periphyton. Interacting microorganisms on rocks, plants, and other surfaces on the bottoms of streams and lakes; communities of organisms surrounding submerged vegetation or roots of vegetation in shallow freshwater environments; see *microbenthos*

autecological Pertaining to ecological studies dealing with a single species and its relationship to the biological and physicochemical aspects of its environment

autecology See *autecological*

autocolony (pl. **autocolonies**) Offspring colonies, miniatures of parent colonies, formed by reproduction (multiple fission) of parental colonies (usually in algae, for example, the chlorophyte *Pediastrum*, Pr-28)

autogamous Self-fertilization; a type of karyogamy characterized by the union of two nuclei both derived from a single parent nucleus

autogamy See *autogamous*

autolysin Substance that enzymatically degrades glycoprotein-type cell walls

autopoiesis Self-maintenance; set of principles defining life and pertaining to membrane-bounded, self-limited, internally organized systems that dynamically maintain their identity in the face of external fluctuations and limitations. Autopoietic entities have the capability to continually replace and repair their constituent parts, ultimately, at the expense of solar energy; see *metabolism*

autopoietic See *autopoiesis*

autospore (pl. **autospores**) Offspring cell produced within parental cell wall that resembles parent cell at the time of release except that it is smaller (typical of chlorophytes of the genus *Chlorella*, Pr-28)

autotroph Mode of nutrition; pertaining to organisms that synthesize organic compounds using an inorganic source of carbon (for example, carbon dioxide). Strict autotrophs derive energy and electrons also from sources other than organic compounds, that is, from sunlight or from the oxidation of hydrogen, ammonia, or other inorganic compounds (Table 1); see *biotrophy*, *heterotrophy*, *saprotrophy*

autotrophic See *autotroph*

autotrophy See *autotroph*

auxiliary cell (pl. **auxiliary cells**) Cell in rhodophytes (Pr-33) to which the diploid zygote is transferred, and from which the sporophyte is generated

auxiliary zoospore (pl. **auxiliary zoospores**) Zoospore within the zoosporangium bearing apically or subapically inserted undulipodia that are retracted upon encystment; characteristic of some members of the Saprolegniaceae (phylum Oomycota, Pr-21); see *principal zoospore*

auxospore (pl. **auxospores**) Diatom cell released from its rigid siliceous test; often the zygotic product of fertilization

auxotrophic mutant (pl. **auxotrophic mutants**) Microorganism capable of growth only when minimal medium is supplemented with a specific substance (for example, vitamin, amino acid) not required for growth of wildtype strains

axenic Pure, lacking strangers; especially a culture containing only a single identified strain or species of organism

axil (pl. **axils**) The angle between the upper surface of a leaf (or thallus of an alga) and the stem (or main branch) that bears it

axonemal dense substance (pl. **axonemal dense substances**) Axosome. Electron-opaque material in which microtubules are embedded at the base of the axonemes of undulipodia or in the outer sheet of the centrosphere in actinopods

axoneme (pl. **axonemes**) Microtubular axis or shaft, exclusive of the covering membrane, extending the length of an undulipodium (cilium, flagellum, or axopod) composed of the [9(2)+2] arrangement of microtubules. Each of the nine doublets is comprised of a complete A-tubule and an incomplete B-tubule; see *microtubular rod*

axoplast (pl. **axoplasts**) Central granule. In actinopods, microtubule-organizing center from which axonemes of axopods arise; devoid of inner differentiation; see *centroplast*

axopod (pl. **axopods**) Cell process stiffened by a microtubular shaft or axoneme; characteristic of actinopods; used primarily in feeding but also in walking by the heliozoan *Sticholonche zanclea*, Pr-31

axopodium (pl. **axopodia**) See *axopod*

axosome (pl. **axosomes**) Electron-opaque fuzzy structure at the base of the central tubules of an undulipodium; see *axonemal dense substance*

axostylar cap (pl. **axostylar caps**) Proteinaceous material covering anterior end of axostyle

axostyle (pl. **axostyles**) Axial motile organelle of metamonads (pyrsonymphids) and parabasalians composed of a patterned array of microtubules and their cross-bridges that runs from the apical end to (and sometimes through) the posterior pole of the organism

azygospore (pl. **azygospores**) Parthenogenetically produced zygospore; characteristic of endomycorrhizal symbionts

B form (pl. **B forms**) In foraminifera (Pr-3) with alternation of generations, the asexual organism, the agamont

backing membrane (pl. **backing membranes**) Part of the endoplasmic reticulum in the blastocladialean (Pr-34) zoospores that extends partway around the side body complex

bacterium (pl. **bacteria**) Microorganism with prokaryotic cell organization

bacterized medium Nutritional fluid or agar containing bacteria (living or dead) as a food source

bactivore Mode of nutrition; referring to organisms that feed on bacteria (Table 2)

bactivorous See *bactivore*

bactivory See *bactivore*

ballistospore (pl. **ballistospores**) Propagule (spore) that is violently discharged for long distances (up to several meters) from its point of origin (for example, in protostelid plasmodial slime molds, Pr-23)

banded root (pl. **banded roots**) Periodically striated, longitudinally orientated, subpellicular fiber (or component fibrils) arising close to the base of a somatic kinetosome (posterior one, if paired), near its microtubular triplets 5–8 and extending anteriorly toward or parallel to the organism's pellicular surface and always on the right side of its kinety ("law of desmodexy"). Structure is diagnostic for ciliates (Pr-6). Striated fibers showing exception to these characteristics and orientation are not true kinetodesmata. Kinetodesmata of a length greater than the interkinetosomal distance along the kinety overlap, shingle-fashion, producing a bundle of fibers. These are well developed in apostome, hymenostome, and scuticociliate ciliates. They are present as large and heavy bundles in certain astomous ciliates; see *kinetodesma*

banded root fiber system (pl. **banded root fiber systems**) See *banded root*

basal apparatus (pl. **basal apparatuses**) Kinetosomes and their associated tubules and fibers present in all undulipodiated cells; unit of organization of the ciliate cortex. The functional organellar complex, including undulipodia, is usually responsible for locomotion. Synonyms include basal apparatus, flagellar apparatus, flagellar root system, proboscis root, root fiber system, undulipodial apparatus, kinetosomal territory, and ciliary corpuscle. Kinetids always consist of at least one kinetosome, but may have pairs or occasionally more than two kinetosomes (for example, they may be dikinetids or polykinetids). Structures associated with the kinetosomes of ciliates usually include cilia, unit membranes, alveoli, kinetodesmata, and various ribbons, bands, or bundles of microtubules (for example, postciliary microtubules and some nematodesmata). Root microtubules of kinetids may be laterally associated microtubules that originate at kinetosomes in definite numbers and follow a defined path within the cell (for example, ciliates). Some kinetids are also comprised of microfibrils, myonemes, parasomal sacs, mucocysts, or trichocysts. Details of the kinetid are essential for taxonomic and evolutionary studies of motile protoctists; see *kinety, monokinetid, oral kinetid, polykinetid, somatic kinetid*

basal body (pl. **basal bodies**) Kinetosome; intracellular organelle not membrane-bounded, characteristic of mastigotes and all other undulipodiated cells. Microtubule structures, cylinders about 0.25 μm in diameter and up to 4 μm long. Their microtubules are organized in the [9(3)+0] array; all undulipodia are underlain by kinetosomes. These basal organelles are necessary for the formation of all undulipodia; kinetosomes differ from centrioles (which share the cross section characteristic of a circle of nine triplets of microtubules) in that from them extend [9(2)+2] axonemes. The term kinetosome, because of its precision, is preferable to basal body

basal disk (pl. **basal disks**) Any plate-shaped structure at the base of a cell process; see *adhesive disk*

basal plate (pl. **basal plates**) Electron-dense, platelike kinetid component positioned at the proximal end of and perpendicular to the kinetosome

basal swelling (pl. **basal swellings**) Enlargement of volume at the base of a structure, often applied to peduncles, undulipodia, or other vertically extended structures

base-plate scale (pl. **base-plate scales**) Base of spined scale on an organic portion of a coccolith

basipetal Proceeding from the apex toward the base

basipetal development Process in which sporangia are made in basipetal sequence from an undifferentiated hypha terminated by a sporangium (for example, oomycotes, Pr-21); see *determinate sporangium, percurrent development*

basiphyte (pl. **basiphytes**) Plant on which an epibiont or epiphyte lives

bathyal Upper part of an aphotic benthic zone, generally the continental slope, at depths from 1000 to 3000 meters, in which algae and plants are excluded because solar radiation cannot penetrate

bathyl See *bathyal*

benthic Community of organisms near the bottom or attached to the bottom of an ocean, sea, lake, or other aquatic environment

benthonic See *benthic*

benthos See *benthic*

beta spore (pl. **beta spores**) Small, colorless spermatia in sexual species of the rhodophyte *Porphyra*, Pr-33

biflagellate Cell possessing two undulipodia, one of which may be nonemergent; adjective referring to such a cell; see *nonemergent flagellum*

bifurcate Having two branches or peaks; forked

bifurcation See *bifurcate*

biliprotein (pl. **biliproteins**) Complex of phycobilins with protein found in cyanobacteria (B-6), rhodophytes (Pr-33), glaucocystophytes, and some cryptophytes (Pr-26); see *phycobilins*, *phycocyanin*

biloculine Describing a foraminiferan test in which each chamber is added to the previous chamber so that only two final chambers are externally visible

bimastigote See *biflagellate*

binary fission Mode of reproduction; division of parent prokaryotic or eukaryotic cell into two roughly equal-sized offspring cells

binucleate Containing two nuclei

bioassay Determination of an unknown concentration of a substance, such as a drug, by comparing its effect on a test organism with the effect of a known standardized concentration

biogenic Pertaining to a structure (for example, stromatolite), substance (for example, amino acid), or pattern (for example, laminated sediment) produced by organisms

bioluminescence Emission of light by living organisms (for example, some marine dinomastigotes, Pr-5)

biomass (pl. **biomasses**) Total weight of all organisms at a given time in a particular area, volume, or habitat, generally expressed in units such as grams/square meter, pounds/acre, or kilograms/hectare

biomineralization Formation of minerals by living organisms. Two kinds are known (1) biologically controlled or matrix-mediated biomineralization, that is, intracellular precipitation of a given mineral type under genetic control of the cell (magnetite in magnetotactic bacteria, calcite by Coleps or coccolithophorids, Pr-25) and (2) biologically induced biomineralization, that is, production of acid, which changes local pH or other environmental alterations that in turn causes potentially mineralizable material to precipitate (for example, extracellular precipitation of iron and manganese oxides by *Leptothrix*, *Bacillus*, or other bacteria; precipitation of amorphous calcium carbonates in lakes due to algal activity); see *mineral*

biostratigraphy Study of the geological arrangement of sedimentary layers (strata), or the origin, composition, distribution, and succession of strata that contain fossils or remnants of fossils. Biostratigraphy, which especially employs fossil foraminifera and coccolithophorids (Pr-3, -25), is exceedingly important in relative dating, reconstruction of environments of deposition, and thus in petroleum exploration

biota Sum of animals (fauna), plants (flora), and microbiota on Earth. The term microbiota is preferable to microfauna (for example, in reference to intestinal symbionts, ciliates, motile bacteria) or microflora (for example, for bacteria)

biotope (pl. **biotopes**) Environment surrounding a community of organisms

biotroph Mode of nutrition; pertaining to organisms that derive carbon and energy from living food sources (Table 2); for example, many symbionts and pathogens are biotrophs; see *autotrophy*, *saprotrophy*, *symbiotrophy*

biotrophic See *biotroph*

biotrophy See *biotroph*

biozone (pl. **biozones**) Biostratigraphic unit; biochron; range zone (for example, sedimentary rock deposits formed during the life span of a certain fossil form); rocks identified by the occurrence of a specific kind of fossil in them; valuable for establishing intercontinental geological correlations

biphasic medium Culture medium that has two phases (for example, agar overlain by a liquid medium); see *culture medium*

bipolar body (pl. **bipolar bodies**) Xenosomes or organelles found in the cytoplasm of kinetoplastid mastigotes; in bodonids, they appear as encapsulated gram-negative bacteria; in trypanosomatids, they appear to be derived from Gram-negative bacteria that have lost their characteristic cell walls

biraphid Having a raphe running along the apical axis on both the epivalve and hypovalve; descriptive of diatoms

birefringence Splitting of a light beam into two components, which travel at different velocities. The principle of birefringence is employed in differential interference, polarizing, and phase-contrast microscopy

biserial Organized in two rows or series (for example, foraminiferal (Pr-3) test with this organization)

biseriate See *biserial*

bisporal Pertaining to structure or organism making two kinds of spores

bisporangium (pl. **bisporangia**) Sporangium, the contents of which divide to form two spores

bispore See *biosporal*

bisporic generation (pl. **bisporic generations**) Two-spored generation; a generation marked by the production of two types of spores

bladder (pl. **bladders**) Saclike or vesicular structure

blade (pl. **blades**) Flat part of algal thallus (for example, of kelp or other foliaceous algae)

blastular embryo (pl. **blastular embryos**) Diploid product of fertilization that forms a hollow ball (blastula); defining characteristic of members of the animal kingdom

bleached mutant (pl. **bleached mutants**) Altered photosynthetic organism (for example, *Euglena gracilis*, Pr-12) that has permanently lost its chloroplasts and accompanying plastid deoxyribonucleic acid

blepharoplast (pl. **blepharoplasts**) Term for kinetosome or other conspicuous microtubule-organizing center involved in cell division as determined by light microscopic observations of live cells (for example, in *Stephanopogon*, other mastigotes, cycads, and ferns)

bloom (pl. **blooms**) Dense growth of a population in aqueous media, aquaria, or nature; characteristic of certain species of planktonic algae, dinomastigotes, ciliates; often detected by discoloration of water; usually self-limiting and of short duration

blue-green algae Cyanobacteria (B-6); Cyanophyceae. The terms Cyanophyceae, cyanophytes, and blue-green algae have been replaced by the term cyanobacteria, which recognizes the fundamental bacterial (prokaryotic) nature of these organisms

boreal Northern; pertaining to the forest areas and tundras of the northern temperate zone and arctic region

bothrosome (pl. **bothrosomes**) Organelle on cell surface limited to phylum Labyrinthulata (Pr-19; labyrinthulids and thraustochytrids) from which the ectoplasmic network arises. In the sagenogen, an electron-dense plug separates the cell cytoplasm from the matrix of the ectoplasmic network

brackish Water with a salinity intermediate between that of seawater (3.4%) and of standard freshwater

bradyzoites Zoites in latent phase; slowly developing merozoites of apicomplexans (Pr-7); see *merozoite, zoite*

brevetoxin complex Fish toxin produced by the dinomastigote *Ptychodiscus (Gymnodinium) brevis* (Pr-5)

brine Seawater that, due to evaporation or freezing, contains dissolved salts in concentrations higher than 3.4%

brittleworts Calcareous charophytes, diatoms (Pr-18); obsolete name

B-tubule (pl. **B-tubules**) Tubule (subfiber) of doublet of axoneme of undulipodium; the outermost microtubule of axoneme doublets or the central tubule of centriolar triplets is called the B-tubule

buccal cavity (pl. **buccal cavities**) Ingestion apparatus; mouth; oral apparatus; peristome. Pouch or depression toward the apical end of the cell and/or on the ventral side containing compound ciliary organelles that lead to the cytopharyngeal/cytostomal area (for example, ciliates)

budding Mode of reproduction by outgrowth of a protrusion (one or more buds) smaller than the parental cell or body that only slowly reaches parental size; see *exogenesis*

bulbil (pl. **bulbils**) Asexual reproductive organ that forms on the rhizoids of some species of charophytes (phylum Chlorophyta, Pr-28), appearing as a white star or sphere

bulla (pl. **bullae**) Blisterlike structure or large vesicle. In foraminifera, it may partially or completely cover the primary or secondary aperture(s)

caducous Becoming detached; falling off prematurely, used originally for floral organs; see *deciduous*

calcareous Containing calcium, usually in the form of $CaCO_3$

calcite Mineral made of calcium carbonate ($CaCO_3$), crystallized in hexagonal form; the major component of common limestone, chalk, and marble; the material from which foraminiferal tests and coccoliths are composed; see *aragonite*

calyptrolith (pl. **calyptroliths**) Callote (skull-cap)-shaped coccolith; holococcolith having the form of an open cap or basket (for example, in *Sphaerocalypta* and *Calyptosphaera*, Pr-25)

canal (pl. **canals**) Channel-shaped or tubular structure; in euglenoids, the tubelike feature connecting the reservoir or anterior invagination to the outside, open only at its anterior end; see *furrow*

canal raphe (pl. **canal raphae**) Bars running at intervals beneath the raphe on the inside of the valve in diatoms

canaliculate Channeled or grooved longitudinally

cancellate Chambered; reticulate

caneolith (pl. **caneoliths**) Elliptical discoid heterococcolith with petal-shaped upper and lower rims and a central area filled with slatlike elements (for example, *Syracosphera*, Pr-25)

cap (pl. **caps**) Reproductive structure in life cycle stage of acetabularians (Pr-28) that becomes filled with nuclei

cap ray (pl. **cap rays**) Chamber in acetabularian reproductive structure (Pr-28)

capillitium (pl. **capillitia**) Anterior part of a myxomycote sporophore that consists of non-protoplasmic threadlike structures

capitulum (pl. **capitula**) Cell in the antheridium of charophytes (Pr-28) from which the antheridial filaments arise; amorphous material capping proximal ends of nematodesmata in some hypostome ciliates

capsalean A loose grouping of spherical cells of colonial coccoid cyanobacteria (B-6) or algae

capsular wall (pl. **capsular walls**) Walls of spherical or nearly spherical structures. In acantharian actinopods (Pr-31), a perforated, fibrillar cover that limits the endoplasm and through which ectoplasm is emitted; in myxozoan spores, wall of the polar capsule consisting of two layers, the inner is electron-lucent and alkaline-hydrolysis resistant, whereas the outer layer is electron-dense and proteinaceous

capsule (pl. **capsules**) Apical, thick-walled vesicle of a myxosporan spore (A-2; one to seven per spore) containing spirally coiled, extrusible polar filament; in heliozoan actinopods (Pr-31), regions of dense cytoplasm at opposite sides of the nucleus during mitosis

capsulogenesis Process of formation of capsulogenic cell that gives rise to multicellular capsule in myxospora; see *polar capsule*

capsulogenic cell (pl. **capsulogenic cells**) In myxosporan sporoblasts, the cell that produces the polar capsule in its cytoplasm

carbon fixation Uptake and conversion of carbon dioxide (CO_2) into organic compounds

carina (pl. **carinas or carinae**) Keel-shaped structure or process (for example, foraminiferan test, Pr-3)

carinal See *carina*

carinal band (pl. **carinal bands**) Foraminiferal shell (Pr-3) having a keel or flange at the margin

carinate See *carinal band*

carnivore An organisms that are heterotrophic and often predatory (for example, ciliates (Pr-6) and other protoctists that feed on zoomastigotes or metazoa); generally refers to a holozoic and predatory, rather than necrotrophic or histophagous, mode of nutrition; see *osmotrophy*, *phagotrophy*

carnivorous See *carnivore*

carnivory The carnivorous mode of nutrition

carotenoids Generally yellow, orange or red isoprenoid (C40) pigments (for example, carotene, fucoxanthin) found in the plastids (and often the cytoplasm) of virtually all phototrophic organisms as part of their photosynthetic apparatus and in many heterotrophs

carpogonial Female gametangium in rhodophytes (Pr-33); the flask-shaped egg-bearing portion of the female reproductive branch; a carpospore-containing oogonium, usually with a trichogyne

carpogonium (pl. **carpogonia**) See *carpogonial*

carposporangium (pl. **carposporangia**) Sporangium derived directly or indirectly from the zygote nucleus produced in the carposporophyte generation in rhodophytes (Pr-33). Can release diploid carpospores (products of mitosis) or haploid carpotetraspores (products of meiosis)

carpospore (pl. **carpospores**) See *alpha spore*

carposporophyte (pl. **carposporophytes**) Diploid red algal organism (Pr-33) produced after fertilization, a phase characterized by the presence of carposporangia (that is, composed of gonimoblast filaments bearing carpospores in florideophycidean rhodophytes)

carpotetraspores Meiotic products formed in carposporangia; carpotetraspores germinate to give rise to gametophyte thalli in some rhodophytes (Pr-33)

carrageenan Sulfated polymer of a-1,3- and §-1, 4-linked D-galactopyranose units; type of phycocolloid produced by some rhodophytes (Pr-33) and marketed commercially for the production of ice cream and other products

cartwheel structure (pl. **cartwheel structures**) Portion of kinetosome; refers to the appearance in the ultrastructural cross section of the microtubular wheel, radial spokes, axle, and dynein arms

catenate Description of cells or other structures arranged end-to-end like beads in a chain

caudal appendage (pl. **caudal appendages**) Tail-end structure (for example, caudal cilium, caudal undulipodium); in ciliates (Pr-6), distinctly longer somatic cilium (occasionally more than one) at or near the posterior or antapical pole, sometimes used in temporary attachment to the substratum

caudate Having a tail or a caudal appendage

caudo-frontal association (pl. **caudo-frontal associations**) See *syzygy*

caulerpicin Toxin produced by *Caulerpa*

C-axis (pl. **C-axes**) Longest axis of the hexagonal pattern, perpendicular to the surface of some foraminiferan tests (Pr-3); see *A-axis*

cDNA Complementary DNA; DNA sequence manufactured from a messenger RNA using the viral enzyme reverse transcriptase. Such a copy lacks the introns (intervening sequences) of the natural gene, since the mRNA sequences corresponding to the introns have been removed by splicing following transcription

celestite A usually white mineral made of strontium sulfate ($SrSO_4$) comprising the spines of some acantharian actinopods (Pr-31)

cell cycle (pl. **cell cycles**) Repeating sequence of growth and division of a cell consisting of interphase, G1 (growth phase 1); S (DNA synthesis); G2 (growth phase 2); and M (mitosis); characteristic of plants, animals, fungi, and some protoctists. Extreme variation in cell cycle theme occurs in protoctists

cell division (pl. **cell divisions**) Division of cell to produce two or more offspring cells; see *cytokinesis, karyokinesis*

cell envelope (pl. **cell envelopes**) Outer membrane, composed of lipids and proteins, that surrounds a cell; regulates exchange of material between cell and environment; universal structure of cells

cell junction (pl. **cell junctions**) Any of a number of connections between cells in multicellular organisms (for example, desmosomes, septa, plasmodesmata, pit connections, and others). Especially developed in animals (for example, gap junctions and septate junctions)

cell membrane (pl. **cell membranes**) See *cell envelope*

cell plate (pl. **cell plates**) Phragmoplast. Collection of vesicles that forms between telophase nuclei, oriented by microtubules, in the development of a new cell wall; the phragmoplast is characteristic of some taxa of chlorophytes and of all plants

cellular slime mold (pl. **cellular slime molds**) Any member of the phylum Acrasea or Dictyostelida (Pr-2); heterotrophic protoctists which during the course of their life cycle move from independently feeding and dividing amebas into a slimy mass that eventually transforms into a stalked structure, which produces cysts capable of germinating into amebas

cellulose plate (pl. **cellulose plates**) Surface covering on dinomastigotes; see *armored plates*

central capsule (pl. **central capsules**) Double or single membranous structure that delimits the ectoplasm from the endoplasm in actinopods

central granule (pl. **central granules**) See *axoplast*

centric Description of diatoms with radially symmetrical valves

centrifugal cleavage The progressive development of cleavage furrows from the central region of the body toward the periphery; usually refers to algal thalli; see *centripetal cleavage*

centriolar plaque (pl. **centriolar plaques**) Flattened microtubule-organizing centers (MTOCs) at the spindle poles to which the spindle microtubules attach, associated with the nuclear membrane; on the ultrastructural level, they are observed to reproduce by extension and duplication (for example, in yeast)

centriole (pl. **centrioles**) Barrel-shaped cell organelle 0.25 μm (diameter) $\times$ 4 μm (length). Kinetosome lacking an axoneme; a [9(3)+0] microtubular structure that forms at each pole of the mitotic spindle during division in most animal cells. Observed to reproduce by a developmental cycle (for example, in which new centriole appears at right angles to the parental one)

centripetal cleavage The progressive development of cleavage furrows from the peripheral regions of the body toward the center; usually refers to algal thalli; see *centrifugal cleavage*

centrocone (pl. **centrocones**) Division center; cone-shaped extranuclear microtubular bundle, at the apex of which is a [9(1)+1] centriole (nine singlet microtubules surrounding a single axial tubule); formed during mitosis in apicomplexans and probably arising from a microtubule-organizing center

centromere (pl. **centromeres**) Structure attaching chromosomes to microtubules of mitotic spindle. Microtubule-organizing center located on chromosomes. Centromeric connections to the spindle are required for chromatid segregation. The centromere, as a region of the chromosome deduced from genetic behavior, is sometimes distinguished from kinetochore (a structure observable in the electron microscope). Some authors consider centromere synonomous with kinetochore. In some parabasalians, centromeres are embedded in the nuclear membrane; see *kinetochore*

centroplast (pl. **centroplasts**) Microtubule-organizing center from which axonemes of axopods arise in actinopods (Pr-31); it is a tripartite disk consisting of an electron-lucent exclusion zone and interaxonemal substance sandwiched between two caps of electron-dense material (for example, centroaxoplasthelid heliozoa); see *axoplast*

centrosphere (pl. **centrospheres**) Translucent, spherical area in which a centroplast resides; the centrosphere is divided into two sheets: (1) a clear exclusion zone and (2) an interaxonemal zone containing material (axonemal dense substance or interaxonemal substance) in which the axonemes are rooted

CER Plastid endoplasmic reticulum of some algae; an extra layer of ribosome-studded membrane surrounding the plastid

Chagas disease South American human trypanosomiasis; disease found in Central and South America caused by infection with *Trypanosoma cruzi* carried by "kissing bugs" (*Triatomine hemipterano*, Pr-11)

chalk Limestone (which is mostly $CaCO_3$) consisting largely of microscopic coccolith blades and spines

chamber (pl. **chambers**) Portion or subdivision (for example, of a test of foraminifera, Pr-3)

chasmolith (pl. **chasmoliths**) Ecological term referring to microorganisms living in rock crevices produced by erosion or by endolithic organisms; see *endolith, epilith, lithophile*

chasmolithic See *chasmolith*

chemosynthate Any metabolic product of chemoautotrophy (chemosynthesis); total chemosynthate contains sugars, amino acids, and other products of metabolism

chemotactic Movement either toward or away from a chemical stimulus (chemotactic agent)

chemotaxis See *chemotactic*

chemotaxonomy Grouping into higher taxa of organisms based on their chemical characteristics

chert Siliceous rock (including flint) of microcrystalline quartz; the embedding matrix for many well-preserved microfossils. Material of which radiolarite is composed

chiasma Region of contact between homologous chromatids when crossing over has occurred during meiosis; these regions resemble the letter chi ("X")

chiasmata See *chiasma*

chitin Hard organic polysaccharide composed of §-1,4-linked acetylglucosamine units. Chitin is found in cell walls of some rhodophytes (Pr-33), chlorophytes (Pr-28), and chytridiomycotes (Pr-35), and in threads secreted by diatoms (Pr-18) and other protoctists

chitinozoa An extinct group, probably protoctists, which left organic microfossil remains in rocks of Proterozoic and early Paleozoic age

chitinozoans See *chitinozoa*

chlamydospore (pl. **chlamydospores**) Asexual spherical structure of fungi or fungal-like protoctists originating by differentiation of a hyphal segment (or segments) used primarily for perennation, not dissemination (for example, monoblepharidalean chytridiomycotes, Pr-35)

chlorophyll Green lipid-soluble pigments required for photosynthesis; all are composed of closed tetrapyrrholes (porphyrins or chlorins) chelated around a central magnesium atom; comprise part of thylakoid membrane in all photosynthetic plastids

chloroplast (pl. **chloroplasts**) Green plastid; membrane-bounded cell organelle containing lamellae (thylakoid membranes), chlorophylls *a* and *b*, usually carotenoids and other pigments, proteins, and nucleic acids in a nucleoid and ribosomes

chloroplast endoplasmic reticulum See *CER*

chloroplast lamellae Thylakoid membranes in chloroplasts, some of which stack to form grana (in some algae and most plants)

chloroxybacterium (pl. **chloroxybacteria**) Prochlorophyta. Chlorophyll *a*, chlorophyll *b*-containing oxygenic phototrophic prokaryotes that are not cyanobacteria (B-6) because they lack phycobiliproteins (for example, *Prochloron*, *Prochlorothrix*, and an open-ocean dwelling marine coccoid)

choanomastigote (pl. **choanomastigotes**) Choanoflagellate; that is, phylum of undulipodiated protoctists consisting of unimastigotes or colonial organisms; primarily marine heterotrophs enclosed by an organic (theca) or siliceous (lorica) structure with collars of tentacles; also, a stage in the development of trypanosomatid mastigotes in which the kinetoplast lies in front of the nucleus and the associated undulipodium emerges at the anterior extremity by way of an expanded undulipodial pocket (Pr-36)

chondriome (pl. **chondriomes**) Complete set of mitochondria or mitochondrial genetic complement of a cell

chromatic adaptation (pl. **chromatic adaptations**) Alteration in the relative quantities of photosynthetic pigments in response to changes in light quality and intensity (leading to reduction or increase in light absorption) usually observed as color changes in algae and cyanobacteria

chromatic granules Type of microbody; membrane-bounded organelles of archaeprotists (Pr-1), 0.5–2.0 μm; especially in trichomonads and other mastigotes that, under anaerobic conditions, generate H_2. Hydrogenosomes have been called "anaerobic mitochondria."

chromatid (pl. **chromatids**) Half chromosome. Chromatids segregate from each other in late metaphase/early anaphase of mitosis, whereas in meiosis they move jointly to the same pole as entire chromosomes segregate from each other

chromatin Eukaryotic DNA complexed with histone (and/or other basic proteins) to form the nucleosome-studded DNA strands that usually condenses (coils and becomes deeply stainable) to form chromosomes during mitotic cell division

chromatoid body (pl. **chromatoid bodies**) Ribonucleoprotein structures in *Entamoeba* cysts (Pr-2) that form from ribosomes in the trophozoite cytoplasm

chromatophore (pl. **chromatophores**) Pigment-containing structure or organelle; the colored portion of a cell or an organism

chromophilic The tendency of a structure or tissue to become colored by taking up stain in a cytological or histological preparation. Chromophilic bodies are cell structures with affinity for stain

chromophore (pl. **chromophores**) Colored portion of molecule; molecule (purified substance) that is colored

chromophyte algae Algae containing plastids with chlorophylls *a* and *c* (lacking chlorophyll *b*) as well as certain carotenoids as accessory pigments. Chrysoplast-containing algae (for example, chrysophytes (Pr-15), diatoms (Pr-18), xanthophytes (Pr-16), and phaeophytes (Pr-17))

chromosome (pl. **chromosomes**) Intranuclear organelle made of chromatin (DNA, histone, and nonhistone protein) and containing most of the cell's genetic material; usually visible only during mitotic nuclear division. In dinomastigotes (Pr-5), the chromosomes, which have a peculiar composition, tend to be visible throughout the life cycle of the cells

chrysolaminarin Storage polymer, colorless and usually found in membranous vacuoles, composed of β-1,3- or β-1,6-linked glucopyranoside units; found in diatoms (Pr-18), chrysophytes (Pr-15), and phaeophytes (Pr-17)

chrysophytes Informal name of members of the phylum Chrysophyta (Pr-15)

chrysoplast (pl. **chrysoplasts**) Golden-yellow plastid of chromophyte algae (for example, diatoms (Pr-18) or chrysophytes (Pr-15)) that contains chlorophylls *a* and *c*

chrysoplast endoplasmic reticulum Membranes studded with ribosomes, surrounding the plastid of chrysophytes (the chrysoplast; Pr-15)

chute (pl. **chutes**) Canal-like, membranous structure associated with the nematocyst/taeniocyst complex of the dinomastigote *Polykrikos* (Pr-5)

chytrid (pl. **chytrids**) Common name limited to organisms in the order Chytridiales of the phylum Chytridiomycota (Pr-35) (and not the entire phylum referred to as chytridiomycotes)

ciguatera poisoning Illness resulting from human ingestion of marine fish taken from areas of red tide. The toxin is produced by dinomastigotes (Pr-5) comprising the red tide (for example, *Peridinium*, *Gymnodinium*)

ciliary axoneme (pl. **ciliary axonemes**) Shaft of cilium

ciliary necklace Structure of membrane particles seen with the electron microscope at the base of the axonemal membrane. Arranged in single rings, double rings, and other conformations; these necklaces may be of taxonomic significance; see *undulipodial bracelet*

ciliature General term referring to the position or arrangement of undulipodia of ciliates (Pr-6)

cilium (pl. **cilia**) Undulipodium. Organelle of motility that protrudes from the cell, comprised of an axoneme covered by the plasma membrane. The term is used to refer to undulipodia of ciliates and of animal tissue cells. Composed of the [9(2)+2] microtubular configuration

cingular Pertaining to the girdle region of the dinomastigote cell (Pr-5); the constriction running transversely; the girdle region of the frustule connecting the two distal valves in diatoms; see *girdle*

cingulum (pl. **cingula**) See *cingular*

circadian rhythm The occurrence of a phenomenon in live cells (for example, cell division, maximum photosynthetic rate, bioluminescence, or enzyme production) with a periodicity of approximately 24 hours

cirrus (pl. **cirri**) Tuft-shaped organelle formed from bundles of undulipodial axonemes covered by a common membrane in ciliates (Pr-6); functions primarily in locomotion, but also in feeding

cisterna (pl. **cisternae**) Flattened membranous vesicle, such as those comprising the Golgi apparatus or endoplasmic reticulum

cisternal membrane (pl. **cisternal membranes**) Membrane surrounding the cisternae of the endoplasmic reticulum or the Golgi apparatus

clade (pl. **clades**) Branch on a phylogenetic tree consisting of a taxon (or a set of directly related taxa) and its descendants. Also refers to the peripheral bifurcations in ebridian skeletons

cladistic analysis (pl. **cladistic analyses**) Cladistics; a subfield of the biological science of systematics; the formal taxonomic examination of clades or branches on evolutionary trees; a method of arranging taxa by the analysis of primitive and derived characteristics to reflect phylogenetic relationships between organisms

cladogram (pl. **cladograms**) Phylogenetic tree that is derived from cladistic analysis

class (pl. **classes**) More inclusive taxon than order and less inclusive than phylum in the systematic hierarchy. For list of protoctist classes (Tables 6, 7, and 9)

clast (pl. **clasts**) Small piece of rock; small product of crustal erosion; see *sand, silt*

cline (pl. **clines**) Gradient; gradation of morphological differences in a species or population of organisms over a geographic area; see *lysocline, nutricline, thermocline*

clonal culture (pl. **clonal cultures**) Culture of genetically identical offspring organisms produced by cell division of a single parent cell

clone (pl. **clones**) Offspring produced from a single parental individual, in the absence of sexual processes

closed mitosis Cryptomitosis. Any mitosis (karyokinesis) during which the nuclear envelope is preserved intact throughout the process of division

cnidocyst (pl. **cnidocysts**) Nematocyst. Complex extrusome produced by dinomastigotes such as *Nematodinium*, Pr-5, or by coelenterates (A-4)

coated vesicle (pl. **coated vesicles**) Vesicular structures surrounded by a layer of the protein clathrin, arising from endocytosis or by the budding of portions of intracellular membranes (for example, from Golgi apparatus); a function in transport of substances into, out of, and between cells

coccal Spherical structure; spherical bacterium

coccalean Spherical in form; usually refers to algae

coccidian life history Stages in the development of an apicomplexan (Pr-7), a member of a large, economically important group of necrotrophs of animals

coccoid Spherical or approximately spherical in form

coccolith (pl. **coccoliths**) Scale; calcified structure, essentially platelike, but often elaborated, found externally on some prymnesiophyte algae (coccolithophorids, Pr-25); made of $CaCO_3$, usually deposited as calcite on an organic substructure or matrix; often abundant as fossil remains of coccolithophorids in chalk; see *calyptrolith, caneolith, cricolith, crystallolith, discoaster, helicoid placolith, heterococcolith, holococcolith, lopodolith, pentalith, placolith, rhabdolith, zygolith*

coccolith vesicle (pl. **coccolith vesicles**) Modified endoplasmic reticulum in which the coccoliths of coccolithophorids (Pr-25) form. Coccolith vesicle–reticular body system; see *cv-rb system*

coccolithogenesis Intracellular process of the formation of coccoliths (Pr-25)

coccolithophorid (pl. **coccolithophorids**) Haptomonad alga bearing coccoliths (Pr-25)

coccolithosome (pl. **coccolithosomes**) Granular particle 25 nm in diameter located in the Golgi apparatus of coccolithophorids (Pr-25); precursor of the coating that surrounds the coccoliths

coccosphere (pl. **coccospheres**) Total coccolith covering of a coccolithophorid (Pr-25); a cell covering of coccoliths in which the coccoliths hold together to form an intact shell of scales

coccus (pl. **cocci**) See *coccal*

coelopodium (pl. **coelopodia**) Structure involved in the capture of prey in polycystine actinopods (Pr-31), consisting of thickened envelopes of cytoplasm; this serves to enclose the appendages of such larger prey as copepods

coelozoic parasite (pl. **coelozoic parasites**) Parasite (necrotroph) of the coelom or body cavity of metazoans; see *histozoic parasite*

coenobial Colony containing a fixed number of cells prior to its release from the parent colony (for example, *Volvox*, Pr-28)

coenobic See *coenobial*

coenobium (pl. **coenobia**) See *coenobial*

coenocyst (pl. **coenocysts**) Multinucleate thick-walled algal cyst; propagule resistant to desiccation

coenocyte (pl. **coenocytes**) Plasmodium; syncitium. Multinucleate structure (thallus) lacking septa or walls; thallus with siphonous, syncitial, or plasmodial organization

coenocytic See *coenocyte*

coevolution The simultaneous development of morphological or physiological features in two or more populations or species that, by their close interaction, exert selective pressures on each other

collar (pl. **collars**) Inverted cone-shaped structure at cell apex; may be protoplasmic (for example, in choanomastigotes, Pr-36) or mineralized

colonial Group of cells or organisms of the same species, derived from the same parent(s) and living in close association as a unit, each member capable of further reproduction

colony (pl. **colonies**) See *colonial*

colony inversion (pl. **colony inversions**) The turning inside out, from undulipodia facing inward, to undulipodia facing outward, of a colony; characteristic of the coenobia of volvocalean chlorophytes (Pr-28)

columnella Structure arising from the stalk of myxomycote sporangium (Pr-23) and extending into the spore

commensalism Ecological term referring to facultative associations between members of different species, in which one associate obtains nutrients or other benefits from the other without damaging or benefitting it

community (pl. **communities**) Interacting populations of organisms of different species, found in the same place at the same time (for example, termite hindgut protist communities)

competitive exclusion principle A principle that the degree of niche overlap of two species will influence the domination of one species by the other

complexity Measure of the amount of DNA that is present in a single copy (unique sequences rather than repeat DNA). This is determined by the kinetics of reassociation of denatured double-stranded DNA and represented as the combined length in nucleotide pairs of all unique DNA fragments

compound rootlet (pl. **compound rootlets**) Rootlet made of microtubules that extends laterally from the rhizostyle in cryptomonads (Pr-26). Complex proximal structure of kinetids

concentric fibril (pl. **concentric fibrils**) Small, solid, long, thin structures arranged as one circle inside another

conceptacle (pl. **conceptacles**) External cavity visible with the naked eye as a receptacle. Contains reproductive cells, usually on the surface of algal thalli; found in phaeophytes (Pr-17) such as the Fucales and Ascoseirales and rhodophytes (Pr-33) such as the Corallinales. The receptacle contains conceptacles

conchocelis phase (pl. **conchocelis phases**) Microscopic, branched, filamentous, endolithic, sporophytic phase of conchospores in the life history of the rhodophyte *Porphyra* and other Bangiales (Pr-33)

conchosporangium (pl. **conchosporangia**) Type of enlarged sporangium, usually produced in series by the conchocelis phase of some rhodophytes (for example, Bangiales, Pr-33)

conchospore (pl. **conchospores**) Spores produced during the conchocelis phase in bangiophycidean rhodophytes (Pr-33); spores produced and released singly by a conchosporangium

confluent Growing, running, or flowing together, as in the intermingling of the mucilagenous sheaths of certain algae or the growth of cells on nutrient agar plates

congener Members of the same genus

congeneric (adj.) See *congener*

conglomerate Course-grained sedimentary rock, composed of rock fragments larger than 2 mm embedded in a fine-grained sand or silt matrix

conidiogenesis Process by which individual conidia form

conidiophore (pl. **conidiophores**) Spore-bearing structure, usually of fungi; the subtending hypha or stalk to a conidium or group of conidia

conidium (pl. **conidia**) Exogenously produced spore, usually deciduous; in oomycotes (Pr-21), equivalent to a caducous sporangium

conjugation Copulation; mating. In prokaryotes, cell contact during the transmission of genetic material from donor to recipient; in eukaryotes, the fusion of nonundulipodiated gametes or gamete nuclei or the fusion of structures leading to fusion of gametes or gamete nuclei; see *lateral conjugation, scalariform conjugation*

conjugation tube Joined outgrowths in conjugating green algae from adjacent cells in which gametes fuse or through which they move prior to fusion

connecting fiber (pl. **connecting fibers**) Fibrillar or amorphous structure linking triplets of different kinetosomes with each other; any filament (long solid structure) connecting other entities in cells or between cells

conoid (pl. **conoids**) Apical cone-shaped structure made up of several spirally arranged microtubules; part of the apical complex in apicomplexans

conspecific Members of the same species

contamination (pl. **contaminations**) Presence in growth medium of organisms other than those desired

continental shelf (pl. **continental shelves**) That part of the edge or margin of a continent between the shoreline and the continental slope; characterized by a very gentle slope of 0.1°

continental slope (pl. **continental slopes**) That part of the edge or margin of a continent between the continental shelf and the continental rise (or oceanic trench); characterized by a greater angle to the horizontal than the continental shelf

continuous culture (pl. **continuous cultures**) Cultivation of organisms or cells in which the growth rate is maintained constant through continuous addition of fresh medium and continuous removal of cell- or organism-containing spent medium

contophora A large group crossing taxonomic boundaries encompassing all algae in which the thylakoids are assembled in groups (grana); that is, all algae except rhodophytes (Pr-33)

contractile vacuole (pl. **contractile vacuoles**) Vacuole in the cortex or ectoplasm of protoctists that functions in osmoregulation of the cytoplasm by alternately dilating and contracting to excrete water from the cell against an osmotic gradient

convergent evolution The independent development of similar structures or behaviors in populations that are not directly related but have been subjected to the same selection pressures (for example, the evolution of cysts in response to desiccation)

coprolite (pl. **coprolites**) Fossil of lithified feces (animal excrement)

coprophile Ecological term referring to organisms that live on or are attached to dung or fecal pellets

coprophilic See *coprophile*

coprophilous See *coprophile*

coprozoic Organisms living in feces

copula Band-shaped midlatitude structure in spherical structures. Silica band in diatoms, an overlapping series of which intervene between the epivalve and hypovalve; sometimes used synonomously with girdle (cingulum) of dinomastigote tests (Pr-5); see *girdle lamella*

copulate Mating, the fusion of gamonts or gametes; see *conjugation*

copulation (pl. **copulations**) See *copulate*

core (pl. **cores**) Core sample; generally refers to a cylindrical section of rock or sediment collected with a coring device

corona (pl. **coronas, coronae**) Crown or crown-shaped structure

coronula (pl. **coronulae**) Little crown-shaped structure (for example, charophytes, Pr-28)

cortex (pl. **cortices**) Morphological descriptive term referring to the outer layer of a cell, organism, or organ; usually made of proteinaceous or polysaccharide complexes; in ciliates (Pr-6), highly structured fibrillar outer covering, one to several micrometers thick, in which the undulipodia are embedded; in algae, tissue underlying the epidermis

corticating Cortex-forming. Cortication refers to the secondarily formed outer cellular covering of algal thalli (for example, charophytes, phaeophytes, rhodophytes, Pr-28, -17, -33)

cortication See *corticating*

corticolous Pertaining to organisms living on the bark of trees

cosmopolitan The growth or occurrence of organisms in all or most parts of the world; widely distributed

costa (pl. **costae**) Highly motile nonmicrotubular intracellular rod in Archaeprotista (for example, parabasalians, Pr-1); elongated solid thickening (fibula) of the valve in a diatom frustule (Pr-18); attachment band, connected at both ends to coiled filaments that confer elasticity to the cortex in acantharian actinopods. Rib or ridge (for example, foraminifera); see *costal strip, subraphe costa*

costal strip (pl. **costal strips**) Siliceous strips which join to form costae, which in turn make up a basketlike lorica in some choanomastigotes (Pr-36)

costate See *costa*

crampon (pl. **crampons**) Branched stalk base in dictyostelids (Pr-2)

craticulum (pl. **craticula**) Irregular siliceous plate forming an internal shell in certain pennate diatom frustules (Pr-18)

crenulate Wavy, ruffled; describing a surface with notches or small waves

cresta (pl. **crestae**) Fibrillar, noncontractile structure, found below the basal portion of the trailing undulipodium in devescovinid mastigotes (Parabasalia, Pr-1)

cribrate Sievelike, profusely perforated; having a cribrum (for example, a closing plate (velum) of the pores (areolae) of a diatom wall with regularly arranged perforations in the silica; Pr-18); aperture composed of many rounded holes grouped together over a defined area

cricolith (pl. **cricoliths**) Elliptical heterococcolith with the elements arranged peripherally on a base-plate scale. Coccolith with $CaCO_3$ elements stacked to form a simple tube (for example, *Hymenomas carterae*; Pr-25)

crista (pl. **cristae**) Tubular or pouchlike and inwardly directed fold of the inner membrane of a mitochondrion; the site of ATP production during aerobic metabolism; rich in respiratory enzymes, cristae may be discoid, platelike, tubular, or vermiform; see *tubular crista*, *vermiform crista*, *vesicular crista*

cross-banded root (pl. **cross-banded roots**) Basal part of kinetid structure; undulipodial rootlets with a striated appearance

crown cell (pl. **crown cells**) Cells that make up the coronula (corona) (for example, in charophytes)

cruciate Cross-shaped; as in the microtubules of the kinetid structure of some chlorophytes or the contents of a tetrasporangium that are oriented at right angles to each other

cruciform mitosis Cross-shaped appearance of the nucleus in metaphase; characterized by an elongated nucleolus arranged perpendicularly to the chromosomes at the equatorial plate; cruciform nuclear division (for example, in plasmodiophorids, Pr-20)

crude culture (pl. **crude cultures**) See *agnotobiotic culture*

crustlike See *crustose*

crustose Growing hyphae or trichomes, usually of alga or lichen body (F-6), that together form a crust

crustose thallus (pl. **crustose thalli**) See *crustose*

cryophile Organism that grows well and completes its life cycle at low temperatures (that is, near 0°C)

cryophilic See *cryophile*

cryoplankton Plankton of polar or other cold regions

cryopreservation Viable preservation of organisms, tissues, or cells by suspension in appropriate solutions and storage at extremely cold temperatures

cryptobiosis Suspended or deathlike condition generally brought on by desiccation or freezing, reversible by anabiosis

cryptobiotic cyst (pl. **cryptobiotic cysts**) Cysts capable of resuscitation; "suspended life" in which respiration and other metabolic activities are scarcely discernible but reversible by anabiosis; see *anabiosis*, *cryptobiosis*

cryptomitosis Closed mitosis. Mitosis in which the nuclear membrane remains intact

cryptomonad (pl. **cryptomonads**) Informal name of cryptophytes, or members of the phylum Cryptophyta (Pr-26)

cryptopleuromitosis Descriptive term referring to mitosis in which a bilaterally symmetric mitotic spindle is located entirely outside the nucleus and the nuclear membrane remains intact. Characteristic of some parabasalians (Pr-1), diatoms (Pr-18), etc

cryptostoma Small cavities on the surface of the thallus containing rows of sterile hairs in phaeophytes (for example, *Adenocystis*, *Scytothamnus*, *Splachnidium*, Pr-17)

cryptostomata See *cryptostoma*

crystallolith (pl. **crystalloliths**) Coccolith type made of disk-shaped rhombohedrons (for example, *Crystallolithus*); holococcoliths with the crystals deposited on the distal surface of an organic scale (Pr-25)

C-tubule (pl. **C-tubules**) One of three tubules forming the kinetosome; the incomplete microtubule comprising the outermost kinetosomal (or centriolar) triplet; see *A-tubule*, *B-tubule*

culture (pl. **cultures**) Laboratory-maintained population of organisms that survives on culture medium and is transferred by inoculation

culture medium (pl. **culture mediums, culture media**) Liquid or solid material providing nutrients for the survival in laboratory culture of protoctists or other organisms

cumatophyte (pl. **cumatophytes**) Alga, usually brown or red, living exposed to surf (for example, the phaeophyte *Postelsia*, Pr-17)

cuneate Narrowly triangular with the acute angle toward the base; wedge-shaped

curved vane assembly (pl. **curved vane assemblies**) Cytoskeletal support element for the ingestion apparatus of phagotrophic euglenoids (Pr-12); four long equidistantly spaced sheets, "j"-shaped in cross section, radiating out from four microtubules immediately adjacent to the cytopharynx

cuticle (pl. **cuticles**) Waxy or fatty layer on the outer wall of epidermal cells. In protests, sometimes synonymous with cortex

cv-rb system (pl. **cv-rb systems**) Coccolith vesicle–reticular body system. Membrane system associated with coccolith formation; includes Golgi and vacuoles (Pr-25)

cyanelle (pl. **cyanelles**) Endocyanome. Intracellular structures considered by some to be cyanobacterial symbionts and by others to be an organelle derived from symbiotic cyanobacteria, active in oxygenic photosynthesis (for example, in glaucocystophytes). Cyanelles are distinguished from rhodoplasts by possessing at least remnants of cell wall material

cyanobacterium (pl. **cyanobacteria**) Phylum B-6: Chlorophyll *a*, phycobiliprotein-containing, oxygenic photosynthetic bacteria; formerly called blue-green algae; phototrophic prokaryotes that use water (some may use sulfide) as an electron donor in the reduction of CO_2, produce oxygen in the light, have paired thylakoids, and are unicellular, or form filaments or thalli. Some filamentous cyanobacteria differentiate specialized cells (heterocysts) for nitrogen fixation; some have gliding motility. The most widespread phylum of phototrophic aerobic prokaryotes, cyanobacteria initiated the rise of gaseous oxygen in Earth's atmosphere some 2 billion years ago

Cyanophyceae Class in the botanical division Cyanophyta of the Plant kingdom; obsolete term for cyanobacteria (B-6)

cyclical transmission Cycle of development of a heteroxenous parasite in which the parasite undergoes a cycle of development in one host before it infects and develops in the alternative host (for example, *Trypanosoma brucei* (Pr-11) undergoes cycle of development in tsetse flies before infecting mammalian host)

cyclosis Protoplasmic streaming. Circulation of cell cytoplasm, characteristic of eukaryotes; internal cell motility based on nonmuscle actinomyosin fibrous protein complexes

cymose renewal Hyphae or sporangial hyphae produced in a cymose arrangement; that is, each main axis is terminated by a single sporangium; secondary and tertiary axes may also end in sporangia (for example, oomycotes, Pr-21); see *sequential zoosporangium formation*

cyrtos Microtubular apparatus surrounding the cytopharynx (for example, hypostome ciliates, Pr-6)

cyst (pl. **cysts**) Kind of propagule; morphological manifestation of "resting state" in protoctist life cycles; formation of structures may or may not be associated with sexual phenomena. Resistant, sporelike, frequently thick-walled structure independently evolved in many protoctists. Nonmotile, dehydrated, usually resistant to environmental change and inactive. In the life cycle of many protoctists, the cyst is generally considered to serve an important role in either protection or dispersal. Cysts are often formed in response to extreme environmental conditions especially desiccation. The organism typically rounds up and becomes surrounded by one or more layers of secreted cystic envelopes or walls, which may be sculptured on the outside and with or without an emergence pore; see *ectocyst, endocyst, gametocyst, gamontocyst, macrocyst, mesocyst, microcyst, multiplicative cyst, oocyst, pansporoblast, resistant cyst, sclerotium, sorocyst, sporocyst, stomatocyst, temporary cyst, trophocyst*

cystocarp (pl. **cystocarps**) Carposporophyte and surrounding tissue or cells provided by the gametophyte in rhodophytes (Pr-33); reproductive structure on the spore-forming female gametophyte

cystogenesis Process by which cysts are formed

cystogenous plasmodium (pl. **cystogenous plasmodia**) Plasmodium that forms cysts (for example, in plasmodiophorids, Pr-20)

cystosorus (pl. **cystosori**) Structures into which cysts may be united, the presence and morphology of which are of taxonomic significance (for example, in plasmodiophorids, Pr-20)

cytobiont (pl. **cytobionts**) Cellular symbionts; see *endocytobiont*

cytochrome (pl. **cytochromes**) Low-molecular-weight proteins conjugated to iron-chelated tetrapyrrholes (for example, iron porphyrins), chromophores often yellow in color; cytochromes act as electron carriers in aerobic respiration and photosynthesis

cytokinesis Cytoplasmic division, exclusive of nuclear division (karyokinesis); also used as synonym of cell division

cytolysis Rupturing of cells (for example, toxicysts induce cytolysis)

cytopharynx (pl. **cytopharynges**) Cell "throat"; region through which particulate food travels after passing through cytostome (for example, in ciliates, Pr-6)

cytoplasm (pl. **cytoplasms**) Fluid portion of cell containing enzymes and metabolites in solution

cytoplasmic inheritance Non-Mendelian (non-nuclear, nonchromosomal) inheritance of distinctive genetic traits. Often associated with the inheritance of plastids or mitochondria, or correlated with the presence of viral, bacterial, or other endocytobionts

cytoplasmic membrane (pl. **cytoplasmic membranes**) See *cell envelope*

cytoplasmic streaming See *cyclosis*

cytoproct (pl. **cytoprocts**) Cell "anus"; anal pore; generally permanent (when present) slitlike opening (though actually usually closed) near the posterior end of the cell, through which egesta may be discharged. In some ciliate species (Pr-6), located in or just to the left of the posterior portion of kinety number one, the cytoproct is a portion of the cortex with taxonomic significance. Its edges, resembling a kind of pellicular ridge and reinforced with microtubules, are argentophilic (e.g., take up silver stain in Protargol and other microscopic slide preparations)

cytopyge See *cytoproct*

cytoskeleton (pl. **cytoskeletons**) Asymmetric scaffolding, often associated with cell motility inside eukaryotic cells. Microfilaments and microtubules and their associated proteins provide a dynamic framework, which influences the shape of protoctists. Secreted organic or inorganic materials in, on, or below the surface of a protoctist may also contribute to the cytoskeleton

cytosome (pl. **cytosomes**) Ingestive apparatus of euglenids (Pr-12)

cytostomal groove (pl. **cytostomal grooves**) Depression or opening of cell through which food particles pass

cytostome (pl. **cytostomata**) Cell "mouth." A two-dimensional, usually permanently open aperture (for example, in *Noctiluca*, Pr-5). In ciliates (Pr-6), the cytostome may open directly to the exterior or be sunken into a cavity such as an atrium, vestibulum, buccal, or peristomal cavity; the end of the ribbed wall in the ciliate cortex, that is, the level in the ciliate cortex at which pellicular alveolar sacs are no longer present

cytotomy "Cell cutting," multiple fission; cytokinesis delayed with respect to karyokinesis resulting in the formation of several offspring simultaneously. A subcategory of plasmotomy. In some monothalamic, multinucleate foraminifera (Pr-3) with organic tests, the whole cell divides unequally by binary fission to form multiple buds

dactylopodium (pl. **dactylopodia**) Digitiform (finger-shaped) determinate pseudopods, typical of some *Mayorella* spp. (phylum Rhizopoda, Pr-2)

DAP pathway Biosynthetic metabolic pathway forming the amino acid lysine; pathway characteristic of bacteria, some protoctists, and plants; see *alpha aminoadipic acid pathway*

dasmotrophy Feeding strategy, which might be called remote, or necrotrophy auxotrophy following induced osmosis, in which an organism obtains essential nutrients by extracting them from other organisms. The method suggested for inducing excess osmosis is by increasing the permeability of the membranes of surrounding cells, which leak nutrients into the medium where *Chrysochromulina* (Haptomonada, Pr-25) can find them

DBV Membrane-bound vesicle, associated with phosphoglucan metabolism, found in heterokont protoctists (for example, oomycotes). Its appearance in thin section using the transmission electron microscope changes with metabolic activity. Sometimes the DBV is electron-translucent with one or more central or eccentrically placed electron-opaque zones, sometimes it has close-packed lamellar formations between electron-opaque and electron-translucent zones. At oospore formation, DBVs coalesce to form a single, large, membrane-bound inclusion known as the ooplast

deciduous Becoming detached when fully developed; see *caducous*

decomposer (pl. **decomposers**) Osmotrophic organism that converts polymeric organic material into monomers by secretion of extracellular digestive enzymes

defined medium (pl. **defined mediums, defined media**) Culture medium in which the precise chemical nature of the ingredients and their starting concentrations have been identified

definitive host (pl. **definitive hosts**) Host in which a symbiont attains sexual maturity (for example, the coccidian *Aggregata eberthi* in cuttle fish); see *intermediate host*

dehiscence Opening of a structure by drying or programmed death of certain structures or cells (for example, to allow the escape of reproductive bodies contained within)

dendrogram (pl. **dendrograms**) Branching graphic representation of taxonomic arrangement; "family tree" based on numerical relationships (that is, derived from quantification of the similarities and differences among organisms)

dendroid Shaped like a tree; treelike

dense body vesicle (pl. **dense body vesicles**) See *DBV*

desert (pl. **deserts**) An area of low moisture due to low rainfall (that is, fewer than 25 cm annually), high evaporation, or extreme cold, and which supports only specialized vegetation; wind often produces distinctive erosional features (for example, dunes)

desmid (pl. **desmids**) Unicellular or filamentous conjugating green alga (Pr-32) of the families Mesotaeniaceae or Desmidiaceae in which amastigote ameboid gametes conjugate

desmodexy, Law of The invariant position of the kinetodesma to the right (not the left) of its kinety in ciliates (Pr-6)

desmokont (pl. **desmokonts**) Member of a subgroup (Desmophyceae) of the dinomastigotes (Pr-5) characterized by two apically inserted undulipodia

desmoschisis Cell division in which the parental wall forms part of the wall of the progeny (for example, thecate dinomastigotes, chlorosarcinalean chlorophytes, Pr-5, -28); see *eleutheroschisis*

desmose (**desmos**) Part of kinetid structure. Connecting fiber; composite fibrillar connection of unknown nature or function between two adjacent kinetosomes or among several kinetosomes that form a localized group (for example, in the blepharoplast complex of many mastigotes); absent in ciliates (Pr-6)

desmosome (pl. **desmosomes**) Type of cell junction, especially in animal tissues; morphologically and compositionally distinct area of cell membrane at which tissue cells of animals, or regions of same cell, adhere firmly together

determinate growth Pertaining to a growth style, like that of a chytrid thallus (Pr-35), a heterotrichous ciliate (Pr-6), or volvocalean chlorophyte (Pr-28), in which growth stops after reaching a determined size; see *indeterminate growth*

determinate sporangium (pl. **determinate sporangia**) Sporangium that terminates the axis (for example, oomycotes, Pr-21); see *basipetal development, percurrent development*

deuteromerite (pl. **deuteromerites**) Posterior portion of the trophozoite in some gregarine apicomplexans (Pr-7) that is separated by a transverse septum from the nucleus-containing protomerite (the anterior cell)

diadinoxanthin Carotenoid found in the plastids of several types of algae (for example, euglenoids, xanthophytes, and eustigmatophytes, Pr-11, -16, -27)

diagenesis Geological term for physical and chemical alterations in sediments after their deposition and prior to their lithification

diagenetic See *diagenesis*

diakinesis Last stage of meiotic prophase I in which bivalents and chiasma disappear as homologs begin to segregate; see *meiosis*

diaminopimelic acid pathway See *DAP pathway*

diapause Temporary suspension in growth and development in insects (A-21) and other animals

diatom (pl. **diatoms**) Any member of the phylum Bacillariophyta (Pr-18); unicellular and colonial aquatic protoctists renowned for their siliceous tests (frustules).

diatomite Sedimentary rock formed from diatom frustules; when poorly lithified, it is equivalent to diatomaceous earth

diatoxanthin Carotenoid found in the plastids of several protoctists (for example, euglenoids, xanthophytes, eustigmatophytes, and diatoms; Pr-12, -16, -27, -18)

DIC Nomarski differential interference contrast light microscopy

dichotomous Pertaining to the branching into two equal or nearly equal parts

dichotypical The condition in desmids (conjugating green algae, Pr-32) in which one semicell resembles members of one species and the other semicell resembles members of a different species

diclinous Refers to antheridia and oogonia on separate hyphae (for example, the oomycote *Pythium lutarium*, Pr-21); see *monoclinous*

dictyosome (pl. **dictyosomes**) Golgi apparatus; Golgi body. Botanical term for this elaboration of the endomembrane system. Portion of the endomembrane system of nearly all eukaryotic cells visible with the electron microscope as membranous structure of flattened saccules, vesicles, or cisternae, often stacked in parallel arrays; involved in elaboration, storage, and secretion of products of cell synthesis; prominent in many protoctists (for example, parabasalians) and less prominent in others (for example, ciliates). Cis Golgi refers to the face of the membrane where vesicles coalesce to form the cisterna; trans Golgi refers to the secreting (maturing face) of the

Golgi apparatus. Cis and trans cisternae contain different enzymes

diel movement (pl. **diel movements**) Locomotion that follows a 24-hour cycle; see *circadian rhythm*

diffuse growth Generalized, indeterminate growth, characteristic of protoctists such as the plasmodial stage of myxomycotes (Pr-23), labyrinthulids (Pr-19), some chrysophytes (Pr-15), etc

digenetic Descriptive of symbiotrophs with development in their life history in two different types of host; see *homoxenous parasite, polyxenous parasite*

dikaryon (pl. **dikarya**) Cell or organism with cells containing a pair of nuclei (fungi); typically each is derived from a different parent

dikaryotic See *dikaryon*

dikinetid (pl. **dikinetids**) Kinetid composed of two kinetosomes and associated structures

dimorphic Two forms; an organism that, during the course of its life cycle, develops two different types of normal morphologies. Two genetic types of individuals in a population (for example, sexual dimorphism or seasonal dimorphism); see *polymorphism*

dimorphism See *dimorphic*

dinokaryon (pl. **dinokarya**) Unique nuclei of dinomastigotes (Pr-5) characterized by their densely packed chromosomes that persist during interphase. The atypical chromosomes contain DNA with small (25 nm) unit fibrils and lack the conventional histone protein that makes up nucleosomes, distinguishing dinokarya from the nuclei of other protoctists

dinokaryotic See *dinokaryon*

dinokont Organism with one undulipodium located in a transversely aligned groove, the other undulipodium beating in a longitudinally aligned groove. Characteristic of a subgroup of the dinomastigotes (Pr-5)

dinomastigote (pl. **dinomastigotes**) Dinoflagellate. Member of the phylum Dinomastigota (Pr-5)

dinomastigote life history Stages in development of dinomastigotes (Pr-5) correlating environment and morphology

dinomitosis Closed extranuclear pleuromitosis; the characteristic mitosis of dinomastigotes (Pr-5); see *dinokaryon*

dinonucleus See *dinokaryon*

dinospore (pl. **dinospores**) Dinomastigote propagule (Pr-5); spore issued from successive multiple fissions, especially in symbiotrophic dinomastigotes

dioecious Descriptive term referring to organisms having male and female structures on different individual members of the same species; see *diclinous, monoecious*

diphasic life cycle Life cycle with two distinct parts; in symbionts, it can refer to two distinct hosts or tissues of attachment

diplobiontic Having two free-living phases in the life history of an organism; see *haplobiontic*

diplohaplontic In algae, an organism with separate haploid and diploid stages in its life history that may or may not be morphologically distinguishable

diploid Eukaryotic cells in which the nucleus contains two complete sets of chromosomes, abbreviated as 2N; see *euploid, haploid, polyploid*

diplokaryon (pl. **diplokarya**) Two diploid nuclei inside single cells characteristic of some microsporans (F-1; *Thelohania, Pleistophora, Tuzetia*); microsporans with such a nuclear arrangement

diplokaryotic See *diplokaryon*

diplontic Pertaining to the life cycles of organisms in which individual cells are diploid throughout the life history. Organisms that undergo gametic meiosis such that haploidy is limited to the gamete stage; see *haplontic*

diplophase A part of the life cycle in which organisms are diploid, each of their cells containing two complete sets of chromosomes

diplosomes See *bipolar body*

diplotene Diplonema, stage in meiosis just prior to diakinesis in which doubled bivalents become clearly visible; see *meiosis*

diplozoic Having a double body form, for example, as a result of incomplete cell division, as in diplomonads (Pr-1); see *monozoic*

discharge vesicle (pl. **discharge vesicles**) Membrane, usually continuous with the inner zoosporangium wall and papilla in chytridiomycotes (Pr-35), that is laid down during or after sporangial discharge; zoospore delimitation is completed within it

discoaster Star-shaped coccolith (Pr-25)

discobolocyst (pl. **discobolocysts**) Ejectile organelle originating in the Golgi that on discharge forms a firm ring with a gelatinous head; function unknown; restricted to mastigotes, especially chrysomonads (Pr-15)

disporous Cells in microsporans (F-1) and myxozoans (A-2), in which two spores have been produced within a single pansporoblast

distromatic Description of a thallus only two cell layers thick

division (pl. **divisions**) Botanical term for a taxonomic group equivalent to phylum; see *fission*

division center (pl. **division centers**) Microtubule-organizing center of mitosis; centriole, centriolar plaque, centrocone, or any one of a number of structures found at the poles of mitotically dividing cells

dixenous See *digenetic*

DNA complexity See *complexity*

DNA hybridization *In vitro* analytical tool involving pairing of complementary DNA and RNA strands to produce a DNA–RNA hybrid or the partial pairing of complementary DNA strands from different genetic sources. Can be used to determine genetic relatedness between organisms and for purification of messenger RNA

dormancy Resting stage; stage in propagule development of lowered metabolism and resistance to environmental extremes of temperature, desiccation, etc

dorsiventral Pertaining to structures or tendencies (for example, flattening) that extend from the dorsal toward the ventral side; also, having distinct dorsal and ventral surfaces

dourine Venereally transmitted disease of horses caused by *Trypanosoma equiperdum*, Pr-11

dual nuclear apparatus Dimorphic nuclei of heterokaryotic cells (for example, ciliates or foraminifera, Pr-6, -3)

dune (pl. **dunes**) A low mound, ridge, bank or hill of loose, windblown granular material (generally sand, sometimes volcanic ash), either bare or with vegetation, capable of movement but always retaining its characteristic shape

dysaerobic Geological or ecological term referring to aquatic environments with low oxygen, or transition zones between oxic and totally anoxic sediments

dyskinetoplastic Pertaining to members of the kinetoplastids (Pr-11) grown in culture in which the kinetoplast has become unstainable and invisible (either because its contents have become dispersed throughout the mitochondrion or because the structure has been lost as a result of faulty kinetoplast reproduction)

dystrophic Ecological term (meaning "bad nourishment") referring to lakes with very low lime content and containing very high quantities of humus (organic matter). Also refers to bay lakes with colored water and limited inorganic nutrient composition

ecad Genetic race, strain, or variety of organisms that has developed an identifiable morphological response to its environment

ecdysis The act of shedding an outer cuticular layer; in dinomastigotes (Pr-5), shedding of theca prior to division

echinate Spiny

ecosystem (pl. **ecosystems**) Communities of plants, animals, and microorganisms together with their immediate environment, capable of the complete cycling of the biological elements (C, N, O, P, S) (for example, forests, deserts, or ponds). The metabolism and community interactions in an ecosystem are such that cycling within an ecosystem is more rapid than between ecosystems

ecotype (pl. **ecotypes**) See *ecad*

ectocarpin Chemotactic pheromone produced by the female gametes of the phaeophyte *Ectocarpus* (Pr-17) causing accumulation of sperm at the source of the pheromone

ectocyst (pl. **ectocysts**) Outermost of the three layers surrounding a cyst; see *endocyst, mesocyst*

ectoparasite (pl. **ectoparasites**) Ecological term referring to the topology of symbiotrophs and hosts: a symbiotroph that lives upon the surface of its host; see *endoparasite, epibiont*

ectoparasitic See *ectoparasite*

ectoplasm (pl. **ectoplasms**) Outermost, relatively rigid and transparent, granule-free layer of the cytoplasm of many cells (for example, amebas); see *stereoplasm*

ectoplasmic network Extracellular matrix; branching and anastomosing, hyaline, membrane-bounded network of ectoplasmic filaments devoid of cytoplasmic organelles that function as an attachment and absorbing structure and is produced by specialized organelles, called sagenogens, on the cell surface of labyrinthulomycotes (Pr-19). In labyrinthulids, the ectoplasmic network completely surrounds the cells and joins them in a common network through which the cells move by a gliding locomotion. In thraustochytrids, the ectoplasmic network arises from one side of each cell and does not surround it

edaphic Referring to soil

egg (pl. **eggs**) Female gamete, nonmotile and usually larger than the male gamete; see *oosphere*

ejectile body Any organelle forcibly ejected from a cell (for example, trichocysts); in cryptomonads, the ribbonlike extrusome that is coiled and contained in a vesicle

ejectisome See *ejectile body*

ejectosome See *ejectile body*

elastic junction (pl. **elastic junctions**) Layer of the periplasmic cortex of chaunacanthid acantharians (phylum Actinopoda, Pr-31) consisting of microfibrils interconnected in very precise patterns

ELC Duplicate copy of a trypanosome variant surface glycoprotein DNA sequence (gene) expressed when transposed to a telomeric site on the chromosome

electrolyte (pl. **electrolytes**) Salt. A substance that dissociates into its constituent ions in aqueous solution

electrophoresis Technique for separating large DNA molecules, including small chromosomes. DNA from lysed organisms is subjected to pulsed or steady electrical current in agarose gel. The longer the linear molecule, the longer it takes to traverse the gel, thus providing a basis for separation

eleutheroschisis Cell division in which the walls of offspring cells are entirely new and free from parental walls (for example, thecate dinomastigotes, Pr-5); see *desmoschisis*

embryo (pl. **embryos**) Early developmental stage of a plant or animal individual that develops by cell and nuclear divisions from a zygote (fertilized egg); see *blastular embryo*

embryophyte (pl. **embryophytes**) Plant. Embryo enclosed in maternal tissue; usually developing into an adult. Phototrophic organism; sporophytes growing from embryos that are dependent for their nutrition on parental tissue during development. Includes all bryophytes and tracheophytes

emergent flagellum (pl. **emergent flagella**) See *emergent undulipodium*

emergent undulipodium (pl. **emergent undulipodia**) Undulipodium in a bimastigote that protrudes (for example, the undulipodium which extends beyond the canal in euglenids); see *nonemergent undulipodium*

encyst (**v.**) To form or become enclosed in a cyst

encystation See *encyst*

encystment See *encyst*

endemic Pertaining to populations of organisms or viruses, including disease agents, constantly present (often in low numbers) in a limited geographical area

endobiont (pl. **endobionts**) Endosymbiont. Ecological term describing the topology of partners in an association in which one partner lives within the other partner (the host); may be intra- or extracellular; see *epibiont*

endobiotic See *endobiont*

endochite Innermost layer of the fucalean oogonium in phaeophytes (Pr-17; for example, *Fucus*)

endocyanome Cyanelle. All the connected cyanelles of a glaucocystophyte

endocyst (pl. **endocysts**) Innermost of the three layers surrounding a heliozoan oocyst; composed of a layer of fibers and Golgi membranes; see *ectocyst, mesocyst*

endocytic Pertaining to what is inside a cell; intracellular; pertaining to topological relations between associates in which one organism lives inside of another cell. Endocytobiotic

endocytobiology The study of intracellular symbionts and cell organelles. *Endocytobiology and Cell Research* is the journal of the International Society of endocytobiology, published by Tübingen University Press, Germany

endocytobiont (pl. **endocytobionts**) Intracellular symbiont

endocytosis Intake of extracellular material through invagination and pinching off of the plasma membrane; includes intake of fluid (pinocytosis), particulate matter (phagocytosis), and neighboring cell material in tissues (endocytosis *sensu stricto*)

endocytotic vesicle (pl. **endocytotic vesicles**) Cell membranes involved in particle uptake

endodyogeny Endogenesis resulting in the production of two offspring cells within the parent cell (for example, in coccidian apicomplexans, Pr-7); see *endopolygeny*

endogenesis Process by which offspring cells are formed inside the parent cell; see *endodyogeny, endopolygeny*

endogenous See *endogenesis*

endogenous budding See *endogenesis*

endogenous cleavage See *endogenesis*

endogenous multiplication See *endogenesis*

endogeny See *endogenesis*

endolith (pl. **endoliths**) Ecological term describing microorganisms living in tiny openings in rocks or rock crevices that have been produced by the metabolic activities of the endolithic organisms themselves; see *chasmolith, epilith, lithophile*

endolithic See *endolith*

endomitosis Endoreplication. Duplication of chromosomes in the absence of karyokinesis and not followed by chromatid segregation; the process thus leads to polytene chromosomes rather than polyploidy

endomycorrhizal fungus (pl. **endomycorrhizal fungi**) Fungal symbionts (usually zygomycotes) of plants that penetrate tissues of the roots and form a specialized swollen type of root tissue that augments nutrient uptake from the soil. V–A (vesicular–arbuscular) mycorrhizae; F-3

endoparasite (pl. **endoparasites**) Ecological term describing the topology of symbiotrophs and hosts in which the symbiotroph lives within its host, either extra- or intracellularly; endobiotic symbiotroph; see *ectoparasite*, *endobiont*

endoparasitic See *endoparasite*

endophyte (pl. **endophytes**) Ecological term referring to the topology of symbiotic associates with plants. Fungi, protoctists, or bacteria living within the tissue of plants or other photosynthetic organisms. Since "-phyte" may refer to fungi, protoctists, and bacteria, which are not plants, the term should be replaced with endobiont, endosymbiotic bacteria, or other specific name

endophytic See *endophyte*

endoplasm Inner central portion of the cytoplasm of cells (such as amebas, Pr-2), more fluid than the ectoplasm; see *rheoplasm*

endoplasmic reticulum Extensive endomembrane system found in most protoctist, plant, fungal, and animal cells, in places continuous with the nuclear membrane, Golgi apparatus, outer membranes of other organelles, and plasma membrane; called rough (RER) if coated with ribosomes, and smooth (SER) if not; see *chloroplast endoplasmic reticulum*

endopolygeny Endogenesis characterized by the production of several offspring cells within the parent cell (for example, in *Toxoplasma*, *Chlorella*, Pr-7, -28); see *endodyogeny*

endoreplication Endomitosis. In ciliates, may refer to the reproduction of nuclei inside other nuclei

endosome (pl. **endosomes**) Nucleolus. Body, or bodies, into which the nucleolar material is organized and which contains ribosomal precursors. Also, a vesicle resulting from endocytosis

endospore (pl. **endospores**) Spore formed by successive cell divisions within a parent wall. In Actinosporea (phylum Myxospora, A-2), an envelope of one or two modified cells, housing the sporoplasm within the sporal cavity between the episporal cells; in dinomastigotes, the thick inner wall of the three-layered cell wall of a hypnozygote; in conjugating green algae, the inner layer of the zygospore wall; see *exospore*, *mesospore*

endospore cell (pl. **endospore cells**) Parent cell inside which spore(s) form; referring primarily to bacteria. Cells that make up the covering in which sporoplasms of actinosporean myxosporans (A-2) originate; may persist in mature spores in some genera

endosymbiont (pl. **endosymbionts**) Endobiont. Ecological term referring to the topology of association of partners, a member of one species living inside a member of a different species. May be intracellular or extracellular

endosymbiosis Ecological term referring to the topology of an association of partners; the condition of one organism living inside another. Includes intracellular symbiosis (endocytobiosis) and extracellular symbiosis

endosymbiotic See *endosymbiont*

endozoic Ecological term referring to any organism that lives inside an animal

endozoite (pl. **endozoites**) Zoite. Usually in apicomplexans (Pr-7); trophic, motile individual formed by endogenesis

entosolenian tube (pl. **entosolenian tubes**) In some foraminifera (Pr-3), an internal tubelike extension from the aperture

entozoic Endozoic. Ecological term referring to any organism that lives in an animal (may also refer specifically to the gut)

enucleation Anucleation; removal of the nucleus from a cell

envelope (pl. **envelopes**) See *lorica*, *plasma membrane*

epibiont (pl. **epibionts**) Ecological term describing the topology of association of partners in which one organism lives on the surface of another organism; see *endobiont*

epibiotic See *epibiont*

epicone (pl. **epicones**) Episome or upper body. Upper surface or hemisphere of a dinomastigote cell, anterior to the cingulum

epicontinental Pertaining to extensive marine environments, that is, inland seas, formed on the surface of continental masses. Characteristic of the lower Paleozoic Era and later

epilimnion In a thermally stratified lake or body of freshwater, the zone between the surface and the thermocline; upper zone of a body of water characterized by having more or less equal distribution of oxygen and in which the temperature is uniform; see *hypolimnion*

epilith (pl. **epiliths**) Ecological term referring to the biota living on the surface of rocks and stones; see *chasmolith*, *endolith*, *lithophile*, *saxicolous*

epilithic See *epilith*

epimastigote Stage in development of kinetoplastids in which the kinetoplast lies anterior to the nucleus and the associated undulipodium emerges laterally to form an undulating membrane along the anterior part of the body, usually becoming free at its anterior end

epimerite (pl. **epimerites**) Anchoring organelle in the anterior region of septate gregarine apicomplexans (Pr-7), set off from the rest of the body by a septum; see *mucron*

epipelic Ecological term referring to biota living attached to the surface of marine or freshwater mud or sand

epipelon See *epipelic*

epiphyte (pl. **epiphytes**) Ecological term referring to the topology of association of partners, one of which is a plant (or traditionally an alga). The second partner grows on the plant using it for support but not nutrition; epiphyton also refers to communities of microbes growing on algae in aquatic environments. The term is only appropriate if host is member of Plant kingdom

epiphytic See *epiphyte*

epiphyton See *epiphyte*

epiplasm Fibrous or filamentous layer of cytoplasm closely applied to the innermost plasma membrane; in ciliates (Pr-6), a layer under the pellicle, comprising a part of the cortex

epipsammic Ecological term referring to biota living on or in fine interstices of sand grains (from psammon, meaning sand)

epipsammon See *epipsammic*

episeme Change in a seme; evolutionary alteration in a trait; see *seme*

episome (pl. **episomes**) Small genome; genetic element (stretch of DNA sequence capable of coding for a product), usually of bacteria; may be integrated or attached to genophore or replicate independently of the genophore (even at rates faster than the genophore). In dinomastigotes (Pr-5), an ill-advised synonym for epicone; see *epicone*

episporal cell (pl. **episporal cells**) Modified valve cell, three of which make up the epispore that houses the sporoplasm in actinosporean myxosporans (A-2; for example, *Tetractinomyxon*)

epispore Spore wall in actinosporean myxosporans, consisting of three valves or episporal cells often bearing long posterior processes; anucleate half of diploid sporoblast that encloses the sporoplasm in haplosporidians (Pr-29)

epithallus (pl. **epithallium, epithalli**) Part of the growing thallus in which the cells or filaments are developed outwardly from an intercalary meristem (for example, coralline rhodophytes, Pr-33)

epitheca (pl. **epithecae**) Epivalve and adjacent portion of girdle in dinomastigotes (Pr-5); anterior portion of a thecate (armored) dinomastigote; a covering for the epicone

epithelium (pl. **epithelia**) Type of animal tissue that lines the surface of kidneys, or other organs. Epidermis of plants

epitope See *antigenic determinant*

epivalve (pl. **epivalves**) Upper test or shell, found opposite to and usually larger than the hypovalve in diatom frustules (Pr-18)

epizoic Ecological term referring to the topology of association of partners in which an organism lives on the surface of an animal; see *epibiont, epiphyte*

epizoon See *epizoic*

epizootic Pertaining to a widespread occurrence of an infectious disease of animals other than people

equatorial groove (pl. **equatorial grooves**) Cingulum. Midlatitude feature in a spherical organism or structure (for example, portion of a diatom frustule between the valves, space between the hypocone and epicone of a dinomastigote test; Pr-18, -5)

equatorial plate (pl. **equatorial plates**) Transient structure observable in many but not all dividing protoctist cells; plane in the equatorial region of the mitotic spindle at which the chromosomes align by way of their movement during the metaphase stage of mitosis or meiosis

equipotential genomes Genomes (total genetic material of cell) resulting from cell division of a parent cell in which both offspring cells are capable of the same extensive further development

ER See *endoplasmic reticulum*

estuary (pl. **estuaries**) The seaward end or the widened funnel-shaped tidal mouth of a river valley where freshwater mixes with and measurably dilutes seawater and where tidal effects are evident

etiolation Bleached condition of photosynthetic eukaryotic organisms growing in the dark characterized by poorly developed plastids and their lack of chlorophyll. Stem elongation and poor leaf development accompanies etiolation in plants

etiological agent (pl. **etiological agents**) Causative agent (for example, of a disease)

eucarpic Referring to development in certain protoctists (for example, oomycotes, Pr-21) and fungi that form reproductive structures on limited portions of the thallus, such that the residual nucleate protoplasm remains capable of further mitotic growth and regeneration; see *holocarpic*

eucaryote (pl. **eucaryotes**) See *eukaryote*

euglenid (pl. **euglenids**) Any member of the phylum Euglenida, Pr-12. Euglenoid refers to Euglena-like features

euglenoid (pl. **euglenoids**) See *euglenid*

euglenoid motion Peculiar flowing, contracting, expanding ("crawling") movement on surfaces displayed by euglenids capable of changing shape, that is, those not restricted by too rigid a pellicle

euhaline Ecological term referring to salinity of water in the normal oceanic range, that is, between 3.3 and 3.8 percent salt as sodium chloride; see *hyperhaline, oligohaline*

eukaryote (pl. **eukaryotes**) Organism comprised of cell(s) with membrane-bounded nuclei. Most contain microtubules, membrane-bounded organelles (that is, mitochondria and plastids) and chromatin organized into more than a single chromosome

eukinetoplastic Pertaining to kinetoplastid mastigotes (Pr-11) in which the DNA of the kinetoplast (kDNA) forms a single stainable mass located close to the kinetosome(s)

eulittoral See *intertidal, littoral*

euphotic zone Ecological term referring to the illuminated portion of a water column, soil profile, microbial mat, etc.; the layer in which, because of the penetration of light, photosynthesis can occur

euplankton Ecological term referring to aquatic organisms that spend their entire lives suspended in a water column

euploid Possessing a chromosome set that is either the haploid complement or an exact multiple of the haploid complement (for example, diploid and triploid); not aneuploid

euryhaline Ecological term referring to organisms that tolerate and grow under wide ranges of salinity; see *stenohaline*, *euhaline*

eurythermal Ecological term referring to an organism that tolerates and grows under a wide range of temperatures; see *stenothermal*

eurythermic See *eurythermal*

eutrophic Ecological term referring to waters rich in dissolved nutrients (for example, nitrate, phosphate) for phototrophs; see *oligotrophy*

eutrophy See *eutrophy*

evaporite flat (pl. **evaporite flats**) Open area covered with nonclastic sedimentary rocks composed primarily of minerals produced from saline solution that became concentrated by evaporation of the solvent

evolute test (pl. **evolute tests**) Foraminiferal test (Pr-3) in which each whorl does not embrace earlier whorls, such that all chambers are visible

exclusion zone (pl. **exclusion zones**) Layer of the centroplast of heliozoan actinopods (Pr-31)

excyst (**v.**) Process of leaving the cyst stage; cyst germination

excystation See *excyst*

excystment See *excyst*

exocytosis Cell secretion; process of eukaryotic cells involving intracellular motility in which substances are eliminated to the exterior by emptying them from a vesicle that fuses with the plasma membrane, forming a cuplike depression; see *endocytosis*

exogenesis Production of smaller cells at the periphery of the parent cell; a type of budding (for example, in suctorian ciliates, Pr-6)

exogenous Origin or development on or from the outside

exon (pl. **exons**) Segment of DNA that is both transcribed to RNA and translated into protein. Gene or part of gene; see *intron*

exospore (pl. **exospores**) Externally borne reproductive cell; not necessarily heat or desiccation resistant; in dinomastigotes (Pr-5), the thick outer layer of the triple-layered cell wall of the hypnozygote; in conjugating green algae, the outermost layer of the zygospore wall; see *endospore*, *mesospore*

exotoxin Soluble poisonous substance passing into the host or the environment during growth of an organism (for example, red-tide dinomastigotes, Pr-5)

expression-linked copy See *ELC*

expulsion vacuole (pl. **expulsion vacuoles**) See *contractile vacuole*

expulsion vesicle (pl. **expulsion vesicles**) See *contractile vacuole*

extant Living; still in existence

extinct No longer existing

extrinsic encystment Formation of cysts during the exponential phase of population growth; sexual resting cysts are produced (for example, in the chrysophyte *Dinobryon cylindricum*, Pr-15); see *intrinsic encystment*

extrusive organelle (pl. **extrusive organelles**) Membrane-bounded structure the contents of which are extruded by protoctists in response to a variety of stimuli, for example, predators, prey, and changes in acidity. Extrusomes are derived from vesicles of the Golgi system and are anchored to the cell membrane by proteinaceous particles; a generalized term referring to various, probably nonhomologous, structures; see *cnidocyst*, *discobolocyst*, *ejectosome, kinetocyst, mucocyst, nematocyst, polar capsule, taeniocyst, toxicyst, trichocyst*

extrusome (pl. **extrusomes**) See *extrusive organelle*

eyespot (pl. **eyespots**) Stigma. Small, pigmented and probably light-sensitive structure in certain undulipodiated protists (for example, euglenids, eustigmatophytes, labyrinthulid zoospores, Pr-12, -27, -19)

facies Part of a sedimentary rock unit characterized by lithological and biological features and segregated from other parts of the unit, usually seen in the field as a coherent rock layer

facultative Optional; for example, a facultative autotroph is an organism that, depending on conditions, can grow either by autotrophy or by heterotrophy; see *obligate*

falx (pl. **falces**) Sickle-shaped structure of opalinids. Specialized area of the cortex along the front edge of the body; a region of kinetidal proliferation that results in the increased length of the kineties; the falx is usually bisected during the symmetrogenic fission of the organisms

fascicle (pl. **fascicles**) Bundles (for example, oomycote mastigonemes or suctorian ciliate tentacles, Pr-6)

fathom (pl. **fathoms**) Unit of water depth (1 fathom = 2 meters)

fauna (pl. **faunas**) Animal life. Inappropriate for protoctists and bacteria

feeding veil (pl. **feeding veils**) Cytoplasmic sheet extended from the sulcus of some non-photosynthetic dinomastigotes (Pr-5) during extracellular digestion when feeding on diatoms (Pr-18) or other dinomastigotes

female (pl. **females**) Gamont. Gender of individual that produces ovaries, eggs, or other sexual organs and receives the male sperm

fenestra (pl. **fenestrae**) Foramen. Opening in a surface; small "window" (for example, lesion in nuclear membrane in essentially closed mitosis); see *polar fenestra*

fermentation Nutritional mode: enzyme-mediated pathway of catabolism of organic compounds in which other organic compounds serve as terminal electron acceptors (a process that yields energy and organic end products in the absence of oxygen)

ferruginous Made of or containing iron; having the reddish brown color of iron rust

fertile sheet (pl. **fertile sheets**) Cell layers lining the inside of the conceptacle from which the reproductive structures, antheridia and oogonia, are produced in fucalean phaeophytes (Pr-17)

fertilization Syngamy or karyogamy. Fusion of two haploid cells, gametes, or gamete nuclei to form a diploid nucleus, diploid cell, or zygote

fertilization cone (pl. **fertilization cones**) Cytoplasmic cone originating at the posterior end of the female gamete of hypermastigotes (for example, the parabasalian *Trichonympha*, Pr-1)

fertilization tube (pl. **fertilization tubes**) Structure facilitating fertilization; structure forming in laterally fused, mating dinomastigote gametes beneath the kinetosomes into which nuclei migrate and fuse during fertilization

Feulgen stain Red stain requiring hydrolysis of deoxyribose and formation of a Schiff base that is quantitatively specific for chromatin DNA. Named after R. Feulgen, a German cytologist at the beginning of the twentieth century

fibril (pl. **fibrils**) General descriptive term for thread-shaped solid structure (for example, 2–3 nm filaments lacking actin seen in motile systems of various protoctists including acantharians and heliozoan actinopods, dinomastigotes, and ciliates, Pr-31, -5, -6); in vorticellid ciliate fibrils, a 20,000 dalton protein called spasmin has been identified which corresponds to the 3 nm microfilaments of their contractile stalk

fibrillar kinetosome props Fibrillar, often coarse structures of kinetids of chytridiomycote zoospore (Pr-35) that connect at an angle of about 45° to the nine C-tubules of the kinetosome triplet tubules and extend to the plasma membrane; see *transition fiber*

fibrous lamina (pl. **fibrous laminae**) Thick microfibrillar network coating the inner surface of the nuclear membrane in the acantharian actinopod *Haliommatidium* (Pr-31)

fibula (pl. **fibulae**) Clasp of buckle-shaped, elongated structure (for example, bar running beneath the displaced raphe on the side of valve in diatoms such as *Nitzschia* and *Hantzschia*; Pr-18); see *subraphe costa*

filament (pl. **filaments**) See *fibril*

filopodium (pl. **filopodia**) Very thin pseudopods that may be stiffened by one or very few microtubules (for example, those of desmothoracid heliozoa (Pr-31) that pass through openings in the central capsule)

filose Terminating in a threadlike process

filose pseudopod (pl. **filose pseudopods**) Cell protrusion or retractile process ending in a filamentous wisp, especially the motile organelles of amebae (Pr-2)

fimbriate Bordered by or decorated with tiny fibers or fibrils

fission Division of any cell or organism; reproduction by division of cells or organisms into two or more parts of equal or nearly equal size. Longitudinal fission: division through long axis (for example, most mastigotes); transverse fission: division through small equatorial plane of ovoid organisms (for example, all ciliates); see *binary fission, homothetogenic fission, interkinetal fission, multiple fission, perkinetal fission, polytomic fission, symmetrogenic fission*

Fjord (pl. **Fjords**) A long, narrow, winding, V-shaped and steep-walled, generally deep inlet or arm of the sea between high cliffs or slopes along a mountainous coast, typically with a shallow sill or threshold of solid rock or earth material submerged near its mouth and becoming deeper farther inland

flabelliform Fan-shaped

flagellar apparatus (pl. **flagellar apparati**) See *basal apparatus*

flagellar bracelet Structure composed of intramembrane particles occurring at the junction between the undulipodium and the cell body; in green algae some consist of two or three closely associated rings of intramembrane particles; possibly homologous to ciliary necklace

flagellar groove (pl. **flagellar grooves**) Invagination of a cell from which undulipodia emerge; see *gullet*

flagellar hair (pl. **flagellar hairs**) Filamentous appendages at right angles to the axoneme and arranged in one or more rows, associated with or coating the undulipodia of many phototrophic mastigotes ("phytoflagellates"). May be simple, nontubular structures or tubular hairs consisting of at least two distinct regions; see *anisokont, mastigoneme, tinsel flagellum,*

flagellar pocket (pl. **flagellar pockets**) Invagination of the cell surface to form a pit or deep pocket from which the undulipodia emerge in euglenids (Pr-12) and kinetoplastids (Pr-11); see *undulipodial groove*

flagellar rootlet (pl. **flagellar rootlets**) Portion of kinetid. Microtubular, fibrous, or amorphous structure originating at kinetosomes, extending proximally into the cell, and terminating somewhere in the cytoplasm, but not at other kinetosomes; see *rhizoplast, rhizostyle, system I fiber, system II fiber*

flagellar transition zone (pl. **flagellar transition zones**) Part of a kinetid, region of the undulipodium at its proximal (basal) end adjacent to the kinetosome displaying cytological characteristics of diagnostic and phylogenetic interest in the systematics of undulipodiated organisms

flagellate (pl. **flagellates**) Eukaryotic microorganism motile via undulipodia (see Introduction)

flagellum (pl. **flagella**) Bacterial flagellum; prokaryotic extracellular structure composed of homogeneous protein polymers, members of a class of proteins called flagellins; moves by rotation at the base; relatively rigid rod driven by a rotary motor embedded in the cell membrane that is intrinsically nonmotile and sometimes sheathed. Undulipodium by contrast, an intrinsically motile intracellular structure used for locomotion and feeding in eukaryotes; composed of a standard arrangement of nine doublet microtubules and two central microtubules composed of tubulin, dynein, and approximately 200 other proteins, none of them flagellin; no flagellum (but every undulipodium) is underlain by a kinetosome. See Introduction for an explanation of the restriction of the term flagellum

flange (pl. **flanges**) Projecting rim that provides strength or support to a structure

flimmer Fine hairlike projections that extend laterally from undulipodia; mastigonemes differ in detail in various protoctist groups; are probably formed from proteins synthesized on the ribosomes of the outer nuclear membrane

flora (pl. **floras**) Plant life. Innappropriate for protoctists, fungi, and bacteria

flotation chamber (pl. **flotation chambers**) Gas-filled portion of a cell lending buoyancy (for example, the final chamber of the foraminiferan *Rosalina bulloides*, Pr-3, which adds buoyancy so that it floats among the plankton)

foliose Leafy; pertaining to leaflike growth. Growing hyphae or trichomes, usually of lichens (F-6) or algae, which together form a leafy or leaflike structure

fomite (pl. **fomites**) Inanimate object that transmits infective stages of necrotrophs or pathogens

foramen (pl. **foramena**) See *fenestra*

foraminiferan test (pl. **foraminiferan tests**) Shell or covering of members of the phylum Foramenifera (Pr-3)

fragmentation Means of asexual reproduction in which the breakup of a parental thallus or filament gives rise to a new individual (for example, some conjugating green algae, large foraminifera, Pr-3)

freshwater (pl. **freshwaters**) Water containing only small quantities of dissolved salts or other materials, such as the water of streams and inland lakes

frond (pl. **fronds**) Leaflike structure; any divided thallus (or leaf)

front (pl. **fronts**) See *cline*

fructification Sorocarp; sporocarp. Structure that contains spores, cysts, or other propagules. This term, derived from botany and ambiguously applied, should be replaced with appropriate protoctistological alternatives

fruit (pl. **fruits**) Botanical term describing structures of angiosperm plants, that is, matured ovary or ovaries of one or more flowers and their associated structures

fruiting body (pl. **fruiting bodies**) See *fructification*

frustule (pl. **frustules**) Siliceous cell wall or test of a diatom, composed of two valves

fucan Sulfated polysaccharides found in phaeophytes (Pr-17), containing l-fucose

fucoidan See *fucan*

fucosan vesicle (pl. **fucosan vesicles**) Small colorless vesicle occurring in cells of phaeophytes (Pr-17) containing fucosan and certain tannins and terpenes

fucoserraten Pheromone produced by the eggs of *Fucus* (Pr-17), which attracts sperm

fucoxanthin Carotenoid, usually in chrysoplasts such as those of diatoms (Pr-18), phaeophytes (Pr-5), and some dinomastigotes (Pr-17)

fultoportula (pl. **fultoportulae**) Organelle of some centric diatoms surrounded by basal pores and buttresses; may bear an external tube that continues internally

fungal Members of the kingdom of Fungi; osmotrophic, chitinous-walled eukaryotic organisms that develop from spores; they lack both embryos and undulipodia at all stages of their life cycle; see *higher fungus*, *lower fungi*

fungi imperfecti Fungi in which sexual stages have not been observed; form "phylum Deuteromycota." Often closely related to identifiable asco- or basidiomycota (F-4, -5)

fungus (pl. **fungi**) See *fungal*

funis Ribbon of microtubules paralleling the recurrent undulipodium (or its intracellular axoneme) to the posterior end of the cell in diplomonads (Pr-1)

furcellaran Sulfated polysaccharide phycocolloid produced by the rhodophyte furcellaria (Pr-33); wall component with mucilaginous properties, similar to carrageenan

furrow (pl. **furrows**) Long narrow structure that differs from a canal in that it is open along its length

fusiform Spindle-shaped; tapering at each end

fusion cell (pl. **fusion cells**) Cell produced by the union of the protoplasts of two or more cells

fusion competence State of a gamete that is capable of undergoing sexual fusion; exposure of one or both gametes to pheromones may be required

fusule (pl. **fusules**) Complex structure perforating the skeleton and through which pass axopodial axonemes; a strand of cytoplasm that connects the region inside the capsule to that outside in polycystine actinopods (Pr-31)

G1 Growth phase 1 or gap 1; a stage of interphase of mitotic cell cycle preceding DNA synthesis ("S" phase), during which growth occurs

G2 Growth phase 2 or gap 2; stage in cell cycle, following DNA synthesis but before mitosis, during which growth occurs. During this stage, protein synthesis and an increase in organelle number are observed, chromatin condenses, and microtubules are polymerized from tubulin prior to spindle formation

gall (pl. **galls**) Hypertrophy, often spherical or irregular-shaped; growth on plants caused by penetration of plant tissues by xenogenous organisms (for example, insects, fungi, protoctists, or bacteria)

gametangium (pl. **gametangia**) Any structure in which gametes or gametic nuclei are generated and from which they are released; see *sporangium*

gamete (pl. **gametes**) Mature haploid reproductive cell or nucleus capable of fusion with another gamete, usually of a different mating type, to form a diploid zygote nucleus; see *anisogametes*, *anisoplanogametes*, *hologamete*, *isogametes*, *isoplanogametes*, *macrogamete*, *microgamete*

gametic meiosis Descriptive of life cycle in which meiosis immediately precedes gamete formation (for example, most animals and protoctists such as diatoms); see *zygotic meiosis*

gametocyst (pl. **gametocysts**) Cyst of gamete or cyst forming immediately after syngamy (for example, in the hypermastigote *Trichonympha*, Pr-1)

gametocyte (pl. **gametocytes**) Gamont composed of a single cell

gametocytotomont (pl. **gametocytotomonts**) Cell whose multiple division product is a gametocyte

gametogamy Fusion of gametes; see *syngamy*

gametogenesis Production of gametes by cell differentiation

gametogony Formation of gametes by multiple fission; in apicomplexans (Pr-7), often as a result of schizogony

gametophyte generation Gametophyte; individual plant or alga composed of haploid cells; gamete-producing generation. Characteristic of all plants, many rhodophytes (Pr-33), and phaeophytes (Pr-17) with life cycles having alternation of generations. The gametophyte generation usually begins with the germination of spores that were produced by meiosis; it terminates with fertilization and diploid zygote formation; see *sporophyte generation*

gamogony Gamogonic process; sexual phase in which gametes are eventually produced; series of karyokineses and/or cytokineses leading to gamonts, individuals that produce gamete nuclei, or gametes capable of fertilization

gamont (pl. **gamonts**) Reproducing organism or cell at a stage in its life cycle during which it produces gametes or other sexual structures; see *agamont*

gamontocyst (pl. **gamontocysts**) Cyst formed around gamonts; when gamonts are single-celled, gamontocysts are gametocysts. In gregarine apicomplexans (Pr-7), a cyst forms around two conjugating gamonts (engaging in syzygy) and fertilization of two ameboid gametes (products of the gamonts) takes place within the gamontocyst

gamontogamous Pertaining to mating of gamonts, copulation, sexual intercourse, and conjugation or to fusion of two or more gamonts followed by gametogamy (for example, in foraminifera and some gregarines, Pr-3, -7)

gamontogamy See *gamontogamous*

Gause's Law See *competitive exclusion principle*

gel electrophoresis See *electrophoresis*

generative cell (pl. **generative cells**) Cell capable of further growth (for example, free, uninucleate cell within a large myxosporan (A-2) trophozoite (plasmodium), which gives rise to a pansporoblast); cell capable of further division or fertilization followed by further division

generative nuclei Nuclei capable of further growth and karyokinesis (for example, small compact nuclei in heterokaryotic foraminifera (Pr-3), which are the antecedents of nuclei of the next generation)

genetic locus (pl. **genetic loci**) Position on a linkage group that can be determined by recombination analysis of inherited traits displaying distinguishable genetic alternatives (alleles)

genetic marker (pl. **genetic markers**) Gene determining a distinguishable phenotype that can be used to identify a cell or individual that carries it; may also be used to identify a nucleus, chromosome, or locus

geniculate Uncalcified portion of a thallus between segments of articulated coralline rhodophytes; see *intergeniculum*

geniculum (pl. **genicula**) See *geniculate*

genome (pl. **genomes**) Sum of all genes of an organism or organelle

genomic complexity See *complexity*

genophore (pl. **genophores**) Gene-bearing structure of prokaryotes and certain organelles ("bacterial chromosome," a term to be avoided) (for example, DNA-containing nucleoid of bacteria, mitochondria, or plastids). Nucleoids are structures visible by microscopy, whereas genophores are their equivalents that are inferred from genetic investigation; see *nucleoid*

genotype Genetic makeup of an organism with respect to specific traits, in contrast to the physical appearance of those traits (phenotype)

genotypic See *genotype*

geological time scale See Introduction p. 24

geosynclinal Very large (hundreds of kilometers long) troughlike depression in the Earth's surface filled with layered sedimentary rocks and produced by orogeny

Geologic Time Scale

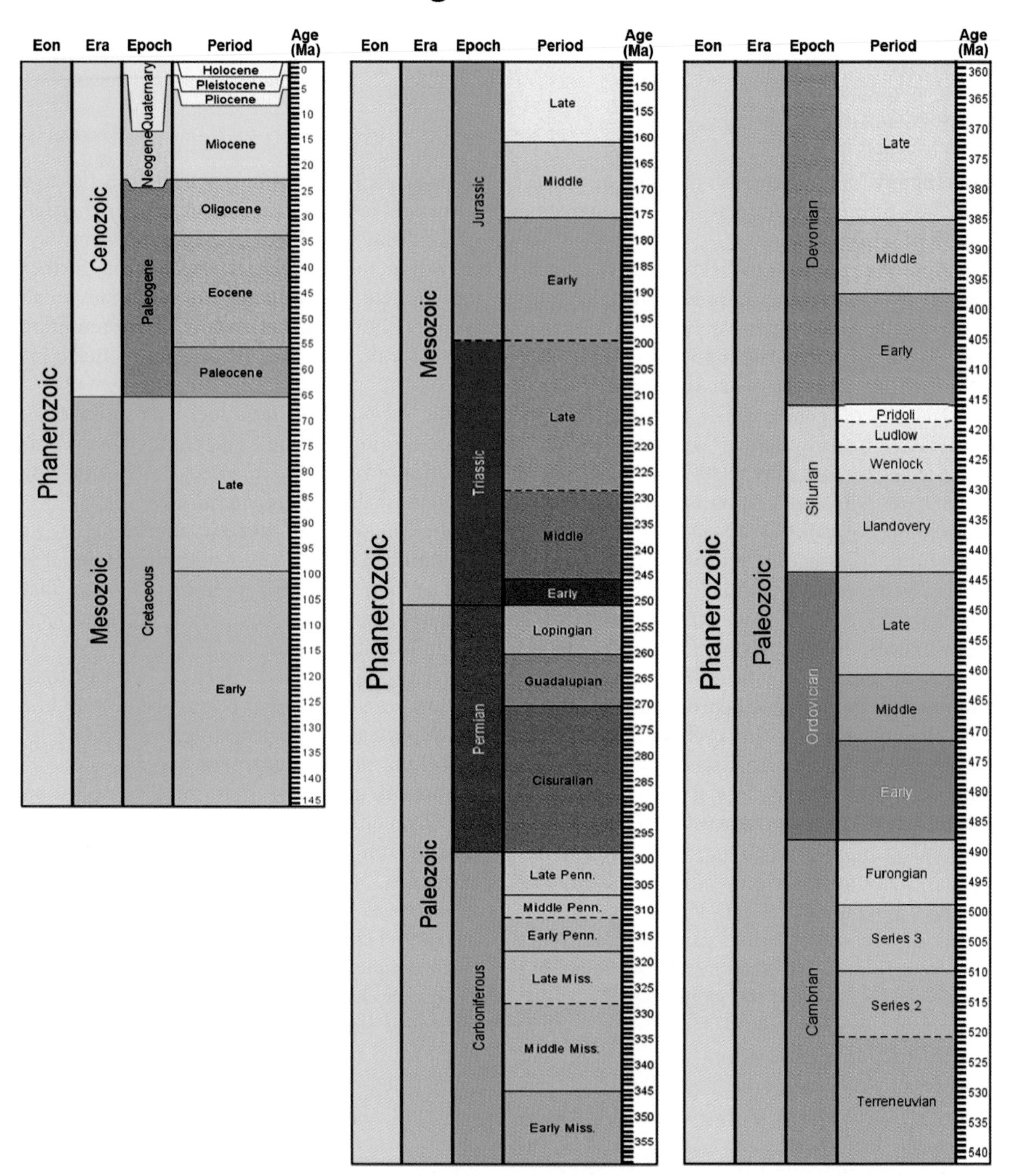

Figure Pr-iii-1 Geologic time scale. Subdivisions are standardized by the International Commission on Stratigraphy (ICS) and the color scheme of the Commission for the Geological Map of the World (CGMW) is shown for these divisions. This GTS shows the "time-rock divisions" for four thousand million years of Earth's natural history. The "relative chronology" is subdivided into time intervals correlated with fossiliferous rock units identified by geographic "type localities" worldwide. Over the past half-century, fossil occurrences have been integrated with paleomagnetic and radiometric age data to calibrate the time scale producing an "absolute chronology" or "geochronologic" time scale based on absolute rather than relative age. This Geologic time scale was generated with TSCreator, a user-defined public JAVA package developed by Adam Lugowski and James Ogg. The program is available from the ICS website (www.stratigraphy.org) and the TSCreator website (www.tscreator.com).

	Eon	Era	Period
Precambrian	Proterozoic	Neoproterozoic	Ediacaran
			Cryogenian
			Tonian
		Mesoproterozoic	Stenian
			Ectasian
			Calymmian
		Paleoproterozoic	Statherian
			Orosirian
			Rhyacian
			Siderian

Age (Ma): 550, 600, 650, 700, 750, 800, 850, 900, 950, 1,000, 1,050, 1,100, 1,150, 1,200, 1,250, 1,300, 1,350, 1,400, 1,450, 1,500, 1,550, 1,600, 1,650, 1,700, 1,750, 1,800, 1,850, 1,900, 1,950, 2,000, 2,050, 2,100, 2,150, 2,200, 2,250, 2,300, 2,350, 2,400, 2,450

	Eon	Era	Period
Precambrian	Archean	Neoarchean	
		Mesoarchean	
		Paleoarchean	
		Eoarchean	
	Hadean (informal)	Hadean (informal)	

Age (Ma): 2,500, 2,550, 2,600, 2,650, 2,700, 2,750, 2,800, 2,850, 2,900, 2,950, 3,000, 3,050, 3,100, 3,150, 3,200, 3,250, 3,300, 3,350, 3,400, 3,450, 3,500, 3,550, 3,600, 3,650, 3,700, 3,750, 3,800, 3,850, 3,900, 3,950, 4,000, 4,050, 4,100, 4,150, 4,200, 4,250, 4,300, 4,350, 4,400, 4,450

geosyncline (pl. **geosynclines**) See *geosynclinal*

geotaxis Directed locomotion toward the gravitational center of Earth. Gravitational response dependent on gravitational sensor (for example, barium sulfate crystals in *Chara*, Pr-28)

geotropism (pl. **geotropisms**) Directed growth toward the center of gravity of Earth. Gravitational response dependent on gravitational receptor (for example, in modified plastids in grasses)

germ cell (pl. **germ cells**) Cell requiring fertilization before it can grow or reproduce (for example, ovum (egg), spermatozoan (sperm), gamete); see *somatic cell*

germ tube (pl. **germ tubes**) Tube-shaped structure capable of further growth (for example, in chytridiomycotes, hyphochytrids, and oomycotes, Pr-35, -14, -21, a short, hyphalike process that develops upon germination of a spore; usually gives rise to more hyphae)

germinate (v.) Begin to grow (for example, from a spore or cyst)

germination chamber (pl. **germination chambers**) Structure in which growth begins (for example, in the chrysophyte *Dinobryon*, Pr-15, a chamber formed from the porus of a germinating stomatocyst, into which go the four offspring protoplasts and from which they eventually emerge)

germling (pl. **germlings**) Bud or newly attached developing propagule capable of growing into an adult at some stage

giant cell (pl. **giant cells**) Usually large cell among normal-sized ones (for example, in dictyostelids, zygote that engulfs and digests other cells to attain a large size; in dasycladalean chlorophytes (Pr-28), large cells produced prior to the division of the primary nucleus)

girdle (pl. **girdles**) See *equatorial groove*

girdle band (pl. **girdle bands**) See *copula*

girdle groove (pl. **girdle grooves**) Surface groove (girdle) in which the transverse undulipodium lies in many dinomastigotes (Pr-5); see *cingulum*

girdle lamella (pl. **girdle lamellae**) Band of thylakoids just inside the plastid membrane and arranged peripherally in some classes of algae (for example, phaeophytes; girdle band of xanthophytes, Pr-16)

glabrous zone (pl. **glabrous zones**) Hairless zone (for example, nonciliated region used for ingestion in some karyorelictid ciliates, Pr-6)

gliding motility Motility of cell or organism always in contact with a solid surface (for example, glass, rocks, conspecifics) in the absence of external appendages; occurs in both prokaryotes and eukaryotes (for example, diatoms, labyrinthulids, Pr-18, -19) but differs in mechanisms. Mechanisms of bacterial gliding (for example, myxobacteria, filamentous cyanobacteria, B-12, -6) are unknown. In the labyrinthulids (Pr-19), motility is thought to be related to the presence of a calcium-dependent contractile system of actinlike proteins; in diatoms, actin microfibrils lying in the cytoplasm beneath the raphe slits have been implicated

glycocalyx (pl. **glycocalyces, glycocalyxes**) Covering; sheath; coat or wall (for example, surface coating secreted by many of the "naked" rhizopod amebas (Pr-2) that cover the plasma membrane; polysaccharide components found outside the bacterial inner lipoprotein membrane)

glycogen body (pl. **glycogen bodies**) Structure composed of the carbohydrate glycogen

glycolipid (pl. **glycolipids**) Class of organic compounds composed of a mixture of small carbohydrate and lipid molecules; lipids with sugar esters

glycosome (pl. **glycosomes**) Organelle; peroxisome-like microbody peculiar to kinetoplastid mastigotes (Pr-11) that lacks peroxidase but contains enzymes of the glycolytic metabolic pathway

glycostyle (pl. **glycostyles**) Flexible surface projection arising from the cell membrane (for example, glycocalyx of some amebas (Pr-2), measuring 110–120 nm in length, may facilitate ingestion of food particles, including bacteria, due to its stickiness)

glyoxylate shunt Biochemical pathway of photorespiration in which organic carbon is converted to amino acids via glyoxylate

glyoxysome (pl. **glyoxysomes**) Organelles; membrane-bounded microbodies harboring the enzymes of glyoxylate metabolism

gnotobiotic Term denoting that the biological composition of a preparation or medium is known; germ free; see *agnotobiotic culture*

golden-yellow algae Phyla Pr-15 or -25. Chrysophytes (algae classified as Chrysophyta or Chrysophyceae) or haptophytes (prymnesiophytes)

Golgi apparatus See *dictyosome*

Golgi body See *dictyosome*

gonimoblast (pl. **gonimoblasts**) In rhodophytes (Pr-33), a filament bearing one or more carpospores or the collection of these filaments that make up the carposporophyte

gonimoblast filament (pl. **gonimoblast filaments**) See *gonimoblast*

gonocyte (pl. **gonocytes**) Dividing cell yielding, by multiple fission, offspring cells capable of propagation (for example, parasitic colonial dinomastigote cell (Pr-5) that gives rise to dinospores during palisporogenesis; in apostome ciliates (Pr-6), offspring cells produced by palintomy)

gonomere (pl. **gonomeres**) Terminal, globular reproductive segments borne on branches called trophomeres in ellobiopsids (Pr-5)

gonospore (pl. **gonospores**) Germ cell; sex or reproductive cell; see *somatic cell*

granellae Crystals of barium sulfate (barite) found in large numbers in the cytoplasm of xenophyophorans (Pr-4)

granellare The plasma body ("protoplasm") of a xenophyophoran (Pr-4) together with its surrounding tubes, which are yellowish and branched in varying degrees

granule (pl. **granules**) Small spherical structure, often unidentified

granuloplasm Granular endoplasm

granuloreticulopodium (pl. **granuloreticulopodia**) Anastomosing pseudopods (for example, in phylum Granuloreticulosa, Pr-3)

granum (pl. **grana**) Stack of thylakoids inside plastids, formed by fusion of membranes of adjacent thylakoids

grex Pseudoplasmodium that, when mature, leaves a trail of slime and migrates toward a dry area before growing upward to form sporophore; a stage in the life cycle of dictyostelids (Pr-2) characterized by the appearance of a translucent migrating structure resembling a tiny shell-less snail

gross culture (pl. **gross cultures**) Agnotobiotic culture; crude culture; culture containing other organisms in addition to the one of interest

growth Increase in size and volume of a cell or of an organism by a number of processes alone or in combination (for example, increase in size, uptake of water, and increase in number of cells). Apical growth refers to growth at tip; basal growth to growth at base; intercalary growth to growth localized at points between base and tip or between two other nongrowing points

gullet (pl. **gullets**) Oral cavity (for example, canal and reservoir of euglenoids or cryptomonads, Pr-12, -26)

guluronic acid Carbohydrate component of alginate

gyrogonite (pl. **gyrogonites**) Whorled, ovate carbonate fossils interpreted to be remains of the female gametangia of charophytes (phylum Chlorophyta, Pr-28). Most Devonian period in age, approximately 420 million years old

habit (pl. **habits**) Descriptive term referring to morphology of growth (for example, bushy, capsalean, crustose, filamentous, foliose, filamentous, single-celled, and viney)

habitat (pl. **habitats**) Immediate surroundings of a population or community. Habitats for protoctists are shown in the illustration "Habitats"; see *biotope*

hadal See *abyssal*

halophil Halophile

halophile (**n**) Ecological term referring to organisms requiring high salt concentrations for growth, including those flourishing in saline environments (for example, salt-requiring bacteria and protists such as *Dunaliella* and *Tetramitus*)

halophilic Halophile

haplobiontic Term referring to the life cycle of organisms that possess only one morphologically distinct stage; life cycle in which there is only one growing phase; cells in this phase usually have haploid nuclei (for example, oomycotes, Pr-21); see *diplobiontic*

haploid Term describing eukaryotic cells or organisms composed of cells in which the nucleus contains one single complete set of chromosomes, abbreviated 1N; see *diploid*

haplomitotic See *haplobiontic*

haplontic Term describing life cycle of organisms in which individual cells are haploid throughout their history. Diploidy is limited to stage immediately preceding meiosis; see *diplontic*

haplophase That part of the life cycle in which organisms are haploid, their cells each containing a single complete set of chromosomes

haplosporosome (pl. **haplosporosomes**) Electron-dense unit-membrane bounded organelle with unknown function; generally spherical, but sometimes having profiles which are oblate, spheroidal, vermiform, pyriform, or cuneiform. Another unit membrane (which distinguishes the organelle from other membrane-bounded, electron-dense inclusions in other eukaryotic cells) is found internally in various configurations free of the delimiting membrane (for example, of haplosporidians and possibly myxozoans and paramyxeans, Pr-29, A-2, Pr-30)

haplostichous Thallus composed of free or consolidated filaments lacking a true parenchymatous organization (for example, typical of some phaeophyte orders, Pr-17)

hapteron (pl. **hapterons**) Kelp holdfast; multicellular attaching organ

haptomonad (pl. **haptomonads**) Member of phylum Haptomonada (Pr-25); protist cell attached to any substratum by modified undulipodium.

haptonema (pl. **haptonemata**) Microtubular appendage, cell organelle, usually coiled, often used as a holdfast; associated with the undulipodia in prymnesiophytes (Pr-25). Structure reveals outer sheath of three concentric membranes and an axoneme of [6(1) + 0]: an inner circle of six or seven microtubules surrounded by a cylinder of endoplasmic reticulum; haptonemata may be long and coiling, or reduced in length and substructure

haptonematal root (pl. **haptonematal roots**) Kinetid of haptonema of prymnesiophytes (Pr-25); root fiber system of haptonema

haptonematal scale (pl. **haptonematal scales**) Scales of the axonemal membrane of haptonema

haptonemid (pl. **haptonemids**) See *haptonemid*

haptophyte (pl. **haptophytes**) Prymnesiophytes (Pr-25) that form haptonemata: coccolithophorids and their relatives; prymnesiophytes with or without coccoliths. Those golden-yellow algae that produce coccoliths, which may fossilize as marine calcium carbonate sediments

haustorium (pl. **haustoria**) Absorbing organelle of osmotrophic protoctists and fungi formed from a projection of a hypha; haustoria penetrate plant cell walls invaginating, but not puncturing, the cell membrane

helicoid placolith (pl. **helicoid placoliths**) Lopodolith. Coccolith subtype; placolith with helical shape; heterococcolith composed of two plates or shields interconnected by a tube (for example, *Helicosphaera*, Pr-25); see *placolith*

hematochrome (pl. **haematochrome**) Astaxanthin or 3,3′ diketo 4,4′-diketo- β-carotene; red to orange colored pigment found in some euglenoids and some chlorophytes (Pr-12, -28)

hematozoic Mode of heterotrophic nutrition; blood-eating (for example, vertebrate bloodstream necrotrophs)

hemiautospore (pl. **hemiautospores**) See *aplanospore*

herbivore See *herbivory*

herbivorous See *herbivory*

herbivory Mode of nutrition referring to organisms feeding on plants (Table 2)

hermaphrodite (pl. **hermaphrodites**) See *hermaphroditic*

hermaphroditic Monoecious. Descriptive term referring to organisms that have both male and female structures on the same individual

heterococcolith (pl. **heterococcoliths**) Coccolith (see Pr-25) composed of distinguishable subelements; heterococcoliths include placoliths and cricoliths and are made up of morphologically diverse calcite structures; coccoliths in which the crystals show a variety of form and modification; see *holococcolith*

heterodynamic undulipodia (pl. **heterodynamic undulipodium**) Undulipodia on the same cell but with different patterns of beating; see *homodynamic undulipodia*

heterogamy State in which morphologically distinguishable gametes are produced by members of a single species (for example, anisogamy and oogamy)

heterogeneric Organisms from different genera

heterogenome (n) From different genomes (for example, microsporidians (A-2) are heterogenomic relative to their host tissue)

heterogenomic See *heterogenome*

heterokaryon (pl. **heterokarya**) Cell or organism with cells containing a pair of nuclei in which it can be shown that each is derived from a genetically distinct parent; see *dikaryon*

heterokaryosis Exhibiting nuclear dimorphism or two or more genetically different nuclei in common cytoplasm (for example, ciliates and foraminifera (Pr-6, -3) with their large and small nuclei and basidiomycete fungi (F-5)); see *homokaryotic*

heterokaryotic See *heterokaryosis*

heterokont See *anisokont*

heterokontimycotina Proposed subdivision to include fungi-like protoctists possessing or thought to be derived from ancestors which once possessed heterokont undulipodia at some stage in their life cycle, that is, oomycotes, labyrinthulids, and hyphochytrids (Pr-21, -19, -14)

heteromorphic life cycle Life cycle in which the different phases are morphologically distinct; for example, alternation of generations in which the diplophase and haplophase are morphologically distinguishable; see *isomorphic life cycle*

heteroplastidy Simultaneous occurrence in one cell or organism of two kinds of plastids (for example, chloroplasts and starch-storing leucoplasts)

heteroside (pl. **heterosides**) Chemical compound: type of complex carbohydrate composed of a monosaccharide (hexose) and a nonsugar compound (organic acid, polyol)

heterospecific Organisms from different species

heterothallic Descriptive term of protoctist life cycle in which two different clones are required for sexual fusion. Single propagules give rise to individuals of a single mating type; the condition of species in which the sexes (mating types) are segregated in separate clones or thalli, two different clones or thalli of compatible mating types are required for fertilization; see *homothallism*

heterothallism See *heterothallic*

heterotrich Filamentous, hairy, or undulipodiated cell or structure with hairs, filaments, or undulipodia of more than a single type (for example, heterotrichous ciliates; filamentous algal morphology composed of both an erect and prostrate portion)

heterotrichous See *heterotrich*

heterotrichy See *heterotrich*

heterotroph (pl. **heterotrophs**) See *heterotrophy*

heterotrophic See *heterotrophy*

heterotrophy Mode of nutrition in which organisms obtain carbon, electrons, and energy from preformed organic compounds (Tables 1 and 2). Examples of heterotrophs include algivores, biotrophs, carnivores, necrotrophs, osmotrophs, parasites, phagotrophs, and saprobes

heterotype (pl. **heterotypes**) Minor variable antigen type (VAT) (that is, trypanosomatid parasites); see *homotype*

heteroxenous See *digenetic*

higher fungus (pl. **higher fungi**) Ascomycetes (F-4), basidiomycetes (F-5), and deuteromycetes (term defined by omission: all fungi except protoctistan "lower fungi"). It is inadvisable to use the terms "higher" and "lower" because of their ambivalence and anthropocentrism

hilar Areola. Spore-connection scar that appears to be composed of numerous globular particles (for example, protostelids and acrasids). Scar on a seed marking point of attachment to ovule in plants

hilum (pl. **hila**) See *hilar*

histone (pl. **histones**) Lysine and arginine-rich protein that complexes with nuclear DNA in eukaryotes to form the nucleosome substructure and therefore components of chromatin. The quantity and quality of this class of basic nucleoproteins varies greatly in protoctists, whereas histones of plants, animals, and fungi are very similar to each other

histophagous See *histophagy*

histophagy Mode of nutrition; heterotrophy of microorganisms that ingest tissues of animals (Table 2); see *carnivory*, *holozoic*, *osmotrophy*, *phagotrophy*

histozoic parasite (pl. **histozoic parasites**) Ecological term referring to symbiotrophs that live in animal tissues (for example, myxosporans, A-2); see *coelozoic parasite*

holdfast (pl. **holdfasts**) Peduncle. Attachment structure, which may be an organ or organelle

holocarpic Pertaining to a mode of development in which the thallus is entirely converted into one or more reproductive structures; the entire cell is used for the production of spores, normally simultaneously but occasionally sequentially (for example, oomycotes, Pr-21); see *eucarpic*

holococcolith (pl. **holococcoliths**) Coccolith type (see Pr-25) not composed of subelements; holococcoliths include calyptroliths, crystalloliths, and zygoliths, and are composed of homogeneous microcrystals; coccoliths in which the $CaCO_3$ is deposited as uniform rhombohedral or hexagonal crystals showing little modification; see *heterococcolith*

hologamete (pl. **hologametes**) Gamete of the same size and structural features as growing cells of the same species

hologenous sperm formation Spermatogenesis in which the entire protoplasm of a microspore converts to form a sperm (for example, in centric diatoms)

holotrophic Phagotrophic. Pertaining to a mode of heterotrophic nutrition involving motile pseudopods in which food is obtained by ingestion of relatively large, solid organic particles (for example, live bacteria or protists); see *carnivory, histophagy, osmotrophy, phagotrophy*

holozoic See *holotrophic*

homodynamic undulipodia Undulipodia on the same cell with the same pattern of beating; see *heterodynamic undulipodia*

homogenomic From the same genome (for example, heterokarya of ciliates, Pr-6)

homokaryosis Possessing only a single kind of nucleus as determined by genetics and morphology (the number of nuclei may be greater than one per cell); see *heterokaryotic*

homokaryotic See *homokaryosis*

homologous Pertaining to a structure or physiology of common evolutionary origin, but not necessarily identical in present structure and/or function (for example, haptonemata and sperm tails are homologous structures); term in molecular biology referring to the degree of sequence similarity of DNA from different sources

homology See *homologous*

homothallic Descriptive term of protoctist life cycle in which members of a single clone are adequate to ensure fertilization. Homothallism describes the sexual system in protoctists and fungi in which single propagules give rise to individuals of compatible mating types such that the sexual process of gamontogamy and/or gametogamy can occur by mating of members of a clone or cells from a single thallus. Homothallic clones or thalli are therefore self-compatible; see *heterothallism*

homothallism See *homothallic*

homothetogenic fission Type of transverse binary fission such that a point-to-point correspondence (homothety) is maintained between structures in both progeny (for example, most ciliates, Pr-6); see *perikinetal fission, symmetrogenic fission*

homotype (pl. **homotypes**) Major variable antigen type (trypanosomatid mastigotes, Pr-11); as homotype is destroyed by host vertebrate antibody response heterotype multiplies to become dominant, forming the new homotype; see *heterotype*

homoxenous parasite (pl. **homoxenous parasites**) Ecological term describing a symbiotroph that completes its life history in a single host; see *heteroxenous*

hormogonium (pl. **hormogonia**) Type of propagule; short filaments that break off from parent organism, disperse, and are capable of further growth (for example, some cyanobacteria (B-6) and algae)

host (pl. **hosts**) Organism that provides nutrition or lodging for symbionts or parasites. The larger member of a symbiotrophic association

H-pieces Cell wall units in some filamentous xanthophytes (for example, *Tribonema*) composed of the joined half-walls of adjacent cells and forming structures that are H-shaped in optical section

humic acids Alkaline- or water-soluble compounds extractable from humus

humus Layer of loose organic debris on land composed of sufficiently decayed organic materials such that their origins are obscured; organic-rich soil

hyaline Glassy, translucent, or transparent; descriptive of ameba cytoplasm or tests of foraminifera (Pr-2, -3). Hyaline tests may be subdivided into subtypes (radial, oblique, compound) depending on the orientation of the crystal laths

hyaloplasm Clear, organelle-free cytoplasm (for example, ectoplasm of ameba pseudopods, Pr-2)

hyalosome (pl. **hyalosomes**) Lens portion of a dinomastigote ocellus (for example, some members of the Warnowiaceae, Pr-5)

hybridization See *DNA hybridization*

hydrogenosomes See *chromatic granules*

hyperhaline Describes water of higher than typical marine salinity; that is, greater than about 3.5% sodium chloride and other salts; see *euhaline, euryhaline, oligohaline*

hyperparasitism Ecological term describing the topology of symbiotrophs in which a symbiont itself maintains a second symbiont; three-way symbiosis (for example, a microsporan parasite of a myxosporan parasite of a fish; F-1, A-2, A-37)

hypersaline See *hyperhaline*

hyperseme Increase in number or size of a trait of evolutionary significance (for example, number of undulipodia in parabasalians, Pr-1); see *seme*

hypertrophy Abnormal enlargement of a body part or structure (for example, a gall)

hypha (pl. **hyphae**) Long slender threadlike cells, walled syncytia, or parts of cells comprising the body of most fungi and many protoctists (for example, oomycotes or elongate cells of the medulla of kelps and fucoids; Pr-21, -17)

hyphal See *hypha*

hyphochytrid (pl. **hyphochytrids**) Informal name of members of the Phylum Hyphochytriomycota (Pr-14)

hypnocyst (pl. **hypnocysts**) Resting cyst of a dinomastigote (Pr-5)

hypnospore (pl. **hypnospores**) Thick-walled aplanospore (for example, in dinomastigotes (Pr-5))

hypnozygote (pl. **hypnozygotes**) Thick-walled zygote; fossilized form of dinomastigote (Pr-5) encountered in core samples; interpreted to be nonmotile zygote of dinomastigotes with a three-layered outer wall

hypocingulum (pl. **hypocingula**) Lower portion of the girdle (cingulum) adjacent to the hypotheca (for example, in dinomastigotes, Pr-5)

hypocone (pl. **hypocones**) Lower surface or hemisphere posterior to the girdle (cingulum) (for example, in dinomastigotes, Pr-5)

hypogynous Botanical term meaning below the ovary or female reproductive structure; as applied to protoctists, refers to antheridia of oomycote *Apodachlya* (Pr-21); in rhodophytes (Pr-33), hypogynous cell subtends a carpogonium in a carpogonial filament. Term should be avoided when referring to protoctists

hypolimnion Zone of water below the thermocline (metalimnion) where the temperature is uniformly low; lower water mass of thermally stratified lakes; see *epilimnion*

hypolith (pl. **hypoliths**) Ecological term referring to organisms dwelling on the undersides of rocks; see *epilith*

hypolithic See *hypolith*

hyposeme Decrease in number or size of a trait of evolutionary significance (for example, reduction of oral apparatus of *Mesodinium rubrum* (Pr-6) relative to other mesodinia); see *seme*

hyposome (pl. **hyposomes**) See *hypocone*

hypothallial Thin, often transparent deposit at the base of sporocarps (for example, myxomycotes, Pr-23); lowermost tissue in crust on which one or more layers of filaments are oriented parallel to the substrate in rhodophytes (Pr-33); medulla

hypothallus (pl. **hypothalli**) See *hypothallial*

hypotheca (pl. **hypothecae**) Hypovalve and hypocingulum portion of diatom frustule; posterior portion of a thecate dinomastigote cell (Pr-18, -5)

hypovalve (pl. **hypovalves**) Lower shell, opposite to and usually smaller than the epivalve (for example, diatom frustules, Pr-5)

hystrichosphere (pl. **hystrichospheres**) Fossil dinomastigote cyst (Pr-5); thick-walled, nearly spherical structure bearing characteristic projections and markings including an apparent excystment aperture (archeopyle). Fossil objects (Proterozoic to Recent) lacking enough detail to be identified as hystrichospheres are classified as acritarchs

ichnofossil (pl. **ichnofossils**) Mark left by extinct organisms resulting from their life activities (for example, *Scolithus*, a vertical tube, burrow, or trail). Geologists often give these genus and species names

ichthyotoxin (pl. **ichthyotoxins**) Any substance toxic to fish

imbrication pattern Overlapping pattern

immunofluorescence Visual detection by fluorescence microscopy of the presence and distribution of specific antigens on or in cells using antibodies bound to fluorescent molecules

imperfect fungi (pl. **imperfect fungus**) See *fungi imperfecti*

imperforate Lacking opening or aperture

inbreeding Mating and production of offspring by organisms known to be derived from recorded common ancestors

indeterminate growth Pertaining to growth that continues indefinitely under optimum conditions (for example, acellular slime molds (Pr-23), labyrinthulids (Pr-19), and mycelial fungi); see *determinate growth*

indeterminate sporangial renewal Continued cell division such that growth and development are not necessarily terminated by the appearance of a sporangium (for example, oomycotes (Pr-21)); see *cymose renewal*, *internal renewal*

inducing factors Pheromones (for example, substances that stimulate the sexual maturation of immature individuals)

infaunal Ecological term referring to organisms covered by sand, mud, or other sediment

infection Initiation of symbiotrophic (including necrotrophic) relationship among organisms of different species

infectious germ (pl. **infectious germs**) Ecological term describing the stage in which a symbiotroph is capable of continuing growth; propagule stage in which parasite is infectious (for example, apicomplexans (Pr-7), motile zoite; myxosporans (A-2), ameboid sporoplasms within the spore)

infraciliature Kinetids taken together; that layer of cortex containing undulipodial substructure. Assemblage of all kinetosomes and associated subpellicular, microfibrillar, and microtubular structures (that is, ciliates, opalinids; Pr-6)

ingestatory apparatus Oral apparatus. Entire complex of structures and organelles involved in or directly related to the mouth sensu lato and its ingestatory function (for example, multiple in suctorians and absent in astomatous ciliates and opalinids; cryptomonad gullet)

ingestion apparatus See *ingestatory apparatus*

inner membrane complex Cellular structure: flattened vesicles forming a double membrane lining the plasma membrane

inoculum (pl. **inocula**) Starter; a subpopulation, usually of microorganisms, used to transfer a culture for continued growth on fresh culture medium

insolation Incoming solar radiation; solar radiation received at the Earth's surface; the flux of direct solar radiation incident on a horizontal surface

interaxonemal substance (pl. **interaxonemal substances**) See *axonemal dense substance*

intercalary Between nodes; see *growth*

intercalary band (pl. **intercalary bands**) Zone, often striated, around margins of dinomastigote (Pr-5) thecal plates where cell growth occurs by addition of wall material

intercameral Between chambers (for example, tests of foraminifera, Pr-3)

intergeniculum (pl. **intergenicula**) Between genicula; calcified section between uncalcified joints; inflexible region of the axis of segmented thalli (for example, coralline rhodophytes or calcified chlorophytes; Pr-33, -28); see *geniculum*

interiomarginal aperture (pl. **interiomarginal apertures**) Opening in a final chamber bounded in part by the wall of an earlier chamber in tests of foraminifera (Pr-3)

interkinetal fission Fission between the kineties; descriptive of longitudinal cell division (for example, mastigotes and opalinids); see *perkinetal fission*

interkinetal space (pl. **interkinetal spaces**) Nonciliated region of a ciliate cortex that lies between kineties

intermediate host (pl. **intermediate hosts**) Animal host in which only the asexual or immature stages of the symbiont occur (for example, coccidian *Aggregata eberthi* (Pr-7) in crab); see *definitive host*

internal renewal Proliferation through the sporangial septum of the sporangiophore producing a new hypha, a sporangium of undetermined size, or a combination of these (for example, oomycotes, Pr-21); see *indeterminate sporangial renewal, percurrent development*

internal toothplate (pl. **internal toothplates**) Projection into the aperture of foraminiferal test (Pr-3). The internal portion of the toothplate usually extends as far as the previous foramen did through the chamber

internode (pl. **internodes**) Portion of stem or thallus lying between the nodes or joints (for example, in the chlorophyte *Chara* or the rhodophyte *Ceramium*, Pr-28, -33)

interphase Growth stage in the cell cycle of eukaryotes between successive mitoses in which the processes of mRNA transcription and protein synthesis are most active. Chromatin is uncondensed and invisible or difficult to see and stain

interseptar Between septae or partitions, referring to the position of a structure

interstitial Existing in small or narrow spaces between things or parts; ecological term referring to organisms or material between sand grains or mud particles; see *psammophile*

intertidal Eulittoral; littoral. Ecological term referring to the areas situated between the tides and therefore covered with seawater at high tide and exposed at low tide. Synonym of littoral in one of its senses: the benthic ocean environment or depth zone between high water and low water

intervening sequence (pl. **intervening sequences**) Untranslated sequence of DNA that forms part of a gene and is removed by splicing of the corresponding mRNA after transcription; see *exon*

interzonal spindle (pl. **interzonal spindles**) Array of microtubules extending from one end of a cell to the other, that is, those microtubules extending in telophase between the two offspring nuclei; distinguished from kinetochoric or chromosomal microtubules that only extend from the kinetochores to the spindle poles

intracristal filament (pl. **intracristal filaments**) Filament occurring within the inner membrane (cristae) of mitochondria

intraerythrocytic Ecological term referring to the topology of symbiotrophs in red blood cells (erythrocytes) of vertebrates (for example, the intraerythrocytic stage of apicomplexans (Pr-7) involves multiple fissions which lead to destruction of the red blood cells and poisoning by the breakdown products of erythrocytes)

intratissular Topological term referring to endosymbionts located within a tissue; see *pseudointratissular*

intrazoic Topological term referring to organisms located inside animals; see *endozoic, entozoic, holozoic*

intrinsic encystment Formation of cysts induced by nutrient depletion (for example, in the chrysophyte *Dinobryon cylindricum*, Pr-15)

intron (pl. **introns**) See *intervening sequence*

intussusception Invagination; the assimilation into a structure of new material and its dispersal among pre-existing material

involucre Nongrowing (sterile) group of cells or filaments that form envelopes around growing (fertile, reproductive) structures

involute Pertaining to tests that are curled spirally; having the whorls closely coiled; curled inward; having the edges rolled over the upper surface toward the midrib. Each whorl may completely embrace and cover earlier whorls so that only the final whorl is externally visible (for example, tests of foraminifera, Pr-3)

iodine test Test for the presence of starch: treatment of cells or tissues with weak aqueous solution of iodine–potassium iodide (Lugol's solution) in which colorless or white starch grains turn blue to black

ion exchange chromatography Technique for separating and identifying the components from mixtures of molecules using a resin that has a higher affinity for some charged organic ions than it has for others

ionophores Class of bacterially derived compounds including antibiotics that facilitate the movement of mono- and divalent cations across biological membranes

isoaplanogametes Nonmotile gametes of equal size (for example, hyphochytrid *Anisolpidium ectocarpii*, Pr-14)

isoenzyme (pl. **isoenzymes**) Variants of a given enzyme occurring within a single organism, having the same affinity for a substrate, but differing sufficiently in the molecular structure so that their separation is possible (usually by electrophoresis)

isofilar Description of filamentous structure composed of stretches of equal or nearly equal width along its length (for example, microsporan polar tube in the everted state or myxozoan polar filament);. see *anisofilar*

isogametes Gametes similar in size and morphology to the corresponding gametes of the opposite mating type; see *anisogametes*

isogametous See *isogametes*

isogamontous See *isogamonts*

isogamonts Gamonts of a given species that are the same in size or form; see *anisogametes*

isogamous Pairing of gametes alike in morphology and size opposite mating types (isogametes); see *anisogamy*

isogamy See *isogamous*

isokont An undulipodiated cell bearing undulipodia of equal length (for example, *Chlamydomonas*, Pr-28); see *anisokont*

isolate (pl. **isolates**) Population or strain of organisms under investigation in the laboratory

isomorphic life cycle Life cycle having alternation of generations in which individuals (for example, gametophyte, sporophyte) are morphologically similar; see *heteromorphic life cycle*

isoplanogametes Undulipodiated gametes (swarmers) of equal size destined for sexual reproduction; see *anisoplanogametes*, *isoaplanogametes*

isoprenoids Class of organic compounds synthesized from multiples of a ubiquitous five carbon compound precursor (isopentenyl pyrophosphate). Includes carotenoids, phytol, terpenes, steroids, and many other important biochemicals

isotherm (pl. **isotherms**) A line drawn on a map or chart linking all points with the same mean temperature for a given period or the same temperature at a given time

isotope fractionation Any process leading to the selective incorporation of certain isotopes of an element (for example, photosynthetic organisms fractionate carbon: they preferentially incorporate a ratio of 12C to 13C greater than that present in the atmosphere)

isozyme (pl. **isozymes**) See *isoenzyme*

K In population biology, the carrying capacity: the density of organisms at which population growth no longer occurs, that is, the reproduction rate equals the mortality rate

Kappa particle (pl. **Kappa particles**) Cytoplasmic particles correlated with the capacity of *Paramecium* (Pr-6) to kill conspecifics. Bacteria (*Caedibacter*, B-3) of *Paramecium*; see *xenosome*

karyogamy Fusion of nuclei; usually follows syngamy and leads to zygote production

karyokinesis Division of the nucleus to form two offspring nuclei

karyomastigont system Mastigont system with its associated nucleus (or nuclei) (for example, diplomonads); see *mastigont system*

karyonide (pl. **karyonides**) Clonal population of organisms, maintained in laboratories, in which no nuclear reorganization has been allowed to occur

karyosome (pl. **karyosomes**) See *endosome, nucleolus*

karyotype (pl. **karyotypes**) Total chromosome complement of an animal, plant, fungus, or protoctist as seen in fixed and stained preparations using a light microscope; karyotyping is a fixation and staining procedure used to determine characteristic morphology and the number of chromosomes for a species

kDNA Circular DNA molecules of trypanosomatids of two size classes, maxicircles and minicircles; usually catenated together in a network; see *kinetoplast*

keel (pl. **keels**) Longitudinal plate or timber extending along the center of the bottom of a ship and projecting outward into the ocean; in protoctists, any projection resembling a keel (for example, peripheral thickening of a foraminiferal test or a ridge on the valves of some pennate diatoms; Pr-3, -18)

kelp Large phaeophytes (Pr-17) that are members of the Laminariales and *Durvillaea* spp

kinete Motile form of the zygote of hematozoan apicomplexans (Pr-7); the kinete has zoitelike features such as pellicle, subpellicular microtubules, rhoptries, and micronemes

kinetid (pl. **kinetids**) See *basal apparatus*

kinetid root (pl. **kinetid roots**) See *banded root*

kinetochore (pl. **kinetochores**) Centromere. Microtubule-organizing center usually located at a constricted region of a chromosome that holds chromatids together. Kinetochores, morphologically visible manifestations of centromeres, are the site of attachment of microtubules forming the spindle fibers during nuclear division (mitosis and meiosis). In general, kinetochores reproduce in synchrony with the chromosomes and divide into two at metaphase, one new centromere segregating with each chromatid to the poles of the mitotic spindle. "Centromere" is a synonym, or if distinguished, centromeres are deduced from genetic behavior, whereas kinetochores are directly visible by electron microscopy

kinetocyst (pl. **kinetocysts**) Extrusome with a complex substructure of "missilelike differentiations" embedded in a fuzzy material (for example, heliozoan and polycystine actinopods, Pr-31)

kinetodesma (pl. **kinetodesmata**) See *banded root*

kinetodesmal fiber (pl. **kinetodesmal fibers**) See *banded root*

kinetodesmal fibril (pl. **kinetodesmal fibrils**) Portion of kinetids of ciliates; striated rootlet fibril originating near triplets 5–8 of the ciliate kinetosome and extending into the cortex; see *banded root, postciliary ribbon*

kinetodesmos See *banded root*
kinetoplast (pl. **kinetoplasts**) Modified mitochondrion; intracellular DNA-containing structure, often near a kinetosome, characteristic of the kinetoplastids (Phylum Kinetoplastida, Pr-11); the mitochondrial DNA that characterizes the kinetoplast is usually associated with the mitochondrial envelope apposed to the kinetosome(s). Kinetoplasts reproduce prior to the nuclei in cell division
kinetoplast DNA See *kDNA*
kinetosome (pl. **kinetosomes**) See *basal body*
kinetosome props See *fibrillar kinetosome props*
kinety (pl. **kineties**) Structure of the ciliate cortex (Pr-6); a row of kinetids. Kineties, typically oriented longitudinally, are composed of kinetids (single, paired, or occasionally several kinetosomes, their axonemes, and other associated cortical structures). These rows are bipolar (though some may be interrupted, fragmented, intercalated, partial, shortened, etc.), with an asymmetry allowing recognition of anterior and posterior poles of the organism; kinety sometimes also refers to linearly aligned buccal infraciliar structures
Knob scale (pl. **knob scales**) Scale; type of undulipodial surface structure in a subgroup of the prymnesiophytes
Koch's postulates Criteria for proving that a specific type of microorganism causes a specific disease; formulated by Robert Koch. The postulates state that the microbe should be found in diseased animals, but not healthy ones; that the organism must be grown in pure culture away from the animal; that this culture should cause the disease when injected into a healthy animal; and that the organism must be able to be reisolated from the experimental animals, recultured and be identified as the original organism
Kofoid system System of thecal plate designation in thecate dinomastigotes (Pr-5), devised by Charles Kofoid
kombu Edible seaweed, commonly prepared from kelp (Pr-17)

L zone Acantharian actinopod myoneme structure; L zones, as seen by electron microscopy, consist of repeated clear areas separated by thin dark transversal lines called T bands. L zone length varies with the degree of contraction of the myoneme
labiate process Lip-shaped structure. Siliceous tube or opening, which projects inward, or even outward, from the valve surface of diatoms (Pr-18). The labiate process terminates in a longitudinal slit surrounded by two liplike structures
lacuna (pl. **lacunae**) Cell structures; layers of flat vesicles underneath the cell membrane (plasmalemma) forming the lacunar system
Lacustrine Ecological term pertaining to lakes
lacustrine ooze Loose sediment, usually rich in small clasts, microbes in suspension, and organic matter. Slimy or muddy material at the bottom of lakes
lag phase The period just after inoculation of a microbial population prior to the detection of exponential growth rate
lageniform Morphological descriptive term meaning flask-shaped
lake (pl. **lakes**) An inland area of open, relatively deep water (fresh or saline) whose surface dimensions are sufficiently large to sustain waves
lamella (pl. **lamellae**) Morphological descriptive term referring to a flat thin scale or flattened saclike structure; see *thylakoid*
lamellate See *lamella*
lamellopodium (pl. **lamellopodia**) Ameboid cell process: broad, flat pseudopodium
lamina (pl. **laminae**) Morphological descriptive term referring to a thin plate or scale; layer

laminarase Enzyme that degrades laminarin

laminarin Carbohydrate stored as food in phaeophytes; polymer of glucose and mannitol with β-1,3 (and some β-1,6) glycoside linkages

lanceolate Morphological descriptive term meaning shaped like a lance head, that is, tapering to a point at the apex and sometimes also at the base

lapidolith (pl. **lapidoliths**) Coccolith type in which the layers of elements are parallel to the coccolith base; holococcolith (for example, *Laminolithus hellenicus*, Pr-25)

latent form (pl. **latent forms**) Stages in the life cycle of an organism that is more or less dormant, cysts or resting spores (for example, the sporocyst or other stages of apicomplexans (Pr-7) that develop slowly and persist for some time without growth)

lateral conjugation Gamontogamy in which conjugation tubes link gametes from adjacent cells in the same filament (for example, the conjugating green alga *Spirogyra* spp., Pr-32)

lateral crest (pl. **lateral crests**) Ridge that supports the adhesive disk (for example, diplomonads, Pr-1)

lateral renewal See *cymose renewal*

laver (pl. **lavers**) Dried edible preparations of algae such as *Ulva* (chlorophyte) and *Porphyra* (rhodophyte, Pr-33)

lectin Protein capable of agglutinating certain cells by binding to specific carbohydrate receptors on the surface of these cells

Leishmaniasis Infection by the genus *Leishmania* (trypanosomatids, Pr-11) that inhabit macrophages of vertebrate blood

lens (pl. **lenses**) Translucent structure capable of light refraction

lentic Ecological term referring to organisms inhabiting standing water

lenticular Morphological descriptive term referring to the shape of a double-convex lens

leptonema First stage of meiotic prophase I, in which chromosomes begin to condense and form threads; see *meiosis*

leptotene See *leptonema*

leucoplast (pl. **leucoplasts**) Cell organelle of algae; colorless or white, often starch-storing plastid

leucosin See *chrysolaminarin*

life See *autopoiesis*

life cycle (pl. **life cycles**) Events throughout the development of an individual organism correlating environment and morphology with genetic and cytological observations (for example, ploidy of the nuclei, fertilization, meiosis, karyokinesis, and cytokinesis)

life history (pl. **life histories**) Events throughout the development of an individual organism correlating environment with changes in external morphology, formation of propagules, and other observable aspects

ligand (pl. **ligands**) Molecule that binds to a complementary site

lignophile Ecological term referring to organisms living in or on the surface of wood of shrubs or trees

lignophilic See *lignophile*

ligula (pl. **ligulas**) Tonguelike extension of the girdle band, which fits beneath the split in the adjacent band in diatoms

limacine movement Monopodial locomotion; limax movement. Sluglike pattern of locomotion produced by single eruptive anterior ectoplasmic pseudopod of an ameba (Pr-2)

limax Morphological descriptive term meaning "shaped like a slug."

limestone Sedimentary rock consisting primarily of calcite (calcium carbonate, $CaCO_3$) with or without magnesium carbonate. Limestone is the most widely distributed of the carbonate rocks

limax ameba Elongate, usually monopodial morphotype typical of the amebas of amoebomastigotes and acrasids (Pr-22, -2)

limax amoeba See *limax ameba*

linella (pl. **linellae**) Long, thin threads composed of a cementlike matter found outside the granellare of xenophyophorans (Pr-4); regarded as an organic part of the test

linkage Linkage group; genetic term describing the condition in which traits are inherited together and thus the genes for these traits are inferred to be physically close together on the same chromosome (linked). In genetically well-mapped eukaryotic organisms, the number of linkage groups corresponds to the number of chromosomes, since linked genes tend to segregate together. In viruses and prokaryotes, the single linkage group corresponds to the genophore

lipid (pl. **lipids**) One of a class of organic compounds soluble in organic, but not aqueous solvents; includes fats, waxes, steroids, phospholipids, carotenoids, and xanthophylls

list Cellulosic extension of the cell wall in some armored dinomastigotes (Pr-5) usually extending out from the cingulum and/or sulcus

lithology The study of rocks on the basis of such characteristics as color, grain size, and mineralogical composition

litholophus Modification of the cell shape in an order of chaunacanthids (phylum Actinopoda, Pr-31). Initially spherical with radial spicules, the acantharian progressively takes the shape of a closed umbrella with all the spicules lying parallel preparatory to cyst formation

lithophile Ecological term referring to organisms dwelling on stones and rocks; see *chasmolith, endolith, epilith, saxicolous*

lithophilic See *lithophile*

lithosome (pl. **lithosomes**) Organelle; vesicular, membrane-bounded cytoplasmic inclusion, comprised of inorganic material laid down in concentric layers (for example, the prostomate ciliates (Pr-6); acantharian actinopods (Pr-31), in which it is composed of $SrSO_4$)

litter Layer of loose organic debris on land, composed of freshly fallen leaves or only slightly decayed materials in which the remains of organisms are detectable

littoral Eulittoral; intertidal. Ecological term referring to that portion of sandy, muddy, and rocky coasts that lies between high- and low-water marks; see *sublittoral, supralittoral*

lobocyte (pl. **lobocytes**) Free cell with a supposedly phagocytic, scavenger function inside large myxosporean trophozoite (plasmodium) of the genus *Sphaeromyxa* (A-2)

lobopodium (pl. **lobopodia**) Lobular, more or less rounded or cylindroid pseudopod; used in both locomotion and feeding

lobose Having many or large lobes; especially in reference to the pseudopods of amebas in the order Lobosa (phylum Rhizopoda, Pr-2)

loculate Chamber having a constricted opening on one side and a velum on the other side (for example, diatom frustule, Pr-18; compartment of a reproductive organ in algae)

locule See *loculate*

longitudinal fission Cell division along the longitudinal axis of an asymmetric cell; see *interkinetal fission, symmetrogenic fission*

longitudinal flagellum (pl. **longitudinal flagella**) Longitudinally aligned undulipodium of dinomastigotes (Pr-5) that originates and partially lies within the sulcus

longitudinal undulipodium (pl. **longitudinal undulipodia**) See *longitudinal undulipodium*

lopodolith (pl. **lopodoliths**) Helicoid placolith; see *coccolith*

lorica (pl. **loricae**) General term for structure external to the cell membrane in many protoctists: envelope; test; shell; valve; sheath; protective covering secreted and/or assembled which may be calcareous, proteinaceous, chitinous, pseudochitinous, siliceous, or tectinous in nature, or made up of foreign matter such as siliceous sand grains, diatom frustules (Pr-5), coccoliths (Pr-25), or debris

loricate See *lorica*

lower fungi Term used to group three mastigote protoctistan groups (chytridiomycetes, hyphochytridiomycetes, and oomycetes, Pr-35, -14, -21) with the zygomycete amastigote fungi (F-2) to distinguish them from the "higher" fungi (asco- and basidiomycetes, F-4, -5). It is inadvisable to use such nontaxonomic general terms as "higher" and "lower" because of their ambivalence and anthropocentrism

luciferase An enzyme that catalyzes the oxidation of luciferins in reactions that generate visible light (bioluminescence). Many types of enzymes and substrates exist. Eukaryotic luciferases always require ATP and oxygen; prokaryotic luciferases may not be ATP dependent

luciferin Any of a number of organic compounds of luminescent organisms that are substrates in luciferase reactions

Lugol's solution See *iodine test*

luminescence See *bioluminescence*

lyophilization Freeze-drying; method for preservation of resistant protoctists

lysis Cell disintegration following rupture of the cell membrane

lysocline Depth in the ocean below which calcium carbonate ($CaCO_3$) skeletons dissolve because of hydrostatic pressure. Neither living nor dead foraminifera (Pr-3) or other protoctists with calcareous tests are found beneath this depth

lysogenic conversion Change in phenotype of a bacterium that accompanies lysogeny, that is, the process in which the genetic material of a virus is incorporated into the genetic material of its host bacterium (for example, in some bacteria, toxin production only occurs when the appropriate virus is incorporated into the genophore)

lysosome (pl. **lysosomes**) Membrane-bounded organelle containing releasable hydrolytic enzymes

lysozyme (pl. **lysozymes**) Enzyme hydrolyzing the peptidoglycans of cell walls and hence used to break open bacteria (for example, in egg white, tears)

macroalgae Algae visible to the naked eye; large algae in contrast to microscopic algae; see *macrophyte*

macrocyst (pl. **macrocysts**) Large cyst (for example, multicellular, irregularly circular or ellipsoidal resting structure about 25–50 μm in size with three distinct walls formed during the sexual cycle in some dictyostelids (Pr-2); protoplasmic, walled, usually multinucleate portions of a myxomycote (Pr-23) sclerotium)

macrogametangium (pl. **macrogametangia**) Algal sexual structure (for example, gametangium containing relatively large locules and thus producing the larger (macro-) gametes)

macrogamete (pl. **macrogametes**) Large, usually female gamete (for example, apicomplexans, Pr-7); see *microgamete*

macrogametocyte (pl. **macrogametocytes**) Descriptive term for the larger of the two gamonts in anisogamontous gamontogamy. Female gamont that transforms into a single macrogamete (for example, in coccidian apicomplexans, Pr-7); see *microgamont*

macrogamont (pl. **macrogamonts**) See *macrogametocyte*

macronucleus (pl. **macronuclei**) Larger of the two kinds of nuclei in ciliate cells (Pr-6); site of messenger RNA synthesis; containing more than two (and often hundreds of) copies of genes, it is required for growth and division; see *micronucleus*

macrophagy Mode of heterotrophic nutrition in which organisms feed on food particles large with respect to their own size

macrophyte (pl. **macrophytes**) Literally "large plant" but inadvisably used for large algae. Usually refers to phaeophytes, rhodophytes, and large chlorophytes (Pr-17, -33, -28)

macrophytic See *macrophyte*

macroschizozoite Life cycle stage of apicomplexans (Pr-7); zoite produced by large schizont

macrosclerotium (pl. **macrosclerotia**) Large sclerotia formed by phaneroplasmodia or aphanoplasmodia in myxomycotes (Pr-23)

macrosporangium (pl. **macrosporangia**) Sporangia containing large spores (macrospores) as opposed to sporangia of similar dimensions containing small spores

macrospore (pl. **macrospores**) Large spore, usually in contrast to microspores made by the same species

macrostome Inducible morph of certain *Tetrahymena* ciliates (Pr-6) in which cell develops a large oral apparatus correlated with carnivorous feeding; see *microstome*

macrothallus (pl. **macrothalli**) Large conspicuous flat-structured morph of relatively large thallus (for example, rhodophytes, chlorophytes, or phaeophytes, Pr-33, -28, -17); see *microthallus*

macrozoospore (pl. **macrozoospores**) Zoospore of large size relative to others produced by the same organism (for example, in the prymnesiophyte *Phaeocystis*, Pr-25); see *microzoospore*

Maerl Unconsolidated poorly lithified clastic sediment, a mixture of clay and calcium carbonate, usually including shells and sometimes including living fragments of deep-water crustose coralline rhodophytes (Pr-33); formed under marine and especially freshwater conditions

magnetotactic Pertaining to directed locomotion in a magnetic field toward a magnetic pole (for example, south- or north-seeking; as in magnetite-containing bacteria or *Chlamydomonas*, Pr-28)

magnetotaxis See *magnetotactic*

maintenance culture Collection of mixed microorganisms (for example, a polyxenic culture of live protoctists maintained in the laboratory by periodic addition of water or new medium); see *culture*

maintenance culture medium (pl. **maintenance culture mediums, maintenance culture media**) Liquid or solid material providing nutrients and osmotic conditions for the maintenance (but not necessarily the growth) of microorganisms in the laboratory

male (pl. **males**) Gamont. Gender of individual that produces sperm or other usually motile gamete and donates it to the female

mannan Polysaccharide component of walls of some rhodophytes and chlorophytes (Pr-33, -28), which yields mannose upon hydrolysis; often a β-1,4 mannopyranoside

mannitol A polyol, that is, a 6-carbon sugar alcohol storage product of phaeophytes (Pr-17)

mannuronic acid Algal metabolite; acid derivative of the sugar mannose

mantle (pl. **mantles**) Outermost portion of valve that is bent at approximately 90° and connected to the girdle (for example, diatoms). Inner shell lining of bivalve molluscs (A-26)

manubrium (pl. **manubria**) Columnar cell connecting the pedicel to the shield cell in the antheridia of charophytes (Pr-28)

marine Of the sea or ocean; inhabitating, found in, or formed in the sea

marine snow Irregularly shaped particulate matter, up to several millimeters in diameter, that precipitates in the ocean, falling toward the bottom. Includes living colonies of protoctists and organic remains of protoctists

marker species Fossil species found in sedimentary facies and used to correlate (and date) that facies over relatively long distances

Marl See *Maerl*

mastigonemate See *flimmer*

mastigoneme (pl. **mastigonemes**) See *flimmer*

mastigont system Intracellular organellar complex found in many mastigotes (for example, parabasalians, diplomonads, retortamonads, Pr-1). Organelles associated with undulipodia, the mastigont system may include the kinetids with their undulipodia, undulating membrane, costa, parabasal bodies, and axostyle; see *karyomastigont system*

mastigote (pl. **mastigotes**) See *flagellate*

mastigote division Karyokinesis and cytokinesis of undulipodiated protists (for example, cryptomonads, euglenids, proteromonads, Pr-26, -12, -1)

mating Syngamy; see *conjugation*

mating type (pl. **mating types**) Strain of organisms incapable of sexual fusion with each other but capable of sexual reproduction with members of another strain of the same organism

maxicircle (pl. **maxicircles**) Large circular DNA molecule (20–38 kb) in the kinetoplast of trypanosomatids (Pr-11), corresponding to mitochondrial DNA of other eukaryotes; held together in a network by minicircle DNA

MC Granulofibrosal material visible with the electron microscope from which microtubules arise and grow. May be associated with centrioles, centriolar plaques, kinetosomes, or other intracellular organelles

mechanical transmission Transmission from one host to another of a symbiont that does not undergo a cycle of development in the vector but retains its morphological and physiological state (for example, *Trypanosoma evansi* (Pr-11) that moves between vampire bats and ungulates, A-37)

median body (pl. **median bodies**) Structures composed of cytoskeletal proteins; incipient adhesive disk of an offspring mastigote formed prior to mitosis (for example, diplomonads: distinguishes species of *Giardia*, Pr-1)

mediocentric Referring to chromosomes with centrally, or nearly centrally, located kinetochores (centromeres)

medium (pl. **mediums, media**) See *culture medium*

medulla Morphological term referring to the central region of an organ (for example, adrenal) or organism (for example, thallus of lichen or alga); see *cortex*

megacytic zone (pl. **megacytic zones**) Region of expansion between adjacent plates where new thecal material is added to allow enlargement of the cell in armored dinomastigotes (Pr-5)

megalospheric test (pl. **megalospheric tests**) Gamont generation test with a large initial chamber in foraminifera (Pr-3)

megasporangium (pl. **megasporangia**) See *macrosporangium*

meiocyte (pl. **meiocytes**) Cell destined to undergo meiosis

meiosis One or two successive divisions of a diploid nucleus that result in the production of haploid nuclei. In organisms with gametic meiosis, meiotic divisions precede the formation of gametes (for example, diatoms). In organisms with zygotic meiosis the zygote undergoes meiosis immediately after it forms (for example, volvocalean algae, Pr-28). Prophase is much longer than in mitosis and can be divided into five consecutive stages: leptonema, zygonema, pachynema, diplonema, and diakinesis. During most meioses homologous chromosomes pair forming the synaptonemal complex. Meiosis is found in all plants and animals, most if not all sexual fungi, and in many, but by no means all, protoctists. Protoctists display extremely varied patterns of life cycles that involve meiosis and fertilization

meiosporangium (pl. **meiosporangia**) Sporangium in which meiosis occurs

meiospore (pl. **meiospores**) Spore produced by meiosis

meiotic See *meiosis*

melanosome (pl. **melanosomes**) Black- or brown-pigmented body containing the tyrosine-derived polymer melanin. Light-sensitive portion of the ocellus onto which the hyalosome focuses (for example, pigmented portion of certain dinomastigote cells); also, any of the melanogenic granules of pigment-producing cells from their earliest recognizable (unpigmented) stage to their completely or partially electron-dense and definitively patterned stage

meristem (pl. **meristems**) Undifferentiated tissue composed of rapidly growing cells; region on protoctist thallus at which new cells arise

meristoderm Superficial layer of rapidly dividing and growing cells covering the thalli of laminarialean phaeophytes (Pr-17)

merogenous sperm formation Spermatogenesis in which a portion of the protoplasm of a microspore converts to form a sperm and the remainder is discarded (for example, centric diatoms, Pr-18)

merogony Multiple fission of apicomplexans, (Pr-7); schizogony resulting in merozoites which themselves undergo multiple cell divisions

meront Life history stage in which active cells grow and divide to form merozoites (for example, apicomplexans and microsporans; Pr-7, F-1)

meroplankton Ecological term referring to neritic organisms that spend part of their life history as plankton and part in benthic communities

meroplanktonic See *meroplankton*

merotomy Division of cells ("cutting up") into portions with or without nuclei

merozoite (pl. **merozoites**) Mitotic product of trophozoites; life history stage. Merozoites may differentiate into meronts or gamonts (for example, *Plasmodium*, Pr-7)

mesocyst (pl. **mesocysts**) Middle of three layers surrounding a heliozoan oocyst (Pr-31); the mesocyst is surrounded by siliceous scales; see *ectocyst*, *endocyst*

mesokaryote Dinokaryotic. Referring to dinomastigote with nuclei that lack conventional histones and have permanently condensed chromosomes. Literally "between prokaryotic and eukaryotic"; see *noctikaryotic*

mesokaryotic See *mesokaryote*

mesomitosis Dinomastigote (Pr-5) karyokinesis; mitosis of the mesokaryotic nucleus in which chromosomes remain condensed and attached to nuclear membrane, breakdown of the nuclear envelope and nucleolus is delayed, and centrioles are lacking; see *metamitosis*, *promitosis*

mesophotic zone Ecological term referring to the region in the water column between the compensation depth (at which the rates of respiration and photosynthesis in phytoplankton are equal over a 24-hour period) and the depth to which no surface light reaches. Dimly lit lower portion of the photic zone

mesosaprobic Ecological term referring to an aquatic environment containing a moderate amount of dissolved organic matter (moderately polluted); see *eutrophy*, *oligotrophy*

mesospore (pl. **mesospores**) Thin middle wall of the three-layered cell wall of hypnozygote (dinomastigote); the middle layer of the zygospore wall (conjugating green algae, Pr-32); see *endospore*, *exospore*

mesotrophic See *mesosaprobic*

metabolism The sum of enzyme-mediated biochemical reactions that continually occurs in cells and organisms and provides the material basis of autopoiesis

metaboly See *euglenoid motion*
metacentric Morphological term pertaining to a mitotic spindle that radiates from centrioles, which lie in the same plane as the metaphase chromosomes (for example, in some chlorophytes, Pr-28); also, referring to chromosomes with centromeres that lie exactly or nearly half way between the chromosome ends or telomeres
metachronal waves See *antiplectic*
metacyclic stage The stage in the development of a parasitic protoctist in its invertebrate host (vector) just before transfer to the vertebrate; this stage is normally infective to the vertebrate host
metacyclogenesis Formation of metacyclic stage
metagenesis Term used for alternation of generations, especially of a sexual and an asexual generation, in heterotrophic organisms
metagenic See *metagenesis*
metalimnion Water at thermocline in lake or other thermally stratified body of water; see *epilimnion, hypolimnion*
metamitosis Conventional metazoan mitosis in which there are centrioles or other conspicuous microtubule-organizing centers at the poles and loss of the nucleolus and nuclear envelope; see *mesomitosis, promitosis*
metaphase plate (pl. **metaphase plates**) See *equatorial plate*
metaphyton Ecological term referring to the biota, especially the microbiota, surrounding plants (metaphytes)
metazoa (pl. **metazoans**) Members of the kingdom Animalia. All organisms developing from a blastular embryo, itself derived from an egg usually fertilized by a sperm. Metazoan bodies are made of cells differentiated into tissues and organs and usually have a digestive cavity with specialized cells. Excludes all "protozoa" a term that includes many different protoctist phyla
microalgae Microscopic algae (as opposed to large algae)
microbe (pl. **microbes**) Any live being not visible to the naked eye and thus requiring visualization by microscopy
microbenthos Bottom-dwelling microbes, small animals, or microbial communities in fresh or marine waters; see *Aufwuchs community, periphyton, plankton, seston*
microbial mat (pl. **microbial mats**) Laminated organo-sedimentary structure composed of stratified communities of microorganisms, usually dominated by phototrophic bacteria, especially cyanobacteria (B-6). Types range from soft, brightly colored layered sandy sediment to lithified carbonates; living precursors of stromatolites
microbiota Sum of microorganisms in a given habitat (for example, termite intestinal microbiota); term preferable to microflora, which implies plants, or microfauna, which implies animals
microbody (pl. **microbodies**) Small intracellular structure; any of a number of organelles of eukaryotic cells bounded by a single membrane and containing a variety of enzymes. Microbodies, usually associated with one or two cisternae of the endoplasmic reticulum, include glycosomes, glyoxysomes, hydrogenosomes, and peroxisomes
Microbody–lipid globule complex Conspicous structure, of unknown function, near the kinetosome of zoospores of chytridiomycotes (Pr-35)
microcyst (pl. **microcysts**) Type of cyst or spore. Microcysts are encysted myxamebas (for example, myxomycotes and acrasids, Pr-23, -2). A two-layered fibrillar wall composed mostly of cellulose comprises the microcyst of dictyostelids, particularly *Polysphondylium* and most of the smaller species of *Dictyostelium* (but not *D. discoideum* and other large species). The wall is secreted in response to adverse environmental conditions, especially the presence of ammonia
microfauna See *microbiota*

microfilament (pl. **microfilaments**) Very small filament or microfibril; general term describing any solid, thin, fibrous proteinaceous structure, generally those in the cytoplasm of eukaryotic cells, some of which are composed of actin and participate in motility

microflora See *microbiota*

microgamete (pl. **microgametes**) Descriptive term for small gamete (for example, in apicomplexans the male gamete); see *macrogamete*

microgametocyte (pl. **microgametocytes**) The smaller of the two gamonts in anisogamontous gamontogamy (for example, in apicomplexans, the gamont that produces microgametes); see *macrogamont*

microgamont (pl. **microgamonts**) See *microgametocyte*

microgranular test (pl. **microgranular tests**) Test of foraminifera (Pr-3) composed of equidimensional subspherical granules of calcite packed closely together without detectable cement. In many forms, there are two layers: an outer layer of irregularly packed granules and an inner, highly ordered, packed layer

microheterotroph (pl. **microheterotrophs**) Term used by oceanographers and ecologists that lumps prokaryotes with nonphotosynthetic or mixotrophic protists to form an arbitrary grouping of diverse facultatively heterotrophic microorganisms fewer than 8 μm in diameter

micrometer (pl. **micrometers**) A millionth of a meter; linear unit of measurement (Table 3)

micron (pl. **microns**) See *micrometer*

micronemes Dense bodies in apicomplexan zoites (Pr-7), most abundant in apical complex area and probably corresponding to secretions of the Golgi apparatus

micronucleus (pl. **micronuclei**) Small nucleus; the smaller of the two types of nuclei in ciliates (Pr-6). The ciliate micronucleus does not synthesize messenger RNA; it is usually diploid and may undergo meiosis prior to syngamy and autogamy. The ciliate micronucleus, required for all sexual processes, is not always necessary for growth or cell division; see *macronucleus*

micronutrient (pl. **micronutrients**) Mineral or element required only in minute quantities for microbial growth (for example, iron, magnesium, cobalt, or zinc)

microorganism (pl. **microorganisms**) See *microbe*

micropaleontology Subdiscipline of geology: study of fossil microbes and the microscopic parts of fossil organisms (for example, pollen and spores)

micropore (pl. **micropores**) Small opening

microschizozoite (pl. **microschizozoites**) Life cycle stage of apicomplexans (Pr-7); zoite produced by small schizont

microsource (pl. **microsources**) Brightly fluorescent spherical bodies, about 0.5 μm in diameter, from which light flashes emanate, distributed primarily in the cortical cytoplasmic region of bioluminescent dinomastigote cells (Pr-5); see *scintillons*

microspheric test (pl. **microspheric tests**) Foraminiferal test (Pr-3) with a small initial chamber. The overall size of the test is generally larger than a test of megalospheric generation. Commonly part of the agamont generation

microspine (pl. **microspines**) Small pointed structure such as that found decorating protist tests

microsporangium (pl. **microsporangia**) Structure that harbors microspores

microspore (pl. **microspores**) Small spore; haploid spore that develops from the microspore parent cell and develops into a male gametophyte (for example, rhodophytes (Pr-33), plants); product of division of a cell that undergoes meiosis giving rise to four sperm cells (for example, diatoms, Pr-18); reproductive structure formed in sporophores (for example, slime molds, Pr-2, -23)

microstome (pl. **microstomes**) Inducible morph of *Tetrahymena* ciliates (Pr-6) in which the cell develops a small oral apparatus correlated with bactivorous feeding; see *macrostome*

microthallus (pl. **microthalli**) Small, inconspicuous phase in the life history of some rhodophytes, phaeophytes, or chlorophytes (Pr-33, -17, -28) that alternates with the macrothallus; see *macrothallus*

microtrophic Nutritional mode referring to heterotrophic organisms (that is, animals, protists) that feed on microbes

microtrophy See *microtrophic*

microtubular bundles Fibrous structures (on inspection by electron microscopy) composed of longitudinally aligned 24 nm microtubules

microtubular Slender, hollow structure primarily made of tubulin proteins (α-tubulin and β-tubulin) each with a molecular weight of about 50 kDa, arranged in a heterodimer. Microtubules are of varying lengths but usually invariant in diameter at 24–25 nm; substructure of axopods, mitotic spindles, kinetosomes, undulipodia, haptonemata, nerve cell processes, and many other intracellular structures; their formation is often inhibitable by colchicine, vinblastine, podophyllotoxin, and other microtubule polymerization-inhibiting drugs

microtubular fiber (pl. **microtubular fibers**) Thin structure (as seen by light microscopy). Electron microscopy reveals it to be a microtubule bundle. Fiber associated with microtubule

microtubular rod (pl. **microtubular rods**) Axoneme. Bundle of parallel microtubules that stiffens axopods of actinopods (Pr-31) or tentacles of suctorian ciliates (Pr-6)

microtubule (pl. **microtubules**) See *microtubule*

microtubule-organizing center See *MC*

microvillus (pl. **microvilli**) Cytoplasmic projection from epithelial cells; may contain microfibrils or microtubules

microzooplankton Ecological term grouping small motile heterotrophs passively moved by currents in aquatic environments. Term should be restricted to microscopic organisms in the kingdom Animalia. Unidentified suspended protoctists are more accurately referred to by size categories such as microplankton or nanoplankton

microzoospore (pl. **microzoospores**) Zoospore of small size relative to others produced by same organism; see *macrozoospore*

mictic Pertaining to two-parent sex; to syngamy or karyogamy leading to fertilization to form an individual with two different parents; see *apomixis*

midlittoral Ecological term referring to the central region of the intertidal zone

mineral (pl. **minerals**) Naturally formed chemical element or compound having a definite chemical composition (for example, calcite, strontium sulfate, silica); see *biomineralization*

minicircles One type of organization of DNA of the kinetoplastids (Pr-11); small circular kinetoplast DNA molecules (0.46–2.5 kb) of unknown function composing the bulk of kDNA and catenated with maxicircles to form a network

mitochondriome (pl. **mitochondriomes**) See *chondriome*

mitochondrion (pl. **mitochondria**) Membrane-bounded intracellular organelles containing enzymes and electron transport chains for oxidative respiration of organic acids and the concomitant production of ATP. Mitochondria have DNA, messenger RNA, and small ribosomes and are thus capable of protein synthesis; they are nearly universally distributed in protoctists but notably absent in *Pelomyxa*, some rhizopods (Pr-2), parabasalians (Pr-1), and certain other protist taxa

mitosis Nuclear division; karyokinesis; although protists vary widely in details of the mitotic process generally four stages are recognizable: prophase, in which the centriole divides and the attached pairs of duplicate chromosomes condense; metaphase, in which the chromosomes move and align at the equatorial plane of the nucleus; anaphase, in which the chromatids separate at their kinetochores and move to opposite poles; and telophase, in which the chromosomes return to their extended state; the result is two separate, identical groups; since the chromosomes replicate once before mitosis, and only one division occurs, the ploidy of the nucleus is unaltered by mitosis; see *cryptomitosis*, *cryptopleuromitosis*, *cytokinesis*, *mesomitosis*, *metamitosis*, *pleuromitosis*, *promitosis*

mitosporangium (pl. **mitosporangia**) Sporangium in which spores are produced by mitotic cell divisions

mitotic See *mitosis*

mitotic apparatus Microtubules, kinetochores, centrioles, centrosomes, and any other transient proteinaceous structures associated with mitotic cell division

mitotic oscillator (pl. **mitotic oscillators**) Hypothetical regulator of synchronous nuclear division in plasmodia of plasmodial slime molds (Pr-23)

mitotic spindle (pl. **mitotic spindles**) Transient microtubular structure that forms between the poles of nucleated cells and is responsible for chromosome movement

mixis See *mictic*

mixotrophic Nutritional mode: facultative chemoheterotrophy in a photoautotrophic organism (Table 2)

mixotrophy See *mixotrophic*

MLC See *microbody–lipid globule complex*

MLS Type of kinetid characteristic of chlorophytes, Pr-28 (for example, charophyte motile cells) containing a band of microtubules that overlies several layers of parallel plates; the proximal portion of the kinetid consists of several layers of which one layer has regularly spaced lamellae oriented perpendicularly to the overlying rootlet microtubules

MOC See *MC*

monad (pl. **monads**) Single cell; free-living, unicellular, usually undulipodiated organism or stage of an organism; mastigote

monadoid See *monad*

moniliform Descriptive morphological term referring to components arranged in a linear order like beads on a string

monocentric Term describing development of thallus in funguslike protoctists (for example, oomycotes, chytridiomycotes, Pr-21, -35) and algae. Thalli with a single central structure into which nutrients flow and from which reproductive structures are initiated; a monocentric thallus may be holocarpic or eucarpic; see *polycentric*

monoclinous Refers to having antheridia and oogonia originating from the same hypha (for example, the oomycote *Aphanomyces stellatus*, Pr-21); see *diclinous*, *monoecious*

monoclonal antibody (pl. **monoclonal antibodies**) Antibodies derived from a single clone of vertebrate plasma cells that have highly specific antigen-binding properties

monodisperse Descriptive term referring to polymers that are homogeneous in molecular weight

monoecious Hermaphroditic; monoclinous. Descriptive term referring to organisms that have male and female structures on the same individual; see *dioecious*

monogenetic The life cycles of symbiotrophic protoctists that occur in only one kind of host; see *heteroxenous*

monokinetid (pl. **monokinetids**) Kinetid containing one kinetosome and one each of associated structures; see *dikinetid*

monolamellar Description of form: structure with one lamella or layer

mononucleate A cell containing a single nucleus

monophyletic Evolutionary term referring to a trait or group of organisms that evolved directly from a common ancestor; see *polyphyletic*

monoploid See *haploid*

monopodial Increasing in length by apical growth; in algae, a type of growth in which the primary axis is maintained as the main line of growth and secondary laterals (offshoots) are produced from the primary axis; see *limacine movement, limax ameba*

monoraphid A diatom frustule (Pr-18) with a single raphe

monospecific Belonging to a single species (for example, a monospecific bloom consists of a single species)

monosporangium (pl. **monosporangia**) Sporangium that produces a single spore, that is, a monospore

monospore (pl. **monospores**) Nonmotile spore produced by mitosis and cytokinesis one at a time from a sporangium; asexual, naked spores of bangiophycidean rhodophytes (Pr-33)

monostromatic Morphological term describing a structure composed of a single layer of cells

monothalamic Bearing only a single chamber; refers to foraminifera (Pr-3) that form a test with only one chamber

monothalamous See *monothalamic*

monoxenic Pertaining to laboratory growth of two species of organisms, one of which is usually studied from a biochemical or ecological viewpoint. A monoxenic culture may involve, for example, a ciliate plus one "stranger" bacterium, alga, yeast, or other ciliate species; the second organism may be unwanted (a contaminant) or may be present in the medium to serve as food for the first; see *axenic*

monoxenous See *monogenetic*

monozoic Possessing a single body form (for example, resulting from complete cell division); see *diplozoic*

morph (pl. **morphs**) Form. Organism or structure with distinguishable size and shape. Environmentally induced form of an organism

morphology Study of form or results from a study of form

morphometrics Subfield of morphology; quantitative study of form, size, and shape variation within a species or strain of organism

morphotype (pl. **morphotypes**) Typical morph. Also term used when taxonomic identification is temporarily uncertain

mouth (pl. **mouths**) See *buccal cavity, oral region*

MTOC See *MC*

muciferous body (pl. **muciferous bodies**) Mucilage-containing extrusome; subpellicularly located, saccular or rod-shaped organelle or a paracrystalline structure, dischargeable through an opening in the pellicle as an amorphous, mucuslike mass (for example, dinomastigotes, pseudociliates, raphidophytes, Pr-5, -24, -15); in ciliates (Pr-6), probably involved in cyst formation (in some species), among other possible functions; not a trichocyst although sometimes used synonomously; see *mucocyst*

mucilage Mucous material, generally composed of polysaccharides

mucocyst (pl. **mucocysts**) Intracellular organelles (less complex than trichocysts) producing or filled with mucilage or mucus. Seen in euglenoids, dinomastigotes, chrysophytes, and prymnesiophytes (Pr-12, -5, -15, -25). May be extrusomes; see *muciferous body*

mucron (pl. **mucrons**) Anteriorly located attachment organelle not separated from the rest of the body by a septum (for example, in gregarine apicomplexan families Ganymedidae and Lecudinidae, Pr-7); analogous structures are found in some symbiotic ciliates (Pr-6) and mastigotes; see *epimerite*

mucus body (pl. **mucus bodies**) See *mucocyst, muciferous body*

mucus trichocyst (pl. **mucus trichocysts**) See *muciferous body*

mud A slimy and sticky or slippery mixture of water, slime, and finely divided particles (silt size or smaller) of aluminosilicate clays or other minerals

multilayered structure See *MLS*

multilocular Foraminifera (Pr-3) characterized by many cells or chambers

multiloculate See *multilocular*

multinucleate Descriptive of cells or tissue containing more than a single nucleus in a membrane-bounded space; see *coenocyte, plasmodium, syncytium*

multiple fission Karyokinesis followed by a delay in cytokinesis such that when cytokinesis occurs, 2N offspring are produced at once, where N represents the number of generations cytokinesis was delayed; see *progressive cleavage*

multiplicative cyst (pl. **multiplicative cysts**) Cyst in which multiple fissions (mitotic cell divisions) occur

multipolar nucleus (pl. **multipolar nuclei**) Dividing nucleus containing spindle microtubules oriented toward more than two poles

multiseriate Morphological term referring to structures (for example, trichomes, filaments, algal "hairs") composed of more than a single row of cells

mural pore (pl. **mural pores**) Minute openings in tests of many foraminiferans (Pr-3)

mutant (pl. **mutants**) Organism bearing an altered gene expressed in its phenotype; organism demonstrating a heritable, detectable, structural or chemical change

mutualism Ecological term referring to associations between organisms that are members of different species such that the associated partners leave more offspring per unit time when together than when they are growing separately

mutualistic See *mutualism*

mycelial Threadlike material (hyphae) that together forms a matted tissuelike structure that makes up the body of most fungi and some protoctists (for example, chytridiomycotes, oomycotes, Pr-35, -21)

mycelium (pl. **mycelia**) See *mycelial*

mycology Study of fungi. Subfield of biology that traditionally included study of fungal-like protoctists (for example, chytrids, plasmodiophorids Pr-35, -20)

mycophagy Mode of nutrition; organisms feeding on fungi

mycosis (pl. **mycoses**) Disease caused by a fungus

mycovirus (pl. **mycoviruses**) Virus of a fungus

Müller's Law Law describing the unique radially symmetrically ordered skeleton of acantharians (phylum Actinopoda, Pr-31) in which the cell may be conceived as a globe from whose center spicules radiate and pierce the surface at fixed latitudes and longitudes. If there are 20 spicules, then there are five quartets, one equatorial, two polar, and two tropical, piercing the globe at latitudes 0°, 30°N, 30°S, 60°N, and 60°S. Longitudes of the piercing points are 0°, 90°W, 90°E, 180°, 45°W, 45°E, 135°W, and 135°E for their respective quartets. Variations in shape of cell, thickness, length or number of spicules are still grouped by some elaboration of Müller's Law

myonemes "Muscle threads." Motile ribbonlike or cylindrical organelles found in acantharian actinopods and some dinomastigotes and ciliates (Pr-31, -5, -6). Consist of densely packed 2–3 nm microfibrils, exhibiting long clear zones cross-striated by thin dark bands (for example, ciliates). Myonemes may play a part in buoyancy regulation (for example, acantharians) Myonemes are responsible for cell contraction (for example, in dinomastigotes and some ciliates, Pr-5, -6). The term is applied to distinguishable cell structures that are probably unrelated

myxameba (pl. **myxamebas**) Ameboid stage of plasmodial (myxomycete, Pr-23) slime molds in which cell lacks cell walls and feeds by phagocytosis; this stage gives way to the formation of a plasmodium and later a stalked sporocarp

myxamoeba (pl. **myxamoebas**) See *myxameba*

naked Wall-less; lacking a cell wall, scales, or decorations; also, ciliates denuded of cilia

nanofossil (pl. **nanofossils**) Microfossils of the smallest kind, usually 1–20 μm in size

nanoplankton Planktonic protists in the 1–20 μm size range; plankton with dimensions of fewer than 70–75 μm that tend to pass through plankton nets

nanoplanktonic See *nanoplankton*

NAOs Microtubule-organizing center just outside, on, or associated with the nuclear membrane of some protoctists (for example, rhodophytes, Pr-33) and fungi. One is found at each of the poles during mitotic division

Nebenkörper German, meaning neighboring body Feulgen-positive body found alongside the nucleus in *Paramoeba* (Pr-2), possibly xenosome of symbiotic origin from bacteria

necrosis Death of cells, a piece of tissue, or an organ, in an otherwise living organism

necrotic See *necrosis*

necrotroph See *Necrotrophy*

necrotrophic See *Necrotroph*

necrotrophy Nutritional mode in which a symbiotroph damages or kills its host; parasitism or pathogenesis (Table 2)

nectomonad (pl. **nectomonads**) Free-swimming (as opposed to attached) stage in the life cycle of trypanosomatids (Pr-11)

negative staining Technique in electron microscopy in which a sample is mixed with a stain (for example, phosphotungstic acid) and sprayed onto a grid; because the stain enters the contours of the sample, objects appear light against a dark background

nemathecium (pl. **nemathecia**) Raised or wart-like area on the surface of the thallus of some florideophycidean rhodophytes (Pr-33); contains reproductive organs

nematocyst (pl. **nematocysts**) Cnidocyst. Modified cell with a capsule containing a threadlike stinger used for defense, anchoring, or capturing prey; some contain poisonous or paralyzing substances (for example, in all coelenterates and ctenophores (A-4, -5); analogous organelles found in some dinomastigotes and some karyorelictid and suctorian ciliates (Pr-5, -6))

nematodesma (pl. **nematodesmata**) Parts of kinetids of certain ciliates (Pr-6). Bundle of microtubules, usually hexagonally packed, that originates in association with the kinetosome and forms part of the wall of the cytopharyngeal apparatus

nematogene (pl. **nematogenes**) Organelle giving rise to the nematocyst in dinomastigotes (Pr-5)

neontology The study of extant species, in contrast with paleontology

neoseme Appearance of a new trait of evolutionary importance (seme); see *seme*

neritic Ecological term referring to the region of shallow water along a seacoast; also refers to organisms in communities near the shoreline (edge) of an ocean; see *pelagic*

neuston Ecological term referring to the surface biota of aquatic environments; those dwelling at the interface between atmosphere and water

neustonic See *neuston*

niche (pl. **niches**) Role performed by members of a species in a biological community

noctikaryotic Refers to dinomastigotes (for example, *Noctiluca*, Pr-5) in which the nucleus changes from the usual mesokaryotic condition (dinokaryotic) to a conventional eukaryotic appearance during the life cycle

node (pl. **nodes**) Protruberance found on the umbilical surfaces of certain foraminifera (Pr-3, for example, Glabratellidae); sites on an algal axis from which new growth arises (for example, charophytes, Pr-28)

nonclastic Pertaining to sediment that is chemically precipitated in place (for example, halite); see *clast*

nonemergent flagellum (pl. **nonemergent flagella**) Undulipodium lacking an emergent axoneme, reduced to a kinetosome only in extreme cases; short undulipodium (for example, in euglenoids, Pr-12); see *emergent undulipodium*

nonemergent undulipodium (pl. **nonemergent undulipodia**) See *nonemergent flagellum*

nongeniculate Descriptive term referring to a structure not formed by joints; nonarticulated; lacking segmentation; also, not bent abruptly at an angle

nori Edible, dried preparation of the rhodophyte Porphyra (Pr-33)

NP Abbreviation of nucleoprotein

nuclear cap (pl. **nuclear caps**) Crescent-shaped, membrane-bounded sac surrounding a third or more of the zoospore nucleus of phylum Blastocladiomycota (Pr-34); the nuclear cap apparently contains all the cell's ribosomes

nuclear cyclosis Intranuclear movement by means of filaments which apparently use the proteins associated with intracellular motility: actin, myosin, and tubulin; the enlargement and slow swiveling movement and rotation of the nucleus is associated with the first meiotic division (for example, some suctorian ciliates and dinomastigotes, Pr-6, -5)

nuclear dualism Heterokaryosis. Possessing two functionally different nuclei in the same cell; characteristic of ciliate cells and a few foraminifera (Pr-6, -3)

nuclear envelope (pl. **nuclear envelopes**) Double-membrane structure, often containing many pores, surrounding the nucleoplasm. Structural criterion defining eukaryotes

nuclear membrane (pl. **nuclear membranes**) See *nuclear envelope*

nuclear plate (pl. **nuclear plates**) See *equatorial plate*

nucleogony Multiple karyokineses to produce many small nuclei at once

nucleoid (pl. **nucleoids**) DNA-containing structure of prokaryotes, not bounded by a membrane; see *genophore*

nucleolar substance (pl. **nucleolar substances**) Stainable material present during or after mitosis and derived from the nucleolus

nucleolar-organizing center (pl. **nucleolar-organizing centers**) Chromosome or chromatin with long secondary constrictions (nucleolar-organizing regions); site of formation of new nucleoli that are precursors to RNA subunits of ribosomes

nucleolar-organizing chromosome (pl. **nucleolar-organizing chromosomes**) See *nucleolar-organizing center*

nucleolus (pl. **nucleoli**) Endosome; karyosome. Structure in the cell nucleus composed of RNA and protein, precursor material to the ribosomes

nucleomorph (pl. **nucleomorphs**) Organelle surrounded by a double-membrane resembling a small nucleus, lying between the plastid ER and plastid membrane in cryptomonads (Pr-26); a membrane-bounded nucleic acid–containing organelle in the periplastidial compartment, thought to be the remnant nucleus of an eukaryotic photosynthetic endosymbiont

nucleonema Network of strands consisting of granular material, located at nucleolar surface (for example, *Pelomyxa palustris*, Pr-2)

nucleoplasma Fluid contents of the nucleus of any eukaryote

nucleus (pl. **nuclei**) Membrane-bounded, spherical, DNA-containing organelle, universal in protictists. Chromatin (DNA, protein) organized into chromosomes; site of DNA synthesis and RNA transcription. Nuclear membranes bear pores. Definitional for eukaryotes

nucleus-associated organelle (pl. **nucleus-associated organelles**) See *NAOs*

nudiform replication Term applied to loricate choanomastigotes (Acanthoecidae, Pr-35) indicating the absence of bundles of component costal strips in one of the two offspring cells resulting from a cell division; see *tectiform replication*

nutricline (pl. **nutriclines**) Ecological term referring to gradients of nutrient concentration in aquatic environments

obligate Compulsory or mandatory as opposed to optional or facultative, for example, an obligate anaerobe can survive and grow only in the absence of gaseous oxygen; see *facultative*

oceanic Pertaining to those areas of the ocean deeper than the littoral and neritic; open-ocean depths

ocelloid Ocelloid. Complex light-perceiving organelle in a few dinomastigote genera (Pr-5), consisting of a large refractive lens (hyalosome) and a pigment-containing cup (melanosome) (for example, *Warnowia*, *Erythropsidinium*); slightly raised areas of a valve that are externally rimmed and enclose an area of fine pores (porelli) in diatoms (Pr-18)

ocellus (pl. **ocelli**) See *ocelloid*

ocular chamber (pl. **ocular chambers**) Component of dinomastigote ocellus (Pr-5); chamber with a canal extending to the sulcus

offspring Filial products. "Daughter cells" should be referred to as offspring cells. "Daughter nuclei" should be referred to as offspring nuclei (Term "daughter" should be avoided in cases where the female gender of the offspring has not been established.)

oligohaline Ecological term referring to marine environments with low salinities, that is, less than about 3.3% salt; see *euhaline*, *euryhaline*, *hyperhaline*

oligophotic zone The region in the aquatic environments below the mesophotic zone, in which the organisms are limited by insufficient sunlight for optimal growth but in which sufficient incident radiation penetrates so that some photosynthesis is possible

oligosaline See *oligohaline*

oligotrophic Ecological term referring to clear water, that is, an aquatic environment deficient in inorganic and organic nutrients, and usually containing high concentrations of dissolved oxygen; see *eutrophy*, *polysaprobic*

oligotrophy See *oligotrophic*

omnivore Practitioner of the heterotrophic mode of nutrition; ingestor of plant, fungal, and/or animal food

omnivorous See *omnivore*

omnivory Mode of nutrition; See *omnivore*

ontogeny Development of an individual organism (for example, animal from fertilized egg to death)

oocyst (pl. **oocysts**) Encysted zygote (for example, coccidian apicomplexans and heliozoan actinopods, Pr-7, -31)

oogamous Fusion of a nonmotile large egg (female gamete) with a small motile sperm (male gamete); extreme form of anisogamy (for example, some protoctists, most animals)

oogamy See *oogamous*

oogenesis Development of ova (egg cells) (for example, animal eggs prior to fertilization)

oogonial cavity (pl. **oogonial cavities**) Space enclosed by the oogonial wall; may be completely filled by an oospore (plerotic) or only partially filled by one or more oospores (aplerotic) (for example, oomycotes, Pr-21)

oogonioplasm Cytoplasm of the oogonium

oogonium (pl. **oogonia**) Uninucleate or coenocytic cell (or the cell wall) that generates female gamete(s) (for example, oomycotes, Pr-21)

ookinete Motile zygote

oomycote (pl. **oomycotes**) Informal name for members of the phylum Oomycota, Pr-21

ooplast (pl. **ooplasts**) Organelle formed in the oospore as a result of coalescence of the dense body vesicles (for example, oomycotes, Pr-21)

oosphere (pl. **oospheres**) Egg. Unfertilized and unpenetrated female gamete containing a single, haploid nucleus (for example, oomycotes, Pr-21)

oospore (pl. **oospores**) Thick-walled spherical structure developing from an oosphere after fertilization in oomycotes, Pr-21; see *aplerotic, plerotic*

oosporogenesis oospore formation

operculum (pl. **opercula**) Lid, covering, or flap of an aperture

opisthe Posterior offspring of transverse binary fission of the parental organism (for example, ciliates); see *proter*

opisthokont Term descriptive of the morphology of a posteriorly undulipodiated mastigote (for example, chytrid zoospores, some dinomastigotes, Pr-35, -5); pertaining to the insertion of the undulipodium at the posterior pole of the cell (in relation to movement)

opisthomastigote (pl. **opisthomastigotes**) Stage in development of a trypanosomatid (Pr-11) in which the kinetoplast lies behind the nucleus and the associated undulipodium emerges at the anterior extremity from a long, narrow undulipodial pocket

oral apparatus See *buccal cavity, ingestion apparatus, oral region*

oral cavity (pl. **oral cavities**) Mouth opening

oral kinetid (pl. **oral kinetids**) Kinetid within the oral region of the ciliate cortex

oral opening (pl. **oral openings**) Mouth cavity

oral region Oral area; mouth; ingestion apparatus; oral apparatus. General term for that part of a protist cell bearing the ingestion apparatus; usually used in a nonspecific way; see *somatic region*

organelle (pl. **organelles**) Distinctive structure detected by microscopy inside a cell. Some, such as mitochondria, nuclei, and plastids, are double membrane bounded and capable of division. Others, such as carboxysomes, ribosomes, and liposomes, are visualizable as locally high concentrations of certain enzymes and other macromolecules

organic test (pl. **organic tests**) Covering or shell of an organism composed of organic materials (for example, chitin, cellulose, or the complex of protein and mucopolysaccharide of foraminifera known as tectin)

orogeny Mountain-building processes

orthomitosis Karyokinesis; mitotic cell division in which spindle tubules are parallel to each other

osmiophilic Osmium-loving. Tendency to stain black with osmium tetroxide, especially characteristic of electron microscopy preparations

osmoregulation Maintenance of constant internal salt and water concentrations in an organism, requiring the input of energy

osmotroph Mode of heterotrophic nutrition; organisms taking in soluble organic compounds; osmotrophic organisms absorb food in the dissolved state from the surrounding medium directly by osmosis, active transport, or pinocytosis (for example, fungi, some protoctists) (Table 2); see *carnivory, histophagy, holozoic, macrophagy, phagotrophy*

osmotrophic See *osmotroph*

osmotrophy Osmotroph

outbreeding Mating and production of offspring by organisms not known to have a traceable common ancestor (genetically unrelated organisms)

outgroup (pl. **outgroups**) Cladistic concept; species or higher monophyletic taxon that is examined in a phylogenetic study to determine which of two homologous characters may be inferred to be symplesiomorphic (apomorphic). One or several out-groups may be examined for each decision. The most critical out-group comparisons involve the sister group of the taxon studied

out-group (pl. **out-groups**) Outgroup

ovary (pl. **ovaries**) Multicellular sex organ of female animals and plants. In flowering plants, an enlarged basal portion of a carpel or of a gynoecium composed of fused carpels that becomes the fruit. Term not appropriate for protoctists

oyster spat Juvenile oyster

pachynema Stage in prophase of meiotic cell divisions in which chromosomes are tightly packed

pachytene Pachynema; stage in meiotic prophase I in which pairs of homologous chromosomes shorten and thicken; see *meiosis*

paedogamy Autogamy. Fusion of two uninucleate sporoplasms and their haploid nuclei in myxosporans (A-2)

paleoecology Subfield of paleontology: attempt to reconstruct past communities of organisms and their environments by study of their fossil remains

paleontology Study of past life on earth primarily by investigation of fossil remains; subfield of geology, essential to evolutionary biology

palintomy Rapid sequence of binary fissions, typically within a cyst and with little or no intervening growth resulting in production of numerous, small offspring cells. Common in various parasitic protists (for example, ciliates and dinomastigotes, Pr-6, -5); produces tomites in apostome ciliates

palisporogenesis Specialized type of division in some blastodinian dinomastigotes (Pr-5) in which the first division results in an individual that continues to feed on the host (trophocyte) and one that is responsible for the subsequent division (gonocyte)

pallium (pl. **pallia**) See *feeding veil*

palmelloid Colonial morphology characteristic of many algae in which nonmotile cells are encased in mucus as a gelatinous mass

panacronematic Description of mastigotes having undulipodia with two rows of mastigonemes (flimmer, fibrils) and a terminal fiber; see *acronematic, pleuronematic*

pankinetoplastic Pertaining to a morphology of kinetoplastids (Pr-11) in which the kDNA is not localized in one or more discrete bodies but is irregularly distributed as stainable masses throughout the kinetoplast mitochondrion

pansporoblast (pl. **pansporoblasts**) Synctium that undergoes cytokinesis to yield parasites. Gives rise to two sporoblasts contained within a single membrane in some apicomplexans (Pr-7). In actinosporean myxosporans (A-2), two- to four-celled envelope-containing groups of eight spores, and sometimes called a pansporocyst; in myxosporans, a thick envelope around one or more spores consisting of two degraded pansporoblast (or pericyte) cells; in microsporans (F-1), obsolete term for the subpersistent membrane of the sporophorous vesicle

pansporoblast envelope (pl. **pansporoblast envelopes**) See *pansporoblast*

pansporocyst (pl. **pansporocysts**) Pansporoblast, usually of actinosporeans (A-2)

pantacronematic See *panacronematic*

pantonematic Description of mastigotes having undulipodia with two rows of mastigonemes, but no terminal filament or fiber; see *acronematic, pantacronematic*

papilla (pl. **papillae**) Small bump or projection. Specialized structure found on the periphery of mature sporangia that is enzymatically degraded at the time of discharge of the sporangial contents, thereby allowing their escape (for example, in *Blastocladiella*, Pr-34)

papillate See *papilla*

PAR Paraflagellar rod. Paraxial rod. Intraundulipodial structure in euglenoids, dinomastigotes, and kinetoplastids (Pr-12, -5, -11); elaborate cross-striated structure of unknown function that extends nearly the entire length of the undulipodium between the membrane and the axoneme

parabasal apparatus Parabasal body plus a single (or pair of) parabasal filament(s)

parabasal body (pl. **parabasal bodies**) Modified Golgi apparatus anterior in the cell which defines the class Parabasalia (Pr-1). Located near the kinetosomes and their associated structures, the structure probably has a secretory function

parabasal filaments Microfibrillar, striated, often paired organelles of parabasalians (Pr-1) arising from a complex kinetid and intimately associated with parabasal bodies

parabasal fold (pl. **parabasal folds**) Bent structure of Golgi limited to members of the phylum Parabasalia (Pr-1)

paracostal granules See *chromatic granules*

paracrystalline Pertaining to cellular inclusions of many types that exhibit a crystal-like organization as seen in the light or electron microscope

paradesmose (pl. **paradesmoses**) Cell structure that links two sets of polar kinetosomes during mitosis (for example, paradesmose in some prasinophytes, Pr-28 composed of a microtubular bundle)

paraflagellar body (pl. **paraflagellar bodies**) Undulipodial swelling; photoreceptor; lateral swelling near the base of the emergent undulipodium; in euglenoids (Pr-12), it is adjacent to the eyespot and presumably carries the photoreceptor for phototaxis; similar in appearance to a paraflagellar rod although much smaller in diameter

paraflagellar rod (pl. **paraflagellar rods**) See *PAR*

paraflagellate (pl. **paraflagellates**) Opalinid; member of the ciliate class Opalina (Pr-6)

parallel evolution See *convergent evolution*

paralytic shellfish poisoning Toxic response due to dinomastigote (Pr-5) bloom in which the toxins do not kill many organisms, but are concentrated within the siphons or digestive glands of filter-feeding bivalve molluscs (A-26)

paramylon Cytoplasmic carbohydrate; the nutritional reserve of euglenoids and prymnesiophytes (Pr-12, -25); β-1,3-glucose polymer, a glucan

paranuclear body (pl. **paranuclear bodies**) Cytoplasmic organelle found in the thraustochytrid labyrinthulomycotes (Pr-19) located adjacent to the nuclei of developing thalli and consisting of a compact mass of inflated smooth endoplasmic reticulum cisternae containing a fine granular material

paranucleus (pl. **paranuclei**) See *Nebenkörper*

paraphyletic group (pl. **paraphyletic groups**) Taxon. Group that includes a common ancestor and some but not all of its descendants

paraphyletic taxa (pl. **paraphyletic taxon**) See *paraphyletic group*

paraphysis (pl. **paraphyses**) Structure of algae: sterile hair growing among reproductive structures (for example, in *Fucus*, Pr-17)

parapyla (pl. **parapylae**) Secondary openings in the central capsule of phaeodarian actinopods (Pr-31); see *astropyle*

parasexuality Any process bypassing standard meiosis and fertilization that forms an offspring cell from more than a single parent (for example, recovery of resistant recombinants in dictyostelids (Pr-2)); see *sex*

parasite (pl. **parasites**) Ecological term referring to organisms that live associated with members of different species as obligate or facultative symbiotrophs that tend toward necrotrophy; see *mutualism*, *necrotrophy*, *pathogen*, *symbiotrophy*

parasitemia level See *parasitemia*

parasitemia Measure of parasites in the circulating blood of vertebrate hosts

parasitic See *parasite*

parasitism Ecological association between members of different species in which one partner (usually the small form) is obligately or facultatively symbiotrophic and tends toward necrotrophy (Table 2)

parasitophorous vacuole (pl. **parasitophorous vacuoles**) Membranous vacuole containing intracellular parasite; originally derived from the host plasma membrane during phagocytosis of the parasite (for example, in the microsporan *Encephalitozoon* and in apicomplexans, F-1, Pr-7); its composition may subsequently be altered by the parasite

parasomal sac (pl. **parasomal sacs**) Structure found associated with each kinetid of the cortex of ciliates (Pr-6); small invagination of the plasma membrane adjacent to kinetosomes

parasome (pl. **parasomes**) See *Nebenkörper*

parasporangium (pl. **parasporangia**) Algal sporangium producing many spores

paratabulation Numbering system for dinomastigote pellicle plates (Pr-5)

paraxial rod (pl. **paraxial rods**) See *PAR*

paraxostylar granules See *chromatic granules*

paraxostyle (pl. **paraxostyles**) Structure found alongside the axostyle (for example, in some Archaeprotista, Pr-1)

parenchyma Tissue made of thin-walled cells that actively grow in any of three dimensions (for example, thalli of large algae)

parenchymatous See *parenchyma*

parenchymous See *parenchyma*

parietal Position of an organ or organelle: near or alongside a wall

paroral kinety Row of kinetosomes around mouth region. In ciliates, zigzag row of kinetosomes on right side of mouth, which form the paroral membrane

parthenogenesis Development of an unfertilized egg into an organism

parthenosporangia (pl. **parthenosporangium**) Receptacles bearing parthenospores

parthenospore (pl. **parthenospores**) Thick-walled spore developing from an unfertilized gamete (for example, conjugating green algae, Pr-32); undulipodiated reproductive cells produced without conjugation (apomictically) (for example, in phaeophytes, Pr-17); haploid dinomastigotes that are morphologically similar or identical to planozygotes but are formed by mitosis instead of syngamy

pathogen (pl. **pathogens**) Ecological term referring to organism that is an obligate or facultative symbiotroph that tends toward necrotrophy and causes symptoms in its host. Disease-causing organism; see *parasite*

pathogenic See *pathogen*

PC Phase-contrast microscopy

pectin Complex polysaccharide extractable from cell walls of plants and some algae

pedicel (pl. **pedicels**) Attachment stalk, holdfast (for example, in some chonotrich ciliates and choanomastigotes, Pr-6, -35); elongated protrusion from the posterior end of a cell; basal portion of a charophyte antheridium (Pr-28)

pedogamy See *autogamous*, *paedogamy*

peduncle (pl. **peduncles**) Holdfast, stalk, base, or stemlike structure; projection from sulcal region used to suck up food during heterotrophic feeding in dinomastigotes (Pr-5; for example, *Katodinium (Gymnodinium) fungiforme*)

pedunculate See *peduncle*

pelagic Ecological term referring to organisms dwelling in open waters of the ocean (as opposed to benthic or neritic)

pellicle Cortex. Outermost living layer of a protoctist, lying beneath any nonliving secreted material; pellicle contains the typical plasma membrane plus the pellicular alveoli or an underlying epiplasm or other membranes (in ciliates, dinomastigotes, and a few others, Pr-6, -5 etc.) and sometimes exhibits ridges, folds, or distinct crests; portion surrounding the cell after the theca is shed by ecdysis in armored dinomastigotes; proteinaceous ridged structure in euglenoids (Pr-12)

pellicular See *pellicle*

pellicular fold (pl. **pellicular folds**) Wrinkles on surface; crenulations of pellicle

pellicular lacunae system System of flat membranous vesicles just beneath the pellicle of cells; micromorphological character typical of glaucocystophytes

pellicular microtubular armature Microtubules located beneath the cell membrane that form a cytoskeleton involved in maintenance of the cell shape (for example, chrysophytes (Pr-15)); subpellicular microtubular cytoskeleton (for example, euglenoids (Pr-12))

pellicular striae Striations, ridges, or striped markings in or on the pellicle

pelliculate See *pellicle*

pelta Crescent-shaped microtubular structure associated with the anterior portion of the axostyle (for example, archaeprotists such as pyrsonymphids or oxymonads, Pr-1)

peneropliform A test that initially grows by adding chambers in a coiled single plane (planispirally) and then adds later chambers in a straight line (rectilinearly) (for example, foraminifera such as *Peneroplis*, Pr-3)

pennate Morphological descriptive term for structure resembling a feather, especially in having similar parts arranged on opposite sides of an axis such as the barbs on the rachis of a feather; refers to shape of some diatoms (Pr-18)

pentalith (pl. **pentaliths**) Coccolith of five identical single calcite crystals; the cleavage plane of the crystals is in the plane of the pentalith (for example, *Braarudosphaera*, Pr-25)

peptide mapping Technique used to compare a given protein from different organisms in which the protein is enzymatically cleaved, the resulting peptides are separated on a gel, and the peptides are identified by staining or reaction with a specific antibody. The similarity between the peptide patterns is related to amino acid sequence similarity

peptidoglycan Glycan tetrapeptide. Rigid layer of bacterial cell walls consisting of *N*-acetylglucosamine and *N*-acetylmuramic acid attached to a few amino acid residues that form a repeating peptide

per os Latin locution meaning orally; by mouth.

PER Specialized layer of endoplasmic reticulum that closely surrounds the plastid and is usually continuous with the nuclear membrane; ribosomes are present on the membrane facing the cytoplasm, but not the membrane facing the periplastidial compartment; see *chloroplast endoplasmic reticulum*

percurrent development Proliferation through the sporangial septum of the sporangiophore, producing a new sporangium of similar dimensions; sporangia produced by limited internal renewal such that successive sporangial septa are formed at approximately the same point on the axis (for example, oomycotes); see *basipetal development*, *internal renewal*

perennation Overwintering in plants; in protoctists, refers to survival of harsh conditions (for example, seasonal desiccation)

perennial Descriptive of organism that lives for more than a year and produces a sexual phase annually or semiannually

perforatorium (pl. **perforatoria**) Reinforced tip of mature microgamete that probably aids in penetration of macrogamete in coccidian apicomplexans (Pr-7)

periaxostylar Material parallel to and surrounding the axostyle

pericarp Sterile layer of cells that surrounds the carposporophyte in some rhodophytes (Pr-33)

perichloroplastic compartment (pl. **perichloroplastic compartments**) Perichloroplastic compartment. Space between the plastid membrane and the plastid endoplasmic reticulum

pericyte (pl. **pericytes**) Outer of two generative cells in myxosporean plasmodium that unite pairwise to produce a pansporoblast; stage in actinosporean myxosporans (A-2) containing two nuclei that arises from the sporoplasm and envelopes the sporogonic cell

peridium (pl. **peridia**) Structure of myxomycete (Pr-23) sporophores consisting of a membranous surface layer

perioral kinety Rows of cilia around the mouth derived from modified somatic kineties (Pr-6)

periphyton See *microbenthos*

periplasm Peripheral cytoplasm. In prokaryotes, the space between the inner plasma membrane and the peptidoglycan layer of the cell wall

periplasmic cortex Surface layer of the protoplasm in the sexual organs remaining after the differentiation of the sexual cells (for example, peronosporean oomycotes, Pr-21); outer pellicle in acantharian actinopods (Pr-31)

periplast Part of eukaryotic cell that lies external to the cell membrane; sometimes composed of elements such as scales, coccoliths, and plates and including specialized structures such as cell walls, pellicles, and thecae; a complex, ornamented plasma membrane

periplastidial compartment (pl. **periplastidial compartments**) See *perichloroplastic compartment*

periplastidial reticulum (pl. **periplastidial reticula**) System of vesicles and tubules located in the periplastidial compartment; membranous reticulum in continuity with the inner membrane of the plastid endoplasmic reticulum, lying within the periplastidial compartment

perispicular cone (pl. **perispicular cones**) Region in acantharian actinopods (Pr-31) in which the capsular wall is connected to the periplasmic cortex making a sleeve around each spicule. The sleeve defines a conical space containing the axoneme, dense granules, vesicles, and myonemes

perispicular vacuole (pl. **perispicular vacuoles**) Spicular vacuole. Large vacuoles in acantharian actinopods (Pr-31); structures in which the spicules are enclosed

peristome (pl. **peristomes**) See *buccal cavity*

perithallial Portion of the growing thallus in which the cells or filaments are developed inwardly from the intercalary meristem (for example, coralline rhodophytes, Pr-33)

perithallium (pl. **perithalli**) See *perithallial*

peritrichous Ciliates that bear an oral ring of ciliature; bacteria flagellated around their periphery

peritrichs See *peritrichous*

perizonium Outer membrane derived from the fertilization membrane after zygote (auxospore) formation in diatoms

perkinetal fission Fission across or through the kineties or rows of cilia; most common type of homothetogenic fission; transverse fission of ciliates, as opposed to the longitudinal fission of mastigotes. Typical of ciliates; see *interkinetal fission*

peroxisome (pl. **peroxisomes**) Organelles containing enzymes, including catalase and peroxidase; site of the oxidation of a variety of substrates to form hydrogen peroxide (H_2O_2), using molecular oxygen as the oxidizing agent

petrographic thin section Slices of rock polished and thin enough to allow light to pass through them; used to detect microfossils in a cryptocrystalline matrix (chert)

PFB See *paraflagellar body*

PFR See *PAR*, *paraflagellar rod*, *paraxial rod*

pH Scale for measuring acidity of aqueous solutions; $pH = -\log[H^+]$; pure water has a pH of

7 (neutral); solutions having a pH greater than 7 are alkaline; less than 7 are acidic

phaeodium (pl. **phaeodia**) Pigmented mass consisting primarily of waste products around the astropyle of the central capsule of phaeodarian actinopods (Pr-31)

phaeophyte (pl. **phaeophytes**) Informal name of members of the phylum Phaeophyta, Pr-17

phaeoplast (pl. **phaeoplasts**) Brown chlorophyll *c*-containing plastid; photosynthetic organelle of phaeophytes, Pr-17

phaeosome (pl. **phaeosomes**) Brown body, may be excretory products (for example, those produced by many actinopods, Pr-31); surface-associated, ectosymbiotic, coccoid cyanobacteria (B-6), mostly *Synechococcus* spp. occurring in association with dinophysoid dinomastigotes (Pr-5)

phaeosome chamber (pl. **phaeosome chambers**) Chamber of actinopods (Pr-31) that harbor phaeosomes; chamberlike modification of the girdle of some complex dinophysoid genera (for example, *Histioneis*, *Citharistes*), in which symbiotic cyanobacteria usually occur

phage (pl. **phages**) Virus of bacteria

phagocytic See *phagocytosis*

phagocytosis Mode of heterotrophic nutrition and immunological defense involving ingestion, by a cell, of solid particles in which pseudopods flow over and engulf particulates

phagocytotic See *phagocytosis*

phagotrophic See *phagotrophy*

phagotroph See *phagotrophy*

phagotrophy Mode of nutrition referring to heterotrophic protoctists or tissue cells that ingest solid food particles by phagocytosis (Table 2); see *carnivory*, *histophagy*, *holozoic*, *osmotrophy*

phaneroplasmodium (pl. **phaneroplasmodia**) Largest and most conspicuous of the three types of plasmodia formed by myxomycotes (Pr-23, primarily of the order Physarales); plasmodium consisting of thin, fanlike advancing regions and a branching network of veins; the veins consist of an outer gel zone of protoplasm and an inner fluid zone, in which protoplasmic streaming occurs; see *aphanoplasmodium*, *protoplasmodium*

phenetic taxonomy Classification of organisms based on their visible, measurable (phenotypic) characteristics without regard to evolutionary (phylogenetic) relationships

phenological Ecological term referring to seasonal variation

pheromone (pl. **pheromones**) Ecological term referring to a chemical substance that when released into the surroundings of organisms influences the behavior or development of other individuals of the same species. If produced by one sex and responded to by the other sex, the substance is called a sex pheromone; see *allelochemic*

phialine lip (pl. **phialine lips**) Flask or cup-shaped outgrowth; in foraminifera (Pr-3), everted rim of aperture, common on neck

phialopore (pl. **phialopores**) Intercellular space in certain volvocalean chlorophytes (Pr-28) through which the colony everts

phlebotominae Dipteran insect family (A-21); sand flies

phoront (pl. **phoronts**) Stage in a polymorphic life cycle during which the protoctist is carried about (generally on or in the integument of) another (generally metazoan) organism. Stage typically preceded by a tomite and followed by a trophont (for example, polymorphic apostome ciliates such as *Hyalophysa*, Pr-6)

photic zone See *euphotic zone*

photoautotrophy Mode of nutrition in which light provides the source of energy. An obligately photoautotrophic organism uses light energy to synthesize cell material from inorganic compounds (carbon dioxide, nitrogen salts) (Table 1)

photoauxotroph See *photoauxotrophy*

photoauxotrophic See *photoauxotrophy*

photoauxotrophy Mode of nutrition, usually of algal mutants that grow phototrophically except for the requirement of a vitamin, amino acid, or other identifiable growth factor

photoheterotroph See *photoheterotrophy*

photoheterotrophic See *photoheterotrophy*

photoheterotrophy Mode of nutrition, limited to bacteria, in which light is used as a source of energy (to generate ATP and osmotic gradients) but organic compounds are used as carbon sources (Table 1)

photoinhibition Physiological response of algae or plants referring to the inhibition of photosynthesis at high light intensities

photokinesis The effect of light intensity on the speed of movement

photoperiodic response (pl. **photoperiodic responses**) Behavioral or growth response of an organism to changes in day length; a mechanism for measuring seasonal time

photoreceptor (pl. **photoreceptors**) Cell structure in which a specialized aggregate of pigments mediates a behavioral reaction to light stimuli (for example, eustigmatophytes, euglenoids, *Paramecium bursaria*, and many other protoctists; Pr-27, -2, -6 etc.)

photoresponse (pl. **photoresponses**) Cell or organismal growth or behavioral response to light stimuli (for example, positive and negative phototaxis; phototropism)

photosensory transduction Reaction chain of light-induced motor responses, that is, the connecting link between photoreceptor and cell motility, consisting of stimulus transformation (conversion of one form of energy to another) and signal transmission (that is, all steps in the reaction chain that cause signal transport)

photosynthate (pl. **photosynthates**) Any metabolic product of photosynthesis; total photosynthate contains sugars, amino acids, organic acids and differs in exact composition in different phototrophs

photosynthesis See *photoautotrophy*

photosystem Functional light-trapping unit; an organized collection of chlorophyll and other pigments embedded in the thylakoids of plastids which trap photon energy and channel it in the form of energetic electrons to the thylakoid membrane

phototactic Movement toward (positive) or away (negative) from a light source

phototaxis See *phototactic*

phototroph See *photoautotrophy*

phototrophic See *photoautotrophy*

phototrophy See *photoautotrophy*

phragmoplast (pl. **phragmoplasts**) Cell plate. System of fusing vesicles guided by microtubules that form perpendicular to the spindle axis at telophase in the plane of division during cytokinesis (for example, in plants and some chlorophytes); see *phycoplast*

phycobilins Class of protein-linked open tetrapyrrhole pigments, water soluble, and generally bluish or red in color (for example, in cyanelles of glaucocystophytes, plastids of rhodophytes, some cryptophytes, and thylakoids of cyanobacteria; Pr-33, -26; B-6)

phycobiliprotein (pl. **phycobiliproteins**) See *biliprotein*

phycobilisome (pl. **phycobilisomes**) Cellular structure containing phycobilin pigments and arranged as protrusions on the surface of the thylakoids of cyanobacteria, rhodophytes, and glaucocystophytes, but within the thylakoids (between membranous stacks) in the plastids of cryptophytes (B-6; Pr-33, -26)

phycobiont (pl. **phycobionts**) Algal symbiotic partner of a lichen (F-6)

phycocolloids Complex polysaccharides produced by algae, the detailed structures of which are largely unknown (for example, agarose, carrageenan)

phycocyanin Type of phycobiliprotein; water-soluble extract is blue; found in cyanobacteria, rhodophytes, and cryptomonads (B-6, Pr-33, -26)

phycoerythrin Type of phycobiliprotein; water-soluble extract is red; found in cyanobacteria, rhodophytes, and cryptomonads (B-6, Pr-33, -26)

phycology See *algology*

phycoma Whole algal body; nonmotile, unicellular, spherical stage in the life history of some prasinophytes (Pr-28; Pterospermataceae, Halosphaeraceae) characterized by a thick, ornamented wall which may contain sporopollenin

phycomycete (pl. **phycomycetes**) Lower fungi. Term for a class of fungi that is obsolete because it grouped zygomycotes (F-2) with unrelated taxa (that is, chytridiomycotes, oomycotes, and other "algal-like" fungi (Pr-35, -21 etc))

phycophage (pl. **phycophages**) Algal virus

phycoplast (pl. **phycoplasts**) System of fusing vesicles guided by microtubules that form parallel to the spindle axis at mitosis and in the plane of division in some algae; see *phragmoplast*

phyllae (pl. **phyllae**) Flat ribbons of microtubules found in the oral region of some ciliates (Pr-6)

phylogenetic tree (pl. **phylogenetic trees**) Graphic or diagrammatic representation of a partial phylogeny (for example, ribosomal RNA or protein sequences) or complete phylogeny (for example, family tree)

phylogeny (pl. **phylogenies**) Hypothesized sequence of ancestor/descendant relationships of groups of organisms as reflected by their evolutionary history

physode (pl. **physodes**) See *fucosan vesicle*

phytoalexin (pl. **phytoalexins**) Compounds of various kinds (some antimicrobial) induced by stress (infection, wound, etc.) in plants in direct response to injury; often are secondary metabolites

phytochrome (pl. **phytochromes**) Pigment associated with the absorption of light found in plants and some algae (for example, conjugating green algae, Pr-32); photoreceptor for red to far-red light; involved in the control of certain developmental processes

phytoflagellate (pl. **phytoflagellates**) Mastigote alga; any swimming protist with at least one undulipodium and one plastid

phytophagy Nutritional mode; organisms that feed on plants (and algae)

phytopicoplankton Ecological term referring to small photosynthetic microbes suspended in the water column, primarily in the ocean; cyanobacteria, chloroxybacteria, and the smallest plastidic protists (for example, *Micromonas*, B-6, -7; Pr-28)

phytoplankton Ecological term referring to aquatic free-floating algae and cyanobacteria (if motile they are unable to swim against the current); see *picoplankton, plankton*

phytotoxic Chemical substances that are poisonous to plants

PI electrophoresis Biochemical technique for separating proteins according to their electrically charged residues, that is, their isoelectric point. A potential difference is applied across a system in which pH increases from anode to cathode. Proteins or peptides present in the system accumulate on a band in the region of the gradient corresponding to their isoelectric point, the point at which their total charge is neutralized

picoplankton Ecological term referring to microorganisms found suspended in aquatic media, especially the ocean. The planktonic cells in the 0.2–2.0 μm size range are dominated by prokaryotes but include small eukaryotes, both with and without plastids. Term refers to size, not to nutritional mode or cell structure; see *phytopicoplankton*, *phytoplankton*, *plankton*

pinnate See *pennate*

pinocytic See *pinocytosis*

pinocytosis Type of eukaryotic intracellular motility process which uses microfibrils for cell "drinking;" endocytosis of liquid, dissolved solutes, and protein-sized particles through formation of membrane tunnels called pinocytotic vesicles; see *phagocytosis*

pinocytotic See *pinocytosis*

pit areas Areas of pit connections

pit connections Type of cell junction: protoplasmic connections joining cells by perforations in the cell wall, which may or may not be plugged; typical of rhodophytes (Pr-33) and fungi. For rhodophytes, pit connection is a misnomer because they are not connections between cells, but rather plugs of proteinaceous material deposited in the pores that result from incomplete wall formation

pit field (pl. **pit fields**) Collection of plasmodesmata at the center of the cross wall between cells in certain chlorophytes (for example, Trentepohliales, Pr-28)

pit plugs See *pit connections*

placoderm desmid (pl. **placoderm desmids**) Kind of conjugating green alga (Pr-32); desmid composed of two semicells that are usually joined by an isthmus and with pores usually present in cell walls. Walls of the two semicells are of different ages; see *saccoderm desmid*

placolith (pl. **placoliths**) Coccolith subtype with upper and lower "shields" composed of radial segments; heterococcolith composed of two plates or shields interconnected by a tube (for example, *Coccolithus*, Pr-25)

plakea Developmental stage in colonial volvocalean chlorophytes (Pr-28); curved plate of cells

planispiral A test of foraminifera (Pr-3) coiled in a single plane

plankton Ecological term referring to suspended, free-floating microscopic or small aquatic organisms in either marine or freshwater environments whose transport is subject to wave movements. Refers to size and passive motility, not to taxonomic affiliation

planktonic See *plankton*

planont Sporoplasm from a freshly germinated spore (for example, myxosporan, A-2)

planozygote (pl. **planozygotes**) Motile zygote of dinomastigotes; enlarged, undulipodiated, and sometimes thick-walled mastigote formed just after fusion

plant (pl. **plants**) Multicellular, diploid organism that develops from an embryo supported by maternal tissue, generally photoautotrophic

plasma membrane (pl. **plasma membranes**) See *cell envelope*

plasmalemma (pl. **plasmalemmas**) See *cell envelope*, *plasma membrane*

Plasmid (pl. **plasmids**) Small piece of naked DNA; small replicon

plasmodesma (pl. **plasmodesmata**) Structural term referring to cell junctions, that is, the tiny cytoplasmic threads that extend through openings in cell walls and connect the protoplasts of adjacent living cells especially in algae (for example, trentepohlialean chlorophytes, Pr-28) and plants

plasmodium (pl. **plasmodia**) Coenocyte; syncytium. Multinucleate mass of cytoplasm lacking internal cell membranes or walls. Multinucleate cell generally has from two to over a dozen nuclei, while plasmodia have over a dozen and up to millions of nuclei per cell

plasmodiocarp Sporophore resembling thickened plasmodial veins or modifications of portions of veins in myxomycotes (Pr-23)

plasmodial See *plasmodium*

plasmogamy Fusion of two cells or plasmodial cytoplasms without karyogamy (fusion of nuclei); cytoplasmic fusion, which may or may not be the first step in the fertilization process. Syngamy without karyogamy that may produce dikarya or heterokarya

plasmotomy Form of binary or occasionally multiple fission of a plasmodium or multinucleate protoctistan cell; division of a plasmodium; characteristic of large, multinucleate amebas (Pr-2), opalinids, myxosporeans, and others (Pr-6, -23, etc), in which nuclei exhibit mitosis following (rather than during or immediately preceding) the process of somatic fission, or in which nuclei may undergo divisions asynchronously. Some mitoses may be found at any time in the two (or more) separable multinucleate masses

plastid (pl. **plastids**) Generic term for photosynthetic organelle in plants and protoctists (all algae). Bounded by double membranes, plastids contain the enzymes and pigments for photosynthesis, ribosomes, nucleoids, and other structures; see *chloroplast, chrysoplast, phaeoplast, rhodoplast*

plastid endoplasmic reticulum (pl. **plastid endoplastic reticula**) See *PER*

plastid matrix (pl. **plastid matrices**) Fluid contents of plastid

plastidic nanoplankton Nanoplanktonic algae; phototrophic nanoplankton; tiny phytoplankton; ecological term specifying certain small plankton; planktonic protists in the 2–20 μm size range that possess plastids; see *picoplankton, phytoplankton, plankton*

plastidic protests Unicellular algae

plastoglobulus (pl. **plastoglobuli**) Lipid droplets usually randomly distributed through the plastid matrix; sometimes seen concentrated at the periphery of the pyrenoid in xanthophytes (Pr-16)

plate formula (pl. **plate formulas**) System of labelling dinomastigote (Pr-5) thecal plates

plate scale (pl. **plate scales**) Oval or circular-shaped flat scales lacking superstructure (for example, chrysophytes) as opposed to spiny scales (for example, prymnesiophytes or prasinophytes, Pr-25, -28)

playa Dry, barren area in the lowest part of an undrained basin (for example, southwestern United States); also, small, sandy land area at the mouth of a stream or along a bay shore; beach (Spanish, meaning shore, beach)

plectenchyma Interwoven tissues comprised of mycelial mass. Structural term designating mycelial tissues found in some heterotrophic protoctists and fungi

plectenchymatous See *plectenchyma*

plectenchymous See *plectenchyma*

pleiomorphic Exhibiting several forms or shapes; many and variable expressions of shape in a genetically uniform population (for example, organisms such as amebas (Pr-2) or other protoctists that display changing form)

pleomorphic See *pleiomorphic*

plerotic Descriptive of oogenesis in oospores (for example, oomycotes, Pr-21); clearly filling the oogonial cavity; see *aplerotic*

plesiomorphic See *plesiomorphy*

plesiomorphy Ancestral, generalized or primitive taxonomic character (seme) present in ancestor at the bifurcation of the lineage; see *symplesiomorphy, synapomorphy*

plethysmothallus (pl. **plesthysmothalli**) Diploid microscopic life cycle phase of some phaeophytes (Pr-17) in which reproduction is by zoospores that transform into diploid thalli (resembling *Ectocarpus* or *Streblonema*) capable of producing more zoospores (in the absence of sexual processes)

pleura (pl. **pleurae**) See *girdle*

pleuromitosis Cryptopleuromitosis. Closed mitosis (nuclear membrane remains intact) with an extranuclear spindle lateral to the nucleus, in which no equatorial plate forms; mitosis with a sharply asymmetrical intranuclear spindle

pleuronematic Descriptive of mastigotes having an undulipodium with one or more rows of mastigonemes; may be panacronematic, pantonematic, or stichonematic

ploidy The number of sets of chromosomes; see *aneuploid, diploid, euploid, haploid, polyploid*

plurilocular sporangium (pl. **plurilocular sporangia**) Sporangium composed of a multicellular structure in which each cell produces a single reproductive cell and spores are produced in several cavities; see *unilocular sporangium*

pluriseriate See *multiseriate*

pnano See *plastidic nanoplankton*

poikilotherm Descriptive term referring to organisms whose body temperatures are very similar to those of their external environment; that is, organisms unable to regulate their body temperature. Characteristic of all protoctists

poikilothermic See *poikilotherm*

polar body (pl. **polar bodies**) One of two cells divided off from ovum during maturation, before gametic nuclei fuse

polar cap (pl. **polar caps**) Chromophilic body beneath the anterior spore wall contained in the polar sac in myxosporean spores

polar capsule (pl. **polar capsules**) See *capsule*

polar fenestra (pl. **polar fenestrae**) Gaps in the nuclear membrane associated with semiopen mitosis (mitosis in which the nuclear membrane dissolves only at the poles of the spindle)

polar filament (pl. **polar filaments**) Distally closed, tubelike structure coiled within the polar capsule of myxosporans (A-2). When everted it has a sticky surface and possibly serves to anchor the hatching spore to the surface of the intestine of its host; "hairpoint" on the terminus of protistan undulipodia; see *anisofilar, isofilar*

polar gaps See *polar fenestra*

polar ring (pl. **polar rings**) Part of apical complex of apicomplexans (Pr-7), typical of sporozoites and merozoites, probably a microtubule-organizing center

polar sac (pl. **polar sacs**) See *anchoring disk*

polar tube (pl. **polar tubes**) Tubular extrusome of microsporan spores (F-1) serving for injection of the sporoplasm into the host cell; see *anisofilar, isofilar*

polarizing microscopy Microscopy in which a specimen is between a polarizer and an analyzer such that if regular features of that specimen lead to alterations in the path of polarized light they are detectable. Useful for analysis of petrographic thin sections and longitudinally aligned microtubules (for example, of axopods) or microfibrils (for example, cellulose walls of charophytes, Pr-28)

polaroplast (pl. **polaroplasts**) Structure consisting of a series of flattened sacs and vesicles, thought to be involved in polar tube extrusion in microsporan spores (F-1)

polycentric (adj.) Descriptive of algal thallus radiating from many centers at which reproductive organs (sporangia or resting spores) are formed; descriptive of cells or organisms demonstrating a number of centers of growth and development and more than one reproductive structure (for example, oomycotes, chytridiomycotes, hyphochytrids, Pr-21, -35, -14); descriptive of chromosomes or chromatids with more than one kinetochore, leading to parallel (rather than V-shaped) segregation of chromatids during anaphase; see *monocentric*

polycomplex (pl. **polycomplexes**) Structures formed by the fusion of components from synaptonemal complexes that have detached from diplotene chromosomes (for example, insects, the haplosporidian *Minchinia louisiana*, A-21, Pr-29)

polyeder (pl. **polyeders**) Polyhedral cell; angular cell formed by zoospores in some chlorophytes (for example, *Pediastrum*, *Hydrodictyon*, Pr-28)

polyenergid (pl. **polyenergids**) Cell containing multiple genomes either within one nucleus or within several nuclei; state of having either multiple nuclei and/or multiple ploidy in a nucleus within a single cell (for example, some radiolarian actinopods and ciliate macronuclei, Pr-31, -6)

polygenomic Having multiple genomes (for example, as in an endosymbiotic association); may also refer to polyploidy

polyglucan granule (pl. **polyglucan granules**) Storage bodies in the cytoplasm of some algal cells; darkstaining polymers of glucose resembling animal glycogen

polykinetid (pl. **polykinetids**) See *cirrus*

polykinetoplastic Stage in trypanosome development (Pr-11) in which the kDNA is present as several distinct kinetoplasts in the mitochondrion

polykinety Row of polykinetids (for example, cirrus). Infraciliary bases, with or without their cilia, of the buccal membranelles sensu lato of certain groups of ciliates (Pr-6) having more than two kinetosomes per unit kinetid (for example, scuticociliates); oral membranelles of the peritrichous ciliates

polymorphic Morphological or genetic differences seen in normal wildtype individuals that are members of the same species and same population; see *dimorphism*

polymorphism See *polymorphic*

polyphyletic Referring to a trait or group of organisms derived by parallel (convergent) evolution from different ancestors; see *monophyletic*

polyploid Descriptive of cells in which a number of sets of chromosomes exceeds two; that is, a multiple of the haploid number of chromosomes greater than diploid (for example, triploid (3N), or hexaploid (6N)); see *polygenomic*

polypodial ameba (pl. **polypodial amebas**) Ameba (Pr-2) that moves by means of several pseudopods that are extended simultaneously; see *monopodial*

polysaprobic Ecological term referring to an aquatic environment rich in dissolved organic material and low in dissolved oxygen; see *oligotrophy*

polyseme Evolutionary change in trait (seme) by varied repetition of that seme (for example, segmentation in worms, increase in kinetids in ciliates, Pr-6); see *seme*

polysiphonous Descriptive of an algal thallus composed of vertically aligned tubes composed of parallel cells (for example, *Polysiphonia*, Pr-33)

polyspore (pl. **polyspores**) See *gonospore*

polysporous See *gonospore*

polystichous Descriptive of an algal thallus that has parenchymatous organization and hence is multicellular in cross section (for example, phaeophytes, Pr-17)

polystromatic Descriptive of algal thallus composed of many cell layers

polytenic Condition in cells in which chromosomes have many times the normal (1x) quantity per length of DNA as a result of repeated replication without division so that the many (poly), threadlike (tenon) chromatids lie side-by-side. Whereas in polyploidy the number of chromosome sets augment, in polyteny the number of chromosomes stays constant but the quantity of DNA per set increases. Polyteny is characteristic of certain stages of macronuclear maturation in hypotrichous ciliates such as *Stylonychia* (Pr-6)

polyteny See *polytenic*

polythalamic Test of foraminifera (Pr-3) having several chambers or cells; multichambered, multilocular

polythalamous See *polythalamic*

polytomic fission Multiple fission; mode of reproduction involving division of a single individual into numerous offspring products; see *merogony, progressive cleavage, schizogony*

polyxenic Pertaining to cultures containing more than one type of unknown (and undesired) organism; descriptive of a culture with many contaminants; see *axenic, monoxenic*

polyxenous parasite (pl. **polyxenous parasites**) Parasite (necrotroph) requiring more than two different hosts for completion of its life cycle; see *heteroxenous*

pond (pl. **ponds**) A body of standing freshwater occupying a small surface depression, usually too small to sustain waves, that is, smaller than a lake and larger than a puddle or pool

population (pl. **populations**) Individuals, members of the same species, found in the same place at the same time

porcellaneous test (pl. **porcellaneous tests**) Test that is white, opaque, or slightly translucent in reflected light (for example, foraminifera, Pr-3)

pore (pl. **pores**) Openings. Minute rounded openings in the chamber wall, usually covered by an internal membrane or sieve (for example, in foraminifera, Pr-3)

pore apparatus Complex pore organ; openings through secondary wall in some desmids (Pr-32) that consist of a lined pore channel and a web of fibrous material at the inner opening

pore plate (pl. **pore plates**) Structure of diatom frustule (Pr-18): fine plate of lightly silicified material with small pores that stretches across the areola of many diatoms; see *rica, velum*

porelli Small, regularly arranged pores in the ocellus of a diatom (Pr-18)

porphyran Sulfated storage carbohydrate of *Porphyra*, composed of galactose units

porus (pl. **pori**) Opening in chrysophyte (Pr-15) stomatocysts (statospores) that is closed by a pectic plug at maturity

postciliary microtubular ribbon (pl. **postciliary microtubular ribbons**) Part of kinetid structure in the ciliate subphylum Postciliodesmatophora (Pr-6); ribbon of microtubules associated with a kinetosome, originating in the right-posterior part at triplet 9 (by convention); see *transverse ribbon*

postciliary ribbon (pl. **postciliary ribbons**) See *postciliary microtubular ribbon*

postciliodesmata Bundle of overlapping postciliary ribbons found in a large group of ciliates (Pr-6). The basis for classification at the level of subphylum (that is, Postciliodesmatophora)

postcingular Descriptive of cell-covering plates on hypotheca in contact with the cingulum in certain dinomastigotes (Pr-5); see *precingular*

posterosome (pl. **posterosomes**) Posterior vacuole formed from coalescence of Golgi vacuoles and involved in polar tube formation in microsporans (F-1)

preapical platelet (pl. **preapical platelets**) Small thecal plate that occurs between the first apical plate and the apical pore complex (APC) in some peridinioid dinomastigotes (Pr-5)

precingular Descriptive of cell-covering plates on the epitheca in contact with cingulum in some dinomastigotes (Pr-5); see *postcingular*

predation Mode of nutrition in which an organism hunts, attacks, and digests other heterotrophic organisms for food (for example, *Didinium* that seizes *Paramecium*, Pr-6)

predator See *predation*

preoral crest (pl. **preoral crests**) Part of the oral apparatus; ridge reinforced by band of microtubules (for example, in kinetoplastid mastigotes, Pr-11)

prespore cell (pl. **prespore cells**) Cell of a slug that will ultimately develop into spores in dictyostelids (Pr-2); earliest stage of sporocarp development in protostelids in which an ameboid cell begins to round up and secrete a slime sheath before the stalk is produced

presporogonic Pertaining is that a part of a life history that precedes the sporogonic one, that is, during which the spores are formed; see *sporogonic*

primary cell (pl. **primary cells**) Pseudoplasmodium enclosing one generative cell (for example, myxosporans, A-2)

primary metabolite (pl. **primary metabolites**) Organic compound produced metabolically and essential for completion of the life cycle of the organism that produces it (for example, any of the 20 protein amino acids or nucleotides in RNA and DNA). Chemical component required for autopoiesis; see *secondary metabolite*

primary nucleus (pl. **primary nuclei**) Large diploid nucleus that undergoes meiosis to give rise to secondary nuclei in certain chlorophytes (for example, Dasycladales, Pr-28)

primary plasmodium (pl. **primary plasmodia**) Sporangial plasmodium. Plasmodium that develops into thin-walled sporangium in plasmodiophorids; see *secondary plasmodium*

primary production Primary productivity; productivity. The production of reduced carbon (organic) compounds by autotrophs (Tables 1 and 2)

primary productivity See *Primary production*

primary rhizoid (pl. **primary rhizoids**) First rootlike protoplasmic extension (rhizoid) that develops from the encysted zoospore in chytridiomycotes (Pr-35)

primary zoospore (pl. **primary zoospores**) Zoospore that germinates directly from a cyst (for example, in plasmodiophorids, Pr-20); see *Secondary zoospore*

principal zoospore (pl. **principal zoospores**) First-formed zoospore, which has laterally inserted undulipodia that are shed on encystment (for example, the oomycote *Phytophthora*, Pr-21); see *Auxiliary zoospore*

proboscis Emergent process on the anterior end of the spermatozoids that contains a band of eight or nine microtubules originating near the kinetosomes (for example, in the xanthophyte *Vaucheria* (Pr-16)); structure thought to facilitate attachment to the egg in the phaeophyte *Fucus* (Pr-17); trunklike extension emerging from the oral area at anterior of certain ciliates (Pr-6; for example, *Dileptus*)

procaryote (pl. **procaryotes**) See *prokaryote*

procentriole (pl. **procentrioles**) Cell organelle; thin, circular, electron-dense, granular structure about 240 nm in diameter with a core of nine radiating spokes and no microtubular elements; found in the trophic cells of *Labyrinthula* (Pr-19), these structures appear to arise *de novo* prior to each mitotic division

procyclic stage (pl. **procyclic stages**) Stage in life cycle that represents the beginning of development in an invertebrate host (for example, trypanosomatid mastigotes, Pr-11)

productivity See *primary production*

progamic fission Binary fission occurring within a gamontocyst and resulting in the formation of two gamonts (for example, heliozoan actinopods, Pr-31)

progeny See *offspring*

progressive cleavage Multiple fission; cytokinesis of multinucleate protoplasm to form uninucleate cells. Phycological term for process comparable to schizogony

prokaryote (pl. **prokaryotes**) Bacterium; member of the kingdom Monera (Kingdom Procaryotae); cell or organism composed of cells lacking a membrane-bounded nucleus

proloculum (pl. **prolocula**) First chamber formed during development of the test of an adult gamont (for example, foraminifera, Pr-3)

promastigote (pl. **promastigotes**) Stage in trypanosomatid development (Pr-11) in which the kinetoplast lies in front of the nucleus and the associated undulipodium emerges laterally to form an undulating membrane along the anterior part of the body, usually becoming free at its anterior end

promitochondrion (pl. **promitochondria**) Structures that develop into mature, cristate mitochondria

promitosis Mitosis in which the nuclear envelope remains closed throughout the process and the nucleolus pinches in two (for example, in many amoebomastigotes, Pr-22); see *mesomitosis, metamitosis*

promitotic See *promitosis*

propagule (pl. **propagules**) Generative structure; any unicellular or multicellular structure produced by organisms, and capable of survival, dissemination, and further growth. Examples include cysts, spores, some kinds of eggs, seeds, and akinetes. Phycologists restrict the term to refer to hormogonia or other multicellular structures that function in asexual reproduction

prophase First stage of mitosis or meiosis in which chromosomes condense and nucleolus and nuclear membrane may begin to disappear. In meiosis, prophase is broken down into five substages: leptotene, zygotene, pachytene, diplotene, and diakinesis; see *meiosis, mitosis*

prosporangium (pl. **prosporangia**) Structure from which a sporangium develops

prostomial Oral region that develops in the anterior of an organism (that is, especially ciliates, Pr-6)

proter Anterior offspring of transverse binary fission of the parental organism; it often retains the mouthparts of the parent (for example, ciliates, Pr-6); see *opisthe*

protist (pl. **protists**) Single-celled (or very few celled and, therefore, microscopic) protoctists

protocentriole (pl. **protocentrioles**) According to the serial endosymbiotic theory, free-living bacterial ancestor of the centriole; see *protomitochondrion, protoplastid, serial endosymbiosis theory*

protoctist (pl. **protoctists**) Eukaryotic microorganisms (the single-celled protists and their multicellular descendants). All eukaryotic organisms with the exception of animals (developing from diploid blastulas), plants (developing from embryos supported by maternal tissue), and fungi (developing from zygo-, asco-, or basidiospores) are protoctists. Protoctists include two-kingdom system "protozoans" and all "fungi" with mastigote stages as well as all algae (including kelps), slime molds, slime nets, and other obscure eukaryotes

protomerite (pl. **protomerites**) Anterior part separated from the deuteromerite by a transverse septum (for example, trophozoites of some gregarine apicomplexans, Pr-7)

protomite (pl. **protomites**) Separate form between the tomont and the tomite; a relatively rare stage in the polymorphic life cycle of a few ciliates (Pr-6, for example, some apostomes)

protomitochondrion (pl. **protomitochondria**) According to the serial endosymbiotic theory, immediate free-living bacterial ancestor to mitochondria (for example, oxygen respirer such as *Paracoccus* or *Daptobacter*, B-3)

protomitosis See *promitosis*

protomont (pl. **protomonts**) Separate form between the feeding trophont and the often encysted true tomont (dividing) stage; a relatively rare stage in the polymorphic life cycle of a few ciliates (Pr-6, for example, some apostomes)

protonema (pl. **protonemata**) Thread-shaped structure developing from a spore (for example, algae and plants) or the product of zygote germination (charophytes, Pr-28)

protoplasm Fluid contents of cells, that is, cytoplasm and nucleoplasm

protoplasmic streaming See *cyclosis*

protoplasmodium (pl. **protoplasmodia**) Smallest and simplest type of myxomycote plasmodium

(Pr-23); microscopic in size, they lack any system of veins or protoplasmic streaming and usually give rise to a single sporophore; see *aphanoplasmodium, phaneroplasmodium*

protoplast Actively metabolizing membrane-bounded part of a cell as distinct from the cell wall. Cells that, after treatment to remove them, lack cell walls

protoplastid (pl. **protoplastids**) According to the serial endosymbiotic theory, immediate free-living bacterial ancestor to chloroplast, rhodoplast, and other plastids (for example, *Prochloron* or cyanobacterium)

protoseptate Descriptive of rudimentary, incomplete, or partial internal walls (septa) separating successive growth stages (for example, foraminifera, superfamily Astrorhizacea; Pr-3)

protosphere (pl. **protospheres**) Phase in the life history of certain chlorophytes (Pr-28); the lobose cell developing from a germinating zygote that will next form a siphonous juvenile

protozoa (pl. **protozoans**) Obsolete term referring, in the two-kingdom classification, to a phylum in the Animal kingdom consisting of large numbers of primarily heterotrophic, microscopic eukaryotes. Traditionally, the smaller heterotrophic protoctists and their immediate photosynthetic relatives (for example, phytomastigotes); see *metazoa*

proximal sheath (pl. **proximal sheaths**) Wedge-shaped or bilobed component of kinetids associated with proximal ends of uppermost kinetosome (for example, ulvophycean chlorophytes, Pr-28)

psammolittoral Ecological term referring to the sandy environment along marine coasts

psammophile Ecological term referring to organisms that live in sandy environments, especially in the spaces between sand grains (for example, many karyorelictid ciliates, Pr-6)

psammophilic See *psammophile*

pseudocapillitium (pl. **pseudocapillitia**) Structure of myxomycote sporophores (Pr-23) consisting of irregularly shaped thread- or plate-like fragments dispersed among the spores (for example, in the order Liceales)

pseudocilium (pl. **pseudocilia**) Nonmotile undulipodium (for example, of the chlorophyte *Tetraspora gelatinosa*, Pr-28)

Type of mastigoneme in glaucocystophytes; protoplasmic protrusion of a cell containing microtubules, derived from the typical axoneme but immotile; see *pseudoflagellum*

pseudocrystalline See *paracrystalline*

pseudoflagellum (pl. **pseudoflagella**) See *pseudocilium*

pseudogene (pl. **pseudogenes**) A nonfunctional gene closely resembling a known gene of a different locus

pseudointracellular Descriptive of position relative to a cell; appearing intracellular but topologically extracellular because of failure to cross the plasma membrane (for example, symbiotrophs contained in parasitophorous vacuoles)

pseudointratissular Descriptive of position relative to tissue; surrounded by tissue and appearing to be inside tissue but topologically external to the tissue because of failure to penetrate into or between cells; see *intratissular*

pseudoparenchyma Thallus construction; contiguous filaments rather than true parenchymatous cells capable of three-dimensional growth; parenchymalike

pseudoplasmodium (pl. **pseudoplasmodia**) Structure resembling a multinucleate plasmodium that has retained its cell membrane boundaries. An aggregate of amebas, especially that constituting the initial stage of sorocarp formation in the cellular slime molds (dictyostelids, acrasids; Pr-2); uninucleate trophozoite cell containing one to several generative cells (myxosporan life cycle stage; A-2); see *slug*

pseudopod (pl. **pseudopods**) Temporary cytoplasmic protrusion of an ameboid cell used for locomotion or phagocytotic feeding

pseudopodium (pl. **pseudopodia**) See *pseudopod*

pseudospore (pl. **pseudospores**) Nonmotile wall-less spore (for example, in some acrasids, Pr-2)

pseudostome (pl. **pseudostomes**) "False mouth"; aperture through which a testate ameba projects its pseudopods

PSP See *paralytic shellfish poisoning*

psychrophile See *cryophile*

pulsed-field gradient See *electrophoresis*

punctum (pl. **puncta**) Pore containing smaller pores (for example, diatom wall-markings, Pr-18)

pustule (pl. **pustules**) Blisterlike, frequently eruptive spot or spore mass (for example, fungi, foraminifera, Pr-3)

pusules Fluid-filled intracellular sacs responsive to changes in pressure. Specialized vacuole-like organelles, presumably osmoregulatory. Usually two per cell and consisting of two closely appressed membranes that bound a vesicle, they open by canals to the kinetosomes and thence to the outside of the cell

pycnosis Darkly staining chromosomes or nuclei; moribund nuclei in cells (for example, nongenerative nuclei of foraminifera and degenerating fragments of ciliate macronuclei; Pr-3, -6)

pycnotic See *pycnosis*

pyrenoid (pl. **pyrenoids**) Proteinaceous structure associated with plastids serving as the center of starch formation or glucan deposits in some algae

pyrenoid cap (pl. **pyrenoid caps**) Starchy structure surrounding specialized region of plastid (pyrenoid)

pyriform Any structure with the form of a tear or pear (for example, *Tetrahymena pyriformis*)

quadraflagellate Quadriundulipodiated; referring to mastigote cell bearing four undulipodia (for example, some trichomonads or chlorophytes, Pr-1, -28)

quadramastigote See *quadraflagellate*

quinqueloculine Foraminiferan test (Pr-3) in which five chambers are visible and each chamber is angled 144° from previous chamber

r Potential maximal rate of increase of a population

radial fibrils Fibers arranged in a spokelike array such as those seen in many thin sections of axonemes

radial wall (pl. **radial walls**) Foraminiferan test (Pr-3) wall composed of calcite or aragonite crystals oriented with their C-axis perpendicular to the surface

radiolarite Rock made of chert (siliceous microcrystalline quartz) composed of radiolarian tests (Pr-31) that have undergone diagenetic alteration

raphe The slit, elongate cleft, groove, or pair of grooves through the valve of most pennate diatoms (Pr-18) that facilitates gliding cell motility

raphe fiber (pl. **raphe fibers**) Structure immediately below the forming raphe, thought to be responsible for the curve of the raphe slit in some diatoms (Pr-18; for example, *Navicula* spp. and *Pinnularia* spp.)

raphe slit (pl. **raphe slits**) See *raphe*

raphe system (pl. **raphe systems**) See *raphe*

R-body (pl. **r-bodies**) Body found inside the kappa particles in the cytoplasm of killer paramecia in some members of the *Paramecium aurelia* complex (Pr-6). Ribbon-shaped body of the kappa extrusome, viruslike in appearance

receptacle (pl. **receptacles**) Swollen structure containing conceptacles on the thalli of phaeophytes (Pr-17) on which reproductive organs (that is, gametangia or sporangia) are borne; see *conceptacle*

rectilinear test (pl. **rectilinear tests**) Test in which chambers accumulate by growth in a straight line (for example, foraminifera, Pr-3)

recurrent flagellum (pl. **recurrent flagella**) Recurrent flagellum. Undulipodium that does not lead an organism but adheres to it; trailing undulipodium of heterokont mastigotes

recurrent undulipodium (pl. **recurrent undulipodia**) See *recurrent flagellum*

red tide (pl. **red tides**) Seawater discolored by the presence of large numbers of dinomastigotes (Pr-5, especially of the genera *Peridinium* and *Gymnodinium*); blooms of some chrysophytes, euglenids, and the ciliate *Mesodinium rubrum* (Pr-6) have also been correlated with red tides

refringent The ability to refract (break up) (for example, rays of light)

regolith Loose, rocky surface materials (boulders, gravel, silt, sand, etc.) covering a planet

replication Process that augments the number of DNA or RNA molecules. Molecular duplication process requiring copying from a template

reproduction Process that augments the number of individuals. A single parent is sufficient for the increase in numbers of individuals in asexual reproduction whereas two parents are required in sexual reproduction. Requires at least one autopoietic entity

reservoir (pl. **reservoirs**) Holding structure or vestibule; deep part of the oral region of some protoctists; the base of the flask-shaped invagination of euglenids (Pr-12)

reservoir host (pl. **reservoir hosts**) Ecological terms primarily used by parasitologists, for habitats of symbiotrophs in which infected species of animals serve as a source from which other species of animals can become infected (for example, antelopes are reservoir hosts for *Trypanosoma rhodesiense* (Pr-11), the causative agent of African sleeping sickness in humans)

residual body (pl. **residual bodies**) Residuum. That which exists after the formation of offspring cells (gametes or zoites); in dinomastigotes (Pr-5), dark brown body left in empty cyst; in apicomplexans (Pr-7), residual cytoplasm and nuclei of the parent cell

residuum (pl. **residua**) See *residual body*

resistant cyst (pl. **resistant cysts**) Resting cyst; resting spore; dormant propagule of many different kinds of protoctists, equivalent to protoctist spore. Stage surrounded by a wall protecting it from desiccation or other physical injuries; thick-walled, uni- or multinucleate cell that can remain dormant for periods of time under adverse environmental conditions; see *spore, statospore, stomatocyst*

resistant sporangium (pl. **resistant sporangia**) Resting spore or covering of many spores (for example, in chytridiomycotes (Pr-35), a zoosporangium with a thickened wall formed in response to desiccation and capable of extended survival)

resting cyst (pl. **resting cysts**) Resistant cyst or protoctist spore; dormant life cycle stage, equivalent to aplanospores or hypnospores of dinomastigotes (Pr-5) and other protoctists

resting spore (pl. **resting spores**) See *resistant cyst, resistant sporangium*

restriction enzyme digestion The use of endonucleases, enzymes that cleave foreign DNA molecules at specific recognition sites, to generate DNA fragments

reticular body (pl. **reticular bodies**) Net-shaped body; any structure that is netlike or covered with netlike ridges

reticulate Referring to any arrangement in a network; netted

reticulopod (pl. **reticulopods**) Very slender, anastomosing pseudopod that is part of a reticulopodial network (for example, phylum Granuloreticulosa-Foraminifera, Pr-3)

reticulopodial network Network of cross-connected pseudopods through which a two-way flow of cytoplasm and food particles is detectable; functions more often in food capture than in locomotion (for example, phylum Granuloreticulosa-Foraminifera, Pr-3)

reticulopodium (pl. **reticulopodia**) See *reticulopod*

retinoid Light-sensing component located within the melanosome of the dinomastigote ocellus (Pr-5); pigment layer in dinomastigote cells that produce ocelli

reversion Change from mastigote to ameba form

rhabde Main branch of the siliceous skeleton of ebridians from which clades branch

rhabdolith (pl. **rhabdoliths**) Heterococcolith bearing a stem or club-shaped extension on its distal face (for example, *Rhabdosphaera*)

rheoplasm More fluid exterior of reticulopodia; see *stereoplasm*

rheotaxis Directed growth in response to flow of current (for example, algae)

rhizoid (pl. **rhizoids**) Rootlike structure usually with anucleate filaments that anchor and absorb (for example, of chytridiomycotes, Pr-35); see *primary rhizoid, rhizomycelium,*

rhizomycelial Delicate rootlike (rhizoidal) system extensive enough to resemble superficially the mycelia of fungi; nucleated rhizoids having the potential for unlimited growth under favorable environmental conditions (for example, of chytridiomycotes, Pr-35); see *rhizoid*

rhizomycelium (pl. **rhizomycelia**) See *rhizomycelial*

rhizoplast (pl. **rhizoplasts**) Fibrillar rhizoplast. Cross-banded microtubular ribbon extending from the bases of kinetosomes and directed toward the nucleus or to cytoplasmic microtubule-organizing centers; in chytridiomycotes (Pr-35), fibrillar structure in the zoospore connecting the kinetosomes (at its proximal face) with the nuclear envelope. A rhizoplast is a type of kinetid

rhizopod (pl. **rhizopods**) See *reticulopod*

rhizopodium (pl. **rhizopodia**) See *reticulopod*

rhizosphere (pl. **rhizospheres**) Root zone of plants

rhizostyle (pl. **rhizostyles**) Kinetid of cryptomonads (Pr-26); posteriorly directed microtubular undulipodial rootlet; the microtubules have winglike projections in some species

rhodomorphin Hormone isolated from the red alga *Griffithsia* (Pr-33) that can induce cell division and is thus involved in processes of cell repair

rhodoplast (pl. **rhodoplasts**) Red plastid; photosynthetic membrane-bounded organelle of red algae (Pr-33) containing chlorophyll *a* and phycobiliproteins

rhoptry Part of the apical complex of some apicomplexans (Pr-7); dense body extending back from the anterior region of the zoite; may be tubular, saccular, or club-shaped (pedunculate); believed to release secretions facilitating entry of the zoite into its hosts' cells

ribosome (pl. **ribosomes**) Organelle composed of protein and ribonucleic acid; site of protein synthesis

rica (pl. **ricae**) Structure of diatom frustules; thin closing plate of silica usually with circular perforations across the areolae of some biraphid pennate diatoms (Pr-18); type of pore plate; see *velum*

rimoportule (pl. **rimoportules**) Diatom organelle (Pr-18) forming a tubular passage through the siliceous wall, infrequently extended into an external tube. Internally, a slitlike aperture surrounded by a stalked ridge, giving the appearance of two lips

rock (pl. **rocks**) Any naturally formed, consolidated (lithified), loosely consolidated (friable), or unconsolidated (for example, sand or gravel) material (but not soil) composed of two or more minerals or occasionally of one mineral and having some degree of chemical or mineralogic constancy

rohr Extracellular infection apparatus of plasmodiophorids (Pr-20); long, tubular cavity; see *schlauch*, *stachel (German, meaning pipe, tube)*

root fiber system (pl. **root fiber systems**) Proximal portion of kinetid; portion of kinetid below kinetosome; see *kinetid*

root microtubule (pl. **root microtubules**) Rootlet microtubules; part of kinetid structure. Microtubules attached proximally to kinetosomes; see *kinetid*

rootlet (pl. **rootlets**) Any small structure extending vertically (or proximally into cells) and resembling a taproot of plants (for example, portion of kinetid, undulipodial (flagellar) rootlet)

rostrum (pl. **rostra**) General term describing the apical end of a cell when it is beak-shaped or when there is a protuberance (especially ciliates or mastigotes). Head. Usually less conspicuous than a proboscis

ruderal Ecological term referring to the habitat of rubbish, waste, or disturbed places; an organism that grows in such a habitat

rumposome (pl. **rumposomes**) Intracellular structure; honeycomb-like organelle of unknown function consisting of regularly fenestrated cisternae in zoospores (chytridiomycote orders Chytridiales and Monoblepharidales; Pr-35)

S phase Phase in the mitotic cell cycle of eukaryotes during which DNA synthesis occurs; see *mitosis*

saccate Any pouched or bag-shaped structure

saccoderm desmid (pl. **saccoderm desmids**) Conjugating green alga (desmid; Pr-32) lacking semicells and pitted walls; see *placoderm desmid*

sagenogen (pl. **sagenogens**) See *bothrosome*

sagenogenetosome (pl. **sagenogenetosomes**) See *bothrosome*

sagittal ring (pl. **sagittal rings**) Ring-shaped component of radiolarian skeletons (Pr-31) that lies in a medial sagittal plane separating the skeleton into fragments

sagittal suture (pl. **sagittal sutures**) Thecal plate boundary between left and right halves of many dinomastigotes (for example, *Prorocentrum*, *Gymnodinium*; Pr-5)

salt marsh (pl. **salt marshes**) Flat, poorly drained land that is subject to periodic or occasional overflow by salt water, containing water that is brackish to strongly saline and usually covered with a thick mat of grassy halophytic plants

saltatory motion Jumping motion, usually intracellular motility (for example, that exhibited by mitochondria and refractile granules in actinopod cytoplasm as a result of cyclosis)

sand A tract or region of rock fragments or detrital particles smaller than pebbles and larger than coarse silt; usually composed of silica but occasionally of carbonate, gypsum, or other composition

saprobe Saprophyte; saprotroph. Organism utilizing a type of heterotrophy in which it obtains food from dead organic matter; organism feeding by osmotrophy, the mode of nutrition involving the absorption of soluble organic nutrients

saprobic See *saprobe*

saprophyte Saprobe. Heterotrophic organism living on and deriving its nutrition from dead organic matter. Obsolete term for bacteria and fungi (for example, fungi living on dead animals). Term to be avoided meaning "plant feeding on dead matter"; refers to osmotrophy of bacteria and fungi

saprophytic See *saprophyte*

saprotroph See *saprotrophy*

saprotrophic See *saprotrophy*

saprotrophy Mode of nutrition of a saprobe; heterotrophic nutrition obtained from a once-living, still recognizable organism (Table 2); see *autotrophy*, *biotrophy*

sarcinoid A growth habit in which a cubical cell packet arises because the component cells divide in successive perpendicular planes (for example, bacteria, algae)

saturation density See *K*

saxicolous Epilithic; lithophilic. Organisms dwelling on the surface of rocks (for example, algae, cyanobacteria)

saxitoxin complex Group of toxins produced by the dinomastigotes *Protogonyaulax* and *Pyrodinium* (Pr-5) that cause paralytic shellfish poisoning; includes saxitoxins, neosaxitoxin, and gonyautoxins

scalariform conjugation Sexual process in conjugating green algae (Pr-32) involving exchange of gametes through conjugation tubes between cells of parallel filaments. The filaments and conjugation tubes form a ladderlike structure

scale (pl. **scales**) Organic or mineralized structures of specific shape deposited on the cell surface. Organic or mineralized platelets forming part of a scaly envelope (or scale case) surrounding a cell; cell structures produced endogenously usually within cisternae of the Golgi apparatus and then deposited on the cell surface through vesicle exocytosis, usually in ordered arrays; often with elaborate surface decoration and sometimes with an outer deposit of $CaCO_3$ (as in coccoliths). Scales may be disklike (plate scales) or elaborated to form cup scales, spine scales, or small dense bodies (knob scales)

scale reservoir (pl. **scale reservoirs**) Invagination of cell surface harboring scales in scaly protoctists. Scales are deposited into the scale reservoir by exocytosis from the Golgi where they are produced. Production is periodic and apparently synchronized with cell division (for example, coccolithophorids, Pr-25)

schizodeme (pl. **schizodemes**) Strain or variety; ecological term referring to a population of kinetoplastids that display similarities in patterns of kDNA as determined by electrophoresis

schizogony Type of multiple fission; formation of offspring cells in apicomplexans, microsporans, and myxosporans by multiple fission; if the products are merozoites, the process can be subtermed merogony, if gametes, gametogony, if sporozoites, sporogony; in the past, some workers have equated schizogony with merogony only. Process comparable to progressive cleavage of algal plasmodia

schizont Multinucleate organism that will undergo schizogony (for example, apicomplexans)

schizozoite (pl. **schizozoites**) See *merozoite*

schlauch Narrow, open-ended extension of the rohr that is oriented toward the cytoplasm of the encysted zoospore of plasmodiophorids (Pr-20). German, meaning hose; see *rohr*, *stachel*

scintillons Particles isolated from cell extracts of luminescent dinomastigotes (Pr-5) that bioluminesce *in vitro*; see *microsource*

sclerotic Type of propagule; darkened amorphous cystlike material derived from desiccated plasmodia of myxomycotes (Pr-23), desiccation-resistant and capable of germination into viable slime mold

sclerotium See *sclerotic*

secondary cytoskeletal microtubule (pl. **secondary cytoskeletal microtubules**) Cytoplasmic microtubules originating at microtubule-organizing centers close to but not directly attached to kinetosomes (for example, in ciliates, euglenids, Pr-6, -12)

secondary metabolite (pl. **secondary metabolites**) Organic compound, produced metabolically, not essential for completion of the life cycle of the organism that produces it (for example, alkaloids, flavonoids, and tannins). They seem primarily to play ecological roles; may serve as pheromones, phytoalexins; see *primary metabolite*

secondary pit connection (pl. **secondary pit connections**) Pit connection developed between two adjacent cells by the cutting off of a small cell from one of the pair of adjacent cells, and the fusion of that small cell with the other member of the pair

secondary plasmodium (pl. **secondary plasmodia**) Plasmodium of plasmodiophorids (Pr-20) that develops into thick-walled resting cysts; see *primary plasmodium*

secondary zoospore (pl. **secondary zoospores**) Zoospore of zoosporangial origin (for example, plasmodiophorids, Pr-20)

sedimentation coefficient (pl. **sedimentation coefficients**) Rate at which a given solute molecule suspended in a less dense solvent sediments in a field of centrifugal force; given in Svedberg units, abbreviated S (for example, ribosomal subunits 23S, 16S; transfer RNA 5S)

segregation Movement to opposite poles of chromatids (mitosis) or chromosomes (meiosis)

seirosporangium (pl. **seirosporangia**) Algal sporangia produced in series at the termini of thalli. Such rows of sporangia may be either branched or unbranched (for example, the rhodophyte *Seirospora seirosperma*, Pr-33)

seirospore (pl. **seirospores**) Spore produced by a seirosporangium

SEM Scanning electron microscope

seme Complex trait of identifiable selective advantage, and therefore of evolutionary importance, resulting from evolution of an interacting set of genes. Unit of study by evolutionary biologists (for example, nitrogen fixation, cell motility, eyes); see *aposeme*, *hyperseme*, *hyposeme*, *neoseme*

semicell (pl. **semicells**) One of a pair, usually mirror-image halves, that forms the cell of placoderm desmids

semiconservative replication Method of DNA replication in which the molecule splits, each half being conserved and acting as a template for the formation of a new strand

septate Partition; cell wall separating constituent cells of multicellular organisms

septate junction (pl. **septate junctions**) Type of cell junction in animal tissues; specialized area of adjoining cell membranes showing partitions (that is, epithelial cells)

septum (pl. **septa**) See *septate*

sequential zoosporangium formation Process in which zoosporangia are formed over a period of time on the same subtending hypha, either by means of regrowth through the sporangial septum, cymose renewal of the hypha below the base of the zoosporangial septum, or basipetal, retrogressive zoosporangium delimitation (for example, oomycotes, Pr-21)

serial endosymbiosis theory Theory that mitochondria, plastids, and undulipodia began as free-living bacteria that established symbioses with other bacterial hosts, that is, that these organelles began as xenosomes

serodeme (pl. **serodemes**) Populations of for example, trypanosomes (clones or strains) related by descent and capable of expressing the same variable antigen type repertoire

sessile Attached, referring to any organism not free to move about because of attachment to other organisms or to rocks; see *vagile*

seston Ecological term for microbial communities or populations of particulate matter (including organisms) suspended in the water column in aquatic environments

SET See *serial endosymbiosis theory*

seta (pl. **setae**) Stiff bristle, hair, or other elongate immotile process (for example, mastigonemes); common in chlorophytes (for example, *Coleochaete*, Pr-28). Hollow projection of the frustule that extends beyond the valve margin (for example, diatoms, Pr-18)

sex Process of formation of new organism containing genetic material from more than a single parent. Minimally involves uptake of genetic material from solution and DNA recombination by at least one autopoietic entity; mode of reproduction involving the formation of haploid nuclei in eukaryotes (meiosis) and fertilization (karyogamy, syngamy) to form zygotes. Sexuality; see *parasexuality*

sex cell (pl. **sex cells**) See *germ cell*

sex pheromone (pl. **sex pheromones**) See *sex pheromone*

SGO Invaginations in the cell membrane where organic substances (for example, spicules) are deposited to form the skeleton (for example, heliozoan actinopods, Pr-31)

shadow casting Technique used in transmission electron microscopy in which a coating of a heavy metal is deposited on a sample at an angle such that the metal builds up on one side, creating a shadow image. The shape and length of the shadow allows calculation of the dimensions of the sample

sheath (pl. **sheaths**) Mucopolysaccharide periplast; extracellular, noncellular matrix produced by cells; thought to protect cells from desiccation (for example, made by pseudoplasmodia of dictyostelids, developing sorocarps of acrasids, sporocarps of protostelids, Pr-2, or by trichomes or coccoid cells of algae and cyanobacteria, B-6)

shield cell (pl. **shield cells**) Wall cell of the antheridium in charophyte chlorophytes Pr-28

shock reaction Behavioral response to a sudden change in environmental conditions; in euglenids, the cell halts, spins, or turns *in situ* end-over-end for a second or more, then proceeds to swim in a random direction

shuttle streaming Protoplasmic streaming in which there is a rapid flow of protoplasm in one direction, a gradual decrease in the flow rate until it ceases, and then a resumption of flow in the opposite direction (for example, myxomycotes, Pr-23)

side body complex Collective name for cistema, microbody, and lipid globules in the zoospores of Phylum Blastocladiomycota (Pr-34; for example, *Blastocladiella*)

sieve area (pl. **sieve areas**) Pr-17: Field of pores lined by plasma membrane through which products of photosynthesis are translocated (for example, in cells of large algae). The pores may be numerous and small (for example, *Laminaria*) or few and large (for example, *Macrocystis*)

sieve element (pl. **sieve elements**) Pr-17: Cells with sieve areas. Sieve elements may be randomly oriented or superimposed in longitudinal series constituting sieve tubes

sieve tube (pl. **sieve tubes**) Pr-17: Longitudinal series of sieve elements that form tubes for translocation of photosynthate (for example, in *Nereocystis* and *Macrocystis*)

silicalemma (pl. **silicalemmata**) Intracellular membranous vesicle derived from Golgi in silica-depositing algae (for example, membrane upon which opaline silica of the diatom frustule, (Pr-18) is deposited). Silicalemma, to which silica adheres tightly, is found associated with microtubule-organizing center in central region between offspring cells just inside cell membrane

silicoflagellate (pl. **silicoflagellates**) Undulipodiated photosynthetic marine protoctists with siliceous tests. Members of the phylum Chrysophyta (Pr-15), they are partially responsible for the depletion of dissolved silica from surface waters

silicoflagellite (pl. **silicoflagellites**) Chert rock composed of accumulated silicomastigote skeletons that were sedimented and diagenetically altered; see *chert*, *diagenesis*

silicomastigote (pl. **silicomastigotes**) See *silicoflagellate*

silt Clastic sediment composed of particles from 60 to 200 μm(grains are larger than clay and smaller than sand)

sinus (pl. **sinuses**) Invaginated region at the isthmus in certain desmids (conjugating green algae, Pr-32)

siphon (pl. **siphons**) General term referring to cell or structure in the shape of a pipe or tube; in algae, multinucleate, without crosswalls, that is, coenocytic, syncytial; see *coenocyte, plasmodium, syncytium, tubular ingestion apparatus*

siphonaceous See *siphon, siphonous*

siphonaxanthin Carotenoid pigment of the chloroplasts of some chlorophytes (for example, some Caulerpales, Siphonocladales, Pr-28)

siphonein Carotenoid pigment of chloroplasts of some chlorophytes, primarily members of the Caulerpaceae (Pr-28)

siphoneous See *siphon*

siphonous See *siphon*

slime molds See *acellular slime mold* (Pr-23), *cellular slime mold* (Pr-2)

slime net (pl. **slime nets**) Members of the phylum Labyrinthulomycota (Pr-19); labyrinthulids and thraustochytrids

slug (pl. **slugs**) See *grex*

soil (pl. **soils**) Regolith or loose, rocky, organic-rich surface cover of planet Earth; area of unconsolidated material over bedrock; usually supporting or capable of supporting growth of plants

soma General term referring to the body (soma) of an organism, especially the parts not involved in reproduction or germination

somatic See *soma*

somatic cell (pl. **somatic cells**) Differentiated cell comprising the tissues of soma; any body cell except germ cells; see *germ cell*

somatic kinetid (pl. **somatic kinetids**) Body kinetid (for example, kinetid of the ciliate cortex (Pr-6), usually not of the oral region)

somatic nucleus (pl. **somatic nuclei**) See *macronucleus*

somatic region Body region (for example, in ciliates (Pr-6), the body of the cell exclusive of the oral region)

somatoneme (pl. **somatonemes**) Tubular hairs on the cell surface that are products of the Golgi apparatus, associated with subpellicular microtubules (for example, proteromonads, Pr-1)

sonication Method for breaking cells open or homogenizing a mixture of particles by use of ultrahigh frequency vibration

sorocarp (pl. **sorocarps**) Multicellular, aerial, stalked structure derived from the aggregation of many individual cells. Often called fructification or fruiting body; ambiguous botanical terms that should be avoided. Applies to dictyostelids (Pr-2), the ciliate *Sorogena* (Pr-6), and acrasids (Pr-2) but not to protostelids; see *sporocarp*

sorocyst (pl. **sorocysts**) Cyst in sorus of cellular slime molds (Pr-2); sorocysts are virtually identical to ameba cysts, and can also be considered spores

sorogen (pl. **sorogens**) Culminating stage of cellular slime mold sorocarp (Pr-2)

sorogenesis Sorocarp development (Pr-2); formation of the stalked structure that bears the propagules

sorophore (pl. **sorophores**) See *sorocarp* (Pr-2)

sorus (pl. **sori**) Cluster of spores, sporangia, or similar structures in which spores are formed (for example, in cellular slime molds (Pr-2))

sperm Male gamete; motile and generally smaller than the female gamete. Zoospore-like structure requiring fertilization for further growth

spermary (pl. **spermaries**) Sperm storage organ

spermatangium (pl. **spermatangia**) Cell that produces spermatia (for example, rhodophytes, Pr-33)

spermatium (pl. **spermatia**) Minute, coccoid, colorless, male gamete released from a spermatangium; spermatia are never undulipodiated in rhodophytes (Pr-33)

spermatozoa See *sperm*

spermatozoid (pl. **spermatozoids**) Anisogamete, protoctist sperm; undulipodiated reproductive cell functioning as a sperm, that is, gamete fertilizing a much larger nonmotile gamete (egg)

sphaerocyst (pl. **sphaerocysts**) Rough-walled, pigmented, spherical cyst that may result from cell fusions in the acrasid *Copromyxa protea* (Pr-2)

sphaeromastigote (pl. **sphaeromastigotes**) Rounded-up cell; developmental stage in kinetoplastids (Pr-11) in which the anterior end cannot be identified, although an undulipodium is present

spherule (pl. **spherules**) Prominent convoluted mass of cisternae at the anterior end of the sporoplasm in haplosporidians (Pr-29), possibly a modified Golgi body. Macrocyst, that is, dormant, usually multinucleate, walled plasmodial segment in myxomycotes

spicular Slender, typically needle-shaped process (for example, biogenic crystals emerging from the siliceous tests of actinopods, Pr-31); small spine; see *spine*

spicular vacuole (pl. **spicular vacuoles**) Vacuole in actinopods in which spicules lie; see *perispicular vacuole*

Spicule (pl. **spicules**) See *spicular*

spicule-generating organelle (pl. **spicule-generating organelles**) See *SGO*

spindle (pl. **spindles**) See *mitotic spindle*

spindle pole body (pl. **spindle pole bodies**) Nucleus-associated organelle (NAO). Granulofibrosal and microtubular material found at the poles of mitotic spindles; type of microtubule-organizing center. Many variations on NAOs exist in organisms that do not form [9(3)+0] kinetosomes (centrioles); see *NAOs*

spine (pl. **spines**) Slender needle-shaped protrusions (for example, actinopods, Pr-31); skeletal projections; defined differently by different authors as either a major rodlike projection from the skeleton or a minor barblike emanation on the skeleton; in the latter case the major projection is a spicule

spirochete (pl. **spirochetes**) Helically shaped bacterium with flagella in the periplasm

sporangial plasmodium (pl. **sporangial plasmodia**) See *primary plasmodium*

sporangial plug (pl. **sporangial plugs**) Solid deposit of acellular cell wall-like callous material that separates the sporangial protoplasm from the protoplasm of the rest of the thallus; expelled prior to sporangial release

sporangiogenesis Formation of sporangium

sporangiophore (pl. **sporangiophores**) Subtending stalk to a sporangium

sporangium (pl. **sporangia**) Hollow unicellular or multicellular structure in which propagules (cysts or spores) are produced and from which they are released; see *gametangium*

spore (pl. **spores**) Type of propagule; small or microscopic agent of reproduction. Some are desiccation- and heat-resistant propagules capable of development into mature or active organisms. Spores are seldom homologous, sometimes even within a single taxon (for example, coccidians, Pr-7). There is little, if any, difference between the spores of the acellular slime molds (Pr-23) and the cysts of amoebomastigotes (Pr-20). Yet the term spore is widely used for the nonresistant propagules developing from the sporangia of free-living myxomycote groups (Pr-23) and the clearly nonhomologous resistant spores of all microsporan and myxosporan groups (F-1, A-2). The use of spore is controversial for the oocysts of gregarines (Pr-7) or the sporocysts of coccidians (Pr-7), even though these stages are both resistant and infective; for some earlier authors, the sporozoites themselves were the "naked spores," similar as they are to the spore stage of myxomycotes (and various nonprotoctist) species; even cyst (for example, oocyst) and spore have sometimes been confounded. Investigators working with apicomplexans suggest replacement of the term spore with specific terms in the life cycle stages of the organisms. Also called vegetative resting state, an ambiguous botanical term to be avoided; see *aplanospore, autospore, auxospore, azygospore, ballistospore, dinospore, endospore, epispore, exospore, hypnospore, macrospore, meiospore, mesospore, microspore, oospore, resting spore, statospore, zoospore, zygospore*

spore morphogenesis Developmental process resulting in formation of a spore

sporelings Growths resulting from germinated spores

sporoblast (pl. **sporoblasts**) Structures giving rise to spores (for example, myxosporans, A-2). Elliptical, nucleated structures pointed at the ends, the result of a process of segmentation undergone by the protoplasm in apicomplexans (Pr-7)

sporoblastic See *sporoblast*

sporocarp (pl. **sporocarps**) Usually stalked spore-bearing structure in which one initial cell is the source of all the spores (for example, myxomycotes and protostelids, Pr-23, -2). Also called fruiting body, an ambiguous botanical term that should be avoided; see *sorocarp*

sporocyst (pl. **sporocysts**) Cyst formed within the divided oocyst that will contain the sporozoites (for example, coccidians, Pr-7); cyst containing spores (for example, microsporans, F-1); sometimes the oocyst itself in gregarine apicomplexans, which actually have no sporocyst stage

sporocyte (pl. **sporocytes**) Diploid (2N) cell that undergoes meiosis to form haploid (lN) spores; aggregations of cells that divide to produce heterokont bimastigote zoospores (for example, in the labyrinthulomycotes *Labyrinthula vitelline* and *L. algeriensis*, Pr-19); product of division of the gonocyte in necrotrophic dinomastigotes (Pr-5); see *palisporogenesis*

sporoduct (pl. **sporoducts**) Tubular expansion of a cyst wall allowing the escape of mature sporocysts in coccidians (Pr-7)

sporogen Stage of sporocarp development in which stalk is being formed; in protostelids (Pr-2), the stage of sporocarp development in which the cell that will ultimately differentiate into a spore or spores is rising off the substrate and depositing the microfibrillar stalk

sporogenesis Sporulation. Formation of spores; reproduction by spores; see *presporogonic*

sporogenic See *sporogenesis*

sporogenous See *sporogenesis*

sporogonial plasmodium (pl. **sporogonial plasmodia**) Structure that undergoes sporogony in apicomplexans (Pr-7)

sporogonic Pertaining to a kind of multiple fission; to multiple mitoses of a spore or zygote without increase in cell size; to zygotic production of haploid sporozoites; to production of sporoblasts by schizogony

sporogony See *sporogonic*

sporont (pl. **sporonts**) Stage in the life cycle that will form sporocysts (for example, in coccidians (Pr-7), zygote within the oocyst wall), sporoblasts (haplosporidians; Pr-29), or spores (paramyxeans; Pr-30)

sporophore (pl. **sporophores**) Any structure that bears spores, usually a multicellular or noncellular stalked aerial structure bearing spores at the apex (for example, myxomycotes, Pr-23); fruiting body, an ambiguous botanical term that should be avoided

sporophorous vesicle (pl. **sporophorous vesicles**) Pansporoblast membrane. Envelope laid down by sporont external to its plasma membrane in microsporans (F-1)

sporophyte generation Life cycle stage in plants and algae: diploid generation that produces spores. The sporophyte is the thallus (body) composed of diploid cells. The sporophyte generation terminates with meiosis, usually during sporogonic processes; see *gametophyte generation*

sporoplasm Ameboid organism within a spore; infective body (for example, in myxosporans, A-2)

sporopollenin Complex, extremely resistant, organic polymer that tends to survive diagenesis in the lithification process. Part of the organic geochemical record of life. Sporopollenin, complex heterogeneous material derived from carotenoids, is found in pollen and some algal cell walls; acid-hydrolysis-resistant material considered the diagenetic product of spore or cyst walls

sporozoa (pl. **sporozoans**) Ambiguous former name for apicomplexans (Pr-7), which also included the spore-forming parasites: myxosporans and microsporans (A-2, F-1)

sporozoite (pl. **sporozoites**) Life cycle stage of apicomplexans (Pr-7); motile product of multiple mitoses (sporogony) of zygote or spores; trophic stage, which is usually infective

sporulation Sporogenesis. Apicomplexan (Pr-7) multiple fission; formation of spores that involves division of a large cell into small spores

stachel Bulletlike structure contained in the rohr whose pointed end is oriented toward the approsorium and the host cell wall in plasmodiophorids. German, meaning stinger or spine; see *rohr*, *schlauch*

stalk (pl. **stalks**) Peduncle; stipe; stem; basal process. Stalk tubes are tubular, microfibrillar components of stalk of the sporocarp in protostelids; outer layer of stalk laid down by prestalk cells in dictyostelids (Pr-2)

stalkless migration Aggregation and migration of slug stage not followed by sorocarp development in dictyostelid cellular slime mold (Pr-2); directional movement of the pseudoplasmodium (slug) in response to environmental stimuli (light, heat, pH, humidity)

statospore (pl. **statospores**) Stomatocyst. Resistant cyst that consists of two pieces in some algae (for example, chrysophytes, xanthophytes, Pr-15, -16); endogenously formed resting stage with a conspicuous plug (for example, in chrysophytes)

stem cell (pl. **stem cells**) Initial cell; cell giving rise by division to identifiable progeny. Ameboid cell located between host cells in which differentiation of the secondary cells occurs (for example, paramyxeans, Pr-30)

stenohaline Ecological term referring to the ability of organisms to tolerate only narrow ranges of salinity; see *euryhaline*

stenothermal Ecological term referring to the ability of organisms to tolerate only limited ranges of temperatures; see *eurythermal*

stenothermic See *stenothermal*

stephanokont A mastigote that bears an anterior ring or crown of undulipodia

stercomares Masses, usually formed as strings, of stercomes lumped together in large numbers and covered by a thin membrane. Products of xenophyophores (Pr-4)

stercomes Fecal pellets of xenophyophores (Pr-4)

stereoplasm Solid axis of reticulopodia (for example, foraminifera, Pr-3); see *rheoplasm*

stichidium (pl. **stichidia**) Specialized branch in rhodophytes (Pr-33) that bears tetrasporangia

stichonematic Mastigote bearing an undulipodium with a single row of mastigonemes; see *pleuronematic*

stigma See *eyespot*

stigmata See *stigma*

stipe (pl. **stipes**) General morphological term referring to slender stalk of an organ or organism
stipitate Stalked; with a stipe or little stalk
stolon system Internal canal system; tubular structure connecting chambers; system of prolonged extensions (for example, tests of foraminifera, Pr-3)
stomatocyst (pl. **stomatocysts**) Statospore. Endogenous silicified resistant cyst produced by chrysophytes (Pr-15)
stomatogenesis Mouth formation, especially in ciliates. In cyrtophoran ciliates (Pr-6), the process involves formation or replacement of all oral kineties, kinetosomes, and the infraciliature plus the associated openings, cavities, etc., in both the proter and opisthe during binary fission. This resorption and reformation provides the basis for classifying the taxon (subphylum Cyrtophora)
strain (pl. **strains**) Population of microorganisms under investigation in the field or taken into the laboratory; see *isolate*
stratum (pl. **strata**) Layer of sedimentary rock
streptospiral Coiled like a ball of wool (for example, foraminiferan test (Pr-3) in which axis of growth and plane of coiling change as it forms)
stria (pl. **striae**) Linear row of alveoli, areolae, or puncta (that is, diatom frustules, Pr-18)
striated (kinetodesmal) fiber (pl. **striated (kinetodesmal) fibers**) See *kinetodesma*
striated disk (pl. **striated disks**) Part of kinetid of zoospores of Monoblepharidales (phylum Chytridiomycota, Pr-35); morphologically distinctive rootlet consisting of a flattened, often fan-shaped assemblage of microtubules and fibrils extending from the side of the kinetosome; see *adhesive disk*
striated fiber (pl. **striated fibers**) See *banded root*
striker (pl. **strikers**) Structure of ejectosome (taeniocyst) that contacts prey
stroma (pl. **stromata**) The fluid contents of an organelle (for example, chloroplast)
stummel Very short or reduced undulipodium in certain prymnesiophytes (haptophytes, Pr-25); the short bulbous haptonema found in some coccolithophorids. German, meaning little stump or butt
stylet (pl. **stylets**) General morphological term for any of several rigid elongated organs or appendages
subaerial Ecological or geological term for processes occurring in the open air on Earth's surface (but not under water) (for example, evaporation on an evaporite flat)
subkinetal microtubule (pl. **subkinetal microtubules**) Portion of cell cortex of ciliates (Pr-6) composed of components derived from many linearly aligned kinetids (for example, set of microtubules that arise from the base of kinetosomes and extend anteriorly or posteriorly beneath a kinety)
sublittoral Ecological term referring to the environment lying below the level of low tide. Subtidal near the shore or just below the shoreline or littoral zone; see *littoral*, *supralittoral*
submetacentric See *mediocentric*
subpseudopodium (pl. **subpseudopodia**) Fine extension at the leading edge of a pseudopodium (for example, amebas, foraminifera, Pr-2, -3)
subraphe costa (pl. **subraphe costae**) Supporting bars in the form of flying buttresses running beneath and at a 90° angle to the raphe of pennate diatoms (Pr-18); they are continuations of the valve costae
subraphe fibula (pl. **subraphe fibulae**) See *subraphe costa*
substrate (pl. **substrates**) Underlayer; carbon source, nitrogen source, food; stable surface to which organisms are attached (for example, rocks); molecule that is acted upon by an enzyme
subtelocentric See *acrocentric chromosome*
succession (pl. **successions**) Ecological term referring to ecosystem change; the more-or-less

regular phenomenon of community replacement though time

sucking disk (pl. **sucking disks**) See *adhesive disk*

sulcal groove (pl. **sulcal grooves**) Groove running from the posterior end anteriorly in dinomastigotes (Pr-5); at the equatorial region it joins the transverse groove; the sulcus contains the insertion and often the proximal part of the longitudinal undulipodium

sulcate See *sulcal groove*

sulcus (pl. **sulci**) See *sulcal groove*

supplementary aperture (pl. **supplementary apertures**) Opening to the exterior, such an aperture is in addition to and independent of the primary aperture (that is, tests of foraminifera, Pr-3)

supralittoral Ecological term referring to the environment of the spray zone lying just above the shore line or littoral zone; that is, above high tide; see *littoral, sublittoral*

surra Disease of camels caused by *Trypanosoma evansi* (Pr-11) and transmitted by biting flies

suture (pl. **sutures**) General morphological term referring to a seam or furrow between adjacent parts (for example, between thecal plates in armored dinomastigotes, Pr-5)

suture line (pl. **suture lines**) Line of adhesion between the two to seven valves of myxosporan spore walls (A-2); contact area between adjacent plates that acts as a line of separation in dinomastigotes (Pr-5); region of discontinuity in the cortex, defined by the end of kineties terminating near or on each other in ciliates (Pr-6)

swarmer (pl. **swarmers**) Zoospore. Mastigote propagule; undulipodiated, dispersive form in the life cycle of protoctists of many different taxa; swarmer cell (for example, some actinopod zoospores, rapidly produced motile cells of chytridiomycotes, dinomastigotes; Pr-31, -35, -5)

symbiont (pl. **symbionts**) Members of a symbiosis, that is, organisms that have an intimate and protracted association with one or more organisms of a different species

symbiosis Prolonged physical association between two or more organisms belonging to different species. Levels of partner integration in symbioses may be behavioral, metabolic, gene product, or genic. For nutritional modes of symbionts (Table 1)

symbiotroph See *symbiotrophy*

symbiotrophic See *symbiotrophy*

symbiotrophy Mode of nutrition involving a heterotrophic symbiont that derives both its carbon and its energy from a living partner (Table 2); see *necrotrophy*

symmetrogenic fission Type of cell division, generally longitudinal, of a parent such that the two offspring are mirror images of one another with respect to principal structures (for example, opalinids, pseudociliates, Pr-6, -24). Typically occurs in nonciliate protoctists; see *homothetogenic fission*

symplectic See *antiplectic*

symplesiomorphic Term derived from cladistics that refers to an ancestral, homologous trait (seme) that arose prior to the bifurcation of the lineages of organisms; see *plesiomorphy, synapomorphy*

symplesiomorphy See *symplesiomorphic*

sympodial Pertaining to a mode of development in which the primary axis is continually replaced by lateral axes, which become dominant but are soon replaced by their own laterals (for example, sympodial branching in phaeophytes, sympodial renewal in oomycotes, Pr-21)

sympodial zoosporangium formation Term describing morphogenesis in chytridiomycotes (Pr-35) in which the zoosporangium forms on an apparent main axis derived from successive secondary axes

synapomorphic Pertaining to an homologous taxonomic character (seme) that arose in the ancestral species with the bifurcation of the lineage; see *plesiomorphy, symplesiomorphy*

synapomorphy See *synapomorphic*

synaptonemal complex (pl. **synaptonemal complexes**) Complex proteinaceous, longitudinally aligned structure seen with the electron microscope that usually unites homologous chromosomes during the prophase of meiosis

synchronous culture (pl. **synchronous cultures**) Culture in which all cells or organisms are simultaneously in the same stage of growth or reproduction

syncytial See *coenocyte, plasmodium*

syncytium (pl. **syncytia**) See *syncytial*

syngamy Fertilization; gametogamy. Fusion of two cells, usually gametes. The nuclear fusion process that often follows syngamy is called karyogamy

synkaryon (pl. **synkarya**) Fusion nucleus; zygotic nucleus; product of fusion of two haploid gametic nuclei or pronuclei

synzoospore (pl. **synzoospores**) Compound zoospore. Multiple zoospore with two to many sets of undulipodia and equivalent multiples of other organelles; usually the result of incomplete cleavage during zoospore formation in multinucleate xanthophytes (Pr-16; for example, *Botrydium*). Huge synzoospores, each forming hundreds of biundulipodiated zoospores, are characteristic of *Vaucheria*

system I fiber (pl. **system I fibers**) Part of kinetid structure; striated rootlet (not consisting of a bundle of 5–8 nm filaments) often associated with rootlet microtubules and exhibiting a narrow (25–35 nm) repeat of cross-striations (that is, pedinomonadalean chlorophytes, Pr-28); see *system II fiber*

system II fiber (pl. **system II fibers**) Part of kinetid structure; rootlet consisting of a bundle of 5–8 nm filaments, often cross-striated (that is, chlorophycean chlorophytes, Pr-28); see *system I fiber*

systematics A biological science; that subfield of evolutionary science that deals with naming, classifying, and grouping organisms on the basis of their evolutionary relationships

syzygy Association side-by-side or end-to-end (frontal syzygy or in caudo-frontal association) of gamonts (especially of gregarine apicomplexans, Pr-7) prior to formation of gametocysts and gametes

T band (pl. **t bands**) Morphological feature seen with an electron microscope in the myonemes of acantharian actinopods (Pr-31). Thin, dark transverse lines separating repeated clear areas known as L zones

T joint (pl. **t joints**) Morphological feature of the loricae of choanomastigotes (Pr-35). Longitudinal costae joined midway along the anterior costal strips

tabular A laminar form, that is, having a flat surface

tabulation (pl. **tabulations**) System of classifying dinomastigote envelope plates (Pr-5)

tactic Movement of an organism or organelle toward or away from a stimulus (for example, geotaxis, phototaxis, magnetotaxis, thigmotaxis)

taeniocyst (pl. **taeniocysts**) Extrusome with a complex structure characteristic of some dinomastigotes (Pr-5)

taeniogene (pl. **taeniogenes**) Organelle that gives rise to the taeniocyst in dinomastigotes (Pr-5)

tannins Brown polyphenolic compounds that yield tannic acid on hydrolysis. Characteristic of phaeophytes (Pr-17) and plants

taxis See *tactic*

taxon (pl. **taxa**) Any formally named and recognized group of organisms. Unit in the hierarchy of biology that classifies all living organisms (for example, in order of descending inclusiveness, taxa include kingdom, phylum, class, order, family, genus, and species)

Taylor–Evitt System System of thecal plate or cyst paraplate designation used in dinomastigote taxonomy (Pr-5)

tectiform replication Asexual reproduction in loricate choanomastigotes (Pr-35) in which the offspring cell may have component costal strips when it departs from the parent lorica (for example, Acanthoecidae); see *nudiform replication*

tectin Complex of protein and mucopolysaccharides comprising some tests (for example, foraminifera, Pr-3); see *organic test*

tectinous See *tectin*

telocentric Referring to chromosomes with centromeres (kinetochores) at the ends (telomeres) of the structure. Terminal (the very end) or subtelocentric chromosomes with very small quantities of chromatin lying distal to the centromere are termed acrocentric; see *acrocentric chromosome*

telomere (pl. **telomeres**) Chromosome end, usually composed of highly repetitious DNA sequences

telophase (pl. **telophases**) Stage in mitosis in which chromosomes are at opposite ends of the spindle, chromatin begins to uncoil, and cytokinesis occurs. Nucleolus and nuclear membrane often reform in telophase; see *mitosis*

TEM Transmission electron microscope

temporary cyst (pl. **temporary cysts**) Cyst produced directly and reversibly from trophic cell in rapid response to feeding or unfavorable conditions (formed by amoebomastigotes, some dinomastigotes, and ciliates such as *Colpoda*, Pr-22, -5, -6)

tentacle (pl. **tentacles**) General term for long protrusion. In suctorian ciliates (Pr-6) they are protoplasmic processes, underlain by microtubules, bearing missilelike projectiles (extrusive organelles) that attack prey. Tentacles are distinguished from undulipodia, haptonemes, stalks, and pseudopodia by their substructure and aggressive function

teratological Monstrous; referring to the formation of abnormal growths (for example, tumors)

terete Cylindrical

terminal cap (pl. **terminal caps**) Component of kinetid associated with the proximal end of the uppermost kinetid; more-or-less electron-dense flap at the anterior end of the kinetosome (ulvophycean and trentepohlialean chlorophytes, Pr-28)

terminal nodule (pl. **terminal nodules**) Diatom (Pr-18) valve structure; site of the terminal pore of raphe on a motile pennate diatom

terminal plate (pl. **terminal plates**) Kinetid substructure (for example, chytridiomycotes and hyphochytrids, Pr-35, -14); structure just proximal to where the axoneme contacts the kinetosome and thus the cytoplasm of the rest of the zoospore

test (pl. **tests**) Cell covering; hardened, continuous periplast; general descriptive term for any of a large number of shells, hard coverings, valves, or thecae; see *agglutinated test, biserial, calcareous, evolute test, hyaline, megalospheric test, microgranular test, microspheric test, organic test, peneropliform, planispiral, porcellaneous test, quinqueloculine, rectilinear test, streptospiral, triserial, trochospiral test, uniserial, valve*

testis (pl. **testes**) Sperm production and storage organ

tethyan realm See *tethys*

tethys The elongated east-west seaway that separated Eurasia from Gondwanaland from at least the early Paleozoic to late Cretaceous Period

tetrapyrrholes Class of carbon compounds formed from four heterocyclic pyrrhole rings linked by single carbon bridges and often chelated with metal ions (for example, Fe^{++} in heme, Mg^{++} in chlorophyll)

tetrasporangium (pl. **tetrasporangia**) Cell in which a diploid nucleus undergoes meiosis to form four haploid spores (tetraspores) in rhodophytes (Pr-33)

tetraspores Spores formed in a tetrasporangium (for example, rhodophytes, Pr-33)

tetrasporoblastic Referring to a life cycle in some rhodophytes (Pr-33) in which carpospores germinate to produce a diploid tetrasporophyte that is borne on the gametophyte

tetrasporophyte (pl. **tetrasporophytes**) Diploid thallus in rhodophytes (Pr-33) that produces tetrasporangia

thallophyte (pl. **thallophytes**) Literally "flat plants"; obsolete term for bacteria, fungi, and other nonvascular photosynthetic and heterotrophic organisms

thallus (pl. **thalli**) General descriptive term, derived from botany, referring to body type in plants and algae. Thalli are flat, leaflike structures undifferentiated into organs and lacking vascular tissue characteristic of tracheophytes, that is, lacking roots, stems, and leaves

thanosis A process by which selected cells are programmed to die as a normal component of development. This loss of cells plays a role in sculpting the structure of an organism during morphogenesis. Thanosis is initiated by specific signals and requires de novo gene expression. (After the Greek god of death, Thanatos.)

theca (pl. **thecae**) General descriptive term used for many unrelated structures; coat, periplast, test, valve, shell, hard covering, enveloping sheath, or case. Total cell wall, composed of many closely fitting cellulose plates (sometimes used equivalently to the amphiesma) in dinomastigotes (Pr-5)

thecal See *theca*

thecal plate (pl. **thecal plates**) Component of cell coat, or hardened structure, external to the outer plasma membrane (for example, dinomastigotes, Pr-5)

thecate See *theca*

thermocline (pl. **thermoclines**) Ecological term referring to a sharp temperature gradient; the zone of water in which temperature decreases rapidly with depth; in lakes, zone between the epilimnion and hypolimnion

thigmotactic Pertaining to organisms that are touch-sensitive or adherent. Thigmotaxis leads to production of structures functioning as holdfasts (for example, certain somatic cilia of some epibiotic ciliates, Pr-6)

thigmotaxis See *thigmotactic*

thorotrast Electron-dense substance that when added to a sample becomes trapped inside phagocytic vesicles; used in electron microscopy to identify such vesicles

thylakoid (pl. **thylakoids**) Photosynthetic membrane, lamella, or sac; photosynthetic membrane bearing chlorophylls, carotenoids, and their associated proteins usually stacked in layers; photosynthetic membranes in bacteria and in plastids

thylakoid doublet (pl. **thylakoid doublets**) Paired thylakoids, the outer surface of which in cyanobacteria (B-6) and rhodophytes (Pr-33) bears phycobilisomes

tight junction (pl. **tight junctions**) Type of cell junction in animal tissues. Continuous bandlike junction between epithelial cells and, rarely, other cells

tinsel See *flimmer*

tinsel flagellum (pl. **tinsel flagella**) Undulipodium bearing mastigonemes; see *whiplash undulipodium*

tinsel undulipodium (pl. **tinsel undulipodia**) See *tinsel undulipodium*

tomite (pl. **tomites**) Stage in the polymorphic life cycle of histophagous ciliates (Pr-6) in which organisms are small, free-swimming, and nonfeeding; one of two or more fission products of a tomont (or sometimes a protomite)

tomont Pre-fission or dividing stage in the polymorphic life cycle of a number of histophagous ciliates (for example, apostomes and some hymenostomes). A large form, typically encysted. Tomont may undergo multiple fission (for example, divide a number of times in quick succession to yield tomites)

totipotency Developmental term referring to propagule or growing cell that is capable of repeating all steps of development and giving rise to all cell types

toxicyst (pl. **toxicysts**) Type of extrusome; slender tubular structure that probably contains both paralytic and proteolytic enzymes helping to penetrate, immobilize, and cytolyze prey

trace element (pl. **trace elements**) See *micronutrient*

trace fossil (pl. **trace fossils**) See *ichnofossil*

transcription Synthesis of messenger RNA from a DNA template with a sequence determined directly by the base pair sequence of the DNA template

transduction The transfer of small replicons (for example, viral or plasmid DNA) from an organelle or bacterium to another organelle or bacterium usually mediated by bacteriophage. Change of energy from one form to another (for example, light to chemical or mechanical energy to heat)

transfection Natural genetic change in bacteria and eukaryotic cells in culture induced by uptake of DNA from aqueous medium

transformation The process of conversion of an ameba to a mastigote by the production of undulipodia or the reverse transformation of a mastigote to an ameba by active absorption of the undulipodia. Characteristic of amoebomastigotes, myxomycotes, phaeophytes, some actinopods, and other organisms (Pr-22, -23, -17, -31 etc). The process is probably of evolutionary significance; whether it is monophyletic is unknown. *Also:* uptake, incorporation, and inheritance of exogenous genetic material (for example, transforming principle DNA of *Hemophilus* bacteria, B-3)

transition fiber (pl. **transition fibers**) Transition zone fibers; part of a kinetid; fine, fibrillar elements connecting the undulipodial membrane in the transition zone with the undulipodial axoneme at a point between the A- and B-tubules

transition region (pl. **transition regions**) See *flagellar transition zone*

transition zone (pl. **transition zones**) See *flagellar transition zone*

transitional helix (pl. **transitional helices**) Coiled fiber. Helical structure, probably composed of ribonuclear protein, in transition zone of undulipodia of most heterokont groups (for example, xanthophytes, eustigmatophytes, proteromonads, chrysophytes; Pr-16, -27, -1, -15); called "Spiralkörper" in chrysophytes

transitional region (pl. **transitional regions**) See *flagellar transition zone*

translation Synthesis of protein on ribosomes from activated amino acids using messenger RNA (mRNA) transcripts as templates

transverse fission See *homothetogenic fission*, *perkinetal fission*

transverse flagellum (pl. **transverse flagella**) Undulipodium that wraps around the cell and lies in the equatorial groove in dinomastigotes (Pr-5)

transverse microtubular ribbon (pl. **transverse microtubular ribbons**) Transverse fiber; part of kinetid structure characteristic of ciliates (Pr-6); ribbon of microtubules associated with kinetosomes that originate near triplets 3, 4, and 5 and extend laterally; see *postciliary ribbon*

transverse ribbon (pl. **transverse ribbons**) See *transverse microtubular ribbon*

transverse undulipodium (pl. **transverse undulipodia**) See *transverse flagellum*

triaene Arrangement in an ebridian skeleton in which the initial branching point is of four branches (for example, in *Hermesinum*)

triatomine bug (pl. **triatomine bugs**) Bloodsucking insects (A-21) of the order Hemiptera, family Reduviidae (subfamily Triatominae), which defecate while feeding; they transmit *Trypanosoma cruzi*, infecting the host via their contaminated fecal material

trichocyst (pl. **trichocysts**) Extrusome underlying the surface of many ciliates (Pr-6) and some mastigotes; capable of sudden discharge to sting prey; probably nonhomologous structures (for example, dinomastigotes, prasinophytes, raphidophytes, Pr-5, -28)

trichocyst pore (pl. **trichocyst pores**) Aperture in the thecal plate through which trichocysts are discharged in armored dinomastigotes (Pr-5)

trichogyne (pl. **trichogynes**) Receptive protuberance or threadlike elongation of a female gametangium to which male gametes become attached (for example, rhodophytes, Pr-33, and many fungi)

trichome (pl. **trichomes**) Morphological term referring to filamentous or threadlike shape (for example, single row of cells of filament, exclusive of sheath, of cyanobacteria (B-6) or algae)

trichothallic growth Mode of cell division in phaeophyte tissue in which active cell division occurs at the base of a filament or group of filaments

triode (pl. **triodes**) Arrangement in an ebridian skeleton in which the initial branching point is of three branches (for example, *Ebria*)

triphasic life cycle Three-part life history displaying three distinct types of morphology. Sequential polymorphism

triserial General morphological term for structures organized in three rows or series (for example, tests of foraminifera, Pr-3)

triseriate See *triserial*

trisomic Karyotype (2N + 1) of a diploid organism with one extra chromosome. The extra chromosome is homologous with one of the existing pairs; one chromosome is present in triplicate

trisomy See *trisomic*

trochospiral test (pl. **trochospiral tests**) Helicoid spiral test. Coiled test in which the pattern of growth involves the addition of chambers in a spiral coil; the hollow or depressed side of the cone-shaped test is the involute side; the higher opposite side is known as the evolute side (for example, foraminifera, Pr-3)

trophic cell (pl. **trophic cells**) Trophic stage; trophont. General term for a heterotrophic cell that feeds and grows, common in the life cycle of many protoctists (for example, apicomplexans and ciliates, Pr-7, -6). Also called vegetative cell, an ambiguous botanical term that should be avoided; see *trophont, trophozoite*

trophic stage (pl. **trophic stages**) See *trophic cell, trophont*

trophocyst (pl. **trophocysts**) Enlarged cell capable of feeding by osmotrophy

trophocyte (pl. **trophocytes**) Feeding cell (for example, in multicellular symbiotrophic dinomastigotes, Pr-5, the cell that attaches the host to the colony)

trophomere (pl. **trophomeres**) Proximal section of the body or thallus of ellobiopsids that carries terminal reproductive structures, the gonomeres

trophont (pl. **trophonts**) Trophic stage. Trophic cell or organism; feeding and growing stage. Adult stage in the life cycle in ciliates (Pr-6). An interfissional form; form that shows a preceding tomite and a succeeding tomont stage, as in the polymorphic life cycles of various symbiotrophic apostome and hymenostome species. Also called vegetative cell, an ambiguous botanical term that should be avoided; see *trophozoite*

trophozoite (pl. **trophozoites**) Motile trophont stage of symbiotrophic protists (primarily apicomplexans, microsporans, and myxosporans, Pr-7, F-1, A-2)

tropism Morphogenetic movement or growth toward or away from an external stimulus (for example, phototropism, geotropism)

trypanosomatid (pl. **trypanosomatids**) Informal name of members of the trypanosome kinetoplastids (Pr-11)

trypomastigote (pl. **trypomastigotes**) Stage in trypanosome (Pr-11) development in which the kinetoplast lies behind the nucleus and the associated undulipodium emerges laterally to form an undulating membrane along the length of the body, usually becoming free at its anterior end

tubular crista (pl. **tubular cristae**) Descriptive term for the morphology of mitochondrial membranes. Cristae that are finger-shaped, circular in transverse section, and round rather than flattened. Characteristic of ciliates, dinomastigotes, and other protoctists; see *vermiform crista, vesicular crista*

tubular ingestion apparatus Siphon. General descriptive term for an oral apparatus that has the form of a long tube

tubulus (pl. **tubuli**) Small tubes (for example, organic tubules that transverse the micrgranular calcareous walls of certain foraminifera, Pr-3); see *microtubule*

tufa Porous, sedimentary rock composed of calcium carbonate formed by evaporation or by precipitation from spring water or seeps

tundra (pl. **tundras**) Treeless area of arctic regions that has a permanently frozen subsoil (permafrost) and low-growing vegetation (for example, lichens, mosses, and stunted shrubs)

tychoplankton Collective term for benthic organisms that become temporarily suspended in water column by turbulence or other disturbance

ultrasonication See *sonication*

ultrastructural Fine structure. The appearance of the cell and/or cell organelles as seen in the transmission electron microscope

ultrastructure See *ultrastructural*

umbilical Morphological term generally meaning navel or button. Refers to a depressed region of a trochospirally coiled foraminifera (Pr-3) surrounded by all the chambers of the last formed whorl

umbilicus (pl. **umbilici**) See *umbilical*

undefined medium (pl. **undefined media**) Culture medium with one or more components, the exact chemical nature of which is unknown; see *defined medium*

undulating membrane (pl. **undulating membranes**) Waving membrane; refers to several kinds of nonhomologous structures: (1) the parallel membrane, an organelle on the right side of the buccal cavity in ciliates with a compound ciliary apparatus; (2) in symbiotic mastigotes, an extension of the plasma membrane combined with the undulipodial membrane so that the axoneme of the undulipodium is attached to the body by a thin fold; or (3) a membranous fibrillar structure not underlain by undulipodia

undulipodial apparatus (pl. **undulipodial apparatuses**) See *basal apparatus*

undulipodial bracelet (pl. **undulipodial bracelets**) See *flagellar bracelet*

undulipodial groove (pl. **undulipodial grooves**) See *flagellar groove*

undulipodial hair (pl. **undulipodial hairs**) See *flagellar hair*

undulipodial pocket (pl. **undulipodial pockets**) See *flagellar pocket*

undulipodial pore (pl. **undulipodial pores**) Opening through which undulipodium protrudes

undulipodial root (pl. **undulipodial roots**) See *flagellar rootlet*

undulipodial rootlet (pl. **undulipodial rootlets**) See *flagellar rootlet*
undulipodial swelling (pl. **undulipodial swellings**) Photoreceptor; lateral swelling near the base of the emergent undulipodium. In euglenids it is adjacent to the eyespot and presumably carries the photoreceptor for phototaxis; similar in appearance to a paraflagellar rod, although much smaller in diameter
undulipodial transition region (pl. **undulipodial transition regions**) See *flagellar transition zone*
undulipodium (pl. **undulipodia**) Cilium. Sperm tail; cell-membrane-covered motility organelle sometimes showing feeding or sensory functions; composed of at least 400 proteins. [9(2)+2] microtubular axoneme usually covered by plasma membrane; limited to eukaryotic cells. Includes cilia and eukaryotic "flagella." Each undulipodium invariably develops from its kinetosome. Contrasts in every way with the prokaryotic motility organelle or flagellum, a rigid structure composed of a single protein (which belongs to the class of proteins called flagellins). Undulipodia in the cell biological literature are often referred to by the outmoded term flagella or euflagella; see *introduction for discussion of these terms*
unialgal culture (pl. **unialgal cultures**) Culture containing only one species of algae; other protoctists, fungi, and/or bacteria may be present; see *axenic*, *monoxenic*
unikaryon (pl. **unikarya**) Organism with a single nucleus
unikaryotic See *unikaryon*
unilocular sporangium (pl. **unilocular sporangia**) Sporangium in which all spores are produced in a single cavity; see *plurilocular sporangium*
unimastigote (pl. **unimastigotes**) Cell with a single undulipodium
uniporate Referring to a structure with a single pore
uniserial Uniseriate. Any of several structures arranged in or consisting of one series or row of structures; descriptive of cells characterized by such an arrangement (for example, foraminiferan tests, Pr-3)
uralga (pl. **uralgae**) Hypothetical common ancestor of all algae (for example, given the direct filiation (monosymbiotic) theory of the origin of the chloroplasts; the uralga is that common ancestral organism thought to combine the features of phototrophic bacteria (including cyanobacteria, algae, and plants)
uroid (pl. **uroids**) Descriptive morphological term for the tail-like protuberance at the posterior end of a moving lobose ameba (Pr-2). Structure is active in pinocytosis and possibly in defecation and water expulsion
uroid region Region in the ameba (Pr-2) opposite locomotory end (posterior); sometimes distinct from rest of body by constriction; see *uroid*
utricle (pl. **utricles**) General morphological term for "little bladder" (for example, the swollen terminus of a filament of the green alga *Codium*, Pr-28)

vacuole (pl. **vacuoles**) A small space or cavity in the protoplasm of a cell containing fluid or air and surrounded by a membrane
vacuome (pl. **vacuomes**) Morphological term referring to the complete system of vacuoles in a cell (analogous to genome or chondriome)
vagile Referring to behavior of a cell or organism; free to move about; see *sessile*
vagility See *vagile*
valve (pl. **valves**) Opposite faces, or distal plates of a diatom frustule (Pr-18) or dinomastigote theca (Pr-5), typically flattened or somewhat convex. Portion of myxosporan (A-2) spore wall, formed by a specialized (valvogenic) cell during sporogenesis; two or more such valves adhere

together along suture line, composing the spore wall (or shell); see *epivalve, hypovalve*

valve cell (pl. **valve cells**) Cell that forms part of the valve in myxozoans

valve face (pl. **valve faces**) Structure of diatoms (Pr-18); the surface of a valve

valve mantle (pl. **valve mantles**) Structure of diatoms (Pr-18); marginal part of valve differentiated by slope, sometimes also by structure, from the valve face

valve view Front view of a diatom valve (Pr-18)

valvogenic cell (pl. **valvogenic cells**) Cell of myxosporans (A-2) in the sporoblast that gives rise to the valve of mature spores

variable antigen type Antigen type (serotype) expressed by a trypanosome as a consequence of having a variant-specific glycoprotein on its surface

variant surface glycoprotein Surface macromolecules of trypanosomes: the glycoprotein is present as a monomolecular layer on the surface of bloodstream trypanosomes (Pr-11) and constitutes the surface coat; its exposed epitope determines the variable antigen type of the organism

VAT See *variable antigen type*

vector (pl. **vectors**) Motile organism (for example, insect, mammal) that transmits symbionts to other organisms. Parasitologists sometimes limit vector to mean an essential intermediate host in which a parasite undergoes a significant life cycle change

vegetative cell (pl. **vegetative cells**) Growing cell; trophont; trophic cell; trophozoite. The term vegetative, borrowed from growing plants, should be avoided

vegetative resting state See *cyst, spore*

vegetative state (pl. **vegetative states**) Trophic state; see *vegetative cell*

velum (pl. **vela**) Veil; in diatoms (Pr-18), thin, perforated layer of silica over an areola, that is, a type of pore plate; known as rica in many biraphid diatoms. In some species of Dasycladales (for example, the chlorophyte *Acetabularia*, Pr-28), a protective covering over emergent lateral branches

ventral cortex Portion of a ciliate (Pr-6) that contains the oral region in cases where the oral region is not at the anterior pole

ventral disk (pl. **ventral disks**) See *adhesive disk*

ventral skid Recurrent undulipodium; usually nonmotile relative to the cell, which serves as a "runner" upon which a mastigote glides over the substrate

ventral sucker (pl. **ventral suckers**) See *adhesive disk*

ventrostomial Morphological term referring to an area on the ciliate cortex (Pr-6); around the oral region, on the ventral side

vermiform crista (pl. **vermiform cristae**) Descriptive term of intramitochondrial membranes; "wormlike" cristae; those occurring as a pancake-shaped flattened plate; see *tubular crista, vesicular crista*

vermifuge Biologically active substance (for example, antihelminthics) having the power or property of expelling worms or other parasites from the intestines of people or domestic animals (for example, reported to occur in some ulvophytes, Pr-28)

verruca (pl. **verrucae**) General term referring to wartlike thickenings

verrucose See *verruca*

vesicle (pl. **vesicles**) General structural term for a membranous sac. Usually refers to a cell organelle in protoctists

vesicular crista (pl. **vesicular cristae**) Descriptive term of intramitochondrial membranes; vesicles or sac-shaped cristae; see *tubular crista, vermiform crista*

vestibulum (pl. **vestibula**) Morphological term referring to different oral structures in protoctists; subapical depression from which undulipodia emerge in cryptomonads (Pr-26); depression of the body at or near the apical end leading to the cytostome–cytopharyngeal complex and adorned with undulipodia in some ciliates (Pr-6); in other protoctists, intracellular compartment containing oral and lateral apertures for nutrient passage and waste disposal

vitreous Glassy; referring to hyaline wall in which the crystals of calcite have their C-axes optically aligned normal to the surface of the shell (for example, the Archaediscidae foraminifera, Pr-3)

VSG See *variant surface glycoprotein*

wall-forming bodies Inclusions that give rise to the oocyst wall after fertilization in coccidian apicomplexans(Pr-7);wall-formingbodiesIaremore-or-less dense granules; wall-forming bodies II have a spongelike appearance

water molds Common name for several unrelated groups of hyphae-forming organisms found in damp or aquatic environments. White rusts; downy mildews; symbiotrophic or osmotrophic funguslike protoctists most of which are members of the phylum Oomycota (Pr-21) in the five-kingdom system

whiplash flagellum (pl. **whiplash flagella**) Undulipodium lacking mastigonemes; see *tinsel undulipodium*

whiplash undulipodium (pl. **whiplash undulipodia**) See *whiplash flagellum*

whorl (pl. **whorls**) General descriptive term for coiled form or radial structures emerging from a common axis (for example, term is applied to a group of chambers which collectively make up a 360° turn of the test in coiled foraminifera, Pr-3, and to the disposition of long cells around the nodal cells in charatean chlorophytes such as *Nitella*, Pr-28)

whorled vesicle (pl. **whorled vesicles**) Intracellular membranous sacs disposed in a whorled conformation (for example, arrangement characteristic of the contractile vacuole of paramecia, Pr-6); see *whorl*

window (pl. **windows**) Opening (for example, between the branches in the siliceous skeleton of ebridians)

wrack Tangled mass of fucalean seaweeds (Pr-17) on the seashore

xanthophylls Class of plastid pigments; oxygenated carotenoids

xanthosome (pl. **xanthosomes**) Yellow bodies (for example, reddish brown or yellowish, rounded, and often aggregated bodies found between the stercomes in the stercomare of xenophyophores, Pr-4)

xenogenous Of alien or foreign origin. Organism of a different species; heterospecific; heterogeneric

xenoma Symbiotic aggregate formed by multiplying intracellular symbiotrophs within their growing host cells, the whole structure increasing in size, as in the single-celled tumors formed by microsporans (F-1)

xenophya (pl. **xenophyae**) Foreign bodies of which the inorganic part of xenophyophoran tests (Pr-4) is composed

xenosome (pl. **xenosomes**) Intracellular structures. Literally, "alien bodies," referring to micrometer-size bodies found in the cytoplasm and nuclei of protoctists of all kinds. Growth in the absence of the host provides the definitive proof that a structure is a xenosome. These may be foreign infective agents but are easily confused with natural components of the organism when their physiological and even genetic incorporation into the life of the host cell has occurred in the remote past. Endosymbiotic entities such as the (bacterial) kappa and omikron (and other Greek letters) particles of *Paramecium*, *Euplotes*, as well as *Holospora*, etc., zooxanthellae, and cyanelles are xenosomes. Most of the Greek-letter particles (cytoplasmic genes of *Paramecium*) are now classified as Gram-negative bacteria in the genus *Caedibacter*. From an evolutionary point of view, the serial endosymbiosis theory claims that plastids and mitochondria began as xenosomes as well

xylan Xylose polymer

zerfall Break up of nuclear material prior to schizogony (for example, agamont stage of the foraminiferan *Allogromia laticollaris*, Pr-3). German, meaning disintegration, breaking up

zoid (pl. **zoids**) See *monad*

zoite (pl. **zoites**) Endozoite. Trophic cell produced by multiple fission (for example, the infective motile stage of apicomplexans, Pr-7, whether of sexual or asexual origin); see *merozoite*, *sporozoite*

zonate Structure that is zoned; marked with zones, bands, rings, or zones of color

zoobenthos Ecological term referring to heterotrophic protoctists and animals that comprise the biota of the benthos

zoochlorella (pl. **zoochlorellae**) Green photosynthetic symbionts found in protoctists and animals. Although many belong to the genus *Chlorella* (Pr-28; for example, algae of *Coleps hirtus*, *Hydra viridis*, and *Paramecium bursaria*, Pr-6, A-4), others belong to the prasinophytes or other taxa; often symbionts are unidentified to genus

zoocyst (pl. **zoocysts**) Undulipodiated propagule

zoonosis Ecological term referring to specific protoctist (occasionally bacterial or viral) symbiotrophs that have animals including people as their hosts. Infection naturally transferable between animals other than humans

zoophagy Mode of heterotrophic nutrition displayed by organisms that feed on animals

zoosporangium (pl. **zoosporangia**) Sporangium that produces zoospores

zoospore (pl. **zoospores**) Swarmer. Mastigote propagule, undulipodiated motile reproductive cell capable of transformation into a different developmental stage but incapable of sexual fusion. Although spermlike in appearance, they are not sperm; see *auxiliary zoospore*, *macrozoospore*, *microzoospore*

zoosporic fungus (pl. **zoosporic fungi**) Protoctists, primarily osmotrophic, that are capable of forming hyphae and have undulipodiated stages in their life cycle. Outmoded term referring to members of the protoctist phyla Chytridiomycota, Hyphochytriomycota, Oomycota, and sometimes Plasmodiophoromycota (Pr-35, -14, -21, -20)

zoosporogenesis Process by which zoospores are formed

zooxanthella (pl. **zooxanthellae**) Yellowish or yellow-brown photosynthetic symbiont found in protoctists and animals. Although many belong to the dinomastigote group *Symbiodinium* (*Gymnodinium*; Pr-5), others belong to diatom (Pr-18) or other taxa; often the symbionts are unidentified to genus

zygocyst (pl. **zygocysts**) Encysted zygote (for example, structure of opalinids (Pr-6) usually found in intestines or feces of anuran amphibians, A-37)

zygolith (pl. **zygoliths**) Dome-shaped coccolith subtype (for example, *Homozygosphaera*, *Periphyllophora*, Pr-25); holococcoliths with arched crossbow(s)

zygospore (pl. **zygospores**) Resistant structure formed by conjugation; thick-walled zygote of the conjugating green algae (Pr-32); large, multinucleate resting spore in zygomycote fungi (F-2)

zygote (pl. **zygotes**) Diploid (2N) nucleus or cell produced by the fusion of two haploid nuclei or cells. In animals, plants, and some protoctists (those undergoing gametic meiosis) the zygote is destined to develop into a new organism. In fungi and protoctists undergoing zygotic meiosis, the zygote stage is unstable and haploid nuclei or cells are formed as soon as the zygote resumes activity

zygotene Zygonema; stage in meiotic prophase I in which homologous chromosomes pair; see *meiosis*

zygotic meiosis Life cycle in which meiosis immediately follows zygote formation as in most fungi and some algae (for example, conjugating green algae, Pr-32); see *gametic meiosis*

zymodeme (pl. **zymodemes**) Strain or variety; ecological term referring to a population of organisms members of which display similarities in patterns of isoenzymes as determined by electrophoresis, that is, individuals having similar zymograms are said to belong to the same zymodeme

zymogram (pl. **zymograms**) Stained gel (starch or agarose) that shows isoenzyme banding patterns following electrophoresis of a cell lysate; method by which zymodemes are established

ORGANISM GLOSSARY

Introduction

Who is this program for?

This glossary incorporates an abbreviated form of the glossary of *The Handbook of Protoctista*, (Margulis *et al.*, 1993). It is based on the *Illustrated Glossary of Protoctista*, (Margulis *et al.*), and created using ETI's Linnaeus II Software for Biodiversity Documentation. Designed for all investigators, instructors, and students who deal with protoctists, the eukaryotic microorganisms, and their descendants (exclusive of the animals, fungi, and plants), it contains the latest information. Understanding the relationships among living organisms is essential for biochemists, botanists, ecologists, cell and molecular biologists, medical researchers, microbiologists, mycologists, parasitologists, phycologists, protozoologists, and zoologists.

How many protoctists are there?

We estimate that there are more than 100,000 species of described, extant protoctists and that many more thousands await discovery. Probably the number in each category is even greater for extinct forms. On encountering the hypertrophied intestines of the East African rhinoceros, Van Hoven (1987) discovered a new world of symbiotic eukaryotes using scanning electron microscopy. A wood-ingesting termite may contain as many as 30 different protist species. Beavers and cervids enjoy diets extremely rich in cellulose. Who can even predict the protoctistan populations residing in these and so many other animals?

No animals or plants

The bewildering diversity of protoctists, so much of it unknown, must be organized on a rational basis. All earlier schemes conceived of members of the protoctists as tiny animals, tiny plants, and later fungi (water molds or aquatic fungi). Even today, many scientists (for example, especially cell biologists, plankton ecologists, and geologists) routinely write about Protozoa and Algae as if they were phyla in the animal and plant kingdoms, respectively. These organisms are no more "one-celled animals and one-celled plants" than people are shell-less multicellular amebas. Indeed, since animals and plants always develop from embryos, neither one-celled animals nor one-celled plants even exist. Unlike all previous works on protoctists (including the Illustrated Guide to the Protozoa by Lee *et al.*, 1985), the *Handbook of Protoctista* does not operate from "the top down," imposing an obsolete two-kingdom view on the unaccommodating Protoctista. Rather, we have attempted to respect this great realm in its own right, conscious of its legacy from the prokaryotes. All scientists agree that protoctists originated by symbiotic mergers of bacteria.

All protoctists are composites

All protoctists are coevolved symbionts; chimeras with multiple ancestry. They have all evolved from more than a single type of microbial symbiont. Within the perspective of formal divisions of the biological sciences—botany, zoology, and mycology— we find ourselves in a period comparable to that of lichenology in the late nineteenth century. The realization that all lichens (superficially "primitive plants") are symbionts of algae or cyanobacteria with fungi was jarring; the implications of the symbiotic nature of lichens for their systematics and taxonomy were profound.[1]

Like lichens, all algae have secondarily and, in some cases, independently, acquired photosynthetic symbionts. Analogous to the fungi of lichens, the heterotrophic components of algae, rather than the phototrophic plastids, tend to be diverse. In all cases plastids in whatever their glorious colors (green—chloroplasts, red—rhodoplasts, blue-green—cyanelles, etc.) are not directly related to their "hosts" (the rest of the heterotrophic cytoplasm in which the plastids reside). Indeed, in many groups of algae (for example, euglenids, prasinophytes, chlorophytes, and conjugating green algae) it is questionable that the heterotrophic hosts are directly related to each other.

Given their symbiotic nature, in spite of botanical tradition, we can no longer tolerate classification of protoctists on the basis of the colors of their coevolved phototrophic symbionts, organelles derived from undigested food. Mitochondria, like plastids, originated from respiring bacteria by

[1]Serious scholars such as Dr. W. Nylander (1867 cited in Abbayes, 1954) denounced derisively the lichen symbiosis concept, which he named "Schwendenerisme" after Schwendener, who articulated the "theorie algo-lichenique" of the symbiotic nature of all lichens. Lichenologists all accept "Schwendenerisme"; most now agree that lichens need to be named and classified with their heterotrophic fungal partners. Whereas in the 25,000 or so species of lichens the diversity in the phycobiont (algal or cyanobacterial symbiont) is relatively limited, lichen fungi are profoundly diverse; the lichen symbiosis is highly polyphyletic.

several independent acquisitions. Therefore, mitochondrial characteristics in different protoctist lineages cannot be used as the basis for classification until details of mitochondrial polyphyly are available.

Taking our cues from lichenologists (Hawksworth and Hill, 1984), in this program we consider the ultrastructure and sexual patterns of the cytoplasmic (heterotrophic) components of protoctist cells to be of paramount importance for the determination of phyla, classes, and other higher taxa. Cell structure and developmental patterns, exclusive of the mitochondria, plastids, and other xenosomal organelles (Corliss, 1987) provide the primary basis of our classification.

Abadehellidae Family in phylum Foramenifera (Pr-3)

Acanthamoebidae Family in phylum Rhizopoda (Pr-2)

Acantharia Class in phylum Actinopoda (Pr-31). Phagotrophic unicells in which strontium sulfate skeletons underlie the periplasmic cortex (outer layer of each cell). Each possesses numerous ribbonlike or cylindrical motile organelles called myonemes. May form mononucleate bimastigote reproductive cells. No sexuality known. Generally marine planktonic

Acantharian actinopods Adjective referring to organisms in the phylum Actinopoda (Pr-31), class Acantharia, a marine class of planktonic microbes. Generally spherical organisms, with a unique radially symmetrical skeleton composed of rods of crystalline strontium sulfate ($SrSO_4$). The skeleton usually has 10 diametrical (20 radial) spines, called spicules, inserted according to a precise rule; known as Müller's law

Acanthoceraceae Family in phylum Bacillariophyta (Pr-18)

Acanthochiasmidae Family in phylum Actinopoda (Pr-31)

Acanthocystidae Family in phylum Actinopoda (Pr-31)

Acanthoecidae Family in phylum Choanomastigota (Pr-35)

Acanthometridae Family in phylum Actinopoda (Pr-31)

Acanthoplegmidae Family in phylum Actinopoda (Pr-31)

Acanthopodina Suborder in phylum Rhizopoda (Pr-2)

Acervulinacea Superfamily in phylum Foramenifera (Pr-3)

Acervulinidae Family in phylum Foramenifera (Pr-3)

Achnanthaceae Family in phylum Bacillariophyta (Pr-18)

Achnanthales Order in phylum Bacillariophyta (Pr-18)

Achnanthidiaceae Family in phylum Bacillariophyta (Pr-18)

Aconchulinida Order in phylum Rhizopoda (Pr-2)

Acrasea Acrasids; phylum of cellular (pseudoplasmodial) slime molds. Phagotrophic, ameboid organisms formed by aggregation of amebas to directly produce multicellular aerial, spore-bearing structures (sorocarps). Damp soil habitats (for example, dead plant parts, soil, or dung). Feed on bacteria; see also *Dictyostelida* (Pr-2)

Acrasid Informal name of cellular slime molds (Pr-2) in the phylum Acrasea, a phylum of microorganisms that has plant, animal, and fungal characteristics. This phylum is a small, probably polyphyletic group of ameboid organisms characterized by the aggregation of amebas and the formation of sorocarps

Acrasida Order in phylum Acrasea (Pr-2)

Acrasidae Family in phylum Acrasea (Pr-2)

Acrochaetiales Order in phylum Rhodophyta (Pr-33)

Acroseiraceae Family in phylum Phaeophyta (Pr-17)

Acroseirales Order in phylum Phaeophyta (Pr-17)

Acrosiphoniaceae Family in phylum Chlorophyta (Pr-28)

Acrotrichaceae Family in phylum Phaeophyta (Pr-17)

Actiniscaceae Family in phylum Dinomastigota (Pr-5)

Actiniscales Order in phylum Dinomastigota (Pr-5)

Actinocephalidae Family in phylum Apicomplexa (Pr-7)

Actinomyxida Order in phylum Myxospora (A-2)

Actinophryida Suborder in phylum Actinopoda (Pr-31)

Actinophryidae Family in phylum Actinopoda (Pr-31)

Actinopoda Pr-31: Phylum of protists, primarily large marine, heterotrophic unicells having long processes called axopods, which develop from axoplasts; See *Acantharia, Heliozoa, Phaeodaria, Polycystina, Radiolaria*

Actinopods Species of the phylum Actinopoda (Pr-31) are heterotrophic protoctists; their cells bear long processes called axopods, which develop from specialized structures called axoplasts

Actinoptychaceae Family in phylum Bacillariophyta (Pr-18)

Actinosporea Class in phylum Myxospora (A-2)

Actinosporeans Informal name of class of organisms in phylum Myxospora (A-2), class Actinosporea. The phylum Myxospora is a vast assemblage of microscopic heterotrophic animals, formerly considered protoctists, that form intricate multicellular spores with nematocyst-like structures called polar capsules. They are symbiotrophs of invertebrates and vertebrates such as fish, amphibians, and reptiles

Acytosteliaceae Family in phylum Dictyostelida (Pr-2)

Adeleida Order in phylum Apicomplexa (Pr-7)

Adeleidae Family in phylum Apicomplexa (Pr-7)

Adinomonadaceae Family in phylum Dinomastigota (Pr-5)

Aggregatidae Family in phylum Apicomplexa (Pr-7)

Alabaminidae Family in phylum Foramenifera

Alariaceae Family in phylum Phaeophyta (Pr-17)

Alatosporidae Family in phylum Myxospora (A-2)

Albuginaceae Family in phylum Oomycota (Pr-21)

Alfredinidae Family in phylum Foramenifera (Pr-3)

Allogromida Order in phylum Foramenifera (Pr-3)

Allogromiidae Family in phylum Foramenifera (Pr-3)

Almaenidae Family in phylum Foramenifera (Pr-3)

Amblyosporidae Family in phylum Microspora (F-1)

Amebomastigota Pr-22: Phylum consisting of heterotrophic, unicellular mastigotes, which, during their life history, reversibly transform to monopodial, uninucleate, or multinucleate amebas

Ammodiscacea Superfamily in phylum Foramenifera (Pr-3)

Ammodiscidae Family in phylum Foramenifera (Pr-3)

Ammosphaeroidinidae Family in phylum Foramenifera (Pr-3)

Amoebida Order in phylum Rhizopoda (Pr-2)

Amoebidae Family in phylum Rhizopoda (Pr-2)

Amoebophryaceae Family in phylum Dinomastigota (Pr-5)

Amphilithidae Family in phylum Actinopoda (Pr-31)

Amphipleuraceae Family in phylum Bacillariophyta (Pr-18)

Amphiroeae Tribe in phylum Rhodophyta (Pr-33)

Amphiroideae Subfamily in phylum Rhodophyta (Pr-33)

Amphisoleniaceae Family in phylum Dinomastigota (Pr-5)

Amphisteginidae Family in phylum Foramenifera (Pr-3)

Amphoraceae Family in phylum Bacillariophyta (Pr-18)

Amphorales Order in phylum Bacillariophyta (Pr-18)
Anadyomenaceae Family in phylum Chlorophyta (Pr-28)
Anaulaceae Family in phylum Bacillariophyta (Pr-18)
Anaulales Order in phylum Bacillariophyta (Pr-18)
Ancistrocomina Order in phylum Ciliophora (Pr-6)
Angeiocystidae Family in phylum Apicomplexa (Pr-7)
Anisolpidiaceae Family in phylum Hyphochytriomycota (Pr-14)
Annulopatellinidae Family in phylum Foramenifera (Pr-3)
Anomoeoneidaceae Family in phylum Bacillariophyta (Pr-18)
Apansporoblastina Suborder in phylum Microspora (F-1)
Aphanochaetaceae Family in phylum Chlorophyta (Pr-28)
Apicomplexa Pr-7: Phylum of protists parasitic on animals defined by a life history including a motile infective form (zoite), which possesses an apical complex. Life history generally has three phases: growth phase (by merogony or endogeny) during which the host is infected by the zoite; a sexual phase with gamete production and fertilization to form zygotes enclosed in oocysts; and a sporogenesis phase, during which the sporoplasm within the oocysts divides successively to form sporozoites, the new infective form. May be monoxenous or heteroxenous
Apicomplexan Informal name of organisms in the phylum Apicomplexa (Pr-7)
Apoaxoplastidiata Superfamily in phylum Actinopoda (Pr-31)
Apodachlyellaceae Family in phylum Oomycota (Pr-21)
Apodiniaceae Family in phylum Dinomastigota (Pr-5)
Apostomatia Subclass in phylum Ciliophora (Pr-6)
Apostomatida Order in phylum Ciliophora (Pr-6)
Arachnoidiscaceae Family in phylum Bacillariophyta (Pr-18)
Arachnoidiscales Order in phylum Bacillariophyta (Pr-18)
Arcellidae Family in phylum Rhizopoda (Pr-2)
Arcellinidae Order in phylum Rhizopoda (Pr-2)
Archaediscacea Superfamily in phylum Foramenifera (Pr-3)
Archaediscidae Family in phylum Foramenifera (Pr-3)
Archigregarinida Order in phylum Apicomplexa (Pr-7)
Archistomatina Suborder in phylum Ciliophora (Pr-6)
Ardissoniaceae Family in phylum Bacillariophyta (Pr-18)
Ardissoniales Order in phylum Bacillariophyta (Pr-18)
Armophorida Order in phylum Ciliophora (Pr-6)
Arnoldiellaceae Family in phylum Chlorophyta (Pr-28)
Arthracanthida Order in phylum Actinopoda (Pr-31)
Arthrocladiaceae Family in phylum Phaeophyta (Pr-17)
Aschemonellidae Family in phylum Foramenifera (Pr-3)
Asterigerinacea Superfamily in phylum Foramenifera (Pr-3)
Asterigerinatidae Family in phylum Foramenifera (Pr-3)
Asterigerinidae Family in phylum Foramenifera (Pr-3)

Asterolampraceae Family in phylum Bacillariophyta (Pr-18)
Asterolamprales Order in phylum Bacillariophyta (Pr-18)
Astomatophorida Order in phylum Ciliophora (Pr-6)
Astracanthidae Family in phylum Actinopoda (Pr-31)
Astrephomenaceae Family in phylum Chlorophyta (Pr-28)
Astrolithidae Family in phylum Actinopoda (Pr-31)
Astromatia Subclass in phylum Ciliophora (Pr-6)
Astromatida Order in phylum Ciliophora (Pr-6)
Astrorhizacea Superfamily in phylum Foramenifera (Pr-3)
Astrorhizidae Family in phylum Foramenifera (Pr-3)
Asymmetrinidae Family in phylum Foramenifera (Pr-3)
Ataxophragmiacea Superfamily in phylum Foramenifera (Pr-3)
Ataxophragmiidae Family in phylum Foramenifera (Pr-3)
Athalamea Class in phylum Foramenifera (Pr-3)
Atlanticellidae Family in phylum Actinopoda (Pr-31)
Attheyaceae Family in phylum Bacillariophyta (Pr-18)
Auerbachiidae Family in phylum Myxospora (A-2)
Aulacanthidae Family in phylum Actinopoda (Pr-31)
Aulacodiscaceae Family in phylum Bacillariophyta (Pr-18)
Aulacosiraceae Family in phylum Bacillariophyta (Pr-18)
Aulacosirales Order in phylum Bacillariophyta (Pr-18)
Aulosphaeridae Family in phylum Actinopoda (Pr-31)
Auriculaceae Family in phylum Bacillariophyta (Pr-18)
Aurosphaeraceae Family in phylum Chrysophyta (Pr-15)
Aveolinidae Family in phylum Foramenifera (Pr-3)
Axoplasthelida Suborder in phylum Actinopoda (Pr-31)

Bacillariaceae Family in phylum Bacillariophyta (Pr-18)
Bacillariales Order in phylum Bacillariophyta (Pr-18)
Bacillariophyceae Class in phylum Bacillariophyta (Pr-18)
Bacillariophycidae Subclass in phylum Bacillariophyta (Pr-18)
Bacillariophyta Pr-18: Diatoms. Phylum of diploid, sexual freshwater and marine algae. Cells enclosed by complex siliceous walls consisting of two valves. Unicellular or colonial, diatoms reproduce by mitotic division with periodic formation of haploid, valve-less gametes. Centric, radially symmetric or pennate, bilaterally symmetric forms. Male gametes posteriorly undulipodiated in some centric genera. Cells nonmotile or motile by gliding, accompanied by secretion through slits in cell walls and adhesion of the secreted material. Plastids with chlorophylls *a* and *c*, fucoxanthin, and other minor pigments. Extensive fossil forms, Lower Cretaceous to Holocene
Baculellidae Family in phylum Foramenifera (Pr-3)
Bagginidae Family in phylum Foramenifera (Pr-3)
Bangiaceae Family in phylum Rhodophyta (Pr-33)

Bangiales Order in phylum Rhodophyta (Pr-33)
Bangiophycidae Subclass in phylum Rhodophyta (Pr-33)
Barkerinidae Family in phylum Foramenifera (Pr-3)
Bathysiphonidae Family in phylum Foramenifera (Pr-3)
Batrachospermales Order in phylum Rhodophyta (Pr-33)
Bellerocheaceae Family in phylum Bacillariophyta (Pr-18)
Berkeleyaceae Family in phylum Bacillariophyta (Pr-18)
Bicosoecids Class in phylum Archaeprotista (Pr-1). Free-living, planktonic, bimastigote, heterokont, heterotrophic, solitary or colonial cells, mostly contained in a vaselike shell or lorica composed of organic material
Biddulphiaceae Family in phylum Bacillariophyta (Pr-18)
Biddulphiales Order in phylum Bacillariophyta (Pr-18)
Biddulphiophycidae Subclass in phylum Bacillariophyta (Pr-18)
Biokovinidae Family in phylum Foramenifera (Pr-3)
Biomyxida Order in phylum Foramenifera (Pr-3)
Biseriamminidae Family in phylum Foramenifera (Pr-3)
Bivalvulida Order in phylum Myxospora (A-2)
Blastocladiaceae Family in phylum Chytridiomycota (Pr-35)
Blastocladialean Adjective referring to organisms in the Phylum Blastocladiomycota (Pr-34)
Blastocladiales Order in phylum Chytridiomycota (Pr-35)
Blastodiniaceae Family in phylum Dinomastigota (Pr-5)
Blastodiniales Order in phylum Dinomastigota (Pr-5)
Blastodinian dinomastigotes Adjective referring to organisms in the phylum Dinomastigota, order Blastodiniales (Pr-5)
Blastogregarinida Order in phylum Foramenifera (Pr-3)
Blepharocorythina Suborder in phylum Ciliophora (Pr-6)
Bodonidae Family in phylum Archaeprotista (Pr-1)
Bodonina Suborder in phylum Archaeprotista (Pr-1)
Boldiaceae Family in phylum Rhodophyta (Pr-33)
Bolivinellidae Family in phylum Foramenifera (Pr-3)
Bolivinidae Family in phylum Foramenifera (Pr-3)
Bolivinitidae Family in phylum Foramenifera (Pr-3)
Bolivinoididae Family in phylum Foramenifera (Pr-3)
Bonnemaisoniales Order in phylum Rhodophyta (Pr-33)
Botrydiaceae Family in phylum Xanthophyta (Pr-16)
Botrydiopsidaceae Family in phylum Xanthophyta (Pr-16)
Botryochloridaceae Family in phylum Xanthophyta (Pr-16)
Botryococcaceae Family in phylum Chlorophyta (Pr-28)
Botryoidea Family in phylum Actinopoda (Pr-31)
Brachydiniaceae Family in phylum Dinomastigota (Pr-5)
Brachysiraceae Family in phylum Bacillariophyta (Pr-18)
Bradyinidae Family in phylum Foramenifera (Pr-3)
Bronnimanniidae Family in phylum Foramenifera (Pr-3)

Brown algae Pr-17: Phylum containing some of the largest multicellular protoctists. Algae are reproduced from heterokont mastigotes or zygotes formed by fusion of eggs with heterokont male gametes. Exclusively marine in subtidal and intertidal zones. May alternate diploid and haploid generations. Plastids contain chlorophylls *a*, *c*, and c_1 and fucoxanthin. Laminarin as storage material
Brown seaweeds See *Brown algae*, Pr-17.
Bryophryida Order in phylum Ciliophora (Pr-6)
Bryopsidaceae Family in phylum Chlorophyta (Pr-28)
Bueningiidae Family in phylum Foramenifera (Pr-3)
Buffhamiaceae Family in phylum Phaeophyta (Pr-17)
Buliminacea Superfamily in phylum Foramenifera (Pr-3)
Buliminellidae Family in phylum Foramenifera (Pr-3)
Buliminidae Family in phylum Foramenifera (Pr-3)
Buliminoidiae Family in phylum Foramenifera (Pr-3)
Burenellidae Family in phylum Microspora (F-1)
Burkeidae Family in phylum Microspora (F-1)
Bursariomorphida Order in phylum Ciliophora (Pr-6)
Buxtehudeidae Family in phylum Microspora (F-1)

Cachonellaceae Family in phylum Dinomastigota (Pr-5)
Calcarinidae Family in phylum Foramenifera (Pr-3)
Caligellidae Family in phylum Foramenifera (Pr-3)
Calonymphidae Family in phylum Archaeprotista (Pr-1)
Candeinidae Family in phylum Foramenifera (Pr-3)
Cannosphaeridae Family in phylum Actinopoda (Pr-31)
Carteriaceae Family in phylum Chlorophyta (Pr-28)
Carterinida Order in phylum Foramenifera (Pr-3)
Carterinidae Family in phylum Foramenifera (Pr-3)
Caryosporidae Family in phylum Apicomplexa (Pr-7)
Caryotrophidae Family in phylum Apicomplexa (Pr-7)
Cassidulinacea Superfamily in phylum Foramenifera (Pr-3)
Cassidulinidae Family in phylum Foramenifera (Pr-3)
Cassigerinellidae Family in phylum Foramenifera (Pr-3)
Castanellidae Family in phylum Actinopoda (Pr-31)
Catapsydracidae Family in phylum Foramenifera (Pr-3)
Catenariaceae Family in phylum Chytridiomycota (Pr-35)
Caucasinidae Family in phylum Foramenifera (Pr-3)
Caudosporidae Family in phylum Microspora (F-1)
Caulerpaceae Family in phylum Chlorophyta (Pr-28)
Caulerpales Order in phylum Chlorophyta (Pr-28)
Caulochytriaceae Family in phylum Chytridiomycota (Pr-35)
Cavosteliidae Family in phylum Myxomycota (Pr-23)
Centritractaceae Family in phylum Xanthophyta (Pr-16)

Centroaxoplastidiata Superfamily in phylum Actinopoda (Pr-31)
Centrocollidae Family in phylum Actinopoda (Pr-31)
Centroplasthelida Suborder in phylum Actinopoda (Pr-31)
Centropyxidae Family in phylum Rhizopoda (Pr-2)
Cephaloidophoridae Family in phylum Apicomplexa (Pr-7)
Cephalolobidae Family in phylum Apicomplexa (Pr-7)
Ceramilaes Order in phylum Rhodophyta (Pr-33)
Ceratiaceae Family in phylum Dinomastigota (Pr-5)
Ceratiomyxaceae Family in phylum Myxomycota (Pr-23)
Ceratiomyxales Order in phylum Myxomycota (Pr-23)
Ceratiomyxidae Family in phylum Myxomycota (Pr-23)
Ceratiomyxomycetidae Subclass in phylum Myxomycota (Pr-23)
Ceratobuliminidae Family in phylum Foramenifera (Pr-3)
Ceratocoryaceae Family in phylum Dinomastigota (Pr-5)
Cerelasmidae Family in phylum Xenophyophora (Pr-4)
Chaetocerophycidae Subclass in phylum Bacillariophyta (Pr-18)
Chaetocerotaceae Family in phylum Bacillariophyta (Pr-18)
Chaetocerotales Order in phylum Bacillariophyta (Pr-18)
Chaetochloridaceae Family in phylum Chlorophyta (Pr-28)
Chaetophoraceae Family in phylum Chlorophyta (Pr-28)
Chaetophorales Order in phylum Chlorophyta (Pr-28)
Chaetosiphonaceae Family in phylum Chlorophyta (Pr-28)
Challengeriidae Family in phylum Actinopoda (Pr-31)
Characiaceae Family in phylum Chlorophyta (Pr-28)
Characiachloridaceae Family in phylum Chlorophyta (Pr-28)
Characiasiphonaceae Family in phylum Chlorophyta (Pr-28)
Characidiopsidaceae Family in phylum Xanthophyta (Pr-16)
Characiopsidaceae Family in phylum Xanthophyta (Pr-16)
Charophyceae Class of green algae (phylum Chlorophyta, Pr-28) containing orders Chlorokybales, Klebsormidiales, Coleochaetales, and Charales. Members of first three orders produce undulipodiated swarmer cells with an intracellular multilayered structure associated with kinetids and typically covered with small, square scales on the cell body. Since Coleochaete (and several other species) has a phragmoplast and forms bimastigote zoospores that resemble plant spermatozoids, some members of these groups are thought to resemble the ancestors of land plants. Members of the order Charales are large, submerged, phragmoplast-forming freshwater algae with multicellular sex organs, consisting of large thalli with an erect main axis with regularly placed whorls of lateral branches with limited growth. The egg cell is enclosed within sterile (nondividing) tissue.
Charophyte Informal name of green algae (phylum Chlorophyceae, Pr-28) in the class Charophyceae
Chaunacanthida Order in phylum Actinopoda (Pr-31)

Chiloguembelinidae Family in phylum Foramenifera (Pr-3)

Chilostomellacea Superfamily in phylum Foramenifera (Pr-3)

Chilostomellidae Family in phylum Foramenifera (Pr-3)

Chlamydodontina Suborder in phylum Ciliophora (Pr-6)

Chlamydomonadaceae Family in phylum Chlorophyta (Pr-28)

Chlamydomonadales Order in phylum Chlorophyta (Pr-28)

Chloramoebaceae Family in phylum Xanthophyta (Pr-16)

Chloramoebales Order in phylum Xanthophyta (Pr-16)

Chlorarachnid Informal name of organisms in the phylum Chlorarachnida

Chlorarachnida Phototrophic marine organisms in which an ameboid plasmodium contains individual green cells linked by a network of reticulopodia. Monospecific phylum (one species *Chlorarachnion reptans*). Cells contain plastids with chlorophylls *a* and *b*. Life history incompletely known but contains a spherical, walled stage, and unimastigote zoospores. Extensive periplastidial compartment indicates origin from symbiosis between amebas and green algae

Chlorarachniophyceae Class in phylum Chlorarachnida (Pr-28)

Chlorobotryaceae Family in phylum Eustigmatophyta (Pr-27)

Chlorochytriaceae Family in phylum Chlorophyta (Pr-28)

Chlorococcaceae Family in phylum Chlorophyta (Pr-28)

Chlorococcales Order in phylum Chlorophyta (Pr-28)

Chlorodendraceae Family in phylum Chlorophyta (Pr-28)

Chlorodendrales Order in phylum Chlorophyta (Pr-28)

Chloromonads Phylum of wall-less heterokont mastigote algae. Solitary cells distinguished by large Golgi apparatus extending over the anterior surface of the single nucleus. Plastids contain chlorophylls *a* and *c*. Found as motile or palmelloid cells in freshwater and marine habitats; sexuality is unknown

Chloromyxidae Family in phylum Myxospora (A-2)

Chloropediaceae Family in phylum Xanthophyta (Pr-16)

Chlorophyceae Class of green algae (phylum Chlorophyta, Pr-28). Mainly freshwater; mastigotes covered by a cell wall (theca) or naked. Cell division characterized by phycoplast and a collapsing telophase spindle

Chlorophyta Pr-28: Green algae; phylum of cosmopolitan, unicellular, or multicellular photosynthetic organisms that form mastigote stages as spores or gametes. Cells possess plastids surrounded by a double membrane. Thylakoid membranes contain chlorophylls *a* and *b*; primary storage material is starch. The colorless or amastigote immediate descendants of these algae are included in phylum; see *Charophyceae, Chlorophyceae, Microthamniales, Pedinomonadales, Prasinophyceae, Prasiolales, Trentepohliales, Ulvophyceae*

Chlorophytes Informal name of organisms in the phylum Chlorophyta, Pr-28. The green algae, a highly diverse division characterized by chloroplasts, having chlorophylls *a* and *b* as the predominating pigments

Chlorarachnid The monospecific phylum Chlorarachnida contains only *C. reptans*. The arrangement of the membranes surrounding the chloroplast appears to be unique. The combination of characters found in *C. reptans* suggests that it may have evolved from an original symbiosis between a colorless eukaryote and a chlorophyll-*b*-containing protoctist

Choanomastigota Pr-36: Choanomastigotes: a phylum marine heterotrophic mastigotes or sessile colonial organisms. Cells enclosed by an organic (theca) or siliceous (lorica) structure with collars of tentacles; also, in phylum Kinetoplastida (Pr-11), a term for a stage in the development of trypanosomatid mastigotes in which the kinetoplast lies anterior to the nucleus and the associated undulipodium emerges

Choanomastigotes Informal name of organisms in the phylum Choanomastigota (Pr-36). They can be distinguished by having a lorica, a hard structure from which a single undulipodium emerges. It is generally thought that choanomastigotes are direct ancestors of the sponges

Chonotrich ciliates Informal name of ciliates (Pr-6) belonging to the subclass Chonotrichia. The ciliates are undoubtedly one of the easiest groups of protoctists to identify since their typical feature is the presence of files of cilia, known as kinities, on the cell surface

Chonotrichia Subclass in phylum Ciliophora (Pr-6)

Chordaceae Family in phylum Phaeophyta (Pr-17)

Chordariaceae Family in phylum Phaeophyta (Pr-17)

Chordariales Order in phylum Phaeophyta (Pr-17)

Choreotrichia Subclass in phylum Ciliophora (Pr-6)

Choreotrichida Order in phylum Ciliophora (Pr-6)

Choristocarpaceae Family in phylum Phaeophyta (Pr-17)

Chromulinaceae Family in phylum Chrysophyta (Pr-15)

Chrysalidinidae Family in phylum Foramenifera (Pr-3)

Chrysamoebaceae Family in phylum Chrysophyta (Pr-15)

Chrysamoebales Order in phylum Chrysophyta (Pr-15)

Chrysanthemodiscaceae Family in phylum Bacillariophyta (Pr-18)

Chrysanthemodiscales Order in phylum Bacillariophyta (Pr-18)

Chrysapiaceae Family in phylum Chrysophyta (Pr-15)

Chrysocapsaceae Family in phylum Chrysophyta (Pr-15)

Chrysocapsales Order in phylum Chrysophyta (Pr-15)

Chrysochaetaceae Family in phylum Chrysophyta (Pr-15)

Chrysococcaceae Family in phylum Chrysophyta (Pr-15)

Chrysomeridaceae Family in phylum Chrysophyta (Pr-15)

Chrysophyceae Class in phylum Chrysophyta (Pr-15)

Chrysophyta Pr-15: Golden-yellow algae. Phylum of photosynthetic and related colorless organisms, single cells or colonial, primarily freshwater plankton. Includes the class Dictyochophyceae (silicomastigotes or silicomastigotes). Plastids contain chlorophylls *a* and *c*; chrysolaminarin as storage product. Form swarmers with heterokont undulipodia. Fossil silicified cysts (stomatocysts) of class Chrysophyceae common, from Upper Cretaceous to Holocene

Chrysosaccaceae Family in phylum Chrysophyta (Pr-15)

Chrysosphaeraceae Family in phylum Chrysophyta (Pr-15)

Chrysosphaerales Order in phylum Chrysophyta (Pr-15)

Chytrid Pr-35: Informal name of water molds belonging to the phylum Chytridiomycota, class Chytridiomycetes. A group of simple nonphotosynthetic, chiefly aquatic, saprobic, or parasitic eukaryotic microorganisms, usually considered as fungi or sometimes protist. They are unicellular or coenocytic, with cell walls containing chitin. Sexual reproduction is oogamous and they also reproduce asexually by motile zoospores. Sperm and zoospores all bear a singular posterior whiplash undulipodium distinguishing them from true fungi

Chytridiaceae Family in phylum Chytridiomycota (Pr-35)

Chytridiales Order in phylum Chytridiomycota (Pr-35)

Chytridiomycetes Phylum Chytridiomycota (Pr-35) forms a group of simple nonphotosynthetic chiefly aquatic, saprobic, or parasitic eukaryotic microorganisms, usually considered as fungi or sometimes protists. They are unicellular or coenocytic, with cell walls containing chitin. Sexual reproduction is oogamous and they also reproduce asexually by motile zoospores. Sperm and zoospores all bear a single posterior whiplash undulipodium, distinguishing them from other fungi

Chytridiomycetes Single class in phylum Chytridiomycota (Pr-35)

Chytridiomycota Pr-35: Phylum of chitinous-walled, heterotrophic aquatic, and soil protoctists that form undulipodiated zoospores and display absorptive nutrition. Filamentous or thalloid organisms form sporangia, which release undulipodiated propagules (zoospores), some of which may behave as gametes and fuse. Zoospores may transform into or fuse with the developing sporangium. Cells contain microbody–lipid globule complex (MLC). Some are necrotrophs in plants

Chytridiomycotes Informal name of water molds belonging to the phylum Chytridiomycota (Pr-35)

Chytridiopsidae Family in phylum Microspora (F-1)

Chytriodiniaceae Family in phylum Dinomastigota (Pr-5)

Chytriodiniales Order in phylum Dinomastigota (Pr-5)

Cibicididae Family in phylum Foramenifera (Pr-3)

Ciliates Informal name of organisms belonging to the phylum Ciliophora (Pr-6)

Ciliophora Pr-6: Phylum of dikaryotic, heterotrophic ciliates, primarily single motile cells with dimorphic nuclei (at least one macro- and one micronucleus, but often more) and complex cortices. The cortex, approximately 1 μm at the outer surface of the ciliate wall, is composed of precisely patterned tubules, fibers, and membranes. Files of kinetosomes, known as kineties, from which cilia extend, comprise a major portion of the cortex. Physiologically active macronucleus divides amitotically, whereas smaller, inactive, diploid, mitotic micronucleus undergoes meiosis and reciprocal transfer in sexuality. Synkaryotes are formed in conjugation and autogamy. Both types of nuclei lack centrioles and divide by closed karyokinesis (no nuclear membrane breakdown). Mostly phagotrophic on bacteria or other protists. Cosmopolitan in aqueous habitats. Some are secondarily photosynthetic by acquisition of algae or plastids, others heterotrophic symbiotrophs, which display dimorphic life cycles

Ciliophryida Suborder in phylum Actinopoda (Pr-31)

Ciliophryidae Family in phylum Actinopoda (Pr-31)

Circoporidae Family in phylum Actinopoda (Pr-31)

Cladochytriaceae Family in phylum Chytridiomycota (Pr-35)

Cladophoraceae Family in phylum Chlorophyta (Pr-28)

Cladopyxidaceae Family in phylum Dinomastigota (Pr-5)

Cladostephaceae Family in phylum Phaeophyta (Pr-17)

Clastodermataceae Family in phylum Myxomycota (Pr-23)

Clathrulinidae Family in phylum Actinopoda (Pr-31)

Clevelandellida Order in phylum Ciliophora (Pr-6)

Climacospheniaceae Family in phylum Bacillariophyta (Pr-18)

Climacospheniales Order in phylum Bacillariophyta (Pr-18)

Coccidia Class in phylum Apicomplexa (Pr-7)

Coccidian apicomplexan Informal name or adjective referring to organisms in the phylum Apicomplexa (Pr-7), class Coccidia. Heterotrophic microbes, spore-forming symbiotrophs of animals. The coccidians are perhaps the best-known group of apicomplexans because many of them cause serious and even fatal diseases in their vertebrate and invertebrate hosts. The major symptoms of coccidian disease are diarrhea and dysentery

Coccidiniaceae Family in phylum Dinomastigota (Pr-5)

Coccomyxaceae Family in phylum Chlorophyta (Pr-28)

Cocconeidaceae Family in phylum Bacillariophyta (Pr-18)

Coccosphaerales Order in phylum Prymnesiophyta (Haptomonada; Pr-25)

Codiaceae Family in phylum Chlorophyta (Pr-28)

Codonosigidae Family in phylum Choanomastigota (Pr-36)

Coelocladiaceae Family in phylum Phaeophyta (Pr-17)

Coelodendriae Family in phylum Actinopoda (Pr-31)

Coelomomycetaceae Family in phylum Chytridiomycota (Pr-35)

Coelotrophidae Family in phylum Apicomplexa (Pr-7)

Coelotrophiida Order in phylum Apicomplexa (Pr-7)

Coilodesmaceae Family in phylum Phaeophyta (Pr-17)

Colaniellacea Superfamily in phylum Foramenifera (Pr-3)

Colaniellidae Family in phylum Foramenifera (Pr-3)

Coleitidae Family in phylum Foramenifera (Pr-3)

Coliphorina Suborder in phylum Ciliophora (Pr-6)

Colpodea Class in phylum Ciliophora (Pr-6)

Colpodida Order in phylum Ciliophora (Pr-6)

Compsopogonaceae Family in phylum Rhodophyta (Pr-33)

Compsopogonales Order in phylum Rhodophyta (Pr-33)

Conaconidae Family in phylum Actinopoda (Pr-31)

Concharidae Family in phylum Actinopoda (Pr-31)

Conjugaphyta Pr-32: Phylum of primarily freshwater filamentous zygonemalean (conjugacean) and desmid green algae distinguished from other chlorophytes by their isogamontous conjugating sexuality and their lack of undulipodia at all stages of development. Reproduction by mitotic division, sexuality by conjugation involving fusion of ameboid gametes to form synkaryon, which develops into resistant spores

Conjugatophyceae Single class in phylum Conjugaphyta (Pr-32)

Conopodina Suborder in phylum Rhizopoda (Pr-3)

Conorbinidae Family in phylum Foramenifera (Pr-3)
Conorboididae Family in phylum Foramenifera (Pr-3)
Copromyxidae Family in phylum Rhizopoda (Pr-2)
Corallinales Order in phylum Rhodophyta (Pr-33)
Corallineae Tribe in phylum Rhodophyta (Pr-33)
Corallinoideae Subfamily in phylum Rhodophyta (Pr-33)
Corethraceae Family in phylum Bacillariophyta (Pr-18)
Corethrales Order in phylum Bacillariophyta (Pr-18)
Corethronophycidae Subclass in phylum Bacillariophyta (Pr-18)
Cornuspiracea Superfamily in phylum Foramenifera (Pr-3)
Cornuspiridae Family in phylum Foramenifera (Pr-3)
Coscinodiscaceae Family in phylum Bacillariophyta (Pr-18)
Coscinodiscales Order in phylum Bacillariophyta (Pr-18)
Coscinodiscophyceae Class in phylum Bacillariophyta (Pr-18)
Coscinodiscophycidae Subclass in phylum Bacillariophyta (Pr-18)
Coscinophragmatacea Superfamily in phylum Foramenifera (Pr-3)
Coscinophragmatidae Family in phylum Foramenifera (Pr-3)
Coskinolinidae Family in phylum Foramenifera (Pr-3)
Cougourdellidae Family in phylum Microspora (F-1)
Cribrariaceae Family in phylum Myxomycota (Pr-23)
Cribratinidae Family in phylum Foramenifera (Pr-3)
Cryptaxohelida Order in phylum Actinopoda (Pr-31)
Crypthecodiniaceae Family in phylum Dinomastigota (Pr-5)
Cryptoaxoplastidiata Superfamily in phylum Actinopoda (Pr-31)
Cryptogemmida Order in phylum Ciliophora (Pr-6)
Cryptomonads Pr-26: Phylum of asymmetric flattened, mastigote algae with distinctive swimming motion or derived palmelloid forms. Vestibular depression from which undulipodia emerge is the anterior portion of the crypt lined with refractile ejectosomes. Periplast, formed by organic plates, is internal to the plasma membrane, rather than external to the cell wall. Plastids contain chlorophyll *c* and phycobilins. Based on the presence of nucleomorph, the phylum is thought to have evolved from the symbiosis between heterotrophic mastigotes and red algae that retain remnant nuclei
Cryptonemiales Order in phylum Rhodophyta (Pr-33)
Cryptophyceae Single class in phylum Cryptophyta (Pr-26)
Cryptophyta See *Cryptomonads*(Pr-26)
Cryptosporidae Family in phylum Apicomplexa (Pr-7)
Culicosporidae Family in phylum Microspora (F-1)
Cuneolinidae Family in phylum Foramenifera (Pr-3)
Cutleriaceae Family in phylum Phaeophyta (Pr-17)
Cutleriales Order in phylum Phaeophyta (Pr-17)
Cyanophoraceae Family in phylum Glaucocystophyta

Cyanophorales Order in phylum Glaucocystophyta
Cyclamminidae Family in phylum Foramenifera (Pr-3)
Cyclolinacea Superfamily in phylum Foramenifera (Pr-3)
Cyclolinidae Family in phylum Foramenifera (Pr-3)
Cyclosporidae Family in phylum Apicomplexa (Pr-7)
Cyclotellaceae Family in phylum Bacillariophyta (Pr-18)
Cymatosiraceae Family in phylum Bacillariophyta (Pr-18)
Cymatosirales Order in phylum Bacillariophyta (Pr-18)
Cymatosirophycidae Subclass in phylum Bacillariophyta (Pr-18)
Cymbaloporidae Family in phylum Foramenifera (Pr-3)
Cymbellaceae Family in phylum Bacillariophyta (Pr-18)
Cymbellales Order in phylum Bacillariophyta (Pr-18)
Cyrtoidea Family in phylum Actinopoda (Pr-31)
Cyrtolophosidida Order in phylum Ciliophora (Pr-6)
Cyrtophora Subphylum in phylum Ciliophora (Pr-6)
Cyrtophorida Order in phylum Ciliophora (Pr-6)
Cystoseiraceae Family in phylum Phaeophyta (Pr-17)

Dactylophoridae Family in phylum Apicomplexa (Pr-7)
Dariopsidae Family in phylum Foramenifera (Pr-3)
Dasycladaceae Family in phylum Chlorophyta (Pr-28)
Dasycladales Order in phylum Chlorophyta (Pr-28)
Delamareaceae Family in phylum Phaeophyta (Pr-17)
Delosinacea Superfamily in phylum Foramenifera (Pr-3)
Delosinidae Family in phylum Foramenifera (Pr-3)
Dermatolitheae Tribe in phylum Rhodophyta (Pr-33)
Desmarestiaceae Family in phylum Phaeophyta (Pr-17)
Desmarestiales Order in phylum Phaeophyta (Pr-17)
Desmidiaceae Family in phylum Conjugaphyta (Pr-32)
Desmocapsaceae Family in phylum Dinomastigota (Pr-5)
Desmocapsales Order in phylum Dinomastigota (Pr-5)
Desmomonadaceae Family in phylum Dinomastigota (Pr-5)
Desmomonadales Order in phylum Dinomastigota (Pr-5)
Desmothoracida Suborder in phylum Actinopoda (Pr-31)
Devescoviidae Family in phylum Archaeprotista (Pr-1)
Diadesmidaceae Family in phylum Bacillariophyta (Pr-18)
Dianemaceae Family in phylum Myxomycota (Pr-23)
Diatom Any member of the phylum Bacillariophyta (Pr-18); unicellular and colonial aquatic protoctists renowned for their two-valved siliceous tests (frustules); see *Bacillariophyta*
Dichotomosiphonaceae Family in phylum Chlorophyta (Pr-28)
Dictyacanthidae Family in phylum Actinopoda (Pr-31)

Dictyochaceae Family in phylum Chrysophyta (Pr-15)

Dictyochales Order in phylum Chrysophyta (Pr-15)

Dictyochophyceae Class in phylum Chrysophyta (Pr-15)

Dictyoneidaceae Family in phylum Bacillariophyta (Pr-18)

Dictyoneidales Order in phylum Bacillariophyta (Pr-18)

Dictyopsellidae Family in phylum Foramenifera (Pr-3)

Dictyosiphonaceae Family in phylum Phaeophyta (Pr-17)

Dictyosiphonales Order in phylum Phaeophyta (Pr-17)

Dictyosphaeriaceae Family in phylum Chlorophyta (Pr-28)

Dictyosteliaceae Family in phylum Rhizopoda, Pr-2

Dictyostelida Dictyostelids (Pr-2); class of cellular (pseudoplasmodial) slime molds. Ameboid amastigote cells aggregate to form sorocarps. Damp soil, freshwater habitats. Differentiated from acrasids by cytology of the myxameba (slime ameba), production of well-differentiated stalk and spore cells, formation of more complex sorocarps and alignment of aggregating myxamebas into streams that form motile pseudoplasmodia (for example, "slugs" or "Mexican hat stage," cf. *Dictyostelium discoideum*). Sexual fusion of compatible myxamebas occurs in plasmodium formation

Dictyostelids Class in the phylum Rhizopoda (Pr-2). The cellular slime molds. Informal name of cellular slime molds in the phylum Dictyostelida

Dictyotaceae Family in phylum Phaeophyta (Pr-17)

Dictyotales Order in phylum Phaeophyta (Pr-17)

Dictyotopsidaceae Family in phylum Phaeophyta (Pr-17)

Dicyclinidae Family in phylum Foramenifera (Pr-3)

Didymiaceae Family in phylum Myxomycota (Pr-23)

Difflugiidae Family in phylum Rhizopoda (Pr-2)

Diffusilinidae Family in phylum Foramenifera (Pr-3)

Dimorphyidae Family in phylum Actinopoda (Pr-31)

Dinobryaceae Family in phylum Chrysomonada (Pr-15)

Dinomastigota Pr-5: Phylum of bimastigotes, usually with one girdle and one transverse undulipodium. Amphiesmal plates form intramembranous wall-like structures; some lack walls. Primarily marine plankton, solitary or colonial cells. Distinctive chromatin organization: nucleus (dinokaryon or mesokaryon) has permanently condensed and visible chromosomes lacking nucleosomes and the histones that comprise them. Substitution of much thymine in DNA by 5-hydroxymethyl uracil. Photosynthetic forms contain plastids with chlorophylls *a* and c_2 and a unique xanthophyll, peridinin; many lack plastids. Resistant cysts fossilize as hystrichospheres

Dinomastigotes Also Dinomastigotes. Informal name of organisms in the phylum Dinomastigota (Pr-5). Dinomastigotes are essentially biundulipodiated, photosynthetic or nonphotosynthetic, walled or naked, unicells. At least 30 species of marine dinomastigotes are luminescent. They are one of the most common sources of luminescence in seawater

Dinophysiaceae Family in phylum Dinomastigota (Pr-5)

Dinophysiales Order in phylum Dinomastigota (Pr-5)

Dinophysid dinomastigotes Adjective referring to organisms in the phylum Dinomastigota (Pr-5), order Dinophysiales

Diploconidae Family in phylum Actinopoda (Pr-31)

Diplocystidae Family in phylum Apicomplexa (Pr-7)

Diplomonad Class in the phylum Archaeprotista (Pr-1). Mastigotes with distinctive karyomastigont systems, lacking mitochondria and Golgi apparatus. All heterotrophic; sexuality unknown. Free-living freshwater or symbiotic, including necrotrophic forms. Cysts found only in symbiotrophic species

Diplomonadida See *Diplomonad*

Diploneidaceae Family in phylum Bacillariophyta (Pr-18)

Diploneidineae Suborder in phylum Bacillariophyta (Pr-18)

Diplosporidae Family in phylum Apicomplexa (Pr-7)

Discamminidae Family in phylum Foramenifera (Pr-3)

Discocephalina Suborder in phylum Ciliophora (Pr-6)

Discocyclinidae Family in phylum Foramenifera (Pr-3)

Discoidae Family in phylum Actinopoda (Pr-31)

Discorbacea Superfamily in phylum Foramenifera (Pr-3)

Discorbidae Family in phylum Foramenifera (Pr-3)

Discorbinellidae Family in phylum Foramenifera (Pr-3)

Discospirinidae Family in phylum Foramenifera (Pr-3)

Dobelliidae Family in phylum Apicomplexa (Pr-7)

Dorataspidae Family in phylum Actinopoda (Pr-31)

Dorisiellidae Family in phylum Apicomplexa (Pr-7)

Dorothiidae Family in phylum Foramenifera (Pr-3)

Dryorhizopsidae Family in phylum Foramenifera (Pr-3)

Duboscqiidae Family in phylum Microspora (F-1)

Duboscquellaceae Family in phylum Dinomastigota (Pr-5)

Dunaliellales Order in phylum Chlorophyta (Pr-28)

Duostominacea Superfamily in phylum Foramenifera (Pr-3)

Duostominidae Family in phylum Foramenifera (Pr-3)

Durvillaeaceae Family in phylum Phaeophyta (Pr-17)

Durvillaeales Order in phylum Phaeophyta (Pr-17)

Dusenburyinidae Family in phylum Foramenifera (Pr-3)

Dysteriina Suborder in phylum Ciliophora (Pr-6)

Earlandiacea Superfamily in phylum Foramenifera (Pr-3)

Earlandiidae Family in phylum Foramenifera (Pr-3)

Earlandinitidae Family in phylum Foramenifera (Pr-3)

Ebridians Coastal, marine, free-living bimastigote having solitary cells with basketlike internal skeletons consisting of siliceous rods. Reproduction by simple fission; sexuality unknown; taxonomy *incertae sedis*. Fossil record from Lower Cenozoic to present with greatest diversity in the Miocene

Echinamoebidae Family in phylum Rhizopoda (Pr-2)

Echinosteliaceae Family in phylum Myxomycota (Pr-23)

Echinosteliales Order in phylum Myxomycota (Pr-23)
Echinosteliopsidae Family in phylum Myxomycota (Pr-23)
Ectocarpaceae Family in phylum Phaeophyta (Pr-17)
Ectocarpales Order in phylum Phaeophyta (Pr-17)
Eggerellidae Family in phylum Foramenifera (Pr-3)
Eimeriida Order in phylum Apicomplexa (Pr-7)
Eimeriidae Family in phylum Apicomplexa (Pr-7)
Elachistaceae Family in phylum Phaeophyta (Pr-17)
Elaeomyxaceae Family in phylum Myxomycota (Pr-23)
Elhasaellidae Family in phylum Foramenifera (Pr-3)
Ellobiopsida Heterotrophic, coenocytic symbiotrophs including necrotrophs especially of planktonic marine anthropods. Larger members arborescent with their absorptive bases anchored in the host nerve tissue. The trunk breaches the cuticle and then divides dichotomously into branches (trophomeres) carrying terminal reproductive segments (gonomeres) that form bimastigote zoospores; taxonomy *incertae sedis*
Elphidiidae Family in phylum Foramenifera (Pr-3)
Endictyaceae Family in phylum Bacillariophyta (Pr-18)
Endochytriaceae Family in phylum Chytridiomycota (Pr-35)
Endogenida Order in phylum Ciliophora (Pr-6)
Endonucleoaxoplasthelida Suborder in phylum Actinopoda (Pr-31)
Endothyracea Superfamily in phylum Foramenifera (Pr-3)
Endothyridae Family in phylum Foramenifera (Pr-3)
Entamoebidae Family in phylum Rhizopoda (Pr-2)
Enteridiaceae Family in phylum Myxomycota (Pr-23)
Enterocystidae Family in phylum Apicomplexa (Pr-7)
Enteromonadidae Family in phylum Archaeprotista (Pr-1)
Entodiniomorphida Order in phylum Ciliophora (Pr-6)
Entodiniomorphina Suborder in phylum Ciliophora (Pr-6)
Entomoneidaceae Family in phylum Bacillariophyta (Pr-18)
Entopylaceae Family in phylum Bacillariophyta (Pr-18)
Entopylales Order in phylum Bacillariophyta (Pr-18)
Eocristellariidae Family in phylum Foramenifera (Pr-3)
Eoglobigerinidae Family in phylum Foramenifera (Pr-3)
Eouvigerinacea Superfamily in phylum Foramenifera (Pr-3)
Eouvigerinidae Family in phylum Foramenifera (Pr-3)
Epistomariidae Family in phylum Foramenifera (Pr-3)
Epistominidae Family in phylum Foramenifera (Pr-3)
Epithemiaceae Family in phylum Bacillariophyta (Pr-18)
Epithemiales Order in phylum Bacillariophyta (Pr-18)
Eponididae Family in phylum Foramenifera (Pr-3)
Eremosphaeraceae Family in phylum Chlorophyta (Pr-28)

Erythropeltidaceae Family in phylum Rhodophyta (Pr-33)
Ethmodiscaceae Family in phylum Bacillariophyta (Pr-18)
Ethmodiscales Order in phylum Bacillariophyta (Pr-18)
Eucomonymphidae Family in phylum Archaeprotista (Pr-1)
Euglenales Order in phylum Euglenida (Pr-12)
Euglenamorphales Order in phylum Euglenida (Pr-12)
Euglenida Pr-12: Phylum of mastigotes with one or two anterior undulipodia. Unilateral hairs present on the emergent portion of the locomotory undulipodium and paramylon as storage material. Most euglenids are freshwater or soil phagotrophs or osmotrophs. Approximately one-third are photosynthetic with plastids that contain chlorophylls *a* and *b* in which photosynthesis supplements heterotrophy; no fully autotrophic species known. Stigma is outside the chloroplasts. Many have a flexible pellicle and move by metaboly. Cells, solitary or colonial, display a characteristic type of closed mitosis
Euglenids See *Euglenida* (Pr-12)
Euglenophyceae Single class in phylum Euglenida (Pr-12)
Euglyphidae Family in phylum Rhizopoda (Pr-2)
Eugregarinida Order in phylum Apicomplexa (Pr-7)
Eunotiaceae Family in phylum Bacillariophyta (Pr-18)
Eunotiales Order in phylum Bacillariophyta (Pr-18)
Eunotiophycidae Subclass in phylum Bacillariophyta (Pr-18)
Euplotida Order in phylum Ciliophora (Pr-6)
Euplotina Suborder in phylum Ciliophora (Pr-6)
Eupodiscaceae Family in phylum Bacillariophyta (Pr-18)
Eupodiscales Order in phylum Bacillariophyta (Pr-18)
Eustigmataceae Family in phylum Eustigmatophyta (Pr-27)
Eustigmatales Single order in phylum Eustigmatophyta (Pr-27)
Eustigmatophyceae Single class in phylum Eustigmatophyta (Pr-27)
Eustigmatophyceae Single class in phylum Eustigmatophyta (Pr-27)
Eustigmatophyta Pr-27: Phylum of mastigote algae that form zoospores with prominent red eyespot at the extreme anterior end of their single anteriorly inserted undulipodium. Reproducing by autospores or zoospores; no sexuality known. Plastids contain chlorophyll *a*, beta-carotene, and violoxanthin
Eustigmatophytes See *Eustigmatophyta* (Pr-27)
Eutreptiales Order in phylum Euglenida (Pr-12)
Evaginogenida Order in phylum Ciliophora (Pr-6)
Exoaxoplastidiata Superfamily in phylum Actinopoda (Pr-31)
Exocryptoaxoplastidiata Superfamily in phylum Actinopoda (Pr-31)
Exogemmida Order in phylum Ciliophora (Pr-6)
Exogenida Order in phylum Ciliophora (Pr-6)
Exonucleoaxoplasthelida Suborder in phylum Actinopoda (Pr-31)

Fabesporidae Family in phylum Choanomastigota (Pr-36)
Fabulariidae Family in phylum Foramenifera (Pr-3)
Favusellidae Family in phylum Foramenifera (Pr-3)
Filosea Class in phylum Rhizopoda (Pr-2)
Fischerinidae Family in phylum Foramenifera (Pr-3)

Flabellina Suborder in phylum Rhizopoda (Pr-2)
Flabellulidae Family in phylum Rhizopoda (Pr-2)
Florideophycidae Subclass in phylum Rhodophyta (Pr-33)
Fonticulidae Family in phylum Rhizopoda (Pr-2), but with uncertain affinity
Foraminiferea Class in phylum Foramenifera (Pr-3)
Fragilariaceae Family in phylum Bacillariophyta (Pr-18)
Fragilariales Order in phylum Bacillariophyta (Pr-18)
Fragilariophyceae Class in phylum Bacillariophyta (Pr-18)
Fragilariophycidae Subclass in phylum Bacillariophyta (Pr-18)
Frontoniina Suborder in phylum Ciliophora (Pr-6)
Fucaceae Family in phylum Phaeophyta (Pr-17)
Fucales Order in phylum Phaeophyta (Pr-17)
Fursenkoinacea Superfamily in phylum Foramenifera (Pr-3)
Fursenkoinidae Family in phylum Foramenifera (Pr-3)
Fusulinacea Superfamily in phylum Foramenifera (Pr-3)
Fusulinida Order in phylum Foramenifera (Pr-3)
Fusulinidae Family in phylum Foramenifera (Pr-3)

Ganymedidae Family in phylum Apicomplexa (Pr-7)
Gavelinellidae Family in phylum Foramenifera (Pr-3)
Geinitzinacea Superfamily in phylum Foramenifera (Pr-3)
Geinitzinidae Family in phylum Foramenifera (Pr-3)
Gelidiales Order in phylum Rhodophyta (Pr-33)
Giardiinae Family in phylum Archaeprotista (Pr-1)
Gigartaconidae Family in phylum Actinopoda
Gigartinales Order in phylum Rhodophyta (Pr-31)
Giraudiaceae Family in phylum Phaeophyta (Pr-33)
Glabratellacea Superfamily in phylum Foramenifera (Pr-17)
Glabratellidae Family in phylum Foramenifera (Pr-3)
Glandulinidae Family in phylum Foramenifera (Pr-3)
Glaucocystaceae Family in phylum Glaucocystophyta
Glaucocystales Order in phylum Glaucocystophyta
Glaucocystophyceae Single class in phylum Glaucocystophyta
Glaucocystophyta Phylum of miscellaneous blue-green nucleated algae (not presented in this volume). Photosynthetic, freshwater organisms containing cyanelles, intracellular organelles interpreted to be modified cyanobacterial symbionts (with chlorophyll *a* and phycobiliproteins) that retain remnants of cell walls
Glaucocystophytes Cyanelle-containing algae. Informal name of organisms in the phylum Glaucocystophyta (not presented in this volume). The phylum Glaucocystophyta comprises a small group of unicellular mastigotes. They live photoautotrophically with the aid of endosymbionts, named cyanelles. All members are freshwater organisms living in the plankton or benthos of lakes, ponds, or ditches
Glaucophyta See *Glaucocystophyta*
Glaucosphaeraceae Family in phylum Glaucocystophyta

Globanomalinidae Family in phylum Foramenifera (Pr-3)
Globigerinacea Superfamily in phylum Foramenifera (Pr-3)
Globigerinelloididae Family in phylum Foramenifera (Pr-3)
Globigerinida Order in phylum Foramenifera (Pr-3)
Globigerinidae Family in phylum Foramenifera (Pr-3)
Globigerinitidae Family in phylum Foramenifera (Pr-3)
Globorotaliacea Superfamily in phylum Foramenifera (Pr-3)
Globorotaliidae Family in phylum Foramenifera (Pr-3)
Globorotalitidae Family in phylum Foramenifera (Pr-3)
Globotextulariidae Family in phylum Foramenifera (Pr-3)
Globotruncanacea Superfamily in phylum Foramenifera (Pr-3)
Globotruncanidae Family in phylum Foramenifera (Pr-3)
Gloeobotrydaceae Family in phylum Xanthophyta (Pr-16)
Gloeochaetaceae Family in phylum Glaucocystophyta
Gloeochaetales Order in phylum Glaucocystophyta
Gloeodiniaceae Family in phylum Dinomastigota (Pr-5)
Gloeopodiaceae Family in phylum Xanthophyta (Pr-16)
Glugeidae Family in phylum Microspora (F-1)
Gomphonemataceae Family in phylum Bacillariophyta (Pr-18)
Gonapodyaceae Family in phylum Chytridiomycota (Pr-35)
Gonyaulacaceae Family in phylum Dinomastigota (Pr-5)
Gonyaulacales Order in phylum Dinomastigota (Pr-5)
Gossleriellaceae Family in phylum Bacillariophyta (Pr-18)
Granuloreticulosa (Foraminifera) Pr-3: Phylum of marine protists having granular reticulopods that form anastomosing networks with distinctive two-way streaming. Mostly enclosed by calcareous or agglutinated tests characteristic of the major class: Foraminiferea. Naked forms in class Athalamea. Possesses single, dimorphic, or many nuclei. Many contain photosynthetic symbionts. Some have complex sexual life cycles, others undulipodiated gametes; diploid asexual reproducing phase (agamont) alternating with a haploid sexually reproducing phase (gamont), or with only one phase (apogamic or apoagamic). Extremely useful as stratigraphic markers because of abundance and diversity in Paleozoic and more recent marine sediment
Green seaweeds Class in phylum Chlorophyta (Pr-28) containing predominantly marine, sessile algae with multicellular or coenocytic, walled growing cells. Heteromorphic life history with multicellular or reduced diploid sporophyte. Mitosis with centrioles, cytokinesis by an in-growing cleavage furrow. Fertilization followed by zygote formation occurs when bi- or quadrimastigote isogamous cells fuse
Gregarine apicomplexan Informal name of organisms in the phylum Apicomplexa (Pr-7), class Gregarinia. Heterotrophic microbes, spore-forming symbiotrophs of animals. Although the gregarines apparently do not cause serious damage to their invertebrate hosts, other apicomplexans are often pathogens
Gregarinia Class in phylum Apicomplexa (Pr-7)
Gregarinidae Family in phylum Apicomplexa (Pr-7)
Gurleyidae Family in phylum Microspora (F-1)

Guttulinopsidae Family in phylum Rhizopoda (Pr-2)

Gymnamoebia Subclass in phylum Rhizopoda (Pr-2)

Gymnidae Family in phylum Actinopoda (Pr-31)

Gymnodiniaceae Family in phylum Dinomastigota (Pr-5)

Gymnodiniales Order in phylum Dinomastigota (Pr-5)

Haddoniidae Family in phylum Foramenifera (Pr-3)

Haematococcaceae Family in phylum Chlorophyta (Pr-28)

Haemogregarinidae Family in phylum Apicomplexa (Pr-7)

Haemosporida Order in phylum Apicomplexa (Pr-7)

Haliommatidae Family in phylum Actinopoda (Pr-31)

Hantkeninacea Superfamily in phylum Foramenifera (Pr-3)

Hantkeninidae Family in phylum Foramenifera (Pr-3)

Haplophragmiidae Family in phylum Foramenifera (Pr-3)

Haplophragmoididae Family in phylum Foramenifera (Pr-3)

Haplosporea Class in phylum Haplosporidia (Pr-29)

Haplosporida Order in phylum Haplosporidia (Pr-29)

Haplosporidia (Haplospora) Pr-29: Phylum of unicellular amastigote symbiotrophs including necrotrophs (pathogens), primarily histozoic or coelozoic in marine animals. Form plasmodia with dense organelles called haplosporosomes in host tissue and produce unicellular, typically uninucleate propagules ("spores" that lack polar capsules and polar filaments)

Haplosporidian Informal name of organisms in the phylum Haplosporidia (Pr-29)

Haplosporidiidae Family in phylum Haplosporidia (Pr-29)

Haplozoaceae Family in phylum Dinomastigota (Pr-5)

Haptoria Subclass in phylum Ciliophora (Pr-6)

Haptorida Order in phylum Ciliophora (Pr-6)

Harpochytriaceae Family in phylum Chytridiomycota (Pr-35)

Hartmannellidae Family in phylum Rhizopoda (Pr-2)

Hastigerinidae Family in phylum Foramenifera (Pr-3)

Hedbergellidae Family in phylum Foramenifera (Pr-3)

Hedraiophryidae Family in phylum Actinopoda (Pr-31)

Heleochloridaceae Family in phylum Chlorophyta (Pr-28)

Heliozoa Class in phylum Actinopoda (Pr-31). Axopods, used for locomotion or predaceous feeding, spherical, free-living, heterotrophic, primarily freshwater unicells that lack central capsules and which radiate from naked, siliceous-coated bodies. Some species also produce pseudopods, filopods, or undulipodia. Autogamy reported in two species

Heliozoan Informal name of organisms belonging to the phylum Actinopoda (Pr-31), class Heliozoa. Small group of predaceous heterotrophic actinopods.

Hematozoa Class in phylum Apicomplexa (Pr-7)

Hematozoan apicomplexan Informal name of organisms belonging to class Hematozoa in the phylum Apicomplexa (Pr-7). Heterotrophic microbes, spore-forming symbiotrophs of animals. Many of them cause serious and even fatal diseases in their vertebrate and invertebrate hosts. The haemosporidia include Plasmodium, which is an agent of malaria

Hemiaulaceae Family in phylum Bacillariophyta (Pr-18)
Hemiaulales Order in phylum Bacillariophyta (Pr-18)
Hemidiscaceae Family in phylum Bacillariophyta (Pr-18)
Hemisphaeramminidae Family in phylum Foramenifera (Pr-3)
Hesseidae Family in phylum Microspora (F-1)
Heterochordariaceae Family in phylum Phaeophyta (Pr-17)
Heterodendraceae Family in phylum Xanthophyta (Pr-16)
Heterodiniaceae Family in phylum Dinomastigota (Pr-5)
Heterogloeaceae Family in phylum Xanthophyta (Pr-16)
Heterogloeales Order in phylum Xanthophyta (Pr-16)
Heterohelicacea Superfamily in phylum Foramenifera (Pr-3)
Heterohelicidae Family in phylum Foramenifera (Pr-3)
Heteronematales Order in phylum Euglenida (Pr-12)
Heteropediaceae Family in phylum Xanthophyta (Pr-16)
Heterophryidae Family in phylum Actinopoda (Pr-31)
Heterotrichia Subclass in phylum Ciliophora (Pr-6)
Heterotrichida Order in phylum Ciliophora (Pr-6)
Heterotrichina Suborder in phylum Ciliophora (Pr-6)
Hexacapsulidae Family in phylum Myxospora (A-2)
Hexactinomyxidae Family in phylum Myxospora (A-2)
Hexalaspidae Family in phylum Actinopoda (Pr-31)
Hexamitidae Family in phylum Archaeprotista (Pr-1)
Hexamitinae Subfamily in phylum Archaeprotista (Pr-1)
Hildenbrandiales Order in phylum Rhodophyta (Pr-33)
Himanthaliaceae Family in phylum Phaeophyta (Pr-17)
Hippocrepinellidae Family in phylum Foramenifera (Pr-3)
Hirmocystidae Family in phylum Apicomplexa (Pr-7)
Holacanthida Order in phylum Actinopoda (Pr-31)
Holomastigotidae Family in phylum Archaeprotista (Pr-1)
Homotrematidae Family in phylum Foramenifera (Pr-3)
Hoplonymphidae Family in phylum Archaeprotista (Pr-1)
Hormosinacea Superfamily in phylum Foramenifera (Pr-3)
Hormosinidae Family in phylum Foramenifera (Pr-3)
Hormosiraceae Family in phylum Phaeophyta (Pr-17)
Hormotilaceae Family in phylum Chlorophyta (Pr-28)
Hospitellidae Family in phylum Foramenifera (Pr-3)
Hyalodiscacea Family in phylum Bacillariophyta (Pr-18)
Hydrodictyaceae Family in phylum Chlorophyta (Pr-28)
Hydruraceae Family in phylum Chrysophyta (Pr-15)
Hymenostomatida Order in phylum Ciliophora (Pr-6)

Hymenostomia Subclass in phylum Ciliophora (Pr-6)

Hyperamminacea Superfamily in phylum Foramenifera (Pr-3)

Hyperamminidae Family in phylum Foramenifera (Pr-3)

Hyperamminoididae Family in phylum Foramenifera (Pr-3)

Hypermastigotes Order in phylum Archaeprotista (Pr-1)

Hyphochytriaceae Family in phylum Hyphochytriomycota (Pr-14)

Hyphochytrids Class in phylum Hyphochytriomycota (Pr-14)

Hyphochytriomycetes See *Hyphochytrids*

Hyphochytriomycota Pr-14: Phylum of osmotrophic or necrotrophic soil and water organisms that reproduce via zoospores. Zoospores, motile by a single, anteriorly directed undulipodium with mastigonemes, form from a multinucleate thallus by reduction of cleavage vesicles. Growth as a heterotrophic thallus follows germination of an encysted zoospore. Autogamy reported in one species, *Anisolpidium ectocarpii.* "Funguslike protoctists."

Hypocomatina Suborder in phylum Ciliophora (Pr-6)

Hypotrichia Subclass in phylum Ciliophora (Pr-6)

Involutinida Order in phylum Foramenifera (Pr-3)

Involutinidae Family in phylum Foramenifera (Pr-3)

Ishigeaceae Family in phylum Phaeophyta (Pr-17)

Islandiellidae Family in phylum Foramenifera (Pr-3)

Isochrysidales Order in phylum Haptomonada, Pr-25

Janiae Tribe in phylum Rhodophyta (Pr-33)

Joeniidae Family in phylum Archaeprotista (Pr-1)

Karreriddae Family in phylum Foramenifera (Pr-3)

Karyoblastea Class of giant, free-living, microaerophilic, multinucleate, algivorous, freshwater amebas (Pr-2). Monospecific: *Pelomyxa palustris.* Each ameba harbors three different morphotypes of endosymbiotic bacteria in proportions that change with conditions; at least one type is methanogenic and a second, perinuclear. Lack mitochondria and possibly Golgi bodies; have nonmotile surface projections that seem to be extreme variations on standard axonemal morphology

Karyorelictea Class in phylum Ciliophora (Pr-6)

Keramosphaeridae Family in phylum Foramenifera (Pr-3)

Kinetoplastida Pr-11: Phylum of free-living or symbiotrophic mastigotes with one or two undulipodia associated with a conspicuous intracellular stainable structure: the kinetoplast. Masses of small and large circular DNA including mitochondrial DNA sequences, within a single differentiating and dedifferentiating mitochondrion renders the kinetoplast nearly as large and just as stainable as nucleic acids of the nucleus

Klossiidae Family in phylum Apicomplexa (Pr-7)

Kofoidiidae Family of hypermastigotes in phylum Archaeprotista (Pr-1)

Kofoidiniaceae Family in phylum Dinomastigota (Pr-5)

Kolkwitziellaceae Family in phylum Dinomastigota (Pr-5)

Kolkwitziellales Order in phylum Dinomastigota (Pr-5)

Komokiacea Superfamily in phylum Foramenifera (Pr-3)
Komokiidae Family in phylum Foramenifera (Pr-3)
Kudoidae Family in phylum Myxospora (A-2)
Kybotiaceae Family in phylum Chrysophyta (Pr-15)

Labyrinthomycotes Informal name of slime net organisms in the phylum Labyrinthulata (Pr-19). These unique organisms produce globose or colonial structures associated with wall-less ectoplasmic networks, which absorb nutrients and attach the cells to surfaces
Labyrinthulea Single class in phylum Labyrinthulata (Pr-19)
Labyrinthulida Single order in phylum Labyrinthulata (Pr-19)
Labyrinthulidae Family in phylum Labyrinthulata (Pr-19)
Labyrinthulids Pr-19: Heterotrophic protoctists that produce an extracellular matrix (a wall-less ectoplasmic network), called a slime network, which absorbs nutrients and attaches the cells within it to surfaces. Ectoplasmic networks are devoid of cytoplasmic constituents; they are produced by cell organelles called sagenogens. Cells divide within the network; in some genera, cells show gliding motility. Reproduction by break up of the net or by propagules (heterokont bimastigote zoospores). The slime net in thraustochytrids is reduced, and the extracellular material is hardened into a structure that resembles superficially a chytrid thallus. Meiotic sexuality observed in at least one species. Saprotrophic to weakly symbiotrophic. Found in marine and estuarine environments
Labyrinthulata See *Labyrinthulids*
Lacosteinidae Family in phylum Foramenifera (Pr-3)
Lagenida Order in phylum Foramenifera (Pr-3)
Lagenidiaceae Family of phylum Oomycota (Pr-21)
Lagynidae Family in phylum Foramenifera (Pr-3)
Lagyniina Suborder in phylum Foramenifera (Pr-3)
Laminariaceae Family in phylum Phaeophyta (Pr-17)
Laminariales Order in phylum Phaeophyta (Pr-17)
Lankesterellidae Family in phylum Apicomplexa (Pr-7)
Larcoidae Family in phylum Actinopoda (Pr-31)
Lasiodiscidae Family in phylum Foramenifera (Pr-3)
Lauderiaceae Family in phylum Bacillariophyta (Pr-18)
Leathesiaceae Family in phylum Phaeophyta (Pr-17)
Lecudinidae Family in phylum Apicomplexa (Pr-7)
Legerellidae Family in phylum Apicomplexa (Pr-7)
Lepidocyclinidae Family in phylum Foramenifera (Pr-3)
Lepidorbitoididae Family in phylum Foramenifera (Pr-3)
Leptocylindraceae Family in phylum Bacillariophyta (Pr-18)
Leptocylindrales Order in phylum Bacillariophyta (Pr-18)
Leptodiscaceae Family in phylum Dinomastigota (Pr-5)
Leptolegniellaceae Family of phylum Oomycota (Pr-21)
Leptomitaceae Family of phylum Oomycota (Pr-21)
Leptomitales Order of phylum Oomycota (Pr-21)
Lessoniaceae Family in phylum Phaeophyta (Pr-17)

Liceaceae Family in phylum Myxomycota (Pr-23)
Liceales Order in phylum Myxomycota (Pr-23)
Licmophoraceae Family in phylum Bacillariophyta (Pr-18)
Licnophorida Order in phylum Ciliophora (Pr-6)
Linderinidae Family in phylum Foramenifera (Pr-3)
Lithodesmiaceae Family in phylum Bacillariophyta (Pr-18)
Lithodesmiales Order in phylum Bacillariophyta (Pr-18)
Lithodesmiophycidae Subclass in phylum Bacillariophyta (Pr-18)
Lithophylleae Tribe in phylum Rhodophyta (Pr-33)
Lithophylloideae Subfamily in phylum Rhodophyta (Pr-33)
Lithopteridae Family in phylum Actinopoda (Pr-31)
Lithothamnieae Tribe in phylum Rhodophyta (Pr-33)
Lithothamnoideae Subfamily in phylum Rhodophyta (Pr-33)
Lithotricheae Tribe in phylum Rhodophyta (Pr-33)
Litostomatea Class in phylum Ciliophora (Pr-6)
Lituolacea Superfamily in phylum Foramenifera (Pr-3)
Lituolidae Family in phylum Foramenifera (Pr-3)
Lituoliporidae Family in phylum Foramenifera (Pr-3)
Lituotubidae Family in phylum Foramenifera (Pr-3)
Lobosea Class in phylum Rhizopoda (Pr-2)
Loeblichiidae Family in phylum Foramenifera (Pr-3)
Loftusiacea Superfamily in phylum Foramenifera (Pr-3)
Loftusiidae Family in phylum Foramenifera (Pr-3)
Lophodiniaceae Family in phylum Dinomastigota (Pr-5)
Lophomonadidae Family in phylum Archaeprotista (Pr-1)
Loxodida Order in phylum Ciliophora (Pr-6)
Loxostomatidae Family in phylum Foramenifera (Pr-3)
Lyrellaceae Family in phylum Bacillariophyta (Pr-18)
Lyrellales Order in phylum Bacillariophyta (Pr-18)

Mackinnoniidae Family in phylum Apicomplexa (Pr-7)
Mallodendraceae Family in phylum Xanthophyta (Pr-16)
Mallomonadaceae Family in phylum Chrysophyta (Pr-15)
Mallomonadales Order in phylum Chrysophyta (Pr-15)
Mamiellaceae Family in phylum Chlorophyta (Pr-28)
Mamiellales Order in phylum Chlorophyta (Pr-28)
Mantonellidae Family in phylum Apicomplexa (Pr-7)
Marteiliidea Class in phylum Paramyxa (Pr-30)
Mastogloiaceae Family in phylum Bacillariophyta (Pr-18)
Mastogloiales Order in phylum Bacillariophyta (Pr-18)
Mastophoreae Tribe in phylum Rhodophyta (Pr-33)
Mastophoroideae Subfamily in phylum Rhodophyta (Pr-33)
Maylisoriidae Family in phylum Foramenifera (Pr-3)
Meandropsinidae Family in phylum Foramenifera (Pr-3)

Medusettidae Family in phylum Actinopoda (Pr-31)

Melobesioideae Subfamily in phylum Rhodophyta (Pr-33)

Melonidae Family in phylum Foramenifera (Pr-3)

Melosiraceae Family in phylum Bacillariophyta (Pr-18)

Melosirales Order in phylum Bacillariophyta (Pr-18)

Merocystidae Family in phylum Apicomplexa (Pr-7)

Merogregarinidae Family in phylum Apicomplexa (Pr-7)

Mesostigmataceae Family in phylum Chlorophyta (Pr-28)

Mesotaeniaceae Family in phylum Conjugaphyta (Pr-32)

Metchnikovellida Order in phylum Microspora (F-1)

Metchnikovellidae Family in phylum Microspora (F-1)

Micractiniaceae Family in phylum Chlorophyta (Pr-28)

Micromonadaceae Family in phylum Chlorophyta (Pr-28)

Microspora F-1: Phylum of minute unicellular fungal symbiotrophs causing single-cell tumors in a vast array of insects and other animals. All lack mitochondria. Propagules are spores that produce a polar tube deployed in the inoculation of the host with no damage to the host cell membrane. Penetration of the animal tissue is through this unprecedented inoculation device. Sexual fusion is reported in some species

Microsporea Class in phylum Microspora (F-1)

Microsporida Order in phylum Microspora (F-1)

Microthamniales Order in phylum Chlorophyta (Pr-28) of exclusively freshwater chlorophytes including several common phycobionts of lichens. Occur as solitary cell packets or branched filaments. Propagate by autospores, aplanospores (especially symbiotic taxa), or naked, bimastigote zoospores. Sexuality is unknown

Microthoracida Order in phylum Ciliophora (Pr-6)

Miliolacea Superfamily in phylum Foramenifera (Pr-3)

Miliolida Order in phylum Foramenifera (Pr-3)

Miliolidae Family in phylum Foramenifera (Pr-3)

Milioliporidae Family in phylum Foramenifera (Pr-3)

Millettiidae Family in phylum Foramenifera (Pr-3)

Minisporida Order in phylum Microspora (F-1)

Miogypsinidae Family in phylum Foramenifera (Pr-3)

Mischococcaceae Family in phylum Xanthophyta (Pr-16)

Mischococcales Order in phylum Xanthophyta (Pr-16)

Mississippinidae Family in phylum Foramenifera (Pr-3)

Mobilida Order in phylum Ciliophora (Pr-6)

Monoblepharidaceae Family in phylum Chytridiomycota (Pr-35)

Monoblepharidales Order in phylum Chytridiomycota (Pr-35)

Monocercomonae Family in phylum Archaeprotista (Pr-1)

Monocystidae Family in phylum Apicomplexa (Pr-7)

Monodopsidaceae Family in phylum Eustigmatophyta (Pr-27)

Monoductidae Family in phylum Apicomplexa (Pr-7)

Monostromataceae Family in phylum Chlorophyta (Pr-28)

Moravamminacea Superfamily in phylum Foramenifera (Pr-3)
Moravamminidae Family in phylum Foramenifera (Pr-3)
Mrazekiidae Family in phylum Microspora (F-1)
Multivalvulida Order in animal phylum Myxospora (A-2)
Myrionemataceae Family in phylum Phaeophyta (Pr-17)
Myriosporidae Family in phylum Apicomplexa (Pr-7)
Myriotrichiaceae Family in phylum Phaeophyta (Pr-17)
Myxidiidae Family in phylum Myxospora (A-2)
Myxobolidae Family in phylum Myxospora (A-2)
Myxochloridaceae Family in phylum Xanthophyta (Pr-16)
Myxochrysidaceae Family in phylum Chrysophyta (Pr-15)
Myxomycota (Pr-23) Phylum of plasmodial slime molds. Phagotrophic bacterivorous organisms form plasmodium. Propagation is by spores shed by sporophore (stalked spore-bearing structure). Spores germinate to form ameboid (myxameba) or undulipodiated cells (mastigote swarmers); each type is a potential gamete or can develop into plasmodia, in many species by synchronous division of plasmodial nuclei. Plasmodium has a reversible type of protoplasmic streaming and ability to increase in size by coalescing with other compatible plasmodia
Myxomycote Myxomycetes. Informal name of organisms belonging to the phylum Myxomycota (Pr-23), Myxomycota
Myxosporan Member of animal phylum Myxospora (A-2), a vast assemblage of microscopic, heterotrophic organisms that form intricate multicellular spores with nematocyst-like structures called polar capsules. Newly reassigned to Kingdom Animalia from Protoctista, they are necrotrophs of invertebrates and vertebrates such as fish, amphibians, and reptiles
Myxosporea Class in phylum Myxospora (A-2)
Myxosporean Adjective referring to organisms in the phylum
Myxospora A-2: Phylum of obligate symbiotrophs that produce multicellular spores, deploy polar capsules that penetrate and attach to animal tissue (for example, oligochaetes, sipunculids, fish, and other vertebrates). Ameboid cells are released through valves. Symbiotrophic (including necrotrophic) heterotrophy is by ameboid cells or plasmodia. Plasmodia are formed by buds (internal or external) or binary or multiple karyokinesis. The two classes are Myxosporea and Actinosporea
Myxosporans Informal name of organisms belonging to the phylum Myxospora (A-2). The phylum Myxospora (formerly Myxospora (A-2)) is a vast assemblage of microscopic heterotrophic protoctists that form intricate multicellular spores with nematocyst-like structures called polar capsules. They are symbiotrophs of invertebrates and vertebrates such as fish, amphibians, and reptiles

Naegeliellaceae Family in phylum Chrysophyta (Pr-15)
Nassellarida Order in phylum Actinopoda (Pr-31)
Nassophorea Class in phylum Ciliophora (Pr-6)
Nassophoria Subclass in phylum Ciliophora (Pr-6)
Nassulida Order in phylum Ciliophora (Pr-6)
Nassulina Suborder in phylum Ciliophora (Pr-6)
Nautococcaceae Family in phylum Chlorophyta (Pr-28)
Naviculaceae Family in phylum Bacillariophyta (Pr-18)

Naviculales Order in phylum Bacillariophyta (Pr-18)
Naviculineae Suborder in phylum Bacillariophyta (Pr-18)
Neidiaceae Family in phylum Bacillariophyta (Pr-18)
Neidiineae Suborder in phylum Bacillariophyta (Pr-18)
Nemaliales Order in phylum Rhodophyta (Pr-33)
Nematochrysidaceae Family in phylum Chrysophyta (Pr-15)
Neocallimasticaceae Family in phylum Chytridiomycota (Pr-35)
Neogoniolithoneae Tribe in phylum Rhodophyta (Pr-33)
Neogregarinida Order in phylum Apicomplexa (Pr-7)
Neonemataceae Family in phylum Xanthophyta (Pr-16)
Neoschwagerinidae Family in phylum Foramenifera (Pr-3)
Nephroselmidaceae Family in phylum Chlorophyta (Pr-28)
Nezzazatidae Family in phylum Foramenifera (Pr-3)
Nivalidae Family in phylum Actinopoda (Pr-31)
Noctilucaceae Family in phylum Dinomastigota (Pr-5)
Noctilucales Order in phylum Dinomastigota (Pr-5)
Nodosariacea Superfamily in phylum Foramenifera (Pr-3)
Nodosariidae Family in phylum Foramenifera (Pr-3)
Nodosinellacea Superfamily in phylum Foramenifera (Pr-3)
Nodosinellidae Family in phylum Foramenifera (Pr-3)
Nonionacea Superfamily in phylum Foramenifera (Pr-3)
Nonionidae Family in phylum Foramenifera (Pr-3)
Nosematidae Family in phylum Microspora (F-1)
Notheiaceae Family in phylum Phaeophyta (Pr-17)
Notodendrodidae Family in phylum Foramenifera (Pr-3)
Nouriidae Family in phylum Foramenifera (Pr-3)
Nubeculariidae Family in phylum Foramenifera (Pr-3)
Nummilitidae Family in phylum Foramenifera (Pr-3)
Nummulitacea Superfamily in phylum Foramenifera (Pr-3)

Oberhauserellidae Family in phylum Foramenifera (Pr-3)
Ochromonadaceae Family in phylum Chrysophyta (Pr-15)
Ochromonadales Order in phylum Chrysophyta (Pr-15)
Odontostomatida Order in phylum Ciliophora (Pr-6)
Oedogoniomycetaceae Family in phylum Chytridiomycota (Pr-35)
Oligohymenophorea Class in phylum Ciliophora (Pr-6)
Oligotrichida Order in phylum Ciliophora (Pr-6)
Olpidiaceae Family in phylum Chytridiomycota (Pr-35)
Oocystaceae Family in phylum Chlorophyta (Pr-28)
Oodiniaceae Family in phylum Dinomastigota (Pr-5)
Oomycetes Informal name of water molds and other protoctists that superficially resemble fungi and belong to the phylum Oomycota (Pr-21)
Oomycota Pr-21: Phylum of conjugating anisogamontous protoctists. Heterotrophic or

osmotrophic in freshwater environments, or symbiotrophic on plants. Uninucleate or coenocytic with haplomitotic ploidy cycle. Undulipodiated heterokonts are zoospores, not gametes; sexuality is by conjugation of nonmotile differentiated (male and female) hyphae

Oomycotes See *Oomycetes*

Opalinata Class in the phylum Archaeprotista (Pr-1). Large heterotrophic protists, mastigotes motile by numerous surface undulipodia. Symbiotrophic in the digestive system (cloaca) of poikilotherm vertebrates (Pr-37), mostly anuran amphibians, for example, frogs. Opalinid cells contain two or many homokaryotic nuclei, lack cytostomes, and display fertilization and cyst formation

Opalinida Order in phylum Archaeprotista (Pr-1)

Opalinidae Family in phylum Archaeprotista (Pr-1)

Opalinids See *Opalinata*

Ophiocytaceae Family in phylum Xanthophyta (Pr-16)

Ophryoglenina Suborder in phylum Ciliophora (Pr-6)

Ophthalmidiidae Family in phylum Foramenifera (Pr-3)

Orbitoclypeidae Family in phylum Foramenifera (Pr-3)

Orbitoidacea Superfamily in phylum Foramenifera (Pr-3)

Orbitoididae Family in phylum Foramenifera (Pr-3)

Orbitolinacea Superfamily in phylum Foramenifera (Pr-3)

Orbitolinidae Family in phylum Foramenifera (Pr-3)

Orbitopsellidae Family in phylum Foramenifera (Pr-3)

Oridorsalidae Family in phylum Foramenifera (Pr-3)

Ormieractinomyxidae Family in phylum Myxospora (A-2)

Ortholineidae Family in phylum Myxospora (A-2)

Osangulariidae Family in phylum Foramenifera (Pr-3)

Ostreobiaceae Family in phylum Chlorophyta (Pr-28)

Ostreopsidaceae Family in phylum Dinomastigota (Pr-5)

Oxinoxisidae Family in phylum Foramenifera (Pr-3)

Oxyphysaceae Family in phylum Dinomastigota (Pr-5)

Oxyrrhinaceae Family in phylum Dinomastigota (Pr-5)

Oxyrrhinales Order in phylum Dinomastigota (Pr-5)

Oxytoxaceae Family in phylum Dinomastigota (Pr-5)

Ozawainellidae Family in phylum Foramenifera (Pr-3)

Pachyphloiidae Family in phylum Foramenifera (Pr-3)

Palaeospiroplectamminidae Family in phylum Foramenifera (Pr-3)

Palaeotextulariacea Superfamily in phylum Foramenifera (Pr-3)

Palaeotextulariidae Family in phylum Foramenifera (Pr-3)

Palmariales Order in phylum Rhodophyta (Pr-33)

Palmellopsidaceae Family in phylum Chlorophyta (Pr-28)

Palmodictyaceae Family in phylum Chlorophyta (Pr-28)

Pannellainidae Family in phylum Foramenifera (Pr-3)

Pansporoblastina Suborder in phylum Microspora (F-1)

Parabasalia Class in phylum Archaeprotista (Pr-1). Uninucleate, cells symbiotrophic in animals, containing one or more parabasal bodies (modified Golgi apparatus) usually associated with nuclei and kinetids. Heterotrophic mastigotes with few to hundreds of thousands of undulipodia. All lack mitochondria. Some have microtubular axostyles and distinctive undulating membranes. Includes three orders: trichomonads, polymonads, and hypermastigotes

Parabasalians See *Parabasalia*

Parahymenostomatina Suborder in phylum Ciliophora (Pr-6)

Paraliaceae Family in phylum Bacillariophyta (Pr-18)

Paraliales Order in phylum Bacillariophyta (Pr-18)

Parameciina Suborder in phylum Ciliophora (Pr-6)

Paramoebidae Family in phylum Rhizopoda (Pr-2)

Paramyxa Pr-30: Phylum of amastigote unicellular symbiotrophs of marine animals. Form propagules (spores) consisting of several cells enclosed inside each other arising from a process of internal cleavage or endogenous budding within an ameboid stem cell

Paramyxidea Class in phylum Paramyxa (Pr-30)

Paraphysomonadaceae Family in phylum Chrysophyta (Pr-15)

Parathuramminacea Superfamily in phylum Foramenifera (Pr-3)

Parathuramminidae Family in phylum Foramenifera (Pr-3)

Parathuramminiina Suborder in phylum Foramenifera (Pr-3)

Paratikhinellidae Family in phylum Foramenifera (Pr-3)

Parrelloididae Family in phylum Foramenifera (Pr-3)

Partisaniidae Family in phylum Foramenifera (Pr-3)

Parvicapsulidae Family in phylum Myxospora (A-2)

Patellinidae Family in phylum Foramenifera (Pr-3)

Pavlovales Order in phylum Haptomonada (Pr-25)

Pavoninidae Family in phylum Foramenifera (Pr-3)

Pedinellaceae Family in phylum Chrysophyta (Pr-15)

Pedinellales Order in phylum Chrysophyta (Pr-15)

Pedinellophyceae Class in phylum Chrysophyta (Pr-15)

Pedinomonadalean chlorophytes Adjective referring to green algae (phylum Chlorophyta, Pr-28) in the order Pedinomonadales. The Pedinomonadales with one genus (Pedimonas) and 13 species occupy an isolated position within the Chlorophyta (Pr-28). They consist of small, naked unimastigoted mastigotes. Pedimonas cells are flattened and asymmetric with lateral to subapical ciliary insertion. The undulipodium propels the cell so that it swims "backward."

Pedinomonadales Order in phylum Chlorophyta (Pr-28). Small, naked mastigotes composed of flattened and asymmetric cells with one undulipodium laterally to subapically inserted, a nonfunctional kinetosome, and chloroplasts. Marine or freshwater; two species symbiotic. Sexuality is unknown

Pegidiidae Family in phylum Foramenifera (Pr-3)

Peneroplidae Family in phylum Foramenifera (Pr-3)

Peniculida Order in phylum Ciliophora (Pr-6)

Pentacapsulidae Family in phylum Myxospora (A-2)

Perenosporean oomycotes Family in the phylum Oomycota (Pr-21). Physiologically and morphologically the Oomycota (Pr-21) are fungi. Most species of Oomycotes are freshwater or terrestrial, a few are oligohaline or marine

Pereziidae Family in phylum Microspora (F-1)

Periaxoplastidiata Superfamily in phylum Actinopoda (Pr-31)

Peridiniaceae Family in phylum Dinomastigota (Pr-5)

Peridiniales Order in phylum Dinomastigota (Pr-5)

Peridinoid dinomastigotes Adjective referring to organisms in the phylum Dinomastigota (Pr-5), order Peridiniales. Dinomastigotes are essentially biundulipodiated, photosynthetic or nonphotosynthetic, walled or naked, unicells. At least 30 species of marine dinomastigotes are luminescent. They are one of the most common sources of luminescence in seawater

Peritrichia Subclass in phylum Ciliophora (Pr-6)

Peroniaceae Family in phylum Bacillariophyta (Pr-18)

Peronosporaceae Family of phylum Oomycota (Pr-21)

Peronosporales Order of phylum Oomycota (Pr-21)

Peronosporean Adjective referring to organisms in the phylum Oomycota (Pr-21), order Peronosporales

Peronosporomycetidae Class (subclass) of phylum Oomycota (Pr-21)

Perryaceae Family in phylum Bacillariophyta (Pr-18)

Pfeifferinellidae Family in phylum Apicomplexa (Pr-7)

Pfenderinidae Family in phylum Foramenifera (Pr-3)

Phacodiniida Order in phylum Ciliophora (Pr-6)

Phacotaceae Family in phylum Chlorophyta (Pr-28)

Phaeocalpida Order in phylum Actinopoda (Pr-31)

Phaeoconchia Order in phylum Actinopoda (Pr-31)

Phaeocystida Order in phylum Actinopoda (Pr-31)

Phaeodaria Class in phylum Actinopoda (Pr-31). Large spherical solitary cells with siliceous skeletons consisting of isolated pieces or numerous hollow tubes. Some lack skeletons. Spheres in ectoplasm develop into polynucleated ameboids and eventually lead to bimastigote propagule formation. Marine planktonic radiolaria-like protists

Phaeodarian actinopods Informal name or adjective referring to organisms in the phylum Actinopoda (Pr-31), class Phaeodaria. Species of the phylum Actinopoda (Pr-31) are heterotrophic protoctists; their cells bear long processes called axopods, which develop from specialized structures called axoplasts. Members of the class Phaeodaria sometimes lack skeletons; when present, the skeleton consists of isolated pieces or numerous hollow tubes

Phaeodendrida Order in phylum Actinopoda (Pr-31)

Phaeodermatiaceae Family in phylum Chrysophyta (Pr-15)

Phaeodinidae Family in phylum Actinopoda (Pr-31)

Phaeogromida Order in phylum Actinopoda (Pr-31)

Phaeogymnocellida Order in phylum Actinopoda (Pr-31)

Phaeophyceae Single class in phylum Phaeophyta (Pr-17)

Phaeophyta (Pr-17) See *Brown algae*

Phaeoplacaceae Family in phylum Chrysophyta (Pr-15)

Phaeosacciaceae Family in phylum Chrysophyta (Pr-15)

Phaeosphaerida Order in phylum Actinopoda (Pr-31)

Phaeosphaeridae Family in phylum Actinopoda (Pr-31)

Phaeothamniaceae Family in phylum Chrysophyta (Pr-15)

Phaeothamniales Order in phylum Chrysophyta (Pr-15)

Phaneraxohelida Order in phylum Actinopoda (Pr-31)

Pharactopeltidae Family in phylum Actinopoda (Pr-31)

Pharyngophorida Order in phylum Ciliophora (Pr-6)

Philasterina Suborder in phylum Ciliophora (Pr-6)

Phragmonemataceae Family in phylum Rhodophyta (Pr-33)

Phthanotrochidae Family in phylum Foramenifera (Pr-3)

Phyllacantha Suborder in phylum Actinopoda (Pr-31)

Phyllopharyngea Class in phylum Ciliophora (Pr-6)

Phyllopharyngia Subclass in phylum Ciliophora (Pr-6)

Phyllostauridae Family in phylum Actinopoda (Pr-31)

Phymatolitheae Tribe in phylum Rhodophyta (Pr-33)

Physaraceae Family in phylum Myxomycota (Pr-23)

Physarales Order in phylum Myxomycota (Pr-23)

Physematidae Family in phylum Actinopoda (Pr-31)

Physodermataceae Family in phylum Chytridiomycota (Pr-35)

Phytodiniaceae Family in phylum Dinomastigota (Pr-5)

Phytodiniales Order in phylum Dinomastigota (Pr-5)

Pilisuctorida Order in phylum Ciliophora (Pr-6)

Pinnulariaceae Family in phylum Bacillariophyta (Pr-18)

Piroplasmida Order in phylum Apicomplexa (Pr-7)

Placentulinidae Family in phylum Foramenifera (Pr-3)

Placopsilinidae Family in phylum Foramenifera (Pr-3)

Plactorecurvoididae Family in phylum Foramenifera (Pr-3)

Plagiogrammaceae Family in phylum Bacillariophyta (Pr-18)

Plagiogrammales Order in phylum Bacillariophyta (Pr-18)

Plagiopylia Subclass in phylum Ciliophora (Pr-6)

Plagiopylida Order in phylum Ciliophora (Pr-6)

Plagiotomida Order in phylum Ciliophora (Pr-6)

Plagiotropidaceae Family in phylum Bacillariophyta (Pr-18)

Planomalinacea Superfamily in phylum Foramenifera (Pr-3)

Planomalinidae Family in phylum Foramenifera (Pr-3)

Planorbulinacea Superfamily in phylum Foramenifera (Pr-3)

Planorbulinidae Family in phylum Foramenifera (Pr-3)

Planulinidae Family in phylum Foramenifera (Pr-3)

Planulinoididae Family in phylum Foramenifera (Pr-3)

Myxomycota Pr-23: Phagotrophic bacterivorous, soil, dung, and plant debris organisms that develop from spores borne in sporophores.

Spores germinate to form amebas that develop into plasmodia (rate of karyokinesis exceeds that of cytokinesis). Conspicuous cyclosis in plasmodium. Can form mastigote and ameba stages as well as sclerotia (dry propagules); See *Myxomycota, Protostelida*

Plasmodiophoraceae Single family in phylum Plasmodiophoromycota (Pr-20)

Plasmodiophorales Single order in phylum Plasmodiophoromycota (Pr-20)

Plasmodiophorids Informal name of zoosporic plant symbiotrophs in the phylum Plasmodiophora (Pr-20). The phylum Plasmodiophora comprises endosymbiotrophs of plants, algae, other aquatic protoctists, and fungi. Plasmodiophorids are microscopic, obligate endobiotic necrotrophs of protoctists and plants; the growing form is a multinucleate protoplast lacking walls

Plasmodiophoromycetes Single class in phylum Plasmodiophora (Pr-20)

Plasmodiophora Pr-20: Phylum of soil and freshwater, obligate symbiotrophs (including necrotrophs) of many plants, fungi, and other protoctists. Multinucleate unwalled protoplasts (plasmodia) develop either into sporangia, which produce zoospores with two anteriorly directed whiplash undulipodia, or cystosori, which form resting bodies that are aggregations of thick-walled, uninucleate cells. Cells show cruciform division. Meiosis is thought to occur based on the presence of synaptonemal complexes in some species

Platysporina Suborder in phylum Myxospora (A-2)

Plectoidea Family in phylum Actinopoda (Pr-31)

Pleurochloridaceae Family in phylum Xanthophyta (Pr-16)

Pleurochloridellaceae Family in phylum Xanthophyta (Pr-16)

Pleuronematina Suborder in phylum Ciliophora (Pr-6)

Pleurosigmataceae Family in phylum Bacillariophyta (Pr-18)

Pleurostomatida Order in phylum Ciliophora (Pr-6)

Pleurostomellacea Superfamily in phylum Foramenifera (Pr-3)

Pleurostomellidae Family in phylum Foramenifera (Pr-3)

Podolampaceae Family in phylum Dinomastigota (Pr-5)

Polycystina Class in phylum Actinopoda (Pr-31). Large solitary cells having regularly perforated silica skeletons with radial axopods emerging among fine ramified pseudopods. Mastigote propagules form having two undulipodia, one emergent, and characterized by intracellular strontium sulfate crystal. No sexuality known. Extant organisms are marine; fossil record of polycystines dates from late Proterozoic eon. Their skeletal debris may be the basis of formations of extensive marine silica deposits (radiolarite)

Polycyttaria Family in phylum Actinopoda (Pr-31)

Polykrikaceae Family in phylum Dinomastigota (Pr-5)

Polymonad Order in phylum Archaeprotista (Pr-1)

Polymonadida See *Polymonad.*

Polymorphinidae Family in phylum Foramenifera (Pr-3)

Polypyramidae Family in phylum Actinopoda (Pr-31)

Polysaccamminidae Family in phylum Foramenifera (Pr-3)

Porospathidae Family in phylum Actinopoda (Pr-31)

Porosporidae Family in phylum Apicomplexa (Pr-7)

Porphyridiaceae Family in phylum Rhodophyta (Pr-33)

Porphyridiales Order in phylum Rhodophyta (Pr-33)

Postciliodesmatophora Subphylum in phylum Ciliophora (Pr-6)

Praebuliminidae Family in phylum Foraminifera (Pr-3)

Prasinophyceae Class in phylum Chlorophyta (Pr-28). Motile solitary green algae; their cell bodies and undulipodia are covered by nonmineralized organic scales. Undulipodia originate from a groove, and Golgi apparatus is in a parabasal position. Reproduction by binary division; no sexuality known

Prasinophytes Informal name of green algae (phylum Chlorophyta (Pr-28)) in the class Prasinophyceae. They have motile chlorophytes that are covered on their cell body and flagella by nonmineralized organic scales

Prasiolaceae Family in phylum Chlorophyta (Pr-28)

Prasiolales Order in phylum Chlorophyta (Pr-28). Multicellular flattened algae composed of walled, uninucleate cells. Reproduction primarily by aplanospores. Oogamous sexual reproduction known in a few species in which a bimastigote sperm has one undulipodium absorbed by the egg resulting in the formation of posteriorly unimastigote planozygote. Marine or freshwater

Proaxoplastidiata Superfamily in phylum Actinopoda (Pr-31)

Progonoiaceae Family in phylum Bacillariophyta (Pr-18)

Proheterotrichida Order in phylum Ciliophora (Pr-6)

Propeniculida Order in phylum Ciliophora (Pr-6)

Prorocentraceae Family in phylum Dinomastigota (Pr-5)

Prorocentrales Order in phylum Dinomastigota (Pr-5)

Prorodontida Order in phylum Ciliophora (Pr-6)

Proschkiniaceae Family in phylum Bacillariophyta (Pr-18)

Prostomatea Class in phylum Ciliophora (Pr-6)

Prostomatida Order in phylum Ciliophora (Pr-6)

Proteromonadida Class in phylum Archaeprotista (Pr-1). Small, symbiotrophic, nonmastigonemate, heterokont mastigotes in which a rhizoplast is associated with the Golgi apparatus and nucleus. Reproduction by multiple fission occurs in some species. Form resistant fecal cysts in the intestinal tract of many amphibians, reptiles, and mammals

Proteromonadidae Family in phylum Archaeprotista (Pr-1)

Proteromonads Informal name of mastigotes in phylum Archaeprotista (Pr-1), class Proteromonadida

Proteromonads See *Proteromonadida*

Protocruziida Order in phylum Ciliophora (Pr-6)

Protoodiniaceae Family in phylum Dinomastigota (Pr-5)

Protosiphonaceae Family in phylum Chlorophyta (Pr-28)

Protostelida Class of Myxomycota (Pr-23). Sporocarp consists of a small delicate stalk bearing one to four spores. Growing stage ameboid; may also possess mastigote and plasmodial stages. Life history may be simple with one type of trophic cell, or complex, with several types. Found worldwide in soil, dung, or on living or dead plant parts

Protostelids Informal name of organisms in phylum Myxomycota (Pr-23), class Protostelida. Protostelids are recognized by the sorocarp, consisting of a delicate stalk bearing one to four spores. The sporocarps rest on basal disks

Protosteliidae Family in phylum Myxomycota (Pr-23)

Protostomatida Order in phylum Ciliophora (Pr-6)

Prunoidae Family in phylum Actinopoda (Pr-31)

Prymnesiales Order in phylum Haptomonada (Pr-25)

Prymnesiophyceae Class in phylum Haptomonada (Pr-25)

Prymnesiophyta (Haptomonada) Pr-25: Phylum of yellow-brown algae, many covered with scales of varying degrees of complexity that may be unmineralized or calcified. Includes coccolithophorids, unicellular organisms with calcified plates (coccoliths). Many possess haptonemas, typically a filiform structure associated with the undulipodia. Generally marine; fossil coccolithophorids dating to the Jurassic Period

Prymnesiophytes Yellow-brown algae including coccolithophorids. Informal name of organisms in the phylum Haptomonada (Pr-25). This phylum comprises a group of algae generally found in marine habitats. The most familiar examples are the coccolithophorids, unicellular organisms with an investiture of calcified plates (coccoliths) often with complex ornamentation

Psammettidae Family in phylum Xenophyophora (Pr-4)

Psamminida Class in phylum Xenophyophora (Pr-4)

Psamminidae Family in phylum Xenophyophora (Pr-4)

Psammodiscaceae Family in phylum Bacillariophyta (Pr-18)

Psammosphaeridae Family in phylum Foramenifera (Pr-3)

Pseudoammodiscidae Family in phylum Foramenifera (Pr-3)

Pseudobolivinidae Family in phylum Foramenifera (Pr-3)

Pseudocharaciopsidaceae Family in phylum Eustigmatophyta (Pr-24)

Pseudociliata (Pr-24) Phylum of marine, benthic organisms with 2–16 homokaryotic nuclei and distinctive kineties, kinetosomes connected by a desmose; cytostome (mouth)–cytopharyngeal apparatus supported by complex fibrillar system for active phagocytosis; reproduction by multiple cell division inside cyst. Formerly classified as ciliates, but lack the infraciliary features characteristic of ciliates. Feed on diatoms, small mastigotes, and bacteria.

Pseudoendothyridae Family in phylum Foramenifera (Pr-3)

Pseudohimantidiaceae Family in phylum Bacillariophyta (Pr-18)

Pseudohimantidiales Order in phylum Bacillariophyta (Pr-18)

Pseudolithidae Family in phylum Actinopoda (Pr-31)

Pseudoparrellidae Family in phylum Foramenifera (Pr-3)

Pseudopleistophoridae Family in phylum Microspora (F-1)

Pseudorbitoididae Family in phylum Foramenifera (Pr-3)

Pseudoscourfieldiaceae Family in phylum Chlorophyta (Pr-28)

Pseudoscourfieldiales Order in phylum Chlorophyta (Pr-28)

Pseudotaxidae Family in phylum Foramenifera (Pr-3)

Psycheneidaceae Family in phylum Bacillariophyta (Pr-18)

Psyedoklossiidae Family in phylum Apicomplexa (Pr-7)

Pterospermataceae Family in phylum Chlorophyta (Pr-28)

Ptychocladiacea Superfamily in phylum Foramenifera (Pr-3)

Ptychocladiidae Family in phylum Foramenifera (Pr-3)
Pulleniatinidae Family in phylum Foramenifera (Pr-3)
Punctariaceae Family in phylum Phaeophyta (Pr-17)
Pyramimonadaceae Family in phylum Chlorophyta (Pr-28)
Pyramimonadales Order in phylum Chlorophyta (Pr-28)
Pyrobotryaceae Family in phylum Chlorophyta (Pr-28)
Pyrocystaceae Family in phylum Dinomastigota (Pr-5)
Pyrocystales Order in phylum Dinomastigota (Pr-5)
Pyrophacaceae Family in phylum Dinomastigota (Pr-5)
Pyrsonymphida Class in phylum Archaeprotista (Pr-1). Heterotrophic mastigotes symbiotrophic in the hindguts of wood-eating cockroaches and termites; 4, 8, or 12 undulipodia and an intrinsically motile longitudinally aligned axostyle composed of laterally connected microtubules. Since all lack mitochondria, the group is presumed anaerobic
Pyrsonymphidae Family in phylum Archaeprotista (Pr-1)
Pyrsonymphids See *Pyrsonymphida*
Pythiaceae Family of phylum Oomycota (Pr-21)
Pythiales Order of phylum Oomycota (Pr-21)

Quadrimorphinidae Family in phylum Foramenifera (Pr-3)

Radiolaria Common name of polycystine and phaeodarian marine actinopods (Pr-31); See *Polycystina, Phaeodaria*
Ralfsiaceae Family in phylum Phaeophyta (Pr-17)
Raphidiophryidae Family in phylum Actinopoda (Pr-31)
Raphidophyceae Class in phylum Chlorophyta (Pr-28)
Raphidophyta See *Chloromonads*
Red algae Informal name of organisms in the phylum Rhodophyta (Pr-33). Phylum of primarily marine, photosynthetic protoctists. Life history involves alternation of generations, which may include two free-living generations and a dependent generation. Sexuality via nonmotile male gametes that penetrate female tissue. Plastids contain chlorophyll *a* and the water-soluble accessory pigments allophycocyanin, phycocyanin, and phycoerythrin localized in phycobilisomes; thylakoids present as single lamellae. Undulipodia absent at all stages
Remaneicidae Family in phylum Foramenifera (Pr-3)
Retortamonadida Class in phylum Archaeprotista (Pr-1). Small, symbiotrophic mastigotes with twisted cell bodies bearing a cytostome in which a trailing undulipodium beats. Symbiotrophic usually found in digestive tract of insects, amphibians, reptiles, rodents, and other animals. Lack mitochondria and Golgi; presumed anaerobes
Retortamonadidae Family in phylum Archaeprotista (Pr-1)
Retortamonads See *Retortamonadida.*
Reussellidae Family in phylum Foramenifera (Pr-3)
Rhabdomonadales Order in phylum Euglenida (Pr-12)
Rhabdonemataceae Family in phylum Bacillariophyta (Pr-18)
Rhabdonematales Order in phylum Bacillariophyta (Pr-18)
Rhabdophora Subphylum in phylum Ciliophora (Pr-6)

Rhaphoneidaceae Family in phylum Bacillariophyta (Pr-18)
Rhaphoneidales Order in phylum Bacillariophyta (Pr-18)
Rhapydioninidae Family in phylum Foramenifera (Pr-3)
Rhipidiaceae Family of phylum Oomycota (Pr-21)
Rhipidiales Order of phylum Oomycota (Pr-21)
Rhizamminidae Family in phylum Foramenifera (Pr-3)
Rhizidiomycetaceae Family in phylum Hyphochytriomycota (Pr-14)
Rhizochloridaceae Family in phylum Xanthophyta (Pr-16)
Rhizochloridales Order in phylum Xanthophyta (Pr-16)
Rhizochrysidaceae Family in phylum Chrysophyta (Pr-15)
Rhizonymphidae Family in phylum Archaeprotista (Pr-1)
Rhizopoda Pr-2: Phylum of amastigote soil, freshwater, and marine amebas. Typically single-celled uninucleate organisms motile by pseudopods, feeding by phagotrophy. Body naked or bears tests of silica, carbonate sand grains or organic materials; many form resistant cysts. Some have two or more nuclei; reproduction by binary fission only. Sexuality is unknown. Cosmopolitan distribution in aquatic or terrestrial habitats; some symbiotrophic to necrotrophic
Rhizosoleniaceae Family in phylum Bacillariophyta (Pr-18)
Rhizosoleniales Order in phylum Bacillariophyta (Pr-18)
Rhizosoleniophycidae Subclass in phylum Bacillariophyta (Pr-18)
Rhodochaetaceae Family in phylum Rhodophyta (Pr-33)
Rhodochaetales Order in phylum Rhodophyta (Pr-33)
Rhodochytriaceae Family in phylum Chlorophyta (Pr-28)
Rhodophyceae Class in phylum Rhodophyta (Pr-33)
Rhodophyta See *Red algae*
Rhodophytes The phylum Rhodophyta (Pr-33) is a well-characterized and morphologically diverse taxon of eukaryotic photosynthetic protists, red algae, that are primarily marine. These seaweeds are cultivated for commercial utilization and consumption for hundreds of years; see *Red algae*
Rhodymeniales Order in phylum Rhodophyta (Pr-33)
Rhoicospheniaceae Family in phylum Bacillariophyta (Pr-18)
Rhynchodida Order in phylum Ciliophora (Pr-6)
Rhynchodina Suborder in phylum Ciliophora (Pr-6)
Riveroinidae Family in phylum Foramenifera (Pr-3)
Robertinacea Superfamily in phylum Foramenifera (Pr-3)
Robertinida Order in phylum Foramenifera (Pr-3)
Robertinidae Family in phylum Foramenifera (Pr-3)
Robuloidacea Superfamily in phylum Foramenifera (Pr-3)
Robuloididae Family in phylum Foramenifera (Pr-3)
Rotaliacea Superfamily in phylum Foramenifera (Pr-3)
Rotaliellidae Family in phylum Foramenifera (Pr-3)
Rotaliida Order in phylum Foramenifera (Pr-3)

Rotaliidae Family in phylum Foramenifera (Pr-3)

Rotaliporacea Superfamily in phylum Foramenifera (Pr-3)

Rotaliporidae Family in phylum Foramenifera (Pr-3)

Rudimicrosporea Class in phylum Microspora (F-1)

Rugoglobigerinidae Family in phylum Foramenifera (Pr-3)

Rzehakinacea Superfamily in phylum Foramenifera (Pr-3)

Rzehakinidae Family in phylum Foramenifera (Pr-3)

Saccamminidae Family in phylum Foramenifera (Pr-3)

Sagosphaeridae Family in phylum Actinopoda (Pr-31)

Salpingoecidae Family in phylum Archaeprotista (Pr-1)

Saprolegniales Order of phylum Oomycota (Pr-21)

Saprolegniomycetidae Class (subclass) of phylum Oomycota (Pr-21)

Sarcinochrysidaceae Family in phylum Chrysophyta (Pr-16)

Sarcinochrysidales Order in phylum Chrysophyta (Pr-16)

Sarcocystidae Family in phylum Apicomplexa (Pr-7)

Sargassaceae Family in phylum Phaeophyta (Pr-17)

Sceletonemataceae Family in phylum Bacillariophyta (Pr-18)

Scenedesmaceae Family in phylum Chlorophyta (Pr-28)

Schackoinidae Family in phylum Foramenifera (Pr-3)

Schizamminidae Family in phylum Foramenifera (Pr-3)

Schizomeridaceae Family in phylum Chlorophyta (Pr-28)

Schnellaceae Family in phylum Myxomycota (Pr-23)

Schubertellidae Family in phylum Foramenifera (Pr-3)

Schwagerinidae Family in phylum Foramenifera (Pr-3)

Sciadiaceae See *Ophiocytaceae*

Sclerosporaceae Family of phylum Oomycota (Pr-21)

Sclerosporales Order of phylum Oomycota (Pr-21)

Scolioneidaceae Family in phylum Bacillariophyta (Pr-18)

Scoliotropidaceae Family in phylum Bacillariophyta (Pr-18)

Scuticociliatida Order in phylum Ciliophora (Pr-6)

Scytosiphonaceae Family in phylum Phaeophyta (Pr-17)

Scytosiphonales Order in phylum Phaeophyta (Pr-17)

Seirococcaceae Family in phylum Phaeophyta (Pr-17)

Selenidiidae Family in phylum Apicomplexa (Pr-7)

Sellaphoraceae Family in phylum Bacillariophyta (Pr-18)

Sellaphorineae Suborder in phylum Bacillariophyta (Pr-18)

Semitextulariidae Family in phylum Foramenifera (Pr-3)

Septemcapsulidae Family in phylum Myxospora (A-2)

Sessilida Order in phylum Ciliophora (Pr-6)

Silicomastigotes See *Dictyochophyceae.*

Silicoloculinida Order in phylum Foramenifera (Pr-3)

Silicoloculinidae Family in phylum Foramenifera (Pr-3)

Silicotubidae Family in phylum Foramenifera (Pr-3)
Sinuolineidae Family in phylum Myxospora (A-2)
Siphogenerinoididae Family in phylum Foramenifera (Pr-3)
Siphoninacea Superfamily in phylum Foramenifera (Pr-3)
Siphoninidae Family in phylum Foramenifera (Pr-3)
Siphonocladaceae Family in phylum Chlorophyta (Pr-28)
Siphonocladales Order in phylum Chlorophyta (Pr-28)
Slime nets See *Labyrinthulids*
Soritacea Superfamily in phylum Foramenifera (Pr-3)
Soritidae Family in phylum Foramenifera (Pr-3)
Sorocarpaceae Family in phylum Phaeophyta (Pr-17)
Spermatochnaceae Family in phylum Phaeophyta (Pr-17)
Sphacelariaceae Family in phylum Phaeophyta (Pr-17)
Sphacelariales Order in phylum Phaeophyta (Pr-17)
Sphaenacantha Suborder in phylum Actinopoda (Pr-31)
Sphaeractinomyxidae Family in phylum Myxospora (A-2)
Sphaeramminidae Family in phylum Foramenifera (Pr-3)
Sphaerellarina Suborder in phylum Actinopoda (Pr-31)
Sphaeridiothricaceae Family in phylum Chrysophyta (Pr-15)
Sphaeriparaceae Family in phylum Dinomastigota (Pr-5)
Sphaerocollina Suborder in phylum Actinopoda (Pr-31)
Sphaeroidinidae Family in phylum Foramenifera (Pr-3)
Sphaeromyxidae Family in phylum Myxospora (A-2)
Sphaeromyxina Suborder in phylum Myxospora (A-2)
Sphaerosporidae Family in phylum Myxospora (A-2)
Sphenomonadales Order in phylum Euglenida (Pr-12)
Spirillinida Order in phylum Foramenifera (Pr-3)
Spirillinidae Family in phylum Foramenifera (Pr-3)
Spirocyclinidae Family in phylum Foramenifera (Pr-3)
Spiroplectamminacea Superfamily in phylum Foramenifera (Pr-3)
Spiroplectamminidae Family in phylum Foramenifera (Pr-3)
Spirotectinidae Family in phylum Foramenifera (Pr-3)
Spirotrichea Class in phylum Ciliophora (Pr-6)
Spirotrichonymphidae Family in phylum Archaeprotista (Pr-1)
Spirotrichosomidae Family in phylum Archaeprotista (Pr-1)
Spizellomycetaceae Family in phylum Chytridiomycota (Pr-35)
Spizellomycetales Order in phylum Chytridiomycota (Pr-35)
Splachnidiaceae Family in phylum Phaeophyta (Pr-17)
Sporadotrichina Suborder in phylum Ciliophora (Pr-6)
Sporochnaceae Family in phylum Phaeophyta (Pr-17)
Sporochnales Order in phylum Phaeophyta (Pr-17)
Sporolitheae Tribe in phylum Rhodophyta (Pr-33)

Sporolithoideae Subfamily in phylum Rhodophyta (Pr-33)

Spraguidae Family in phylum Microspora (F-1)

Spumellarida Order in phylum Actinopoda (Pr-31)

Spyroidea Family in phylum Actinopoda (Pr-31)

Squamulinacea Superfamily in phylum Foramenifera (Pr-3)

Squamulinidae Family in phylum Foramenifera (Pr-3)

Staffellidae Family in phylum Foramenifera (Pr-3)

Stainforthiidae Family in phylum Foramenifera (Pr-3)

Stannomida Class in phylum Xenophyophora (Pr-4)

Stannomidae Family in phylum Xenophyophora (Pr-4)

Stauraconidae Family in phylum Actinopoda (Pr-31)

Staurojoenidae Family in phylum Archaeprotista (Pr-1)

Stauroneidaceae Family in phylum Bacillariophyta (Pr-18)

Stemonitaceae Family in phylum Myxomycota (Pr-23)

Stemonitales Order in phylum Myxomycota (Pr-23)

Stemonitomycetidae Subclass in phylum Myxomycota (Pr-23)

Stenophoridae Family in phylum Apicomplexa (Pr-7)

Stephanopyxidaceae Family in phylum Bacillariophyta (Pr-18)

Stephoidea Family in phylum Actinopoda (Pr-31)

Stichogloeaceae Family in phylum Chrysophyta (Pr-15)

Sticholonchidae Family in phylum Actinopoda (Pr-31)

Stichotrichida Order in phylum Ciliophora (Pr-6)

Stichotrichina Suborder in phylum Ciliophora (Pr-6)

Stichtrichia Subclass in phylum Ciliophora (Pr-6)

Stictocyclaceae Family in phylum Bacillariophyta (Pr-18)

Stictocyclales Order in phylum Bacillariophyta (Pr-18)

Stilostomellidae Family in phylum Foramenifera (Pr-3)

Stipitococcaceae Family in phylum Xanthophyta (Pr-4)

Striariaceae Family in phylum Phaeophyta (Pr-17)

Strobilidiina Suborder in phylum Ciliophora (Pr-6)

Strombidinopsina Suborder in phylum Ciliophora (Pr-6)

Stylocephalidae Family in phylum Foramenifera (Pr-3)

Stylococcaceae Family in phylum Chrysophyta (Pr-15)

Stypocaulaceae Family in phylum Phaeophyta (Pr-17)

Suctoria Subclass in phylum Ciliophora (Pr-6)

Surirellaceae Family in phylum Bacillariophyta (Pr-18)

Surirellales Order in phylum Bacillariophyta (Pr-18)

Symphiacanthid Informal name of organisms in the phylum Actinopoda (Pr-31), order Symphiacanthida. These Symphiacanthida have 20 radial spicules united in the center of the endoplasm into a dense sphere that cannot be dissociated with sulfuric acid

Symphiacanthida Order in phylum Actinopoda (Pr-31)

Synactinomyxidae Family in phylum Myxospora (A-2)

Synchytriaceae Family in phylum Chytridiomycota (Pr-35)

Syndiniaceae Family in phylum Dinomastigota (Pr-5)

Syndiniales Order in phylum Dinomastigota (Pr-5)

Synhymeniida Order in phylum Ciliophora (Pr-6)

Syringamminidae Family in phylum Xenophyophora (Pr-4)

Syringodermataceae Family in phylum Phaeophyta (Pr-17)

Syringodermatales Order in phylum Phaeophyta (Pr-17)

Syzraniidae Family in phylum Foramenifera (Pr-3)

Tabellariaceae Family in phylum Bacillariophyta (Pr-18)

Tawitawiacea Superfamily in phylum Foramenifera (Pr-3)

Tawitawiidae Family in phylum Foramenifera (Pr-3)

Taxopodida Suborder in phylum Actinopoda (Pr-31)

Telomyxidae Family in phylum Microspora (F-1)

Teratonymphidae Family in phylum Archaeprotista (Pr-1)

Testaceafilosida Order in phylum Rhizopoda (Pr-2)

Testacealobosa Subclass in phylum Rhizopoda (Pr-2)

Tetractinomyxidae Family in phylum Myxospora (A-2)

Tetradimorphyidae Family in phylum Actinopoda (Pr-31)

Tetrahymenina Suborder in phylum Ciliophora (Pr-6)

Tetrasporaceae Family in phylum Chlorophyta (Pr-28)

Tetrasporales Order in phylum Chlorophyta (Pr-28)

Tetrataxacea Superfamily in phylum Foramenifera (Pr-3)

Tetrataxidae Family in phylum Foramenifera (Pr-3)

Textulariacea Superfamily in phylum Foramenifera (Pr-3)

Textulariellidae Family in phylum Foramenifera (Pr-3)

Textulariida Order in phylum Foramenifera (Pr-3)

Textulariidae Family in phylum Foramenifera (Pr-3)

Textulariopsidae Family in phylum Foramenifera (Pr-3)

Thalassicollidae Family in phylum Actinopoda (Pr-31)

Thalassionemataceae Family in phylum Bacillariophyta (Pr-18)

Thalassionematales Order in phylum Bacillariophyta (Pr-18)

Thalassiophysaceae Family in phylum Bacillariophyta (Pr-18)

Thalassiosiraceae Family in phylum Bacillariophyta (Pr-18)

Thalassiosirales Order in phylum Bacillariophyta (Pr-18)

Thalassiosirophycidae Subclass in phylum Bacillariophyta (Pr-18)

Thalassophysidae Family in phylum Actinopoda (Pr-31)

Thalicolidae Family in phylum Apicomplexa (Pr-7)

Thecadiniaceae Family in phylum Dinomastigota (Pr-5)

Thecamoebidae Family in phylum Rhizopoda (Pr-2)

Thecina Suborder in phylum Rhizopoda (Pr-2)

Thelohaniidae Family in phylum Microspora (F-1)

Thigmotrichina Suborder in phylum Ciliophora (Pr-6)
Thomasinellidae Family in phylum Foramenifera (Pr-3)
Thoracosphaeraceae Family in phylum Dinomastigota (Pr-5)
Thoracosphaerales Order in phylum Dinomastigota (Pr-5)
Thraustochytrids See *Labyrinthulids*
Thraustochytriidae Family in phylum Labyrinthulata (Pr-19)
Tilopteridales Order in phylum Phaeophyta (Pr-17)
Tintinnina Suborder in phylum Ciliophora (Pr-6)
Tiplopteridaceae Family in phylum Phaeophyta (Pr-17)
Tournayellacea Superfamily in phylum Foramenifera (Pr-3)
Tournayellidae Family in phylum Foramenifera (Pr-3)
Toxariaceae Family in phylum Bacillariophyta (Pr-18)
Toxariales Order in phylum Bacillariophyta (Pr-18)
Thraustochytrid Informal name of slime net organisms belonging to the phylum Labyrinthulata (Pr-19). These unique organisms (slime nets) produce colonial structures associated with wall-less ectoplasmic networks, which absorb nutrients and attach the cells to surfaces
Tremachoridae Family in phylum Foramenifera (Pr-3)
Trentepohliaceae Family in phylum Chlorophyta (Pr-28)
Trentepohliales Order in phylum Chlorophyta (Pr-28). Microscopic, branched filamentous chlorophytes, usually with differentiated reproductive cells: quadrimastigote zoospores and bimastigote isogametes. Cells walled, usually uninucleate. Often occur in subaerial habitats. Some are plant symbiotrophs, and at least one species is lichen phycobiont
Treubariaceae Family in phylum Chlorophyta (Pr-28)
Triactinomyxidae Family in phylum Myxospora (A-2)
Triadiniaceae Family in phylum Dinomastigota (Pr-5)
Tribonemataceae Family in phylum Xanthophyta (Pr-16)
Tribonematales Order in phylum Xanthophyta (Pr-16)
Trichiaceae Family in phylum Myxomycota (Pr-23)
Trichiales Order in phylum Myxomycota (Pr-23)
Trichohyalidae Family in phylum Foramenifera (Pr-3)
Trichomonadida Order in phylum Archaeprotista (Pr-1)
Trichomonadidae Family in phylum Archaeprotista (Pr-1)
Trichomonads Informal name of mastigotes (phylum Archaeprotista (Pr-1)) belonging to the class Parabasalia, order Trichomonadida. These heterotrophic protists have only been found in association with animals. They are characterized by the presence of one or more parabasal bodies, which are Golgi complexes associated with the kinetosomes
Trichomonads See *Trichomonadida*
Trichonymphidae Family in phylum Archaeprotista (Pr-1)
Trichosida Order in phylum Rhizopoda (Pr-2)
Trichostomatia Subclass in phylum Ciliophora (Pr-6)
Trilosporidae Family in phylum Myxospora (A-2)
Trimosinidae Family in phylum Foramenifera (Pr-3)
Trochamminacea Superfamily in phylum Foramenifera (Pr-3)

Trochamminidae Family in phylum Foramenifera (Pr-3)
Trocholonidae Family in phylum Foramenifera (Pr-3)
Trypanochloridaceae Family in phylum Xanthophyta (Pr-16)
Trypanosomatids Trypanosomes. Informal name of organisms in the phylum Archaeprotista (Pr-1). They are eukinetoplastic, hemomastigotes, invariably necrotrophic. May be spread by biting flies. For example, transmission by tsetse flies causes sleeping sickness
Trypanosomatidae Family in phylum Archaeprotista (Pr-1)
Trypanosomatina Suborder in phylum Archaeprotista (Pr-1)
Tuberitinidae Family in phylum Foramenifera (Pr-3)
Tubulina Suborder in phylum Rhizopoda (Pr-2)
Turrilinacea Superfamily in phylum Foramenifera (Pr-3)
Turrilinidae Family in phylum Foramenifera (Pr-3)
Tuscaroridae Family in phylum Actinopoda (Pr-31)
Tuzetiidae Family in phylum Microspora (F-1)

Udoteaceae Family in phylum Chlorophyta (Pr-28)
Ulotrichaceae Family in phylum Chlorophyta (Pr-28)
Ulotrichales Order in phylum Chlorophyta (Pr-28)
Ulvaceae Family in phylum Chlorophyta (Pr-28)
Ulvales Order in phylum Chlorophyta (Pr-28)
Ulvellaceae Family in phylum Chlorophyta (Pr-28)
Ulvophyceae See *Green seaweeds*
Ulvophycean Adjective referring to green algae (phylum Chlorophyta (Pr-28)) in the class Ulvophyceae. Predominantly marine organisms, nearly all the benthic green algae known from saline habitats worldwide are referable to this class
Unikaryonidae Family in phylum Microspora (F-1)
Uradiophoridae Family in phylum Apicomplexa (Pr-7)
Urophlyctaceae Family in phylum Chytridiomycota (Pr-35)
Urosporidae Family in phylum Apicomplexa (Pr-7)
Urostylina Suborder in phylum Ciliophora (Pr-6)
Uvigerinidae Family in phylum Foramenifera (Pr-3)

Vaginulinidae Family in phylum Foramenifera (Pr-3)
Vahlkampfidae Family in phylum Rhizopoda (Pr-2)
Valoniaceae Family in phylum Chlorophyta (Pr-28)
Valvulinellidae Family in phylum Foramenifera (Pr-3)
Valvulinidae Family in phylum Foramenifera (Pr-3)
Variisporina Suborder in phylum Myxospora (A-2)
Vaucheriaceae Family in phylum Xanthophyta (Pr-16)
Vaucheriales Order in phylum Xanthophyta (Pr-16)
Verbeekinidae Family in phylum Foramenifera (Pr-3)
Verneuilinacea Superfamily in phylum Foramenifera (Pr-3)
Verneuilinidae Family in phylum Foramenifera (Pr-3)

Verrucalvaceae Family of phylum Oomycota (Pr-21)

Vestibuliferida Order in phylum Ciliophora (Pr-6)

Victoriellidae Family in phylum Foramenifera (Pr-3)

Virgulinellidae Family in phylum Foramenifera (Pr-3)

Volvocaceae Family in phylum Chlorophyta (Pr-28)

Volvocalean Adjective referring to green algae (phylum Chlorophyta (Pr-28)) in the order Volvocales. They are coenobic colonial organisms with biundulipodiated cells arranged in multiples of two. The undulipodial apparatus was transformed during evolution from two undulipodia beating in opposite directions to two undulipodia beating in approximately the same direction

Volvocales Order in phylum Chlorophyta (Pr-28)

Warnowiaceae Family in phylum Dinomastigota (Pr-5)

Wenyonellidae Family in phylum Apicomplexa (Pr-7)

Xanthophyceae Single class in phylum Xanthophyta (Pr-16)

Xanthophyta Pr-16: Phylum of primarily freshwater, yellow-green, heterokont mastigote algae. Coccoid unicells and multicellular descendants; double-membrane bounded plastids contain chlorophylls *a* and *c*. Plastids, which store fat or oil, not starch, are surrounded by plastid endoplasmic reticulum. Reproduction by zoospores or their amastigote equivalent (hemiautospores). Sexual fusion of egg and sperm reported

Xanthophytes Yellow-green algae. Informal name of organisms in the phylum Xanthophyta (Pr-16). This is a phylum of phototrophic protoctists possessing green parietal plastids. They are distinguished from members of the Chlorophyta (Pr-28) by the absence of chlorophyll *b*, the presence of chlorophyll *c*, and the fact that xanthophytes never store starch

Xenophyophora Pr-4: Phylum of heterotrophic protoctists, all of which (except one group in shallow water) live in the abyssal marine benthos. Large ameboid organisms organized as plasmodia enclosed by a branched, tubelike organic cement. Tests patched from hard parts of skeletons, sponges, or foraminifera, radiolaria spicules, and mineral grains. Life history is not completely known.

Yamikovellidae Family in phylum Apicomplexa (Pr-7)

Zooxanthellaceae Family in phylum Dinomastigota (Pr-5)

Zygnemataceae Family in phylum Conjugaphyta (Pr-32)

Zygnematales Order in phylum Conjugaphyta (Pr-32)

References

Key Words:
Introduction
Abbayes, H. Des, "Histoire de la Botanique en France." In: D. de Virille, ed., Paris: Eighth International Botanical Congress Paris-Nice, pp.235–241; 1954.

Key Words:
Introduction
Balows, A., H. G. Truper, M. Dworkin, W. Harder, and K.-H. Schleifer, *The prokaryotes*, Vols I–IV, 2d ed. Springer-Verlag; New York; 1992.

Key Words:
Introduction
Barnes, R. S. K., ed., *A synoptic classification of living organisms*. Oxford Blackwell Scientific Publications and Sinauer Associates; Sunderland, MA; 1984.

Key Words:
Chytridiomycota
Barr, D. J. S., "Phylum Chytridiomycota." pp. 454–466 In: L. Margulis, J. O. Corliss, M. Melkonian, and D. J. Chapman, eds., *Handbook of protoctista*. Jones & Bartlett; Boston; 1990.

Key Words:
Introduction
Barr, D. J. S., and P. M. E. Allan. Zoospore ultrastructure of Polymyxa graminis (Plasmodiophoromycetes) *Canadian Journal of Botany* 60:2496–2504; 1982. In: L. Margulis, J. O. Corliss, M. Melkonian, and D. J. Chapman, eds., *Handbook of protoctista*. Jones & Bartlett; Boston; 1990.

Key Words:
Chytridiomycota
Bates, L., and J. A. Jackson eds., *Dictionary of geological terms*, 3rd ed. Anchor Press/Doubleday; Garden City, NY; 1984.

Key Words:
Glossaries
Bates, L., and Jackson, J. A., eds., *Dictionary of geological terms*, 3rd ed. Anchor Press/Doubleday; Garden City, NY; 1984.

Key Words:
Introduction
Bermudes, D., and L. Margulis. "Symbiont acquisition as neoseme: Origin of species and higher taxa." *Symbiosis* 4:185–198; 1987.

Key Words:
Introduction
Blackmore, S., and E. Tootill, eds., *The Penguin dictionary of botany*. Penguin Books; New York; 1984.

Key Words:
Glossaries
Blanton, R. L. The spore hilium of *Acrasis rosea. Journal of the Elisha Mitchell Scientific Society* 97:95–100; 1981. In: L. Margulis, J. O. Corliss, M. Melkonian, and D. J. Chapman, eds., *Handbook of protoctista*. Jones & Bartlett; Boston; 1990.

Key Words:
Acrasea
Bold, H. C., and M. J. Wynne, *Introduction to the algae: Structure and reproduction*. Prentice-Hall, Inc.; Englewood Cliffs, NJ; 1978.

Key Words:
Glossaries
Brown, R. W., *Composition of scientific words*. Smithsonian Institution Press; Washington, DC; 1956.

Key Words:
Glossaries
Brugerolle, G., and J. P. Mignot, "Phylum Proteromonadida." In: L. Margulis, J. O. Corliss,

M. Melkonian, and D. J. Chapman, eds., *Handbook of protoctista.* Jones & Bartlett; Boston; 1990.

Key Words:
Proteromonadida
Brugerolle, *et al.*, "Comparison et evolution des structures cellulaires chez plusiers éspeces de Bodonides et Cryptobiides appartenant genres Bodo, Cryptobia et Trypanoplasma (Kinetoplastida, Mastigophora)." *Protistologica* 15:197–221; 1979.

Key Words:
Zoomastigina
Kinetoplastida
Buck, K. R., "Phylum Choanomastigotes." In: L. Margulis, J. O. Corliss, M. Melkonian, and D. J. Chapman, eds., *Handbook of protoctista.* Jones & Bartlett; Boston; 1990.

Key Words:
Choanomastigotes
Cachon, J., M. Cachon, and K. W. Estep, "Phylum Actinopoda, classes: Polycystina and Phaeodaria." pp. 334–346. In: L. Margulis, J. O. Corliss, M. Melkonian, and D. J. Chapman, eds., *Handbook of Protoctista.* Jones & Bartlett; Boston; 1990.

Key Words:
Actinopoda
Canning, E. U., *et al.*, *Systematic Parasitology* 5:147–159; 1988.

Key Words:
Table 6 Classes of the Phyla of the Kingdom Protoctista
Cavalier-Smith, T. "Eukaryote kingdoms seven or nine?" *BioSystems* 14:461–481; 1981. In: L. Margulis, J. O. Corliss, M. Melkonian, and D. J. Chapman, eds., Handbook of *Protoctista.* Jones & Bartlett; Boston; 1990.

Key Words:
Zoomastigina
Pseudociliata
Clayton, M. N., "Phylum Phaeophyta." In: L. Margulis, J. O. Corliss, M. Melkonian, and D. J. Chapman, eds., *Handbook of protoctista.* Jones & Bartlett ; Boston; 1990.

Key Words:
Phaeophyta
Copeland, H. F., *Classification of the Lower Organisms.* Pacific Books; Palo Alto; 1956.

Key Words:
Introduction
Corliss, J. O., The ciliated protozoa: Characterization, classification and guide to the literature, 2d ed. Pergamon Press; Oxford-New York; 1979.

Key Words:
Introduction
Corliss, J. O., "Class Pseudociliata." In: L. Margulis, J. O. Corliss, M. Melkonian, and D. J. Chapman, eds., *Handbook of protoctista.* Jones & Bartlett; Boston; 1990.

Key Words:
Pseudociliata
Corliss, J. O., "Problems in cytoterminology and nomenclature for the protists." In: L. H. Huang, ed., *Advances in culture collections,* Vol. 1. USFCC/ASM; pp. 23–37; 1991.

Key Words:
Glossaries
Deasey, M. C., and L. S. Olive, Role of Golgi apparatus in sorogenesis by the cellular slime mold *Fonticula alba, Science.* 213:561–563. In: L. Margulis, J. O. Corliss, M. Melkonian, and D. J. Chapman, eds., *Handbook of protoctista.* Jones & Bartlett; Boston; 1990.

Key Words:
Acrasea
Desportes, I., Annales des Sciences Naturelles, Zoologie et Biologie Animale 12eme Serie 17:215–228; 1975.

Key Words:
Oomycota
Dick, M. W., "*Phylum Oomycota.*" In: L. Margulis, J. O. Corliss, M. Melkonian, and D. J. Chapman, eds., *Handbook of protoctista.* Jones & Bartlett; Boston; 1990.

Key Words:
Amebomastigota
Dyer, B. D., "Class Bicoecids." In: L. Margulis, J. O. Corliss, M. Melkonian, D. J. Chapman, eds., *Handbook of protoctista.* Jones & Bartlett; Boston; 1990.

Key Words:
Oomycota
Dyer, B. D., "Class Amebomastigota." In: L. Margulis, J. O. Corliss, M. Melkonian, D. J. Chapman, eds., *Handbook of protoctista.* Jones & Bartlett; Boston; 1990.

Key Words:
Bicoecids
Dylewski, D. P., "Phylum Plasmodiophoromycota." In: L. Margulis, J. O. Corliss, M. Melkonian, and D. J. Chapman, eds., *Handbook of protoctista.* Jones & Bartlett; Boston; 1990.

Key Words:
Plasmodiophoromycota
Estep, K. W., and F. MacIntyre, "Taxonomy, life cycle, distribution and dasmotrophy of Chrysochromulina: A theory accounting for scales, haptonema, muciferous bodies and toxicity." *Marine Ecology Progress Series* 57:11–21; 1989.

Key Words:
Dasmotrophy
Febvre, J., "Class Acantharia." In: L. Margulis, J. O. Corliss, M. Melkonian, and D. J. Chapman, eds., *Handbook of protoctista.* Jones & Bartlett; Boston; 1990.

Key Words:
Acantharia
Febvre-Chevalier, C., "Class Heliozoa." In: L. Margulis, J. O. Corliss, M. Melkonian, and D. J. Chapman, eds., *Handbook of protoctista.* Jones & Bartlett; Boston; 1990.

Key Words:
Heliozoa
Floyd, G. L., and C. J. O'Kelly, "Ulvophyceae." In: L. Margulis, J. O. Corliss, M. Melkonian, and D. J. Chapman, eds., *Handbook of protoctista.* Jones & Bartlett; Boston; 1990.

Key Words:
Ulvophyceae
Fox, G. E., R. B. Stackebrandt, R. B. Hespell, J. Gibson, J. Maniloff, T. A. Dyer, R. S. Wolfe, W. E. Balch, R. Tanner, L. Magrum, L. B. Zablen, R. Blakemore, R. Gupta, L. Bonen, B. J. Lewis, D. A. Stahl, K. R. Luehrsen, K. N., Chen, and C.R. Woese, "The phylogeny of prokaryotes." *Science* 209:457–463; 1980.

Key Words:
Introduction
Frederick, L. "Class Myxomycota." In: L. Margulis, J. O. Corliss, M. Melkonian, D. J. Chapman, eds., *Handbook of protoctista.* Jones & Bartlett; Boston; 1990.

Key Words:
Myxomycota
Gabrielson, P. W., D. J. Garbary, M. R. Sommerfeld, R. A. Townsend, and P. L. Tyler, "Phylum

Rhodophyta." In: L. Margulis, J. O. Corliss, M. Melkonian, and D. J. Chapman, eds., *Handbook of protoctista*. Jones & Bartlett; Boston; 1990.

Key Words:
Cryptophyta
Cryptomonads
Gillott, M. A. "Phylum Cryptophyta." In: L. Margulis, J. O. Corliss, M. Melkonian, and D. J. Chapman, eds., *Handbook of protoctista*. Jones & Bartlett; Boston; 1990.

Glaessner, M. F., *Dawn of animal life*. Cambridge University Press; Cambridge, England; 1984.

Key Words:
Rhodophyta
Gillott, M. A., and S. P. Gibbs, "Comparison of the ciliary rootlets and periplast in two marine cryptomonads." *Canadian Journal of Botany* 61:1964–1978; 1983.

Key Words:
Introduction
Graham, L., "Charophyceae, (orders Chlorokybales, Klebsormidiales, Coleochaetales)." In: L. Margulis, J. O. Corliss, M. Melkonian, and D. J. Chapman, eds., *Handbook of protoctista*. Jones & Bartlett; Boston; 1990.

Key Words:
Charophyceae
Chlorokybales
Klebsormidiales
Coleochaetales
Grant, M. C., "Charophyceae (order Charales)." In: L. Margulis, J. O. Corliss, M. Melkonian, and D. J. Chapman, eds., *Handbook of protoctista*. Jones & Bartlett; Boston; 1990.

Key Words:
Charophyceae
Charales
Gray, M., "The bacterial ancestry of mitochondria and plastids." *BioScience* 33:693–699; 1984.

Key Words:
Introduction
Green, J. C., K. Perch-Nielsen, and P. Westbroek, "Phylum Prymnesiophyta." In: L. Margulis, J. O. Corliss, M. Melkonian, and D. J. Chapman, eds., *Handbook of protoctista*. Jones & Bartlett; Boston; 1990.

Key Words:
Prymnesiophyta
Prymnesiophytes
Griffin, J. L., "Fine structure and taxonomic position of the giant amoeboid mastigote *Pelomyxa palustris*." *Journal of Protozoology* 35:300–315; 1988.

Key Words:
Table 6 Classes of the Phyla of the Kingdom Protoctista
Haines, K. C., K. D. Hoagland, and G. A. Fryxell, "A preliminary list of algal culture collections of the world." In: J. R. Rosowski, and B. C. Parker, eds., *Selected Papers in Phycology II*. Phycological Society of America; Lawrence, KS; pp. 820–826; 1982. [Provides list of algal culture collections with addresses and phone numbers, fee and restriction informatin, cross-refernced taxonomically.]

Key Words:
Table 4 Sources of Living Protoctists and Their Culture
Hawksworth, D. L., D. J. V. Hill, "The lichen-forming fungi." Chapman and Hall; New York; 1984.

Key Words:
Introduction
Heywood, P., Ultrastructure of mitosis in the chloromonadophycean alga *Vacuolaria Virescens, Journal of Cell Science* 31:37–51; 1978.

Key Words:
Introduction
Hibberd, D. J. "The ultrastructure and taxonomy of the Chrysophyceae and Prymnesiophyceae (Haptophyceae): A survey with some new observations on the ultrastructure of the Chrysophyceae." *Botanical Journal of the Linnean Society. Linnean Society of London* 72:55–80; 1976.

Key Words:
Chrysophyceae
Prymnesiophyceae
Hibberd, D. J., "Phylum Eustigmatophyta." In: L. Margulis, J. O. Corliss, M. Melkonian, and D. J. Chapman, eds., *Handbook of protoctista.* Jones & Bartlett; Boston; 1990.

Key Words:
Eustigmatophyta
Hindák, Frantisek, "Culture collection of algae at Laboratory of Algology in Trebon." *Archiv für Hydrobiologie,* Algological Studies 2/3(suppl. 39): 86–126; 1970.

Key Words:
Table 4 Sources of Living Protoctists and Their Culture
Hogg, J. "On the distinctions between a plant and an animal, and on a fourth kingdom of nature." *The Edinburgh New Philosophical Journal (new series)* 12:216–225; 1860.

Key Words:
Introduction
Hoshaw, R. W., R. M. McCourt, and J. C. Wang, "Phylum Conjugaphyta." In: L. Margulis, J. O. Corliss, M. Melkonian, and D. J. Chapman, eds., *Handbook of protoctista.* Jones & Bartlett; Boston; 1990.

Key Words:
Conjugaphyta
Kendrick, B., *The fifth kingdom.* Mycologue Publications; Waterloo, Ontario; 1985.

Key Words:
Introduction
Kies, L., and B. P. Kremer, "Phylum Glaucocystophyta." In: L. Margulis, J. O. Corliss, M. Melkonian, and D. J. Chapman, eds., *Handbook of protoctista.* Jones & Bartlett; Boston; 1990.

Key Words:
Glaucocystophyta
King, R. C., and W. D. Stansfield, *A dictionary of genetics,* 3rd ed. Oxford University Press; New York-Oxford; 1985.

Key Words:
Glossaries
Kivic, P. A., and P. L. Walne. *Origins of life* 13:269–288; 1984. In: L. Margulis, J. O. Corliss, M. Melkonian, and D. J. Chapman, eds., *Handbook of protoctista.* Jones & Bartlett; Boston; 1990.

Key Words:
Zoomastigina
Pseudociliata
Krieg, N. R., J. G. Holt, eds. *Bergey's manual of systematic bacteriology.* Vol. I. Williams and Wilkins; Baltimore; 1984.

Key Words:
Introduction
Kristiansen, J. "Phylum Chrysophyta." In: L. Margulis, J. O. Corliss, M. Melkonian, and

D. J. Chapman, eds. *Handbook of protoctista.* Jones & Bartlett; Boston; 1990.

Key Words:
Chrysophyta
Kulda, J., and E. Nohynkovà, "Giardia and Giardiasis." In: J. P. Kreier, ed., *Parasitic Protozoa*, Vol. 2. Academic Press; New York; pp. 2–138; 1978.

Key Words:
Zoomastigina
Diplomonadida
Lake, J. A., "Origin of the eukaryotic nucleus determined by rate-invariant analysis of rRNA sequences." *Nature* 331:184–186; 1988.

Key Words:
Introduction
Lazcano, A., "RNA world and molecular phylogeny." In: S. Bengtson, ed., *Evolution on the early Earth.* Columbia University Press; New York; 1993.

Key Words:
Introduction
Lee, J. J., "Phylum Foramenifera." In: L. Margulis, J. O. Corliss, M. Melkonian, and D. J. Chapman, eds., *Handbook of protoctista.* Jones and Bartlett; Boston, MA; 1990.

Key Words:
Foraminifera
Foramenifera
Introduction
Lee, J. J., S. H. Hutner, and E. C. Bovee, *An illustrated guide to the Protozoa.* Society of Protozoologists; Lawrence, KS; 1985.

Key Words:
Introduction
Lom, J., "Phylum Myxozoa." In: L. Margulis, J. O. Corliss, M. Melkonian, and D. J. Chapman, eds., *Handbook of protoctista.* Jones and Bartlett; Boston, MA; 1990.

Key Words:
Myxozoa
Lynn, D. H., and E. B. Small, "Phylum Ciliophora." In: L. Margulis, J. O. Corliss, M. Melkonian, and D. J. Chapman, *Handbook of protoctista.* Jones and Bartlett; Boston, MA; 1990.

Key Words:
Introduction
Margulis, L., "Undulipodia, flagella and cilia." *BioSystems* 12:105–108; 1980.

Key Words:
Introduction
Margulis, L. *Early life.* Jones and Bartlett; Boston, MA; 1982.

Key Words:
Introduction
Margulis, L., "Undulipodiated cells." *BioScience* 35:333; 1985.

Key Words:
Introduction
Margulis, L., "Systematics: The view from the origin and early evolution of life. Secession of the protoctista from the animal and plant kingdoms." In: D. Hawksworth and R. G. Davies, eds., *Prospects in systematics.* Clarendon Press; Oxford, UK; pp. 430–443; 1988.

Key Words:
Introduction
Margulis, L., Symbiosis in cell evolution: Microbial communities in the archean and proterozoic eons, 2d ed. W.H. Freeman and Company; New York; 1992.

Key Words:
Introduction
Margulis, L., and D. Bermudes, "Symbiosis as a mechanism of evolution: Status of cell symbiosis theory." *Symbiosis* 1:101–124; 1985.

Key Words:
Introduction
Margulis, L., and D. Sagan, "Order amongst animalcules: The protoctista kingdom and its undulipodiated cells." *BioSystems* 18:141–147; 1985.

Key Words:
Ciliates
Ciliophora
Margulis, L., and D. Sagan, *Origins of sex*. Yale University Press; New Haven, CT; 1986.

Key Words:
Introduction
Margulis, L., and K. V. Schwartz, *Five kingdoms*. W.H. Freeman and Company; New York; 1988.

Key Words:
Introduction
Margulis, L., J. O. Corliss, M. Melkonian, and D. J. Chapman, *Handbook of protoctista*. Jones and Bartlett; Boston, MA; 1990.

Key Words:
Introduction
Margulis L., H. I. McKhann, L. Olendzenski, *Illustrated Glossary of Protoctista* Jones and Bartlett Publishers. Boston; 1993.

Key Words:
Introduction
McEnery, M., and J. J. Lee, "*Allogromia laticollaris*: A foraminiferan with unusual apogamic metagenic life cycle." *Journal of Protozoology* 23:94–108; 1976.

Key Words:
Foraminifera
Foraminiferan
Foramenifera
Mehlhorn, H., A. O. Heydorn, J. Senaud, and E. Schein, "La modalités de la transmission des protozoaires symbiotrophs des genres *Sarcocystis* et *Theileria* agents de graves malladies." *L'Année Biologique* 18:97–120; 1979.

Key Words:
Apicomplexa
Melkonian, M., "Class Chlorophyceae." In: L. Margulis, J. O. Corliss, M. Melkonian, and D. J. Chapman, eds., *Handbook of protoctista*. Jones and Bartlett; Boston, MA; 1990.

Key Words:
Chlorophyceae
Melkonian, M., "Class Microthamniales." In: L. Margulis, J. O. Corliss, M. Melkonian, and D. J. Chapman, eds., *Handbook of protoctista*. Jones and Bartlett; Boston, MA; 1990.

Key Words:
Microthamniales
Melkonian, M., "Class Pedinomonadales." In: L. Margulis, J. O. Corliss, M. Melkonian, and D. J. Chapman, eds., *Handbook of protoctista*. Jones and Bartlett; Boston, MA; 1990.

Key Words:
Pedinomonadales
Melkonian, M., "Class Prasinophyceae." In: L. Margulis, J. O. Corliss, M. Melkonian, and D. J. Chapman, eds., *Handbook of protoctista*. Jones and Bartlett; Boston, MA; 1990.

Key Words:
Prasinophyceae
O'Kelly, C. J., and G. L. Floyd, "Class Prasiolales." In: L. Margulis, J. O. Corliss, M. Melkonian,

and D. J. Chapman, eds., *Handbook of protoctista.* Jones and Bartlett; Boston, MA; 1990.

Key Words:
Prasiolales
O'Kelly, C. J., and G. L. Floyd, "Class Trentepohliales." In: L. Margulis, J. O. Corliss, M. Melkonian, and D. J. Chapman, eds., *Handbook of protoctista.* Jones and Bartlett; Boston, MA; 1990.

Key Words:
Trentepohliales
Patterson, D. J., "The fine structure of *Opalina ranarum* (family Opalinidae): Opalinid phylogeny and classification." *Protistologica* 21:413–428; 1986.

Key Words:
Table 6 Classes of the Phyla of the Kingdom Protoctista
Perkins, F. O., "Phylum Haplosporidia." In: L. Margulis, J. O. Corliss, M. Melkonian, and D. J. Chapman, eds., *Handbook of protoctista.* Jones and Bartlett; Boston, MA; 1990.

Key Words:
Haplosporidia
Pirozynski, K., and D. Malloch, "The origin of land plants: A matter of mycotrophism." *BioSystems* 6:153–164; 1975.

Key Words:
Introduction
Poindexter, J., *Microbiology: An introduction to protists.* Macmillan; New York; 1971.

Key Words:
Introduction
Porchet-Henneré, E., and A. Richard, "La schizogonie chez *Aggregata eberthi*, etude en microscope electronique." *Protistologica* 7:227–259; 1971.

Key Words:
Apicomplexa
Porter, D., "Phylum Labyrinthulata." In: L. Margulis, J. O. Corliss, M. Melkonian, and D. J. Chapman, eds., *Handbook of protoctista.* Jones and Bartlett; Boston, MA; 1990.

Key Words:
Labyrinthulata
Raikov, I. B., *The protozoan nucleus.* Springer-Verlag; Vienna; 1982.

Key Words:
Karyoblastea
Raper, K. B., A. C. Worley, and T. A. Kurzynski, "*Copromyxella*: A new genus of Acrasidae." *American Journal of Botany* 65:1011–1026; 1978.

Key Words:
Acrasea
Raven, P. H., and G. B. Johnson, *Biology.* Times Mirror/Mosby College Publishing; St. Louis–Toronto–Santa Clara; 1986.

Key Words:
Glossaries
Richardson, T. B., "Origins and evolution of the earliest land plants." In: J. W. Schopf, ed., *Major events in the history of life.* Jones and Bartlett; Boston, MA; pp. 95–118; 1992.

Key Words:
Introduction
Roberts, K. R., K. D. Stewart, and K. R. Mattox, "The ciliary apparatus of *Chilomonas paramecium* (Cryptophyceae) and its comparison with certain zoomastigotes." *Journal of Phycology* 17:159–167; 1981.

Key Words:
Cryptophyta
Round, F. E., and R. M. Crawford, "Phylum Bacillariophyta." In: L. Margulis, J. O. Corliss, M. Melkonian,

and D. J. Chapman, eds., *Handbook of protoctista.* Jones and Bartlett; Boston, MA; 1990.

Key Words:
Bacillariophyta
Diatoms
Schlosser, U. G., "Sammlung von Algenkulturen." *Berichte der Deutschen Botanischen Gesellschaft* 95:181–287; 1982.

Key Words:
Table 4 Sources of Living Protoctists and Their Culture
Schopf, J. W., ed., *Major events in the history of life.* Jones and Bartlett; Boston, MA; 1992.

Key Words:
Introduction
Schuster, F. L., "Phylum Rhizopoda." In: L. Margulis, J. O. Corliss, M. Melkonian, and D. J. Chapman, eds., *Handbook of protoctista.* Jones and Bartlett; Boston, MA; 1990.

Key Words:
Amoeba
Rhizopoda
Schwemmler, W., Reconstruction of cell evolution: A periodic system. CRC Press; Boca Raton, FL; 1984.

Key Words:
Introduction
Schwemmler, W., *Symbiogenesis: A macromechanism of evolution.* Walter de Gruyter; Berlin; 1989.

Key Words:
Introduction
Senaud, J., H. Augustin, and B. Doens-Juteau, "Observations ultrastructurales sur la development sexual de la coccide *Eimeria acervulina* (Tyzzer, 1929) dans l'epithelium intestinal du poulet: La microgametogenese et la macrogametogenese." *Protistologica* 16:241–257; 1980.

Key Words:
Apicomplexa
Seravin, L. N., and A. V. Goodkov, "The flagella of the freshwater amoeba *Pelomyxa palustris.*" *Tsitoligya* 29:721–724 (in Russian); 1987.

Key Words:
Introduction
Table 6 Classes of the Phyla of the Kingdom Protoctista
Shmagina, A. P., *Mertsatel' Noe Dvizhenie. (Ciliary movement.)* Medgiz; Moscow (in Russian); 1948.

Key Words:
Introduction
Sieburth, J. McN., and K. Estep, "Precise and meaningful terminology in marine microbial ecology." *Marine Microbial Food Webs* 1:1–16; 1985.

Key Words:
Introduction
Sneath, P. H. A., N. S. Mair, M. E. Sharpe, and J. G. Holt, eds., *Bergey's manual of systematic bacteriology*, Vol. 2. Williams and Wilkins; Baltimore, MD; 1986.

Key Words:
Introduction
Sonea, S., and M. Panisset, *A new bacteriology.* Jones and Bartlett; Boston, MA; 1983.

Key Words:
Acrasea
Spiegel, F. W., "Class Protostelida." In: L. Margulis, J. O. Corliss, M. Melkonian, and D. J. Chapman, eds., *Handbook of protoctista.* Jones and Bartlett; Boston, MA; 1990.

Key Words:
Introduction
Spiegel, F. W., and L. S. Olive, "New evidence for the validity of *Copromyxa protea*." *Mycologia* 70:843–847; 1978.

Key Words:
Protostelida
Starr, R. C., "The culture collection of algae at the University of Texas at Austin." *Journal of Phycology* 14(suppl. 47):47–100; 1978. [Provides listing of cultures held, cross-referenced to CCAP1, along with media recipes for growth of algae and information about ordering.]

Key Words:
Table 4 Sources of Living Protoctists and Their Culture
Taylor, F. J. R., "Phylum Dinoflagellata (Dinomastigota)." In: L. Margulis, J. O. Corliss, M. Melkonian, and D. J. Chapman, eds., *Handbook of protoctista*. Jones and Bartlett; Boston, MA; 1990.

Key Words:
Dinoflagellata
Dinomastigota
Van Hoven, W., "Isolated cilioprotistan evolution in African rhino intestines." International Society for Evolutionary Protistology, Royal Holloway and Bedford New College; Egham, England; July 1987. [Oral communication.]

Key Words:
Introduction
Vavra, J., and V. Spraque, "Glossary for the microsporidia." In: L. A. Bulla and T. C. Cheng, eds., *Biology of the microsporidia, comparative pathobiology*, Vol. 1. Plenum Press; New York–London; 1976.

Key Words:
Glossaries
Vickerman, K. "Class Kinetoplastida." In: L. Margulis, J. O. Corliss, M. Melkonian, and D. J. Chapman, eds., *Handbook of protoctista*. Jones and Bartlett; Boston, MA; 1990.

Key Words:
Kinetoplastida
Vidal, G., "The oldest eukaryotic cells." *Scientific American* 250:48–57; 1984.

Key Words:
Introduction
Vivier, E., and I. Desportes, "Phylum Apicomplexa." In: L. Margulis, J. O. Corliss, M. Melkonian, and D. J. Chapman, eds., *Handbook of protoctista*. Jones and Bartlett; Boston, MA; 1990.

Key Words:
Apicomplexa
Walne, P. L., and P. A. Kivic, "Phylum Euglenida." In: L. Margulis, J. O. Corliss, M. Melkonian, and D. J. Chapman, eds., *Handbook of protoctista*. Jones and Bartlett; Boston, MA; 1990.

Key Words:
Euglenida
Euglenophyta
Whisler, H. C., "Ellobiopsida." In: L. Margulis, J. O. Corliss, M. Melkonian, and D. J. Chapman, eds., *Handbook of protoctista*. Jones and Bartlett; Boston, MA; 1990.

Key Words:
Ellobiopsida
Whittaker, R. H., "On the broad classification of organisms." *Quarterly Review of Biology* 34:210–226; 1959.

Key Words:
Introduction
Woese, C.R., "Bacterial evolution." *Microbial Reviews* 51:221–271; 1987.

Key Words:
Introduction
Zahn, R. K., "A green alga with minimal eukaryotic features: *Nanochlorum eukaryotum*." *Origins of Life* 13:289–303; 1984.

INDEX

Note: Page references followed by the letter "f" refer to a figure; by the letter "t" to a table, and by "phy" to the phylogenetic diagrams at the beginning of each chapter. Letters and numbers in parentheses after an entry refer to the kingdom and phylum. See the Table of Contents or the Appendix for the entire list. Examples of genera from the heading of each phylum do not appear in the Index but are listed in the Appendix.